Algebra – Abstract and Modern

AF647675

Dr U. M. Swamy
Dean, Faculty of Science (Retired)
Department of Mathematics
Andhra University
Visakhapatnam

Dr A. V. S. N. Murty
Professor of Mathematics
Srinivasa Institute of Engineering and Technology
Amalapuram

PEARSON

Chennai • Delhi • Chandigarh

Associate Editor–Acquisitions: King D Charles Fenny
Associate Editor–Production: M. R. Ramesh

Copyright © 2012 Dorling Kindersley (India) Pvt. Ltd

This book is sold subject to the condition that it shall not, by way of trade or otherwise, be lent, resold, hired out, or otherwise circulated without the publisher's prior written consent in any form of binding or cover other than that in which it is published and without a similar condition including this condition being imposed on the subsequent purchaser and without limiting the rights under copyright reserved above, no part of this publication may be reproduced, stored in or introduced into a retrieval system, or transmitted in any form or by any means (electronic, mechanical, photocopying, recording or otherwise), without the prior written permission of both the copyright owner and the publisher of this book.

ISBN 978-81-317-5892-2

First Impression

Published by Dorling Kindersley (India) Pvt. Ltd, licensees of Pearson Education in South Asia.

Head Office: 7th Floor, Knoweldge Boulevard, A-8(A), Sector 62,
Noida 201 309, UP, India.
Registered Office: 11 Community Centre, Panchsheel Park,
New Delhi 110 017, India.

Compositor: Q CON Technologies Pvt. Ltd, Chennai.
Printer: Saurabh Printers Pvt. Ltd.

Dedicated to my wife

LAKSHMI

and to my dearest daughters

MYTHRI and SOWMYA

– U. M. Swamy

Contents

Preface

This book is designed for a two-semester sequence as a first course in abstract algebra for advanced undergraduate and junior post-graduate students. A glance at the table of contents will reveal the scope of the book; the range of topics covered is reasonably standard, with no major surprises. Our intention is to present a text that is logically developed, precise, and in keeping with the spirit of the times. Guided by the principle that a routine diet of definitions, theorems and results soon becomes unpalatable, we have concentrated on supplementing the concepts with examples and counter-examples and on establishing the important and fruitful results in a formal, rigorous fashion. En route, we have tried to showcase the power and elegance of the abstract – modern approach in mathematics, particularly in algebra, and chosen the title 'Algebra – Abstract and Modern' for this book.

The reader is not presumed to possess any previous knowledge of the concepts of modern algebra, except certain mathematical maturity and a will to learn abstract thinking. Consequently, the book's initial chapters are somewhat elementary, with the exposition proceeding at a leisurely pace, filling in the details of proofs, particularly of basic results. To smoothen the approach, we have devoted Part I to preliminaries consisting of two chapters, one on sets, relations, function, partitions and the cardinality of a set and the other on number systems, matrices and determinants. This part also serves as a vehicle for introducing some of the notation and terminology concerning the language of basic mathematics to be used in the later parts. Proofs of most of the results in Part I are skipped and given as exercises to encourage interested readers to work on them.

There are three parts in the main text of the book, Part II (Chapter 3–8), Part III (Chapter 9–13) and Part IV (Chapters 14–16) covering Group Theory, Ring Theory and Field Theory respectively. Each chapter is divided into a suitable number of sections in which definitions of the various concepts are immediately followed by a sufficient number of examples and counter-examples. Worked exercises are included in each section in addition to a set of exercises of varying levels of difficulty at the end of each section. These exercises are an integral part of the book and require the reader's active participation. Some of them introduce a variety of ideas not treated in the body of the text and impart certain additional information about concepts discussed in chapters. We have given a brief introduction of vector spaces and linear transformations to the extent necessary for a discussion on Galois Theory. We have resisted the temptation to use Exercises, except those in Part I, to develop results that will be needed thereafter. As a result, the reader does not need to work on all the exercises to assimilate the ideas presented in the rest of the book. However, for the benefit of slow learners, answers/hints for all even-numbered exercises have been provided.

When the publishers approached us with a proposal to take up the project of writing a text book on Algebra, we considered various opinions on what should be attempted within the framework of a first course in algebra. In selecting textual material, we have followed, to a considerable extent, our own interests, condensing or omitting altogether a number of topics that other authors might have pursed more vigorously. The measure of success of our efforts in writing this book is directly proportional to the number of readers stimulated to expand their horizons in the realms of algebra. Comments and suggestions for the improvement of the quality of the book are most welcome and will be acknowledged in later editions. We may be excused for any possible typos.

We profusely thank all persons who directly or indirectly helped us in bringing out this book. We are grateful to the people at Pearson Education, to Mr. King D Charles Fenny in particular, for their encouragement and help in completing this project. A special word of appreciation and thanks goes to my wife, Lakshmi, who patiently helped me in the early morning hours on the days when I was writing this book.

U. M. Swamy
A. V. S. N. Murty

PART I

Preliminaries

1

Sets and Relations

The concept of a set was used even by the ancient mankind without having an exact idea of what it was. In modern mathematics, the notion of a set is most basic. In fact, almost all the mathematical systems are certain collection of sets and their theories can be categorised as parts of set theory. We do not intend to discuss axiomatic development of set theory. But, any person with an intention of starting to learn the present day algebra must necessarily possess certain elementary knowledge of set theory. This chapter provides a fairly good platform to refresh with those elementary notions of sets, relations, functions and the cardinality of a set.

1.1 SETS AND SUBSETS

A set is usually defined as a well-defined collection of objects, in the sense that, given any object we must be in a position to decide whether the object belongs to the collection or not. First, let us take up two examples.

Example 1.1.1. Let us call a positive integer, a prime number if it has exactly two positive divisors, namely 1 and itself. Clearly, 1 is not a prime number, since 1 has only one positive divisor. Let C be the collection of all prime numbers. We shall argue that C is a well-defined collection of objects. Let a be any object. If a is not a positive integer, then we can immediately say that a does not belong to the collection C. Suppose that a is a positive integer, we can evaluate all the positive divisors of a and see whether these are two in number. For example, let $a = 123456789$, on simple examination, we can say that 3 divides a (since the sum of the digits of a is $\frac{9(9+1)}{2} = 45$) and $3 \neq 1$ and $3 \neq a$ and hence a is not a prime number, so that a does not belong to the collection C. On the other hand, let $b = 123457687$. It may be difficult for us to decide whether b is prime or not. However, one thing is certain, it is either a prime or not a prime, but not both. Therefore, C is a well-defined collection of objects.

Example 1.1.2. Let C be the collection of all sets A satisfying the property that A is not an object in A (or A does not belong to A). We shall argue that C is not a well-defined collection. Suppose on the contrary that C is a well-defined collection, that is, C is a set. Then, if C is an object in C, it follows that C is not an object in C. On the other hand, if C is not an object in C, then it follows that C is an object in C. Either way, it leads to a contradiction. Therefore, we cannot decide whether C is an object in C. Therefore, C is not a well-defined collection.

Definition 1.1.1. A well-defined collection of objects is called a *set.* If S is a set, then the objects in S are called *elements* of S. We write $a \in S$ and read 'a belongs to S', when a is an object in S. We write $a \notin S$ to say that a does not belong to S.

Sets are usually denoted by uppercase letters, such as A, B, X, Y, etc. and the elements of sets are denoted by lowercase letters, such as a, b, x, y, etc.

Example 1.1.3

1. The collection of all intelligent persons in India is not a set, since, if we select a person from India, we cannot say with certainty whether he/she belongs to the collection or not, as there is no standard scale for the evaluation of intelligence.
2. For a similar reason, as detailed above, the collection of all tall persons in India is not a set.
3. The collection of all prime numbers is a set, as discussed in Example 1.1.1.
4. The collection of all positive integers, which are not prime, is a set.

In this book, it is convenient to represent a set with the help of certain property or properties satisfied only by the elements of the set. In order to represent a set by this method, we write between the brackets { } a variable x which stands for each of the set followed by the property or properties possessed by each element of the set and these two are separated by a symbol ':' or '|', read as 'such that'. Therefore, we write

$$\{x : p(x)\} \quad \text{or} \quad \{x \mid p(x)\}$$

to represent the set of all objects x that satisfy the statement $p(x)$. For example, the set of all prime numbers is represented by

$$\{x : x \text{ is a prime number}\}.$$

The set of all positive odd integers is represented by

$$\{x : x \text{ is a positive integer and } x \text{ is odd}\}$$

which is same as the set $\{1, 3, 5, 7, \ldots\}$.

Definition 1.1.2. A collection having no objects is clearly a set and is called the *empty set* or *null set* and is denoted by the symbol $\varnothing$.

Example 1.1.4. The set $\{x : x \text{ is an even integer and } 2 < x < 4\}$ is the empty set, since there is no even integer x, such that $2 < x < 4$. Similarly,

$$\{x : x \text{ is an integer and } x^2 + 2 = 0\}$$

is the empty set.

Notation 1.1.1. The implication symbol $\Rightarrow$ will be read as 'implies'. If P and Q are statements, then $P \Rightarrow Q$ stands for the statement 'the truth of P implies the truth of Q' or simply 'P implies Q'. The symbol $\Leftrightarrow$ is read as 'implies and implied by'. For any statements P and Q, $P \Leftrightarrow Q$ stands for 'P implies and implied by Q' or 'P if and only if Q'.

Example 1.1.5

1. Let P be the statement, 'x is an integer and $x^2 = 0$' and Q be the statement, '$x = 0$'. Then, we have $P \Leftrightarrow Q$ since, for any integer x, $x^2 = 0$ if and only if $x = 0$.
2. Let P be the statement, 'x is a real number and $x^2 = x$' and Q be the statement, '$x = 0$ or $x = 1$'. Then, $P \Leftrightarrow Q$ since, for any real number x, $x^2 = x$ if and only $x = 0$ or $x = 1$.

Definition 1.1.3. Let A and B be two sets. Then, we say that

1. A is *equal* to B and express this by $A = B$ if, for any object x,

$$x \in A \Leftrightarrow x \in B.$$

2. A is a *subset* of B (or A is contained in B) and express this by $A \subseteq B$ if, for any object x,

$$x \in A \Rightarrow x \in B.$$

For any two sets A and B, clearly $A = B$ if and only if $A \subseteq B$ and $B \subseteq A$. Whenever we are required to prove that two given sets A and B are equal, we usually prove that $A \subseteq B$ and $B \subseteq A$, that is, for any object x,

$$x \in A \Rightarrow x \in B \quad \text{and} \quad x \in B \Rightarrow x \in A.$$

Sometimes, we say that a set A is *smaller than* B (or B is larger than A) if A is a subset of B. A is said to be a *proper subset* of B and write $A \subset B$ if $A \subseteq B$ and $A \neq B$. Also, instead of writing $A \subseteq B$ or $A \subset B$, some times we write $B \supseteq A$ or $B \supset A$ and say that B contains A (or B is a superset of A) or B properly contains A, respectively. We write $A \nsubseteq B$ if A is not a subset of B.

Definition 1.1.4. For any set S, the collection of all subsets of S is again a set and is called the *power set of S* and is denoted by $\mathbb{P}(S)$.

Note that the power set $\mathbb{P}(S)$ of any set S is always nonempty, since the empty set $\varnothing$ is a subset of every set S. In fact, if S is the empty set $\varnothing$, then

$$\mathbb{P}(\varnothing) = \{\varnothing\},$$

a set consisting of only one element. It can be easily proved that, for any non-negative integer n, a set S has exactly n elements if and only if the power set $\mathbb{P}(S)$ has exactly 2^n elements.

Definition 1.1.5. A set whose element are sets is called a *class of sets* or *family of sets*.

Class of sets will be usually denoted by script letters, such as $\mathcal{A}$, $\mathcal{B}$, $\mathcal{C}$, etc. For any set S, the power set $\mathbb{P}(S)$ is a class of sets. A class $\mathcal{C}$ of sets is called an *indexed class* if there exists a set I such that, for each $i \in I$, there is a unique member A_i in $\mathcal{C}$ associated with i and the class $\mathcal{C}$ is equal to the class of all A_i, $i \in I$; in this case, we write

$$\mathcal{C} = \{A_i : i \in I\} \quad \text{or} \quad \mathcal{C} = \{A_i\}_{i \in I}$$

and I is called the *index set*.

Example 1.1.6. For any positive integer n, let

$$A_n = \left\{x : x \text{ is a real number and } 0 \leq x \leq \frac{1}{n}\right\}.$$

Then, $\{A_n\}_{n\in\mathbb{Z}^+}$ is an indexed class of sets and the set $\mathbb{Z}^+$ of positive integers is the index set.

Definition 1.1.6. For any indexed class of sets $\{A_i\}_{i\in I}$, we define the set as

$$\bigcap_{i\in I} A_i = \{a : a \in A_i \text{ for all } i \in I\}.$$

This set is called the *set intersection* of A_i's, $i \in I$. In particular, if $A_1, A_2, \ldots, A_n$ are sets, we define

$$\bigcap_{i=1}^{n} A_i = \left\{a : a \in A_i \text{ for } i = 1, 2, \ldots, n\right\}$$

and is also denoted by $A_1 \cap A_2 \cap \ldots \cap A_n$. For any sets A and B, we have

$$A \cap B = \{x : x \in A \text{ and } x \in B\}.$$

Two sets A and B are said to be *disjoint* if $A \cap B = \varnothing$, that is, there are no common elements of A and B.

Definition 1.1.7. For any indexed class $\{A_i\}_{i\in I}$ of sets, we define the set as

$$\bigcup_{i\in I} A_i = \{a : a \in A_i \text{ for some } i \in I\}.$$

This set is called the *set union* of A_i's, $i \in I$. In particular, for any sets $A_1, A_2, \ldots, A_n$, we define

$$\bigcup_{i=1}^{n} A_i = \{a : a \in A_i \text{ for some } 1 \le i \le n\}$$

and this is also denoted by $A_1 \cup A_2 \cup \ldots \cup A_n$. For any sets A and B, we have

$$A \cup B = \{x : x \in A \text{ or } x \in B\}.$$

Example 1.1.7. For any positive integer n, let

$$A_n = \left\{x : x \text{ is a real number and } 0 \le x \le \frac{1}{n}\right\}.$$

Then, we have $A_n \supset A_{n+1}$ for any n and therefore

$$[0, 1] = A_1 \supset A_2 \supset A_3 \supset \ldots \supset A_n \supset A_{n+1} \supset \ldots$$

$$\bigcap_{n \in \mathbb{Z}^+} A_n = \{x : x \text{ is a real number and } 0 \le x \le \frac{1}{n} \text{ for all } n \in \mathbb{Z}^+\} = \{0\}$$

and $\bigcup_{n \in \mathbb{Z}^+} A_n = A_1 = [0, 1]$.

The following theorems can be easily proved by straight-forward verifications.

Theorem 1.1.1. The following holds good for any sets A, B and C.

1. $A \cup B \subseteq C \Leftrightarrow A \subseteq C$ and $B \subseteq C$
2. $A \subseteq B \cap C \Leftrightarrow A \subseteq B$ and $A \subseteq C$
3. $A \cap B \subseteq A \subseteq A \cup B$
4. $A \cup A = A = A \cap A$
5. $A \cup B = B \cup A$ and $A \cap B = B \cap A$
6. $(A \cup B) \cup C = A \cup (B \cup C)$ and $(A \cap B) \cap C = A \cap (B \cap C)$
7. $A = A \cap B \Leftrightarrow A \subseteq B \Leftrightarrow A \cup B = B$
8. $A \cap (A \cup B) = A = A \cup (A \cap B)$
9. $A \cap (B \cup C) = (A \cap B) \cup (A \cap C)$
10. $A \cup (B \cap C) = (A \cup B) \cap (A \cup C)$
11. $A \cap (\bigcup_{i \in I} A_i) = \bigcup_{i \in I} (A \cap A_i)$ for any indexed class $\{A_i\}_{i \in I}$ of sets.
12. $A \cup (\bigcap_{i \in I} A_i) = \bigcap_{i \in I} (A \cup A_i)$ for any indexed class $\{A_i\}_{i \in I}$ of sets.
13. $A \subseteq B \Rightarrow A \cap C \subseteq B \cap C$ and $A \cup C \subseteq B \cup C$
14. $A \cap B \subseteq A \cap C$ and $A \cup B \subseteq A \cup C \Leftrightarrow B \subseteq C$

Definition 1.1.8. For any two sets A and B, the *difference* of A with B is defined as

$$A - B = \{x : x \in A \text{ and } x \notin B\}.$$

Theorem 1.1.2 (De Morgan Laws). For any indexed class $\{B_i\}_{i \in I}$ of sets and for any sets A, B and C, the following holds good.

1. $A - (\bigcup_{i \in I} B_i) = \bigcap_{i \in I} (A - B_i)$
2. $A - (\bigcap_{i \in I} B_i) = \bigcup_{i \in I} (A - B_i)$
3. $B \subseteq C \Rightarrow A - C \subseteq A - B$ and $B - A \subseteq C - A$
4. $(\bigcup_{i \in I} B_i) - A = \bigcup_{i \in I} (B_i - A)$

5. $(\bigcap_{i \in I} B_i) - A = \bigcap_{i \in I} (B_i - A)$
6. $(A \cup B) - C = (A - C) \cup (B - C)$
7. $(A \cap B) - C = (A - C) \cap (B - C)$
8. $A - (B \cup C) = (A - B) \cap (A - C)$
9. $A - (B \cap C) = (A - B) \cup (A - C)$
10. $(A - B) - C = A - (B \cup C) = (A - C) - B$
11. $A - (B - C) = (A - B) \cup (A \cap C)$
12. $A \cap B = \varnothing \Leftrightarrow A \subseteq A - B \Leftrightarrow B \subseteq B - A$
13. $A - \varnothing = A$
14. $\varnothing - A = \varnothing$

Definition 1.1.9. For any sets A and B, the *symmetric difference* of A and B is defined as

$$A \oplus B = (A - B) \cup (B - A).$$

That is, $A \oplus B = \{x : x \text{ belongs to exactly one of } A \text{ and } B\}$.

Theorem 1.1.3. The following holds good for any sets A, B and C.

1. $A \oplus B = B \oplus A$
2. $(A \oplus B) \oplus C = A \oplus (B + C)$
   $$= (A \cap B \cap C) \cup ((A - B) - C) \cup ((B - C) - A) \cup ((C - A) - B)$$
3. $A \oplus \varnothing = A$
4. $A \oplus A = \varnothing$

Theorem 1.1.4. For any sets A, B and C,

$$A \cap (B \oplus C) = (A \cap B) \ominus (A \cap C).$$

EXERCISE 1(A)

1. Express each of the following sets in the form $\{x : P(x)\}$ and specify the property $P(x)$.
   (i) The set of all rational numbers, whose denominators are not divisible by 5.
   (ii) The set of all integer multiples of 5 in between -96 and 96.
   (iii) The set of all points in the three-dimensional Euclidean space, whose distance from (0, 0) is a rational number.

(iv) The set of all pairs of real numbers, whose sum of their squares is nonzero.

(v) The set of all even primes.

(vi) The set of all subsets of $\{1, 2, 3, 4\}$ not containing 3.

2. Write explicitly all elements in each of the following sets

    (i) $\{a : a \text{ is an integer and } 0 \leq a^2 \leq 26\}$

    (ii) $\{\frac{r}{s} : r \text{ and } s \text{ are nonzero integers and } -1 < \frac{r}{s} < 1\}$

    (iii) $\{A : A \subseteq \{a, b, c\} \text{ and } b \notin A\}$

    (iv) The set of all three-digit positive integers, whose all the digits are even and are in strictly increasing order.

    (v) The set of all pairs of integers, whose sum of the squares is zero.

    (vi) The set of all integers, whose squares are in between 10 and 15.

3. Let $A = \{a \in \mathbb{Z}^+ : 3 \text{ divides } a\}$ and

    $B = \{a \in \mathbb{Z}^+ : \text{The sum of the digits in } a \text{ is divisible by } 3\}$.

    Prove that $A = B$.

4. Describe $\mathbb{P}(X)$ if $X = \{1, 2, 3\}$.

5. Let $X = \{a \in \mathbb{R} : -1 \leq a \leq 1\}$ and

    $Y = \{r \in \mathbb{R} : r = \sin t - \cos t \text{ for some } t \in \mathbb{R}\}$.

    Is $X = Y$?

6. Let $X = \{1, 2, 3, \ldots, 100\}$, $A = \{a \in X : a = b^2, b \in \mathbb{Z}\}$,

    $B = \{a \in X : a \text{ is odd}\}$ and for each $1 \leq i \leq 96$,

    $C_i = \{i, i + 1, i + 2, i + 3, i + 4\}$. Write explicitly all elements in each of the following sets.

    (i) $A \cap B$

    (ii) $A \cup B \cup C_2$

    (iii) $(\bigcup_{i=1}^{96} C_i) \cap A$

    (iv) $B \cap (\bigcup_{i=20}^{25} C_i)$

    (v) $X - (A \cup B)$

    (vi) $X - (\bigcup_{i=6}^{90} C_i)$

    (vii) $A - (\bigcup_{i=1}^{96} C_i)$

    (viii) $A - B$

7. For any two sets A and B, prove that

$$A = A \cap B \Leftrightarrow A \subseteq B \Leftrightarrow A \cup B = B.$$

8. Prove Theorem 1.1.1.

9. Prove Theorem 1.1.2.

10. Prove Theorem 1.1.3.

11. Prove Theorem 1.1.4.

12. Prove or disprove each of the following for any sets X and Y.
    (i) $\mathbb{P}(X \cap Y) = \mathbb{P}(X) \cap \mathbb{P}(Y)$
    (ii) $\mathbb{P}(X \cup Y) = \mathbb{P}(X) \cup \mathbb{P}(Y)$
    (iii) $\mathbb{P}(X - Y) = \mathbb{P}(X) - \mathbb{P}(Y)$
    (iv) $\mathbb{P}(X) = \mathbb{P}(Y) \Leftrightarrow X = Y$

1.2 RELATIONS AND FUNCTIONS

Consider the set A of all points in the plane and the set B of all straight lines in the plane. For any $x \in A$ and $L \in B$, let us write

$$x \; R \; L \text{ if } x \text{ lies on } L \text{ (or } L \text{ passes through } x).$$

Then, R is a relation between the elements of A and the elements of B. Here, $x\,R\,L$ can be read as 'x is related to L' and R denotes the relation 'lies on'. We can also consider R as the set of ordered pairs (x, L) such that x lies on L. This pair is ordered in the sense that x and L cannot be interchanged, because the first component of the pair is a point of A and the second component is a point of another set B and because the statement 'L lies on x' has no meaning. Therefore, we can consider R as a set of ordered pairs (x, L) satisfying the property that x lies on L. This concept is formalised in this section by introducing an abstract concept of a relation and by discussing the various properties of relations.

Definition 1.2.1. A pair of elements (not necessarily in the same set) written in a particular order is called an *ordered* pair and is written by listing its elements in a particular order, separated by a comma, and enclosing the pair in brackets. In the ordered pair (x, L), x is called the first component (or first coordinate) and L is called the second component (or second coordinate).

The ordered pairs (x, L) and (L, x) are different even though they consist of the same pair of elements. For example, the pairs $(2, 5)$ and $(5, 2)$ represent two different points in the plane.

Definition 1.2.2. Let A and B be any two sets. Then, the set of all ordered pairs (a, b) with $a \in A$ and $b \in B$ is called the *Cartesian product* of A and B and is denoted by $A \times B$. That is,

$$A \times B = \{(a, b) : a \in A \text{ and } b \in B\}.$$

Example 1.2.1. If $A = \{1, 2\}$ and $B = \{a, b, c\}$, then

$$A \times B = \{(1, a), (1, b), (1, c), (2, a), (2, b), (2, c)\}$$
$$A \times A = \{(1, 1), (1, 2), (2, 1), (2, 2)\}$$
$$B \times B = \{(a, a), (a, b), (a, c), (b, a), (b, b), (b, c), (c, a), (c, b), (c, c)\}.$$

Note 1.2.1. For any sets A and B,

$$A \times B = \varnothing \Leftrightarrow A = \varnothing \text{ or } B = \varnothing$$
$$A \times B = B \times A \Leftrightarrow A = B$$

Definition 1.2.3. For any sets $A_1, A_2, \ldots, A_n$, we define the Cartesian product of $A_1, A_2, \ldots, A_n$ as the set

$$A_1 \times A_2 \times \ldots \times A_n = \{(a_1, a_2, \ldots, a_n) : a_i \in A_i \text{ for all } 1 \le i \le n\}.$$

In particular, for any set A and for any positive integer n, we define

$$A^n = \{(a_1, a_2, \ldots, a_n) : a_i \in A \text{ for all } 1 \le i \le n\}.$$

Definition 1.2.4. Let A and B be any sets. Then, any subset of $A \times B$ is called a *relation* from A to B. For any relation R from A to B (that is, $R \subseteq A \times B$), if $(a, b) \in R$, then we say that '*a is R-related to b*' or '*a is related to b with respect to R*' or '*a and b have relation with R*' and is usually denoted by $a\ R\ b$.

Definition 1.2.5. Let A be any nonempty set. A relation from A to itself is called a '*binary relation on A*'.

Example 1.2.2. Let $\mathbb{Z}$ be the set of all integers and n a positive integer. Define

$$R = \{(a, b) \in \mathbb{Z} \times \mathbb{Z} : n \text{ divides } a - b\}$$
$$S = \{(a, b) \in \mathbb{Z} \times \mathbb{Z} : a = nb\}.$$

Then, both R and S are binary relations on $\mathbb{Z}$.

Definition 1.2.6. Let A, B, and C be sets, R be a relation from A to B and S be a relation from B to C, that is, $R \subseteq A \times B$ and $S \subseteq B \times C$. Define

$$S \circ R = \{(a, c) \in A \times C : a\ R\ b \text{ and } b\ S\ c \text{ for some } b \in B\}.$$

In other words, for any $a \in A$ and $c \in C$,

$$(a, c) \in S \circ R \Leftrightarrow \text{There exists } b \in B \text{ such that } (a, b) \in R \text{ and } (b, c) \in S.$$

Note that, when $S \circ R$ is defined, $R \circ S$ may not be defined. Even when $S \circ R$ and $R \circ S$ are both defined, they may not be equal. $S \circ R$ is called the *composition* of R with S.

Definition 1.2.7. For any relation R from a set A to a set B, the *inverse* of R is defined by

$$R^{-1} = \{(b, a) : (a, b) \in R\}.$$

Note that R^{-1} is a relation from B to A, $R \circ R^{-1}$ is a binary relation on B and $R^{-1} \circ R$ is a binary relation on A. It can be easily verified that $(R^{-1})^{-1} = R$ and $(S \circ R)^{-1} = R^{-1} \circ S^{-1}$ for any relation R from A to B and any relation S from B to C.

Definition 1.2.8. A relation R from a set A to a set B is called a *function (or a mapping) of A into B*, if for each $a \in A$, there exists unique $b \in B$ such that $(a, b) \in R$. Usually, functions are denoted by lowercase letters f, g, h, etc. If f is a function of A into B, then $f \subseteq A \times B$ satisfying the following conditions:

(i) For each $a \in A$, there exists $b \in B$ such that $(a, b) \in f$.

(ii) If $(a, b) \in f$ and $(a, b_1) \in f$, then $b = b_1$.

If f is a function of A into B and $(a, b) \in f$, then we write $a f b$ or $(a)f = b$ or $f(a) = b$. More popular convention is writing $f(a) = b$. This is reasonable, since b corresponds to a uniquely. In this case, b is called '*the image of a under f*' and a is called 'a pre-image of b under f'. We write simply $f: A \to B$, to denote that f is a function from A into B. If $f: A \to B$, any element a of A will have exactly one image $f(a)$ in B, while an element b of B may have any number of pre-images in A or may not have any pre-image at all. These circumstances lead to the following.

Definition 1.2.9. Let $f: A \to B$ be a function.

1. f is said to be an *injection* (or *a one-one function*) if each element of B has at most one pre-image in A; or, equivalently, for any $a_1, a_2 \in A$,

$$f(a_1) = f(a_2) \Rightarrow a_1 = a_2$$

2. f is said to be a *surjection* (or an *onto function*) if each element of B has atleast one pre-image in A; or, equivalently,

$$b \in B \Rightarrow f(a) = b \quad \text{for some } a \in A.$$

3. f is said to be a *bijection* (or *a one-to-one function*) if each element of B has exactly one pre-image in A, or, equivalently, f is both an injection as well as a surjection.

Note that, to describe a function $f : A \to B$, it is enough if we prescribe the image $f(a)$ of each element a in A.

Example 1.2.3. Let $\mathbb{Z}$ be the set of all integers.

1. Define $f : \mathbb{Z} \to \mathbb{Z}$ by $f(a) = a^2$ for any $a \in \mathbb{Z}$. Then, f is a function of $\mathbb{Z}$ into itself and f is neither an injection (since $f(-1) = 1 = f(1)$) nor a surjection (since there is no pre-image for -1).
2. Define $g : \mathbb{Z} \to \mathbb{Z}$ by $g(a) = 4a$ for any $a \in \mathbb{Z}$. Then, g is an injection, but not a surjection.
3. Define $h : \mathbb{Z} \to \mathbb{Z}$ by $h(a) = a + 2$ for any $a \in \mathbb{Z}$. Then, h is a bijection.
4. Let N be the set of all nonnegative integers and define $m : \mathbb{Z} \to N$ by $m(a) = |a|$ for any $a \in \mathbb{Z}$, where $|a| = a$ or $-a$ depending upon a is positive or not. Then, m is a surjection, but not an injection.

Let $f : A \to B$ be a function. Then, A is called '*the domain of f*' and is denoted by Dom(f) and B is called '*the codomain of f*' and is denoted by Codom(f). The set $\{f(a) : a \in A\}$ is called '*the image of f*' and is denoted by Im(f). Note that Im(f) is a subset of the codomain B and is not necessarily equal to B. In fact, f is a surjection if and only if Im(f) = Codom(f).

Definition 1.2.10. Let $f : A \to B$ and $g : B \to C$ be functions. Then, the composition $g \circ f$ is also a function from A to C. Recall from Definition 1.2.6 that $g \circ f$ is defined by

$$\begin{aligned} g \circ f &= \{(a, c) \in A \times C : (a, b) \in f \quad \text{and} \quad (b, c) \in g \quad \text{for some } b \in B\} \\ &= \{(a, c) \in A \times C : f(a) = b \quad \text{and} \quad g(b) = c\} \\ &= \{(a, c) \in A \times C : g(f(a)) = c\}. \end{aligned}$$

Therefore, $g \circ f : A \to C$ is a function defined by

$$(g \circ f)(a) = g(f(a)) \quad \text{for any } a \in A.$$

Note that $g \circ f$ is defined only when Codom(f) = Dom(g) or Im(f) $\subseteq$ Dom(g). In fact, we have

$$\text{Dom}(g \circ f) = \text{Dom}(f)$$

and

$$\text{Codom}(g \circ f) = \text{Codom}(g).$$

Two functions f and g are said to be equal if their domains are equal and $f(a) = g(a)$ for all the elements a in the common domain. For two functions f and g, both $f \circ g$ and $g \circ f$ may be defined but still they may not be equal, consider the following example.

Example 1.2.4. Define $f : \mathbb{Z} \to \mathbb{Z}$ and $g : \mathbb{Z} \to \mathbb{Z}$ by $f(a) = a + 1$ and $g(a) = a^2$ for any $a \in \mathbb{Z}$.

Then, $(f \circ g)(a) = f(g(a)) = f(a^2) = a^2 + 1$
and $(g \circ f)(a) = g(f(a)) = g(a + 1) = (a + 1)^2 = a^2 + 2a + 1$.

Therefore, $(f \circ g)(a) \neq (g \circ f)(a)$ for $0 \neq a \in \mathbb{Z}$ and hence $f \circ g$ and $g \circ f$ are not equal.

Note that, if f and g are injections (surjections, bijections), then so is $f \circ g$. Further, we have $(f \circ g) \circ h = f \circ (g \circ h)$ for any functions f, g and h, whenever the compositions are defined.

Definition 1.2.11. Let A be any nonempty set and define a function $I_A : A \to A$ by $I_A(a) = a$ for all $a \in A$. Then, I_A is called the '*identity function on A*'. I_A will also be denoted by Id_A or Id on X.
For any function $f : A \to B$, it can be seen that

$$f \circ I_A = f = I_B \circ f.$$

In the following, we give certain characterisation properties for injections, surjections and bijections.

Theorem 1.2.1. Let $f : A \to B$ be a function, then

1. f is an injection if and only if there exists a function $g : B \to A$ such that
$$g \circ f = I_A.$$
2. f is a surjection if and only if there exists a function $h : B \to A$ such that
$$f \circ h = I_B.$$

Theorem 1.2.2. A function $f : A \to B$ is a bijection if and only if there exists a function $g : B \to A$ such that

$$f \circ g = I_B \quad \text{and} \quad g \circ f = I_A.$$

In this case, g is unique and is called the *inverse* of f and is denoted by f^{-1}. Note that, for any $a \in A$ and $b \in B$,

$$f(a) = b \Leftrightarrow a = f^{-1}(b)$$

and that f^{-1} is also a bijection.

If $f: A \to B$ is a surjection, then usually we say that f is a surjection of A onto B, instead of A into B, just to mention that f is an onto function (or surjection).

Definition 1.2.3. Let A and B be two sets. A is said to be *equivalent* or *equipotent* with B if there exists a bijection of A onto B; in this case, we denote it by $A \simeq B$.

If $f: A \to B$ is a bijection, then $f^{-1}: B \to A$ is also a bijection and therefore we have

$$A \simeq B \Leftrightarrow B \simeq A.$$

Also, since the identity function $I_A : A \to A$ is a bijection, we have

$$A \simeq A \quad \text{for any set } A.$$

Further, if $f: A \to B$ and $g: B \to C$ are bijections, then $g \circ f: A \to C$ is also a bijection and therefore

$$A \simeq B \quad \text{and} \quad B \simeq C \Rightarrow A \simeq C.$$

Example 1.2.5. Let E be the set of all even integers and $\mathbb{Z}$ be the set of all integers. Then, $E = \{2a : a \in \mathbb{Z}\}$ and E is equivalent to $\mathbb{Z}$; for, define $f: E \to \mathbb{Z}$ by

$$f(a) = \begin{cases} 2b & \text{if } a = 4b \\ b & \text{if } a = 2b \text{ and } b \text{ is odd.} \end{cases}$$

Then, f is a bijection. Therefore $E \simeq \mathbb{Z}$.

Definition 1.2.12. Let $f: X \to Y$ be a function, $A \subseteq X$ and $B \subseteq Y$. The image of A under f is defined as the set

$$f(A) = \{f(a) : a \in A\}.$$

The *inverse image* of B under f is defined as the set

$$f^{-1}(B) = \{x \in X : f(x) \in B\}.$$

Then, clearly $f(A)$ is a subset of Y for all $A \subseteq X$ and $f^{-1}(B)$ is a subset of X for all $B \subseteq Y$. In other words, f induces a function from the power set $\mathbb{P}(X)$ into the power set $\mathbb{P}(Y)$ and another function from $\mathbb{P}(Y)$ into $\mathbb{P}(X)$. In this context, we have the following.

Theorem 1.2.3. The following holds good for any function $f: X \to Y$ and subsets A_1 and A_2 of X and B_1 and B_2 of Y.

1. $f(\emptyset) = \emptyset$ and $f(X) = \text{Im}(f)$
2. $f^{-1}(\emptyset) = \emptyset$ and $f^{-1}(Y) = X$
3. $f(A_1 \cup A_2) = f(A_1) \cup f(A_2)$

4. $f^{-1}(B_1 \cup B_2) = f^{-1}(B_1) \cup f^{-1}(B_2)$
5. $f^{-1}(B_1 \cap B_2) = f^{-1}(B_1) \cap f^{-1}(B_2)$
6. $f(A_1 \cap A_2) \subseteq f(A_1) \cap f(A_2)$
7. $A \subseteq f^{-1}(f(A))$ for any $A \subseteq X$
8. $f(f^{-1}(B)) \subseteq B$ for any $B \subseteq Y$.

Notice that there are only one-side inclusions in (6), (7) and (8). In general, these one-side inclusions cannot be replaced by the equality in these. In this context, we have the following.

Theorem 1.2.4. The following are equivalent to each other for any function $f: X \to Y$

1. f is an injection.
2. $A = f^{-1}(f(A))$ for any $A \subseteq X$.
3. $f(A_1 \cap A_2) = f(A_1) \cap f(A_2)$.

Theorem 1.2.5. A function $f : X \to Y$ is a surjection if and only if $f(f^{-1}(B)) = B$ for any subset B of Y.

Definition 1.2.13. Let $f: X \to Y$ be a function and $Z \subseteq X$. Then, $(Z \times Y) \cap f$ is a function of $\mathbb{Z}$ into Y and is called the *restriction of f to Z* and is denoted by $f|Z$. Note that

$$(f|Z)(a) = f(a) \quad \text{for any } a \in Z.$$

EXERCISE 1(B)

1. Prove each of the following for any sets A, B and C.
   (i) $(A \cup B) \times C = (A \times C) \cup (B \times C)$
   (ii) $(A \cap B) \times C = (A \times C) \cap (B \times C)$
   (iii) $(A - B) \times C = (A \times C) - (B \times C)$
   (iv) $A \times B = A \times C \Rightarrow A = \varnothing$ or $B = C$
2. In each of the following cases, find sets A, B, C and D to disprove the statement.
   (i) $A = B \Leftrightarrow A - C = B - C$
   (ii) $A = B \Leftrightarrow A \cap C = B \cap C$
   (iii) $A = B \Leftrightarrow A \cup C = B \cup C$
   (iv) $(A - B) \times (C - D) = (A \times C) - (B \times D)$
   (v) $(A \cup B) \times (C \cup D) = (A \times C) \cup (B \times D)$

3. Prove that $(A \cap B) \times (C \cap D) = (A \times C) \cap (B \times D)$ for any sets A, B, C and D.

4. If A has n elements and B has m elements, then prove that $A \times B$ has nm elements. Determine the number of relations from A to B.

5. State whether each of the following is a function and substantiate your answers.
   (i) $R = \{(a, b) \in \mathbb{Z} \times \mathbb{Z} : b^2 = a\}$
   (ii) $R = \{(a, b) \in \mathbb{Z} \times \mathbb{Z} : a^2 = b\}$
   (iii) $R = \{(a, b) \in \mathbb{R} \times \mathbb{R} : 2a^2 - 1 = b\}$
   (iv) $R = \{(a, b) \in \mathbb{R} \times \mathbb{R} : 2b^2 - 1 = a\}$
   (v) $R = \{(a, b) \in \mathbb{Q} \times \mathbb{Q} : a^2 + b^2 \text{ is an integer}\}$
   (vi) $R = \{(a, b) \in \mathbb{Z} \times \mathbb{Q} : 2b = a\}$.

6. If A has n elements and B has m elements, then determine the number of functions from A into B.

7. Prove that, for any relation R from A to B and any relation S from B to C, $R^{-1} \circ S^{-1} = (S \circ R)^{-1}$.

8. Prove that a function $f: X \rightarrow Y$ is an injection if and only if $f|Z$ is an injection for every subset Z of X. Is this true if we replace injection with surjection?

9. For any function $f: X \rightarrow Y$ and $A \subseteq B \subseteq X$, prove that $(f|B)|A = f|A$.

10. Let $\emptyset = A \subseteq X$ and $f: A \rightarrow Y$ be a function. Does there exist a function $g: X \rightarrow Y$ such that $g|A = f$? If yes, how many such functions g can be found?

11. Let $f: A \rightarrow B$ and $g: B \rightarrow C$ be the functions. If f and g are bijections, prove that $g \circ f$ is a bijection. Is the converse true? Substantiate your answer.

12. If $f: \mathbb{R} \rightarrow \mathbb{R}$ is defined by $f(a) = a^2$ for all $a \in \mathbb{R}$, then determine $f^{-1}[-2, 8)$, $f^{-1}(-\infty, 0]$, $f^{-1}(-1, 1)$ and $f^{-1}(\mathbb{Z})$.

13. For any real number a, let $[a]$ be the largest integer less than or equal to a and define $f: \mathbb{R} \rightarrow \mathbb{Z}$ by $f(a) = [a]$ for any $a \in \mathbb{R}$.

    Then, determine the following sets
   (i) $f((-1, 1))$
   (ii) $f([-1, 1])$
   (iii) $f^{-1}(\{0, 1\})$
   (iv) $f^{-1}(E)$, where E is the set of even integers.
   (v) $f^{-1}\left(-\frac{2}{3}, \frac{2}{3}\right)$
   (vi) $f^{-1}(\{0\})$

14. Prove Theorem 1.2.1.

15. Prove Theorem 1.2.2.

16. Prove Theorem 1.2.3.

17. Give an example of a function $f: X \to Y$ and a subset A of X for which A is properly contained in $f^{-1}(f(A))$.

18. Give an example of a function $f: X \to Y$ and a subset B of Y such that $f(f^{-1}(B))$ is properly contained in B.

19. If X is an n-element set and Y is an m-element set, how many bijections can be there from X onto Y?

20. Let $f: X \to Y$ be a function, $\varnothing \neq A \subseteq X$ and $\varnothing \neq B \subseteq Y$. Then, prove the following.
    (i) $f(f^{-1}(f(A))) = f(A)$
    (ii) $f^{-1}(f(f^{-1}(B))) = f^{-1}(B)$
    (iii) $(f|A)^{-1}(B) = f^{-1}(B) \cap A$.

21. Let $f: X \to Y$ and $g: Y \to X$ be mappings, such that $g \circ f = I_A$. Prove that the following are equivalent to each other.
    (i) f is a surjection.
    (ii) g is an injection.
    (iii) f is a bijection.
    (iv) g is a bijection.

    and that, in this case, $f \circ g = I_B$.

22. Let $f: X \to Y$ and $g: Y \to Z$ be mappings, such that $g \circ f$ is an injection. Then, prove that f is an injection and that, when f is a surjection also, then g is an injection.

23. Let $\mathscr{C}[0, 1]$ be the set of all real-valued continuous functions defined on the closed interval $[0, 1]$ and $\mathscr{C}'[0, 1]$ be the set of all differentiable functions f in $\mathscr{C}[0, 1]$, such that $f(0) = 0$ and the derivative f' is continuous. Prove that the function

$$D : \mathscr{C}'[0, 1] \to \mathscr{C}[0, 1] \text{ defined by } D(f) = f'$$

    is a bijection.

24. Let A be an n-element set and B be an m-element set. Find the number of injections of A into B in each of the following cases.
    (i) $n = m$, (ii) $n > m$, (iii) $n < m$.

25. Define $f: \mathbb{Z}^+ \to \mathbb{Z}^+$ by $f(a) = 2a - 1$. Prove that there exist infinitely many functions $g: \mathbb{Z}^+ \to \mathbb{Z}^+$, such that $g \circ f = I_{\mathbb{Z}^+}$ and there is no function $h: \mathbb{Z}^+ \to \mathbb{Z}^+$ such that $f \circ g = I_{\mathbb{Z}^+}$.

26. Prove Theorem 1.2.4.

27. Prove Theorem 1.2.5.

28. Let $f_1, f_2, \ldots, f_n$ be bijections, such that $f_1 \circ f_2 \circ \ldots \circ f_n$ is defined. Then, prove that $f_n^{-1} \circ f_{n-1}^{-1} \circ \cdots \circ f_1^{-1}$ is defined and is equal to $(f_1 \circ f_2 \circ \ldots \circ f_n)^{-1}$.

29. Let $f: X \to Y$ and $g: Y \to X$ be functions, such that $g \circ f$ is an injection and $f \circ g$ is a surjection. Then, prove that both $g \circ f$ and $f \circ g$ are bijections.

30. Let $f: X \to Y$ be a function. If $g: Y \to X$ is a function, such that $g \circ f = I_X$ ($f \circ g = I_Y$), then g is called a left (respectively, right) inverse of f. Prove that the following are equivalent to each other.
    (i) f has a unique left inverse.
    (ii) f is a bijection.
    (iii) f has a unique right inverse.

31. Let n and m be positive integers greater than 1, such that n and m have no common factors except 1. Let I_n be the set $\{1, 2, \ldots, n\}$. Prove that there is a bijection $f: I_{n+m} \to I_{n+m}$ such that $f(n+m) = n+m$ and $f(i+1) - f(i) \in \{n, -m\}$ for all $1 \le i < m+n$.

32. Let X be a nonempty set. Prove that $f \circ g = g \circ f$ for all bijections f and g of X onto itself if and only if X has almost two elements in X.

33. Let $f: X \to Y$ be a function and $F: \mathbb{P}(Y) \to \mathbb{P}(X)$ be defined by $F(A) = f^{-1}(A)$ for all $A \subseteq Y$. Then, prove the following.
    (i) f is injection if and only if F is surjection.
    (ii) f is surjection if and only if F is injection.
    (iii) f is a bijection if and only if F is a bijection.

34. Let X be a set and define $f: \mathbb{P}(\mathbb{P}(X)) \to \mathbb{P}(X)$ by

$$f(\mathscr{C}) = \bigcup_{A \in \mathscr{C}} A \quad \text{for any } \mathscr{C} \subseteq \mathbb{P}(X).$$

Then, find two distinct right inverses of f.

35. Define $f: \mathbb{Z}^+ \times \mathbb{Z}^+ \to \mathbb{Z}^+$ by $f(a, b) = 2^{a-1}(2b-1)$. Then, prove that f is a bijection and find f^{-1}.

36. Let X and Y be nonempty sets and Y^X be the set of all functions of X into Y. Prove the following for any $\varnothing \neq A \subseteq X$.
    (i) The function $\eta: Y^X \to Y^A$, defined by $\eta(f) = f/A$, is a surjection.
    (ii) If Y has atleast two elements, then η is a bijection $\Leftrightarrow A = X$.

37. Let X be any set and $2 = \{0, 1\}$. Prove that the map $\chi: \mathbb{P}(X) \to 2^X$, defined by

$$\chi(A)(x) = \begin{cases} 1 & \text{if } x \in A \\ 0 & \text{if } x \notin A \end{cases}$$

for any $A \in \mathbb{P}(X)$ and $x \in X$, is a bijection.

38. Deduce from the above that if A is an n-element set, then $\mathbb{P}(A)$ is a 2^n-element set.

1.3 EQUIVALENCE RELATIONS AND PARTITIONS

Dividing a set into disjoint subsets is called a partitioning of the set. In this section, we discuss a special type of binary relations on a set which induce partitions of the set.

Definition 1.3.1. Let S be any nonempty set and R, binary relation on S.

1. R is said to be *reflexive on S* if $(a, a) \in R$ for all $a \in S$.
2. R is said to be *symmetric* if $(a, b) \in R \Rightarrow (b, a) \in R$.
3. R is said to be *transitive* if $(a, b) \in R$ and $(b, c) \in R \Rightarrow (a, c) \in R$.
4. R is said to be an *equivalence relation on S*, if it is reflexive on S, symmetric and transitive.

Example 1.3.1

1. Let X be any nonempty set and

$$\Delta_X = \{(x, x) : x \in X\}.$$

Then Δ_X is an equivalence relation on S and is called the *diagonal* on X. Δ_X can also be defined as

$$\Delta_X = \{(x, y) \in X \times X : x = y\}.$$

2. For any set X, the whole of $X \times X$ is an equivalence relation on X.
3. For any positive integer n, let

$$R_n = \{(a, b) \in \mathbb{Z} \times \mathbb{Z} : n \text{ divides } a - b\}.$$

Then, R_n is an equivalence relation on $\mathbb{Z}$.

4. Let $R = \{(a, b) \in \mathbb{Z} \times \mathbb{Z} : a = 0 = b \text{ or } ab > 0\}$.

Then, R is an equivalence relation on $\mathbb{Z}$.

Definition 1.3.2. Let R be an equivalence relation on a set X and $x \in X$. The *R-equivalence class of x* (or simply, the *R-class* of x) is defined to be the set

$$R(x) = \{y \in X : (x, y) \in R\}.$$

The following can be proved easily.

Theorem 1.3.1. Let R be an equivalence relation on a set X. Then, the following holds good.

1. For any x and $y \in X$,

$$(x, y) \in R \Leftrightarrow R(x) = R(y)$$

and $(x, y) \notin R \Leftrightarrow R(x) \cap R(y) = \emptyset$.

2. Any two R-equivalence classes in X are either equal or disjoint.
3. $\bigcup_{x \in X} R(x) = X$.

Definition 1.3.3. Let X be a nonempty set. A class $\mathscr{C}$ of sets is said to be a *partition* of X if the following conditions are satisfied:

1. Each member of $\mathscr{C}$ is a nonempty subset of X.
2. For any $A, B \in \mathscr{C}$, either $A = B$ or $A \cap B = \varnothing$.
3. The union of all the members of $\mathscr{C}$ is X.

In other words, a class $\mathscr{C}$ of nonempty subsets of a set X is called a partition of X if each element in X is in exactly one member of $\mathscr{C}$.

The following is an immediate consequence of Theorem 1.3.1.

Theorem 1.3.2. For any equivalence relation R on a set X, the class of all R-equivalence classes in X is a partition of X and is denoted by X/R; that is,

$$X/_R = \{R(x) : x \in X\}.$$

$X/_R$ is called the partition on X induced by R or the *quotient* of X by R.

The converse of the above result is also true, in the sense that, for any partition $\mathscr{C}$ of X, there exists an equivalence relation $R_{\mathscr{C}}$ on X such that the partition of X induced by $R_{\mathscr{C}}$ is precisely equal to the given partition $\mathscr{C}$.

Theorem 1.3.3. Let $\mathscr{C}$ be a partition of a nonempty set X. Define

$R = \{(x, y) \in X \times X : \text{both } x \text{ and } y \text{ belong to the same member of } \mathscr{C}\}$.

Then, R is an equivalence relation on X and $X/R = \mathscr{C}$. (In fact, if $x \in A \in \mathscr{C}$, then $A = R(x)$.)

These two processes $R \mapsto X/R$ and $\mathscr{C} \mapsto R_{\mathscr{C}}$ are inverses to each other in the sense that

$$R \mapsto X/R \mapsto R_{X/R} = R \quad \text{and} \quad \mathscr{C} \mapsto R_{\mathscr{C}} \mapsto X/R_{\mathscr{C}} = \mathscr{C}$$

for any equivalence relation R on X and for any partition $\mathscr{C}$ of X. Therefore, we have the following.

Theorem 1.3.4. For any nonempty set X, let $\xi(X)$ be the set of all equivalence relations on X and $\text{Part}(X)$ be the set of all partitions of X. Then,

$$\xi(X) \simeq \text{Part}(X),$$

that is, there is a bijection of $\xi(X)$ onto $\text{Part}(X)$.

Example 1.3.2. Consider the relation R given in Example 1.3.1 (4), we have

$$R = \{(a, b) \in \mathbb{Z} \times \mathbb{Z} : a = 0 = b \text{ or } ab > 0\}.$$

For any $a \in \mathbb{Z}$, $R(a) = \{0\}$ or $\mathbb{Z}^+$ or $\mathbb{Z}^-$ according as $a = 0$ or $a > 0$ or $a < 0$, respectively, where $\mathbb{Z}^-$ stands for the set of all negative integers. Therefore,

$$X/_R = \{\{0\}, \mathbb{Z}^+, \mathbb{Z}^-\}.$$

Example 1.3.3. Let n be a positive integer and R_n be the equivalence relation on $\mathbb{Z}$ defined in Example 1.3.1 (3); that is,

$$R_n = \{(a, b) \in \mathbb{Z} \times \mathbb{Z} : n \text{ divides } a - b\}.$$

Then, for any $a \in \mathbb{Z}$, the R-class of a is given by

$$R_n(a) = \{a + nx : x \in \mathbb{Z}\}.$$

If $a = qn + r$, $q, r \in \mathbb{Z}$ and $0 \le r < n$, then $R_n(a) = R_n(r)$ and hence $R_n(0)$, $R_n(1)$, …, $R_n(n-1)$ are all the distinct R_n-classes in $\mathbb{Z}$. That is, there are exactly n R_n-classes in $\mathbb{Z}$.

Definition 1.3.4. Let $\mathscr{C}_1$ and $\mathscr{C}_2$ be two partitions on a set X. Then, $\mathscr{C}_2$ is said to be a *refinement* of $\mathscr{C}_1$ if every member of $\mathscr{C}_2$ is a union of members of $\mathscr{C}_1$.

Theorem 1.3.5. Let R and S be two equivalence relations on a set X and $X/_R$ and $X/_S$ be partitions corresponding to R and S, respectively. Then, $R \subseteq S$ if and only if $X/_S$ is a refinement of $X/_R$.

Proof: Suppose that $R \subseteq S$, then, for any $x, y \in X$,

$$(x, y) \in R \Rightarrow (x, y) \in S$$

and hence $R(x) \subseteq S(x)$ for all $x \in X$. It can be seen that $S(x) = \bigcup_{y \in S(x)} R(y)$. Therefore, $X/_S$ is a refinement of $X/_R$. Conversely suppose that $X/_S$ is a refinement of $X/_R$. Let $(x, y) \in R$. Then, $S(x)$ is a member of $X/_S$ and hence $S(x)$ is a union of members of $X/_R$. Therefore,

$$S(x) = \bigcup_{z \in Z} R(z) \quad \text{for some } Z \subseteq X.$$

Now, since $x \in S(x)$, we get that $x \in R(z)$ for some $z \in Z$ and hence $(x, z) \in R$. Since $(x, y) \in R$ also, we have that $(y, z) \in R$ so that $y \in R(z) \subseteq S(x)$. Therefore, $(x, y) \in S$. Thus, $R \subseteq S$. ◀

The following theorem is a simple verification.

Theorem 1.3.6. The intersection of any class of equivalence relations on a set X is again an equivalence relation on X. In particular, if R and S are equivalence relations on X and X/R and X/S are the corresponding partitions of X, then $R \cap S$ is also an equivalence relation whose corresponding partition is $\{R(x) \cap S(x) \mid x \in X\}$. In other words,

$$(R \cap S)(x) = R(x) \cap S(x) \quad \text{for any } x \in X.$$

These can be better understood by the following figures showing partitions of R, S and $R \cap S$.

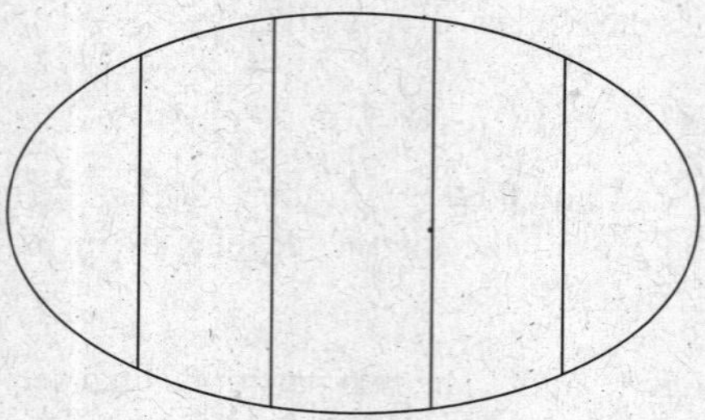

X/R

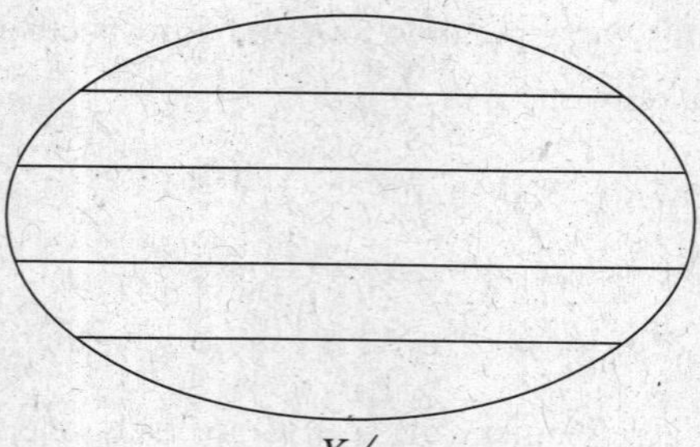

X/S

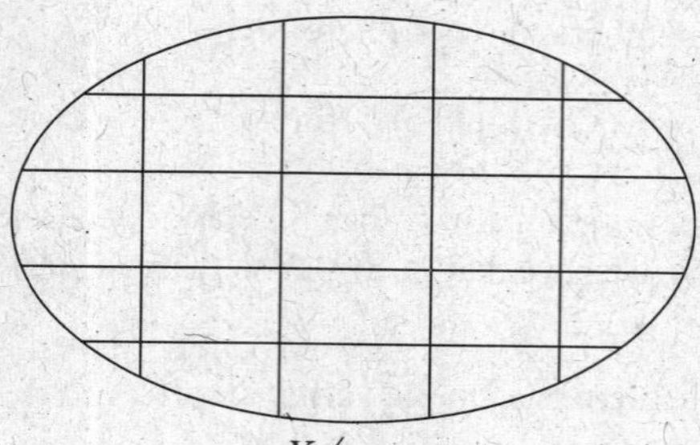

$X/R \cap S$

Recall that a binary relation R on a set X is an equivalence relation on X if and only if $R = R^{-1}$, $R \circ R \subseteq R$ and $\Delta_X \subseteq R$, where Δ_X is the diagonal of X; in fact, in this case, $R \circ R = R$. In general, for any equivalence relations R and S on a set X, the composition $R \circ S$ may not be an equivalence relation. In this context, we have the following.

Theorem 1.3.7. Let R and S be equivalence relations on a set X. Then, the following are equivalent to each other:

1. $R \circ S$ is an equivalence relation on X.
2. $R \circ S$ is symmetric.
3. $R \circ S$ is transitive.
4. $R \circ S \subseteq S \circ R$
5. $S \circ R \subseteq R \circ S$
6. $R \circ S = S \circ R$
7. $S \circ R$ is symmetric.
8. $S \circ R$ is transitive.

Theorem 1.3.8 (Fundamental theorem of functions). Any function $f: X \to Y$ can be expressed as

$$f = g \circ h$$

for some injection g and some surjection h.

Proof: Let $f: X \to Y$ be a function. Define

$$R = \{(a, b) \in X \times X : f(a) = f(b)\}.$$

Then, R is an equivalence relation on X. Consider the partition $X/R = \{R(x) : x \in X\}$ and define

$$h : X \to X/R \text{ by } h(x) = R(x) \quad \text{for any } x \in X.$$

Also, define
$g : X/R \to Y$ by $g(R(x)) = f(x)$. If $R(x) = R(x')$, then $(x, x') \in R$ and hence $f(x) = f(x')$. Therefore, g is a well-defined function and clearly g is an injection. Also, it is clear that $g(h(x)) = f(x)$ for all $x \in X$. Thus, $f = g \circ h$, g is an injection and h is a surjection. ◀

EXERCISE 1(C)

1. Which of the following are equivalence relations?
   (i) $\{(a, b) \in \mathbb{R} \times \mathbb{R} : a - b \text{ is a rational number}\}$
   (ii) $\{(a, b) \in \mathbb{Q} \times \mathbb{Q} : a - b \text{ is an integer}\}$
   (iii) Let $\mathbb{Z}^* = \mathbb{Z} - \{0\}$ and
   $R = \{(a, b) \in \mathbb{Z}^* \times \mathbb{Z}^*: b = 2^n a \text{ for some } n \in \mathbb{Z}\}$
   (iv) $\{(a, b), (c, d) \in \mathbb{R}^2 \times \mathbb{R}^2 : a^2 + b^2 = c^2 + d^2\}$
   (v) $\{(a, b) \in \mathbb{Z} \times \mathbb{Z} : a = nb \text{ for some } n \in \mathbb{Z}\}$
   (vi) $\{(a, b) \in \mathbb{Z}^+ \times \mathbb{Z}^+ : n \text{ divides both } a \text{ and } b \text{ for some } 1 < n \in \mathbb{Z}\}$
   (vii) $\{(a, b) \in \mathbb{R} \times \mathbb{R} : ab \text{ is a rational number}\}$
   (vii) Let $M(\mathbb{R})$ be the set of all mappings of $\mathbb{R}$ into $\mathbb{R}$ and
   $R = \{(f, g) \in M(\mathbb{R}) \times M(\mathbb{R}) : f(a) = g(a) \text{ for some } a \in \mathbb{R}\}$
   (ix) For any set X,
   $$\{(A, B) \in \mathbb{P}(X) \times \mathbb{P}(X) : A \oplus B \text{ is finite}\}$$
   (x) Let $\mathbb{R}^* = \mathbb{R} - \{0\}$: and
   $R = \{(a, b) \in \mathbb{R}^* \times \mathbb{R}^* : 0 < ab^{-1} \in \mathbb{Q}\}$.

2. Give three examples of binary relations showing that a relation can satisfy any one of reflexivity, symmetricity and transitivity without satisfying the other two.

3. Prove that reflexivity, symmetricity and transitivity of a relation are independent in the sense that no two of them imply the other.

4. If R is an equivalence relation on a set X and $\phi \neq Y \subseteq X$, then prove that $R \cap (Y \times Y)$ is an equivalence relation on R.

5. Let $X = \mathbb{Z} \times (\mathbb{Z} - \{0\})$ and
$$\mathbb{R} = \{((a, b), (c, d)) \in X \times X: ad = bc\}.$$
Prove that R is an equivalence relation on X.

6. Let $X = \mathbb{Z}^+ \times \mathbb{Z}^+$ and $R = \{((a, b), (c, d)) \in X \times X : a + d = b + c\}$.
Prove that R is an equivalence relation on X.

7. Describe the partitions corresponding to each of the equivalence relations given in Exercises 5 and 6 above.

8. Prove Theorem 1.3.1.

9. Prove Theorem 1.3.2.

10. Prove Theorems 1.3.3 and 1.3.4 and apply these to the relations given in Exercise 5 above.

11. Prove Theorems 1.3.5 and 1.3.6.

12. Let R be a binary relation on a nonempty set X. Then, prove that R is an equivalence relation on X if and only if R is reflexive on X and

$$(a, b) \in R \quad \text{and} \quad (b, c) \in R \Rightarrow (c, a) \in R.$$

13. Describe the equivalence relations on $\mathbb{Z}$ corresponding to the following partitions of $\mathbb{Z}$:
    (i) $\{\ldots, -5, -1, 3, 7, \ldots\}, \{\ldots, -6, -2, 2, 6, \ldots\}$,
    $\{\ldots, -7, -3, 1, 5, \ldots\}, \{\ldots, -8, -4, 0, 4, \ldots\}$
    (ii) $\{2n : n \in \mathbb{Z}\}, \{2n + 1 : n \in \mathbb{Z}\}$
    (iii) $\mathbb{Z}^-, \{0\}, \mathbb{Z}^+$
    (iv) $\{\ldots, -3, 0, 3, 6, \ldots\}, \{\ldots, -2, 1, 4, 7, \ldots\}, \{\ldots, -1, 2, 5, 8, \ldots\}$.

14. Describe the partitions corresponding to the equivalence relations given in Exercise 1 above.

15. Prove Theorem 1.3.7.

1.4 THE CARDINALITY OF A SET

The concepts of cardinality of a set and of cardinal number are very important in the abstract study of any branch of mathematics and, in particular, in the study of abstract algebra. In this section, we give a brief introduction of these concepts.

Definition 1.4.1. For any set X, let $|X|$ denote the class of all sets that are equivalent to X (that is, bijective with X). Then, $|X|$ is called the *cardinality* of X or the *cardinal number* of X or, simply, a cardinal number.

If we define, for any two sets A and B, $A \simeq B$ whenever there is a bijection of A onto B, then $\simeq$ is actually an equivalence relation on the class of all sets. The following is a direct consequence of the discussion made after Definition 1.2.3.

Theorem 1.4.1. Let A, B and C be any sets. Then, the following holds good.

1. $|A| = |B| \Leftrightarrow A \simeq B \Leftrightarrow A \in |B| \Leftrightarrow B \in |A|$
2. $A \in |B|$ and $B \in |C| \Rightarrow A \in |C|$.

Definition 1.4.2. For any nonnegative integer n, let I_n be the set of positive integers less than or equal to n. That is,

$$I_n = \{1, 2, 3, \ldots, n\}.$$

Note that, if $n = 0$, then I_n is the empty set.

Theorem 1.4.2. The following are equivalent to each other for any nonnegative integers n and m.

1. $|I_n| = |I_m|$
2. $I_n \simeq I_m$
3. $n = m$

In view of the above theorem, we denote the cardinality of I_n by simply n. Note that, for any set A, $|A| = n$ if and only if there is a bijection of A onto the set $\{1, 2, \ldots, n\}$ and, for this reason, we say that A has n-elements or A is an n-element set if $|A| = n$.

Definition 1.4.3. A set A is called a *finite set* if the cardinality of A is a nonnegative integer. A is called an *infinite set* if it is not a finite set.

In other words, a set A is called finite if A is bijective with the set I_n for some nonnegative integer n. A is called infinite if it is not bijective with I_n for any nonnegative integer n.

Definition 1.4.4. A cardinal number is said to be *finite* if any (and hence all) of its members are finite sets.

Example 1.4.1. The set $\mathbb{Z}^+$ of positive integers is an infinite set, for we can easily check that there cannot be a bijection of $\mathbb{Z}^+$ onto I_n for any nonnegative integer n. If $f: I_n \to \mathbb{Z}^+$ is a function, we can choose $m \in \mathbb{Z}^+$ such that $f(a) < m$ for all $a \in I_n$.

Theorem 1.4.3. Let n be a nonnegative integer and X be a set, such that $|X| = n$. Then, for any subset Y of X, $|Y| = m$ for some $0 \leq m \leq n$.

Corollary 1.4.1. Every subset of a finite set is finite. Equivalently, any superset of an infinite set is infinite.

We can identify any nonnegative integer n with the cardinal number $|A|$, where A is a set with n elements. It can be easily seen that, for any nonnegative integers n and m, $n \leq m$ if and only if there is an injection of A into B, where A and B are sets of cardinalities n and m, respectively. This suggests an extension of the usual ordering $\leq$ on the set $\mathbb{N}$ of nonnegative integers to that of cardinal numbers.

Definition 1.4.5. Let α and β be two cardinal numbers and X and Y be sets, such that $|X| = \alpha$ and $|Y| = \beta$. Then, we define α is less than or equal to β (and express this by $\alpha \leq \beta$) if there is an injection of X into Y.

First of all, we have to prove that $\leq$ is a well-defined relation on the cardinals, in the sense of the following.

Theorem 1.4.4. Let X, Y, A and B be sets, such that $|X| = |A|$ and $|Y| = |B|$. Then, there is an injection of X into Y if and only if there is an injection of A into B.

Proof: Since $|X| = |A|$ and $|Y| = |B|$, there are bijections $f : X \to A$ and $g : Y \to B$. If $h : X \to Y$ is an injection, then $g \circ h \circ f^{-1}$ is an injection of A into B. On the other hand, if $p : A \to B$ is an injection, then $g^{-1} \circ p \circ f$ is an injection of X into Y. ◀

Thus, $\leq$ is a well-defined binary operation on the set of cardinals. Since $A \cong A$ for any set A, it follows that $\leq$ is reflexive on the set of cardinals.

Also, since the composition of injections is again an injection, we have that $\leq$ is a transitive relation. In addition to the reflexivity and transitivity of the relation $\leq$, we have another important property, namely the anti-symmetricity; that is, $\alpha \leq \beta$ and $\beta \leq \alpha$ are possible only if $\alpha = \beta$. The proof of this is not that straight forward and requires a skilled proof.

Theorem 1.4.5. (Schroeder–Bernstein Theorem). Let X and Y be sets and $f : X \to Y$ and $g : Y \to X$ be injections, then there exists a bijection of X onto Y.

Proof: Put $Z = g(Y)$. Then, Z is a subset of X.

Define $h : X \to Z$ by $h(x) = g(f(x))$ for any $x \in X$. Then, since g and f are injections, h is also an injection. We define sequences $\{X_n\}$ and $\{Z_n\}$ of sets as follows:

$$X_1 = X \quad \text{and} \quad Z_1 = Z$$

and, for $n > 1$, $X_n = h(X_{n-1})$ and $Z_n = h(Z_{n-1})$. Then, $X_n = h^{n-1}(X)$ and $Z_n = h^{n-1}(Z)$, where $h^{n-1} = h \circ h \circ \ldots \circ h$ ($n - 1$ times) and $h^0 = Id_X$. We have

$$X = X_1 \supseteq Z_1 \supseteq X_2 \supseteq Z_2 \supseteq X_3 \supseteq Z_3 \supseteq X_4 \supseteq \ldots$$

Define $p : X \to Z$ by

$$p(x) = \begin{cases} h(x) & \text{if } x \in X_n - Z_n \text{ for some } n \\ x & \text{otherwise} \end{cases}$$

Then, it can be easily verified that p is a bijection of X onto Z. Now, define $q : X \to Y$ by

$$q(x) = y \quad \text{if } g(y) = p(x).$$

Since g is an injection and $p(x) \in Z = g(Y)$, there will be unique $y \in Y$ such that $p(x) = g(y)$. Therefore, q is a well-defined mapping of X into Y. It can be verified that q is a bijection of X onto Y. ◀

Corollary 1.4.2. The relation $\leq$ is an anti-symmetric, transitive and reflexive relation on the set of cardinals.

One can also define a relation $\geq$ on the set of cardinals as $|Y| \geq |X|$ if there is a surjection of Y onto X. This is also an antisymmetric, transitive and reflexive relation on the set of cardinals. In fact, this is precisely the relation $\leq$ in view of the following.

Theorem 1.4.6. Let X and Y be any nonempty sets. Then, there is an injection of X into Y if and only if there is a surjection of Y onto X.

Proof: Suppose that $f : X \to Y$ is an injection, choose an arbitrary element $x_0 \in X$. Define $g : Y \to X$ by

$$g(y) = \begin{cases} x & \text{if } f(x) = y \\ x_0 & \text{otherwise (that is, } y \notin f(X)) \end{cases}.$$

Since f is an injection, for each $y \in f(X)$, there exists unique $x \in X$ such that $f(x) = y$. Therefore, g is a well-defined function of Y into X. Also, for any $x \in X$, $f(x) \in Y$ and $g(f(x)) = x$ and hence g is a surjection of Y onto X.
Conversely suppose there is a surjection $g : Y \to X$. For each $x \in X$, consider the set

$$A_x = g^{-1}(\{x\}) = \{y \in Y : g(y) = x\}.$$

Since g is a surjection, each A_x, $x \in X$, is a nonempty subset of Y. Therefore, $\{A_x : x \in X\}$ is a nonempty class of nonempty sets. By an important axiom of set theory (known as the *axiom of choice*), there exists a function $c : X \to \cup_{x \in X} A_x$ such that $c(x) \in A_x$ for each $x \in X$ (such a function is called a choice function). Now, define $f : X \to Y$ by $f(x) = c(x)$.
Note that $A_x \cap A_{x'} = \emptyset$ for any $x \neq x' \in X$ and hence $c(x) \neq c(x')$ if $x \neq x'$. Therefore, f is an injection of X into Y. ◀

If A is a finite set and $|A| = n$, then we know that its power set $\mathbb{P}(A)$ (the set of all subsets of A) is of cardinality 2^n and that $n < 2^n$. We can extend this to arbitrary cardinals. First let us have the following.

Definition 1.4.6. If α is the cardinal of a set A, then the cardinal of the power set $\mathbb{P}(A)$ is denoted by 2^α, for the simple reason that $\mathbb{P}(A) \simeq \{0,1\}^A$ under the bijection $B \mapsto \chi_B$, where χ_B is defined by $\chi_B(a) = 1$ or 0 according as $a \in B$ or $a \notin B$. χ_B is called the characteristic map of B.

For any two cardinals α and β, we write $\alpha < \beta$ if $\alpha \leq \beta$ and $\alpha \neq \beta$. Also write $\alpha \nleq \beta$ to say that α is not less than or equal to β.

Theorem 1.4.7. For any cardinal α, $\alpha < 2^\alpha$.

Proof: Let α be a cardinal and A be a set such that $|A| = \alpha$. Since the cardinal of the empty set Ø is 0 and $\mathbb{P}(\emptyset) = \{\emptyset\}$ which is a nonempty set, we get that $|\emptyset| = 0 < 1 = 2^0 = |\mathbb{P}(\emptyset)|$ therefore, we can suppose that A is a nonempty set. Define

$$f: A \to \mathbb{P}(A) \text{ by } f(a) = \{a\} \quad \text{for any } a \in A.$$

Then, clearly f is an injection and hence

$$\alpha = |A| \leq \mathbb{P}(A)| = 2^\alpha$$

Now, we prove that $|A| \neq |\mathbb{P}(A)|$ or, equivalently, $|\mathbb{P}(A)| \nleq |A|$. By Theorem 1.4.6, it is enough if we can prove that there is no surjection of A onto $\mathbb{P}(A)$. Suppose, if possible, that there is a surjection $g : A \to \mathbb{P}(A)$. Then, for each $a \in A$, $g(a)$ is a subset of A and every subset of A is of the form $g(a)$ for some $a \in A$ (since g is a surjection). Now, consider the set B defined by

$$B = \{a \in A : a \notin g(a)\}.$$

Then, B is a subset of A and hence $B = g(a)$ for some $a \in A$. Now,

$$a \in B \Rightarrow a \in g(a) \Rightarrow a \notin B$$
$$\text{and } a \notin B \Rightarrow a \notin g(a) \Rightarrow a \in B,$$

which are contradictions, since exactly one of the statements $a \in B$ and $a \notin B$ must be valid. Therefore, there is no surjection of A onto $\mathbb{P}(A)$ and hence $|\mathbb{P}(A)| \nleq |A|$. Thus, $\alpha = |A| < |\mathbb{P}(A)| = 2^\alpha$. ◀

Next, we have a brief discussion on countable cardinals.

Definition 1.4.7. Let X be any set and $\mathbb{Z}^+$ be the set of positive integers. Then, X is said to be a *countable set* and $|X|$ said to be a *countable cardinal* if $|X| = |\mathbb{Z}^+|$; that is, if X is equipotent with $\mathbb{Z}^+$ and if $f: \mathbb{Z}^+ \to X$ is bijection, then X can be expressed as $X = \{f(1), f(2), \ldots, f(n), \ldots\}$ or, simply $X = \{x_1, x_2, \ldots\}$. If X is not a countable set, then X is called an *uncountable set* and $|X|$ is called an *uncountable cardinal.*

Definition 1.4.8. A set X is called *at most countable* if it is either finite or countable.

Theorem 1.4.8. The following are equivalent to each other for any non-empty set X:

1. There is an injection $f: X \to \mathbb{Z}^+$
2. X is at most countable.
3. X is a subset of a countable set.
4. There is a surjection $g : \mathbb{Z}^+ \to X$.

Proof: (1) $\Rightarrow$ (2): Suppose that there is an injection $f: X \to \mathbb{Z}^+$, put $Y = f(X)$. Then, $X \simeq Y \subseteq \mathbb{Z}^+$. Suppose that X is not finite, then Y is an infinite subset of $\mathbb{Z}^+$. Define $g : \mathbb{Z}^+ \to Y$ as follows.
Let $g(1)$ be the least element in Y (use the well-ordering principle in $\mathbb{Z}^+$). Having defined $g(1), \ldots, g(n-1)$, let $g(n)$ be the least element in $Y - \{g(1), g(2), \ldots, g(n-1)\}$, for any $n > 1$. Since Y is infinite, $Y - \{g(1), \ldots, g(n-1)\} \neq \emptyset$ for any $n > 1$ and hence g is welldefined. Now, we have

$$g(1) < g(2) < \ldots < g(n) < \ldots$$

Clearly, g is an injection of $\mathbb{Z}^+$ into Y. We prove that g is surjection also. Let $y \in Y$. Then, the number of $g(m)$, such that $g(m) \leq y$ is finite and hence we can choose the largest m such that $g(m) \leq y$. Then, $g(m) \leq y < g(m+1)$. If $g(m) < y$, then we get that $g(m+1) \leq y$ (since $g(m+1)$ is the least in $Y - \{g(1), \ldots, g(m)\}$ and $y \in Y - \{g(1), \ldots, g(m)\}$. Therefore, $g(m) = y$ and hence g is a surjection also and hence $\mathbb{Z}^+ \simeq Y$ so that Y is countable. Thus, X is countable ($g^{-1} \circ f: X \to \mathbb{Z}^+$ is a bijection).
(2) $\Rightarrow$ (3): If $X \simeq I_n$ for some n or $X \simeq \mathbb{Z}^+$, in either case, X can be treated as a subset of a countable set.
(3) $\Rightarrow$ (4): If X is a subset of a countable set Y, then there is a bijection $f: Y \to \mathbb{Z}^+$. Then, the restriction of f to X is an injection of X into $\mathbb{Z}^+$. Therefore, by Theorem 1.4.6, there is a surjection of $\mathbb{Z}^+$ into X.
(4) $\Leftrightarrow$ (1) follows from Theorem 1.4.6. ◀

Corollary 1.4.3. The set $\mathbb{R}$ of real numbers is uncountable.

Proof: Let $X = \{0 \cdot x_1 x_2 \ldots : x_i = 0 \text{ or } 1 \text{ for all } i\}$. Then, $X \subseteq \mathbb{R}$. We prove that X is uncountable. Suppose, if possible, that X is countable. Let $f: \mathbb{Z}^+ \to X$ be a bijection. Then, $X = \{f(1), f(2), \ldots\}$. Let

$$f(n) = 0 \cdot x_{n_1} x_{n_2} x_{n_3} \ldots, \text{ where } x_{n_i} = 0 \text{ or } 1.$$

For each $n \in \mathbb{Z}^+$, define $y_n = 1 - x_{n_n}$ and consider

$$y = 0 \cdot y_1 y_2 y_3 \ldots$$

Then, since $y_n \neq x_{n_n}$ for each n, we get that $y \neq f(n)$ for all n and hence $y \notin X$, which is a contradiction. Thus, X is uncountable and thus so is $\mathbb{R}$. ◀

For any cardinal number α, we have proved that $\alpha < 2^\alpha$ (Theorem 1.4.7) and, in particular $|\mathbb{Z}^+| < 2^{|\mathbb{Z}^+|}$ which automatically implies that $2^{|\mathbb{Z}^+|}$ is an uncountable cardinal. In the following, we prove that the cardinal number of the set $\mathbb{R}$ of real numbers is precisely $2^{|\mathbb{Z}^+|}$. The cardinal number of $\mathbb{Z}^+$ will be usually denoted by $\aleph_0$ and that of $\mathbb{R}$ by c.

Theorem 1.4.9

$$c = 2^{\aleph_0}$$

Proof: Recall that $\mathbb{P}(\mathbb{Z}^+) \simeq 2^{\mathbb{Z}^+}$, the set of all mapping of $\mathbb{Z}^+$ into the two-element set $\{1, 0\}$. We prove that there is a bijection of $2^{\mathbb{Z}^+}$ onto $\mathbb{R}$. Define $f : 2^{\mathbb{Z}^+} \to \mathbb{R}$ by

$$f(g) = 0 \cdot g(1)\, g(2) \ldots \quad \text{for any } g \in 2^{\mathbb{Z}^+},$$

where $0 \cdot g(1)\, g(2)\ldots$ is the real number in the interval $[0, 1)$ whose decimal places are $g(1)$, $g(2)$, ... Then, clearly f is an injection. On the other hand, noted that any real number x can be represented in the binary scale in the form

$$x = \ldots x_7 x_5 x_3 x_1 \cdot x_2 x_4 x_6 x_8 \ldots$$

where each $x_n = 0$ or 1 for every $n \in \mathbb{Z}^+$. Now, define $h : \mathbb{R} \to 2^{\mathbb{Z}^+}$ by

$$h(x)(n) = x_n \text{ if } x = \ldots x_5 x_3 x_1 \cdot x_2 x_4 x_6 \ldots$$

for any $n \in \mathbb{Z}^+$ and $x \in \mathbb{R}$. Then, h can be easily verified to be an injection. Therefore, we have two injections $f : 2^{\mathbb{Z}^+} \to \mathbb{R}$ and $h : \mathbb{R} \to 2^{\mathbb{Z}^+}$. By the Schroeder–Bernstein Theorem 1.4.5, it follows that there is a bijection of $\mathbb{R}$ onto $2^{\mathbb{Z}^+}$. Thus,

$$c = |\mathbb{R}| = \left|2^{\mathbb{Z}^+}\right| = 2^{|\mathbb{Z}^+|} = 2^{\aleph_0}.$$ ◀

Theorem 1.4.10. Let $X_1, X_2, \ldots$ be finite sets. Then, $\bigcup_{n=1}^{\infty} X_n$ is at most countable.

Proof: Let $X = \bigcup_{n=1}^{\infty} X_n$. Without loss of generality, we can assume that each X_n is nonempty. Let $|X_n| = m_n$ and $X_n = \{x_{n_1}, x_{n_2}, \ldots, x_{nm_n}\}$. Then,

$$X = \{x_{11}, x_{12}, \ldots, x_{1n_1}, x_{21}, \ldots, x_{2n_2}, x_{31}, \ldots\}$$

which shows that there is a surjection of $\mathbb{Z}^+$ onto X. By Theorem 1.4.8, X is at most countable. ◀

Corollary 1.4.4. If A and B are finite sets, then $A \cup B$, $A \cap B$ and $A \times B$ are all finite sets.

Proof: Clearly, $A \cup B$ and $A \cap B$ are finite. Also, for each $a \in A$, $\{a\} \times B \simeq B$ and hence finite and therefore $A \times B$ is finite, since $A \times B = \cup_{a \in A} \{a\} \times B$. ◀

Theorem 1.4.11. If X and Y are countable sets, then $X \cup Y$ and $X \times Y$ are countable and $X \cap Y$ is at most countable.

Proof: Suppose that X and Y are countable, then we can write

$$X = \{x_1, x_2, x_3, \ldots\} \quad \text{and} \quad Y = \{y_1, y_2, y_3, \ldots\}.$$

Since $X \cap Y$ is a subset of the countable set X, $X \cap Y$ is at most countable (by Theorem 1.4.8). Also, since

$$X \cup Y = \{x_1, y_1, x_2, y_2, x_3, y_3, \ldots\},$$

it follows that $X \cup Y$ is countable. For each $n \in \mathbb{Z}^+$, let

$$A_n = \{(x_i, x_j) : i + j = n + 1\}.$$

Then, each A_n is a finite set and $X \times Y = \bigcup_{n=1}^{\infty} A_n$. By Theorem 1.4.10, $X \times Y$ is at most countable. Further, since $\{x\} \times Y \simeq Y$, Y is infinite and $\{x\} \times Y \subseteq X \times Y$, it follows that $X \times Y$ is infinite. Thus, $X \times Y$ is countable. ◀

Corollary 1.4.5. If $X_1, X_2, \ldots, X_n$ are countable sets, then so are $\bigcup_{i=1}^{n} X_i$ and $X_1 \times X_2 \times \cdots \times X_n$.

Corollary 1.4.6. $\mathbb{Z}$, $\mathbb{Z} \times \mathbb{Z}$ and $\mathbb{Q}$ (the set of rational numbers) are countable.

Corollary 1.4.7. Countable union of countable sets is countable.

Proof: Let $\{X_1, X_2, \ldots, X_n, \ldots\}$ be a countable class of countable sets. Then, for each $n \in \mathbb{Z}^+$, there exists a bijection $f_n : \mathbb{Z}^+ \to X_n$. Now, let $X = \bigcup_{n=1}^{\infty} X_n$ and define $f : \mathbb{Z}^+ \times \mathbb{Z}^+ \to X$ by

$$f(n, m) = f_n(m).$$

Then, clearly f is a surjection. Also, since $\mathbb{Z}^+ \times \mathbb{Z}^+$ is countable, there is a bijection $g : \mathbb{Z}^+ \to \mathbb{Z}^+ \times \mathbb{Z}^+$. Now, $f \circ g$ is a surjection of $\mathbb{Z}^+$ onto X. Therefore, X is at most countable. But, since X is infinite, it follows that X is countable. ◀

Note that countable product of at most countable sets may not be countable. For consider the following.

Example 1.4.2. The set $2 = \{0, 1\}$ is a finite set and $\mathbb{Z}^+$ is a countable set. Here, $2^{\mathbb{Z}^+}$ ($\simeq \mathbb{R}$) is uncountable.

EXERCISE 1(D)

1. Prove that the cardinal numbers of $\mathbb{Z}^+$, $\mathbb{Z}^-$, $\mathbb{Z}$, $\mathbb{Q}^+$ and $\mathbb{Q}$ are all equal to each other.

2. Let $f : X \to Y$ and $g : Y \to Z$ be injections. Then, prove the following:
   (i) $|X| = |Z| \Rightarrow |Y| = |Z|$
   (ii) $|X| = |Y|$ and $f(X) \subseteq A \subseteq Y \Rightarrow |A| = |Y|$
   (iii) $X \subseteq A \subseteq Y$ and $|X| = |Y| \Rightarrow |A| = |Y|$

3. Prove that any infinite subset of a countable set is countable.

4. If X is a countable set and $f : X \to Y$ is a surjection, then prove that Y is at most countable.

5. Prove that a set X is infinite if and only if $|X| = |Y|$ for some proper subset Y of X.

6. For any positive integer n, prove that $(\mathbb{Q}^+)^n$ is countable.

7. If X is a set such that $|X| = |\mathbb{P}(\mathbb{P}(X))|$, then prove that there exists a surjection $f : X \to \mathbb{P}(X)$.

8. Deduce from Exercise 7 above that $|X| < |\mathbb{P}(\mathbb{P}(X))|$ for any set X.

9. If $|X| = |A|$ and $|Y| = |B|$, then prove that $|A^B| = |X^Y|$.

10. For any sets X and Y, prove that $|X| = |Y|$ if and only if $|\mathbb{P}(X)| = |\mathbb{P}(Y)|$.

11. Prove that $|\mathbb{P}(\mathbb{Z})| = |\mathbb{P}(\mathbb{Q})|$.

12. Give an example of a set $\mathscr{C}$ of circles in the plane such that every circle with positive radius properly contains a member of $\mathscr{C}$.

13. Prove that $|(a, b)| = |(c, d)|$ for any intervals (a, b) and (c, d) in $\mathbb{R}$ with $a < b$ and $c < d$.

14. Let $I_n = \{1, 2, \ldots, n\}$ for any $n \in \mathbb{Z}^+$. Prove that $|\mathbb{Z}^+| = |\mathbb{Z}^+ - I_n|$ for any $n \in \mathbb{Z}^+$.

15. Let X be a countable set and $\mathbb{P}_F(X)$ be the set of all finite subsets of X. Then, prove that

$$\left|\bigcup_{n=1}^{\infty} X^n\right| = |X| = |\mathbb{P}_F(X)|.$$

16. Prove that the set of polynomials in the indeterminate x over the set of rational numbers is countable.

17. A real number a is said to be an *algebraic number* if there exists a nonzero polynomial

$$f(x) = a_0 + a_1 x + \cdots + a_n x^n \text{ with } a_i \in \mathbb{Q}$$

such that $f(a) = 0$. Prove that the set of algebraic numbers is countable.

18. A real number is said to be *transcendental* if it is not algebraic. Prove that the set of transcendental numbers is uncountable.

19. Prove that the set of complex numbers is uncountable.

20. Prove that the set of complex numbers, whose real and imaginary parts are rational numbers, is countable.

2

Number Systems

This chapter is meant to review some of the important properties of the set of positive integers, the set of integers, the set of rational numbers, the set of real numbers and the set of complex numbers. We do not discuss any axiomatic development of these systems. We simply assume familiarity with addition and multiplication of these and their usual properties. Also, we briefly discuss the concept of a partial order on a set in general and the usual ordering on the real number system, in particular, these facilitate us in facing several encounters with these throughout this book. Further, we recall the notion of a matrix over the number systems and some important elementary properties of the matrices and their determinants.

2.1 INTEGERS

In this section, we review certain important elementary properties of integers, by assuming familiarity with the addition, subtraction, multiplication and the usual ordering in these (that is, $m \leq n$ if and only if $n - m$ is nonnegative). As mentioned in the beginning of the book, we follow the notations given below.

$\mathbb{Z}$: The set of integers $\{\ldots, -2, -1, 0, 1, 2, \ldots\}$
$\mathbb{Z}^+$: The set of positive integers $\{1, 2, 3, \ldots\}$

$\mathbb{Z}^-$: The set of negative integers $\{\ldots, -3, -2, -1\}$
$\mathbb{N}$: The set of nonnegative integers $\{0, 1, 2, 3, \ldots\}$
$m \le n$ if and only if $n - m \in \mathbb{N}$
$m < n$ if and only if $n - m \in \mathbb{Z}^+$

For any subset S of $\mathbb{Z}$ and $a \in S$, a is called the smallest (or least) member of S if $a \le s$ for all $s \in S$. It is well known that, for any given integers m and n with $m < n$, there are only finitely many (at most $n - m$ number of) integers a, such that $m < a < n$.

Theorem 2.1.1 (Well-ordering Property of $\mathbb{Z}^+$). Any nonempty set of positive integers has the smallest member.

Proof: Let S be a nonempty set of positive integers. That is, $\varnothing \neq S \subseteq \mathbb{Z}^+$. Suppose, if possible, that S has no smallest member. Since S is nonempty, we can choose $a \in S$. Then, a is not smallest in S and hence there exists $a_1 \in S$ such that $a \not\le a_1$; that is, $a_1 < a$. Again since a_1 is not smallest in S, we get $a_2 \in S$ such that $a_2 < a_1$. Continuing this, we get an infinite set of integers such that

$$0 < \cdots < a_n < a_{n-1} < \cdots < a_2 < a_1 < a$$

which is a contradiction, since there can be only a finite number of integers between 0 and a. Thus, S has smallest member.

In fact, the above well-ordered property of $\mathbb{Z}^+$ can be extended to $\mathbb{Z}$ as given below whose proof is similar to the above one.

Theorem 2.1.2. Let S be a nonempty subset of $\mathbb{Z}$ and $b \in \mathbb{Z}$ such that $b < s$ for all $s \in S$. Then, S has smallest member.

Now, we derive some more properties of $\mathbb{Z}^+$ as consequences of the well-ordering property.

Theorem 2.1.3 (First Principle of Induction). Let $b \in S \subseteq \mathbb{Z}^+$ such that

$$b \le n \in S \Rightarrow n + 1 \in S.$$

Then, $m \in S$ for all $m \ge b$.

Proof: Put $T = \{m \in \mathbb{Z}^+ : b \le m \text{ and } m \notin S\}$. It is enough if we can prove that T is the empty set. Suppose, if possible, that T is not empty. Then, $b - 1 < \text{m}$ for all $m \in T$. Therefore, by Theorem 2.1.2, T has smallest member, say m_0. Then, since $m_0 \in T$,

$$b \leq m_0 \quad \text{and} \quad m_0 \notin S.$$

Then, since $b \in S$, we get that $b < m_0$ and hence $b \leq m_0 - 1$ and $m_0 - 1 \notin S$ (otherwise $b \leq m_0 - 1 \in S$, so that $m_0 \in S$). This implies that $m_0 - 1 \in T$, which is a contradiction to the least property of m_0. Thus, T is empty and hence $m \in S$ for all $m \geq b$.

Corollary 2.1.1. Let $S \subseteq \mathbb{Z}^+$ such that $1 \in S$ and $n + 1 \in S$ whenever $n \in S$. Then, $S = \mathbb{Z}^+$.

Corollary 2.1.1 is the actual statement of the first principle of induction. However, this has more general form in Theorem 2.1.3. Often, induction is stated in terms of a proposition on $\mathbb{Z}^+$. A proposition on a set S means that, for each $a \in S$, $P(s)$ is a statement about s and $P(s)$ is either true or false. The above principle of induction is equivalent to the following.

Let P be a proposition on $\mathbb{Z}^+$ such that $P(b)$ is true for some $b \in \mathbb{Z}^+$ and $P(n + 1)$ is true whenever $n \geq b$ and $P(n)$ is true. Then, $P(n)$ is true for all $n \geq b$.

Example 2.1.1. Let us prove that $1 + 2 + \cdots + n = \dfrac{n(n+1)}{2}$ for all $n \in \mathbb{Z}^+$.
Let $P(n)$ be the statement '$1+2+\cdots+n = \dfrac{n(n+1)}{2}$' and $S = \{n \in \mathbb{Z}^+ : P(n) \text{ is true}\}$. Then, clearly $1 \in S \subseteq \mathbb{Z}^+$.

$$\begin{aligned}
n \in S &\Rightarrow P(n) \text{ is true} \\
&\Rightarrow 1 + 2 + \cdots + n = \frac{n(n+1)}{2} \\
&\Rightarrow 1+2+\cdots+n+n+1 = \frac{n(n+1)}{2} + n + 1 = \frac{(n+1)(n+2)}{2} \\
&\Rightarrow P(n + 1) \text{ is true} \\
&\Rightarrow n + 1 \in S.
\end{aligned}$$

Therefore, by the first principle of induction, $S = \mathbb{Z}^+$ and hence $P(n)$ is true for all $n \in \mathbb{Z}^+$.

The following shows the importance of the general version of the first principle of induction given in Theorem 2.1.3.

Example 2.1.2. Let us find all positive integers n for which $3^n < n!$
Let $S = \{n \in \mathbb{Z}^+ : 3^n < n!\}$.
Clearly $3^1 > 1!$, $3^2 > 2!$, $3^3 > 3!$, $3^4 > 4!$, $3^5 > 5!$ and $3^6 = 729 > 720 = 6!$
But $3^7 = 2187 < 5040 = 7!$, $3^8 < 8!$ and $3^9 < 9!$ which suggests that $n \in S$

for all $n \geq 7$. To prove this, let us apply the first principle of induction (Theorem 2.1.3). We have $7 \in S$ and

$$\begin{aligned} 7 \leq n \in S &\Rightarrow 3^n < n! \\ &\Rightarrow 3^{n+1} = 3^n \cdot 3 < n!(n+1) \\ &\Rightarrow 3^{n+1} < (n+1)! \\ &\Rightarrow n+1 \in S. \end{aligned}$$

Thus, by Theorem 2.1.3, $n \in S$ for all $n \geq 7$. Since we have already checked that $n \notin S$ for $0 < n < 7$, we get that

$$S = \{n \in \mathbb{Z}^+ : 7 \leq n\}.$$

The first principle of induction does not work sometimes when we need to know the truth of one or more smaller cases and not necessarily the immediately preceding one. To handle situations like this, we need another form of induction given below.

Theorem 2.1.4 (Second Principle of Induction). Let S be a set of positive integers and $b \in S$ such that, for any $b \leq n \in \mathbb{Z}^+$,

$$b, b+1, \ldots, n \in S \Rightarrow n+1 \in S.$$

Then, $m \in S$ for all $m \geq b$.

Proof: Put $T = \{m \in \mathbb{Z}^+ : b \leq m \text{ and } m \notin S\}$. We need to prove that T is empty. Suppose, if possible, that T is nonempty. Then, by the well-ordering property of $\mathbb{Z}^+$ (Theorem 2.1.1), T has a smallest member, say m_0. Since $m_0 \in T$, we have $b \leq m_0$ and $m_0 \notin S$. But, since $b \in S$, we get that $b < m_0$. Also, by the least property of m_0, any integer less than m_0 cannot be in T. Therefore,

$$b, b+1, \ldots, m_0 - 1 \in S.$$

By the hypothesis, it follows that $m_0 \in S$, which is a contraction. Thus, T is empty and hence $m \in S$ for all $m \geq b$.

Corollary 2.1.2. Let $S \subseteq \mathbb{Z}^+$ such that $1 \in S$ and, for any $n \in \mathbb{Z}^+$,

$$m \in S \quad \text{for all } m < n \Rightarrow n \in S.$$

Then, $S = \mathbb{Z}^+$.

We have actually used the well-ordering property of $\mathbb{Z}^+$ to prove both the first and second principles of induction. However, we can prove the well-ordering property using either of the induction principles. In the following, we prove that all three are equivalent.

Theorem 2.1.5. The following are equivalent to each other:

1. Well-ordering property of $\mathbb{Z}^+$ (Theorem 2.1.1).
2. The first principle of induction (Theorem 2.1.3).
3. The second principle of induction (Theorem 2.1.4).

Proof: The proof of Theorem 2.1.3 is precisely the proof of (1) $\Rightarrow$ (2).
(2) $\Rightarrow$ (3): Assume that the first principle of induction holds.
Let $b \in S \subseteq \mathbb{Z}^+$ and, for any $n \geq b$,

$$b, b + 1, \ldots, n \in S \Rightarrow n + 1 \in S.$$

Put $A = \{a \in \mathbb{Z}^+ : a \geq b \text{ and } b, b + 1, \ldots, a \in S\}$. By our assumption on S, it follows that $b \in A$ and

$$b \leq n \in A \Rightarrow n + 1 \in A.$$

From the first principle of induction, $m \in A$ for all $m \geq b$. In particular, $m \in S$ for all $m \geq b$.
(3) $\Rightarrow$ (1): Assume that the second principle of induction holds. Let A be a nonempty set of positive integers. Suppose, if possible, that A has no smallest member. Put $S = \mathbb{Z}^+ - A$. Then, $1 \notin A$ and hence $1 \in S$. For any $n \geq 1$, if $1, 2, \ldots, n \in S$, then $1, 2, \ldots, n \notin A$ and therefore $n + 1 \notin A$ (otherwise $n + 1$ becomes the smallest member in A) and hence $n + 1 \in S$. From the second principle of induction, it follows that $n \in S$ for all $n \geq 1$; that is, $S = \mathbb{Z}^+$ and hence A is empty, which is a contradiction. Thus, A has a smallest member.

The next result is one of the best applications of the second principle of induction. Before this, let us recall that a positive integer $p > 1$ is called a *prime number* (or simply, *prime*) if 1 and p are the only factors of p (a is said to be a *factor* of b if $ac = b$ for some integer c).

Theorem 2.1.6 (Fundamental Theorem of Arithmetic). Any positive integer greater than 1 can be uniquely expressed as a product of prime numbers.

Proof: Let $S = \{n \in \mathbb{Z}^+ : n > 1 \text{ and } n \text{ is a product of primes}\}$. Then 2, being a prime, is a member of S. Let $2 \leq n \in \mathbb{Z}^+$ such that $2, 3, \ldots, n \in S$. If $n + 1$ is a prime, then clearly $n + 1 \in S$. Suppose that $n + 1$ is not a prime; then there is a factor a of $n + 1$ such that $a \neq 1$ and $a \neq n + 1$. Therefore, there are positive integers a and b such that

$$n + 1 = ab \quad \text{and} \quad 1 < a < n + 1 \quad \text{and} \quad 1 < b < n + 1.$$

From the induction hypothesis, a and $b \in S$ and hence a and b can be expressed as products of primes and therefore so is $n + 1$. Thus, $n + 1 \in S$. By the second principle of induction, it follows that $m \in S$ for all $m \geq 2$. Thus, any $n > 1$ can be expressed as a product of primes.

We prove the uniqueness of the factorization also by using induction principle. Let

$$p_1 p_2 \cdots p_r = n = q_1 q_2 \cdots q_s$$

where p_i's and q_j's are prime numbers. Suppose that $p_i = q_j$ for some i and j. We can suppose, by renumbering of p_i's and q_j's, that $p_1 = q_1$. Then, $p_2 p_3 \cdots p_r = q_2 q_3 \cdots q_s < n$ and hence, by the induction hypothesis, $r = s$ and each p_i is equal to some q_j and vice versa. Next suppose that $p_i \neq q_j$ for all i and j. Without loss of generality, we can suppose that $p_1 > q_1$. Then,

$$\begin{aligned} n > (p_1 - q_1) p_2 \cdots p_r &= p_1 p_2 \cdots p_r - q_1 p_2 \cdots p_r \\ &= q_1 q_2 \cdots q_s - q_1 p_2 \cdots p_r \\ &= q_1 (q_2 \cdots q_s - p_2 \cdots p_r). \end{aligned}$$

Again by the induction hypothesis, $q_1 = p_i$ for some $2 \leq i \leq r$ or q_1 divides $p_1 - q_1$. But $q_1 \neq p_i$ for all i and hence q_1 divides $p_1 - q_1$. Therefore, $q_1 = p_1$, a contradiction to our assumption. Thus, $r = s$ and each p_i is equal to some q_j and vice versa.

The following is an important property of integers which we use throughout this book. The proof of this is again by the well-ordering property of $\mathbb{Z}^+$.

Theorem 2.1.7 (The Division Algorithm in $\mathbb{Z}$). Let a and b be any integers and $b > 0$. Then, there exist unique integers q and r such that

$$a = bq + r \quad \text{and} \quad 0 \leq r < b.$$

(Here q is called the *quotient* and r is called the *remainder* of a modulo b.)

Proof: If $a = 0$, we can take $q = 0 = r$ and, if $b = 1$, then we can take $q = a$ and $r = 0$. Therefore, we can assume that $b > 1$ and $a \neq 0$. Put

$$S = \mathbb{Z}^+ \cap \{a - bx : x \in \mathbb{Z}\}.$$

Since $b > 1$, $a < b|a|$ and hence $a - b(-|a|) \in S$. Therefore, S is a nonempty subset of $\mathbb{Z}^+$. By the well-ordering property of $\mathbb{Z}^+$, S has a smallest member, say m_0.

Let $m_0 = a - bx > 0$ (since $m_0 \in S$). If $b < m_0$, then $0 < m_0 - b = a - b(x+1) \in S$ and $m_0 - b < m_0$ which is a contradiction to the least property of m_0. Therefore, we have $m_0 \leq b$. If $m_0 = b$, then $a - bx = b$ and hence $a = b(1 + x)$, so that we can take $q = x + 1$ and $r = 0$. Thus, $m_0 < b$ and

$$a = bx + (a - bx) = bx + m_0$$

so that we can take $q = x$ and $r = m_0$.
To prove the uniqueness of q and r, let

$$bq + r = a = bq' + r', 0 \leq r < b \quad \text{and} \quad 0 \leq r' < b.$$

Then, $b(q - q') = r' - r$ and hence

$$b|q - q'| = |r' - r| < b.$$

This implies that $|q - q'| = 0$; that is, $q = q'$ and $r = r'$.

Definition 2.1.1. For any m and $n \in \mathbb{Z}^+$, let CD(m, n) be the set of all common divisors (factors) of m and n in $\mathbb{Z}^+$. That is,

$$\text{CD}(m, n) = \{c \in \mathbb{Z}^+ : c \text{ divides both } m \text{ and } n\}.$$

Clearly CD(m, n) is a nonempty subset of $\mathbb{Z}^+$ for any m and $n \in \mathbb{Z}^+$, since 1 is a divisor of any positive integer. Also, for any a and $b \in \mathbb{Z}^+$,

$$a \text{ divides } b \Rightarrow a \leq b$$

and hence every member of CD(m, n) is less than or equal to both m and n. This implies that CD(m, n) is finite and has a largest (greatest) member, which is called the *greatest common divisor* of m and n and is denoted by g.c.d.$\{m, n\}$ or, simply (m, n). The following is an interesting property of the g.c.d.'s.

Theorem 2.1.8. Let m and n be positive integers Then, the following are equivalent to each other for any $d \in$ CD(m, n).

1. $d =$ g.c.d.$\{m, n\}$
2. $d = ma + nb$ for some a and $b \in \mathbb{Z}$
3. Every member of CD(m, n) divides d.

Proof: Let $d \in CD(m, n)$; that is, d is a common divisor of m and n.

(1) $\Rightarrow$ (2): Suppose that d is the greatest member of $CD(m, n)$. Let $S = \mathbb{Z}^+ \cap \{ma + nb : a \text{ and } b \in \mathbb{Z}\}$. Clearly S is a nonempty subset of $\mathbb{Z}^+$ (for example, $m^2 + n^2 \in S$). By the well-ordering property of $\mathbb{Z}^+$, S has a smallest member, say d_0. We prove that $d_0 = d$. Since $d_0 \in S$, we have

$$0 < d_0 = ma + nb \quad \text{for some } a \text{ and } b \in \mathbb{Z}.$$

We first prove that $d_0 \in CD(m,n)$. By the division algorithm (Theorem 2.1.7), we can write

$$m = d_0 q + r, 0 \leq r < d_0, q \quad \text{and} \quad r \in \mathbb{Z}.$$

Now, $r = m - d_0 q = m - (ma + nb)q = m(1 - aq) + n(-bq)$. If $r > 0$, then $r \in S$ and $r < d_0$ which is a contradiction to the least property of d_0 in S. Therefore, $r = 0$ and $m = d_0 q$. Thus, d_0 divides m and, similarly d_0 divides n and hence $d_0 \in CD(m, n)$. From this, it follows that $d_0 \leq d$. Also, since $d \in CD(m, n)$, d divides m and n and hence d divides $ma + nb = d_0$. Therefore, $d \leq d_0$. Thus, $d = d_0 = ma + nb$.

(2) $\Rightarrow$ (3) and (3) $\Rightarrow$ (1) are trivial.

Definition 2.1.2. Two positive integers m and n are said to be *relatively prime* (or, *prime to each other*) if $(m, n) = 1$. This is equivalent to saying that $CD(m, n) = \{1\}$.

Note that m and n are relatively prime if and only if there is no prime number dividing both m and n. The following is an important consequence of Theorem 2.1.8.

Theorem 2.1.9. Let m, n and $r \in \mathbb{Z}^+$ such that m divides nr and $(m, n) = 1$. Then, m divides r.

Proof: By Theorem 2.1.8, there exist integers a and b such that $ma + nb = 1$. Now,

$$r = r1 = r(ma + nb) = mra + nrb.$$

Since m divides mra and m divides nr, it follows that m divides r.

Corollary 2.1.3. Let p be a prime number and m and n be positive integers such that p divides mn. Then, p divides either m or n.

Proof: Suppose that p does not divide m. Then, $(p, m) = 1$ and therefore, by Theorem 2.1.9, p divides n.

Definition 2.1.3. Let X be any set. A function $s : \mathbb{Z}^+ \to X$ is called a *sequence in* X. A sequence s is usually represented by $\{s(1), s(2), \ldots\}$ or $\{s_1, s_2, \ldots\}$.

Quite often a sequence s is described inductively by giving $s(1)$ and a rule to find $s(n + 1)$ from $s(n)$. For example, we define $s(n) = n!$ inductively by

$$1! = 1$$

and $(n + 1)! = (n + 1)n!$ for any $n \geq 1$.

In the next theorem, we prove that this inductive method of defining a sequence works well in the sense that there exists a unique sequence $s : \mathbb{Z}^+ \to X$ satisfying the given conditions for determining $s(n + 1)$ from $s(n)$.

Theorem 2.1.10 (Recursion Theorem). Let X be a set and $x_1 \in X$. Suppose that $f : \mathbb{Z}^+ \times X \to X$ is a mapping. Then, there exists a unique sequence $s : \mathbb{Z}^+ \to X$ such that

$$s(1) = x_1 \text{ and } s(n + 1) = f(n, s(n)) \quad \text{for all } n \in \mathbb{Z}^+.$$

Proof: First we prove the existence of a sequence $s : \mathbb{Z}^+ \to X$ satisfying the required conditions. Let

$$\mathbb{P} = \{S \subseteq \mathbb{Z}^+ \times X : (1, x_1) \in S \quad \text{and} \quad \text{if } (n, x) \in S, \text{ then } (n + 1, f(n, x)) \in S\}.$$

Then, $\mathbb{P}$ is nonempty, since $\mathbb{Z}^+ \times X \in \mathbb{P}$. Let T be the intersection of all members in $\mathbb{P}$. Then, clearly T is a member of $\mathbb{P}$ and is contained in every member of $\mathbb{P}$. Put

$$A = \{n \in \mathbb{Z}^+ : \text{there is a unique } x \in X \text{ such that } (n, x) \in T\}.$$

We prove, by induction principle, that $A = \mathbb{Z}^+$. Suppose, if possible, that $1 \notin A$. Since $(1, x_1) \in T$, there exists $x \neq x_1$ in X such that $(1, x) \in T$. Then, $T - \{(1, x)\}$ is a member of $\mathbb{P}$ and hence it contains T, which is a contradiction. Therefore, $1 \in A$.

Next, let $n \in A$. Suppose, if possible, that $n + 1 \notin A$. Let x_n be the unique element in X such that $(n, x_n) \in T$. Since $T \in \mathbb{P}$, it follows that $(n + 1, f(n, x_n)) \in T$. Since $n + 1 \notin A$, there exists $y \neq f(n, x_n)$ in X such that $(n + 1, y) \in T$. Again, it can be easily verified that $T - \{(n + 1, y)\}$ is a member of $\mathbb{P}$ and hence

$$T \subseteq T - \{(n + 1, y)\}, \text{ which is a contradiction.}$$

Therefore, $n + 1 \in A$. By the first principle of induction, it follows that $A = \mathbb{Z}^+$. Thus, for each $n \in \mathbb{Z}^+$, there exists unique element, say x_n, in X such that $(n, x_n) \in T$. Now, define $s : \mathbb{Z}^+ \to X$ by $s(n) = x_n$. Then,

$$s(1) = x_1, \quad \text{since } (1, x_1) \in T$$

and $s(n+1) = f(n, s(n))$, since $(n, s(n)) \in T$ and hence $(n+1, f(n, s(n))) \in T$. Thus, s is a sequence satisfying the required properties.
Next, we prove the uniqueness of s. Let s and t be sequences such that

$$s(1) = x_1 = t(1)$$
$$\text{and} \quad s(n+1) = f(n, s(n)) \quad \text{and} \quad t(n+1) = f(n, t(n))$$

for all $n \in \mathbb{Z}^+$. Let $B = \{n \in \mathbb{Z}^+ : s(n) = t(n)\}$. Then, $1 \in B$ and, if $n \in B$, then $s(n) = t(n)$ and hence

$$s(n+1) = f(n, s(n)) = f(n, t(n)) = t(n+1),$$

so that $n + 1 \in B$. Again, by the first principle of induction, $B = \mathbb{Z}^+$. Thus, $s(n) = t(n)$ for all $n \in \mathbb{Z}^+$ and hence $s = t$.

Before we close this section, let us recall the concept of absolute value of an integer.

Definition 2.1.4. For any $a \in \mathbb{Z}$, define

$$|a| = \begin{cases} a & \text{if } a \geq 0 \\ -a & \text{if } a < 0 \end{cases}.$$

Note that $|a| \geq 0$ for all $a \in \mathbb{Z}$. The following can be proved by straight forward verification.

Theorem 2.1.11. The following holds for any integers a, b and c.

(i) $|a| = 0 \Leftrightarrow a = 0$
(ii) $|ab| = |a||b|$
(iii) $|a - b| = |b - a|$
(iv) $|a + b| \leq |a| + |b|$
(v) $|a + b| = |a| + |b|$ if and only if either both a and b are nonpositive or nonnegative.
(vi) $|a - b| \leq |a - c| + |c - b|$
(vii) $||a| - |b|| \leq |a - b|$
(viii) $||a| - |b|| = |a - b|$ if and only if $|a + b| = |a| + |b|$.

EXERCISE 2(A)

1. Prove the following for any $n \in \mathbb{Z}^+$:
   (i) $1^2 + 2^2 + 3^2 + \cdots + n^2 = n(n+1)(2n+1)/6$
   (ii) $1^3 + 2^3 + 3^3 + \cdots + n^3 = \left(\frac{n(n+1)}{2}\right)^2$
   (iii) $1 \cdot 1! + 2 \cdot 2! + 3 \cdot 3! + \cdots + n \cdot n! = (n+1)! - 1$
   (iv) If X is a set and $|X| = n$, then $|\mathbb{P}(X)| = 2^n$
   (v) $x^n - 1 = (x-1)(x^{n-1} + x^{n-2} + \cdots + x + 1)$
   (vi) $(x+y)^n = \sum_{r=0}^{n} \binom{n}{r} x^{n-r} y^r$
   (vii) $\sum_{r=1}^{n} \frac{4n}{3^n} = 3 - \frac{2n+3}{3^n}$
   (viii) $\sum_{r=1}^{n} \frac{1}{r(r+1)} = \frac{n}{n+1}$
   (ix) $\sum_{r=1}^{n} r(r+1) = \frac{n(n+1)(n+2)}{3}$
   (x) $1 + 3 + 5 + \cdots + (2n-1) = n^2$.
2. Find all $n \in \mathbb{Z}^+$ for which $2n + 1 < 2^{n-1}$.
3. Let X be a set such that $|X| = n \geq 2$. Prove that these are exactly $\frac{n(n-1)}{2}$ subsets each with exactly two elements.
4. For any n and $r \in \mathbb{Z}^+$ such that $1 \leq r \leq n$, prove that

$$\sum_{i=r}^{n} \binom{i}{r} = \binom{n+1}{r+1}$$

5. Use the Binomial theorem given in Exercise 1 (vi) above to prove the following.
   (i) $\sum_{r=0}^{n} \binom{n}{r} = 3^n$ for all $n \in \mathbb{Z}^+$
   (ii) $(a+b)^n \in a\mathbb{Z}^+ + b^n$ for all a, b and $n \in \mathbb{Z}^-$
   (iii) $\sum_{\substack{r \text{ even} \\ 0 \leq r \leq n}} \binom{n}{r} = \sum_{\substack{r \text{ odd} \\ 0 < r \leq n}} \binom{n}{r}$
6. Prove that the set of prime numbers is infinite.
7. For any $a \in \mathbb{Z}^+$ and for any prime number p, prove that there is a largest nonnegative integer n for which p^n divides a.
8. Prove that any positive integer a can be expressed as

$$a = p_1^{n_1}\, p_2^{n_2} \ldots p_r^{n_r}$$

where $p_1, p_2, \ldots, p_r$ are distinct primes and $n_1, n_2, \ldots, n_r$ are nonnegative integers.

9. Let a, b and $c \in \mathbb{Z}^+$. c is said to be a common multiple of a and b if both a and b divide c. Prove that there is a least common multiple for any a and $b \in \mathbb{Z}^+$. This will be denoted by l.c.m.$\{a, b\}$ or $[a, b]$.

10. Let $a = p_1^{n_1} p_2^{n_2} \ldots p_r^{n_r}$, where p_i's are distinct primes and n_i's are nonnegative integers and let $b \in \mathbb{Z}^+$. Then, prove that b divides a if and only if
$b = p_1^{m_1} p_2^{m_2} \ldots p_r^{m_r}$, where $m_i \in \mathbb{Z}$ and $0 \leq m_i \leq n_i$.

11. Let $a = \prod_{i=1}^{r} p_i^{n_i}$ and $b = \prod_{i=1}^{r} p_i^{m_i}$, where p_i's are distinct primes and n_i's are nonnegative integers. Then, prove that
g.c.d.$\{a, b\} = \prod_{i=1}^{r} p_i^{k_i}$, where k_i = minimum of n_i and m_i
and l.c.m.$\{a, b\} = \prod_{i=1}^{r} p_i^{d_i}$, where d_i = maximum of n_i and m_i.

12. For any positive integers a and b, prove that the product of a and b is equal to the product of their g.c.d. and l.c.m.

13. Let a, b and c be positive integers such that a divides both b and c. Then, prove that a divides $mb + nc$ for any integers m and n.

14. Let $1 < n \in \mathbb{Z}^+$. Prove that either n is a prime or has a prime divisor which is $\leq \sqrt{n}$.

15. Let $a_1, a_2, \ldots, a_r \in \mathbb{Z}^+$ and a = g.c.d.$\{a_1, a_2, \ldots, a_r\}$. Then, prove the following:
    (i) $d = b_1a_1 + b_2a_2 + \cdots + b_ra_r$ for some $b_1, b_2, \ldots, b_r \in \mathbb{Z}$
    (ii) If $S = \{b_1a_1 + b_2a_2 + \cdots + b_ra_r : b_i \in \mathbb{Z}\}$, then
    $S = a\mathbb{Z}$ and a is the least member of $S \cap \mathbb{Z}^+$.
    (iii) If $b \in \mathbb{Z}^+$ and b divides a_i for all $1 \leq i \leq r$, then b divides a.

16. Let $a, b, c, d \in \mathbb{Z}^+$ and $a = bc + d$. Then, prove that
g.c.d.$\{a, b\}$ = g.c.d.$\{b, d\}$

17. For any positive integer a, prove that $1 + a + a^2$ and $1 + a$ are relatively prime.

18. Prove the following for any $a, b, c, d \in \mathbb{Z}^+$.
    (i) g.c.d.$\{a, c\} = 1 =$ g.c.d.$\{b, c\} \Leftrightarrow$ g.c.d.$\{ab, c\} = 1$
    (ii) g.c.d.$\{a, b\} = 1 \Leftrightarrow$ g.c.d.$\{a, b^n\} = 1$ for any $n \in \mathbb{Z}^+$.

19. Establish the *Euclidean Algorithm* given below to find the greatest common divisor of given positive integers a and b.

    * Let $a, b \in \mathbb{Z}^+$ with $a > b$ and b not a divisor of a. Use division algorithm (Theorem 2.1.7) repeatedly as necessary, to write

$$a = bq_1 + r_1$$
$$b = r_1q_2 + r_2$$
$$r_1 = r_2q_3 + r_3$$
$$\vdots$$

$$r_{n-3} = r_{n-2}q_{n-1} + r_{n-1}$$
$$r_{n-2} = r_{n-1}q_n + r_n$$

where r_n is the last nonzero reminder. Then,
g.c.d.$\{a, b\} = r_n = xa + yb$ for some $x, y \in \mathbb{Z}$.

20. Find x and $y \in \mathbb{Z}$ such that
g.c.d.$\{969, 1273\} = 969x + 1273y$.

21. For any positive integers a and b, prove that a divides b if and only if $2^a - 1$ divides $2^b - 1$.

22. For any a and $b \in \mathbb{Z}^+$, prove that a and b are relatively prime if and only if $ma + nb = 1$ for some m and $n \in \mathbb{Z}$ and, in this case, $|m|$ and $|n|$ are relatively prime.

23. Let a and b be relatively prime and $a, b, c \in \mathbb{Z}^+$, then prove that ab divides c if and only if both a and b divide c.

24. If $a, b, c, d \in \mathbb{Z}^+$ such that $a = c(a, b)$ and $b = d(a, b)$, then prove that $(c, d) = 1$.

25. Let $a, b \in \mathbb{Z}^+$ such that $b > 1$ and $(a, b) = 1$. Then, prove that $(a + bc, b) = 1$ for any $c \in \mathbb{Z}^+$ and there is a unique $n \in \mathbb{Z}^+$ such that $1 \leq n < b$ and $(n, b) = 1$ and b divides $an - 1$.

2.2 CONGRUENCE MODULO *n*

Here, we briefly discuss an important equivalence relation, namely congruence modulo n, on the set $\mathbb{Z}$ of integers. The importance of this is due to the fact that it is compatible with addition, subtraction and multiplication in $\mathbb{Z}$. Let us first agree with the following notation.

Definition 2.2.1. For any subsets A and B of $\mathbb{Z}$, let

$$A + B = \{a + b : a \in A \text{ and } b \in B\}$$

and $AB = \{ab : a \in A \text{ and } b \in B\}$.

Note that $A + B = B + A$, $AB = BA$, $(AB)C = A(BC)$ and $(A + B) + C = A + (B + C)$ for any A, B and $C \subseteq \mathbb{Z}$. Also AB is empty if and only if A is empty or B is empty. If A is a single element set $\{a\}$, then we write $a + B$ for $\{a\} + B$ and aB for $\{a\}B$.

Theorem 2.2.1. Let n be any positive integer. Then,

$$n\mathbb{Z}, 1 + n\mathbb{Z}, \ldots, (n - 1) + n\mathbb{Z}$$

form a partition of $\mathbb{Z}$.

Proof: For any i and $j \in \{0, 1, 2, \ldots, n-1\}$,

$$\begin{aligned}(i + n\mathbb{Z}) \cap (j + n\mathbb{Z}) \neq \varnothing &\Rightarrow i + na = j + nb \text{ for some } a \text{ and } b \in \mathbb{Z} \\ &\Rightarrow |i - j| = n|b - a| \\ &\Rightarrow i - j = 0 \text{ (since } |i - j| < n) \\ &\Rightarrow i = j\end{aligned}$$

Therefore, for any $i \neq j$, $i + n\mathbb{Z}$ and $j + n\mathbb{Z}$ are disjoint. Also, for any $a \in \mathbb{Z}$, we have (by the division algorithm, Theorem 2.1.7)

$$a = nq + r$$

for some q and $r \in \mathbb{Z}$ such that $0 \leq r < n$ and hence

$$a \in r + n\mathbb{Z} \quad \text{for some } r = 0, 1, \ldots, n-1.$$

Thus, the sets $r + n\mathbb{Z}$, $0 \leq r \leq n-1$ form a partition of $\mathbb{Z}$.

Theorem 2.2.2. Let $\equiv_n$ be the equivalence relation on $\mathbb{Z}$ corresponding to the partition given in the above theorem. Then, for any a and $b \in \mathbb{Z}$

$$a \equiv_n b \text{ if and only if } n \text{ divides } |a - b|.$$

Proof: For any a and $b \in \mathbb{Z}$. $a \equiv_n b$ if and only if a and b belong to the same set in the partition. Therefore,

$$\begin{aligned}a \equiv_n b &\Leftrightarrow a \quad \text{and} \quad b \in i + n\mathbb{Z} \quad \text{for some } 0 \leq i < n \\ &\Leftrightarrow a = i + nx \quad \text{and} \quad b = i + ny \quad \text{for some } x, y \in \mathbb{Z} \\ &\Leftrightarrow a - b = n(x - y) \\ &\Leftrightarrow |a - b| = n|x - y|, x, y \in \mathbb{Z} \\ &\Leftrightarrow n \text{ divides } |a - b|.\end{aligned}$$

Definition 2.2.2. The equivalence relation $\equiv_n$ obtained above is called the *congruence modulo n*. Quite often, we write

$$a \equiv b \pmod{n} \quad \text{for } a \equiv_n b.$$

That is, for any integers a and b and for any positive integer n,

$$a \equiv b \pmod{n} \Leftrightarrow n \text{ divides } |a - b|.$$

If $\equiv_n (a)$ is the equivalence class containing a corresponding to $\equiv_n$, then we have

$$\equiv_n (a) = a + n\mathbb{Z} = r + n\mathbb{Z}, \quad \text{where } a = nq + r, 0 \leq r < n.$$

Also, for any a and $b \in \mathbb{Z}$

$$\begin{aligned} \equiv_n (a) = \equiv_n (b) &\Leftrightarrow a \equiv b \pmod{n} \\ &\Leftrightarrow a \in b + n\mathbb{Z} \\ &\Leftrightarrow a - b \in n\mathbb{Z} \\ &\Leftrightarrow n \text{ divides } |a - b|. \end{aligned}$$

In the following, we prove that the congruence modulo n is compatible with the usual arithmetical operations.

Theorem 2.2.3. Let $1 < n \in \mathbb{Z}$. Then the following holds for any a, b, c and $d \in \mathbb{Z}$.

(i) $a \equiv b \pmod{n}$ and $c \equiv d \pmod{n} \Rightarrow a + c \equiv b + d \pmod{n}$
(ii) $a \equiv b \pmod{n} \Rightarrow -a \equiv -b \pmod{n}$
(iii) $a \equiv b \pmod{n} \Rightarrow ac \equiv bc \pmod{n}$
(iv) $a \equiv b \pmod{n}$ and $c \equiv d \pmod{n} \Rightarrow ac \equiv bd \pmod{n}$
(v) $a \equiv b \pmod{n} \Rightarrow a^m \equiv b^m \pmod{n}$ for all $m \in \mathbb{Z}^+$
(vi) $ac \equiv bc \pmod{n}$ and $(c, n) = 1 \Rightarrow a \equiv b \pmod{n}$

Proof:

(i) $$\begin{aligned} a \equiv_n b \text{ and } c \equiv_n d &\Rightarrow n \text{ divides } |a - b| \text{ and } |c - d| \\ &\Rightarrow nx = a - b \text{ and } ny = c - d \text{ for some } x, y \in \mathbb{Z} \\ &\Rightarrow n(x + y) = (a + c) - (b + d), x + y \in \mathbb{Z} \\ &\Rightarrow n \text{ divides } |(a + c) - (b + d)| \\ &\Rightarrow (a + c) \equiv_n (b + d) \end{aligned}$$

(ii) and (iii) can be proved similarly.

(iv) $$\begin{aligned} a \equiv_n b \text{ and } c \equiv_n d &\Rightarrow ac \equiv_n bc \text{ and } bc \equiv_n bd \text{ (by (iii))} \\ &\Rightarrow ac \equiv_n bd \text{ (since } \equiv_n \text{ is transitive)} \end{aligned}$$

(v) is a simple consequence of (iv) and the principle of induction.

(vi) $$\begin{aligned} ac \equiv_n bc \text{ and } (c, n) = 1 &\Rightarrow n \text{ divides } |ac - bc| = c|a - b| \\ &\Rightarrow n \text{ divides } |a - b| \text{ (since } (c, n) = 1) \\ &\Rightarrow a \equiv_n b. \end{aligned}$$

Corollary 2.2.1. Let $1 < n \in \mathbb{Z}$ and $a, b, a_1, a_2, \ldots, a_r, b_1, b_2, \ldots, b_r \in \mathbb{Z}$. Then, the following holds:

(i) $a_i \equiv b_i \pmod{n}$ for all $1 \leq i \leq r \Rightarrow \sum_{i=1}^{r} a_i \equiv \sum_{i=1}^{r} b_i \pmod{n}$ and $\prod_{i=1}^{r} a_i \equiv \prod_{i=1}^{r} b_i \pmod{n}$.

(ii) If $f(x)$ is a polynomial with integer coefficients and

$$a \equiv b \pmod{n}, \text{ then } f(a) \equiv f(b) \pmod{n}.$$

Proof: To prove (i), use Theorem 2.2.3 (i) and apply induction on r. (ii) is a consequence of (i), (iii) and (v) of Theorem 2.2.3.

Observe that, when $x \equiv y \pmod{8}$, either of x and y may be replaced by the other in any polynomial congruence modulo n (by Corollary 2.2.1 (ii)) and this idea can be used in solving linear congruences $ax \equiv b \pmod{n}$ when a and n are relatively prime. Before proving this, let us have the following.

Theorem 2.2.4. Let $1 < n \in \mathbb{Z}$ and $a \in \mathbb{Z}$ such that $(|a|, n) = 1$. Then, there exists unique $r \in \mathbb{Z}^+$ such that $1 \leq r < n$, $(r, n) = 1$ and $ar \equiv 1 \pmod{n}$.

Proof: Since $(|a|, n) = 1$, there exists u and $v \in \mathbb{Z}$ such that $u|a| + vn = 1$ (by Theorem 2.1.8) and hence

$$sa + vn = 1 \qquad (*)$$

for some s and $v \in \mathbb{Z}$. We use the division algorithm to get r and $q \in \mathbb{Z}$ such that

$$s = nq + r \quad \text{and} \quad 0 \leq r < n.$$

If $r = 0$, then n divides $sa + vn = 1$, which is a contradiction to the hypothesis that $n > 1$. Therefore, we have $1 \leq r < n$. Also, by substituting $s = nq + r$ in (*), we get that

$$(nq + r)a + vn = 1$$

$$\text{or} \quad ra + (v + qa)n = 1 \qquad (**)$$

which implies that $ra - 1$ is a multiple on n. Thus, $ra \equiv 1 \pmod{n}$ and $1 \leq r < n$. Also, from equation (**), we get that $(r, n) = 1$. If r' is any other integer such that $1 \leq r' < n$ and $r'a \equiv 1 \pmod{n}$, then $(r - r')\,a \equiv 0 \pmod{n}$, so that $|r - r'||a|$ is a multiple of n. Since $(|a|, n) = 1$, it follows that n divides $|r - r'|$. Since $1 \leq r, r' < n$, we get that $r - r' = 0$ or $r = r'$.

Theorem 2.2.5. Let $n \in \mathbb{Z}^+$ and a and $b \in \mathbb{Z}$ such that $(|a|, n) = 1$. Then, the linear congruence equation

$$ax \equiv b \pmod{n}$$

has a unique solution r in $\{0, 1, 2, \ldots, n-1\}$ and the set of all integer solutions of this is precisely equal to the congruence class $r + n\mathbb{Z}$.

Proof: By Theorem 2.2.4, there exists unique s such that $1 \le s < n$, $(s, n) = 1$ and $as \equiv 1 \pmod{n}$. Choose the unique r such that $0 \le r < n$ and $sb \in r + n\mathbb{Z}$ (or, equivalently, $sb \equiv r \pmod{n}$. Then,

$$ar \equiv asb \equiv b \pmod{n}.$$

Therefore, r is a solution of $ax \equiv b \pmod{n}$ in $\{0, 1, \ldots, n-1\}$. We prove that the congruence class $\equiv_n (r) = r + n\mathbb{Z}$ is equal to the set of integer solutions of $ax \equiv b \pmod{n}$. For any $x \in \mathbb{Z}$,

$$\begin{aligned} ax \equiv b \pmod{n} &\Leftrightarrow sax \equiv sb \pmod{n} \\ &\Leftrightarrow x \equiv r \pmod{n} \\ &\Leftrightarrow x \in r + n\mathbb{Z} \end{aligned}$$

The uniqueness of r is clear.

Example 2.2.1. Let us find all integer solutions of $55x \equiv 65 \pmod{80}$. First observe that, if m is a common divisor of a, b and n, then $ax \equiv b \pmod{n}$ if and only if $\frac{a}{m}x \equiv \frac{b}{m}\left(\mathrm{mod}\,\frac{n}{m}\right)$. Since 5 is a common divisor of 55, 65 and 80, $55x \equiv 65 \pmod{80}$ if and only if $11x \equiv 13 \pmod{16}$. Note that $(16, 11) = 1$. A quick check reveals that $3 \cdot 11 \equiv 1 \pmod{16}$. Therefore, the integer solutions of $11x \equiv 13 \pmod{16}$ are all $y \equiv 3 \cdot 13 \pmod{16}$ or, equivalently, all $y \equiv 7 \pmod{16}$ (since $39 \equiv 7 \pmod{16}$). Thus, integer solutions of $55x \equiv 65 \pmod{80}$ are members of $7 + 16\mathbb{Z}$ and those in $\{0, 1, 2, \ldots, 79\}$ are precisely 7, 23, 39, 55, 71.

The method that is followed in Example 2.2.1 above is formalized in the following.

Theorem 2.2.6. Let $1 < n \in \mathbb{Z}^+$ and a and $b \in \mathbb{Z}$. Then, $ax \equiv b \pmod{n}$ has an integer solution if and only if the g.c.d. (a, n) divides b.

Proof: Let $d = (a, n)$. Suppose that $ax \equiv b \pmod{n}$ has an integer solution. Let s be an integer solution of $ax \equiv b \pmod{n}$. Then, $as \equiv b \pmod{n}$ and hence $nr = as - b$ for some $r \in \mathbb{Z}$. Now, $b = as - nr$, d divides both a and n and hence

d divides b. Conversely suppose that d divides b. Put $a' = a/d$, $b' = b/d$ and $n' = n/d$. Then, $(a', n') = 1$ and hence, by Theorem 2.2.5, $a'x \equiv b' \pmod{n'}$ has an integer solution, say x_0. Then, n' divides $|a'x_0 - b'|$ and hence $n'q = a'x_0 - b'$ for some $q \in \mathbb{Z}$. Now

$$dn'q = da'x_0 - db'$$

and hence $nq = ax_0 - b$, which implies that $ax_0 \equiv b \pmod{n}$. Thus, $ax \equiv b \pmod{n}$ has an integer solution.

We close this section by developing tests for the divisibility of integers by various primes. These tests are easy for small primes, but these are not practical for large primes. The following is easy, since any integer is divisible by 2 if and only if the last digit in it is one of 0, 2, 4, 6, and 8.

Theorem 2.2.7. Let $a \in \mathbb{Z}^+$, then a is even if and only if $a \equiv r \pmod{10}$ for some $r \in \{0, 2, 4, 6, 8\}$. Also a is divisible by 5 if and only if $a \equiv 0 \pmod{10}$ or $a \equiv 5 \pmod{10}$ (that is, the last digit in a is 0 or 5).

Theorem 2.2.8. Let $a \in \mathbb{Z}^+$ and $a = a_r a_{r-1} \ldots a_1 a_0 = \sum_{i=0}^{r} a_i 10^i$, where a_i's are integers such that $0 \leq a_i \leq 9$. Then,

$$a \equiv \sum_{i=0}^{r} a_i \pmod{3},\ a \equiv \sum_{i=0}^{r} a_i \pmod{9}$$

and $a \equiv (a_0 + a_2 + a_4 + \cdots) - (a_1 + a_3 + a_5 + \cdots) \pmod{11}$.

Proof: Consider the polynomial given by

$$f(x) = a_0 + a_1 x + a_2 x^2 + \cdots + a_r x^r.$$

Then, $f(10) = \sum_{i=0}^{r} a_i 10^i = a$. Since $10 \equiv 1 \pmod{3}$, it follows from Corollary 2.2.1 (ii) that $f(10) \equiv f(1) \pmod{3}$. Then, $a \equiv \sum_{i=0}^{r} a_i \pmod{3}$. Also, since $10 \equiv 1 \pmod{9}$, we get that $a = f(10) \equiv f(1) = \sum_{i=0}^{r} a_i \pmod{9}$. Similarly, since $10 \equiv -1 \pmod{11}$, we get that $a = f(10) \equiv f(-1) = (a_0 + a_2 + \cdots) - (a_1 + a_3 + \cdots) \pmod{11}$.

Corollary 2.2.2. Let $a \in \mathbb{Z}^+$ and $a = \sum_{i=0}^{r} a_i 10^i$ with $0 \leq a_i \leq 9$. Then,

(i) 3 divides a if and only if 3 divides $\sum_{i=0}^{r} a_i$,

(ii) 9 divides a if and only if 9 divides $\sum_{i=0}^{r} a_i$, and

(iii) 11 divides a if and only if 11 divides $(a_0 + a_2 + \cdots) - (a_1 + a_3 + \cdots)$.

Theorem 2.2.9. Let $p > 5$ be a prime and $a \in \mathbb{Z}^+$ such that $a = 10k + a_0$, where k and a_0 are nonnegative integers and $0 \leq a_0 \leq 9$. Then, there exists unique $m_p \in \mathbb{Z}^+$ satisfying the following:

(i) $1 \leq m_p < p$

(ii) $10m_p \equiv 1 \pmod{p}$

(iii) For any $s \equiv m_p \pmod{p}$, p divides $a \Leftrightarrow p$ divides $k + sa_0$ (in particular, p divides $a \Leftrightarrow p$ divides $k + m_pa_0$).

Proof: Notice that, when a is represented in decimal system $a = \sum_{i=0}^{r} a_i 10^i$ or $a = a_r a_{r-1} \ldots a_1 a_0$ with $0 \leq a_i \leq 9$, then $a = 10k - a_0$ where $k = a_r a_{r-1} \ldots a_1 = \sum_{i=0}^{r} a_i 10^{i-1}$. Since p is prime and $p > 5$, $(p, 10) = 1$. By Theorem 2.2.4, there exists unique $m_p \in \mathbb{Z}^+$ such that $1 \leq m_p < p$, $(m_p, p) = 1$and $10m_p \equiv 1 \pmod{p}$. By Theorem 2.2.3,

$$am_p \equiv 10km_p + a_0m_p \equiv k + a_0m_p \text{ (since } 10m_p \equiv_p 1).$$

Therefore, p divides am_p if and only if p divides $k + a_0m_p$.

Since $(p, m_p) = 1$, it follows that

p divides $a \Leftrightarrow p$ divides am_p

$\Leftrightarrow p$ divides $k+m_pa_0$

$\Leftrightarrow p$ divides $k + sa_0$ for any $s \equiv m_p \pmod{p}$.

The above theorem can be better understood by the following examples, in which we test certain positive integers for their divisibility by a given prime $p > 5$.

Example 2.2.2

1. Let us test the divisibility of 62354 by 7. Here, $a = 62354 = 10k + a_0$, where $k = 6235$ and $a_0 = 4$ since $10.5 \equiv 1 \pmod{7}$, $m_7 = 5$.

   7 divides 62354 $\Leftrightarrow$ 7 divides $k - m_7a_0$

   $\Leftrightarrow$ 7 divides $6235 + 5.4$ (= 6255)

   $\Leftrightarrow$ 7 divides $625 + 5.5$ (= 650)

   $\Leftrightarrow$ 7 divides $65 + 5.0$ (= 65)

   Since 7 does not divide 65, it follows that 7 does not divide 62354.

2. Consider $a = 5876438$ and test its divisibility by 7 we have $10.5 \equiv 1 \pmod{7}$ and hence $m_7 = 5$.

   7 divides $a \Leftrightarrow$ 7 divides $587643 + 5.8$ (= 587683)

   $\Leftrightarrow$ 7 divides $58768 - 5.3$ (= 58783)

$\Leftrightarrow$ 7 divides 5878 + 5.3 (= 5893)
$\Leftrightarrow$ 7 divides 589 + 5.3 (= 604)
$\Leftrightarrow$ 7 divides 60 + 5.4 (=80)
$\Leftrightarrow$ 7 divides 8 + 5.0 (=8)

Since 7 does not divide 8, if follows that 7 does not divide 5876438.

Example 2.2.3

1. Test 7892654 for its divisibility by 11. Since $10.10 \equiv 1 \pmod{11}$, we have $m_{11} = 10$. Also, since $10 \equiv -1 \pmod{11}$, we can take $s = -1$ in Theorem 2.2.9 (iii).Therefore,

   11 divides 7892654 $\Leftrightarrow$ 11 divides 789265 + (−1) 4 (= 789261)
   $\Leftrightarrow$ 11 divides 78926 + (−1)1 (= 78925)
   $\Leftrightarrow$ 11 divides 7892 +(−1)5 (= 7887)
   $\Leftrightarrow$ 11 divides 788 + (−1)7 (= 781)
   $\Leftrightarrow$ 11 divides 78 + (−1)1(= 77),
   which is true.

   Thus, 11 divides 7892654. In this context, note that Corollary 2.2.2 (iii) is a better test for the divisibility by 11.
2. Test 7892654 for the divisibility by 13.
   Since $10.4 \equiv 1 \pmod{13}$, we have $m_{13} = 4$. Therefore,

   13 divides 7892654 $\Leftrightarrow$ 13 divides 789265 + 4.4 (= 789273)
   $\Leftrightarrow$ 13 divides 78927 + 4.3 (= 78939)
   $\Leftrightarrow$ 13 divides 7893 + 4.9 (= 7929)
   $\Leftrightarrow$ 13 divides 792 + 4.9 (= 828)
   $\Leftrightarrow$ 13 divides 82 + 4.8 (= 114)
   $\Leftrightarrow$ 13 divides 11 + 4.4 (= 27)

   Since 13 does not divide 27, it follows that the given number 7892654 is not divisible by 13.
3. Test whether 7892654 is divisible by 23.
   Since $10.7 \equiv 1 \pmod{23}$, $m_{23} = 7$.

   We have 23 divides 7892654 $\Leftrightarrow$ 23 divides 789265 + 7.4 (= 789293)
   $\Leftrightarrow$ 23 divides 78929 + 7.3 (= 78950)
   $\Leftrightarrow$ 23 divides 7895 + 7.0 (= 7895)
   $\Leftrightarrow$ 23 divides 789 + 7.5 (= 824)

$\Leftrightarrow$ 23 divides 82 + 7.4 (= 110)

$\Leftrightarrow$ 23 divides 11 + 7.0 (= 11),

which is not true.

Thus, 23 does not divide 7892654.

EXERCISE 2(B)

1. State whether the following are true or false.
   (i) $10^n \equiv 0 \pmod{2^n}$ for any $n \in \mathbb{Z}^+$
   (ii) $6789453 \equiv 5987654 \pmod 3$
   (iii) $237092 \equiv 236092 \pmod{100}$
   (iv) $13^2 \equiv 31^2 \pmod{23}$
   (v) $786 - 687 \pmod{11}$
2. Find the set of integer solutions for each of the following.
   (i) $15x \equiv 25 \pmod{35}$
   (ii) $21x \equiv 35 \pmod{49}$
   (iii) $25x \equiv 16 \pmod{20}$
   (iv) $27x \equiv 21 \pmod{24}$
   (v) $7x \equiv 16 \pmod{17}$
   (vi) $9x \equiv 14 \pmod{15}$
3. Prove that, for any prime p, $(p - 1)! + 1 \equiv 0 \pmod p$. (This is known as *Wilson's theorem.*)
4. Test the following divisibilities.
   (i) 876453 by 3
   (ii) 746538 by 9
   (iii) 587642 by 7
   (iv) 7896534 by 11
   (v) 87965325 by 17
   (vi) 97865432 by 19
   (vii) 67892345 by 23
   (viii) 79862345 by 29
5. For any prime p and for integer a, prove that $a^p \equiv a \pmod p$. (This is known as *Fermat's theorem.*)
6. Let $n \in \mathbb{Z}^+$. A set $\{a_0, a_1, \ldots, a_{n-1}\}$ of distinct integers is called a *transversal for congruence mod n* if $a_i \in i + n\mathbb{Z}$ for each $0 \leq i \leq n - 1$.

Prove the following for any transversal $\{a_0, a_1, \ldots, a_{n-1}\}$ for congruence modulo n.

(i) $a_i \not\equiv a_j \pmod{n}$ for any $i \neq j$

(ii) $a_i \equiv i \pmod{n}$ for any $0 \leq i \leq n - 1$

(iii) For any $a \in \mathbb{Z}$, $a \equiv a_i \pmod{n}$ for some $0 \leq i \leq n - 1$.

7. Find whether the following are transversals.

   (i) $\{0, 3, 2, 1\}$ for congruence mod 5.

   (ii) $\{-3, -2, -1, 0, 1, 2, 3\}$ for congruence mod 7.

   (iii) $\{0, 2, 2^2, 2^3, \ldots, 2^{10}\}$ for congruence mod 11.

   (iv) $\{2, 4, 6, \ldots, 2n\}$ for congruence mod n, if n is odd.

   (v) $\{a, 2a, 3a, \ldots, na\}$ for congruence mod n if $(a, n) = 1$

   (vi) $\{5, 10, 15, 20, \ldots, 105\}$ for congruence mod 21.

8. For any $n \in \mathbb{Z}^+$, prove that any n consecutive integers form a transversal for congruence mod n.

9. Characterise all n in each of the following cases that satisfy the given condition

   (i) $1 + 2 + 3 + \cdots + (n - 1) \equiv 0 \pmod{n}$

   (ii) $1^2 + 2^2 + 3^2 + \cdots + (n - 1)^2 \equiv 0 \pmod{n}$

   (iii) $1^3 + 2^3 + 3^3 + \cdots + (n - 1)^3 \equiv \pmod{n}$

10. For any $n \in \mathbb{Z}^+$, prove that 3 divides n implies 3 divides m for m any rearrangement of the digits in n.

11. For any $n \in \mathbb{Z}^+$, prove that $10^{3^n} \equiv 1 \pmod{3^{n+2}}$.

12. Find all the digits x $(0 \leq x \leq 9)$ for which 12x, 527, 846, 531 is divisible by 3; 9; or 11.

13. If $a_1, a_2, \ldots, a_n \in \{0, 1, 2, \ldots, 9\}$ and $a_1 \neq 0$, prove that 11 divides $a_1a_2\ldots a_na_na_{n-1}\ldots a_2a_1$.

14. Prove the following for any relatively prime positive integers m and n:

    (i) For any a and $b \in \mathbb{Z}^+$, $a \equiv b \pmod{mn} \Leftrightarrow a \equiv b \pmod{m}$ and $a \equiv b \pmod{n}$

    (ii) If c and d are both integer solutions of $x \equiv a \pmod{n}$ and of $x \equiv b \pmod{m}$, then $c \equiv d \pmod{mn}$.

15. For any $r \in \mathbb{Z}^+$, let $M_r = 1 + 100 + 100^2 + \cdots + 100^{r-1}$.

    (i) Prove that $M_r = 101010\ldots01$ having exactly r ones.

    (ii) Prove that each of 7, 9 and 11 divides M_r for infinitely many r's.

2.3 RATIONAL, REAL AND COMPLEX NUMBERS

Although we assume familiarity with the rational real, and complex numbers, we prefer to give the structure of rational numbers for the simple reason that we imitate this construction later in this book in a more general set up. We do not attempt the construction of real numbers, since it is outside the scope of this book.

Definition 2.3.1. Let $\mathbb{Z}$ be the set of all integers and $\mathbb{Z}^* = \mathbb{Z} - \{0\}$. We define a binary relation R on the set $\mathbb{Z} \times \mathbb{Z}^*$ as follows: For any (a, b) and (c, d) in $\mathbb{Z} \times \mathbb{Z}^*$,

$(a, b)\, R\, (c, d) \Leftrightarrow ad = bc$ (that is, the products ad and bc are equal).

The following is a straight forward verification.

Theorem 2.3.1. R is an equivalence relation on $\mathbb{Z} \times \mathbb{Z}^*$.

Definition 2.3.2. For any $(a, b) \in \mathbb{Z} \times \mathbb{Z}^*$, the equivalence class of (a, b) corresponding to R will be denoted by $\frac{a}{b}$. That is,

$$\frac{a}{b} = R(a, b) = \{(c, d) \in \mathbb{Z} \times \mathbb{Z}^* : (a, b)\, R\, (c, d)\}$$

i.e., $\frac{a}{b} = \{(c, d) \in \mathbb{Z} \times \mathbb{Z}^* : ad = bc\}$.

For example, $\frac{2}{3}$ represents the set of all pairs (c, d) of integers, with $d \neq 0$, such that $2d = 3c$. Note that, for any (a, b) and $(c, d) \in \mathbb{Z} \times \mathbb{Z}^*$,

$$\frac{a}{b} = \frac{c}{d} \Leftrightarrow (a, b)\, R\, (c, d) \Leftrightarrow ad = bc.$$

Definition 2.3.3. For any $(a, b) \in \mathbb{Z} \times \mathbb{Z}^*$, the R-equivalence class $\frac{a}{b}$ is called a *rational number* and the set of all rational numbers is denoted by $\mathbb{Q}$. That is,

$$\mathbb{Q} = \left\{\frac{a}{b} : a \text{ and } b \in \mathbb{Z} \text{ and } b \neq 0\right\}$$

Note that $\mathbb{Q}$ is precisely the quotient of $\mathbb{Z} \times \mathbb{Z}^*$ by R. In the following, we introduce the arithmetical operations addition, subtraction, multiplication and division. First of all, observe that the following holds for any $a, b, c, d \in \mathbb{Z}$ with $b \neq 0$ and $d \neq 0$.

$$\frac{ad}{bd} = \frac{a}{b}$$

$$\frac{0}{b} = \frac{0}{d} = \frac{0}{1}$$

$$\frac{a}{b} = \frac{0}{d} \Leftrightarrow a = 0$$

$$\frac{ab}{b} = \frac{a}{1}$$

Definition 2.3.4. For any $r = \frac{a}{b}$ and $s = \frac{c}{d}$ in $\mathbb{Q}$, we define the following.

$$r + s = \frac{a}{b} + \frac{c}{d} = \frac{ad + bc}{bd}$$

$$-r = -\left(\frac{a}{b}\right) = \frac{-a}{b}$$

$$r \cdot s = \left(\frac{a}{b}\right)\left(\frac{c}{d}\right) = \frac{ac}{bd}$$

$$\frac{1}{r} = \frac{b}{a} \text{ if } a \neq 0$$

$$\frac{r}{s} = \frac{\left(a/b\right)}{\left(c/d\right)} = \frac{ad}{bc}, \text{ if } c \neq 0.$$

It can be easily verified that $r + s$, $-r$, $r \cdot s$ and $\frac{r}{s}$ does not depend on the integers a, b, c and d, but they depend on the classes $\frac{a}{b}$ and $\frac{c}{d}$; that is,

$$\frac{a}{b} = \frac{a'}{b'} \text{ and } \frac{c}{d} = \frac{c'}{d'} \Rightarrow \frac{ad + bc}{bd} = \frac{a'd' + b'c'}{b'd'}, \quad \frac{-a}{b} = \frac{-a'}{b'}$$

$$\frac{ac}{bd} = \frac{a'c'}{b'd'} \text{ and } \frac{ad}{bc} = \frac{a'd'}{b'c'} \quad \text{if } c \neq 0 \text{ and } c' \neq 0.$$

That is, the operations $+$, $-$, $\cdot$ and / are well-defined. Also, it can be easily seen that the following arithmetical laws are satisfied in $\mathbb{Q}$.

1. $r + s = s + r$
2. $r + (s + t) = (r + s) + t$
3. $r + 0 = r$, where $0 = \frac{0}{1}$.

4. $r + (-r) = 0\ (= \frac{0}{1})$
5. $r \cdot (s \cdot t) = (r \cdot s) \cdot t$
6. $r \cdot (s + t) = r \cdot s + r \cdot t$
7. $r \cdot s = s \cdot r$
8. $r \cdot 1 = r$, where $1 = \frac{1}{1}\ (= \frac{b}{b}$ for any $0 \neq b \in \mathbb{Z})$
9. $r \cdot (\frac{1}{r}) = 1$

Definition 2.3.5. Any nonempty set together with the operations $+, -, \cdot$ and / satisfying the properties (1) to (9) above is called a *field*.

We will be discussing about fields in great detail later in Part III and Part IV of this book. We just want to highlight here that the set $\mathbb{Q}$ of rational numbers is a field. For any integer a, consider the rational number $\frac{a}{1}$ and, we can see that the map $a \mapsto \left(\frac{a}{1}\right)$ is an injection of the set $\mathbb{Z}$ of integers into the set $\mathbb{Q}$ of rational numbers. If we identify a with $a/1$, then we can see that $\mathbb{Z}$ is a subset of $\mathbb{Q}$ and, for any a and $b \in \mathbb{Z}$,

$$\frac{a}{1}+\frac{b}{1}=\frac{a+b}{1}, -\left(\frac{a}{1}\right)=\frac{-a}{1} \quad \text{and} \quad \frac{a}{1}\cdot\frac{b}{1}=\frac{ab}{1}.$$

These demonstrate that the usual arithmetical operations addition, subtraction and multiplication on the integers are simply the restrictions of those on rational numbers to $\mathbb{Z}$. Thus, for all practical purposes, we can treat integers as rational numbers by means of the identification of a with $\frac{a}{1}$.

As we have constructed rational numbers from integers, we can construct real numbers from rational numbers. However, the procedure is not as simple as the construction of rational numbers. We need some more techniques from analysis to construct real numbers from rational number. However, for the benefit of an enthusiastic reader, a brief sketch of the construction of real numbers is given in the exercises. The proofs are not very difficult, but require care, attention and some elementary knowledge about sequences, Cauchy sequences, convergent sequences and their limits. The real number system is denoted by $\mathbb{R}$ and it is known that $\mathbb{R}$ is a field.

Next we construct the system of complex numbers. Consider the Cartesian product $\mathbb{R} \times \mathbb{R}$, where $\mathbb{R}$ is the set of real numbers. We define addition, subtraction, multiplication and division to make $\mathbb{R} \times \mathbb{R}$ a field. For any $z = (a, b)$ and $w = (c, d)$ in $\mathbb{R} \times \mathbb{R}$, let us define

$$z + w = (a, b) + (c, d) = (a + c, b + d)$$
$$-z = -(a, b) = (-a, -b)$$
$$z \cdot w = (a, b) \cdot (c, d) = (ac - bd, ad + bc)$$

$$\frac{1}{z}=\frac{1}{(a,b)}=\left(\frac{a}{a^2+b^2},\frac{-b}{a^2+b^2}\right) \quad \text{if} \quad a \neq 0 \text{ or } b \neq 0$$

and
$$\frac{z}{w}=\frac{(a,b)}{(c,d)}=\left(\frac{ac+bd}{c^2+d^2},\frac{bc-ad}{c^2+d^2}\right) \quad \text{if} \quad c \neq 0 \text{ or } d \neq 0.$$

Theorem 2.3.2. $\mathbb{R} \times \mathbb{R}$, together with the operations defined above, is a field.

For any $a \in \mathbb{R}$, consider $(a, 0)$ in $\mathbb{R} \times \mathbb{R}$. Then, $a \mapsto (a, 0)$ is an injection of $\mathbb{R}$ into $\mathbb{R} \times \mathbb{R}$ and satisfies the following for any a and $b \in \mathbb{R}$.

$$(a + b, 0) = (a, 0) + (b, 0)$$
$$(-a, 0) = -(a, 0)$$
$$(ac, 0) = (a, 0)\cdot(c, 0)$$
$$\left(\frac{1}{a}, 0\right)=\frac{1}{(a,\ 0)} \quad \text{if} \quad a \neq 0$$
$$\left(\frac{c}{a}, 0\right)=\frac{(c,\ 0)}{(a,\ 0)} \quad \text{if} \quad a \neq 0.$$

These show that we can identify $\mathbb{R}$ with the subset $\mathbb{R} \times \{0\}$ of $\mathbb{R} \times \mathbb{R}$, by means of the injection $a \mapsto (a, 0)$ and that the usual arithmetical operations on $\mathbb{R}$ coincide with those on $\mathbb{R} \times \mathbb{R}$ restricted to $\mathbb{R}$ $(= \mathbb{R} \times \{0\})$.

Now, let us identify another distinguished element, namely $(0, 1)$, in $\mathbb{R} \times \mathbb{R}$. First observe that

$$(0, 1) \cdot (0, 1) = (-1, 0) = -1$$

since we are identifying $(a, 0)$ with a,

$$\text{Put } (0, 1) = i.$$

Then, $i^2 = -1$ and hence i is a root of the polynomial $x^2 + 1$ over $\mathbb{R}$. Further, any element $z = (a, b)$ in $\mathbb{R} \times \mathbb{R}$ can be expressed as

$$z = (a, 0) + (0, 1)(b, 0) = a + ib$$

by identifying $(a, 0)$ with a, $(b, 0)$ with b and $(0, 1)$ with i. Thus,

$$\mathbb{R} \times \mathbb{R} = \{a + ib : a \text{ and } b \in \mathbb{R}\}$$

where i is the element $(0, 1)$. $a + ib$ is the usual familiar form of complex numbers and let us agree to call any element of $\mathbb{R} \times \mathbb{R}$ as a complex number.

$\mathbb{R} \times \mathbb{R}$, together with the arithmetical operations defined above, is denoted by $\mathbb{C}$. Thus, we have

$$\mathbb{C} = \{a + ib : a \text{ and } b \in \mathbb{R}\}$$

and the arithmetical operations on $\mathbb{C}$ take the following form.

$$(a + ib) + (c + id) = (a + c) + i(b + d)$$
$$-(a + ib) = (-a) + i(-b)$$
$$(a + ib)\cdot(c + id) = (ac - bd) + i(ad + bc)$$
$$\frac{1}{(a+ib)} = \left(\frac{a}{a^2+b^2}\right) + i\left(\frac{-b}{a^2+b^2}\right)$$
$$\frac{(a+ib)}{(c+id)} = \left(\frac{ac+bd}{c^2+d^2}\right) + i\left(\frac{bc-ad}{c^2+d^2}\right)$$

Now, we have the following number systems.

$\mathbb{Z}^+$ = The set of positive integers
$\mathbb{N}$ = The set of nonnegative integers
$\mathbb{Z}$ = The set of integers
$\mathbb{Q}$ = The set of rational numbers
$\mathbb{R}$ = The set of real numbers
$\mathbb{C}$ = The set of complex numbers

These are interrelated by

$$\mathbb{Z}^+ \subset \mathbb{N} \subset \mathbb{Z} \subset \mathbb{Q} \subset \mathbb{R} \subset \mathbb{C}$$

in such a way that the usual arithmetic operations addition, subtraction and multiplication on each of these are precisely restrictions of those on the next system. Moreover, $\mathbb{Q}$, $\mathbb{R}$ and $\mathbb{C}$ are fields while the others are not.

We close this section with an additional operation, namely the complex conjugation, on $\mathbb{C}$.

Definition 2.3.6. For any $z = a + ib \in \mathbb{C}$ with a and $b \in \mathbb{R}$, the *complex conjugate* of z is defined as the complex number.

$$\overline{z} = a - ib\ (= a + i(-b)).$$

If $z = a + ib$, and a and $b \in \mathbb{R}$, then a and b are called *real part* and *imaginary part* of z, respectively.

The following are easy verifications.

Theorem 2.3.3. The following holds for any complex numbers z and z':

1. $\overline{z+z'} = \overline{z} + \overline{z}'$
2. $\overline{\overline{z}} = \overline{z}$
3. $\overline{zz'} = \overline{z}\,\overline{z}'$
4. $\frac{z+\overline{z}}{2}$ = The real part of z.
5. $\frac{z-\overline{z}}{2i}$ = The imaginary part of z
6. $z = \overline{z} \Leftrightarrow z \in \mathbb{R} \Leftrightarrow$ The real part of $z = 0$.

Theorem 2.3.4. For any $z = a + ib$ (a and $b \in \mathbb{R}$) $\in \mathbb{C}$,

$$z\,\overline{z} = a^2 + b^2,$$

The nonnegative square root of $z\overline{z}$ is called the *absolute value* of z and is denoted by $|z|$; That is,

$$|z|^2 = z\,\overline{z} = a^2 + b^2.$$

The map $z \mapsto |z|$ satisfies the following properties:

1. $|z + z'| \leq |z| + |z'|$
2. $|zz'| = |z||z'|$
3. $|z| = 0 \Leftrightarrow z = 0$
4. $|rz| = |r||z|$ for any real number r.
5. For any real number r, $|r| = \begin{cases} r & \text{if } r \geq 0 \\ -r & \text{if } r < 0 \end{cases}$

EXERCISE 2(C)

A sequence $\{a_n\}$ of rational numbers is said to be a *Cauchy sequence* if, for each positive rational number $\in$, there exists $n_0 \in \mathbb{Z}^+$ such that

$$|a_n - a_m| < \in \quad \text{for all } n \text{ and } m \geq n_0.$$

A sequence $\{a_n\}$ in $\mathbb{Q}$ is said to be convergent if there exists $r \in \mathbb{R}$ such that, for each rational $\in > 0$, there exists $n_0 \in \mathbb{Z}^+$ such that $|a_n - r| < \in$ for all $n \geq n_0$ and in this case we write $a_n \to r$ and r = limit of a_n.

Prove the following:

1. $\mathbb{Q}$ is countable and $\mathbb{R}$ and $\mathbb{C}$ are uncountable.

2. For any a and b in $\mathbb{Q}$ such that $a < b$, then the set $\{r \in \mathbb{Q} : a < r < b\}$ is bijective with $\mathbb{Q}$.

3. If a, b, c and $d \in \mathbb{Q}$ with $a < b$ and $c < d$, then $(a, b)_{\mathbb{Q}}$ is bijective with $(c, d)_{\mathbb{Q}}$, where

$$(a, b)_{\mathbb{Q}} = \{r \in \mathbb{Q} : a < r < b\}.$$

4. For any real numbers a, b, c and d with $a < b$ and $c < d$,

$$(a, b) \simeq (c, d) \simeq \mathbb{R} \text{ and } (a, b)_{\mathbb{Q}} \simeq (c, d)_{\mathbb{Q}} \simeq \mathbb{Q}$$

where $(a, b) = \{s \in \mathbb{R} : a < s < b\}$.

5. Between any two real numbers, there is a rational number.

6. $\left\{\frac{1}{n}\right\}$ is a Cauchy sequence.

7. Every convergent sequence in $\mathbb{Q}$ is a Cauchy sequence.

8. Let CS($\mathbb{Q}$) be the set of all Cauchy sequences in $\mathbb{Q}$.

   For any $\{a_n\}$ and $\{b_n\}$ in CS($\mathbb{Q}$), define

$$\{a_n\} \sim \{b_n\} \quad \text{if and only if } |a_n - b_n| \to 0.$$

   Then, $\sim$ is an equivalence relation on CS($\mathbb{Q}$).

9. For each $\{a_n\} \in$ CS($\mathbb{Q}$), there exists $r \in \mathbb{R}$ such that $a_n \to r$.

10. For each $r \in \mathbb{R}$, there exists $\{a_n\} \in$ CS($\mathbb{Q}$) such that $a_n \to r$ and, if $\{b_n\}$ is another Cauchy sequence in $\mathbb{Q}$ such that $b_n \to r$, then $\{a_n\} \sim \{b_n\}$.

11. The quotient CS($\mathbb{Q}$)/$\sim$ is bijective with $\mathbb{R}$.

12. For any $a \in \mathbb{Q}$, the sequence $\{a_n\}$, such that $a_n = a$ for all n, is called a *constant* sequence and is denoted by $\{a\}$. Then, $a \mapsto \widetilde{\{a\}}$ is an injection of $\mathbb{Q}$ into CS($\mathbb{Q}$)/$\sim$.

13. If $\{a_n\}$ and $\{b_n\} \in$ CS($\mathbb{Q}$), then $\{a_n + b_n\}$ and $\{a_n b_n\} \in \mathbb{Q}$.

14. For any $\{a_n\}$, $\{b_n\}$, $\{a_n'\}$ and $\{b_n'\} \in$ CS($\mathbb{Q}$),

$$\{a_n\} \sim \{a_n'\} \text{ and } \{b_n\} \sim \{b_n'\} \Rightarrow \{a_n + b_n\} \sim \{a_n' + b_n'\} \text{ and } \{a_n b_n\} \sim \{a_n' b_n'\}.$$

15. For any $\{a_n\}$ and $\{b_n\} \in$ CS($\mathbb{Q}$), define

$$\widetilde{\{a_n\}} + \widetilde{\{b_n\}} = \widetilde{\{a_n + b_n\}}$$

$$-\widetilde{\{a_n\}} = \widetilde{\{-a_n\}}$$

$$\widetilde{\{a_n\}} \cdot \widetilde{\{b_n\}} = \widetilde{\{a_n b_n\}},$$

where $\widetilde{\{a_n\}}$ is the $\sim$-equivalence class of $\{a_n\}$ in CS($\mathbb{Q}$). Then, the operations $+$, $-$ and $\cdot$ are well-defined on CS($\mathbb{Q}$)/$\sim$.

2.4 ORDERING

The well-ordering property of positive integers is with respect to the natural or usual ordering. This natural ordering is there on the rational number system and the real number system also. However, there is no such ordering on the complex number system. In this section, we introduce the abstract concept of a partial ordering on a given set and discuss its elementary properties in general and those of the natural ordering on $\mathbb{R}$ in particular. Let us begin with the following.

Definition 2.4.1. Let X be a nonempty set. A binary relation θ on X is said to be a *partial order* or a *partial ordering* if θ is reflexive, transitive and anti-symmetric (that is, $a\ \theta\ b$ and $b\ \theta\ a$ only when $a = b$). A pair $(X, \leq)$ is called a *partially ordered set* or, simply, a *poset* if X is a nonempty set and $\leq$ is a partial order on X.

A partial order is usually denoted by the symbol $\leq$ (which is read as 'less than or equal to'). It can be easily verified that, if $\leq$ is partial order on a set X, then the inverse of $\leq$ is also a partial order on X and is denoted $\geq$. That is, for any a and $b \in X$,

$$a \leq b \Leftrightarrow b \geq a.$$

If $a \leq b$ and $a \neq b$, then we write $a < b$.

Example 2.4.1

1. $(\mathbb{Z}^+, \leq)$, $(\mathbb{Z}, \leq)$, $(\mathbb{Q}, \leq)$ and $(\mathbb{R}, \leq)$ are all partially ordered sets, where $\leq$ is the natural ordering.
2. For any nonempty set X, the equality relation is a partial order on X. That is, for any a and $b \in X$, if we define $a \leq b$ if and only if $a = b$, then $\leq$ is a partial order on X. Note that this is the only binary relation on X which is both an equivalence relation and a partial order on X.
3. Let $\mathbb{P}(X)$ be the set of all subsets of a given set X. For any A and $B \in \mathbb{P}(X)$, define

   $$A \leq B \text{ if and only if } A \text{ is a subset of } B.$$

   Then, $\leq$ is a partial order on $\mathbb{P}(X)$.
4. Let X be the set of all real valued functions defined on a set A (that is, $X = \mathbb{R}^A$). For any f and $g \in X$, define

   $$f \leq g \text{ if and only if } f(a) \leq f(b) \text{ in } \mathbb{R} \text{ for all } a \in A.$$

   Then, $\leq$ is a partial order on X.

5. Let $(X_1, \leq), (X_2, \leq), \ldots, (X_n, \leq)$ be posets and $X = X_1 \times X_2 \times \ldots \times X_n$. For any $a = (a_1, a_2, \ldots, a_n)$ and $b = (b_1, b_2, \ldots, b_n)$, define

$$a \leq b \text{ if and only if } a_i \leq b_i \quad \text{for all } 1 \leq i \leq n.$$

Then, $\leq$ is a partial order on X and is called the *coordinate-wise ordering*.

6. In Example (5) above, define $a \leq b$ if and only if either $a = b$ or there is i_0, $1 \leq i_0 \leq n$, such that $a_i = b_i$ for all $i < i_0$ and $a_{i_0} < b_{i_0}$. Then, $\leq$ is a partial order on $X = X_1 \times X_2 \times \cdots \times X_n$ and is called the *lexicographic ordering* or *dictionary ordering*.

Definition 2.4.2. A partial order $\leq$ on a set X is called a *total order* if, for any a and $b \in X$, either $a \leq b$ or $b \leq a$ and, in this case, $(X, \leq)$ is called a *totally ordered set*.

$\mathbb{R}$ together with the natural ordering is totally ordered set. In Examples (2) and (3), the partial orders are not total orders, except when X has at most one element.

Definition 2.4.3. Let $(X, \leq)$ be a poset, $A \subseteq X$ and $x \in X$. Then, x is called a *lower bound* (*upper bound*) of A in X if $x \leq a$ (respectively $a \leq x$) for all $a \in A$. If A has a lower bound (upper bound) in X, then A is said to be *bounded below* (respectively, *bounded above*). A is said to be *bounded* if it is both bounded below and bounded above. A lower bound x of A is called *greatest lower bound* if $y \leq x$ for all lower bounds y of A in X and it is denoted by $\text{glb}_X A$ or, simply, glb A when there is no ambiguity about X. Similarly, an upper bound x of A is called *least upper bound* and denoted by $\text{lub}_X A$ or lub A if $x \leq y$ for all upper bounds y of A in X.

Example 2.4.2

1. In $(\mathbb{R}, \leq)$, $\mathbb{Z}^+$ is bounded below and not bounded above, while the set $\mathbb{Z}^-$ of negative integers is bounded above and not bounded below.
2. 1 is the lub of the interval $(0, 1)$ and 0 is the glb of $(0, 1)$ in $(\mathbb{R}, \leq)$.

Definition 2.4.4. Let $(X, \leq)$ be a poset, $A \subseteq X$ and $a_0 \in A$. a_0 is said to be the *least* (*greatest*) *element* of A if $a_0 \leq a$ is (respectively $a \leq a_0$) for all $a \in A$. a_0 is said to be a *maximal element* of A if there is no element a in A with $a_0 < a$. Similarly, a_0 is said to be a *minimal element* of A if there is no element a in A with $a < a_0$.

Clearly any subset of a poset can have at most one least element and at most one greatest element. Also, the least element (greatest element), if it exists, is minimal (respectively, maximal). However, a subset A of a poset may possess more than one minimal (maximal) elements and any minimal (maximal) element is not necessarily the least (respectively, greatest) element.

Example 2.4.3

1. The interval (0, 1) in ($\mathbb{R}$, $\leq$) has neither a minimal element nor a maximal element.
2. Let X be a set with more than one element and consider the poset ($\mathbb{P}(X)$, $\subseteq$) of all subsets of X. Let Y be the set of all nonempty subsets of X. Then, Y has minimal elements; in fact, for any $x \in X$, $\{x\}$ is a minimal element in Y and is not the least element, since $\{x\} \nsubseteq \{y\}$ for any $y \neq x$ in X. Also, let Z be the set of all proper subsets of X. Then, Z has maximal elements; in fact, for any $x \in X$, $X - \{x\}$ is a maximal element in Z and is not greatest in Z.

Definition 2.4.5. Let $(X, \leq)$ be a poset. Any nonempty subset A in X, such that, for any a and $b \in A$, either $a \leq b$ or $b \leq a$, is called a *chain* in X.

In other words, any totally ordered subset of a poset is called a chain.

Example 2.4.4

1. $\mathbb{Z}$ is a chain in ($\mathbb{R}$, $\leq$). In fact, $\mathbb{R}$ itself is a chain in ($\mathbb{R}$, $\leq$) and hence any nonempty subset of $\mathbb{R}$ is a chain.
2. If $X = \{a, b, c, d\}$, then

$$A = \{\emptyset, \{a\}, \{a, b\}, \{a, b, c\}, X\}$$

is a chain in the poset ($\mathbb{P}(X)$, $\subseteq$).

3. Consider the set $\mathbb{Z}^+$ of positive integers and, for any a and $b \in \mathbb{Z}^+$, define a/b if a divides b (that is, $ac = b$ for some $c \in \mathbb{Z}^+$). Then, | is a partial order on $\mathbb{Z}^+$. For any $a \in \mathbb{Z}^+$,

$$\{a^n : n \in \mathbb{Z}^+\}$$

is a chain in ($\mathbb{Z}^+$, 1).

The following is an important axioms of set theory, though its popular name is Zorn's lemma. It is a lemma used to prove some other equivalent axioms of set theory.

Zorn's lemma 2.4.1. Let $(X, \leq)$ be a poset in which each chain has an upper bound in X. Then, $(X, \leq)$ has a maximal element.

The following is an equivalent form of Zorn's lemma and, in this form only the Zorn's lemma is used several times in this book.

Corollary 2.4.1 (An equivalent form of Zorn's lemma). Let $\mathcal{S}$ be a class of subsets of a given set X satisfying the following.

If $\mathscr{C}$ is a subclass of $\mathscr{S}$ such that, for any A and $B \in \mathscr{C}$, either $A \subseteq B$ or $B \subseteq A$, then $\bigcup_{A \in \mathscr{C}} A \in \mathscr{S}$. (In this case, $\mathscr{S}$ is said to be closed under unions of chains.) Then, $\mathscr{S}$ has a maximal member, in the sense that, there is a member M in $\mathscr{S}$ such that M is not properly contained in any other member of $\mathscr{S}$.

Next, we take up a brief discussion of the well-ordering principle which is also an axiom equivalent to the Zorn's lemma. First, let us have the following.

Definition 2.4.6. Let X be a set. A partial order $\leq$ on X is called a *well-order* if every nonempty subset of X has a least element with respect to $\leq$.

It can be easily seen that any well-order on a set is a total order; for, the set $\{a, b\}$ should possess a least element which must be either a or b. However, there are total orders which are not well-orders. Consider the examples given below.

Example 2.4.5

1. The natural order on $\mathbb{Z}^+$ is a well-order (by Theorem 2.1.1).
2. The natural order on $\mathbb{Q}$ is a total order, but not a well-order; for, the interval $(0, 1) \cap \mathbb{Q}$ has no least member, since for any $0 < a < 1$, there is a rational number r such that $0 < r < a$.
3. The division order | on $\mathbb{Z}^+$ (that $a|b$ if a divides b) is not a total order (for example, if p and q are distinct primes, then $p \nmid q$ and $q \nmid p$) and hence not a well-order.

The Principle of Well-ordering 2.4.1. Any nonempty set can be well-ordered; that is, if X is a given nonempty set, then there is a well-order on X.

We close this section with a mention of another important axiom of set theory, namely the *axiom of choice* which is known to be equivalent to each of the Zorn's lemma and the principle of well-ordering. First, we have the following.

Definition 2.4.7. Let $\{A_i\}_{i \in I}$ be an indexed nonempty class of nonempty sets (that is, I is a nonempty set and, each A_i, $i \in I$, is a nonempty set). Then, any function $c : I \to \bigcup_{i \in I} A_i$ such that $c(i) \in A_i$ for all $i \in I$ is called a *choice function*.

This amounts to saying that a choice function c is simply choosing one element from each A_i, $i \in I$. If the index set I is a finite set, say $I = \{1, 2, \ldots, n\}$, then the choice functions can be easily seen to be just elements of the Cartesian product $A_1 \times A_2 \times \cdots \times A_n$ and, in this case, the existence of choice function is precisely equivalent to say that the Cartesian product $A_1 \times A_2 \times \cdots \times A_n$ is nonempty. This idea can be extended to define the Cartesian product of infinite class of sets. That is, if $\{A_i\}_{i\in I}$ is an infinite class of nonempty sets, then their Cartesian product can be defined as

$$\prod_{i\in I} A_i = \{c : I \rightarrow \bigcup_{i\in I} A_i : c(i) \in A_i \text{ for all } i \in I\}.$$

The axiom of choice, given below, say that the Cartesian product of any nonempty class of nonempty sets is a nonempty set.

The Axiom of Choice 2.4.1. Given any nonempty class $\{A_i\}_{i\in I}$ of nonempty sets, there is a choice function $c : I \rightarrow \bigcup_{i\in I} A_i$ (that is, c is a function such that $c(i) \in A_i$ for all $i \in I$).

EXERCISE 2(D)

1. List all the partial orders on a 2-element set, a 3-element set and a 4-element set.

2. Prove that the number of partial orders on an n-element set is less than or equal to $2^{\frac{n(n-1)}{2}}$.

3. Prove that the lexicographic ordering on $X_1 \times X_2 \times \cdots \times X_n$ is a total ordering if and only if the partial orders on each X_i is a total order.

4. Prove that any well-order on any set is a total order.

5. Give an example of total order which is not a well-order.

6. Let $(X_1, \leq), \ldots, (X_n, \leq)$ be posets and $X = X_1 \times X_2 \times \cdots \times X_n$. Prove that the lexicographic ordering on X is a well-order if and only if the partial order $\leq$ on each of the X_i's is a well-order.

2.5 MATRICES

Though matrices are originated from the study of solutions of certain systems of linear equations and are later found to be in one-to-one correspondence with linear transformations of a finite dimensional linear space into another finite dimensional linear space, but these have acquired an independent status and form one of the most important areas of study in modern abstract algebra. In particular, matrices are a rich source of examples and counter examples of several concepts in noncommutative algebraic structures which we come across throughout this book. Actually, we study later in detail about matrices over an abstract ring. However, in this section, we briefly discuss matrices over the real number system or complex number system. Let us begin with the following.

Definition 2.5.1. For any positive integer n, let I_n denote the set of integers from 1 to n; that is,

$$I_n = \{1, 2, \ldots, n\}.$$

For any m and $n \in \mathbb{Z}^+$, a mapping

$$A : I_m \times I_n \rightarrow \mathbb{R} \text{ (or } \mathbb{C}\text{)}$$

is called an *$m \times n$ matrix over $\mathbb{R}$ or $\mathbb{C}$*, as the case may be or, simply, an *$m \times n$ matrix*, when there is no ambiguity about $\mathbb{R}$ or $\mathbb{C}$. An $m \times n$ matrix A is usually represented by the values $A(i, j)$, which are real or complex numbers, for each $1 \leq i \leq m$ and $1 \leq j \leq n$ and we express the matrix A by writing

$$A = (a_{ij}), \quad \text{where } a_{ij} = A(i, j).$$

Also, we express an $m \times n$ matrix A by an array of mn real or complex numbers a_{ij}, $1 \leq i \leq m$ and $1 \leq j \leq n$, written in m rows and n columns with a_{ij} in the ith row and j^{th} column as exhibited below.

$$A = \begin{pmatrix} a_{11} & a_{12} & a_{13} & \cdots & a_{1n} \\ a_{21} & a_{22} & a_{23} & \cdots & a_{2n} \\ \vdots & & & & \\ \cdots & \cdots & a_{i3} & \cdots & a_{in} \\ \vdots & & & & \\ a_{m1} & a_{m2} & a_{m3} & \cdots & a_{mn} \end{pmatrix}$$

Here, a_{ij} is called the ij^{th} *entry* in the matrix A and $m \times n$ is called the *size* of A. Actually, the size of A is not an integer, but it is a pair (m, n) (which is usually written as $m \times n$) of integers. An $m \times n$ matrix $A = (a_{ij})$ and an $r \times s$ matrix $B = (b_{ij})$ are said to be equal if $m = r$, $n = s$ and $a_{ij} = b_{ij}$ for all $1 \leq i \leq m$ and $1 \leq j \leq n$; that is, A and B have equal number of rows and equal number of columns and have the same ij^{th} entry for each i and j. The n-tuple $(a_{i_1}, a_{i_2}, \ldots, a_{i_n})$ is called the ith *row* and the m-tuple $(a_{1_j}, a_{2_j}, \ldots, a_{m_j})$ is called the jth *column* of the $m \times n$ matrix $A = (a_{ij})$. A $1 \times n$ matrix is called a *row matrix* and an $m \times 1$ matrix is called a *column matrix*.

Definition 2.5.2. An $n \times n$ matrix is called a *square matrix* of order n and the n-tuple $(a_{11}\ a_{22} \ldots a_{nn})$ is called the *diagonal* of a square matrix $A = (a_{ij})$. A is called a *diagonal matrix* if $a_{ij} = 0$ for any $i \neq j$; that is, except the entries on the diagonal, all other entries are 0. A diagonal matrix $A = (a_{ij})$ is called a *scalar matrix* if $a_{11} = a_{22} \cdots = a_{nn}$

Definition 2.5.3. A square matrix $A = (a_{ij})$ is called an *upper triangular matrix* if $a_{ij} = 0$ for all $i > j$ and A is called a *lower triangular matrix* if $a_{ij} = 0$ for all $i < j$.

The set of all $m \times n$ matrices over the real number system (complex number system) is denoted by $M_{m\times n}(\mathbb{R})$ ($M_{m\times n}(\mathbb{C})$, respectively). The set of all square matrices of order n is denoted by $M_n(\mathbb{R})$ or $M_n(\mathbb{C})$ as the case may be.

Example 2.5.1

1. $\begin{pmatrix} 2 & 3 & 1 & -1 \\ 1 & 0 & 2 & 3 \\ 0 & 1 & 1 & 1 \end{pmatrix}$ is a 3×4 matrix over $\mathbb{R}$ (over $\mathbb{C}$ also, since $\mathbb{R} \subseteq \mathbb{C}$).

2. $\begin{pmatrix} 1 & 2 & -1 \\ 2 & 3 & 0 \\ 2 & i & -i \end{pmatrix}$ is a square matrix of order 3 over $\mathbb{C}$.

3. $\begin{pmatrix} 1 & 2 & 0 \\ 0 & 2 & 3 \\ 0 & 0 & -1 \end{pmatrix}$ is an upper triangular matrix of order 3.

4. $\begin{pmatrix} 2 & 0 & 0 & 0 \\ 3 & 1 & 0 & 0 \\ -1 & 2 & 2 & 0 \\ 5 & -2 & 0 & 4 \end{pmatrix}$ is a lower triangular matrix of order 4.

5. $\begin{pmatrix} 2 & 0 & 0 \\ 0 & 1 & 0 \\ 0 & 0 & -1 \end{pmatrix}$ is a diagonal matrix.

6. $\begin{pmatrix} 2 & 0 & 0 \\ 0 & 2 & 0 \\ 0 & 0 & 2 \end{pmatrix}$ is a scalar matrix.

Note that, for any positive integer n, if we define $f: \mathbb{R} \to M_n(\mathbb{R})$ by $f(a) = (a_{ij})$, where

$$a_{ij} = \begin{cases} a \text{ if } i = j \\ 0 \text{ if } i \neq j \end{cases},$$

for any $a \in \mathbb{R}$, then f is an injection of $\mathbb{R}$ into $M_n(\mathbb{R})$ and therefore, we can identify a in $\mathbb{R}$ with the scalar matrix $f(a)$ in $M_n(\mathbb{R})$ and hence $\mathbb{R}$ can be identified with the subset of $M_n(\mathbb{R})$ consisting of all scalar matrices of order n. Similarly $\mathbb{C}$ can be identified with a subset of $M_n(\mathbb{C})$. In the following, we extend the arithmetical operations addition, subtraction and multiplication in the real and complex number systems to $M_n(\mathbb{R})$ or $M_n(\mathbb{C})$. We do this in a more general set up.

Definition 2.5.4. Let m and n be any positive integers and $A = (a_{ij})$ and $B = (b_{ij})$ be any $m \times n$ matrices over $\mathbb{R}$ or $\mathbb{C}$. Then, we define
$A + B = (c_{ij})$, where $c_{ij} = a_{ij} + b_{ij}$ for all $1 \leq i \leq m$ and $1 \leq j \leq n$ and define

$$-A = (-a_{ij})$$

$A + B$ will also be expressed as $A + B = (a_{ij} + b_{ij})$.

Example 2.5.2

Let $A = \begin{pmatrix} 2 & 3 & 1 & -4 \\ -1 & 2 & 0 & -3 \\ 4 & 5 & -2 & 2 \end{pmatrix}$ and $B = \begin{pmatrix} 3 & 1 & 2 & 5 \\ 2 & 3 & -1 & 4 \\ -1 & -2 & 0 & 1 \end{pmatrix} \in M_{3\times 4}(\mathbb{R})$.

Then, $A + B = \begin{pmatrix} 2+3 & 3+1 & 1+2 & -4+5 \\ -1+2 & 2+3 & 0+(-1) & -3+4 \\ 4+(-1) & 5+(-2) & -2+0 & 2+1 \end{pmatrix}$

$$= \begin{pmatrix} 5 & 4 & 3 & 1 \\ 1 & 5 & -1 & 1 \\ 3 & 3 & -2 & 3 \end{pmatrix}$$

and $-A = \begin{pmatrix} -2 & -3 & -1 & 4 \\ 1 & -2 & 0 & 3 \\ -4 & -5 & 2 & -2 \end{pmatrix}$.

Definition 2.5.5. Let m, n and r be any positive integers and $A = (a_{ij}) \in M_{m\times n}(\mathbb{R})$ and $B = (b_{ij}) \in M_{n\times r}(\mathbb{R})$. Then, we define the product AB as an $m \times r$ matrix given by

$$A \cdot B = (c_{ij}) \quad \text{where } c_{ij} = \sum_{k=1}^{n} a_{ik} b_{kj}$$

for all $1 \leq i \leq m$ and $1 \leq j \leq r$. That is,

$$c_{ij} = a_{i1}b_{1j} + a_{i2}b_{2j} + \cdots + a_{in}b_{nr}.$$

Note that for the product AB to be defined it is necessary that the number of columns in A must be equal to the number of rows in B. Therefore, even if AB is defined, BA may not be defined. If we define the dot product of two n-tuples, $a = (a_1, a_2, \ldots, a_n)$ and $b = (b_1, b_2, \ldots, b_n)$ by

$$a \cdot b = a_1b_1 + a_2b_2 + \cdots + a_nb_n,$$

then the ij^{th} entry in the product AB is precisely the dot product of ith row in A and j^{th} column in B. Also, note that the product of any two square matrices of the same order is always defined.

Example 2.5.3. Let $A = \begin{pmatrix} 2 & 1 & 3 & 2 \\ 3 & 2 & -1 & 0 \\ 6 & 4 & 2 & 1 \end{pmatrix} \in M_{3\times 4}(\mathbb{R})$ and $B = \begin{pmatrix} 3 & 2 & -1 \\ 2 & 0 & 1 \\ 1 & -1 & 0 \\ 4 & 2 & 3 \end{pmatrix} \in M_{4\times 3}(\mathbb{R})$

Then, $A \cdot B = (c_{ij})$, where

$C_{11} = 2 \cdot 3 + 1 \cdot 2 + 3 \cdot 1 + 2 \cdot 4 = 19$
$C_{12} = 2 \cdot 2 + 1 \cdot 0 + 3\,(-1) + 2\,2 = 5$
$C_{13} = 2 \cdot (-1) + 1{\cdot}1 + 3{\cdot}0 + 2{\cdot}3 = 5$
$C_{21} = 3 \cdot 3 + 2{\cdot}2 + (-1){\cdot}1 + 0 \cdot 4 = 12$
$C_{22} = 3 \cdot 2 + 2{\cdot}0 + (-1) \cdot (-1) + 0 \cdot 2 = 7$
$C_{23} = 3 \cdot (-1) + 2 \cdot 1 + (-1) \cdot 0 + 0 \cdot 3 = -1$
$C_{31} = 6 \cdot 3 + 4 \cdot 2 + 2 \cdot 1 + 1 \cdot 4 = 32$
$C_{32} = 6 \cdot 2 + 4 \cdot 0 + 2{\cdot}(-1) + 1 \cdot 2 = 12$
and $C_{33} = 6 \cdot (-1) + 4 \cdot 1 + 2 \cdot 0 + 1 \cdot 3 = 1$

and hence $AB = \begin{pmatrix} 19 & 5 & 5 \\ 12 & 7 & -1 \\ 32 & 12 & 1 \end{pmatrix}$. Similarly, we can compute BA and see that $AB \neq BA$. Here, note that AB is a 3×3 matrix and BA is a 4×4 matrix. Even if AB and BA are of same size, they may not be equal; for, consider the matrices

$$A = \begin{pmatrix} 0 & 2 & 1 \\ 0 & 1 & 2 \\ 3 & 0 & 1 \end{pmatrix} \quad \text{and} \quad B = \begin{pmatrix} 1 & 0 & 2 \\ 2 & 1 & 0 \\ 3 & 2 & 1 \end{pmatrix}.$$

Then, $AB = \begin{pmatrix} 7 & 4 & 1 \\ 8 & 5 & 2 \\ 6 & 2 & 7 \end{pmatrix}$ and $BA = \begin{pmatrix} 6 & 2 & 3 \\ 0 & 5 & 4 \\ 3 & 8 & 8 \end{pmatrix}$ and therefore $AB \neq BA$.

Definition 2.5.6

1. For any m and $n \in \mathbb{Z}^+$, the $m \times n$ matrix all of whose entries are zero is called the *zero matrix* and is denoted by $O_{m\times n}$ or, simply O, when there is ambiguity about the size of the matrix.
2. For any $n \in \mathbb{Z}^+$, the square matrix (δ_{ij}) is called the *identity matrix* of order n, where δ_{ij} is defined by

$$\delta_{ij} = \begin{cases} 1 & \text{if } i = j \\ 0 & \text{if } i \neq j \end{cases}.$$

The identity matrix of order n is denoted by $I_{n\times m}$ or, simply I, when there is no ambiguity about the order of the matrix.

Theorem 2.5.1. Let m and $n \in \mathbb{Z}^+$ and A, B and C be $m \times n$ matrices over $\mathbb{R}$ or $\mathbb{C}$. Then, the following holds:

1. $A + B = B + A$
2. $(A + B) + C = A + (B + C)$
3. $A + O_{m\times n} = A$
4. $A + (-A) = O_{m\times n}$

Theorem 2.5.2. The following holds for any matrices A, B and C, in the sense that whenever one side of an equation is defined, then the other side is also defined and both sides of that equation are equal.

1. $A(BC) = (A\,B)C$
2. $A(B + C) = AB + AC$
3. $(A + B)C = AC + BC$
4. $AI = A = IA$, where I is the identity matrix of appropriate order.

In addition to the operations addition and multiplication of matrices, we have yet another operation of matrices, namely the scalar multiplications. The real or complex numbers are called scalars and we multiply any matrix by any scalar as defined below.

Definition 2.5.7. Let $A = (a_{ij})$ be an $m \times n$ matrix and a be a scalar; that is $a \in \mathbb{R}$ or $\mathbb{C}$. Then, the matrix aA is defined as

$$aA = (aa_{ij}).$$

That is, aA is obtained by multiplying each entry a_{ij} of A by a to get ij^{th} entry of aA.

Theorem 2.5.3. The following holds for any matrices A and B and for any scales a and b.

1. $a(A + B) = aA + aB$, whenever A and B are of same size.
2. $(a + b)A = aA + bA$
3. $a(AB) = (aA)B = A = (aB)$, whenever $A \cdot B$ is defined.
4. $a(bA) = b(aA) = (ab)A$

Definition 2.5.8. Let $n \in \mathbb{Z}^+$, $1 \leq i \leq n$ and $1 \leq j \leq n$. Then, the $n \times n$ matrix whose ij^{th} entry is 1 and all other entries are 0 is called a *matrix unit* and is denoted by E_{ij}. Note that each E_{ij} is a square matrix of order n. In general, the order of E_{ij} is not mentioned in the notation of the matrix unit E_{ij} and the order is to be understood as per the context. However, we call E_{ij} as the $n \times n$ *matrix unit,* when it is necessary to mention the order of E_{ij}. Note that, for any scalar a, aE_{ij} is the square matrix whose ij^{th} entry is a and all other entries are 0 and hence we have the following.

Theorem 2.5.4

1. Any $n \times n$ matrix A can be expressed as

$$A = \sum_{i=1}^{n}\sum_{j=1}^{n} a_{ij} E_{ij}$$

2. $E_{ij}E_{rs} = \delta_{jr}E_{is}$, where $\delta_{jr} = 1$ or 0 according as $j = r$ or $j \neq r$.

We close this section by introducing another important operation on matrices.

Definition 2.5.9. For any $m \times n$ matrix $A = (a_{ij})$, the *transpose* of A is defined as the $n \times m$ matrix obtained by interchanging the rows and columns of A. The transpose of A is denoted by A^t; that is,

$$A^t = (a_{ji}).$$

Example 2.5.4

1. If $A = \begin{pmatrix} 1 & 2 & 3 \\ 2 & -1 & 0 \end{pmatrix}$, then $A^t = \begin{pmatrix} 1 & 2 \\ 2 & -1 \\ 3 & 0 \end{pmatrix}$

2. If $A=\begin{pmatrix} 3 & 1 & 2 & 3 \\ -1 & 0 & 2 & -4 \\ 4 & -2 & -3 & 0 \end{pmatrix}$, then $A^t=\begin{pmatrix} 3 & -1 & 4 \\ 1 & 0 & -2 \\ 2 & 2 & -3 \\ 3 & -4 & 0 \end{pmatrix}$.

3. If A is a square matrix, then A^t is also a square matrix of order same as of A.

Theorem 2.5.5. The following holds for any $m \times n$ matrices A and B.

1. $(A^t)^t = A$
2. $(aA)^t = aA^t$
3. $(A + B)^t = A^t + B^t$
4. $(-A)^t = -A^t$

Definition 2.5.10. A square matrix A of order n is said to be *nonsingular* or *invertible* if there exists a square matrix B of order n such that

$$AB = I = BA.$$

A matrix is said to be *singular* if it is not nonsingular.

Theorem 2.5.6

1. If A is an $m \times n$ matrix and B is an $n \times r$ matrix, then $(AB)^t = B^tA^t$.
2. If A is a nonsingular square matrix, then there exists a unique square matrix B such that

$$AB = I = BA$$

and this B is called the inverse of A and is denoted by A^{-1}.

3. If A and B are nonsingular square matrices of the same order, then AB is nonsingular and $(AB)^{-1} = B^{-1}A^{-1}$.
4. A square matrix A is nonsingular if and only if its transpose A^t is nonsingular and, in this case $(A^t)^{-1} = (A^{-1})^t$.

EXERCISE 2(E)

1. Compute the following for the matrices

$$A=\begin{pmatrix} 2 & 3 & 1 \\ -1 & 2 & -3 \\ 1 & 3 & -2 \end{pmatrix},\ B=\begin{pmatrix} 2 & 1 & 0 \\ 0 & 2 & -1 \\ 1 & 0 & -2 \end{pmatrix} \text{ and } C=\begin{pmatrix} 0 & 1 & 3 \\ 2 & 3 & 0 \\ 1 & 0 & 2 \end{pmatrix}$$

(i) $A + B$

(ii) $(A + B) + C$

(iii) $B + C$

(iv) $A + (B + C)$

(v) AB

(vi) A^t

(vii) B^t

(viii) B^tA^t

(ix) BA

(x) A^tB^t

2. For any two matrices A and B, prove that both AB and BA are defined if and only if A and B^t are of the same size and that, in this case, both AB and BA are square matrices.

3. Prove that a square matrix A of order n is a scalar matrix if and only if $AB = BA$ for all square matrices B of order n.

4. Prove Theorem 2.5.2.

5. Prove Theorem 2.5.3.

6. For any scalar a, let S_a be the $n \times n$ scalar matrix in which all the diagonal entries are a and other entries are 0. Prove that $S_aA = aA = AS_a$ for all $n \times n$ matrices A.

7. Prove Theorems 2.5.4 and 2.5.5.

8. Prove Theorem 2.5.6.

9. Prove the following for any integer $n \geq 0$.

(i) $\begin{pmatrix} 1 & 1 \\ 0 & 1 \end{pmatrix}^n = \begin{pmatrix} 1 & n \\ 0 & 1 \end{pmatrix}$

(ii) $\begin{pmatrix} 1 & 1 & 0 \\ 0 & 1 & 1 \\ 0 & 0 & 1 \end{pmatrix}^n = \begin{pmatrix} 1 & n & \frac{n(n-1)}{2} \\ 0 & 1 & n \\ 0 & 0 & 1 \end{pmatrix}$

where, for any square matrix A, A^n is defined inductively by $A^0 = I$ and $A^n = A^{n-1} \cdot A$ for any $n > 0$.

10. Prove that the sum and product of two upper triangular matrices are again upper triangular matrices and that the same statement for lower triangular matrices is also true.

11. A matrix A is said to be *symmetric* if $A = A^t$, and A is called *skew-symmetric* if $A = -A^t$. Prove the following.
    (i) Any symmetric or skew-symmetric matrix is a square matrix.
    (ii) If A and B are symmetric matrices, then so is $rA + sB$ for any scalars r and s.
    (iii) For any matrix A, $A \cdot A^t$ and $A^t \cdot A$ are both symmetric.
    (iv) For any symmetric matrices A and B, the product AB is symmetric if and only if $AB = BA$.
12. Prove that the diagonal entries of a skew-symmetric matrix are all zero.
13. For any square matrix A, prove that $A + A^t$ is symmetric.
14. Prove that any square matrix can be expressed as the sum of a symmetric matrix and a skew-symmetric matrix.

2.6 DETERMINANTS

In this section, we briefly discuss an important function known as determinant function which maps square matrices into scalars. The term 'determinant of A' is conventionally used to call the value of this function at a given square matrix A. Determinants have definite importance as a theoretical tool, besides their effectiveness as a device for computations. For example they provide us with simple criterion for the nonsingularity; namely, a square matrix is nonsingular if and only if its determinant is nonzero.

There are several ways of defining the determinant function. However, we prefer the classical definition which uses permutations. In view of this, we first have a brief discussion on permutations. We begin with the following.

Definition 2.6.1. For any positive integer n, let $I_n = \{1, 2, \ldots, n\}$. Any bijection of I_n onto itself is called a permutation on I_n. The set of all permutations on I_n is denoted by S_n.

Any permutation f on I_n can expressed by means of an array (a $2 \times n$ matrix)

$$\begin{pmatrix} 1 & 2 & 3 & \cdots & n \\ f(1) & f(2) & f(3) & \cdots & f(n) \end{pmatrix}$$

symbolising that each i is mapped to $f(i)$. Note that the order of the columns in this representation of f is immaterial. For example

$$\begin{pmatrix} 1 & 2 & 3 & 4 & 5 \\ 3 & 4 & 1 & 5 & 2 \end{pmatrix} \text{ and } \begin{pmatrix} 3 & 4 & 2 & 1 & 5 \\ 1 & 5 & 4 & 3 & 2 \end{pmatrix}$$

represent the same permutation f which is given by $f(1) = 3, f(2) = 4, f(3) = 1,$ $f(4) = 5$ and $f(5) = 2$.

If f and g are permutations on I_n, the $f \circ g, f^{-1}$ and g^{-1} are also permutations on I_n. The set S_n of all permutations on I_n has the structure of a group, which we thoroughly discuss later in chapter 6. Note that $|S_n| = n!$ for any $n \in \mathbb{Z}^+$. A permutation f is called an *r-cycle* (or *cycle of length r*) if f maps r elements $i_1, i_2, \ldots, i_r$ cyclically, keeping the remaining elements, if any fixed and such an r-cycle will be denoted by

$$f = (i_1\ i_2 \ldots i_r)$$

That is, $f(i_1) = i_2, f(i_2) = i_3, \ldots, f(i_{r-1}) = i_r$ and $f(i_r) = i_1$ and $f(i) = i$ for all $i \in I_n - \{i_1, i_2, \ldots, i_r\}$. Observe that $(i_1\ i_2 \ldots i_r), (i_2\ i_3 \ldots i_r i_1), \ldots, (i_r\ i_1\ i_2 \ldots i_{r-1})$ are all represent the same cycle. A 2-cycle is called a *transposition*. Note that, if f is an r-cycle, the $f^r(= f \circ f \circ \ldots \circ f, r$ times) is the identity map on I_n and r is the least such positive integer. In particular, if f is a transposition, then f^2 is the identity map and f interchanges two elements in I_n and keeps all other elements fixed.

Two cycles $(a_1\ a_2 \ldots a_r)$ and $(b_1\ b_2 \ldots b_s)$ are said to be disjoint if $a_i \neq b_j$ for all $1 \leq i \leq r$ and $1 \leq j \leq s$. It can be easily proved that $f \circ g = g \circ f$ for any disjoint cycles f and g and that any permutation on I_n can be expressed, in an essentially unique way, as a product of disjoint cycles. Further, any cycle is a product of transpositions (since $(a_1\ a_2 \ldots a_r) = (a_1 a_r) \circ (a_1 a_{r-1}) \circ \ldots \circ (a_1 a_2)$) and hence any permutation can be expressed as a product of transpositions, although not necessarily uniquely. For example $(2\ 4) \cdot (4\ 5) \cdot (1\ 3) = (1\ 3) \circ (2\ 4) \circ (1\ 3) \circ (4\ 5) \circ (1\ 3)$. However, it can be proved (see Corollary 6.4.2) that, if a permutation can be expressed as a product of even number of transpositions, then it cannot be expressed as a product of odd number of transpositions. In view of this, a permutation is called an *even (odd) permutation* if it is a product of even (odd, respectively) number of permutations. If f and g are even permutations, then clearly $f \circ g, f^{-1}$ and g^{-1} are also even (since $f = f_1 \circ f_2 \circ \ldots \circ f_r$ implies that $f^{-1} = f_r^{-1} \circ f^{-1}_{r-1} \circ \ldots \circ f_2^{-1} \circ f_1^{-1}$). Note that an r-cycle is even if and only if r is odd.

Definition 2.6.2. For any permutation f on I_n, the *signature* of f, denoted by sgn f, is defined by

$$\operatorname{sgn} f = \begin{cases} 1 & \text{if } f \text{ is even} \\ -1 & \text{if } f \text{ is odd} \end{cases}.$$

It can be easily verified that, for any permutation f and g on I_n,

$$\operatorname{sgn}(f \circ g) = \operatorname{sgn} f \cdot \operatorname{sgn} g \text{ and } \operatorname{sgn} f = \operatorname{sgn} f^{-1}.$$

In the following, we give a formal definition of the determinant function.

Definition 2.6.3. Let $A = (a_{ij})$ be an $n \times n$ matrix. Then the sum

$$\sum_{f \in S_n} (\operatorname{sgn} f)\, a_{1f(1)} a_{2f(2)} \cdots a_{nf(n)}$$

is called the *determinant* of A and is denoted by $\det A$ or $|A|$.

Examples 2.6.1

1. Let $A = \begin{pmatrix} a_{11} & a_{12} \\ a_{21} & a_{22} \end{pmatrix}$ be a 2×2 matrix. Since S_2 has only two elements, namely the identity e which is an even permutation and the transposition $\sigma = (1\ 2)$ which is odd, we have

$$\begin{aligned} \det A &= (\operatorname{sgn} e) a_{11} a_{22} + (\operatorname{sgn} \sigma) a_{12} a_{21} \\ &= a_{11} a_{22} - a_{12} a_{21} \end{aligned}$$

2. Consider a 3×3 matrix $A = \begin{pmatrix} a_{11} & a_{12} & a_{13} \\ a_{21} & a_{22} & a_{23} \\ a_{31} & a_{32} & a_{33} \end{pmatrix}$

S_3 has 3! elements; these are

$$e = \begin{pmatrix} 1 & 2 & 3 \\ 1 & 2 & 3 \end{pmatrix},\ f = \begin{pmatrix} 1 & 2 & 3 \\ 2 & 3 & 1 \end{pmatrix},\ g = \begin{pmatrix} 1 & 2 & 3 \\ 3 & 1 & 2 \end{pmatrix}$$

$$\alpha = \begin{pmatrix} 1 & 2 & 3 \\ 2 & 1 & 3 \end{pmatrix},\ \beta = \begin{pmatrix} 1 & 2 & 3 \\ 1 & 3 & 2 \end{pmatrix},\ \gamma = \begin{pmatrix} 1 & 2 & 3 \\ 3 & 2 & 1 \end{pmatrix}$$

Note that e is the identity, $f = (1\ 2\ 3)$, $g = (1\ 3\ 2)$ $\alpha = (1\ 2)$, $\beta = (2\ 3)$ and $\gamma = (1\ 3)$ and hence e, f and g are even and α, β and γ are odd. Therefore $\det A = a_{11}a_{22}a_{33} + a_{12}a_{23}a_{31} + a_{13}a_{21}a_{32} - a_{12}a_{21}a_{33} - a_{11}a_{23}a_{32} - a_{13}a_{22}a_{31}$

3. The determinant of the zero matrix is 0.
4. $\det I = 1$, where I is the identity matrix of any order.

In the following we prove some results that facilitate the evaluation of the determinant of any square matrix in a less tedious manner. First recall that, if $A = (a_{ij})$ is an $n \times n$ matrix, $R_1, \ldots, R_n$ are the n rows of A and $C_1, C_2, \ldots, C_n$ are the n-columns of A, then we express A as

$$A = \begin{pmatrix} R_1 \\ R_2 \\ \vdots \\ R_n \end{pmatrix} \quad \text{or} \quad A = (C_1, C_2, \ldots, C_n)$$

where $R_i = (a_{i1}\ a_{i2}\ \ldots\ a_{in})$ and $C_j = \begin{pmatrix} a_{1j} \\ a_{2j} \\ \vdots \\ a_{nj} \end{pmatrix}$.

Here, each R_i is a $1 \times n$ matrix and each C_j is a $n \times 1$ matrix.

Theorem 2.6.1. Let $A = (a_{ij})$ be an $n \times n$ matrix and $R_1, R_2, \ldots, R_n$ be the rows of A. For any fixed i, $1 \le i \le n$, let $S_i = (b_{i1}\ b_{i2}\ \ldots\ b_{in})$ be a $1 \times n$ matrix. Then, the following holds.

1. $$\det\begin{pmatrix} R_1 \\ R_2 \\ \vdots \\ R_i + S_i \\ R_{i+1} \\ \vdots \\ R_n \end{pmatrix} = \det\begin{pmatrix} R_1 \\ \vdots \\ \vdots \\ R_i \\ \vdots \\ \vdots \\ R_n \end{pmatrix} + \det\begin{pmatrix} R_1 \\ \vdots \\ R_{i-1} \\ S_i \\ R_{i+1} \\ \vdots \\ R_n \end{pmatrix}$$

2. $$\det\begin{pmatrix} R_1 \\ \vdots \\ R_{i-1} \\ aR_i \\ R_{i+1} \\ \vdots \\ R_n \end{pmatrix} = a\det\begin{pmatrix} R_1 \\ \vdots \\ R_{i-1} \\ R_i \\ R_{i+1} \\ \vdots \\ R_n \end{pmatrix}$$

Proof:

1. The left hand side of the equation is

$$\sum_{f \in S_n} (\operatorname{sgn} f) a_{1f(1)} \ldots a_{i-1f(i-1)} (a_{if(i)} + b_{if(i)}) a_{i+1f(i+1)} \ldots a_{nf(n)}$$

$$= \sum_{f \in S_n} (\operatorname{sgn} f) a_{1f(1)} \cdots a_{i-1f(i-1)} a_{if(i)} \cdots a_{nf(n)}$$

$$+ \sum_{f \in S_n} (\operatorname{sgn} f) a_{1f(1)} \cdots a_{i-1f(i-1)} b_{if(i)} a_{i+1f(i+1)} \cdots a_{nf(n)}$$

$$= \text{The right hand side of (1)}$$

2. This is clear from the definition.

Theorem 2.6.2. For any square matrix A, $\det A = \det A^t$

Proof: Let $A = (a_{ij})$ be an $n \times n$ matrix. Then, $A^t = (a_{ji})$ and

$$\det A^t = \sum_{f \in S_n} (\operatorname{sgn} f)\, a_{f(1)1} a_{f(2)2} \cdots a_{f(n)n}$$

$$= \sum_{f \in S_n} (\operatorname{sgn} f) a_{1f^{-1}(1)} a_{2f^{-1}(2)} \cdots a_{nf^{-1}(n)}$$

$$= \sum_{f \in S_n} (\operatorname{sgn} f^{-1}) a_{1f^{-1}(1)} a_{2f^{-1}(2)} \cdots a_{nf^{-1}(n)}$$

$$= \det A \text{ (since } S_n = \{f^{-1} : f \in S_n\}).$$

Theorem 2.6.3. If two rows of a square matrix A are equal, then $\det A = 0$.

Proof: Let $A = (a_{ij})$ be an $n \times n$ matrix and $R_1, R_2, \ldots, R_n$ be the n rows of A. Suppose that the rth row and s^{th} row are equal; that is, $R_r = R_s$ and $r \neq s$. Without loss of generality, we can assume that $r < s$. Let g be the transposition (r, s). Let

$$A = \{f \in S_n : f(r) < f(s)\} \quad \text{and} \quad B = \{f \in S_n : f(r) > f(s)\}.$$

Then, the map $\alpha : A \to B$ defined by $\alpha(f) = f \circ g$ is a bijection. Also, note that $A \cup B = S_n$ (since $r \neq s, f(r) \neq f(s)$ for any $f \in S_n$) and $A \cap B = \emptyset$. Therefore,

$$\det A = \sum_{f \in A} (\operatorname{sgn} f) a_{1f(1)} \cdots a_{rf(r)} \cdots a_{sf(s)} \cdots a_{nf(n)}$$

$$+ \sum_{f \in B} (\operatorname{sgn} f) a_{1f(1)} \cdots a_{rf(r)} \cdots a_{sf(s)} \cdots a_{nf(n)}$$

$$= \sum_{f \in A} (\operatorname{sgn} f) a_{1f(1)} \cdots a_{rf(r)} \cdots a_{sf(s)} \cdots a_{nf(n)}$$

$$+ \sum_{f \in A} (\operatorname{sgn} f \circ g) a_{1fg(1)} \cdots a_{rfg(r)} \cdots a_{sfg(s)} \cdots a_{nfg(n)}$$

$$= \sum_{f \in A} (\operatorname{sgn} f)[a_{1f(1)} \cdots a_{rf(r)} \cdots a_{sf(s)} \cdots a_{nf(n)} - a_{1fg(1)} \cdots a_{rfg(r)} \cdots a_{sfg(s)} \cdots a_{nfg(n)}]$$

$$= \sum_{f \in A} (\operatorname{sgn} f)[a_{1f(1)} \cdots a_{rf(r)} \cdots a_{sf(s)} \cdots a_{nf(n)} - a_{1f(1)} \cdots a_{rf(s)} \cdots a_{sf(r)} \cdots a_{nf(n)}]$$

$= 0$ (since $a_{rj} = a_{sj}$ for all $1 \le j \le n$).

Theorem 2.6.4. Let $A = (a_{ij})$ be an $n \times n$ matrix and B be the matrix obtained from A by interchanging the rth row and the sth row, then $\det A = -\det B$.

Proof: We can assume that $r < s$. We have $A = \begin{pmatrix} R_1 \\ \vdots \\ R_n \end{pmatrix}$, where R_i is the ith row of A.

Put $C = \begin{pmatrix} R_1 \\ \vdots \\ R_r + R_s \\ \vdots \\ R_s + R_r \\ \vdots \\ R_n \end{pmatrix}$. Then, the rth row of C = s^{th} row of C and all other i^{th} rows of C are same as those of A. Now, by Theorem 2.6.3, we have

$$0 = \det C = \det \begin{pmatrix} R_1 \\ \vdots \\ R_r \\ \vdots \\ R_s \\ \vdots \\ R_n \end{pmatrix} + \det \begin{pmatrix} R_1 \\ \vdots \\ R_r \\ \vdots \\ R_r \\ \vdots \\ R_n \end{pmatrix} + \det \begin{pmatrix} R_1 \\ \vdots \\ R_s \\ \vdots \\ R_s \\ \vdots \\ R_n \end{pmatrix} + \det \begin{pmatrix} R_1 \\ \vdots \\ R_s \\ \vdots \\ R_r \\ \vdots \\ R_n \end{pmatrix}$$

$= \det A + 0 + 0 + \det B$

Thus, $\det A = -\det B$.

The following is an immediate consequence of Theorems 2.6.2 and 2.6.4.

Corollary 2.6.1. Let A be an $n \times n$ matrix and B be the matrix obtained from A by interchanging two columns of A. Then, $\det A = -\det B$.

Corollary 2.6.2. If $R_1, R_2, \ldots, R_n$ are the rows of an $n \times n$ matrix A and f is a permutation on $\{1, 2, \ldots, n\}$, then

$$\det\begin{pmatrix} R_{f(1)} \\ \vdots \\ R_{f(n)} \end{pmatrix} = (\operatorname{sgn} f)\det A$$

Proof: Let f be a product of m transpositions. Then the matrix A can be transformed to $\begin{pmatrix} R_{f(1)} \\ \vdots \\ R_{f(n)} \end{pmatrix}$ by m interchanges of the rows. Therefore, by Theorem 2.6.4,

$$\det\begin{pmatrix} R_{f(1)} \\ \vdots \\ R_{f(n)} \end{pmatrix} = (-1)^m \det\begin{pmatrix} R_1 \\ \vdots \\ R_n \end{pmatrix} = (\operatorname{sgn} f)\det A.$$

The following is one of the most important properties of the determinants of matrices.

Theorem 2.6.5. For any $n \times n$ matrices A and B,

$$\det(A \cdot B) = \det A \cdot \det B$$

Proof: Let $A = (a_{ij})$ and $B = (b_{ij})$ be two $n \times n$ matrices and $AB = (c_{ij})$. Then, $c_{ij} = \sum_{r=1}^{n} a_{ir}\, b_{rj}$ for any $1 \le i, j \le n$.

$$\begin{aligned}
\det(AB) &= \sum_{f \in S_n} (\operatorname{sgn} f)\, c_{1f(1)}\, c_{2f(2)} \ldots c_{nf(n)} \\
&= \sum_{f \in S_n} (\operatorname{sgn} f)\left(\sum_{r_1=1}^{n} a_{1r_1}\, b_{r_1 f(1)}\right)\cdots\left(\sum_{r_n=1}^{n} a_{nr_n}\, b_{r_n f(n)}\right) \\
&= \sum_{1 \le r_1, r_2, \ldots, r_n \le n} (a_{1r_1}\, a_{2r_2} \ldots a_{nr_n})\left(\sum_{f \in S_n} (\operatorname{sgn} f)\, b_{r_1 f(1)} b_{r_2 f(2)} \cdots b_{r_n f(n)}\right)
\end{aligned}$$

In the above summation, if $r_1, r_2, \ldots, r_n$ are not all distinct, then, by Theorem 2.6.3,

$$\sum_{f \in S_n} (\operatorname{sgn} f)\, b_{r_1 f(1)}\, b_{r_2 f(2)} \ldots b_{r_n f(n)} = 0$$

and therefore, we can consider only the summands corresponding to distinct n-tuples $r_1, r_2, \ldots, r_n$. Therefore,

$$\det(AB) = \sum_{g\in S_n} (a_{1g(1)}\, a_{2g(2)} \cdots a_{ng(n)}) \left(\sum_{f\in S_n} (\operatorname{sgn} f)\, b_{g(1)f(1)} \cdots b_{g(n)f(n)} \right)$$

$$= \left(\sum_{g\in S_n} (\operatorname{sgn} g)\, a_{1g(1)} \cdots a_{ng(n)} \right) (\det B) \text{ (by Theorem 2.6.10)}$$

$$= (\det A)\,(\det B).$$

Corollary 2.6.3. If A is a nonsingular $n \times n$ matrix, then $\det A \neq 0$ and

$$\det A^{-1} = \frac{1}{\det A}$$

Proof: If $AA^{-1} = I$ (the identity $n \times n$ matrix), then

$$\det A \cdot \det A^{-1} = \det (AA^{-1}) = \det (I) = 1.$$

Next, we discuss an expansion for the determinant of a matrix which provides us with an inductive algorithm to find the value of det A. First, we have the following.

Definition 2.6.4. Let $A = (a_{ij})$ be an $n \times n$ matrix and, for any $1 \leq i, j \leq n$, let

$$A_{ij} = \sum_{\substack{f\in S_n \\ f(i)=j}} (\operatorname{sgn} f) \prod_{r\neq i} a_{rf(r)}$$

$$= \sum_{\substack{f\in S_n \\ f(i)=j}} (\operatorname{sgn} f)\, a_{1f(1)} \ldots a_{i-1f(i-1)} \cdot a_{i+1f(i+1)} \cdots a_{nf(n)}.$$

Then, A_{ij} is called the cofactor of a_{ij} in det A.

Theorem 2.6.6. The following holds for any $n \times n$ matrix $A = (a_{ij})$.

1. $\displaystyle\sum_{j=1}^{n} a_{rj}\, A_{ij} = \begin{cases} \det A & \text{if } r = i \\ 0 & \text{if } r \neq i \end{cases}$

2. $\displaystyle\sum_{i=1}^{n} a_{ir}\, A_{ij} = \begin{cases} \det A & \text{if } r = j \\ 0 & \text{if } r \neq j \end{cases}$

Proof: Consider all the summands in the sum

$$\det A = \sum_{f \in S_n} (\operatorname{sgn} f)\, a_{1f(1)} \ldots a_{2f(2)} \ldots a_{nf(n)}$$

that contain a given entry a_{ij} as a factor. These are corresponding to those permutations f for which $f(i) = j$. Therefore, the sum of all the summands in the summation for det A involving a_{ij} as a factor is

$$\sum_{\substack{f \in S_n \\ f(i)=j}} (\operatorname{sgn} f)\, a_{1f(1)} \ldots a_{2f(2)} \cdots a_{nf(n)} = a_{ij} A_{ij}$$

and hence

$$\begin{aligned}\det A &= \sum_{f \in S_n} (\operatorname{sgn} f)\, a_{1f(1)}\, a_{2f(2)} \ldots a_{nf(n)} \\ &= \sum_{j=1}^{n} \sum_{\substack{f \in S_n \\ f(i)=j}} (\operatorname{sgn} f)\, a_{1f(1)}\, a_{2f(2)} \ldots a_{nf(n)} \\ &= \sum_{j=1}^{n} a_{ij}\, A_{ij} \text{ for each } 1 \le i \le n.\end{aligned}$$

Similarly, det $A = \sum_{i=1}^{n} a_{ij}\, A_{ij}$ for each $1 \le j \le n$.
Next, let $1 \le r \ne i \le n$ and consider

$$\begin{aligned}\sum_{j=1}^{n} a_{rj}\, A_{ij} &= \sum_{j=1}^{n} a_{rj} \sum_{\substack{f \in S_n \\ f(i)=j}} (\text{sgnf}) \prod_{s \ne i} a_{sf(s)} \\ &= \sum_{f \in S_n} (\operatorname{sgn} f)\, a_{1f(1)} \ldots a_{i-1f(i-1)}\, a_{rf(i)}\, a_{i+1f(i+1)} \ldots a_{nf(n)} \\ &= 0 \text{ (by Theorem 2.6.7)}\end{aligned}$$

Similarly $\sum_{i=1}^{n} a_{ir}\, A_{ij}$ if $r \ne j$.

The equation $\sum_{j=1}^{n} a_{ij}\, A_{ij} = \det A$ given in (1) above is called the *expansion of det A with respect to the* i^{th} *row* and the equation $\sum_{i=1}^{n} a_{ij}\, A_{ij} = \det A$ given in (2) above is called the *expansion of det A with respect to the* j^{th} *column*. The following result provides a method to evaluate the cofactors A_{ij} for the matrix $A = (a_{ij})$.

Theorem 2.6.7. Let $A = (a_{ij})$ be an $n \times n$ matrix and, for any $1 \leq i, j \leq n$, let A^{ij} be the submatrix of A obtained by deleting the ith row and j^{th} column. Then, the cofactor A_{ij} of a_{ij} in det A is given by

$$A_{ij} = (-1)^{i+j} \det A^{ij}$$

Proof: Let us first find the cofactor A_{11} of a_{11} in det A. By Definition 2.6.4, we have

$$A_{11} = \sum_{\substack{f \in S_n \\ f(1)=1}} (\operatorname{sgn} f)\, a_{2f(2)}\, a_{3f(3)} \ldots a_{nf(n)}.$$

If $f \in S_n$ and $f(1) = 1$, then the restriction of f to $\{2, 3, \ldots, n\}$ is a permutation on $\{2, 3, \ldots, n\}$ and any permutation on $\{2, 3, \ldots, n\}$ can be uniquely extended to a permutation f on $\{2, 3, \ldots, n\}$ by defining $f(1) = 1$. Also, for $f \in S_n$ with $f(1) = 1$, f is an even permutation if and only if the restriction of f to $\{2, 3, \ldots, n\}$ is even. Thus, the above equation is precisely same as

$$A_{11} = \det A^{11} = (-1)^{1+1} \det A^{11}.$$

Next, to find the value of a general cofactor A_{ij} of a_{ij}, let us bring a_{ij} to the (1, 1) position by performing some row and column interchanges on A. To bring a_{ij} to the (1, 1) position, we move the jth column to the left to $j-1^{\text{th}}$ column (that is, interchanging j^{th} column and $j-1^{\text{th}}$ column), then to $j-2^{\text{th}}$ column, ..., to 1st column, so that after $j-1$ interchanges of the columns, the jth column becomes the first column. Next, in a similar way, we move the ith row up to the 1st row in $i-1$ interchanges of the rows. Now, we have a matrix B that is obtained from A by $j-1$ interchanges of columns and $i-1$ interchanges of rows. Therefore, by Theorem 2.6.4 and Corollary 2.6.1, the determinant of the new matrix B is $(-1)^{(j-1)+(i-1)} \det A = (-1)^{i+j} \det A$. If $B = (b_{ij})$, then $b_{11} = a_{ij}$ and the matrix obtained by deleting the 1st row and 1st column in B is precisely A^{ij}. Thus,

$$A_{ij} = (-1)^{i+j} \det A^{ij}.$$

Corollary 2.6.4. For any $n \times n$ matrix $A = (a_{ij})$,

$$\det A = \sum_{j=1}^{n} (-1)^{i+j} a_{ij} \det A^{ij}, \quad \text{for each } 1 \leq i \leq n$$

$$= \sum_{i=1}^{n} (-1)^{i+j} a_{ij} \det A^{ij}, \quad \text{for each } 1 \leq j \leq n.$$

Example 2.6.2. Let us determine the determinant of a 3×3 matrix $A = (a_{ij})$, using Theorem 2.6.7 and Corollary 2.6.4. We have

$$A^{11} = \begin{pmatrix} a_{22} & a_{23} \\ a_{32} & a_{33} \end{pmatrix}, \quad A^{12} = \begin{pmatrix} a_{21} & a_{23} \\ a_{31} & a_{33} \end{pmatrix} \quad \text{and} \quad A^{13} = \begin{pmatrix} a_{21} & a_{22} \\ a_{31} & a_{32} \end{pmatrix}.$$

Now,

$$\begin{aligned} \det A &= (-1)^{1+1} a^{11} \det A^{11} + (-1)^{1+2} a^{12} \det A^{12} + (-1)^{1+3} a^{13} \det A^{13} \\ &= a_{11} \begin{vmatrix} a_{22} & a_{23} \\ a_{32} & a_{33} \end{vmatrix} - a_{12} \begin{vmatrix} a_{21} & a_{23} \\ a_{31} & a_{33} \end{vmatrix} + a_{13} \begin{vmatrix} a_{21} & a_{22} \\ a_{31} & a_{32} \end{vmatrix} \\ &= a_{11}(a_{22} a_{33} - a_{32} a_{23}) - a_{12}(a_{21} a_{33} - a_{31} a_{23}) + a_{13}(a_{21} a_{32} - a_{31} a_{22}). \end{aligned}$$

This is the expansion of det A with respect to the 1st row. Notice that we get the same value for det A by expanding it with respect to any other row or any column.

The following result characterizes nonsingular matrices in terms of the value of their determinants.

Theorem 2.6.8. A square matrix A is nonsingular if and only if $\det A \neq 0$ and, in this case, the inverse of A is given by

$$A^{-1} = \frac{1}{\det A} (A_{ij})^t$$

where (A_{ij}) is the matrix whose ij^{th} entry is the cofactor A_{ij} of the ij^{th} entry in A.

Proof: Let $A = (a_{ij})$ be an $n \times n$ matrix and A_{ij} be the cofactor of a_{ij} in det A.
Let B be the transpose of (A_{ij}).
That is, $B = (A_{ij})^t = (b_{ij})$; say.
Then, $b_{ij} = A_{ji}$ for all i and j. By Theorem 2.6.6,

$$\sum_{j=1}^{n} a_{rj}\, b_{ji} = \sum_{j=1}^{n} a_{rj}\, A_{ij} = \delta_{ri} \det A$$

where $\delta_{ri} = 1$ or 0 according as $r = i$ or $r \neq i$. This implies that, the ri^{th} entry in the product matrix AB is det A if $r = i$, and 0 if $r \neq i$. Therefore,

$$AB = \det A \cdot I$$

where I is the $n \times n$ identity matrix. Similarly, $BA = \det A \cdot I$. Thus, $\det A \neq 0$ implies that

$$A \cdot \left(\frac{1}{\det A} B\right) = I = \left(\frac{1}{\det A}\right) B \cdot A$$

and hence A is nonsingular and $A^{-1} = \dfrac{1}{\det A} B = \dfrac{1}{\det A}(A_{ij})'$. Converse follows from Corollary 2.6.3.

Definition 2.6.5. For any $n \times n$ matrix $A = (a_{ij})$, the transpose of the matrix (A_{ij}) is called *adjoint* of A and is denoted by adj A, where A_{ij} is the cofactor of a_{ij} in $\det A$.

Corollary 2.6.5. For any square matrix A,

$$A \cdot (\text{adj } A) = \det A \cdot I = (\text{adj } A) \cdot A.$$

EXERCISE 2(F)

1. Evaluate the determinants of each of the following matrices

(i) $\begin{pmatrix} 2 & 4 & 1 \\ 3 & -1 & 2 \\ -4 & -3 & -2 \end{pmatrix}$

(ii) $\begin{pmatrix} 2 & 1 & 3 & 4 \\ 1 & 4 & 2 & 3 \\ 3 & 4 & 1 & 2 \\ 4 & 3 & 2 & 1 \end{pmatrix}$

(iii) $\begin{pmatrix} 4 & 3 & 2 & 1 \\ 1 & 2 & 3 & 4 \\ 2 & -3 & 1 & -4 \\ -3 & 4 & -4 & -1 \end{pmatrix}$

2. Prove that the determinant of an upper (or a lower) triangular matrix is equal to the product of the diagonal entries.

3. Prove that $\det \begin{vmatrix} 1 & a & a^2 \\ 1 & b & b^2 \\ 1 & c & c^2 \end{vmatrix} = (a-b)(b-c)(c-a)$.

4. Prove that $\det\begin{pmatrix} 1 & x_1 & x_1^2 & \cdots & x_1^{n-1} \\ 1 & x_2 & x_2^2 & \cdots & x_2^{n-1} \\ \vdots & & & & \\ i & x_n & x_n^2 & \cdots & x_n^{n-1} \end{pmatrix} = \prod_{1 \le i \le j \le n} (x_j - x_i).$

5. Prove that $\det\begin{pmatrix} 1 & a & a^3 \\ 1 & b & b^3 \\ 1 & c & c^3 \end{pmatrix} = (b-c)(c-a)(a-b)(a+b+c).$

6. If A is a nonsingular square matrix such that $A^2 = A$, then prove that $\det A = 1$.

7. Prove the following:

(i) $\det\begin{pmatrix} (b+c)^2 & a^2 & a^2 \\ b^2 & (c+a)^2 & b^2 \\ c^2 & c^2 & (a+b)^2 \end{pmatrix} = 2\,abc\,(a+b+c)^3$

(ii) $\det\begin{pmatrix} a-b-c & 2b & 2c \\ 2a & b-c-a & 2c \\ 2a & 2b & c-a-b \end{pmatrix} = (a+b+c)^3$

8. If A is a $n \times n$ matrix such that $A^m = O_{n\times n}$ for some $m \in \mathbb{Z}^+$, then prove that A is singular.

9. Let $A = (a_{ij})$ and $B = (b_{ij})$ be $n \times n$ matrices such that $b_{ij} = (-1)^{i+j} a_{ij}$ for all i and j. Then, prove that $\det A = \det B$.

10. Prove that a square matrix A is nonsingular if and only if A' is nonsingular.

If A is an $n \times n$ skew-symmetric matrix, then prove that $\det A = (-1)^n \det A$?

11. If n is odd, prove that any skew-symmetric $n \times n$ matrix is singular.

12. Prove that for any $n \times n$ nonsingular matrix A, $\det(\text{adj } A) = (\det A)^{n-1}$.

13. Prove the following for any $n \times n$ matrices A and B:

(i) $\det(AB) = \det(BA)$

(ii) $\det(A \cdot A') = (\det A)^2$

(iii) If A is nonsingular, then $\det(ABA^{-1}) = \det B$.

PART II
Group Theory

3

Groups

3.1 BINARY SYSTEMS

It is well known that the product of two integers is again an integer. That is, if a and b are integers, then the product $a \cdot b$ is again an integer. Here, the symbol '$\cdot$' denotes the 'operation of taking product' of a and b, in this order. Similarly, if A and B are two 2×2 matrices over the real number system, then the product $A \cdot B$ is again a 2×2 matrix over $\mathbb{R}$. Here, the symbol '$\cdot$' denotes the 'operation of taking product' of the matrices A and B, in this order. Further, if f and g are two mappings of a set X into itself, then the composition $g \circ f$ is also a mapping of X into itself. Here, the symbol '$\circ$' denotes the 'operation of taking composition' of g and f, in this order. Also, if A and B are subsets of a given set X, then the union $A \cup B$ is also a subset of X. Here again, the symbol $\cup$ denotes the 'operation of taking union' of A and B, in this order.

In each of these cases, from any two elements of a given set, we obtain another element of the same set by performing an operation on the two elements in a specific order. This is formalized in the following definition.

Definition 3.1.1. Let S be a nonempty set and $S \times S$ be the set of all ordered pairs of elements of S. That is,

$$S \times S = \{(a, b) : a \in S \text{ and } b \in S\}.$$

A mapping $f : S \times S \to S$ is called a *binary operation on S*.

If f is a binary operation on a set S and a and b are elements of S, then we write $a\,f\,b$ for $f(a, b)$. This is only for convenience. Recall that we write $a \cdot b$ for the product of two integers a and b. By definition, for any set S, any mapping of $S \times S$ into S is a binary operation on S. Certain special binary operations on certain special sets are important to be mentioned. In the following, we list several binary operations on certain special sets, as examples.

Example 3.1.1

1. The usual multiplication '$\cdot$' is a binary operation on the set $\mathbb{Z}$ of integers. Quite often, we simply write ab for $a \cdot b$.
2. The usual addition $+$ is a binary operation on the set $\mathbb{Z}$ of integers.
3. Let us define the mapping $- : \mathbb{Z} \times \mathbb{Z} \to \mathbb{Z}$ by $-(a, b) = a - b$, the usual difference of b with a, for any integers a and b. Then, $-$ is a binary operation on $\mathbb{Z}$. Note that $-$ is not a binary operation on the set $\mathbb{Z}^+$ of positive integers, since $a - b$ need not be positive for any two positive integers a and b. Likewise, $-$ is not a binary operation on the set $\mathbb{Z}^-$ of negative integers. Note that both the multiplication and addition given in (1) and (2), respectively are binary operations on $\mathbb{Z}^+$. The operation $-$ is called the *difference* operation. Note that each of addition, difference and multiplication is a binary operation on the set $\mathbb{Q}$ of rational numbers and on the set $\mathbb{R}$ of real numbers.
4. Let $\mathbb{R}$ be the set of real numbers and, for any real numbers a and b, define

   $$a \wedge b = \text{The minimum of } a \text{ and } b$$

   and $$a \vee b = \text{The maximum of } a \text{ and } b.$$

   Then, both $\wedge$ and $\vee$ are binary operations on $\mathbb{R}$. In fact, these are binary operations on any nonempty subset of $\mathbb{R}$ and, in particular, on $\mathbb{Q}$, $\mathbb{Z}$ and $\mathbb{Z}^+$.
5. Consider the set $\mathbb{Z}^+$ of positive integers. For any a and b in $\mathbb{Z}^+$, define

   $$a\,g\,b = (a, b), \text{ the greatest common divisor of } a \text{ and } b$$

   and $$a\,\ell\,b = [a, b], \text{ the least common multiple of } a \text{ and } b.$$

   Then, g and ℓ are both binary operations of $\mathbb{Z}^+$. Usually, we write (a, b) and $[a, b]$ to denote respectively the greatest common divisor and the least common multiple of any positive integers a and b.
6. Let X be any set and $\mathbb{P}(X)$, the power set of X; that is, $\mathbb{P}(X)$ is the set of all subsets of X. For any A and $B \in \mathbb{P}(X)$, define

   $$A \cap B = \{x : x \in A \text{ and } x \in B\}$$
   $$A \cup B = \{x : x \in A \text{ or } x \in B\}$$
   $$A - B = \{x : x \in A \text{ and } x \notin B\}$$
   $$A + B = (A - B) \cup (B - A).$$

Then, $\cap$, $\cup$, $-$ and $+$ are all binary operations on the set $\mathbb{P}(X)$ and are respectively called *intersection, union, difference* and *symmetric difference*. Note that $\mathbb{P}(X)$ is not empty even if X is empty.

7. For any positive integers m and n, let $M_{m\times n}(\mathbb{R})$ be the set of all $m \times n$ matrices over the real number system $\mathbb{R}$. For any $A = (a_{ij})$ and $B = (b_{ij})$ in $M_{m\times n}(\mathbb{R})$, define

$$A + B = (c_{ij}),$$

where $c_{ij} = a_{ij} + b_{ij}$ and $+$ is the usual addition of real numbers. Then, $+$ is a binary operation on $M_{m\times n}(\mathbb{R})$ and is called the *addition of matrices* (of same order).

8. For any positive integer n, let $M_n(\mathbb{R})$ be the set of all $n \times n$ matrices (square matrices of order n). For any $A = (a_{ij})$ and $B = (b_{ij})$ in $M_n(\mathbb{R})$, define

$$A \cdot B = (d_{ij}), \quad \text{where } d_{ij} = \sum_{k=1}^{n} a_{ik} b_{kj}$$

for any $1 \leq i, j \leq n$. That is,

$$d_{ij} = a_{i1}b_{1j} + a_{i2}b_{2j} + \cdots + a_{in}b_{nj}.$$

(The operations involved in defining d_{ij} above are the usual addition and multiplication of real numbers.) Then, '$\cdot$' is a binary operation on $M_n(\mathbb{R})$ and is called the *multiplication of square matrices* (of the same order).

Note that, we can define addition and multiplication (as in (7) and (8) above) on the sets $M_{m\times n}(\mathbb{Z})$ and $M_n(\mathbb{Z})$ of matrices over the set $\mathbb{Z}$ of integers or on the sets $M_{m\times n}(\mathbb{Q})$ and $M_n(\mathbb{Q})$ of matrices over the set $\mathbb{Q}$ of rational numbers.

9. Let X be any set and $M(X)$ be the set of all mappings of X into itself. For any mappings f and g in $M(X)$, define

$$f \circ g : X \to X \text{ by } (f \circ g)(x) = f(g(x)) \quad \text{for any } x \in X.$$

Then, $\circ$ is a binary operation on $M(X)$ and is called the *composition of mappings*.

10. Let $\mathbb{C}$ be the set of complex numbers; that is $\mathbb{C}$ is the set of expressions of the form $a + bi$, where a and b are real numbers.

$$\mathbb{C} = \{a + bi : a \text{ and } b \text{ are real numbers}\}.$$

For any $a + bi$ and $c + di$ in $\mathbb{C}$, define

$$(a + bi) + (c + di) = (a + c) + (b + d)i$$

and

$$(a + bi) \cdot (c + di) = (ac - bd) + (ad + bc)i.$$

Then, + and '$\cdot$' are binary operations on $\mathbb{C}$ and are called the *usual addition* and *multiplication of complex numbers*, respectively.

11. Let n be a positive integer and

$$\mathbb{Z}_n = \{0, 1, 2, \ldots, n-1\}.$$

For any a and b in $\mathbb{Z}_n$, define

$$a +_n b = \begin{cases} \text{the usual sum } a+b, & \text{if } a+b<n \\ a+b-n, & \text{if } a+b \geq n \end{cases}.$$

Then, $+_n$ is a binary operation on $\mathbb{Z}_n$ and is called the *addition modulo n*.

12. Let $\mathbb{Z}_n$ be the set considered above. For any a and b in $\mathbb{Z}_n$, define

$$a \cdot_n b = r,$$

where r is the remainder obtained by dividing the usual product ab with n; that is r is the integer such that

$$ab = qn + r, q \text{ and } r \in \mathbb{Z} \quad \text{and} \quad 0 \leq r < n.$$

For example, $7 \cdot_8 6 = 2$, $5 \cdot_{10} 8 = 0$ and $8 \cdot_9 8 = 1$. Then, $\cdot_n$ is a binary operation on $\mathbb{Z}_n$ and is called the *multiplication modulo n*. Note that, in the example given in (11) above, for any a and b in $\mathbb{Z}_n$, $a +_n b$ can also be viewed as the remainder obtained by dividing the usual sum $a + b$ with n, since $0 \leq a + b < 2n$. $a +_n b$ and $a \cdot_n b$ are respectively called the *sum* and *product* of a and b modulo n.

13. Let n be any integer and define, for any integers a and b,

$$a * b = a + b + n.$$

Then, $*$ is a binary operation on the set $\mathbb{Z}$ of integers.

14. Let E be the set of all English words (whether meaningfull or not); that is, E is the set of all finite sequences

$$a_1\, a_2 \ldots a_n,\ n > 0 \text{ and } a_i\text{'s are in alphabet of English.}$$

For any $a = a_1\, a_2 \ldots a_n$ and $b = b_1\, b_2 \ldots b_m$, define

$$a * b = a_1 b_m.$$

Then, $*$ is a binary operation on E. Note that $a * b$ is the two letter word consisting of the first letter of a followed by the last letter of b.

15. Let X be any nonempty set and $\mathbb{R}^X$ be the set of all mappings from X into $\mathbb{R}$. For any f and g in $\mathbb{R}^X$, define $f + g$ and $f \cdot g : X \to \mathbb{R}$ by

$$(f + g)(x) = f(x) + g(x)$$
$$\text{and} \quad (f \cdot g)(x) = f(x) \cdot g(x), \quad \text{for all } x \in X.$$

Note that the $+$ and $\cdot$ on the right hand sides of the above are the usual addition and multiplication in the real number system $\mathbb{R}$. Then, $+$ and $\cdot$ are binary operations on $\mathbb{R}^X$ and are respectively called the *point-wise addition* and *point-wise multiplication*.

16. The above example can be generalized as follows. Let $*$ be a binary operation on a nonempty set S and X be a nonempty set. Let S^X be the set of all mappings of X into S. For any f and g in S^X, define $f * g : X \to S$ by

$$(f * g)(x) = f(x) * g(x) \quad \text{for all } x \in X.$$

Then, $*$ is a binary operation on S^X and is called the *point-wise operation* on S^X with respect to the operation $*$ on S.

Note that, in (16) above (and so is in (15)), we have denoted the operations on S^X and in S with the same symbol $*$. There should not be any confusion. The $*$ on the left sides is the one we are defining on S^X and that on the right sides is the given binary operation on S.

Note 3.1.1. In defining a binary operation on a set S, one should observe the following:

(i) For each ordered pair of elements in S, the element assigned to it must be again an element of S.

(ii) Exactly one element of S must be assigned to each ordered pair of elements in S.

For example, consider the set $\mathbb{R}$ of real numbers and, for any a and b in $\mathbb{R}$, define $a * b = \frac{a}{b}$. Then, $*$ is not a binary operation on $\mathbb{R}$, since $*$ is not defined for all ordered pairs of elements in $\mathbb{R}$. Note that $2 * 0$ is not defined, while $0 * 2$ is defined. However, this $*$ is a binary operation on a smallest set, namely, the set $\mathbb{R} - \{0\}$ of nonzero real numbers.

Let us consider another example. Let S be the set of all people in a particular village and define, for any a and b in S, $a * b = c$ where c is a person whose height is equal to the minimum of the heights of a and b. Then, $*$ is not a binary operation on S, since $a * b$ may not be an unique element in S; there can be more than one person in the village whose height is equal to the minimum of those of a and b.

Note 3.1.2. Let S be a finite set with n elements. Then, the number of elements in $S \times S$ is n^2. Any binary operation S is simply a mapping of $S \times S$ into S; that is, an element of $S^{s \times s}$. Therefore, there are exactly n^{n^2} many binary operations on S.

Definition 3.1.2. A pair $(S, *)$ is said to be a *binary system* if S is a nonempty set and $*$ is a binary operation on S. Here, S is called the *underlying set* in the binary system $(S, *)$.

Definition 3.1.3. A binary system $(S, *)$ is said to be *finite* if the underlying set S is finite.

A finite binary system $(S, *)$ can be represented by means of a table as detailed below. Let $S = \{a_1, a_2, \ldots, a_n\}$. These elements $a_1, a_2, \ldots, a_n$ are to be listed across the top of the table and at the left of the table, both in the same order. The element $a_i * a_j$ is written in the ith row and j^{th} column as given in the table given below and $*$ is to written on the extreme left of the top.

$*$	a_1	a_2	a_3	$\ldots a_j \ldots$	a_n
a_1	$a_1 * a_1$	$a_1 * a_2$	$a_1 * a_3$	$\ldots a_1 * a_j \ldots$	$a_1 * a_n$
a_2	$a_2 * a_1$	$a_2 * a_2$	$a_2 * a_3$	$\ldots a_2 * a_j \ldots$	$a_2 * a_n$
a_3	$a_3 * a_1$	$a_3 * a_2$	$a_3 * a_3$	$\ldots a_3 * a_j \ldots$	$a_3 * a_n$
$\vdots$ a_i $\vdots$	$\vdots$ $a_i * a_1$ $\vdots$	$\vdots$ $a_i * a_2$ $\vdots$	$\vdots$ $a_i * a_3$ $\vdots$	$\vdots$ $\cdots a_i * a_j \cdots$ $\vdots$	$\vdots$ $a_i * a_n$ $\vdots$
a_n	$a_n * a_1$	$a_n * a_2$	$a_n * a_3$	$\ldots a_n * a_j \ldots$	$a_n * a_n$

In the example given below, we shall construct the table representing the binary system $(\mathbb{Z}_9, +_9)$ where $\mathbb{Z}_9 = \{0, 1, 2, \ldots, 8\}$ and $+_9$ is the addition modulo 9 (see Example 3.1.1 (11)).

Example 3.1.2. Let $\mathbb{Z}_9 = \{0, 1, 2, \ldots, 8\} = \{a \in \mathbb{Z} : 0 \leq a < 9\}$ and $+_9$ the addition modulo 9. $+_9$ is the binary operation on $\mathbb{Z}_9$ defined by

$$a +_9 b = \begin{cases} \text{the usual sum } a+b, & \text{if } a+b<9 \\ a+b-9, & \text{if } a+b \geq 9 \end{cases}.$$

for any a and b in $\mathbb{Z}_9$. The following table represents the binary system $(\mathbb{Z}_9, +_9)$.

$+_9$	0	1	2	3	4	5	6	7	8
0	0	1	2	3	4	5	6	7	8
1	1	2	3	4	5	6	7	8	0
2	2	3	4	5	6	7	8	0	1
3	3	4	5	6	7	8	0	1	2
4	4	5	6	7	8	0	1	2	3

5	5	6	7	8	0	1	2	3	4
6	6	7	8	0	1	2	3	4	5
7	7	8	0	1	2	3	4	5	6
8	8	0	1	2	3	4	5	6	7

Here, $2 +_9 7 = 0$, $2 +_9 5 = 7$, $4 -_9 5 = 0$, $5 +_9 7 = 3$, $6 +_9 8 = 5$, $8 +_9 8 = 7$, etc.

Example 3.1.3. Let S be the set of all positive divisors of 36 and, for any a and b in S, define

$$a \, g \, b = (a, b), \text{ the greatest common divisor (GCD) of } a \text{ and } b$$

(see Example 3.1.1 (5)). Then, (S, g) is a binary system which is represented by the table given below. We have

$$S = \{1, 2, 3, 4, 6, 9, 12, 18, 36\}$$

g	1	2	3	4	6	9	12	18	36
1	1	1	1	1	1	1	1	1	1
2	1	2	1	2	2	1	2	2	2
3	1	1	3	1	3	3	3	3	3
4	1	2	1	4	2	1	4	2	4
6	1	2	3	2	6	3	6	6	6
9	1	1	3	1	3	9	3	9	9
12	1	2	3	4	6	3	12	6	12
18	1	2	3	2	6	9	6	18	18
36	1	2	3	4	6	9	12	18	36

Let $(S, *)$ be a binary system. For any elements a, b and c in S, the expression $a * b * c$ has no meaning, since $*$ is a binary operation and hence $*$ is defined for pairs of elements. For example, $1 - 2 - 3$ has no meaning and we should specify whether it is $(1 - 2) - 3$ (this is what we usually take) or $1 - (2 - 3)$. Note that $(1 - 2) - 3 \neq 1 - (2 - 3)$. For arbitrary elements a, b and c in a binary system $(S, *)$, $a * b$ and c are two elements in S and hence $(a * b) * c$ is defined and so is $a * (b * c)$. In general, $(a * b) * c$ and $a * (b * c)$ may be different. In this context, we have the following definition.

Definition 3.1.4. A binary operation * on a set S is said to be *associative* if

$$(a * b) * c = a * (b * c)$$

for all elements a, b and c in S.

Definition 3.1.5. A pair $(S, *)$ is said to be a *semigroup* if S is a nonempty set and * is an associative binary operation on S.

Example 3.1.4. The binary operations given in Example 3.1.1, except those in (3), (6) and (16), are all associative and therefore, these together with the corresponding underlying sets, are semigroups. $(\mathbb{Z}, -)$ is not a semigroup, since $-$ is not associative. For any set X, $(\mathbb{P}(X), \cap)$, $(\mathbb{P}(X), \cup)$ and $(\mathbb{P}(X), +)$, given in Example 3.1.1 (6), are all semigroups. However, $(\mathbb{P}(X), -)$ is not a semigroup; For, consider $X = \{a, b, c, d\}$, $A = \{a, b, c\}$, $B = \{c, d\}$ and $c = \{b\}$. Then, A, B and $C \in \mathbb{P}(X)$ and

$$(A - B) - C = (\{a, b, c\} - \{c, d\}) - \{b\} = \{a, b\} - \{b\} = \{a\}$$

and

$$A - (B - C) = \{a, b, c\} - (\{c, d\} - \{b\}) = \{a, b, c\} - \{c, d\} = \{a, b\}$$

and therefore $(A - B) - C \neq A - (B - C)$, so that $-$ is not associative. Note that the operation * on S^X given in Example 3.1.1 (16) is associative if and only if the operation * on S is associative.

The associativity of a binary operation involves three arbitrary elements in the underlying set. If we take four elements a, b, c and d in a binary system $(S, *)$, then we get several expressions involving * and the elements a, b, c and d, in this order. These are given below.

$$(a * b) * (c * d)$$
$$a * (b * (c * d))$$
$$a * ((b * c) * d)$$
$$((a * b) * c) * d$$
$$(a * (b * c)) * d.$$

One can easily prove that these expressions represent one single element, if * is associative. In fact, we can generalize and extend the associativity for any finite sequence of elements. First, let us have the following definition.

Definition 3.1.6. Let $(S, *)$ be a binary system and $a_1, a_2, \ldots, a_n \in S$. A meaningful expression involving * and $a_1, a_2, \ldots, a_n$, in this order, is called a *meaningful product of $a_1, a_2, \ldots, a_n$, in this order.* The one given in the following definition is a meaningful product.

Definition 3.1.7. Let $(S, *)$ be a binary system and $a_1, a_2, \ldots, a_n \in S$. The standard product $\Pi_{i=1}^{n} a_i$ of $a_1, a_2, \ldots, a_n$, in this order, is defined inductively as follows.

$$\prod_{i=1}^{n} a_i = \begin{cases} a_1 & \text{if } n = 1 \\ \left(\prod_{i=1}^{n-1} a_i\right) * a_n & \text{if } n > 1 \end{cases}$$

For example,

$$\prod_{i=1}^{2} a_i = a_1 * a_2$$

$$\prod_{i=1}^{3} a_i = \left(\prod_{i=1}^{2} a_i\right) * a_3 = (a_1 * a_2) * a_3$$

$$\prod_{i=1}^{5} a_i = (((a_1 * a_2) * a_3) * a_4) * a_5.$$

Theorem 3.1.1 (Generalised Associative Law). Let $(S, *)$ be a semigroup and $a_1, a_2, \ldots, a_n$ elements of S. Then, all meaningful products of $a_1, a_2, \ldots, a_n$, in this order, are equal to each other.

Proof: We shall use induction on n to prove that each meaningful product of $a_1, a_2, \ldots, a_n$ is equal to their standard product. If $n = 1$ or 2, the theorem is trivial. Suppose that $n > 2$ and assume that any meaningful product of $b_1, b_2, \ldots, b_m$, with $m < n$, is equal to the standard product of $b_1, b_2, \ldots, b_m$, in this order. Let x be any meaningful product of $a_1, a_2, \ldots, a_n$, in this order. Then, there exists r such that $1 \leq r < n$ and

$$x = s * t$$

where s and t are meaningful products of $a_1, a_2, \ldots, a_r$ and $a_{r+1}, a_{r+2}, \ldots, a_n$, in these orders, respectively. By the induction hypothesis, we get that

$$s = \prod_{i=1}^{r} a_i \quad \text{and} \quad t = \prod_{j=r+1}^{n} a_j$$

Now, $x = s * t$

$$= \left(\prod_{i=1}^{r} a_i\right) * \left(\prod_{j=r+1}^{n} a_j\right)$$

$$= \left(\prod_{i=1}^{r} a_i\right) * \left(\left(\prod_{j=r+1}^{n-1} a_j\right) * a_n\right)$$

$$= \left[\left(\prod_{i=1}^{r} a_i\right) * \left(\prod_{j=r+1}^{n-1} a_j\right)\right] * a_n \text{ (since } * \text{ is associative)}$$

$$= \left(\prod_{i=1}^{n-1} a_i\right) * a_n \text{ (by the induction hypothesis)}$$

$$= \prod_{i=1}^{n} a_i$$

Thus, x is the standard product of $a_1, a_2, ..., a_n$, in this order. ◀

Definition 3.1.8. A binary operation * on a set S is said to be *commutative* if

$$a * b = b * a \quad \text{for all } a \text{ and } b \text{ in } S.$$

Example 3.1.5. Except in (3), (6), (8), (9), (14) and (16), all other binary operations given in Example 3.1.1 are commutative. The operations $\cap$, $\cup$ and $+$ on $\mathbb{P}(X)$, given in Example 3.1.1 (6), are all commutative and the operation $-$ on $\mathbb{P}(X)$ is not commutative. Also, the operation * on S^X, given in Example 3.1.1 (16) is commutative if and only if the operation * on S is commutative.

Theorem 3.1.2 (Generalised Commutative Law). Let * be a commutative and associative binary operation on a set S and $a_1, a_2, \ldots, a_n \in S$. Then, for any permutation σ of $\{1, 2, \ldots, n\}$, any meaningful product of $a_1, a_2, \ldots, a_n$ is equal to any meaningful product of $a_{\sigma(1)}, a_{\sigma(2)}, \ldots, a_{\sigma(n)}$.

Proof: Since * is associative, it is enough to prove that

$$\prod_{i=1}^{n} a_i = \prod_{i=1}^{n} a_{\sigma(i)}$$

for any permutation σ of $\{1, 2, \ldots, n\}$ (by Theorem 3.1.1). We shall prove theorem using induction on n. If $n = 1$, the theorem is trivial, if $n = 2$, the theorem follows from the commutativity of *. Let $n > 2$ and assume that the theorem is true for any $n - 1$ elements in S. Let σ be a permutation of $\{1, 2, \ldots, n\}$. That is, σ is a bijection of $\{1, 2, \ldots, n\}$ onto itself. Let $\sigma(n) = k$. Then, we have

$$\{\sigma(1), \sigma(2), \ldots, \sigma(n-1)\} = \{1, 2, \ldots, k-1, k+1, k+2, \ldots, n\}.$$

Consider the standard product

$$\begin{aligned}
\prod_{i=1}^{n} a_{\sigma(i)} &= \left(\prod_{i=1}^{n-1} a_{\sigma(i)}\right) * a_{\sigma(n)} \\
&= \left(\left(\prod_{i=1}^{k-1} a_i\right) * \left(\prod_{i=k+1}^{n} a_i\right)\right) * a_k \text{ (by induction hypothesis)} \\
&= \left(\prod_{i=1}^{k-1} a_i\right) * \left(a_k * \prod_{i=k+1}^{n} a_i\right) \text{(by associativity and commutativity of *)} \\
&= \prod_{i=1}^{n} a_i
\end{aligned}$$

◀

Note that the operation * in a finite binary system $(S, *)$ represented by the corresponding table is commutative if and only if the entries in the table are symmetric with respect to the diagonal that starts at the upper left corner of the table and terminates at the lower right corner (let us call this left–right diagonal). There is no such single technique to check the associativity of a given binary operation.

Worked Exercise 3.1.1. Determine the number of commutative binary operations on a set with n elements.

Answer: Let S be a set with n elements. $S \times S$ has n^2 elements. A binary operation on S is just a mapping of $S \times S$ into S. Therefore, there are exactly n^{n^2} number of binary operations on S. As mentioned above, a binary operation is commutative if and only if the entries in the corresponding table are symmetric with respect to the left–right diagonal. The number of pairs (a, b) in $S \times S$ with $a \neq b$, is $n^2 - n$. If * is a commutative binary operation on S, then $a * b = b * a$ for all $a, b \in S$ and hence, we can consider only half of the number of pairs (a, b) with $a \neq b$. These together with the pair (a, a), $a \in S$, constitute a set X consisting of

$$\frac{n^2 - n}{2} - n\left(= \frac{n^2 + n}{2}\right)$$

elements. A commutative binary operation on S can be identified with a mapping of X into S. Therefore, the number of commutative binary operations on S is equal to

$$|S^X| = n^{\frac{n^2+n}{2}}.$$

Worked Exercise 3.1.2. Let X be a nonempty set and $M(X)$ be the set of all mappings of X into itself. Let o be the composition of mappings on $M(X)$. Then prove that o is a commutative operation on $M(X)$ if and only if X has exactly one element.

Answer: If X has exactly one element, then $M(X)$ also has only one element and the result is trivial. Conversely suppose that X has more than one element, choose $a \neq b \in X$ and define f and $g : X \rightarrow X$ by

$$f(x) = a \text{ for all } x \in X$$

and $\quad g(a) = b,\ g(b) = a$ and $g(x) = x$ for all $x \notin \{a, b\}$.

Then, $\quad (f \circ g)(a) = f(g(a)) = f(b) = a$

and $\quad (g \circ f)(a) = g(f(a)) = g(a) = b.$

Therefore, $(f \circ g)(a) \neq (g \circ f)(a)$ and hence $f \circ g \neq g \circ f$.

Worked Exercise 3.1.3. Determine the number of noncommutative binary operations on a 5-element set.

Answer: Let S be a 5-element set. The total number of binary operations on S is 5^{25}. The number of commutative binary operations on S is

$$5^{\frac{5^2+5}{2}} = 5^{15}.$$

Therefore, the number of noncommutative binary operations on S is

$$5^{25} - 5^{15} = 5^{15}(5^{10} - 1).$$

EXERCISE 3(A)

1. Construct tables representing the following binary systems:
   (a) $(\mathbb{Z}_4, +_4)$
   (b) $(\mathbb{Z}_{10}, \cdot_{10})$

(c) $(\mathbb{P}(X), \cup)$, where $X = \{a, b, c\}$

(d) (S, ℓ), where S is the set of positive divisors of 100 and ℓ is defined by

$$a \,\ell\, b = \text{LCM of } \{a, b\}$$

(e) $(M(X), o)$, where $X = \{a, b\}$, $M(X)$ is the set of mappings of X into itself and o is the composition of mappings

(f) $(S, \cdot)$, where $S = \{1, i, -1, -i\}$ and '$\cdot$' is the usual multiplication of complex numbers.

2. Compare the tables in (a) and (f) above.

3. Fill in the blanks in the following table such that the binary operation * represented by the table is commutative.

*	a	b	c	d	e	f
a	d	e		c		b
b		f	d		f	
c	a		c	a	d	
d		c		e	d	
e	b				b	d
f		d	c	b		a

4. Compute the following from the table given in (3) above.

$$(b * (d * a)) * (c * (b * a))$$
$$((a * b) * c) * d$$
$$(a * b) * (c * d)$$

5. Is the operation * given in (3) above associative?

6. Give an example of an associative binary operation which is not commutative.

7. Prove that the associativity and the commutativity are independent of each other.

8. Prove or disprove the statement:

   Every commutative binary operation on a 2-element set is associative.

9. Which of the following binary operations are associative or commutative?

   (a) On the set $\mathbb{Z}$ of integers, $a * b = (a + 3)(b + 2)$.

   (b) On the set $\mathbb{Z}$, $a * b = a^{|b|}$.

   (c) On the set $\mathbb{R}$ of real numbers, $a * b = a - b$.

   (d) On the set $\mathbb{R}$, $a * b = a + ab$.

(e) On the power set $\mathbb{P}(X)$ of a set X, $A * B = (X - A) \cup (X - B)$.

(f) On any set X, $a * b = a$.

(g) On the set $\mathbb{Z}$, $a * b = 0$.

(h) On the set $\mathbb{Z}^+$ of positive integers,

$$a * b = 5^{a+b}$$

10. Prove or disprove the following statement:

    Every binary operation on a set S is both commutative and associative if and only if S has exactly one element.

11. Compute the number of commutative binary operations on a 4-element set.

12. Compute the number of noncommutative binary operations on a 3-element set.

13. Let $(S, *)$ be a binary system and

$$A = \{x \in S : (x * b) * c = x * (b * c)\} \text{ for all } b \text{ and } c \in S\}.$$

    If A is nonempty, prove that $(A, *)$ is a semigroup.

14. Let $(S, *)$ be a semigroup and e be any element not in S and $S' = S \cup \{e\}$. Define a binary operation $+$ on S' as follows. For any a and $b \in S'$, define

$$a + b = \begin{cases} a*b & \text{if both } a \text{ and } b \in S \\ a & \text{if } b = e \\ b & \text{if } a = e \end{cases}$$

    Then prove that $(S', +)$ is a semigroup and $e + x = x = x + e$ for all $x \in S'$.

3.2 GROUPS

The integer 0 has a special property in the binary system $(\mathbb{Z}, +)$ and is unique satisfying this property; namely

$$a + 0 = a = 0 + a \quad \text{for all } a \in \mathbb{Z}.$$

Similarly, the integer 1 is the unique element in the binary system $(\mathbb{Z}, \cdot)$ satisfying the property

$$a \cdot 1 = a = 1 \cdot a \quad \text{for all } a \in \mathbb{Z}$$

Likewise, the identity map I_X, defined on any set X by $I_X(x) = x$ for all $x \in X$, is the unique element in the binary system $(M(X), \text{o})$ satisfying the property

$$f \circ I_X = f = I_X \circ f \quad \text{for all } f \in M(X),$$

where $M(X)$ is the set of mappings of X into itself and o is the composition of mapping. An abstraction of these ideas is made in the following definition.

Definition 3.2.1. Let $(S, *)$ be a binary system and e be an element of S.

1. e is said to be a *right identity* in $(S, *)$ if $a * e = a$ for all $a \in S$.
2. e is said to be *left identity* in $(S, *)$ if $e * a = a$ for all $a \in S$.
3. e is said to be an *identity* in $(S, *)$ if e is both a left identity and a right identity; that is, $a * e = a = e * a$ for all $a \in S$.

Example 3.2.1

1. 0 is the only identity in $(\mathbb{Z}, +)$.
2. 1 is the only identity in $(\mathbb{Z}, \cdot)$, where '$\cdot$' is the usual multiplication of integers.
3. Let X be any nonempty set and $I_X : X \to X$ be defined by $I_X(x) = x$ for all $x \in X$. Then, I_X is the identity in $(M(X), \circ)$.
4. Let m and n be any positive integers and $M_{m\times n}(\mathbb{R})$ be the set of all $m\times n$ matrices over $\mathbb{R}$. Let $O_{m\times n}$ be the $m \times n$ matrix in which each entry is the number 0. Then,
$$O_{m\times n} + A = A = A + O_{m\times n}$$
for all matrices A and hence $O_{m\times n}$ is the identity in $(M_{m\times n}(\mathbb{R}), +)$, where $+$ is the usual addition of matrices.
5. Let S be any nonempty set and define $a * b = b$ for all a and b in S. Then, every element of S is a left identity in the binary system $(S, *)$.
6. Let S be any nonempty set and define $a * b = a$ for all a and $b \in S$. Then, every element of S is a right identity in $(S, *)$.
7. For any set X, the empty set $\varnothing$ is the identity in $(\mathbb{P}(X), \cup)$ and also in $(\mathbb{P}(X), +)$, where $\cup$ is union operation and $+$ is the symmetric difference operation.
8. Also for any set X, the whole set X is the identity in $(\mathbb{P}(X), \cap)$, where $\cap$ is the intersection operation.

From examples (7) and (8) above, the concept of identity is depending on the binary operation of the system. Also, it depends on the underlying set. For example, 0 is the identity in the system $(\mathbb{Z}, +)$ where as it is not the identity in $(\mathbb{Z}^+, +)$, since 0 is not an element in the underlying set $\mathbb{Z}^+$.

Also, from examples (5) and (6), observe that a binary system can possess any number of right identities without having any left identity and vice versa. However, if e is a right identity and f is a left identity, then e must be equal to f. This is proved in the following theorem.

Theorem 3.2.1. Let $(S, *)$ be a binary system and e be a left identity in $(S, *)$. Then, every right identity in $(S, *)$ is an identity and coincides with e.

Proof: Let f be a right identity in $(S, *)$. Then,

$$f = e * f \text{ (since } e \text{ is left identity)}$$
$$= e \text{ (since } f \text{ is right identity)}$$

Therefore, $f = e$ and hence f is a left identity also. Thus, f is an identity and $f = e$.

Corollary 3.2.1. There can be almost one identity in any binary system.

Proof: Let $(S, *)$ be a binary system and e and f be identities in $(S, *)$. Since e is a left identity and f is a right identity, $e = f$ by the above theorem.

Example 3.2.2

1. Let E be the set of even integers and is the usual multiplication of integers. Then, there is no identity in the binary system $(E, \cdot)$.
2. Let S be any set and define $a * b = b$ for all $a, b \in S$. Then, every element of S is a left identity in $(S, *)$. However, $(S, *)$ has no right identity, unless S is a singleton set (one element set).
3. Similarly, if we define $a * b = a$ for all $a, b \in S$ then every element of S is a right identity in $(S, *)$ and there are no left identities in $(S, *)$, unless S is a singleton set.
4. The integer 0 is the only identity in $(\mathbb{Z}, +)$.

Definition 3.2.2. A semigroup $(S, *)$ is called a monoid if it has identity. That is, a binary system $(S, *)$ is called a monoid if $*$ is associative and the identity exists in $(S, *)$.

Example 3.2.3

1. The set $M_n(\mathbb{R})$ of all $n \times n$ square matrices over $\mathbb{R}$ together with the matrix multiplication is a monoid. Here, the matrix I_n, in which all the diagonal entries are 1 and the others are 0, is the identity element in $M_n(\mathbb{R})$. I_n is called the identity matrix. Note that $I_n = (a_{ij})$, where $a_{ij} = 1$ or 0 according as $i = j$ or $i \neq j$.
2. $(\mathbb{Z}, +)$, $(\mathbb{Z}, \cdot)$ and $(\mathbb{Z}^+, \cdot)$ are all monoids, where $+$ and '$\cdot$' are the usual addition and multiplication, respectively. 0 is the identity in $(\mathbb{Z}, +)$ and 1 is the identity in $(\mathbb{Z}, \cdot)$ and in $(\mathbb{Z}^+, \cdot)$.

3. $(\mathbb{Z}^+, +)$ is a semigroup, but not a monoid.
4. The set $M(X)$ of all mappings of a set X into itself together with the composition of mappings is a monoid. The identity in $(M(X), \text{o})$, is precisely the identity map I_X defined by $I_X(x) = x$ for all $x \in X$.

Next, we shall take up the solvability of linear equations of type $a * x = b$, where a and b are given elements in a binary system $(S, *)$. It is well known that, for any real numbers a and b, there is a unique real number x satisfying the equation

$$a + x = b.$$

Our usual procedure of finding x is the following.

Consider $a + x = b$

$$\begin{aligned} (-a) + (a + x) &= (-a) + b && \text{(by adding } -a) \\ (-a + a) + x &= -a + b && \text{(by associativity)} \\ 0 + x &= -a + b && (-a + a = 0) \\ x &= -a + b && \text{(0 is the identity in } (\mathbb{R}, +)). \end{aligned}$$

Also, if we substitute $-a + b$ for x in $a + x = b$, we get that

$$a + (-a + b) = (a + (-a)) + b = 0 + b = b.$$

This is to say that $-a + b$ is the unique real number satisfying the equation a $+ x = b$. In this process finding the unique solution of $a + x = b$, we have skipped one step by not explaining what $-a$ is. It is obvious that $-a$ is the unique real number x satisfying the equation $x + a = 0$. This concept is abstracted in the following definition.

Definition 3.2.3. Let $(S, *)$ be a monoid in which e is the identity and $a \in S$.

1. An element $a^\ell \in S$ is called a *left inverse* of a if
$$a^\ell * a = e$$
2. An element $a^r \in S$ is called a *right inverse* of a if
$$a * a^r = e$$
3. An element $a' \in S$ is called an *inverse* of a if a' is both a left inverse and a right inverse of a; that is,
$$a' * a = e = a * a'.$$
4. a is called *invertible*, if there exists an inverse of a.

Example 3.2.4

1. In the monoid $(\mathbb{R}, +)$, the number 0 is the identity and every element of $\mathbb{R}$ has inverse.
2. In the monoid $(\mathbb{Z}, \cdot)$, 1 is the identity, where '$\cdot$' is the usual multiplication. Here, 1 and -1 are the only elements having inverses.
3. Consider the set $M(\mathbb{Z})$ of all mappings of $\mathbb{Z}$ into itself. Then, $(M(\mathbb{Z}), 0)$ is a monoid, in which the identity map I, defined by $I(x) = x$ for all $x \in \mathbb{Z}$, is the identity element. Define $f : \mathbb{Z} \to \mathbb{Z}$ by $f(x) = 2x$ for all $x \in \mathbb{Z}$. For each integer a, define $g_a : \mathbb{Z} \to \mathbb{Z}$ by

$$g_a(x) = \begin{cases} \dfrac{x}{2} & \text{if } x \text{ is an even integer} \\ a & \text{if } x \text{ is an odd integer.} \end{cases}$$

Then, $(g_a \circ f)(x) = g_a(f(x)) = g_a(2x) = \frac{2x}{2} = x$ for all $x \in \mathbb{Z}$. Therefore,

$$g_a \circ f = I$$

and hence g_a is a left inverse of f, for each integer a.

4. In (3) above, note that f has no right inverse; for, if g is a right inverse of f, then

$$f \circ g = I$$

and, in particular, $(f \circ g)(1) = I(1) = 1$ and hence

$$2 \cdot g(1) = 1$$

which is false, since we cannot get an integer $g(1)$ such that $2g(1) = 1$.

From the examples (3) and (4) above, we have noticed that an element in a monoid can have several left inverses without having any right inverses. However, if an element has both left inverse and right inverse, then they must be equal. This is proved in the following theorem.

Theorem 3.2.2. Let $(S, *)$ be a monoid in which e is the identity and $a \in S$. Let a^ℓ and a^r be left inverse and right inverse of a, respectively. Then, $a^\ell = a^r$ and a is invertible.

Proof: We are given that $a^\ell * a = e = a * a^r$. Now, consider

$$\begin{aligned} a^\ell &= a^\ell * e && \text{(since } e \text{ is the identity)} \\ &= a^\ell * (a * a^r) && \text{(since } a^r \text{ is right inverse of } a\text{)} \end{aligned}$$

$$= (a^\ell * a) * a^r \quad \text{(by associativity)}$$
$$= e * a^r \quad \text{(since } a^\ell \text{ is left inverse of } a)$$
$$= a^r \quad \text{(since } e \text{ is identity)}$$

Therefore, $a^\ell = a^r$ and hence $a^\ell * a = e = a * a^\ell$ so that a^ℓ $(= a^r)$ is inverse of a.

Corollary 3.2.2. Any element in a monoid has at most one inverse.

Proof: Let $(S, *)$ be a monoid in which e is the identity and $a \in S$. Suppose a' and a'' are inverses of a. Then, a' is a left inverse and a'' is a right inverse of a and hence, by the above theorem $a' = a''$.

Note that in a monoid, certain elements may be invertible and other elements may not be invertible. The identity element e in any monoid is always invertible and, since $e * e = e$, e is the inverse of itself. In the following, we give an example of a monoid in which each element is inverse of itself.

Example 3.2.5. Let X be any set and $\mathbb{P}(X)$ be the power set of X. For any A and $B \in \mathbb{P}(X)$, define

$$A + B = (A - B) \cup (B - A).$$

Then, $(\mathbb{P}(X), +)$ is a monoid with the empty set $\emptyset$ as the identity element. Here, for any $A \in \mathbb{P}(X)$,

$$A + A = (A - A) \cup (A - A) = \emptyset \cup \emptyset = \emptyset$$

and hence A is the inverse of itself.

In certain monoids, some elements may have left inverses, some elements may have right inverses, some may have both left and right inverses and some may have neither left inverses nor right inverses. In the following theorem, we give one such example.

Theorem 3.2.3. Let X be any nonempty set and $M(X)$ be the set of all mappings of X into itself. Then, $(M(X), \circ)$ is a monoid in which the following holds for any $f \in M(X)$.

1. f has a left inverse in $M(X)$ if and only if f is an injection.
2. f has a right inverse in $M(X)$ if and only if f is a surjection.
3. f is invertible in $M(X)$ if and only if f is a bijection.

Proof: We know that $(M(X), o)$ is a monoid (see Example 3.2.3 (4)) in which o is the composition of mapping and the map $I_X : X \to X$, defined by $I_X(x) = x$ for all $x \in X$, is the identity. Let f be an arbitrary element of $M(X)$; that is, $f : X \to X$ is a mapping.

1. Suppose that f has a left inverse in $M(X)$. Then, there exists $g \in M(X)$ such that

$$g \circ f = I_X.$$

For any $a, b \in X$, we have

$$\begin{aligned} f(a) = f(b) &\Rightarrow g(f(a)) = g(f(b)) \\ &\Rightarrow (g \circ f)(a) = (g \circ f)(b) \\ &\Rightarrow I_X(a) = I_X(b) \\ &\Rightarrow a = b. \end{aligned}$$

Therefore, f is an injection.

Conversely, suppose that f is an injection. Define

$g : X \to X$ by

$$g(x) = \begin{cases} a & \text{if } x = f(a) \text{ for some } a \in X \\ s & \text{otherwise} \end{cases},$$

where s is an arbitrarily chosen fixed element of X. Note that, since f is an injection, there can be at most one $a \in X$ such that $x = f(a)$ and hence g is welldefined. Now, for any $a \in X$,

$$(g \circ f)(a) = g(f(a)) = a = I_X(a)$$

and hence $g \circ f = I_X$, so that g is a left inverse of f.

2. Suppose that f has a right inverse in $M(X)$, Then, there exists $h \in M(X)$ such that

$$f \circ h = I_X.$$

For any $x \in X$, we have $h(x) \in X$ and

$$f(h(x)) = (f \circ h)(x) = I_X(x) = x$$

and therefore f is a surjection.

Conversely suppose that f is a surjection. Define $h : X \to X$ as follows. For any $x \in X$, choose one element $a_x \in X$ such that $f(a_x) = x$ (since f is a surjection, $f^{-1}\{x\}$ is a nonempty subset of X, for each $x \in X$ and now, we have to the axiom of choice). Now, define

$$h(x) = a_x, \quad \text{for each } x \in X.$$

Since $(f \circ h)(x) = f(h(x)) = f(a_x) = x$ for all $x \in X$, we have $f \circ h = I_X$ and hence h is a right inverse of f.

3. This follows from (1) and (2) and from Theorem 3.2.2.

Definition 3.2.4. A monoid $(G, *)$ is called a *group* if every element of S is invertible.
To be more elaborate, A pair $(G, *)$ is called a group if the following are satisfied:

1. G is a nonempty set and $*$ is a binary operation on G.
2. $a * (b * c) = (a * b) * c$ for all a, b and $c \in G$.
3. There exists $e \in G$ such that

$$a * e = a = e * a \quad \text{for all } a \in G.$$

4. For each $a \in G$, there exists $a' \in G$ such that

$$a' * a = e = a * a'$$

Recall that a pair $(G, *)$ is called a binary system if (1) is satisfied, semigroup if (1) and (2) are satisfied, monoid if (1), (2) and (3) are satisfied and is called a group if all the four (1), (2), (3), and (4) are satisfied.

Also, recall that the element e in (3) is unique and is called the identity in $(G, *)$. Further, the element a' in (4) is unique and is called the inverse of a. The inverse of a is usually denoted by a^{-1}.

Example 3.2.6

1. $(\mathbb{Z}, +)$, $(\mathbb{Q}, +)$, $(\mathbb{R}, +)$ and $(\mathbb{C}, +)$ are all groups in which $+$ is the usual addition, 0 is the identity and $-a$ is the inverse of any element a.
2. $(\mathbb{Q} - \{0\}, \cdot)$, $(\mathbb{R} - \{0\}, \cdot)$ and $(\mathbb{C} - \{0\}, \cdot)$ are all groups in which '$\cdot$' is the usual multiplication, 1 is the identity and $\frac{1}{a}$ is the inverse of any element a.
3. Neither $(\mathbb{Z}, \cdot)$ nor $(\mathbb{Z} - \{0\}, \cdot)$ are groups, since not all elements are invertible.
4. For any set X, $(\mathbb{P}(X), +)$ is a group in which $+$ is the symmetric difference operation, $\emptyset$ is the identity and, every element is inverse of itself.
5. Let X be a nonempty set and $S(X)$ the set of all bijections of X onto itself. Then, $(S(X), \circ)$ is a group in which $\circ$ is the composition of mappings, I_X is the identity and f^{-1} is the inverse of any bijection f. Recall that, for any bijections f and g, the composition $f \circ g$ is also a bijection.
6. The set $M_{m \times n}(\mathbb{R})$ of all $m \times n$ matrices over $\mathbb{R}$ together with the addition of matrices is a group. Here the zero matrix, in which all the entries are 0, is the identity and, for any $A = (a_{ij})$, the matrix $(-a_{ij})$ is the inverse of A.

7. For any positive integer n, $(\mathbb{Z}_n, +_n)$ is a group where $\mathbb{Z}_n = \{0, 1, 2, \dots, n-1\}$ and $+_n$ is the addition modulo n. Here, 0 is the identity and, for any $a \in \mathbb{Z}_n$, $n - a$ is the inverse of a. This group $(\mathbb{Z}_n, +_n)$ is called the *additive group of integers modulo n.*
8. Let n be a positive integer. A $n \times n$ square matrix A is called *nonsingular* if its determinant is not zero. The set $\text{NSM}_n(\mathbb{R})$ of all nonsingular $n \times n$ matrices over $\mathbb{R}$ together with the matrix multiplication is a group. Here, the identity matrix I_n, in which all the diagonal entries are 1 and other entries are 0, is the identity. It is well known that a $n \times n$ square matrix A is nonsingular if and only if there exists $n \times n$ matrix B such that

$$AB = I_n = BA.$$

9. For any points $a = (a_1, a_2)$ and $b = (b_1, b_2)$ in the two-dimensional Euclidean space $\mathbb{R} \times \mathbb{R}$, $d(a, b)$ be the usual Euclidean distance between a and b; that is,

$$d(a, b) = \sqrt{(a_1 - b_1)^2 + (a_2 - b_2)^2}.$$

Let X be the set of all points on a given geometrical figure (in fact, X may be any nonempty subset of $\mathbb{R} \times \mathbb{R}$).

A bijection f of X onto itself is called a *symmetry of X* if
$d(f(a), f(b)) = d(a, b)$ for all $a, b \in X$.

Let Sym(X) be the set of all symmetries of X. Then, (Sym(X), o) is a group in which o is the composition of mappings, I_X is the identity and f^{-1} is the inverse of any f. This group (Sym(X), o) is called the *group of symmetries of X.*
10. Let p be any prime number and $G_p = \{1, 2, \dots, p-1\}$. Then, $(G_p, \cdot_p)$ is a group in which $\cdot_p$ is the multiplication modulo p (see Example 3.1.1 (12)) defined by

$a \cdot_p b$ = the remainder obtained by dividing the usual product ab with p.

Here, 1 is the identity and, for any $0 < a < p$, the GCD of a and p is 1 and hence, there exists integers α and β such that

$$\alpha a + \beta b = 1.$$

If we divide α by p, we get the remainder b. Then, $0 < b < p$ and $b \cdot_p a = 1$and hence b is the inverse of a in $(G_p, \cdot_p)$.

Theorem 3.2.4. Let $(M, *)$ be a monoid and G be the set of all invertible elements in $(M, *)$. Then, $(G, *)$ is a group.

Proof: First we shall observe that G is a nonempty set, since the identity e in $(M, *)$ is always invertible and hence $e \in G$. Also, if a and $b \in G$ and a' and b' are inverses of a and G, respectively, then

$$\begin{aligned}(a * b) * (b' * a') &= a * (b * b') * a' \\ &= a * e * a' \\ &= a * a' = e\end{aligned}$$

and

$$\begin{aligned}(b' * a') * (a * b) &= b' * (a' * a) * b \\ &= b' * e * b \\ &= b' * b = e\end{aligned}$$

and hence $a * b$ is invertible and $b' * a'$ is the inverse of $a * b$, so that $a * b \in G$. Therefore, $*$ becomes a binary operation on G. Also, since $*$ satisfies associativity on M, $*$ is associative on G also. Since $e \in G$, $(G, *)$ is a monoid and clearly, the inverse of any invertible element is also invertible. Therefore, every element in $(G, *)$ is invertible. Thus, $(G, *)$ is a group.

Definition 3.2.5. Let $(M, *)$ be a monoid and $a \in M$. For any nonnegative integer n, define

$$a^n = \begin{cases} e, \text{ the identity} & \text{if } n = 0 \\ a^{n-1} * a & , \text{if } n > 0 \end{cases}.$$

If a is invertible and n is a negative integer, define

$$a^n = (a')^{-n}, \text{ when } a' \text{ is the inverse of } a.$$

Note that $a^0 = e$, $a^1 = a$, $a^2 = a * a$, $a^3 = (a * a) * a$, etc. and, if a is invertible, then

$$\begin{aligned} a^{-1} &= a', \quad \text{the inverse of } a \\ a^{-2} &= (a')^2 \end{aligned}$$

and

$$a^{-n} = (a')^n \quad \text{for any } n \in Z^+.$$

This justifies the notation a^{-1} for the inverse of a.

Worked Exercise 3.2.1. Let a be an invertible element in a monoid $(M, *)$ and a' be the inverse of a. Then prove the following for any integers m and n.

1. $a^{n+m} = a^n * a^m$
2. $(a')^n = a^{-n}$
3. $(a^n)^m = a^{nm} = (a^m)^n$

Answer: We shall fix $n \in \mathbb{Z}$ and use induction on m.

1. Case (i): Suppose that $m \geq 0$.
   If $m = 0$, then $a^{n+m} = a^n = a^n * e = a^n * a^m$.
   Let $m > 0$ and assume that $a^{n+(m-1)} = a^n * a^{m-1}$. Then,
   $$\begin{aligned} a^n * a^m &= a^n * (a^{m-1} * a) && \text{(by definition of } a^m\text{)} \\ &= (a^n * a^{m-1}) * a && \text{(by associativity)} \\ &= (a^{n+(m-1)}) * a && \text{(by induction hypothesis)} \\ &= a^{n+m} \end{aligned}$$
   Thus, $a^{n+m} = a^n * a^m$ for all $n, m \in \mathbb{Z}$ with $m \geq 0$.
   Case (ii): Suppose that $m < 0$.
   Consider, $a^n * a^m = a^n * (a')^{-m}$ (by definition of a^m)
   $$= \begin{cases} a^{n-(-m)} & \text{if } n \geq -m \\ (a')^{-m-n} & \text{if } -m > n \end{cases}$$
   $$= a^{n+m}$$

2. This is trivial if $n = 0$, since $(a')^0 = e = a^0$
   If $n > 0$, then $-n < 0$ and hence, by definition,
   $$a^{-n} = (a')^{-(-n)} = (a')^n.$$
   If $n < 0$, then, again by definition,
   $$(a')^n = a^{-n}, \text{ since } a \text{ is the inverse of } a'.$$

3. This is trivial if $m = 0$ or $n = 0$. Therefore, we can assume that $mn \neq 0$.
   Case (i): Suppose that $mn > 0$.
   If both m and n are positive, then, by (1),
   $$(a^n)^m = a^{nm} = a^{mn} = (a^m)^n.$$
   If both m and n are negative, then
   $$\begin{aligned} (a^n)^m &= ((a^n)')^{-m} \\ &= (((a')^{-n})')^{-m} \\ &= (((a')')^{-n})^{-m} \\ &= (a^{-n})^{-m} \\ &= a^{(-n)(-m)} = a^{nm} \end{aligned}$$

Case (ii): Suppose that $mn < 0$. To be specific, suppose that $n < 0$ and $m > 0$. Then,

$$\begin{aligned} a^{nm} &= (a')^{-nm} \\ &= (a')^{(-n)m} \\ &= ((a')^{-n})^m \text{ (since } -n > 0 \text{ and } m > 0) \\ &= (a^n)^m. \end{aligned}$$

So is the case when $n > 0$ and $m < 0$.

Thus, $(a^n)^m = a^{nm} = a^{mn} = (a^m)^n$.

Worked Exercise 3.2.2. Let G be the set of all rotations of the plane about the origin in the plane and o the composition of mappings. Thus, prove that (G, o) is a group.

Answer: The rotation about the origin through an angle θ can be represented analytically as the map $f_\theta : \mathbb{R} \times \mathbb{R} \to \mathbb{R} \times \mathbb{R}$ defined by

$$f_\theta(x, y) = (x\cos\theta - y\sin\theta,\ x\sin\theta + y\cos\theta)$$

Therefore, $G = \{f_\theta : 0 \leq \theta < 2\pi\}$. Note that

$$f_\theta \text{ o } f_\phi = \begin{cases} f_{\theta+\phi} & \text{if } \theta+\phi < 2\pi \\ f_{\theta+\phi-2\pi} & \text{if } \theta+\phi \geq 2\pi \end{cases}$$

It can be easily verified that o is an associative binary operation on G and that f_0 (= the identity map) is the identity in (G, o). Also for any θ, $f_{2\pi-\theta}$ is the inverse of f_θ (considering $f_{2\pi} = f_0$). Thus, (G, o) is a group.

Worked Exercise 3.2.3. For any real numbers a and b with $a \neq 0$, define

$$f_{a,b} : \mathbb{R} \to \mathbb{R} \text{ by } f_{a,b}(x) = ax + b \text{ for all } x \in \mathbb{R},$$

let $G = \{f_{a,b} : 0 \neq a \in \mathbb{R} \text{ and } b \in \mathbb{R}\}$. Then prove that (G, o) is a group, where o is the composition of mappings.

Answer: Note that, for any a, b, c and $d \in \mathbb{R}$,

$$\begin{aligned} (f_{a,b} \text{ o } f_{c,d})(x) &= f_{a,b}(cx + d) \\ &= a(cx + d) + b \\ &= acx + ad + b \\ &= f_{ac,ad+b}(x) \end{aligned}$$

and hence $f_{a,b} \circ f_{c,d} = f_{ac,\, ad+b}$. Also, if $a \neq 0$ and $c \neq 0$, then $ac \neq 0$. Therefore, 0 is an associative binary operation on G. Further, $f_{1,0}$ is the identity in $(G, \circ)$, since $f_{1,0}(x) = 1 \cdot x + 0 = x$ for all $x \in \mathbb{R}$. Also, $f_{\frac{1}{a}, -\frac{b}{a}}$ is the inverse of $f_{a,b}$. Thus, $(G, \circ)$ is a group.

We shall conclude this section with two more important examples of groups given below.

Example 3.2.7. Let $G = \{f_1, f_2, f_3, f_4, f_5, f_6\}$, where each f_i is a function of $\mathbb{R} - \{0, 1\}$ into itself as defined below.

$$f_1(x) = x, f_2(x) = \frac{1}{x}, f_3(x) = 1 - x,$$

$$f_4(x) = \frac{1}{1-x}, f_5(x) = \frac{x-1}{x} \quad \text{and} \quad f_6(x) = \frac{x}{x-1}.$$

Then, $(G, \circ)$ is a group, where o is the composition of mappings. The following table represents the binary operation o on G.

0	f_1	f_2	f_3	f_4	f_5	f_6
f_1	f_1	f_2	f_3	f_4	f_5	f_6
f_2	f_2	f_1	f_4	f_3	f_6	f_5
f_3	f_3	f_5	f_1	f_6	f_2	f_4
f_4	f_4	f_6	f_2	f_5	f_1	f_3
f_5	f_5	f_3	f_6	f_1	f_4	f_2
f_6	f_6	f_4	f_5	f_2	f_3	f_1

Note that f_1 is the identity and the inverses are given by

$$f_1^{-1} = f_1, f_2^{-1} = f_2, f_3^{-1} = f_3, f_4^{-1} = f_5, f_5^{-1} = f_4, f_6^{-1} = f_6.$$

Also note that $f_2 \circ f_4 = f_3 \neq f_6 = f_4 \circ f_2$.

Example 3.2.8. The group discussed here is called the *group of symmetries of the square* (see Example 3.2.6 (9)). Let X be the set of all points in a square of unit side. Recall that a symmetry of X is a bijection f of X into itself such that

$$d(a, b) = d(f(a), f(b)) \quad \text{for all } a \text{ and } b \in X.$$

Let G be the set of all symmetries of X. Then, (G, o) is a group, where o is the usual composition of mappings Note that G consists of exactly eight symmetries and these are listed below.

e = The identity function.

r_1 = The clock-wise rotation about the centre of the square through an angle $\frac{\pi}{2}$.

r_2 = The clock-wise rotation about the centre through an angle π.

r_3 = The clock-wise rotation about the centre through an angle $\frac{3\pi}{2}$.

h = The reflection about the horizontal line through the centre of the square.

v = The reflection about the vertical line through the centre of the square.

d_1 = The reflection about the diagonal D_1.

d_2 = The reflection about the diagonal D_2.

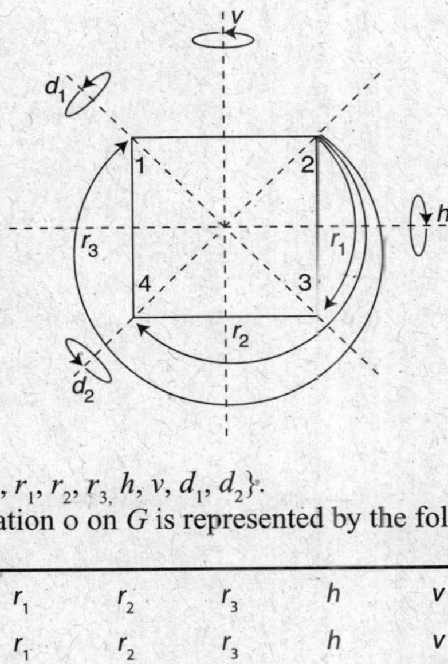

We have $G = \{e, r_1, r_2, r_3, h, v, d_1, d_2\}$.

The binary operation o on G is represented by the following table.

o	e	r_1	r_2	r_3	h	v	d_1	d_2
e	e	r_1	r_2	r_3	h	v	d_1	d_2
r_1	r_1	r_2	r_3	e	d_1	d_2	v	h
r_2	r_2	r_3	e	r_1	v	h	d_2	d_1
r_3	r_3	e	r_1	r_2	d_2	d_1	h	v
h	h	d_2	v	d_1	e	r_2	r_3	r_1
v	v	d_1	h	d_2	r_2	e	r_1	r_3
d_1	d_1	h	d_2	v	r_1	r_3	e	r_2
d_2	d_2	v	d_1	h	r_3	r_1	r_2	e

EXERCISE 3(B)

1. Determine the following in which $+_n$ and $\cdot_n$ are the addition and multiplication modulo n, for a given positive integer n.
   (i) $7 +_{12} 11$
   (ii) $8 +_{10} 7$
   (iii) $7 \cdot_{12} 11$
   (iv) $8 \cdot_{10} 7$
   (v) $4 \cdot_6 5$
   (vi) $(7 +_{10} 6) \cdot_{10} 8$
   (vii) 7^7 in $(\mathbb{Z}_8, \cdot_8)$
   (viii) 5^{-6} in $(G_7, \cdot_7)$
   (ix) 7^{-8} in $(G_{11}, \cdot_{11})$
   (x) 6^8 in $(\mathbb{Z}_9, \cdot_9)$

2. List all the invertible elements in each of the following monoids.
   (i) $(\mathbb{Z}_{10}, +_{10})$
   (ii) $(\mathbb{Z}_{10}, \cdot_{10})$
   (iii) $(\mathbb{Z}_{36}, \cdot_{36})$
   (iv) $(\mathbb{Z}_4, \cdot_4)$
   (v) $(M(X), \text{o})$, where $M(X)$ is the set of all mappings of X into itself
   (vi) $(\mathbb{P}(X), +)$.
   (vii) $(\mathbb{R}^+, \cdot)$
   (viii) $(\mathbb{Z}^+, \cdot)$

3. Determine which of the following gives a group structure on the given set.
   (i) For any $a, b \in \mathbb{Z}$, $a * b = a + b + ab$.
   (ii) For any $a, b \in \mathbb{R}^+$, $a * b = \frac{a}{b}$.
   (iii) $a * b = \frac{ab}{2}$ for any $a, b \in \mathbb{R}^+$.
   (iv) $a * b = |ab|$ for any $a, b \in \mathbb{C}$.
   (v) $a * b = a + b - 2$ for any $a, b \in \mathbb{Z}$.
   (vi) For any $a, b, \in \mathbb{R}^+$, $a * b = 5ab$.
   (vii) For any $a, b \in \mathbb{Q}^+$, $a * b = |ab|$.
   (viii) $a * b = a + b + ab$, for any $a, b \in \mathbb{Z}$.

4. Prove that the matrices

$$\begin{pmatrix} 1 & 0 \\ 0 & 1 \end{pmatrix}, \begin{pmatrix} -1 & 0 \\ 0 & 1 \end{pmatrix}, \begin{pmatrix} -1 & 0 \\ 0 & -1 \end{pmatrix}, \begin{pmatrix} 1 & 0 \\ 0 & -1 \end{pmatrix}$$

form a group under the matrix multiplication. What is identity element? Determine the inverse of each element.

5. State whether each of the following is true (T) or false (F):
   (i) Any semigroup has exactly one left identity.
   (ii) Any semigroup has at most one right identity.
   (iii) Any group has exactly one left identity.
   (iv) In a group, each element has exactly one right inverse.
   (v) There is a group with exactly one element.
   (vi) For each positive integer n, there is a group with exactly n elements.
   (vii) Every semigroup has an identity.
   (viii) For any positive integer n, there is a semigroup with exactly n elements in which every element is a right identity.

6. Let $(G, *)$ be a group. Prove that the identity e is the only element satisfying $x * x = x$.

7. Let n be a positive integer and G be the set of all n^{th} root of unity; that is,

$$G = \{z : z \text{ is a complex number and } z^n = 1\}.$$

Prove that G is group under the usual multiplication of complex numbers.

8. Let $(G, *)$ be a group and X be any nonempty set. Let G^X be the set of all mappings of X into G. For any $f, g \in G^X$, define $f * g : X \to G$ by

$$(f * g)(x) = f(x) * g(x) \text{ for all } x \in X.$$

Prove that $(G^X, *)$ is a group. What is the identity in this group? Determine the inverse of any $f \in G^X$.

9. For any positive integer n, prove that $(\mathbb{Z}_n, \cdot_n)$ is a monoid, where $\mathbb{Z}_n = \{0, 1, 2, \ldots n - 1\}$ and $\cdot_n$ is the multiplication modulo n.

10. For any $n > 1$, prove that $(\mathbb{Z}_n, \cdot_n)$ is never a group.

11. For any $1 \leq a < n$, prove that a is invertible in the monoid $(\mathbb{Z}_n, \cdot_n)$ if and only if a is relatively prime to n.

12. For any prime number p, prove in detail that $(G_p, \cdot_p)$ is a group, where

$$G_p = \mathbb{Z}_p - \{0\}.$$

13. For any positive integer n, give an example of a group having exactly 2^n elements.

14. Is $\{1, 2, 3, 4\}$ a group under multiplication modulo 5?

15. Is $\{1, 2, 3, 4, 5\}$ a group under multiplication modulo 6?

16. Prove that the following are equivalent to each other for any integer $n > 1$.
    (1) $\cdot_n$ is a binary operation on $\mathbb{Z}_n - \{0\}$.
    (2) $(\mathbb{Z}_n - \{0\}, \cdot_n)$ is a group.
    (3) n is a prime number.
    (4) any $1 \leq a < n$ is relatively prime with n.

17. Let G be the set of all rotations about the origin in the plane and reflections in the lines through origin. Then prove that (G, o) is a group, where o is the composition of mappings.

18. Consider the regular n-gon (polygon of n equal sides and equal internal angles) inscribed in the unit circle in the plane, so that one of the vertices is $(1, 0)$. Let R_n be the set of all rotations about the origin which maps this regular n-gon into itself. Prove that (R_n, o) is a group, where o is the composition of mappings. How many elements are there in this group?

19. Let D_n be the set of all rotations and reflections which the regular n-gon, given in 18 above, into itself. Then prove that (D_n, o) is a group, where o is the composition of mappings. How many elements D_n has? The group (D_n, o) is called the *dihedral group of degree n* (see Theorem 6.4.8). The elements of D_n are called the *symmetries* of the regular n-gon.

20. For any rational numbers r and s, define $r \sim s$ if and only if $r - s$ is an integer. Then prove that $\sim$ is an equivalence relation on the set $\mathbb{Q}$ of rational numbers. Let $\mathbb{Q}/\mathbb{Z}$ denote the set of equivalence classes w.r.t. $\sim$ in $\mathbb{Q}$ and, for any classes $\tilde{r}, \tilde{s}$, define $\tilde{r} + \tilde{s} = \widetilde{r+s}$. Then prove that $(\mathbb{Q}/\mathbb{Z}, +)$ is a group.

3.3 ELEMENTARY PROPERTIES OF GROUPS

In this section, we shall derive certain important elementary properties of groups. In particular, we obtain several sets of equivalent conditions for a semigroup to become a group. Let us agree to denote the identity element in an abstract monoid or group by e. We begin with the following theorem.

Theorem 3.3.1. Let $(G, *)$ be a group and a, b and $c \in G$. Then, the following holds.

1. $a * b = e \Leftrightarrow a^{-1} = b \Leftrightarrow b^{-1} = a$, where e denotes the identity in the group.
2. $(a^{-1})^{-1} = a$
3. $(a * b)^{-1} = b^{-1} * a^{-1}$
4. $a * b = c \Leftrightarrow a = c * b^{-1} \Leftrightarrow b = a^{-1} * c$

Proof:

1. $a*b=e \Rightarrow a^{-1}=a^{-1}*e=a^{-1}*(a*b)=(a^{-1}*a)*b=e*b=b$

   $a^{-1}=b \Rightarrow a*b=a*a^{-1}=e$

   $a*b=e \Rightarrow b^{-1}=e*b^{-1}=(a*b)*b^{-1}=a*(b*b^{-1})=a*e=a$

   $b^{-1}=a \Leftrightarrow a*b=b^{-1}*b=e$

2. Since $a^{-1} * a = e$, it follows from (1) that $(a^{-1})^{-1} = a$
3. Since $(a * b) * (b^{-1} * a^{-1}) = a * (b * b^{-1}) * a^{-1} = a * e * a^{-1} = a * a^{-1} = e$, again from (1) it follows that $(a * b)^{-1} = b^{-1} * a^{-1}$
4. $a*b=c \Rightarrow c*b^{-1}=(a*b)*b^{-1}=a*(b*b^{-1})=a*e=a$

   $a=c*b^{-1} \Rightarrow a*b=(c*b^{-1})*b=c*(b^{-1}*b)=c*e=c$

   $a*b=c \Rightarrow a^{-1}*c=a^{-1}*(a*b)=(a^{-1}*a)*b=e*b=b$

   $b=a^{-1}*c \Rightarrow a*b=a*(a^{-1}*c)=(a*a^{-1})*c=e*c=c$

Note that if we take e for c in (4), we get (1).

Let us recall that a semigroup is a pair $(S, *)$ where S is a nonempty set and $*$ is an associative binary operation on S and that a semigroup with identity is called a monoid and also that a monoid is called a group if every each of its elements is invertible.

Theorem 3.3.2. Let $(S, *)$ be a semigroup. Then, $(S, *)$ is a group if and only if the following conditions are satisfied.

1. $(S, *)$ has a right identity e. That is, there exists $e \in S$ such that $a * e = a$ for all $a \in S$.
2. For each $a \in S$, there exists $a' \in S$ such that $a * a' = e$

Proof: If $(S, *)$ is a group, then clearly (1) and (2) are satisfied. Conversely suppose that the conditions (1) and (2) are satisfied. By (1), there exists $e \in S$, such that $a * e = a$ for all $a \in S$. We shall prove that this e is actually the identity in $(S, *)$. Let a be an arbitrary element in S, By (2), there exists a' and x in S such that

$$a*a'=e \quad \text{and} \quad (a'*a)*x=e \tag{i}$$

Consider

$$\begin{aligned}(a'*a)*(a'*a) &= a'*(a*a')*a \\ &= (a'*e)*a \text{ (by (i))} \\ &= a'*a\end{aligned} \tag{ii}$$

Now,

$$\begin{aligned} e &= (a' * a) * x && \text{(by (i))} \\ &= ((a' * a) * (a' * a)) * x && \text{(by (ii))} \\ &= (a' * a) * ((a' * a) * x) && \text{(by associativity)} \\ &= (a' * a) * e && \text{(by (i))} \\ &= a' * a \end{aligned}$$

Thus, for any $a \in S$, there exists $a' \in S$ such that

$$a' * a = e = a * a' \qquad \text{(iii)}$$

Also, $e * a = (a * a') * a = a * (a' * a) = a * e = a$.
Thus, e is the identity in $(S, *)$ and, for any $a \in S$, a' is the inverse of a (by (iii)), Therefore, $(S, *)$ is a group.

On the lines of above proof, one can also prove that a semigroup is a group if and only if it has left identity with respect to which every element has left inverse.

Recall that, for any given real numbers a and b, the equation $a + x = b$ has a unique solution in $\mathbb{R}$. In fact, this is an important defining property of a group as proved in the following theorem.

Theorem 3.3.3. A semigroup $(S, *)$ is a group if and only if, for any elements a and b in S, the equation

$$a * x = b \quad \text{and} \quad y * a = b$$

are solvable in S (in the sense that there are elements x and y in S satisfying these equations).

Proof: Let $(S, *)$ be a semigroup. If $(S, *)$ is a group, then for any $a, b \in S$, we have $a^{-1} * b$ and $b * a^{-1}$ are elements of S such that

$$a * (a^{-1} * b) = (a * a^{-1}) * b = e * b = b$$

and

$$(b * a^{-1}) * a = b * (a^{-1} * a) = b * e = b$$

and therefore the equations $a * x = b$ and $y * a = b$ have solutions in S.
Conversely suppose that these equations have solutions in S for any a and b in S.

Let a be an arbitrary element in S. Then, there exists $e \in s$ such that

$$a * e = a \quad \text{(since } a * x = a \text{ is solvable in } S\text{)}.$$

We shall prove that $b * e = b$ for all elements $b \in S$. To prove this, let $b \in S$. Then, choose an element $s \in S$ such that

$$s * a = b \quad \text{(since } y * a = b \text{ is solvable in } S\text{)}.$$

Now, $b * e = (s * a) * e = s * (a * e) = s * a = b$.
Thus, e is a right identity in $(S, *)$. Also, since $a * x = e$ is solvable in S, we get that, for each $a \in S$, there exists $a' \in S$ such that $a * a' = e$. Thus, by the above Theorem 3.3.2, $(S, *)$ is a group.

Recall that, in the elementary school mathematics, one is used to conclude $b = c$ whenever $a + b = a + c$ for some a and we were used to give reasoning for this by saying 'subtracting a from both sides' which amounts to adding $-a$ both sides.

This is abstracted in the following theorem.

Theorem 3.3.4. Let $(G, *)$ be a group and a, b and $c \in S$. Then,

$$a * b = a * c \Rightarrow b = c \text{ (left cancellation law)}$$

and
$$b * a = c * a \Rightarrow b = c \text{ (right cancellation law)}.$$

Proof: Consider

$$\begin{aligned} a * b = a * c &\Rightarrow a^{-1} * (a * b) = a^{-1} * (a * c) \\ &\Rightarrow (a^{-1} * a) * b = (a^{-1} * a) * c \\ &\Rightarrow e * b = e * c \\ &\Rightarrow b = c \end{aligned}$$

Also, $b * a = c * a \Rightarrow (b * a) * a^{-1} = (c * a) * a^{-1}$
$$\begin{aligned} &\Rightarrow b * (a * a^{-1}) = c * (a * a^{-1}) \\ &\Rightarrow b * e = c * e \\ &\Rightarrow b = c \end{aligned}$$

A semigroup may satisfy both the left and right cancellation laws without being a group. This is to say that the converse of the above theorem is not true. For, consider the following examples.

Example 3.3.1

1. Consider the semigroup $(\mathbb{Z}^+, +)$, where $\mathbb{Z}^+$ is the set of positive integers and $+$ is the usual addition. Since $(\mathbb{Z}, +)$ is a group, $(\mathbb{Z}, +)$ satisfies both the cancellation laws. Since $\mathbb{Z}^+$ is a subset of $\mathbb{Z}$, $(\mathbb{Z}^+, +)$ also satisfies both the cancellation laws. Nevertheless, $(\mathbb{Z}^+, +)$ is not a group, since this has no identity.
2. A monoid may satisfy the cancellation laws without being a group. Consider the set $\mathbb{W}$ of all nonnegative integers. Then, for the same reason given above, $(\mathbb{W}, +)$ is a monoid satisfying both the cancellation laws and this is not a group, since no element, except 0, has inverse.

Even though the converse of Theorem 3.3.4 is not true in general, we prove the converse in the case of finite semigroups. Recall that a semigroup $(S, *)$ is called finite if the underlying set S is finite.

Theorem 3.3.5. Let $(S, *)$ be a finite semigroup satisfying both the cancellation laws. Then, $(S, *)$ is a group.

Proof : Since S is a finite set, we can enumerate the elements of S. Let $a_1, a_2, \ldots, a_n$ be all the distinct elements of S. That is,

$$S = \{a_1, a_2, \ldots, a_n\}.$$

Let a and b be any arbitrary elements in S and let

$$a * S = \{a * a_1, a * a_2, \ldots, a * a_n\}.$$

Then, $a * a_i$'s are all distinct elements in S, since

$$\begin{aligned} a * a_i = a * a_j &\Rightarrow a_i = a_j \text{ (by left cancellation law)} \\ &\Rightarrow i = j \text{ (since } a_i\text{'s are distinct).} \end{aligned}$$

Therefore, $a * S$ is an n-element subset of S and S also has n-elements and hence

$$a * S = S.$$

In particular, $b \in S = a * S$ and hence $b = a * x$ for some $x \in S$.
Similarly, by using the right cancellation law, we can prove that $S * a = S$ and hence $y * a = b$ for some $y \in S$. Therefore, for any elements a and b in S, the equations

$$a * x = b \quad \text{and} \quad y * a = b$$

are solvable in S. Thus, by Theorem 3.3.3, $(S, *)$ is a group.

Definition 3.3.1. A binary operation $*$ on a set S is said to be commutative if $a * b = b * a$ for all a and $b \in S$.

A group $(G, *)$ is said to be a *commutative group* or *abelian group* (in honour of a great algebraist Abel) if $*$ is commutative ; that is,

$$a * b = b * a \quad \text{for all } a \text{ and } b \in G.$$

Example 3.3.2

1. $(\mathbb{Z}, +)$, $(\mathbb{Q}, +)$, $(\mathbb{R}, +)$ and $(\mathbb{C}, +)$ are all abelian groups, since the addition $+$ is commutative.
2. $(\mathbb{Q}-\{0\},\cdot)$, $(\mathbb{R}-\{0\},\cdot)$ and $(\mathbb{C}-\{0\},\cdot)$ are abelian groups, since the multiplication is commutative.
3. For any set X, $(\mathbb{P}(X), +)$ is an abelian group, since, for any A and B in $\mathbb{P}(X)$,
$$A + B = (A - B) \cup (B-A) = (B-A) \cup (A - B) = B + A.$$
4. Let X be a set with atleast three elements and $S(X)$ the set of all bijections of X onto itself. Then, $(S(X), \text{o})$ is a group which is not abelian. For, consider three distinct elements a, b and c in X and define f and $g : X \to X$ by
$$f(a) = b, f(b) = a \quad \text{and} \quad f(x) = x \quad \text{for all } x \neq a, b$$
and $$g(b) = c, g(c) = b \quad \text{and} \quad g(x) = x \quad \text{for all } x \neq b, c.$$
Then, $$(f \text{ o } g)(a) = f(g(a)) = f(a) = b$$
and $$(g \text{ o } f)(a) = g(f(a)) = g(b) = c \neq b = (f\text{o}g)(a).$$
Therefore, $f \text{ o } g \neq g \text{ o } f$. Thus, $(S(X), \text{o})$ is an abelian group.
5. The matrix multiplication is not commutative. Let $NSM_n(\mathbb{R})$ be the set of all nonsingular $n \times n$ matrices over $\mathbb{R}$. Then, $(NSM_n(\mathbb{R}), \cdot)$ is a group which is not abelian if $n > 1$.
6. The addition of matrices is a commutative operation. $(M_{m\times n}(\mathbb{R}), +)$ is an abelian group for any positive integers m and n, where $M_{m\times n}(\mathbb{R})$ is the set of all $m \times n$ matrices over $\mathbb{R}$.

Theorem 3.3.6. The following are equivalent to each other for any group $(G, *)$.

1. $(G, *)$ is an abelian group.
2. $(a * b)^{-1} = a^{-1} * b^{-1}$ for all a and $b \in G$.
3. $(a * b)^2 = a^2 * b^2$ for all a and $b \in G$.

Proof:

(1) ⇒ (2): If $(G, *)$ is an abelian group and a and $b \in G$, then, by Theorem 3.3.1 (1),

$$(a * b)^{-1} = b^{-1} * a^{-1} = a^{-1} * b^{-1}.$$

(2) ⇒(3): Suppose that $(a * b)^{-1} = a^{-1} * b^{-1}$ for all a and $b \in G$. Then, for any a and $b \in G$, we have

$$\begin{aligned}(a * b)^2 &= (a * b) * (a * b)\\ &= (a * (b^{-1})^{-1}) * ((a^{-1})^{-1} * b)\\ &= a * ((b^{-1})^{-1} * (a^{-1})^{-1}) * b\\ &= a * (a^{-1} * b^{-1})^{-1} * b\\ &= a * ((a * b)^{-1})^{-1} * b\\ &= a * (a * b) * b\\ &= (a * a) * (b * b) = a^2 * b^2\end{aligned}$$

(3) ⇒ (1): For any a and $b \in G$

$$\begin{aligned}(a * b)^2 = a^2 * b^2 &\Rightarrow (a * b) * (a * b) = (a * a) * (b * b)\\ &\Rightarrow a * (b * a) * b = a * (a * b) * b\\ &\Rightarrow b * a = a * b \text{ (by cancellation laws)}\end{aligned}$$

Worked Exercise 3.3.1. Let X be any nonempty set and $S(X)$ be the set of all bijections of X onto itself. Then prove that $(S(X), \mathrm{o})$ is an abelian group if and only if $|X| < 3$.

Answer: If $|X| \geq 3$, then we have proved in Example 3.3.2 (4) that the group $(S(X), \mathrm{o})$ is not abelian.

On the other hand, suppose that $|X| < 3$. Then, $|X| = 1$ or 2. If $|X| = 1$, then $S(X)$ has only one element, namely the identity map and $S(X) = \{I_X\}$ is clearly abelian. If $|X| = 2$, say $X = \{a, b\}$, then there are exactly two bijections, namely the identity map I_X and the function $f : X \to X$ defined by $f(a) = b$ and $f(b) = a$ and therefore $S(X) = \{I_X, f\}$ which is clearly an abelian group.

Worked Exercise 3.3.2. Let $(G, *)$ be a group. Then prove that $(G, *)$ is abelian if and only if there exist three consecutive integers n such that $(a * b)^n = a^n * b^n$ for all $a, b \in G$.

Answer: If $(G, *)$ is an abelian group, then for any a and $b \in G$, we have

$$(a * b)^0 = e = e * e = a^0 * b^0$$

$$(a * b)^1 = a * b = a^1 * b^1$$

and

$$\begin{aligned}(a * b)^2 &= (a * b) * (a * b)\\ &= a * (b * a) * b\\ &= a * (a * b) * b\\ &= (a * a) * (b * b) = a^2 * b^2\end{aligned}$$

and hence, for $n = 0, 1, 2$, $(a * b)^n = a^n * b^n$ for all $a, b \in S$. Conversely suppose that there exists an integer n such that

$$(a * b)^{n-1} = a^{n-1} * b^{n-1} \quad \text{(i)}$$

$$(a * b)^n = a^n * b^n \quad \text{(ii)}$$

and

$$(a * b)^{n+1} = a^{n+1} * b^{n+1} \quad \text{(iii)}$$

for all a and $b \in G$. From (i) and (ii) we have

$$\begin{aligned}a * (a^{n-1} * b) * b^{n-1} &= a^n * b^n\\ &= (a * b)^n\\ &= (a * b) * (a * b)^{n-1}\\ &= (a * b) * a^{n-1} * b^{n-1}\\ &= a * (b * a^{n-1}) * b^{n-1}\end{aligned}$$

and hence, from the cancellation laws, we get that

$$a^{n-1} * b = b * a^{n-1} \quad \text{for all } a, b \in G \quad \text{(iv)}$$

Similarly, by using (ii) and (iii) we get that

$$a^n * b = b * a^n \quad \text{for all } a, b \in G \quad \text{(v)}$$

Now, for any a and $b \in G$, consider

$$\begin{aligned}b^{n-1} * (a * b) &= (b^{n-1} * a) * b\\ &= (a * b^{n-1}) * b && \text{(by (iv))}\\ &= a * (b^{n-1} * b)\\ &= a * b^n\\ &= b^n * a && \text{(by (v))}\\ &= (b^{n-1} * b) * a\\ &= b^{n-1} * (b * a)\end{aligned}$$

By the left cancellation law, we get $a * b = b * a$ for all a and $b \in G$, Thus, the group $(G, *)$ is abelian.

Worked Exercise 3.3.3. Let $(G, *)$ be a group such that $a^2 = e$ for all $a \in G$. Then prove that $(G, *)$ is an abelian group.

Answer: For any elements a and $b \in G$, we have $a * a = a^2 = e$ and $b * b = b^2 = e$ and hence $a^{-1} = a$ and $b^{-1} = b$. Now

$$(a * b)^{-1} = a * b = a^{-1} * b^{-1}$$

Thus, by Theorem 3.3.6, $(G, *)$ is an abelian group.

Worked Exercise 3.3.4. Let $(G, *)$ be a group such that $x^2 \neq e$ for all $x \neq e$ in G. Then prove that $(G, *)$ is an abelian group if and only if

$$(a * b)^2 = (b * a)^2 \quad \text{for all } a \text{ and } b \in G.$$

Answer: If $(G, *)$ is an abelian group, then clearly

$$(a * b)^2 = (b * a)^2 \quad \text{for all } a \text{ and } b \in G.$$

Conversely suppose that $(a * b)^2 = (b * a)^2$ for all a and $b \in G$.
Let a and b be arbitrary elements of G and consider

$$\begin{aligned} a^2 = (a * e)^2 &= (a * (b^{-1} * b))^2 \\ &= ((a * b^{-1}) * b)^2 \\ &= (b * (a * b^{-1}))^2 \qquad \text{(by hypothesis)} \\ &= b * a * b^{-1} * b * a * b^{-1} \\ &= b * a * e * a * b^{-1} \\ &= b * a^2 * b^{-1} \end{aligned}$$

Therefore, $a^2 = b * a^2 * b^{-1}$ for all a and $b \in G$ and hence $a^2 * b = b * a^2 * b^{-1} * b = b * a^2$.
Therefore, $a^2 * b = b * a^2$ for all a and $b \in G$.
Now, put $x = (a * b) *(a^{-1} * b^{-1})$ and consider

$$\begin{aligned} x^2 = x * x &= (a * b) * (a^{-1} * b^{-1}) * (a * b) * (a^{-1} * b^{-1}) \\ &= (a * b) * (a^{-1} * b^{-1} * a) * b * (a^{-1} * b^{-1}) \\ &= (a * b) * (a * a^{-2} * b^{-1} * a) * b * (a^{-1} * b^{-1}) \end{aligned}$$

$$
\begin{aligned}
&= (a * b) * (a * b^{-1} * a^{-2} * a) * b * (a^{-1} * b^{-1}) \\
&= (a * b) * (a * b^{-1} * a^{-1}) * b * (a^{-1} * b^{-1}) \\
&= (a * b) * (a * b * b^{-2} * a^{-1}) * (b * a^{-1} * b^{-1}) \\
&= (a * b) * (a * b * a^{-1} * b^{-2}) * (b * a^{-1} * b^{-1}) \\
&= (a * b) * (a * b) * a^{-1} * b^{-1} * a^{-1} * b^{-1} \\
&= (a * b)^2 * (a^{-1} * b^{-1})^2 \\
&= (b * a)^2 * (a^{-1} * b^{-1})^2 \\
&= (b * a)^2 * (b * a)^{-2} = e
\end{aligned}
$$

Therefore, $x^2 = e$ and hence, by hypothesis, $x = e$. From this it follows that $a * b * a^{-1} * b^{-1} = e$ and hence $a * b = (a^{-1} * b^{-1})^{-1} = (b^{-1})^{-1} * (a^{-1})^{-1} = b * a$. This $(G, *)$ is an abelian group.

Worked Exercise 3.3.5. Let $(G, *)$ be a group, a and $b \in G$ and m and n be relatively prime positive integers such that

$$a^m = b^m \quad \text{and} \quad a^n = b^n.$$

Then prove that $a = b$.

Answer: Since m and n are relatively prime there exist integers r and s such that

$$rm + sn = 1$$

Now, consider

$$
\begin{aligned}
a = a^1 &= a^{rm + sn} \\
&= (a^m)^r * (a^n)^s \\
&= (b^m)^r * (b^n)^s \\
&= b^{rm+sn} \\
&= b^1 = b
\end{aligned}
$$

Worked Exercise 3.3.6. Let $(G_1, *), (G_2, *), \ldots, (G_n, *)$ be groups and

$$G = G_1 \times G_2 \times \cdots \times G_n.$$

For any $a = (a_1, a_2, \ldots, a_n)$ and $b = (b_1, b_2, \ldots, b_n) \in G$, define

$$a * b = (a_1 * b_1, a_2 * b_2, \ldots, a_n * b_n).$$

Then prove that $(G, *)$ is a group and $(G, *)$ is abelian if and only if each $(G_n, *)$ is abelian.

Answer: Clearly $*$ is a binary operation on G and, using the associativity of the operations on $G_1, G_2, \ldots, G_n$, we can prove that $*$ is associative on G. Also, if $e_1, e_2, \ldots, e_n$ are identities in $G_1, G_2, \ldots, G_n$, respectively, then the element $e = (e_1, e_2, \ldots, e_n)$ becomes identity in $(G, *)$. Further, for any $a = (a_1, a_2, \ldots, a_n)$ in G, $(a_1^{-1}, a_2^{-1}, \ldots, a_n^{-1})$ is the inverse of a in $(G, *)$. Thus, $(G, *)$ is a group.

If each $(G_i, *)$ is abelian, then

$$\begin{aligned} a * b &= (a_1, a_2, \ldots, a_n) * (b_1, b_2, \ldots, b_n) \\ &= (a_1 * b_1, a_2 * b_2, \ldots, a_n * b_n) \\ &= (b_1 * a_1, b_2 * a_2, \ldots, b_n * a_n) \\ &= (b_1, b_2, \ldots, b_n) * (a_1, a_2, \ldots, a_n) \\ &= b * a \end{aligned}$$

For all $a, b \in G$ and hence $(G, *)$ is abelian. Conversely suppose that $(G, *)$ is abelian. Fix $1 \leq i \leq n$. For any a_i and $b_i \in G_i$, consider

$$a = (e_1, \ldots, e_{i-1}, a_i, e_{i+1}, \ldots, e_n)$$

and $b = (e_1, \ldots, e_{i-1}, b_i, e_{i+1}, \ldots, e_n)$.

Since $(G, *)$ is abelian, we have $a * b = b * a$ and, in particular, their i^{th} coordinates must be equal and therefore $a_i b_i = b_i a_i$. Thus, $(G_i, *)$ is abelian for all $1 \leq i \leq n$.

Worked Exercise 3.3.7. Describe the group of symmetries of the set X of all points on the perimeter of an equilateral triangle.

Answer: Let 1, 2 and 3 be vertices of an equilateral triangle and its altitudes be as shown in the adjacent figure. Let X be the set of all points on the perimeter of the triangle.

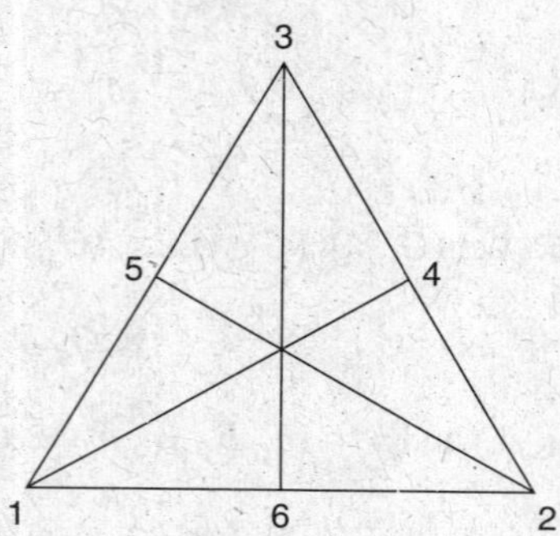

Recall that a symmetry of X is a bijection f of X onto itself such that

$$d(f(a), f(b)) = d(a, b)$$

for all a and $b \in X$, where $d(a, b)$ is the usual Euclidean distance between a and b. Therefore, a symmetry of X should map each vertex to a vertex only and hence we can identify the group (Sym(X), o) with the group ($S(V)$, o), where $S(V)$ is the set of bijections of the set V of vertices onto $V = \{1, 2, 3\}$. It follows that Sym(X) has exactly b elements which are described below. Each of these map each of 1, 2, 3 to the number given vertically below that.

$$e = \begin{pmatrix} 1 & 2 & 3 \\ 1 & 2 & 3 \end{pmatrix}, \; a = \begin{pmatrix} 1 & 2 & 3 \\ 2 & 3 & 1 \end{pmatrix}, \; b = \begin{pmatrix} 1 & 2 & 3 \\ 3 & 1 & 2 \end{pmatrix}$$

$$c = \begin{pmatrix} 1 & 2 & 3 \\ 1 & 3 & 2 \end{pmatrix}, \; d = \begin{pmatrix} 1 & 2 & 3 \\ 3 & 2 & 1 \end{pmatrix} \text{ and } s = \begin{pmatrix} 1 & 2 & 3 \\ 2 & 1 & 3 \end{pmatrix}.$$

These six elements e, a, b, c, d and s are related by the following equations.

$$a \circ a = b; a \circ a \circ a = e = c \circ c = d \circ d = s \circ s = b \circ b \circ b$$
$$c \circ d = a; d \circ c = a \circ a = b; d \circ s = a$$
$$s \circ d = b; c \circ s = b; s \circ c = a$$
$$a \circ c = s; c \circ a = d; a \circ d = c$$
$$d \circ a = s; s \circ a = c; a \circ s = d$$

The following table describes the binary operation o on Sym(X) = $S(V)$

o	e	a	b	c	d	s
e	e	a	b	c	d	s
a	a	b	e	s	c	d
b	b	e	a	d	s	c
c	c	d	s	e	a	b
d	d	s	c	b	e	a
s	s	c	d	a	b	e

Note that this group is not abelian, since $a \circ c \neq c \circ a$.

EXERCISE 3(C)

1. Prove the following for any elements a, b and c in a group G in which e is the identity.

   (i) $a * b = e \Leftrightarrow b * a = e$

   (ii) $(a * b) * c = e \Leftrightarrow (b * c) * a = e$

2. Give an example of a finite semigroup satisfying the left cancellation law, but not satisfying the right cancellation law.

3. Give an example of a finite semigroup satisfying the right cancellation law, which is not a group.

4. Let $(G, *)$ be a semigroup satisfying the following.

   (i) $(G, *)$ has a left identity e.

   (ii) For each $a \in G$, there exists $a' \in G$ such that $a * a' = e$.

   Then prove that $(G, *)$ is a group.

5. Prove that a group $(G, *)$ is abelian if and only if $(a * b)^n = a^n * b^n$ for all a and b in G and for all integers n.

6. In any finite semigroup, prove that there exists an element e such that $e^2 = e$.

7. Let m and n be relatively prime positive integers and $(G, *)$ a group such that

   $$a^m * b^m = b^m * a^m \quad \text{and} \quad a^n * b^n = b^n * a^n$$

   for all a and $b \in G$. Then prove that $(G, *)$ is an abelian group.

8. Let $(G, *)$ be a finite group and suppose that the number of elements in G is even. Then prove that there exists an element a, other than the identity, in G such that $a^2 = e$.

9. For any element a in a finite group $(G, *)$, prove that there exists a positive integer n such that $a^n = e$, the identity in G.

10. For any finite group $(G, *)$, prove that there exists a positive integer n such that

    $$a^n = e \quad \text{for all } a \in G$$

    where e is the identity in $(G, *)$.

11. For any elements a and b in a group $(G, *)$ and for any positive integer n, prove that

    $$(a * b * a^{-1})^n = a * b * a^{-1} \Leftrightarrow b^n = b.$$

12. Let $(G, *)$ be a group and X be any nonempty set. Let G^X be the set of all mappings of X into G. For any f and $g \in G^X$, define $f * g : X \to G$ by

    $$(f * g)(x) = f(x) * g(x) \quad \text{for all } x \in G.$$

Then prove that $(G^X, *)$ is a group which is abelian if and only if so is G.

13. Give an example of a nonabelian group having exactly six elements.

14. Prove that any group with fewer than six elements is abelian.

15. For any integer $n > 1$, prove that the set of all nonsingular $n \times n$ matrices over $\mathbb{R}$ forms a nonabelian group under the multiplication of matrices.

16. Let $(G, *)$ be a group. Define a new binary operation o on G by

$$a \circ b = b * a$$

for all a and b in G. Then prove that $(G, \circ)$ is a group which is abelian if and only if $(G, *)$ is abelian.

3.4 FINITE GROUPS AND GROUP TABLES

We know that a binary operation on a finite set can be represented by means of a table. In this section, we shall take up finite groups and the description of their group structure in terms of the table representing the binary operation. First consider the smallest group. Any group should contain the identity element e and hence $\{e\}$ is the smallest group. Since $e * e = e$, the table for the group $(\{e\}, *)$ is trivial, as given below.

*	e
e	e

Next, we consider a two element group G. Then, there should be only one element in G other than the identity e and therefore $G = \{e, a\}$, where $a \neq e$, we have $a * e = a = e * a$ and $e * e = e$. What could be $a * a$? It cannot be a, for, if $a * a = a$ then $a * a = a * e$ and, by the cancellation law, $a = e$ which is false. Therefore, the only possibility is $a * a = e$. The table for the group $(G, *)$, where $G = \{e, a\}$, is given below.

*	e	a
e	e	a
a	a	e

Next, we shall take up a 3-element group. In this case, there are exactly two elements, say $a \neq b$, in G other than the identity e. That is,

$$G = \{e, a, b\} \text{ and } e, a \text{ and } b \text{ are distinct.}$$

The table representing this group $(G, *)$ should be like the one given below

$*$	e	a	b
e	e	a	b
a	a		
b	b		

Let us search for the possible entries for the vacant places in the table. First observe that $a * b \neq a$ (since $a * b = a = a * e \Rightarrow b = e$, which is false). Similarly, $a * b \neq b$ (since $a \neq e$). Therefore, the only possibility is $a * b = e$ and $b * a = e$.

$*$	e	a	b
e	e	a	b
a	a		e
b	b	e	

Next, we shall search for $a * a$. First of all $a * a \neq a$ (since $a \neq e$). Also,

$$\begin{aligned} a * a = e &\Rightarrow a * a * b = e * b = b \\ &\Rightarrow a * e = b \\ &\Rightarrow a = b, \quad \text{which is false.} \end{aligned}$$

Therefore, $a * a \neq e$ and $a * a \neq a$ and hence the only possibility is $a * a = b$ and similarly $b * b = a$. Now, the table is complete and is given below

$*$	e	a	b
e	e	a	b
a	a	b	e
b	b	e	a

The above procedure for arriving at the full table representing the group $\{e, a, b\}$ yields the fact that the table of any three element group looks like the same, except the interchanging of the elements a and b or relabeling the elements a and b as b and a.

Consider the group $(\mathbb{Z}_n, +_n)$, where $\mathbb{Z}_n = \{0, 1, 2, \ldots, n-1\}$ and $+_n$ is the addition modulo n. For $n = 2$ and $n = 3$, the tables representing $(\mathbb{Z}_2, +_2)$ and

$(\mathbb{Z}_3, +_3)$ look like exactly the above tables representing a 2-element group and a 3-element group. In these cases, 0 is the identity, $\mathbb{Z}_2 = \{0, 1\}$ and $\mathbb{Z}_3 = \{0, 1, 2\}$.

$+_2$	0	1
0	0	1
1	1	0

$(\mathbb{Z}_2, +_2)$

$+_3$	0	1	2
0	0	1	2
1	1	2	0
2	2	0	1

$(\mathbb{Z}_3, +_3)$

Careful examination of these tables reveal certain necessary conditions that a table representing a binary operation on a finite set must satisfy certain properties for the operation to give a group structure on the set. There must be one element of the set which is the identity in the group and is denoted by e. Since $a * e = a$ for all elements a in the set, the column of the table under e at the very top must contain exactly the elements appearing at the extreme left in the same order. Also, since $e * a = a$ for all elements a in the set, the row of the table opposite e at the extreme left must contain exactly the same elements appearing across the very top of table in the same order. Further, since any element of the set has left inverse and right inverse, the row having a at the extreme left must contain e in some place and the column under a at the top must contain e at some place. Therefore, e must appear in each row and in each column. In fact, for any elements a and b in the group, the equations $a * x = b$ and $y * a = b$ have unique solutions in the group. This is equivalent to saying that, for a given element a in the group, every element of the group appears exactly once in the row with a at the extreme left and exactly once in the column with a at the very top. In the following, we have formulated a converse of the above argument.

Theorem 3.4.1. Let $*$ be an associative binary operation on a nonempty finite set G. Then, $(G, *)$ is a group if and only if, for any $a \in G$, every element of G appears in that row with a at the extreme left and in the column with a at the very top.

Proof: Consider the table representing the operation $*$ on G. For any elements a and b in G, the equation $a * x = b$ is solvable in G if and only if b appears in the row with a at the extreme left. Also, the equation $y * a = b$ is solvable in G if and only if b appears in the column, with a at the very top. Now, the theorem is a direct consequence of Theorem 3.3.3.

Example 3.4.1. The table representing the group $(\mathbb{Z}_9, +_9)$ is given below. Recall that $\mathbb{Z}_9 = \{0, 1, 2, 3, 4, 5, 6, 7, 8\}$ and $+_9$ is the addition modulo 9.

$+_9$	0	1	2	3	4	5	6	7	8
0	0	1	2	3	4	5	6	7	8
1	1	2	3	4	5	6	7	8	0
2	2	3	4	5	6	7	8	0	1
3	3	4	5	6	7	8	0	1	2
4	4	5	6	7	8	0	1	2	3
5	5	6	7	8	0	1	2	3	4
6	6	7	8	0	1	2	3	4	5
7	7	8	0	1	2	3	4	5	6
8	8	0	1	2	3	4	5	6	7

Example 3.4.2. Let $X = \{1, 2, 3\}$ and $S(X)$ be the set of all bijections of X onto itself. $S(X)$ has six elements and these are

$$e=\begin{pmatrix}1 & 2 & 3\\ 1 & 2 & 3\end{pmatrix},\ a=\begin{pmatrix}1 & 2 & 3\\ 2 & 3 & 1\end{pmatrix},\ b=\begin{pmatrix}1 & 2 & 3\\ 3 & 1 & 2\end{pmatrix},$$

$$c=\begin{pmatrix}1 & 2 & 3\\ 1 & 3 & 2\end{pmatrix},\ d=\begin{pmatrix}1 & 2 & 3\\ 3 & 2 & 1\end{pmatrix}\ \text{and}\ \ s=\begin{pmatrix}1 & 2 & 3\\ 2 & 1 & 3\end{pmatrix}$$

Then, $(S(X), \text{o})$ is a group, where o is the composition of mappings and its table is given below.

o	*e*	*a*	*b*	*c*	*d*	*s*
e	*e*	*a*		*c*	*d*	*s*
a		*b*	*e*		*c*	*d*
b	*b*	*e*	*a*	*d*		*c*
c	*c*	*d*		*e*	*a*	*b*
d	*d*				*e*	
s	*s*		*d*	*a*	*b*	*e*

The vacancies in the above table can be filled in using Theorem 3.4.1.

Worked Exercise 3.4.1. Let $X = \{1, 2, 3\}$ and $\mathbb{P}(X)$ be the set of all subsets of X. Construct the table representing the group $(\mathbb{P}(X), +)$, where $+$ is the symmetric difference.

Answer: $\mathbb{P}(X)$ has 2^3 ($= 8$) elements since X is a 3-element set. For convenience, we label them as given below.

$$e = \varnothing, \quad a = \{1\}, \quad b = \{2\}, \quad c = \{3\},$$
$$p = \{1, 2\}, \quad q = \{2, 3\}, \quad r = \{1, 3\}, \quad X = \{1, 2, 3\}.$$

+	e	a	b	c	p	q	r	X
e	e	a	b	c	p	q	r	X
a	a	e	p	r	b	X	c	q
b	b	p	e	q	a	c	X	r
c	c	r	q	e	X	b	a	p
p	p	b	a	X	e	r	q	c
q	q	X	c	b	r	e	p	a
r	r	c	X	a	q	p	e	b
X	X	q	r	p	c	a	b	e

$(\mathbb{P}(X), +)$.

Worked Exercise 3.4.2. Let $G_7 = \{1, 2, 3, 4, 5, 6\}$ and $\cdot_7$ be the multiplication modulo 7. Construct the table representing the group $(G_7, \cdot_7)$.

Answer:

$\cdot_7$	1	2	3	4	5	6
1	1	2	3	4	5	6
2	2	4	6	1	3	5
3	3	6	2	5	1	4
4	4	1	5	2	6	3
5	5	3	1	6	4	2
6	6	5	4	3	2	1

$(G_7, \cdot_7)$.

Worked Exercise 3.4.3. Examine whether the table below represent a group structure.

*	*a*	*b*	*c*	*d*	*e*
a	*b*	*c*	*d*	*e*	*a*
b	*a*	*d*	*c*	*a*	*b*
c	*d*	*c*	*b*	*e*	*c*
d	*c*	*a*	*a*	*b*	*d*
e	*a*	*b*	*c*	*d*	*e*

Answer: The underlying set is $X = \{a, b, c, d, e\}$. If the table represents a group $(X, *)$, then the set of entries in each row and in each column must be equal to the set X. The third row opposite to c at extreme left consists of d, c, b, e and a is absent in this and hence the given table does not represent a group structure on X. Note that $c * x = a$ is not solvable in X. However, the element e is the identity in the binary system $(X, *)$.

EXERCISE 3(D)

1. Examine the following tables representing binary systems and determine which of them represent a group structure.

(i)

*	0	1	2	3	4
0	0	0	0	0	0
1	0	1	2	3	4
2	0	2	4	1	3
3	0	3	1	4	2
4	0	4	3	2	1

(ii)

*	*a*	*b*	*c*
a	*a*	*b*	*c*
b	*b*	*c*	*a*
c	*c*	*a*	*b*

(iii)

*	*a*	*b*	*c*	*d*	*e*
a	*b*	*c*	*d*	*e*	*a*
b	*c*	*d*	*e*	*a*	*b*
c	*d*	*e*	*a*	*b*	*c*
d	*e*	*a*	*b*	*c*	*d*
e	*a*	*b*	*c*	*d*	*e*

(iv)

*	1	2	3	4
1	1	2	3	4
2	2	3	4	1
3	3	4	2	3
4	4	3	2	1

(v)

$\cdot$	i	$-i$	1	1
i	-1	1	$-i$	i
$-i$	1	-1	i	$-i$
-1	$-i$	i	1	-1
1	i	$-i$	-1	1

(vi)

*	1	2	3	4
1	1	2	3	4
2	2	3	4	1
3	3	4	1	2
4	4	1	2	3

2. Let $G_{11} = \{1, 2, 3, 4, 5, 6, 7, 8, 9, 10\}$ and $\cdot_{11}$ be the multiplication modulo 11. Then prove that $(G_{11}, \cdot_{11})$ is a group and construct the table representing the group.

3. Let n be a positive integer greater than n and $S = \{1, 2, \ldots, n - 1\}$. Prove that the multiplication modulo n is a binary operation on S if and only if n is a prime number.

4. Construct tables representing all the 2-element groups, 3-elements groups, 4-element groups and 5-element groups.

5. By observing the tables in 4 above, prove that every 2-element groups, 3-element groups, 4-element groups and 5-element groups is abelian.

6. Prove that a finite group $(G, *)$ is not abelian only if $|G| > 5$.

7. Give an example of a nonabelian group with exactly six elements.

8. For any positive integer n, give an example of an abelian group with exactly n elements and construct a table representing it.

9. Let G be the set of all rational numbers with odd denominators. Prove that $(G, +)$ is a group, where $+$ is the usual addition of rational numbers.

10. Let $A = \begin{pmatrix} 0 & 1 \\ -1 & 0 \end{pmatrix}$ and $B = \begin{pmatrix} 0 & i \\ i & 0 \end{pmatrix}$, where i is the complex number such that $i^2 = 1$. Let

$$Q_8 = \{(A^n B^m) : n, m \equiv \mathbb{Z}\}.$$

Prove that $(Q_8, \cdot)$ is a group, where '$\cdot$' is the usual multiplication of complex numbers and that Q_8 has exactly 8 elements. This group is called the *Quaternion group of order 8*.

4

Subgroups and Quotient Groups

A nonempty set together with certain operations on it is called an algebraic system. In the study of any algebraic system, the subsets of the underlying set which are closed under the operations of the algebraic system called subsystems or substructures. In general, we are not interested in arbitrary subsets of the underlying set A in the algebraic system; for, they do not reflect the fact that A has an algebraic structure imposed on it. Whatever subsets we do consider will be those endowed with algebraic properties derived from those of the given algebraic system. In this chapter, we introduce the notion of a subgroup of a group and study various properties of subgroups of a several group.

First let us slightly change our notation followed till now. If $*$ is the binary operation on a given group $(G, *)$ and a and b are arbitrary elements of G, it is cumbersome to write $a * b$. Let us agree to write simply ab for $a * b$, without writing the specific symbol $*$ in between a and b. Already we are practicing this; for example, for any two real numbers a and b, we write ab to denote the product of a and b. Here afterwards, we simply write ab for $a * b$, except on specific occasions where the binary operation in the group is a special one we are familiar with. For example, when we consider the group $(\mathbb{R}, +)$, we write $a + b$.

Also, instead of saying that '$(G, *)$ is a group', we simply say that 'G is a group' or 'G is a group under $*$'. It is not that $*$ is unimportant and need not be mentioned. Actually the binary operation in a group is like a backbone to the group structure. However, for convenience and simplicity, we ignore to mention the binary operation $*$, when only one operation is under consideration and there is no ambiguity. When we consider two binary operations on the same set, then we invariably specify the binary operation which is under consideration. When there is no ambiguity about the binary operation, we use G to denote the group as well as the underlying set of the group.

4.1 SUBGROUPS

Recall that a binary operation $*$ on a set S is a mapping of $S \times S$ into S. If A is a subset of S, then $A \times A$ is a subset of $S \times S$ and if the restriction of $*$ to $A \times A$ is a binary operation on A, then the restriction also will be denoted by $*$ and is called the operation A induced by the operation $*$ on S. In particular, if $(G, *)$ is a group and A is a subset of G such that $a * b \in A$ whenever a and $b \in A$, then $*$ can be treated as a binary on A with respect to which A can be a group and, in this case, we say that A is a group under $*$.

Definition 4.1.1. Let $(G, *)$ be a group. A subset H of G is said to be a *subgroup* of $(G, *)$ if H on its own becomes a group under $*$.

Before going, for example, we obtain equivalent conditions on a subset of a group to be a subgroup. These facilitate us in checking whether a given subset is a subgroup.

Theorem 4.1.1. The following are equivalent to each other for any nonempty subset H of a group G:

1. a and $b \in H \Rightarrow ab \in H$ and $a^{-1} \in H$.
2. a and $b \in H \Rightarrow ab^{-1} \in H$.
3. a and $b \in H \Rightarrow a^{-1}b \in H$.
4. H is a subgroup of G.

Proof: Let G be a group and H be a nonempty subset of G.

(1) $\Rightarrow$ (2): a and $b \in H \Rightarrow a$ and $b^{-1} \in H$ (by (1))

$$\Rightarrow ab^{-1} \in H \quad \text{(by (1))}$$

(2) $\Rightarrow$ (3): Suppose that a and $b \in H \Rightarrow ab^{-1} \in H$.
If $a \in H$, then $e = aa^{-1} \in H$ and hence

$$a \in H \Rightarrow a^{-1} = ea^{-1} \in H \quad \text{(since } e \text{ and } a \in H)$$

$$\begin{aligned}\text{Now, } a \text{ and } b \in H &\Rightarrow a^{-1} \text{ and } b^{-1} \in H \\ &\Rightarrow a^{-1}(b^{-1})^{-1} \in H \\ &\Rightarrow a^{-1}b \in H\end{aligned}$$

(3) $\Rightarrow$ (4): Suppose that a and $b \in H \Rightarrow a^{-1}b \in H$.
Since H is nonempty, we can choose $h \in H$. Then,

$$e = h^{-1}h \in H$$

Also, $a \in H \Rightarrow a^{-1} = a^{-1}e \in H$ (since a and $e \in H$)

$$\begin{aligned}\text{Now, } a \text{ and } b \in H &\Rightarrow a^{-1} \text{ and } b \in H \\ &\Rightarrow (a^{-1})^{-1}b \in H \\ &\Rightarrow ab \in H\end{aligned}$$

That is, $a * b \in H$ whenever a and $b \in H$. Therefore, the operation $*$ on G, when restricted to H, becomes a binary operation on H. Since $*$ is associative on G, so is on H. Also, since $e \in H$ and e is the identity in $(G, *)$, e becomes the identity in $(H, *)$ also. Further, for any $a \in H$, $a^{-1} \in H$ and $aa^{-1} = e = a^{-1}a$ and hence a^{-1} is the inverse of a in $(H, *)$. Thus, $(H, *)$ is a group and hence H is a subgroup of G.

(4) $\Rightarrow$ (1): Suppose H is a subgroup of G. Then, $(H, *)$ is a group and hence $a * b \in H$, whenever a and $b \in H$. Also, for any $a \in H$, the inverse of a exists in H also. Let a' and a^{-1} be inverses of a in H and G, respectively. Then both these are inverses of a in G and hence equal. Therefore, $a^{-1} = a' \in H$. Now,

$$a \text{ and } b \in H \Rightarrow ab \in H \quad \text{and} \quad a^{-1} \in H.$$

If the symbol $+$ is used to denote the binary operation in a group, then we write $-a$ for the inverse a^{-1} of a and write $a - b$ for ab^{-1} (for psychological reasons!). In this case, a nonempty subset H of G is a subgroup of G if and only if $a - b \in H$ whenever a and $b \in H$. Recall that, in the group $(\mathbb{R}, +)$ or $(\mathbb{Z}, +)$ or $(\mathbb{C}, +)$, 0 is the identity and $-a$ is the inverse of an element a. Also, the element a^n defined in the Definition 3.2.5 will be denoted by na, when $+$ is used to denote the binary operation. This is only for not violating our usual practice right from the elementary school stage. Recall that we are habituated to write

$$na \text{ for } a + a + \cdots + a \quad (n \text{ times})$$

$$\text{and} \qquad a^n \text{ for } a \cdot a \cdot a \cdot \ldots \cdot a \quad (n \text{ times})$$

$$-\,a \text{ for the inverse of } a \text{ in } (\mathbb{R}, +)$$

$$a - b \text{ for } a + (-b)$$

$$\frac{1}{b} \text{ for } b^{-1}, \text{ the inverse of } b \text{ in } (\mathbb{R} - \{0\}, \cdot)$$

$$\frac{a}{b} \text{ for } ab^{-1}.$$

When H is a nonempty finite subset of a group G, we get a simpler criterion for H to be a subgroup of G, which gives a simpler procedure to check whether a given finite subset is a subgroup. This criterion is obtained in the following theorem.

Theorem 4.1.2. Let H be a nonempty finite subset of a group G. Then, H is a subgroup of G if and only if a and $b \in H \Rightarrow ab \in H$.

Proof: Suppose that a and $b \in H \Rightarrow ab \in H$. In order to prove that H is a subgroup of G, we have to only prove that $a^{-1} \in H$ whenever $a \in H$ (by Theorem 4.1.1). Now, let $a \in H$. Since H is given to be finite, we can write

$$H = \{a_1, a_2, \ldots, a_n\}, a_i \neq a_j \quad \text{for } i \neq j.$$

Consider the set $aH = \{aa_1, aa_2, \ldots, aa_n\}$.

We have $aa_i = aa_j \Rightarrow a_i = a_j$ (by cancellation law)

$$\Rightarrow i = j$$

and hence $aa_1, aa_2, \ldots, aa_n$ are all distinct. Also, since a and $a_i \in H$, we have $aH \subseteq H$ and H and aH have the same number of elements. From the finiteness of H, it follows that $aH = H$. In particular,

$$a \in H = aH$$

and hence $a = aa_i$ for some $1 \leq i \leq n$.
Since all these are elements in the group G, we get that $e = a_i \in H$. Now $e \in H = aH$ and hence

$$e = aa_j \quad \text{for some } 1 \leq j \leq n.$$

Therefore, $a^{-1} = a_j \in H$. Thus, H is a subgroup of G.

The finiteness of H in the above theorem is necessary; for, consider the set $\mathbb{Z}^+$ of positive integers. Then, $\mathbb{Z}^+$ is a subset of $\mathbb{Z}$, $(\mathbb{Z}, +)$ is a group and $a + b \in \mathbb{Z}^+$ whenever a and $b \in \mathbb{Z}^+$ and still $\mathbb{Z}^+$ is not a subgroup of $(\mathbb{Z}, +)$.

Now, we collect certain examples of subgroups of groups. Some of the subgroups given below are earlier given as examples of groups.

Example 4.1.1

1. If H is a subgroup of a group K and K is a subgroup of a group G, then clearly H is a subgroup of G. If $+$ denotes the usual addition of numbers, $\mathbb{Z}$ is a group of $(\mathbb{Q}, +)$, $\mathbb{Q}$ is a subgroup of $(\mathbb{R}, +)$ and $\mathbb{R}$ is a subgroup of $(\mathbb{C}, +)$ and hence $\mathbb{Z}$, $\mathbb{Q}$ and $\mathbb{R}$ are all subgroups of $(\mathbb{C}, +)$.
2. If e is the identity in a group G, then clearly $\{e\}$ and G are subgroups of G and are called the *trivial subgroups* of G. Any subgroup other than $\{e\}$ and G is called a *nontrivial subgroup*. A group G is called *nontrivial* if $G \neq \{e\}$ and *trivial* if $G = \{e\}$.
3. If $\cdot$ is the usual multiplication of numbers, then $\mathbb{Q} - \{0\}$ is a subgroup of $(\mathbb{R} - \{0\}, \cdot)$ and $\mathbb{R} - \{0\}$ is a subgroup of $(\mathbb{C} - \{0\}, \cdot)$. Therefore, both $\mathbb{Q} - \{0\}$ and $\mathbb{R} - \{0\}$ are subgroups of $(\mathbb{C} - \{0\}, \cdot)$.
4. Let X be any nonempty set and $(S(X), \circ)$ be the group of bijections of X onto itself, where $\circ$ is the composition of mappings. Let x_0 be an arbitrary element of X and

$$H_{x_0} = \{f \in S(X) : f(x_0) = x_0\}.$$

Then, H_{x_0} is a subgroup of $(S(X), \circ)$ Also, for any subset Y of X, the set

$$H_Y = \{f \in S(X) : f(y) = y \quad \text{for all } y \in Y\}$$

is a subgroup of $(S(X), \circ)$.

5. Let n be a positive integer and $\text{NSM}_n(\mathbb{R})$ be the set of all nonsingular $n \times n$ matrices over the real number system $\mathbb{R}$. Then, $(\text{NSM}_n(\mathbb{R}), \cdot)$ is a group where $\cdot$ is the usual multiplication of matrices. Let

$$H = \{(a_{ij}) \in \text{NSM}_n(\mathbb{R}) : a_{ij} = 0 \quad \text{for all } i > j\}$$

and

$$K = \{(a_{ij}) \in \text{NSM}_n(\mathbb{R}) : a_{ij} = 0 \quad \text{for all } i < j\}.$$

Then, H and K are subgroups of $\text{NSM}_n(\mathbb{R})$.

6. Let $A = \left\{ \begin{pmatrix} 1 & a \\ 0 & 1 \end{pmatrix} : a \in \mathbb{R} \right\}$. Then, A is a subgroup of $(\text{NSM}_2(\mathbb{R}), \cdot)$. One can easily verify that, for any a and $b \in \mathbb{R}$,

$$\begin{pmatrix} 1 & a \\ 0 & 1 \end{pmatrix} \cdot \begin{pmatrix} 1 & b \\ 0 & 1 \end{pmatrix} = \begin{pmatrix} 1 & a+b \\ 0 & 1 \end{pmatrix}$$

and $$\begin{pmatrix} 1 & a \\ 0 & 1 \end{pmatrix}^{-1} = \begin{pmatrix} 1 & -a \\ 0 & 1 \end{pmatrix}.$$

7. Consider the additive group $(\mathbb{Z}_4, +_4)$ of integers modulo 4. We have $\mathbb{Z}_4 = \{0, 1, 2, 3\}$ and 0 is the identity. If H is a nontrivial subgroup of $\mathbb{Z}_4$, then $H = \{0, 2\}$ (since $1 \in H \Rightarrow 2 = 1 +_4 1 \in H$ and $3 = 1 +_4 1 +_4 1 \in H$ and $3 \in H \Rightarrow 1 = 3 +_4 3 +_4 3 \in H \Rightarrow 1, 2, 3 \in H$). Therefore, $\{0\}$, $\{0, 2\}$ and $\mathbb{Z}_4$ are the only subgroups of $(\mathbb{Z}_4, +_4)$.
8. The group $(\mathbb{Z}_5, +_5)$ has no nontrivial subgroups; for, if H is a subgroup of $\mathbb{Z}_5$, then

$$\begin{aligned}
&1 \in H \Rightarrow 1, 2, 3, 4 \in H \Rightarrow H = \mathbb{Z}_5 \\
&2 \in H \Rightarrow 2 +_5 2 +_5 2 \in H \Rightarrow 1 \in H \Rightarrow H = \mathbb{Z}_5 \\
&3 \in H \Rightarrow 1 = 3 +_5 3 \in H \Rightarrow H = \mathbb{Z}_5 \\
&4 \in H \Rightarrow 3 = 4 +_5 4 \in H \Rightarrow H = \mathbb{Z}_5.
\end{aligned}$$

Since every subgroup of a group G should contain the identity e in G, it follows that $\{e\}$ is the smallest subgroup of G. Also, clearly G is the largest subgroup of G. In the following theorem, we describe the smallest subgroup of a group G containing a given element of the group.

Theorem 4.1.3. Let G be a group and $a \in G$. Let

$$<a> = \{a^n : n \text{ is an integer}\}.$$

Then, $<a>$ is the smallest subgroup of G containing a.

Proof: Recall that a^n is defined as

$$a^n = \begin{cases} e & \text{if } n = 0 \\ a^{n-1} \cdot a & \text{if } n > 0 \\ (a^{-1})^{-n} & \text{if } n < 0 \end{cases}$$

and hence $a^n \in H$ for all integers n and for all subgroups H of G containing a. Also, by Worked Exercise 3.2.18, $<a>$ is a subgroup of G and contains a. Thus, $<a>$ is the smallest subgroup containing a.

Definition 4.1.2. For any group G and $a \in G$, $<a> = \{a^n : n \text{ is an integer}\}$ is called the *cyclic subgroup generated by a* in G.

Before going for a detailed discussion on cyclic subgroups generated by elements of an arbitrary group, we shall first discuss certain elementary properties of subgroups.

Theorem 4.1.4. The intersection of any class of subgroups of a group G is again a subgroup of G.

Proof: Let $\mathscr{C}$ be a class of subgroups of a group G and let $H = \bigcap_{C \in \mathscr{C}} C = \{a : a \in C \text{ for all } C \in \mathscr{C}\}$.
If $\mathscr{C}$ is empty class, then, by logical convention, $H = G$ and hence H is a subgroup of G. Therefore, we can assume that $\mathscr{C}$ is a nonempty class.
Since the identity element e must be in every subgroup, we get that $e \in H$ and hence H is a nonempty subset of G. Now,

$$\begin{aligned} a \text{ and } b \in H &\Rightarrow a \text{ and } b \in C, \quad \text{for all } C \in \mathscr{C} \\ &\Rightarrow ab^{-1} \in C, \quad \text{for all } C \in \mathscr{C} \\ &\Rightarrow ab^{-1} \in H. \end{aligned}$$

Therefore, H is a subgroup of G.

Definition 4.1.3. For any subset S of a group G, let $<S>$ be the intersection of all subgroups of G containing S. Then, by the above theorem, $<S>$ is a subgroup of G containing S and is called the *subgroup generated by S in G*.

Note 4.1.1

1. For any subset S of a group G, $<S>$ is the smallest subgroup of G containing S.
2. $<\emptyset> = \{e\}$ and $<G> = G$, for any group G.
3. $<\{a\}> = <a> = \{a^n : n \in \mathbb{Z}\}$ for any $a \in G$.
4. For any nonempty subset S of a group G, S is a subgroup of G if and only if $S = <S>$.

In the following, we describe the elements of $<S>$, for any nonempty subset S of a group. The description of elements of $<\{a\}>$ is already given in Theorem 4.1.3. This is generalised in the following theorem.

Theorem 4.1.5. Let S be a nonempty subset of a group G and $<S>$ be the smallest subgroup of G containing S. Then,

$$<S> = \left\{ \prod_{i=1}^{n} s_i : n \in \mathbb{Z}^+ \text{ and, for each } i,\ s_i \text{ or } s_i^{-1} \in S \right\}.$$

Proof: Let A be the set defined on the right side of the required equality. If H is any subgroup of G containing S, then s and $s^{-1} \in H$ for all $s \in S$ and therefore any product of elements of S and their inverses must be in H. This is to say that $A \subseteq H$ for all subgroups H containing S. Also, $\varnothing \neq S \subseteq A$ and, since

$$(s_1\, s_2\, \ldots\, s_n)(t_1\, t_2\, \ldots\, t_m)^{-1} = s_1\, s_2\, \ldots\, s_n t_m^{-1} t_{m-1}^{-1}\, \ldots\, t_2^{-1} t_1.$$

Therefore, A is also a subgroup of G containing S. Thus, $<S> = A$.

Corollary 4.1.1. For any subset S of a group G,

$$<S> = <S^{-1}>, \quad \text{where } S^{-1} = \{s^{-1} : s \in S\}.$$

We have proved in Theorem 4.1.4 that the intersection of any class of subgroups is again a subgroup. A similar statement is not true for unions of subgroups. In this context, we have the following theorem.

Theorem 4.1.6. Let A and B be subgroups of group G. Then, $A \cup B$ is a subgroup of G if and only if either $A \subseteq B$ or $B \subseteq A$.

Proof: If $A \subseteq B$, then $A \cup B = B$ and, if $B \subseteq A$ then $A \cup B = A$ and hence, in this case, $A \cup B$ is a subgroup of G. Conversely suppose that $A \cup B$ is a subgroup of G. Assume that $A \nsubseteq B$. Then, there exists $a \in A$ such that $a \notin B$. Now,

$$\begin{aligned} b \in B &\Rightarrow a \text{ and } b \in A \cup B \quad (\text{since } a \in A) \\ &\Rightarrow ab \in A \cup B \quad (\text{since } A \cup B \text{ is a subgroup}) \\ &\Rightarrow ab \in A \quad \text{or} \quad ab \in B \\ &\Rightarrow ab \in A \quad (\text{since } ab \in B \Rightarrow a = (ab)b^{-1} \in B) \\ &\Rightarrow b = a^{-1}(ab) \in A \end{aligned}$$

Therefore, $B \subseteq A$.

From the above, it follows that, for any subgroups A and B of a group G, $A \cup B$ is not a subgroup in general. However, we have noticed earlier that $A \cap B$ is always a subgroup and this is the largest subgroup contained in both A and B. Also, there is a smallest subgroup containing both A and B (which need not be $A \cup B$). In certain cases, we can describe the elements of this elegantly. First, we have the following definition.

Definition 4.1.4. Let G be a group and A and B be subsets of G. Define

$$AB = \{ab : a \in A \text{ and } b \in B\} \quad \text{and} \quad A^{-1} = \{a^{-1} : a \in A\}.$$

Note that, from the associativity of the operation in G, we get that $(AB)C = A(BC)$ for any subsets A, B and C of G. Also, observe that a nonempty subset A of G is a subgroup of G if and only if $AA^{-1} = A$.

Theorem 4.1.7. Let A and B be subgroups of a group G. Then, AB is a subgroup of G if and only if $AB = BA$ and, in this case, AB is the smallest subgroup of G containing both A and B.

Proof: Suppose that AB is a subgroup of G. Then,

$$AB = (AB)^{-1} = B^{-1}A^{-1} = BA.$$

Conversely, if $AB = BA$, then

$$\begin{aligned}(AB)(AB)^{-1} &= (AB)(B^{-1}A^{-1})\\ &= A(BB^{-1})A^{-1}\\ &= (AB)A^{-1}\\ &= (BA)A^{-1}\\ &= B(AA^{-1}) = BA = AB\end{aligned}$$

and hence AB is a subgroup of G.
Since $e \in A$ and $e \in B$, if follows that

$$A = A\{e\} \subseteq AB \quad \text{and} \quad B = \{e\}B \subseteq AB.$$

If H is any subgroup containing A and B, then clearly $AB \subseteq H$. Thus, when $AB = BA$, AB is the smallest subgroup containing both A and B.

Corollary 4.1.2. For any subgroups A and B of an abelian group G, AB is also a subgroup of G.

Worked Exercise 4.1.1. Prove that any subgroup of the group $(\mathbb{Z}, +)$ is the subgroup generated by a single nonnegative integer.

Answer: Recall that, when the symbol $+$ is used for the binary operation in a group, then we write na for a^n and as such

$$\langle a \rangle = \{na : n \text{ is an integer}\}.$$

Now, let A be a subgroup of $(\mathbb{Z}, +)$. If $A = \{0\}$, then clearly $A = <0>$. Suppose that $A \neq \{0\}$. Then, there exists $a \neq 0$ such that $a \in A$. Since A is a subgroup, $-a$ also is in A. Since a or $-a$ is positive, it follows that $A \cap \mathbb{Z}^+$ is a nonempty subset of $\mathbb{Z}^+$. By the well-ordering principle, $A \cap \mathbb{Z}^+$ has the smallest member, say m. Then, since $m \in A$, we get that $<m> \subseteq A$. On the other hand, let $x \in A$. By the division algorithm, we can write

$$x = qm + r, \quad q, r \in \mathbb{Z} \quad \text{and} \quad 0 \leq r < m.$$

Then, $r = x - qm \in A$ (since x and $m \in A$). Since $r < m$ and since m is the least positive integer in A, it follows that $r = 0$ and hence $x = qm \in <m>$. Thus, $A = <m> = \{nm : m \in \mathbb{Z}\} = n\mathbb{Z}$.

Worked Exercise 4.1.2. Compute all subgroups of the group $(n\mathbb{Z}, +)$ for any positive integer n.

Answer: Let n be a positive integer and A be a subgroup of $(n\mathbb{Z}, +)$. Then, A is a subgroup of $(\mathbb{Z}, +)$ and hence $A = <a>$ for some $a \geq 0$. Since $A \subseteq n\mathbb{Z}$, we get that $a \in n\mathbb{Z}$ and hence $a = nq$ for some $q \in \mathbb{Z}$. Therefore, a is multiple of n. Conversely, if a is a multiple of n, then $a\mathbb{Z}$ is a subgroup of $(n\mathbb{Z}, +)$. Thus, the subgroups of $(n\mathbb{Z}, +)$ are precisely of the form $a\mathbb{Z}$, where a is an integral multiple of n.

Worked Exercise 4.1.3. Compute all the subgroups of $(\mathbb{Z}_n, +_n)$ for any positive integer.

Answer: Let n be a positive integer and A be a subgroup of $(\mathbb{Z}_n, +_n)$. Suppose that $A \neq \{0\}$. Let m be the least positive integer in A. As in Worked Exercise 4.1.1, we can prove that $A = <m>$ and that m is a positive divisor of n. Note that, since $m \in A \subseteq \mathbb{Z}_n$, $m < n$. Therefore, the subgroups of $(\mathbb{Z}_n, +_n)$ are precisely of the form $<m>$, where $m = 0$ or a positive divisor of n. Note that, for $m \neq 0$,

$$<m> = \{0, m, 2m, \ldots, (q-1)m\},$$

where $qm = n$.

EXERCISE 4(A)

1. Determine whether the set given is a subgroup of the group in each of the following.
   (i) $\{0, 1, 2, 3, 4\}$ in $(\mathbb{Z}_8, +_8)$
   (ii) $\{0, 3, 6, 9\}$ in $(\mathbb{Z}_{12}, +_{12})$

(iii) $\mathbb{R}^+$ in $(\mathbb{R}, +)$

(iv) $\mathbb{R}^+$ in $(\mathbb{R}-\{0\}, \cdot)$

(v) $\mathbb{Q}^+$ in $(\mathbb{R}^+, \cdot)$

(vi) $\pi\mathbb{Q}$ in $(\mathbb{R}, +)$

(vii) $\{z \in \mathbb{C} \mid |z| = 1\}$ in $(\mathbb{C} - \{0\}, \cdot)$

(viii) $5\mathbb{Z}$ in $(8\mathbb{Z}, +)$

2. Determine whether the following are subgroups of the group of nonsingular $n \times n$ matrices under the usual multiplication of matrices, where n is a given positive integer.
   (i) The set of all $n \times n$ matrices whose determinant is 2.
   (ii) The set of all $n \times n$ matrices whose determinant is 1.
   (iii) The set of $n \times n$ matrices whose determinant is a positive real number.
   (iv) $\{(a_{ij}) \in NSM_n(\mathbb{R}) : a_{ij} = 0 \text{ for all } i < j\}$.
   (v) $\{(a_{ij}) \in NSM_n(\mathbb{R}) : a_{ij} = 0 \text{ for all } i > j\}$.
   (vi) $\{(a_{ij}) \in NSM_n(\mathbb{R}) : a_{ij} = 0 \text{ for } i \neq j\}$.
   (vii) The set of all $n \times n$ matrices whose determinant is a negative real number.
   (viii) $\{(a_{ij}) \in NSM_n(\mathbb{R}) : a_{ij} = 0 \text{ for } i \neq j \text{ and } a_{ii} = a_{jj} \text{ for all } i \text{ and } j\}$.

3. Let $\mathscr{C}$ be a nonempty class of subgroups of a group G such that, for any A and B in $\mathscr{C}$, there is a member C in $\mathscr{C}$ containing both A and B. Then prove that $\bigcup_{C \in \mathscr{C}} C$ is a subgroup of G.

4. Let G be a group such that $G = \langle a \rangle$ for exactly, one element a in G. Then prove that G has at most two elements.

5. Let G be a group having exactly two subgroups. Then prove that $G = \langle a \rangle$ for some $a \in G$.

6. Let n be positive integer and consider the group $(\mathbb{Z}_n, +_n)$ of integers modulo n. For any $0 < d < n$, prove that $\mathbb{Z}_n = \langle d \rangle$ if and only if d is relatively prime with n.

7. Prove that there is a bijection between the set of subgroups of $(\mathbb{Z}_n, +_n)$ and the set of positive divisors of n.

8. Let G be a group and $a \in G$. Prove that the set

$$C_a = \{x \in G : ax = xa\}$$

is a subgroup of G. C_a is caller the *centralizer* of a.

9. Let S be any subset of G. Then prove that the set

$$C_S = \{x \in G : ax = xa \text{ for all } a \in S\}$$

is a subgroup of G.

10. For any group G, prove that the set

$$Z(G) = \{x \in G : ax = xa \text{ for all } a \in G\}$$

is a subgroup of G. $Z(G)$ is called the *centre* of G.

11. Determine all the subgroups of each of the following:
    (i) $(\mathbb{Z}_{24}, +_{24})$
    (ii) $(\mathbb{Z}_7, +_7)$

12. For any positive integer n, prove that $(\mathbb{Z}_n, +_n)$ has exactly two subgroups if and only if n is prime.

13. For any subgroup H of a group G and $a \in G$, prove that

$$aHa^{-1} = \{axa^{-1} : x \in H\} \text{ is also a subgroup of } G.$$

14. For any finite subgroup H of a group G and $a \in G$ prove that H and aHa^{-1} has equal number of elements.

15. Determine all the subgroups of the group $(S(X), \cdot)$ of bijections of X onto itself, where X is a 3-element set.

16. Prove that a nonempty subset H of a group G is a subgroup of G if and only if $HH^{-1} \subseteq H$.

4.2 CYCLIC GROUPS

The concept of a cyclic group is an important tool in determining the structure of a finite (or finitely generated) abelian group. In fact, we prove later that any finitely generated abelian group is a finite product of cyclic groups. In order to understand the structure of a finitely generated abelian group, one has to understand cyclic group. In this section, we thoroughly discuss the various properties of cyclic groups.

First let us recall that, for any element a of a group G, the smallest subgroup of G containing a is given by

$$<a> = \{a^n : n \text{ is an integer}\}$$

and is called the *cyclic subgroup* generated by a in G.
If one uses + to denote the binary operation in a group, we write na for a^n and $-a$ for the inverse of a.

Definition 4.2.1. A group G is called a *cyclic group* if $G = <a>$ for some $a \in G$; that is, there exists $a \in G$ such that

$$G = \{a^n : n \text{ is an integer}\}.$$

In this case, a is called a *generator* of G.

Example 4.2.1

1. $(\mathbb{Z}, +)$ is a cyclic group, since $<1> = \mathbb{Z}$. In fact, $\mathbb{Z} = <-1>$ also and hence both 1 and -1 are generators of the group $\mathbb{Z}$. Later we will be proving that these are the only generators of $\mathbb{Z}$.
2. For any positive integer n, the group $(\mathbb{Z}_n, +_n)$ of integers modulo n is a cyclic group. Here also, 1 and -1 ($=_n n - 1$) are generators of $\mathbb{Z}_n$. Later, we shall prove that any positive integer less than n and relatively prime with n is a generator of $\mathbb{Z}_n$.
3. Let $G = \{1, -1, i, -i\}$. Then, G is a cyclic group under the usual multiplication of complex numbers. Here, $<i> = G = <-i>$ and hence i and $-i$ are the only generators of G, as we can easily see that $<1> = \{1\} \neq G$ and $<-1> = \{1, -1\} \neq G$.
4. The group $\mathbb{Z}_2 \times \mathbb{Z}_2$ is not cyclic, since, for any element a in this group, $2a = 0$, the identity and hence $<a> = \{0, a\} \neq \mathbb{Z}_2 \times \mathbb{Z}_2$.

The following is a fundamental tool in the study of cyclic groups and we might have used this earlier. Here, we offer a proof. First recall that for any real number a, there exists largest integer, less than or equal to a and it is denoted by $[a]$. That is, $[a]$ is the unique integer such that $[a] \leq a < [a] + 1$. $[a]$ is called *integral part* of a.

Theorem 4.2.1 (The Division Algorithm). Let n be a positive integer and a be any integer. Then, there exist unique integers q and r such that

$$a = qn + r \quad \text{and} \quad 0 \leq r < n$$

q and r are respectively called the *quotient* and the *remainder* obtained by dividing a with n.

Proof: Let $q = [\frac{a}{n}]$, the integral part of $\frac{a}{n}$, and $r = a - qn$. Then, q is an integer such that

$$q \leq \frac{a}{n} < q + 1.$$

Since a, q and n are all integers, r is also an integer. We have

$$qn \leq a < qn+n$$

and hence $0 \leq a - qn < n$.
Therefore, we have $a = qn + r$ and $0 \leq r < n$. For proving the uniqueness, let q_0 and r_0 be any integers such that $a = q_0 n + r_0$ and $0 \leq r_0 < n$.
Then, $\frac{a}{n} = q_0 + \frac{r_0}{n}$ and $0 \leq \frac{r_0}{n} < 1$. Therefore,

$$q_0 \leq \frac{a}{n} < q_0 + 1 \quad \text{and hence } q_0 = \left[\frac{a}{n}\right] = q$$

and $r_0 = a - q_0 n = a - qn = r$.

Example 4.2.2

1. The quotient and remainder when 46 is divided by 8 are respectively 5 and 6, since

$$46 = 5 \cdot 8 + 6 \quad \text{and} \quad 0 \leq 6 < 8.$$

2. The quotient and remainder when 46 is divided by 8 are respectively -6 and 2, since

$$-46 = (-6)8 + 2 \quad \text{and} \quad 0 \leq 2 < 8.$$

We shall make use of the division algorithm in proving the following theorem.

Theorem 4.2.2. Every subgroup of a cyclic group is cyclic.

Proof: Let H be a subgroup of a cyclic group G and let $a \in G$ such that

$$G = \langle a \rangle = \{a^n : n \text{ is an integer}\}.$$

If $H = \{e\}$, we are through, since, $H = \langle e \rangle$. Suppose that $H \neq \{e\}$. Since, for any n,

$$a^n \in H \Leftrightarrow a^{-n} = (a^n)^{-1} \in H,$$

it follows that there exists a positive integer n such that $a^n \in H$. By the well-ordering principle, there exist least positive integer m such that $a^m \in H$. Since H is a subgroup and $a^m \in H$, we have $\langle a^m \rangle \subseteq H$. On the other hand, let $x \in H$.

Then, $x \in G$ and hence $x = a^n$ for some integer n. By the division algorithm, there exist integers q and r such that

$$n = qm + r \quad \text{and} \quad 0 \leq r < m.$$

Now, $a^r = a^{n-qm} = a^n(a^m)^{-q} \in H$ (since $a^n, a^m \in H$). By the least property of m, r should not be positive and hence $r = 0$ and $n = qm$, so that

$$x = a^n = a^{qm} = (a^m)^q \in <a^m>.$$

Therefore, $H \subseteq <a^m>$ and hence $H = <a^m>$. Thus, H is a cyclic group.

Theorem 4.2.3. Every cyclic group is abelian.

Proof: Let G be a cyclic group and $a \in G$ such that

$$G = <a> = \{a^n : n \text{ is an integer}\}.$$

Let x and $y \in G$. Then, there exist integers m and n such that $x = a^m$ and $y = a^n$. Now,

$$xy = a^m a^n = a^{m+n} = a^{n+m} = a^n a^m = yx.$$

Thus, G is an abelian group.

The converse of the above theorem is not true; for example, the group $\mathbb{Z}_2 \times \mathbb{Z}_2$ is an abelian group which is not cyclic (see Example 4.2.1 (4)).

Example 4.2.3. Consider the cyclic group $(\mathbb{Z}, +)$ in which $\mathbb{Z} = <1>$. Following the proof of Theorem 4.2.2, one can prove that any subgroup of $(\mathbb{Z}, +)$ must be of the form $<n> = \{mn : m \in \mathbb{Z}\} = \mathbb{Z}n$ for some nonnegative integer n.

Example 4.2.4. Let n be a positive integer and consider the group $(\mathbb{Z}_n, +_n)$ of integer modulo n. Here again one can prove that any subgroup of $(\mathbb{Z}_n, +_n)$ must be of form $<d> = \{0, d, 2d, 3d, \ldots, (m-1)d\}$, where d is a positive divisor of n and $md = n$ (see Worked Exercise 4.1.3).

Definition 4.2.2. If G is a finite group, then the number of elements in G is called the order of G and is denoted by $|G|$.

Note that $|G|$ is precisely the cardinality of G. If $|G| = n$, then G is called a group of order n. In the following, we define the concept of the order of an element in a group which, in the finite case, turns out to be the order of the cyclic subgroup $<a>$ generated by a.

Definition 4.2.3. Let G be a group and $a \in G$. If $a^n = e$ for some positive integer n, then the least positive integer m such that $a^m = e$ is called the *order of a* and is denoted by $O(a)$. In this case, a is said to be an element of order m. If $a^n \neq e$ for all positive integers n, then the order of a is defined to be infinity and a is said to be an element of infinite order.

In other words, a is said to be an element of order m if m is the least element in the set $\{n \in \mathbb{Z}^+ : a^n = e\}$. If this set is empty, then a is said to be of infinite order.

Theorem 4.2.4. Let a be an element of finite order in a group G, then the following holds:

1. $O(a^{-1}) = O(a)$
2. For any integer n, there exists $0 \leq r < O(a)$ such that $a^n = a^r$.
3. For any integer n, $a^n = e$ if and only if $O(a)$ divides n.
4. $<a> = \{e, a, a^2, \ldots, a^{m-1}\}$, where $m = O(a)$.

Proof:

1. This is a direct consequence of the fact that, for any integer n, $a^n = e$ if and only if $(a^{-1})^n = e$ (since $(a^{-1})^n = a^{-n} = (a^n)^e$).
2. Let n be any integer. By the division algorithm, there exist integers q and r such that

   $$n = qO(a) + r \quad \text{and} \quad 0 \leq r < O(a).$$

   Now, $a^n = a^{qO(a)+r} = a^{qO(a)} \cdot a^r = (a^{O(a)})^q a^r = e^q a^r = ea^r = a^r$.
3. Let n be an integer. If $O(a)$ divides n, then $n = qO(a)$ for some integer q and hence

   $$a^n = a^{qO(a)} = (a^{O(a)})^q = e^q = e$$

   conversely suppose that $a^n = e$. Then, as in (2), we can write $n = qO(a) + r$ for some integers q and r such that $0 \leq r < O(a)$. Then, as in (2),

   $$a^r = a^n = e$$

   Since $0 \leq r < O(a)$ and $O(a)$ is the least positive integer such that $a^{O(a)} = e$, it follows that $r = 0$ and $n = qO(a)$. Thus, $O(a)$ divides n.
4. $$\begin{aligned} <a> &= \{a^n : n \text{ is an integer}\} \\ &= \{a^r : 0 \leq r < O(a)\} \\ &= \{e = a^0, a, a^2, \ldots\ldots, a^{O(a)-1}\}. \end{aligned}$$

Corollary 4.2.1. For any element a of finite order in a group,

$$O(a) = |\langle a \rangle|.$$

That is, the order of the element a is precisely equal to the order of the cyclic group $\langle a \rangle$.

Proof: In the above theorem, we have proved that

$$\langle a \rangle = \{e, a, a^2, \ldots, a^{O(a)-1}\}.$$

Also, if $a^r = a^s$, then $a^{r-s} = e = a^{s-r}$ and hence $O(a)$ divides $r - s$ and $s - r$ and; if $0 \leq r$ and $s < O(a)$, then $r = s$. Thus, $e, a, a^2, \ldots, a^{O(a)-1}$ are distinct. Thus, $|\langle a \rangle| = O(a)$.

It is well known that the greatest common divisor (g.c.d.) of two positive integers m and n can be written as a linear combination m and n. We shall prove this using Theorem 4.2.2.

Theorem 4.2.5. Let m and n be two positive integers and (m, n) be the greatest common division of m and n. Then, there exist integers r and s such that

$$(m, n) = rm + sn.$$

Proof: Let $H = \{am + bn : a \text{ and } b \in \mathbb{Z}\}$. Note that $H = \mathbb{Z}m + \mathbb{Z}n = \langle m \rangle + \langle n \rangle$. Since $(\mathbb{Z}, +)$ is a cyclic group and H is a subgroup of $\mathbb{Z}$ (by Corollary 4.1.2), we get that H is a cyclic subgroup of $(\mathbb{Z}, +)$ (by Theorem 4.2.2). Therefore, there exists a positive integer d such that

$$H = \langle d \rangle = \mathbb{Z}d.$$

In particular, since $d \in H$, $d = rm + sn$ for some integers r and s. We shall prove that d is the g.c.d. of m and n. Since $m = 1m + 0n \in H = \mathbb{Z}d$, d is a divisor of m. Similarly, d is a divisor of n. Also, if q is any common divisor of m and n, then

$$m = qk \quad \text{and} \quad n = qt \quad \text{for some } k, t \in \mathbb{Z}$$

and hence m and $n \in \langle q \rangle$ so that $am + bn \in \langle q \rangle$ for all integers a and b and, in particular, $d = rm + sn \in \langle q \rangle = \mathbb{Z}q$. Therefore, q divides d. Thus, d is the greatest common divisor of m and n.

The converse of the above therefore is also true, in the sense that, if d is a common divisor of m and n and d is of the form $rm + sn$, $r, s \in \mathbb{Z}$, then d is the greatest common divisor of m and n. In particular, m and n are relatively prime if and only if $1 = rm + sn$ for some r and $s \in \mathbb{Z}$.

Theorem 4.2.6. Let G be a group and $a \in G$ such that $O(a) = m < \infty$. Then, for any $0 \leq r < m$,

$$O(a^r) = \frac{m}{(m,\ r)}$$

where (m, r) is the g.c.d. of m and r.

Proof: Let $0 \leq r < m$ be fixed and $d = (m, r)$. Then, by Theorem 4.2.5, there exist integers s and t such that

$$d = sm + tr.$$

Put $b = a^r$. Since d divides both m and r, $\frac{m}{d}$ and $\frac{r}{d}$ are integers and

$$1 = s\left(\frac{m}{d}\right) + t\left(\frac{r}{d}\right)$$

and therefore $\frac{m}{d}$ and $\frac{r}{d}$ are relatively prime.

We have

$$b^{\frac{m}{d}} = (a^r)^{\frac{m}{d}} = a^{\frac{rm}{d}} = (a^m)^{\frac{r}{d}} = e.$$

On the other hand, for any integer q,

$$\begin{aligned} b^q = e &\Rightarrow (a^r)^q = e \\ &\Rightarrow a^{rq} = e \\ &\Rightarrow O(a) \text{ divides } rq \\ &\Rightarrow m \text{ divides } rq \\ &\Rightarrow \frac{m}{d} \text{ divides } \frac{r}{d} \cdot q \\ &\Rightarrow \frac{m}{d} \text{ divides } q \left(\text{since } \left(\frac{m}{d}, \frac{r}{d}\right) = 1\right). \end{aligned}$$

Therefore, $\frac{m}{d}$ is the least positive integer k such that $b^k = e$.

$$\therefore \ \mathrm{O}\left(a^r\right) = \mathrm{O}(b) = \frac{m}{d} = \frac{m}{(m, r)}.$$

Corollary 4.2.2. Let G be a group, $a \in G$ and $\mathrm{O}(a) = m < \infty$. If d is a positive divisor of m, then

$$\mathrm{O}\left(a^d\right) = \frac{m}{d} = \frac{\mathrm{O}(a)}{d}.$$

Proof: This follows from the above theorem and the fact that, for any positive divisor d of m, $(m, d) = d$.

Let us recall that, if $G = \langle a \rangle$, then a is called a generator of the cyclic group G. In the next result, we derive formulae to determine the number of generators of a cyclic group. First, we have the following definition.

Definition 4.2.4. For any positive integer n, let $\phi(n)$ be the number of positive integers less than or equal to n and relatively prime with n. Then, $\phi : \mathbb{Z}^+ \to \mathbb{Z}^+$ is a function and is called the *Euler–Totient function*, which is an important arithmetical function in the theory of numbers.

Note that $\phi(1) = 1 = \phi(2)$, $\phi\ (3) = 2 = \phi(4)$, $\phi(5) = 4$, $\phi(6) = 2$ and $\phi(7) = 6$. In fact, for any prime number p, $\phi(p) = p - 1$, since any positive integer less than p is relatively prime with p.

Theorem 4.2.7. Let G be a cyclic group and $a \in G$ such that $G = \langle a \rangle$. Then, G is infinite if and only if $a^n \neq a^m$ for all $n \neq m \in \mathbb{Z}$ and, in this case, a and a^{-1} are the only generators of G.

Proof: We are given that $G = \langle a \rangle = \{a^n : n \in \mathbb{Z}\}$. Suppose that $a^n = a^m$ for some $n \neq m \in \mathbb{Z}$. We can assume that $n < m$. Then, $m - n$ is a positive integer and $a^{m-n} = a^m(a^n)^{-1} = e$. Therefore, a is an element of finite order and, by Theorem 4.2.4 (4),

$$G = \langle a \rangle = \{e, a, a^2, \ldots, a^{\mathrm{O}(a)-1}\}$$

and hence G is finite.

Conversely suppose that G is finite. Then, since $a^n \in G$ for each $n \in \mathbb{Z}$, $a^n = a^m$ for some $n \neq m \in \mathbb{Z}$. Next, suppose that G is infinite. Then, $a^n \neq a^m$ for all $n \neq m \in \mathbb{Z}$ and, in particular, $a \neq a^{-1}$. Also,

$$G = \langle a \rangle = \langle a^{-1} \rangle$$

and therefore a and a^{-1} are two distinct generators of G. Now, suppose that b is any generator of G. Then,

$$<b> = G = <a>$$

and hence $a = b^n$ and $b = a^m$ for some n and $m \in \mathbb{Z}$. Since $a = b^n = (a^m)^n = a^{mn}$, it follows that $1 = mn$. Since m and n are integers, we get that $m = 1 = n$ or $m = -1 = n$ and hence $b = a$ or $b = a^{-1}$. Thus, a and a^{-1} are the only generators of G.

Note that, for any group G and $a \in G$, a is a generator of G if and only if a^{-1} is a generator of G. Also, a group G can have exactly two generators, but still G may be finite. For consider the following.

Example 4.2.5. In the group $(\mathbb{Z}_3, +_3)$ of integers modulo 3, 1 and 2 $(= -1)$ are the only generators and $(\mathbb{Z}_3, +_3)$ is a finite group.

Theorem 4.2.8. Let G be a finite cyclic group of order n and $a \in G$ such that $G = <a>$.

1. For any $1 \le r < n$, a^r is a generator of G if and only if r is relatively prime with n.
2. G has exactly $\phi(n)$ generators.

Proof: By Corollary 4.2.1, $n = |G| = |<a>|$ and hence $\text{O}(a) = n$ so that

$$G = \{e, a, a^2, \ldots, a^{n-1}\}.$$

1. Let $1 \le r < n$. Then, by Theorem 4.2.6,

$$\begin{aligned} a^r \text{ is a generator of } G &\Leftrightarrow n = \text{O}(a^r) = \frac{n}{(n, r)} \\ &\Leftrightarrow (n, r) = 1. \end{aligned}$$

2. This follows from the definition of $\phi(n)$ and from (1) above. Note that, since $\phi(1) = 1$, (2) is trivial when $n = 1$.

Example 4.2.6

1. $(\mathbb{Z}, +)$ is a cyclic group with 1 and -1 as the only generators.
2. $(\mathbb{Z}_n, +_n)$ is a finite cyclic group with $\phi(n)$ generators.
3. There are exactly two generators in each of the groups $(\mathbb{Z}_3, +_3)$, $(\mathbb{Z}_4, +_4)$ and $(\mathbb{Z}_6, +_6)$, since $\phi(3) = 2 = \phi(4) = \phi(6)$.

4. Since $\phi(8) = 4$, there are four generators for $(\mathbb{Z}_8, +_8)$ and these are 1, 3, 5 and 7.
5. For any prime number p, there are $p - 1$ generators for the group $(\mathbb{Z}_p, +_p)$, since $\phi(p) = p - 1$. That is, any nonzero (nonidentity) element $\mathbb{Z}_p$ is a generator.

Worked Exercise 4.2.1. Compute the order of 16 in $(\mathbb{Z}_{24}, +_{24})$.

Answer: $\mathbb{Z}_{24} = \langle 1 \rangle$ and $O(1) = 24$ in $\mathbb{Z}_{24}$

$$O(16) = \frac{24}{(16, 24)} = \frac{24}{8} = 3.$$

Worked Exercise 4.2.2. For any positive integers a and b, prove that

$$a\mathbb{Z} + b\mathbb{Z} = (a, b)\mathbb{Z} \quad \text{and} \quad a\mathbb{Z} \cap b\mathbb{Z} = [a, b]\mathbb{Z}$$

where (a, b) and $[a, b]$ are respectively the g.c.d. and l.c.m. of a and b.

Answer: These follow from the fact that, for any n and $m \in \mathbb{Z}^+$,

$$n\mathbb{Z} \subseteq m\mathbb{Z} \Leftrightarrow n \in m\mathbb{Z} \Leftrightarrow m \text{ divides } n$$

and that $a\mathbb{Z} + b\mathbb{Z}$ is the smallest subgroup of $\mathbb{Z}$ containing $a\mathbb{Z}$ and $b\mathbb{Z}$. Also $a\mathbb{Z} \cap b\mathbb{Z}$ is the largest subgroup of $\mathbb{Z}$ contained in $a\mathbb{Z}$ and $b\mathbb{Z}$.

Worked Exercise 4.2.3. Determine all the generators of $36\mathbb{Z} + 24\mathbb{Z}$.

Answer: Since $36\mathbb{Z} + 24\mathbb{Z} = (36, 24)\mathbb{Z} = 12\mathbb{Z}$ and $12\mathbb{Z}$ is an infinite cyclic group generated by 12, we get that 12 and -12 are the only generators of $12\mathbb{Z} = 36\mathbb{Z} + 24\mathbb{Z}$.

Worked Exercise 4.2.4. If a cyclic group has exactly two generators, then what can we say about the order of G?

Answer: If G is an infinite cyclic group, then, by Theorem 4.2.7, G has exactly two generators. Suppose that G is a finite cyclic group of order n and that G has exactly two generators. Then, $\phi(n) = 2$ and we have to determine all n for which $\phi(n) = 2$. Clearly $\phi(3) = 2 = \phi(4) = \phi(6)$. If $n > 6$ and n is odd, then $1, 2, n - 1$ are distinct and relatively prime with n. Let $n > 6$ and n be even. Suppose that $2 < p_1 < p_2 < \cdots < p_r$ are all the distinct primes dividing n and $m = p_1 p_2 \ldots p_r - 2$. Then, $(m, n) = 1$ and $1 < m < n - 1$ and hence $\phi(n) > 2$. Therefore, $\phi(n) = 2$ if and only if $n = 3$ or 4 or 6.

EXERCISE 4(B)

1. State whether each of the following are true and substantiate your answer.
   (i) Every finite abelian group is cyclic.
   (ii) An infinite group is abelian if and only if it is cyclic.
   (iii) $(\mathbb{Q}, +)$ is a cyclic group, where $\mathbb{Q}$ is the set of rational numbers.
   (iv) $(\mathbb{R}, +)$ is a cyclic group.
   (v) $(\mathbb{C}, +)$ is a cyclic group.
   (vi) If G_1 and G_2 are cyclic groups, then $G_1 \times G_2$ is also a cyclic group.
   (vii) $(\mathbb{C} - \{0\}, \cdot)$ is an abelian group which is not cyclic.
   (viii) $(\mathbb{Q} - \{0\}, \cdot)$ is a cyclic group.
   (ix) Any group of order 5 or 7 is cyclic.
   (x) The group $S(X)$ of bijections of a set X onto itself is a cyclic group under the composition of mappings.

2. Which of the following are cyclic groups? Substantiate your answers.
   (i) $(\mathbb{R} - \{0\}, \cdot)$
   (ii) $(\mathbb{P}(X), +)$, where X is a set.
   (iii) The group of quaternions which is of order 8.
   (iv) $(\mathbb{Q}^+, \cdot)$
   (v) $(\mathbb{R}^+, \cdot)$
   (vi) For any positive Integer n, the group of all n^{th} roots of unity of under the usual multiplication of complex numbers.

3. What can we say about a cyclic group having exactly one generator?

4. List all the elements in each of the following subgroups of the groups mentioned.
   (i) $<7>$ in $(\mathbb{Z}_{18}, +_{18})$
   (ii) $<5>$ in $(\mathbb{Z}_{20}, +_{20})$
   (iii) $<3>$ in $(\mathbb{Z}_{12}, +_{12})$
   (iv) $<3>$ in $(\mathbb{Z}_{16}, +_{16})$
   (v) $<i>$ in $(\mathbb{C} - \{0\}, \cdot)$
   (vi) $<\sqrt{2}>$ in $(\mathbb{R}^+, \cdot)$
   (vii) $<\sqrt{2}>$ in $(\mathbb{R}, +)$
   (viii) $<\sqrt{-2}>$ in $(\mathbb{C} - \{0\}, \cdot)$
   (ix) $<e>$ in any group G, where e is the identity in G.
   (x) $<12>$ in $(3\mathbb{Z}, +)$.

5. Prove that O(a) is finite for any element a of a finite group and give an example of an infinite group in which every element is of finite order.

6. If G is a group in which O(a) is finite for all $a \in G$, then can G be finite?

7. Let K and H be finite cyclic subgroups of orders m and n respectively in an abelian group G. If m and n relatively prime, prove that G has a cyclic subgroup of order mn.

8. In Exercise 7 above, if the least common multiple of m and n is K, then prove that G has a cyclic subgroup of order K.

9. If A and B are subgroups of a group G and one of A and B is cyclic, then prove that $A \cap B$ is a cyclic subgroup of G.

10. If a and b are elements of a group G, such that O(a) and O(b) are relatively prime positive integers, then prove that $<a> \cap <b> = \{e\}$.

11. Let G be a finite cyclic group of order n and d be a positive divisor of n. Then prove that the equation $x^d = e$ has exactly d solutions in G.

12. Prove that the set $\{4, 8, 12, 16\}$ is a group under the multiplication modulo 20. What is the identity element? Is this a cyclic group? If so, what are its generators?

13. Let $G = \{7, 35, 49, 77\}$. Then prove that $(G, \cdot_{84})$ is a group, where $\cdot_{84}$ is the multiplication modulo 84. What is the identity in G? Is this a cyclic group? If so, what are its generators?

14. Is $\mathbb{Z} \times \mathbb{Z}$ a cyclic group, where the operation is a coordinate-wise addition?

15. Is $\mathbb{Z}_9 \times \mathbb{Z}_{16}$ cyclic? If so, what are the generators?

16. Let G and H be finite cyclic groups of orders m and n, respectively. Then prove that $G \times H$ is cyclic if and only if m and n are relatively prime.

17. If G is a finite cyclic group and H is an infinite cyclic group, then can $G \times H$ be cyclic?

18. Let G be a group and $a \in G$ such that O(a) = 24. Then find a generator of the group $<a^9> \cap <a^{10}>$. In general, find a formula for the generator of $<a^n> \cap <a^m>$ for any $1 \leq n, m < 24$.

19. Let G be a finite cyclic group of order n. Then, for each positive divisor r of n, prove that $<a^{\frac{n}{r}}>$ is the only subgroup of order r and that the map $r \rightarrow <a^{\frac{n}{r}}>$ is a bijection of the set of positive divisors of n onto the set of all subgroups of G.

20. Find all the subgroups of $(\mathbb{Z}_{24}, +_{24})$ and $(\mathbb{Z}_{30}, +_{30})$.

21. Given a positive integer m, give an example of a cyclic group having exactly m subgroups.

22. List all the subgroups of $(\mathbb{Z}_{625}, +_{625})$.

23. Let U_n be the group of all n^{th}-roots of unity under the usual multiplication of complex numbers. Any generator of U_n is called a *primitive n^{th}-root of unity.* Determine all the primitive n^{th}-roots of unity for each of $n = 5, 7$ and 11.

24. Prove that any group having only a finite number of subgroups must be finite.

25. Give an example of a nonabelian group such that every proper subgroup is cyclic.

26. Let G be an abelian group and n be a positive integer. Then prove that the set $\{a \in G : O(a) \text{ divides } n\}$ is a subgroup of G.

27. Let a be an element of order n in a group G. Then prove that $O(a^r) = O(a^{n-r})$ for all $1 \leq r < n$.

28. Let G be a cyclic group of order 24 and $a \in G$ such that $a^8 \neq e$ and $a^{12} \neq e$. Then prove that a is a generator of G.

29. For any elements a and b of a group, prove that $O(ab) = O(ba)$.

30. Let G be an abelian group. Prove that the set of elements of finite order in G is a subgroup of G.

4.3 COSETS OF A SUBGROUP

For a given subgroup of a group G, we consider two important equivalence relations on G and study the relationship between their equivalence classes and the given subgroup. Let us begin with the following theorem.

Theorem 4.3.1. Let H be a subgroup of a group G and define two binary relations L_H and R_H on G as follows:

$$(a, b) \in L_H \quad \text{if } a^{-1}b \in H$$

and

$$(a, b) \in R_H \quad \text{if } ab^{-1} \in H$$

for any a and $b \in G$. Then, L_H and R_H are equivalence relations on G.

Proof: Let us recall that a reflexive, symmetric and transitive relation on G is called an equivalence relation. For any $a \in G$,

$$a^{-1}a = e \in H \quad \text{(since } H \text{ is a subgroup)}$$

and hence $(a, a) \in L_H$. Therefore, L_H is a reflexive relation on G. Also, for any a and $b \in G$,

$$\begin{aligned}(a, b) \in L_H &\Rightarrow a^{-1}b \in H \\ &\Rightarrow (a^{-1}b)^{-1} \in H \quad \text{(since } H \text{ is a subgroup of } G) \\ &\Rightarrow b^{-1}a \in H \\ &\Rightarrow (b, a) \in L_H\end{aligned}$$

and hence L_H is symmetric. Further,

$$\begin{aligned}(a, b) \text{ and } (b, c) \in L_H &\Rightarrow a^{-1}b \quad \text{and} \quad b^{-1}c \in H \\ &\Rightarrow (a^{-1}b)(b^{-1}c) \in H \\ &\Rightarrow a^{-1}c \in H \\ &\Rightarrow (a, c) \in L_H\end{aligned}$$

and hence L_H is transitive. Thus, L_H is an equivalence relation on G. Similarly, R_H is an equivalence relation.

Definition 4.3.1. Let H be a subgroup of a group G and $a \in G$. Define

$$\begin{aligned} aH &= \{ax : x \in H\} \\ \text{and} \quad Ha &= \{xa : x \in H\}. \end{aligned}$$

aH is called a *left coset of H corresponding to a in G* and Ha is called a *right coset of H corresponding to a in G*.

Recall that, for any equivalence relation θ on a set X, the equivalence class of θ containing a given element $x \in X$ is given by

$$\theta(x) = \{y \in X : (x, y) \in \theta\}$$

and that the equivalence class of θ form a partition of X in the sense that any two distinct equivalence classes of θ are disjoint and the union of all equivalence classes of θ is equal to the whole set X. In the following, we prove that the equivalence classes of L_H (respectively R_H) are precisely the left (right) cosets of H. Note that, if G is an abelian group then $aH = Ha$ for all $a \in G$.

Theorem 4.3.2. Let H be a subgroup of a group G and $a \in G$. Then,

$$L_H(a) = aH \quad \text{and} \quad R_H(a) = Ha$$

That is, $\{b \in G : a^{-1}b \in H\} = aH$ and $\{b \in G : ab^{-1} \in H\} = Ha$.

Proof: For any $x \in G$,

$$\begin{aligned} x \in L_H(a) &\Leftrightarrow (a, x) \in L_H \\ &\Leftrightarrow a^{-1}x \in H \\ &\Leftrightarrow x = a(a^{-1}x) \in aH \end{aligned}$$

and hence $L_H(a) = aH$. Similarly,

$$\begin{aligned} x \in R_H(a) &\Leftrightarrow (a, x) \in R_H \\ &\Leftrightarrow ax^{-1} \in H \\ &\Leftrightarrow xa^{-1} = (ax^{-1})^{-1} \in H \\ &\Leftrightarrow x = (xa^{-1})a \in Ha \end{aligned}$$

and hence $R_H(a) = Ha$.

Corollary 4.3.1. Let H be a subgroup a group G and a and $b \in G$. Then,

$$aH = bH \Leftrightarrow a^{-1}b \in H$$

and $Ha = Hb \Leftrightarrow ab^{-1} \in H$.

In particular, $aH = H \Leftrightarrow a \in H \Leftrightarrow Ha = H$.

Corollary 4.3.2. For any subgroup H of a group G, any two left (right) cosets of H in G are either equal or disjoint.

Corollary 4.3.3. For any subgroup H of a group G, the left (right) cosets of H in G form a partition of G, then note that, if we use $+$ to denote the operation on the group G, then write $a + H$ for aH and $H + a$ for Ha.

Example 4.3.1

1. If $H = \{e\}$, then $aH = \{a\} = Ha$ for any $a \in G$ and any subgroup H of G.
2. Consider the group $(\mathbb{Z}, +)$ of integers under the usual addition and let H be a subgroup of $(\mathbb{Z}, +)$. Then, by Worked Exercise 4.1.1, $H = n\mathbb{Z}$ for some nonnegative integer n. If $H = \{0\}$, then for any $a \in \mathbb{Z}$, the left coset $a + H = \{a\} = H + a$. Suppose that $H \neq \{0\}$. Then, $n > 0$. For any $a \in \mathbb{Z}$, choose q and $r \in Z$ such that

   $$a = qn + r \quad \text{and} \quad 0 \leq r < n$$

   and hence $r - a = (-q)n \in \langle n \rangle = H$ and therefore $a + H = r + H$.
   Thus,

   $$0 + H (= H), 1 + H, 2 + H, \ldots, (n - 1) + H$$

   are all the left (right) cosets of H in $\mathbb{Z}$. One can observe that these are all distinct. Thus, there are exactly n cosets of $n\mathbb{Z} = H$ in $\mathbb{Z}$.
3. Let $X = \{1, 2, 3\}$ be a 3-element set and $S(X)$ be the group of all bijections of X onto itself under the composition of mappings. We know from

Example 3.4.2 that $S(X) = \{e, a, b, c, d, s\}$ and the group $(S(X), o)$ is represented by the table given below.

o	e	a	b	c	d	s
e	e	a	b	c	d	s
a	a	b	e	s	c	d
b	b	e	a	d	s	c
c	c	d	s	e	a	b
d	d	s	c	b	e	a
s	s	c	d	a	b	e

Let $H = \{e, s\}$, then H is a subgroup of $(S(X), o)$. The left and right cosets of H in $S(X)$ are given below.

$$\begin{aligned} eH &= \{ee, es\} = \{e, s\} = H \\ aH &= \{ae, as\} = \{a, d\} = dH \\ bH &= \{be, bs\} = \{b, c\} = cH. \end{aligned}$$

Therefore, there are exactly three distinct left cosets of H in $S(X)$ and each of these contain exactly two elements. Also

$$\begin{aligned} He &= \{ee, se\} = \{e, s\} = H \\ Ha &= \{ea, sa\} = \{a, c\} = Hc \\ Hb &= \{eb, sb\} = \{b, d\} = Hd. \end{aligned}$$

Therefore, again there are exactly three right cosets of H in $S(X)$ and each of these contain exactly two elements. Note that, even though the number of left cosets of H is equal to the number of right cosets, a left coset may not be a right coset and vice versa. For example, aH is not equal to any right coset.

4. Consider the group $(\mathbb{Z}_{24}, +_{24})$ of integers modulo 24. Let us compute all the subgroups of $\mathbb{Z}_{24}$ and their cosets. This being an abelian group, any left coset of a subgroup is a right coset. We know that the subgroups of $\mathbb{Z}_{24}$ correspond to the positive divisors of 24 which are precisely 1, 2, 3, 4, 6, 8, 12 and 24.

   For any divisor d of 24, let

$$H_d = \{0, d, 2d, \ldots, (q-1)d\}, \text{ where } q = \frac{24}{d}.$$

Then
$$\begin{aligned} H_1 &= \{0, 1, 2, \ldots, 24-1\} = \mathbb{Z}_{24} \\ H_2 &= \{0, 2, 4, 6, 8, 10, 12, 14, 16, 18, 20, 22\} \\ H_3 &= \{0, 3, 6, 9, 12, 15, 18, 21\} \end{aligned}$$

$$H_4 = \{0, 4, 8, 12, 16, 20\}$$
$$H_6 = \{0, 6, 12, 18\}$$
$$H_8 = \{0, 8, 16\}$$
$$H_{12} = \{0, 12\}$$
$$H_{24} = \{0\}$$

The above eight are all the subgroup of $\mathbb{Z}_{24}$. Coming to the cosets, it is clear that $\mathbb{Z}_{24}$ is the only coset of H_1. There are two cosets of H_2 namely

$$0 +_{24} H_2 (= H_2) \quad \text{and} \quad 1 +_{24} H_2;$$

For any $a \in \mathbb{Z}_{24}$, if a is even, then $a \in H_2$ and hence $a +_{24} H_2 = 0 +_{24} H_2$ and, if a is odd, then $a - 1$ is even and hence $a - 1 \in H_2$ so that $a +_{24} H_2 = 1 +_{24} H_2$.

In general, for any divisor d of 24, there are exactly d cosets of H_d and these are

$$0 +_{24} H_d = H_d,\ 1 +_{24} H_d,\ 2 +_{24} H_d,\ \ldots,\ (d-1) +_{24} H_d.$$

Worked Exercise 4.3.1. Let H be a subgroup of a group G such that there are exactly two left cosets of H in G. Then prove that every left coset of H in G is a right coset and vice versa.

Answer: Since $H(= eH)$ is a left coset, it is given that there is only one more left coset of H in G and let this be aH. Then, $aH \neq H$ and $a \notin H$. Now,

$$aH \cap H = \varnothing \quad \text{and} \quad aH \cup H = G$$

and hence aH and H are complements to each other in G. That is, $G - H = aH$ and $xH = aH$ for all $x \notin H$. If $x \in H$, then $Hx = H = He$. On the other hand, for any $x \notin H$,

$$Hx = H^{-1}x = (x^{-1}H)^{-1} = (aH)^{-1} = H^{-1}a^{-1} = Ha^{-1} = Ha.$$

Thus, Ha and H are the only right cosets of H in G and hence $Ha = G - H = aH$.

EXERCISE 4(C)

1. Determine all the subgroups of each of the following and list their cosets.

   (i) $(\mathbb{Z}_{50}, +_{50})$

(ii) $(\mathbb{Z}_{20}, +_{20})$

(iii) $(\mathbb{Z}_{29}, +_{29})$

(iv) $\mathbb{Z}_2 \times \mathbb{Z}_2$

2. Consider the subgroup $\mathbb{Z}$ of the group $(\mathbb{R}, +)$. Prove that there is a bijection between the set of left cosets of $\mathbb{Z}$ in $\mathbb{R}$ and the interval $[0, 1)$.

3. Let H be a subgroup of a group G such that $aha^{-1} \in H$ for all $h \in H$ and $a \in G$. Then prove that every left coset aH is equal to the right coset Ha.

4. For any subgroup H of a group G, prove that $aH \to Ha^{-1}$ is a bijection of the set of left cosets of H in G onto the set of right cosets of H in G.

5. Let H be a subgroup of a group G, $a \in G$ and $K = aHa^{-1}$. Then prove that K is a subgroup of G and that there is a bijection of the set of left cosets of H in G onto the set of left cosets of K in G.

6. Let $X = \{1, 2, 3\}$ and $(S(X), o)$ be the group of bijections of X onto itself. Let $f \in S(X)$ be such that $f(1) = 2, f(2) = 3$ and $f(3) = 1$. Then prove that $H = \{e, f, f^{-1}\}$ is a subgroup of $S(X)$. Compute all the left and right cosets of H in $S(X)$.

7. Let $A = <2\pi>$ be the cyclic subgroup generated by 2π in the group $(\mathbb{R}, +)$. Prove that the trigonometric functions sine and cosine are constant on any left coset of A in $(\mathbb{R}, +)$.

8. Let $(S(X), o)$ be the group of all bijections of a nonempty set X onto itself and $x \neq y \in X$. Let

$$A_x = \{f \in S(X) : f(x) = x\}$$

and
$$A_{x,y} = \{f \in S(X) : f(x) = y\}.$$

Prove that A_x is a subgroup of $S(X)$ and that $A_{x,y}$ is not a subgroup of $S(X)$.

9. In Exercise 8 above, prove that $A_{x,y}$ is a coset of A_x in $S(X)$. Is it a left coset or a right coset?

10. Let A be a subgroup of a group G and x, y and $z \in G$ such that $xyA = xzA$. Then, prove that $yA = zA$. Also, prove that $Ayx = Azx$ implies $Ay = Az$.

11. Give an example of a subgroup A of a group G such that the product of two left cosets of A in G is not a left coset of A in G.

12. For any subgroup A of a group G, prove that the only left (right) coset of A in G, which is also a subgroup of G, is A itself.

13. Let A and B be two subgroups of a group G and x and $y \in G$ such that $Ax = By$. Then prove that $A = B$.

14. Prove that a subset S of a group G cannot be a left coset of two distinct subgroups of G.

4.4 LAGRANGE'S THEOREM

We have proved in the previous section that the left cosets of a subgroup H in a group G form a partition of the group G. This provides a counting technique that the total number of elements of a finite group G is equal to the sum of those in the cosets of a subgroup. In fact, we prove that the number of elements in any finite subgroup H is equal to that in any of its left or right coset and deduce the famous theorem of Lagrange which states that the order of any subgroup of a finite group is a divisor of the order of the group. We prove this theorem of Lagrange and obtain certain important consequences.

First, let us recall that the number of elements in a finite group G is called the *order* of G and is denoted by $|G|$ and that, if a group G is infinite, we say that the order of G is infinite. In general, for any finite set X, $|X|$ denote the number of elements in X. Note that $|X| = 0$ if and only if X is the empty set.

Theorem 4.4.1. Let H be a subgroup of group G and $a \in G$. Then, H, aH and Ha are all bijective to each other.

Proof: Define $f: H \to aH$ by $f(x) = ax$ for all $x \in H$. Clearly f is a surjection. By the left cancellation law, f is an injection also. Therefore, f is a bijection of H onto aH. Similarly the map $x \mapsto xa$ is a bijection of H onto Ha. Thus, $Ha \simeq H \simeq aH$.

Theorem 4.4.2 (Lagrange's Theorem). Let G be a finite group and H be a subgroup of G. Then, the order of H is a divisor of the order of G. That is, $|H|$ divides $|G|$.

Proof: Since G is a finite set and H is a nonempty subset of G, $|G|$ and $|H|$ are positive integers. Again, since G is finite, the number of left (right) cosets of H in G is also finite. Let $a_1H, a_2H, \ldots, a_nH$ be all the distinct left cosets of H in G. By the above theorem,

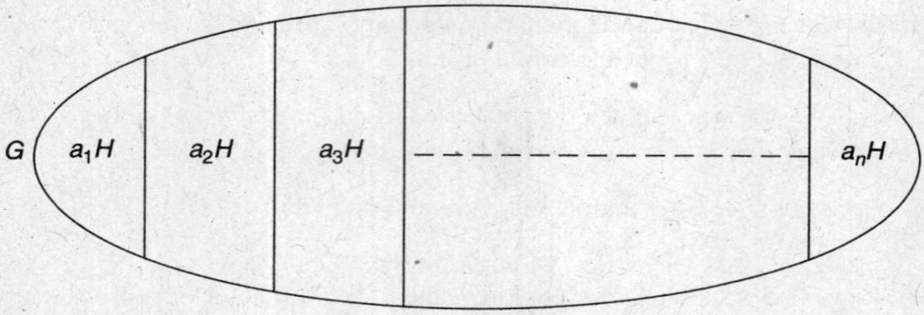

$|a_iH| = |H|$ for all $1 \leq i \leq n$. Also, we have

$$a_iH \cap a_jH = \varnothing \quad \text{for all } i \neq j$$

and $a_1H \cup a_2H \cup \ldots \cup a_nH = G$.

Therefore, $|G| = |a_1H| + |a_2H| + \ldots + |a_nH|$

$$= |H| + |H| + \cdots + |H| \quad (n \text{ times})$$
$$= n|H|.$$

Thus, $|G| = n|H|$ and $|H|$ is a divisor of $|G|$. ◀

Corollary 4.4.1. For any subgroup H of a finite group G, the number of left cosets of H in G is equal to the number of right cosets of H in G which is same as $\frac{|G|}{|H|}$.

Proof: In the proof of the above theorem, we have proved that $|G| = n|H|$ and hence

$$\frac{|G|}{|H|} = n = \text{The number of left cosets of } H \text{ in } G.$$

In the above proof, we can replace left cosets with right cosets and prove similarly that $|G| = m|H|$, where m is the number of right cosets of H in G. Now,

$$n = \frac{|G|}{|H|} = m.$$ ◀

Definition 4.4.1. For any subgroup H of a finite group G, the number of left (right) cosets of H in G is called the *index of H in G* and is denoted by $i_G(H)$.

Corollary 4.4.2. For any subgroup H of a finite group G,

$$i_G(H) = \frac{|G|}{|H|}.$$

Note that, even when G is an infinite group, a subgroup can have only finitely many cosets and one can talk about the index of such a subgroup. Consider the following example.

Example 4.4.2.

1. Let n be a positive integer and $H = n\mathbb{Z} = \{na : a \in \mathbb{Z}\}$. Then, H is a subgroup of $(\mathbb{Z}, +)$ and, by Example 4.3.1 (2),

$$H, 1 + H, 2 + H, \ldots, (n - 1)H$$

are all the cosets of H in $\mathbb{Z}$. Therefore, the index of H in $\mathbb{Z}$ is n.

2. Consider the group $(\mathbb{Z}_{12}, +_{12})$ of integers modulo 12. Let $H = <3> = \{0, 3, 6, 9\}$. Then, $|H| = 4$ and $|\mathbb{Z}_{12}| = 12$ and hence $\frac{12}{4}$ (= 3) is the index of H in $\mathbb{Z}_{12}$. We have $0 +_{12} H = 3 +_{12} H = 6 +_{12} H = 9 +_{12} H = H$,

$$1 +_{12} H = 4 +_{12} H = 7 +_{12} H = 10 +_{12} H = \{1, 4, 7, 10\},$$
$$2 +_{12} H = 5 +_{12} H = 8 +_{12} H = 11 +_{12} H = \{2, 5, 8, 11\}.$$

Therefore, H , $1 +_{12} H$ and $2 +_{12} H$ are all the distinct left cosets of H in $\mathbb{Z}_{12}$.

3. Consider the example given in Example 4.3.1 (3) in which $H = \{e, s\}$ and G = the group $S(X)$ of bijection of a 3-element set X onto itself. Here, $|G| = 3! = 6$, $|H| = 2$ and hence $i_G(H) = \frac{|G|}{|H|} = \frac{6}{2} = 3$ and therefore there are exactly 3 distinct left(right) cosets of H in G.

For any element a in group, recall that the order of a is defined as the least positive integer n such that $a^n = e$ (if all there is one such) and is denoted by $O(a)$.

Theorem 4.4.3. Let a be an element in a finite group G. Then, $O(a)$ divides $|G|$.

Proof: First note that, since G is finite and $a^n \in G$ for all integers n, $a^n = a^m$ for some $n \neq m$ and hence $a^r = e$ for some positive integer r. Therefore, a is an element of finite order. Let $O(a) = n$. Then, by Corollary 4.2.1, $O(a)$ is equal to the order of the subgroup $<a>$ generated by a in G. By the Lagrange's Theorem 4.4.2, $|<a>|$ divides $|G|$. Thus, $O(a)$ divides $|G|$.

Corollary 4.4.3. In a finite group G, $a^{|G|} = e$ for all $a \in G$.

Proof: Let G be a finite group, $a \in G$ and $O(a) = n$. Then, by the above theorem, $nm = |G|$ for some $m \in \mathbb{Z}^+$ and hence

$$a^{|G|} = a^{nm} = (a^n)^m = e^m = e.$$

Theorem 4.4.4. Let G be a group of order a prime number. Then, G is a cyclic group and every nonidentity element in G is a generator of G.

Proof: Let $|G| = p$ be a prime number. For any $a \neq e$ in G, $O(a) > 1$ and $O(a)$ is a divisor of p. Since p is prime, $O(a) = p$ and therefore

$$\langle a\rangle \subseteq G \quad \text{and} \quad |\langle a\rangle| = O(a) = p = |G|.$$

Since G is a finite set, $\langle a\rangle = G$. Thus, G is a cyclic group and every $a \neq e$ in G is a generator of G.

In the next three theorems, we derive two important results in the theory of numbers, using the Lagrange's theorem. First, let us have the following definition.

Definition 4.4.2. Let n be any positive integer. For any integers a and b, a is said to be *congruent to b modulo n* if n divides $a - b$ and we denote this by

$$a \equiv b \ (\text{mod } n).$$

That is, $a \equiv b$ (mod n) if and only if n divides $a - b$ or $a - b \in n\mathbb{Z}$.

Actually, the above is precisely the equivalence relation R_H on $\mathbb{Z}$ defined in Theorem 4.3.1, where $H = n\mathbb{Z}$, the cyclic subgroup generated by n in the group $(\mathbb{Z}, +)$. Besides being an equivalence relation, it has some other properties. Some of these are listed in the following theorem.

Theorem 4.4.5. Let n be a positive integer. The following holds for any integers a, b and c.

1. $a \equiv a$ (mod n)
2. $a \equiv b$ (mod n) $\Rightarrow b \equiv a$ (mod n)
3. $a \equiv b$ (mod n) and $b \equiv c$ (mod n) $\Rightarrow a \equiv c$ (mod n)
4. $a \equiv b$ (mod n) $\Rightarrow a + c \equiv b + c$ (mod n)
5. $a \equiv b$ (mod n) $\Rightarrow ac \equiv bc$ (mod n)
6. $a \equiv 0$ (mod n) $\Leftrightarrow a$ is an integral multiple of n.
7. For each $a \in \mathbb{Z}$, there exists an integer r such that $0 \leq r < n$ and $a \equiv r$ (mod n).
8. For any $0 \leq r \neq s < n$, $r \not\equiv s$ (mod n).

Proof: (1) to (6) are direct implications of the above definition. For any $a \in \mathbb{Z}$, we can use the division algorithm to get integers q and r such that

$$a = qn + r \quad \text{and} \quad 0 \leq r < n$$

and now, $a - r = qn$ and hence $a \equiv r$ (mod n). This proves (7). To prove 8, consider $0 \leq r, s < n$. Then, $|r - s| < n$ and hence $r - s$ cannot be a multiple of n, unless $r = s$.

Recall that, for any $n \in \mathbb{Z}^+$, we have defined $\phi(n)$ to be the number of positive integers which are less than or equal to n and relatively prime with n and that the function $\phi : \mathbb{Z}^+ \to \mathbb{Z}^+$ is called the Euler–Totient function.

Theorem 4.4.6 (Euler's Theorem). Let n be a positive integer. Then,

$$a^{\phi(n)} \equiv 1 \pmod{n}$$

for all integers a which are relatively prime with n, where ϕ is the Euler–Totient function.

Proof: The theorem is trivial if $n = 1$. Therefore, we can assume that $n > 1$. Let S be the set $\{1, 2, \ldots, n - 1\}$. Then, S is a monoid with respect to the multiplication $\cdot_n$ modulo n in which 1 is the identity. We know that an element r in S is invertible if and only if r is relatively prime with n. Let G be the set of all invertible elements in $(S, \cdot_n)$; that is,

$$G = \{r \in \mathbb{Z}^+ \mid r < n \quad \text{and} \quad (r, n) = 1\}.$$

Then, $(G, \cdot_n)$ is a group where $\cdot_n$ is the multiplication modulo n. Also, we know that $|G| = \phi(n)$ (see Definition 4.2.4). Now, let a be any integer which is relatively prime with n. Then, by the division algorithm, we can write

$$a = qn + r, \quad \text{where } q \text{ and } r \in \mathbb{Z} \quad \text{and} \quad 0 \le r < n.$$

Since $(a, n) = 1$, we get that $(r, n) = 1$ and $r > 0$.

Therefore, $r \in G$. Now we have

$$\begin{aligned} a^{\phi(n)} &= (qn + r)^{\phi(n)} \\ &= sn + r^{\phi(n)} \quad \text{for some } s \in \mathbb{Z} \\ &= sn + r^{|G|} \\ &\equiv 1 \pmod{n} \quad \text{(by Corollary 4.4.3).} \end{aligned}$$

Theorem 4.4.7 (Fermat's Theorem). Let p be a prime number and a be any integer. Then,

$$a^p \equiv a \pmod{p}.$$

Proof: We have to prove that p divides $a^p - a = a(a^{p-1} - 1)$.
If p divides a, then clearly p divides $a(a^{p-1} - 1)$. Suppose that p does not divide a. Then, since p is prime and (a, p) is a divisor of p, it follows that $(a, p) = 1$. Also, $\phi(p) = p - 1$. Now, by the Euler's theorem proved above, we get that

$$a^{p-1} = a^{\phi(p)} \equiv 1 \pmod p$$

and hence p divides $(a^{p-1} - 1)$, so that p divides $a(a^{p-1} - 1)$.

In the following, we introduce certain counting techniques which play crucial roles in the proofs of various results in the theory of finite groups. If A and B are subgroups of a group G, recall that the set AB may not be subgroup and that AB is a subgroup if and only if $AB = BA$. However, it is quite natural to think of the number of elements in AB in terms of those in A and B, when A and B are finite subgroups. The following is a precise answer to this.

Theorem 4.4.8. Let A and B be finite subgroups of a group G and $AB = \{ab : a \in A \text{ and } b \in B\}$. Then,

$$|AB| = \frac{|A||B|}{A \cap B} = |BA|.$$

Proof: First note that it is quite possible that $ab = a_1b_1$ for distinct elements a and a_1 in A and distinct elements b and b_1 in B. We shall find out how often does an element ab appear as a product of an element in A and an element in B. For any $x \in A \cap B$, we have

$$ab = (ax)(x^{-1}b)$$

and $ax \in A$ (since a and $x \in A$) and $x^{-1}b \in B$ (since x and $b \in B$). Also, if $ab = a_1b_1$, where a and $a_1 \in A$, and b and $b_1 \in B$, then

$$a^{-1}a_1 = bb_1^{-1} = s, \text{ say} \quad \text{and} \quad s \equiv A \cap B.$$

Also, $a_1 = as$ and $b_1 = s^{-1}b$.
This is to say that any representation of $x = ab$ as a product of an element of A and an element of B must be of the form

$$(as)(s^{-1}b), \quad \text{where } s \in A \cap B.$$

Thus, for each $a \in A$ and $b \in B$, the product ab appears $|A \cap B|$ times as a product of an element of A and an element of B. This implies that the number of distinct elements in AB is the total number in the listing of AB as a product of an element in A and an element in B (that is, $|A\,|B|$) divided by the number of times a given element (that is, $|A \cap B|$). Thus,

$$|AB| = \frac{|A||B}{|A \cap B|} = \frac{|B||A|}{|B \cap A|} = |BA|,$$

since $A \cap B = B \cap A$.

Corollary 4.4.4. Let A and B be subgroups of a finite group G such that $A \cap B = \{e\}$. Then, $|A| \geq \sqrt{|G|}$ or $|B| \geq \sqrt{|G|}$.

Proof: Since $AB \subseteq G$, we have

$$|G| \geq |AB| = \frac{|A||B|}{|A \cap B|} = |A||B| \quad \text{(by the above theorem)}$$

and hence $\sqrt{|G|} \geq |A|$ or $\sqrt{|G|} \geq |B|$.

For any subgroup A of a finite group G, we have $|A|.\, i_G(A) = |G|$ and hence both the order and index of A in G are divisors of the order of $|G|$. In the following, we derive some more properties of the index of a subgroup of a finite group. First note that if A and B are subgroups of a group G and $A \subseteq B$, then A is a subgroup of B also and we can talk of the index of A in B also.

Worked Exercise 4.4.1. Prove that the following holds for any subgroups A and B of a finite group G.

1. If $A \subseteq B$, then $i_G(A) = i_B(A) \cdot i_G(B)$
2. $i_A(A \cap B) \leq i_G(B)$
3. $i_A(A \cap B) = i_G(B)$ if and only if $G = AB$.
4. $i_G(A \cap B) = i_G(A) \cdot i_G(B)$ if and only if $G = AB$.

Answer:

1. Suppose that $A \subseteq B$. Then, by Corollary 4.4.2, we have

$$i_G(A) = \frac{|G|}{|A|} = \frac{|G|}{|B|} \cdot \frac{|B|}{|A|} = i_G(B) \cdot i_B(A)$$

2. $i_A(A \cap B) = \dfrac{|A|}{|A \cap B|} = \dfrac{|AB|}{|B|} \leq \dfrac{|G|}{|B|} = i_G(B)$.

3. Consider $|AB| = \dfrac{|A||B|}{|A \cap B|} = i_A(A \cap B)|B|$. Now

$$i_A(A \cap B) = i_G(B) \Rightarrow |AB| = i_G(B)|B| = |G| \Rightarrow G = AB$$

and, conversely, if $G = AB$, then

$$i_G(B) = \frac{|G|}{|B|} = \frac{|AB|}{|B|} = \frac{|A|}{|A \cap B|} = i_A(A \cap B)$$

4. If $G = AB$, then

$$i_G(A \cap B) = \frac{|G|}{|A \cap B|} = \frac{|G|}{|A||B| \;/|AB|} = \frac{|G|}{|A|}\frac{|G|}{|B|} = i_G(A) \cdot i_G(B)$$

and, conversely, if $i_G(A \cap B) = i_G(A) \cdot i_G(B)$, then

$$\frac{|G|}{|A \cap B|} = \frac{|G|}{|A|}\frac{|G|}{|B|} \quad \text{and hence } |G| = \frac{|A||B|}{|A \cap B|} = |AB|$$

which implies that $G = AB$.

Worked Exercise 4.4.2. Let G be a group of order $p^r n$, where p is a prime not dividing n. Let A and B be subgroups of G of orders p^r and p^s, respectively. If $B \not\subseteq A$, prove that AB is not a subgroup of G.

Answer: We are given that $|A| = p^r$ and $|B| = p^s$. Since $|B|$ divides $|G|$ and p^r is the largest power of p dividing $|G|$, we get that $0 \le s \le r$. Since $|A \cap B|$ is a divisor of $|A|$, $|A \cap B| = p^t$ for some $t \ge 0$. Suppose, AB is a subgroup of G, then $|AB|$ is a divisor of $|G| = p^r n$. But $|AB| = \frac{|A||B|}{A \cap B}$. since $|A|$, $|B|$ and $|A \cap B|$ are all powers of p, $|AB| = p^\alpha$ for some $\alpha \ge 0$ and $\alpha \le r$ (since p^r is the largest power of p dividing $|G|$). Therefore,

$$|AB| = p^\alpha \le p^r = |A| \le |AB| \quad (\text{since } A \subseteq AB)$$

and hence $|AB| = p^r$, so that $A = AB \supseteq B$, which is a contradiction. Thus, AB is not a subgroup of G.

Worked Exercise 4.4.3. Let A and B be finite subgroups of a group such that $|A| = p^r$ and $|B| = q^s$, where p and q are distinct primes and r and s are positive integers. Then prove that $A \cap B = \{e\}$.

Proof: Since $A \cap B$ is a subgroup of A as well as B, $|A \cap B|$ is a common divisor of $|A|$ and $|B|$ and hence $|A \cap B|$ is a divisor of (p^r, q^s), which is equal to 1 since p and q are distinct primes. Thus, $|A \cap B| = 1$ and hence $|A \cap B| = 1$.

EXERCISE 4(D)

1. State whether the following are true or false and justify your answer.
   (i) There is no subgroup of order 9 in $(\mathbb{Z}_{24}, +_4)$.
   (ii) There is a subgroup of order 36 in $(\mathbb{Z}_{120}, +_{120})$.
   (iii) In any group of order 240, there is a subgroup of index 36.
   (iv) If X is a 5-element set, then the group $(S(X), \text{o})$ of bijections of X onto itself has a subgroup of order 24.

(v) For any positive integer n, there is an element of order n in $(\mathbb{R}, +)$.

(vi) If X is a 6-element set, then the group $S(X)$ has an element of order 27.

(vii) The order of any element in a finite group is finite.

(viii) 43 divides $2^{42}-1$.

(ix) 19 divides $9^{18}-1$.

(x) 8 divides 729^4-1.

2. If A and B are subgroups of a group G such that $|A|$ is a prime and $A \cap B \neq \{e\}$, then prove that $A \subseteq B$.

3. If A is a subgroup of index 2 in a group G, then prove that $x^2 \in A$ for all $x \in G$.

4. Prove that the following are equivalent to each other for any subgroup A of a group G.

   (i) $i_G(A) = 2$.

   (ii) $x^{-1}y \in A$ for all x and $y \in G - A$.

   (iii) $xy^{-1} \in A$ for all x and $y \in G - A$.

5. If G is a group having no nontrivial subgroups, then prove that G is a cyclic group of order prime or G is trivial.

6. If A and B are subgroups of finite index in a group G, prove that $A \cap B$ is also of finite index and that

$$i_G(A \cap B) \leq i_G(A) \cdot i_G(B).$$

7. If A and B are subgroups of a group G and a and $b \in G$, then prove that $Aa \cap Bb = \varnothing$ or $Aa \cap Bb = (A \cap B)c$ for some $c \in G$.

8. Let A and B be subgroups of finite index in a group G such that $AB = BA$. Then prove that

$$i_{AB}(A \cap B) = i_{AB}(A) \cdot i_{AB}(B).$$

9. If an abelian group has two subgroups of orders n and m, then prove that it has a subgroup whose order is the least common multiple of n and m.

10. Let a and b be elements of a group such that $a^5 = e$ and $ab^{-1}a = b^2$, then find $O(b)$.

11. Determine all the subgroups of a group of order 137.

12. Let G be an abelian group of order $2n$, where n is an odd positive integer. Prove that G contains exactly one element of order 2.

13. Prove that any group of order 4 is abelian and give an example of a noncyclic group of order 4.

14. Prove that any nonabelian group has atleast six elements and give an example of a 6-element nonabelian group.

4.5 NORMAL SUBGROUPS

For any subgroup A of a group G, the collection of left cosets of A in G is in general not the same as the collection of right cosets of A in G. Subgroups for which these two collections are same are of special importance. In this section, we discuss certain important properties of these special subgroups.

Definition 4.5.1. A subgroup N of a group G is said to be *normal in G* if every left coset of N in G is a right coset of N in G.

Note that if a left coset aN happens to be a right coset, then it must be the coset Na only; for if $aN = Nb$, then $a \in aN = Nb$ and hence $ab^{-1} \in N$ so that $Na = Nb$. In the following, we obtain several other equivalent conditions for a subgroup to be normal.

Theorem 4.5.1. The following are equivalent for any subgroup N of a group G.

1. N is a normal subgroup of G.
2. $aN = Na$ for all $a \in G$.
3. $aNa^{-1} \subseteq N$ for all $a \in G$.
4. Every right coset of N in G is a left coset of N in G.
5. The product of any two left cosets of N in G is a left coset of N in G.
6. $(aN)(bN) = abN$ for all a and b in G.
7. $(Na)(Nb) = Nab$ for all a and b in G.
8. $L_N = R_N$ (see Theorem 4.3.1); that is, for any a and $b \in G$, $a^{-1}b \in N$ if and only if $ab^{-1} \in N$.

Proof:
(1) $\Rightarrow$ (2): Suppose that N is a normal subgroup of G. Then, for any $a \in G$, $aN = Nb$ for some $b \in G$, $aN = Nb$ for some $b \in G$ which implies that $a \in (Nb) \cap (Na)$ and hence $Nb = Na$. Therefore, $aN = Na$.
(2) $\Rightarrow$ (3): If $aN = Na$, then $aNa^{-1} = N$.
(3) $\Rightarrow$ (4): Suppose that $aNa^{-1} \subseteq N$ for all $a \in G$. Then,

$$aN = (aNa^{-1})a \subseteq Na$$

and, since $a^{-1}N(a^{-1})^{-1} \subseteq N$, we get that $a^{-1}N \subseteq Na^{-1}$ and hence $Na = a(a^{-1}N)a \subseteq a(Na^{-1})a = aN$. Thus, $Na = aN$ for all $a \in G$
(4) $\Rightarrow$ (5): Suppose that every right coset of N in G is a left coset of N in G. For any a and $b \in G$, first choose $c \in G$ such that $Nb = cN$ and now

$$(aN)(bN) = a(Nb)N = a(cN)N = acNN = acN.$$

Thus, $(aN)\,(bN)$ is a left coset of N in G.

$(5) \Rightarrow (6)$: Assume (5). For any a and $b \in G$, there exists $x \in G$ such that

$$(aN)(bN) = xN.$$

Now, $ab = ae{\cdot}be \in (aN)(bN) = xN$ and hence $(abN) \cap (xN) \neq \emptyset$ so that $(aN)(bN) = xN = abN$.

$(6) \Rightarrow (7)$: Let a and $b \in G$. Then, a^{-1} and $b^{-1} \in G$ and consider

$$\begin{aligned}(Na)\,(Nb) &= (b^{-1}N^{-1}\,a^{-1}N^{-1})^{-1}\\ &= ((b^{-1}N)\,(a^{-1}N))^{-1}\\ &= (b^{-1}\,a^{-1}\,N)^{-1}\text{ (by (6))}\\ &= N^{-1}ab\\ &= Nab.\end{aligned}$$

$(7) \Rightarrow (8)$: For any a and $b \in G$, consider

$$\begin{aligned}a^{-1}b \in N \Rightarrow b = a(a^{-1}b) \in aN&\\ &\Rightarrow e = ebb^{-1} \in N\,(aN)\,b^{-1} = (Na)(Nb^{-1}) = Nab^{-1}\text{ (by (7))}\\ &\Rightarrow e = xab^{-1},\quad \text{for some } x \in N\\ &\Rightarrow ab^{-1} = x^{-1} \in N.\end{aligned}$$

Similarly $ab^{-1} \in N \Rightarrow a^{-1}b \in N$. Thus, $L_N = R_N$.

$(8) \Rightarrow (1)$: For any $a \in G$, consider

$$\begin{aligned}x \in Na &\Leftrightarrow xa^{-1} \in N\\ &\Leftrightarrow x^{-1}a \in N\text{ (by (8))}\\ &\Leftrightarrow a^{-1}x = (x^{-1}a)^{-1} \in N\\ &\Leftrightarrow x \in aN.\end{aligned}$$

Thus, $aN = Na$ for all $a \in G$ and hence N is a normal subgroup of G. ◀

Example 4.5.1

1. In any group G, the trivial subgroup $\{e\}$ and G are always normal in G.
2. Consider the group $(S(X), \circ)$ of all bijections of a 3-element set X onto itself (see Example 3.4.2). Following the notation given in Example 3.4.2, let $H = \{e, s\}$. Then, H is a subgroup of $S(X)$. Also,

   $$bsb^{-1} = bsa = ca = d \notin H.$$

   Therefore, H is not a normal subgroup of $S(X)$.

Clearly every subgroup of an abelian group is normal. However, the converse of this may not be true in general. That is, a group in which every subgroup is normal may not be abelian; for, consider the following example.

Example 4.5.2. Consider the set $G = \{1, -1, i, -i, j, -j, k, -k\}$. Define the binary operation on G as just multiplication of numbers, treating i, j and k as numbers satisfying the following rules.

$$i^2 = j^2 = k^2 = -1$$
$$ij = k \quad \text{and} \quad ji = -k$$
$$jk = i \quad \text{and} \quad kj = -i$$
$$ki = j \quad \text{and} \quad ik = -j$$

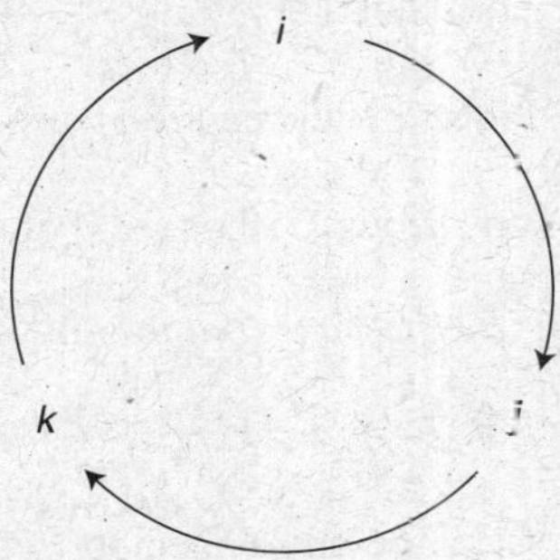

Then, G is a group under this operation. This group is called the *Quaternion group* of 8 elements. Since $ij = k \neq -k = ji$, this group G is not an abelian group. However, every subgroup of G is normal, as proved in Worked Exercise 4.5.2.

Worked Exercise 4.5.1. Prove that any subgroup of index 2 in any group is normal.

Answer: Let A be a subgroup of a group G and $i_G(A) = 2$. Then, there are exactly two left cosets of A in G. By Worked Exercise 4.3.1, A is normal subgroup of G.

Worked Exercise 4.5.2. Prove that every subgroup of the Quaternion group of 8 elements is normal.

Answer: Let G be the 8-element Quaternion group and A be a subgroup of G. Then, the order of A must be a divisor of the order of G (by the Lagrange's Theorem 4.4.2). Therefore, $|A|$ divides $|G| = 8$ and hence $|A| = 1$ or 2 or 4 or 8.

If $|A| = 1$, then $A = \{1\}$ and hence A is normal.
If $|A| = 8$, than $A = G$ and hence A is normal in G.
If $|A| = 4$, than $i_G(A) = \frac{|G|}{|A|} = \frac{8}{4} = 2$ and hence, by Worked Exercise 4.5.1, A is normal.
Finally suppose that $|A| = 2$, then $A = \{1, a\}$, where a is an element of G such that $a^2 = 1$ and $a \neq 1$. The only such element in G is -1 and hence

$$A = \{1, -1\}.$$

Since $(-1)x = -x = x(-1)$ and $1x = x = 1x$ for all $x \in G$, it follows that

$$xA = \{x, -x\} = Ax$$

for all $x \in G$. Thus, A is normal in G.

Worked Exercise 4.5.3. Prove that the centre of any group is normal.

Answer: Recall that, for any group G, the centre of G is defined as the set

$$Z(G) = \{a \in G : ax = xa \text{ for all } x \in G\}.$$

Thus, for any $x \in G$, $xZ(G) = Z(G)x$ and hence $Z(G)$ is a normal subgroup of G.

Worked Exercise 4.5.4. Let X be any nonempty set X and $(S(X), \circ)$ be the group of bijections of X onto itself. For any $x \in X$, let

$$H_x = \{f \in S(X) : f(x) = x\}.$$

Then prove that H_x is a normal subgroup of $S(X)$ for all $x \in X$ if and only if X has at most two elements.

Answer: Suppose that X has three distinct elements, say x, y and z. Let f and $g : X \to X$ be defined by

$$f(x) = y, f(y) = x \quad \text{and} \quad f(s) = s \quad \text{for all } s \in X - \{x, y\}$$

and $$g(y) = z, g(z) = y \quad \text{and} \quad g(s) = s \quad \text{for all } s \in X - \{x, y\}.$$

Then, $g \in H_x$ and $f \in S(X)$ and

$$(fgf^{-1})(x) = f(g(f^{-1}(x)) = f(g(y)) = f(z) = z \neq x$$

and hence $fgf^{-1} \notin H_x$. Therefore, $fH_xf^{-1} \not\subseteq H_x$ and hence H_x is not normal in $S(X)$.

On the other hand, suppose X has at most two elements. If $|X| = 1$, then $S(X) = \{e\} = H_x$ and hence H_x is normal. If $|X| = 2$, then $X = \{x, y\}$, where $x \neq y$, and $H_x = \{e\} = H_y$, which is clearly normal.

EXERCISE 4(E)

1. Let A and B be subgroups of a group G and one of them be normal in G. Then prove that AB is a subgroup of G. Is AB normal in G?

2. For any normal subgroup A of a group G and for any subgroup B of G, prove that $A \cap B$ is a normal subgroup of B.

3. Let G be the set consisting of the following eight matrices.

$$\begin{pmatrix} 1 & 0 \\ 0 & 1 \end{pmatrix}, \begin{pmatrix} -1 & 0 \\ 0 & -1 \end{pmatrix}, \begin{pmatrix} 0 & 1 \\ -1 & 0 \end{pmatrix}, \begin{pmatrix} 0 & -1 \\ 1 & 0 \end{pmatrix}$$

$$\begin{pmatrix} i & 0 \\ 0 & i \end{pmatrix}, \begin{pmatrix} 0 & i \\ i & 0 \end{pmatrix}, \begin{pmatrix} -i & 0 \\ 0 & i \end{pmatrix} \text{ and } \begin{pmatrix} 0 & -i \\ -i & 0 \end{pmatrix}$$

Prove that G is a group under the multiplication of matrices over the complex numbers. Also prove that G is not abelian and that every subgroup of G is normal. Compare this with the example given in Worked Exercise 4.5.2.

4. Determine all the normal subgroups of the group $S(X)$ of bijections of a 3-element set X onto itself.

5. Let G be a group and $n \in \mathbb{Z}^+$. If A is the unique subgroup of order n in G, then prove that A is normal in G.

6. Let A be a subgroup of a group G such that, for any x and $y \in G$,

$$xA = yA \Rightarrow Ax = Ay$$

Then prove that A is normal subgroup of G.

7. Let A and B be two normal subgroups of a group G such that $|A \cap B| = 1$. Then prove that $ab = ba$ for all $a \in A$ and $b \in B$.

8. Give an example of a group G and subgroups A and B such that $A \subseteq B$, A is normal in B, B is normal in G and A is not normal in G.

9. Let G be a group and $a \in G$. Define

$$N(a) = \{x \in G : ax = xa\}.$$

Prove that $N(a)$ is a subgroup of G containing $\langle a \rangle$ as a normal subgroup. $N(a)$ is called the *normaliser of a in G*.

10. Let A be a normal subgroup of a finite group G such that $|A|$ and $i_G(A)$ are relatively prime. For any $x \in G$, prove that $x \in A$ if and only if $x^{|A|} = e$.

11. Let G be a group containing a nontrivial subgroup A which is contained in every nontrivial subgroup of G. Then prove that every element of G is of finite order. What could be the order of A?

12. Let A be a subgroup of a group G and

$$C(A) = \{x \in G : xa = ax \text{ for all } a \in A\}.$$

Prove that $C(A)$ is a subgroup of G containing A as a normal subgroup. $C(A)$ is called the *centralizer* of A in G.

13. For any subgroup A of group G, let

$$N(A) = \{x \in G : xA = Ax\}$$

Prove that $N(A)$ is the largest subgroup of G containing A as a normal subgroup. $N(A)$ is called the *normaliser* of A in G.

14. Compare the normaliser and the centralizer of a subgroup of a group.

15. For any subgroup A of a group G, let

$$N = \bigcap_{a \in G} (a\,A\,a^{-1}).$$

Then prove that N is the largest normal subgroup of G containing A.

16. Let A be a subgroup of a group G such that $x^2 \in A$ for all $x \in G$. Then prove that A is normal in G.

17. For any real numbers a and b, define $T_{ab} : \mathbb{R} \to \mathbb{R}$ by $T_{ab}(x) = ax+b$ for all $x \in \mathbb{R}$. Prove that the set $\{T_{ab} : a \text{ and } b \in \mathbb{R}\}$ is a group under the usual composition of mappings in which $\{T_{1b} : b \in \mathbb{R}\}$ is a normal subgroup.

18. Prove that the intersection of any class of normal subgroups of a group G is again a normal subgroup of G.

19. Prove that, for any group G, the centre $\mathbb{Z}(G) = \bigcap_{a \in G} N(a)$, where $N(a)$ is the normaliser of a in G.

20. For any set X with $|X| \geq 3$, prove that $\mathbb{Z}(S(X)) = \{e\}$.

21. Let G_1 and G_2 be any groups and A_1 and A_2 be subsets of G_1 and G_2 respectively. Then prove that $A_1 \times A_2$ is a (normal) subgroup of the group $G_1 \times G_2$ if and only if A_1 and A_2 are (normal) subgroups of G_1 and G_2, respectively.

22. If G_1 and G_2 are groups, prove that $\{e_1\} \times G_2$ and $G_1 \times \{e_2\}$ are normal subgroups of $G_1 \times G_2$, where e_1 and e_2 are identities in G_1 and G_2, respectively.

4.6 QUOTIENT GROUPS

For any normal subgroup N of a group G, we have proved earlier that the product of any two left cosets of N in G is again a left coset of N in G. That is, taking product can be considered as a binary operation on the set of all left cosets of N in G, which will be actually a group under this operation. This is proved in the following definition.

Definition 4.6.1. Let G be a group and A and B be subsets of G. Then, we define the product $A \cdot B$ as the set

$$A \cdot B = \{ab : a \in A \text{ and } b \in B\}.$$

Then, AB is called the product of A and B, in this order, induced by the binary operation on G. Then, $\cdot$ is a binary operation on the set $\mathbb{P}(G)$ of all subsets of G and is called the *multiplication of subsets of the group G*.

Theorem 4.6.1. Let N be a normal subgroup of a group G and

$$G/N = \{aN : a \in G\}.$$

Then, G/N is a group under the multiplication of subsets of the group; that is, for any aN and $bN \in G/N$,

$$aN \cdot bN = \{xy : x \in aN \text{ and } y \in bN\}.$$

Proof: First of all recall that, $aN \cdot bN = abN$ for all a and $b \in G$ (since N is a normal subgroup of G) and hence the multiplication of subsets of the group G is a binary operation on the set G/N of all left cosets of N in G.
For any a, b and $c \in G$, we have

$$\begin{aligned}(aN \cdot bN) \cdot cN &= (abN) \cdot cN \\ &= ((ab) \cdot c)N \\ &= (a(bc))N \\ &= aN \cdot (bN \cdot cN)\end{aligned}$$

and hence the operation on G/N is associative.
Also, the coset $eN = N$ satisfies the property that

$$(eN) \cdot (aN) = eaN = aN = (ae)N = (aN) \cdot (eN)$$

for all $aN \in G/N$ and hence $eN (= N)$ is the identity in G/N. Further, for any $aN \in G/N$ with $a \in G$, consider

$$(a^{-1}N)\cdot(aN) = a^{-1}a\cdot N = eN = aa^{-1}N = (aN)(a^{-1}N)$$

and therefore $a^{-1}N$ is the inverse of aN in G/N. Thus, G/N is a group under the multiplication of subsets of the group G.

Definition 4.6.2. For any normal subgroup N of a group G, the group G/N constructed above is called the quotient group of G by N. Whenever we refer to a G/N as a group, we only mean the set of left(right) cosets of N in G together with the multiplication of subsets of the group G.

Corollary 4.6.1. For any subgroup N of an abelian group G, G/N is an abelian group.

Theorem 4.6.2. Let N be a normal subgroup of a finite group G. Then,

$$|G| = |N|\,|G/N|$$

Proof: This follows from the facts that $|G/N| = i_G(N)$ and $|G| = |N|\cdot i_G(N)$.

Example 4.6.1. Let $(\mathbb{Z}, +)$ be the group of all integers under the usual addition and let n be any positive integer and $<n> = \{na : a \in \mathbb{Z}\}$ be the subgroup of $(\mathbb{Z}, +)$ generated by n. Since $+$ is the operation on the group $\mathbb{Z}$, the elements of $\mathbb{Z}/<n>$ are of the form $a + <n>$ with $a \in \mathbb{Z}$. From Example 4.3.1 (2), we know that any coset of $<n>$ in $\mathbb{Z}$ is of the form $r + <n>$, where $0 \leq r < n$. Thus,

$$\mathbb{Z}/<n> = \{N, 1 + N, 2 + N, \ldots, (n-1) + N\}, \quad \text{where } N = <n>.$$

Example 4.6.2. Let E be the Euclidean plane and pick any point P in E and fix a coordinate system with P as origin. Then, any point in E can be expressed uniquely as an ordered pair (x, y) of real numbers and P corresponds to $(0, 0)$. For any points $Q_1 = (x_1, y_1)$ and $Q_2 = (x_2, y_2)$, define

$$Q_1 + Q_2 = (x_1 + x_2, y_1 + y_2).$$

The following diagram illustrates this addition of points by the construction of parallelogram with adjacent sides PQ_1 and PQ_2. Then, E is an abelian group with respect to the above operation.

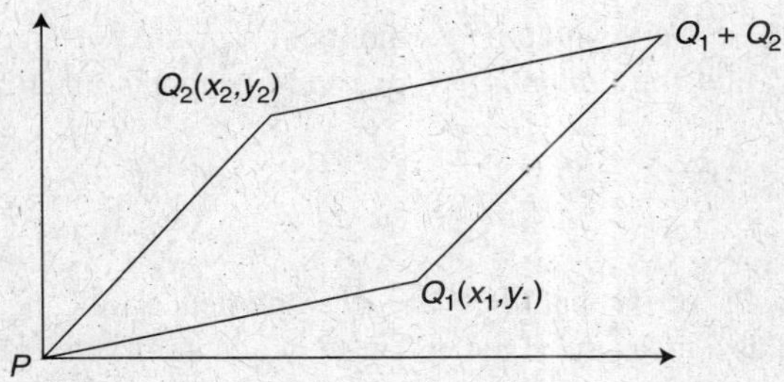

Let L be a straight line in E passing through P. Then, one can easily check that L is a subgroup of E. Let us describe the cosets of L in E. If Q is any point in E, then one can prove that the coset $Q - L$ is precisely the line passing through Q and parallel to L. Therefore, the quotient group E/L consists precisely all the straight lines parallel to L, including L. For any points Q_1 and Q_2 in E, we have

$$(Q_1 + L) + (Q_2 + L) = (Q_1 + Q_2) + L.$$

This is illustrated in the following figure.

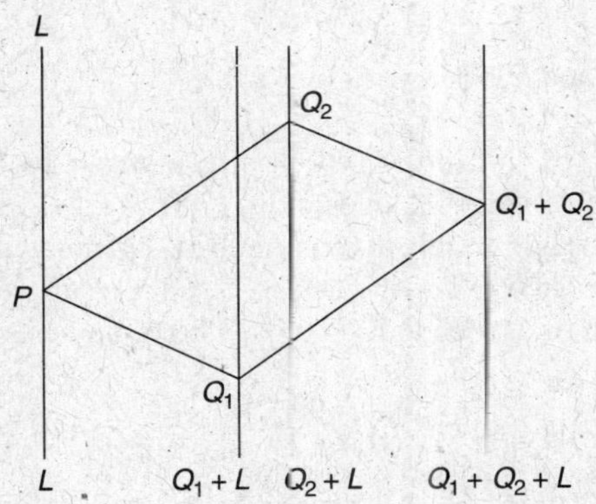

Worked Exercise 4.6.1. Let a be an element of finite order in a group G and N be a normal subgroup of G. Then prove that the order of the element aN in G/N is a divisor of $\mathrm{O}(a)$.

Proof: Let $\mathrm{O}(a) = n$. Then,

$$(aN)^n = a^nN = eN = N \quad (\text{since } \mathrm{O}(a) = n,\ a^n = e)$$

N is the identity in the quotient group G/N. Therefore, aN is of finite order in G/N and $\mathrm{O}(aN)$ is a divisor of $n = \mathrm{O}(a)$ (by Theorem 4.2.4 (3)).

Definition 4.6.3. Let a and b be elements of a group G. Then, the product $aba^{-1}b^{-1}$ is called the *commutator of a and b* and is usually denoted by $[a, b]$; that is,

$$[a, b] = ab\, a^{-1}b^{-1}.$$

One can easily verify that $ab = [a, b]ba$ and hence we can view the commutator $[a, b]$ as a measure of the extent to which ab differs from ba. In fact, the elements a and b commute (that is, $ab = ba$) if and only if $[a, b] = e$, the identity in G. The commutators of elements of a group G may not form a subgroup of G, in general. However, we have the following theorem.

Theorem 4.6.3. Let G be a group and consider the set

$$[G,G]=\left\{\prod_{i=1}^{n}[a_i,b_i] : a_i \text{ and } b_i \in G\right\}.$$

Then, $[G, G]$ is a subgroup of G. Also, any subgroup of G containing $[G, G]$ is normal in G.

Proof: For any a and $b \in G$, we have

$$[a, b]^{-1} = (ab\, a^{-1}b^{-1})^{-1} = bab^{-1}a^{-1} = [b, a]$$

and therefore $[G, G]$ is precisely the subgroup of G generated by the set $\{[a, b] : a \text{ and } b \in G\}$ (see Theorem 4.1.5). Thus, $[G, G]$ is a subgroup of G. Next, let A be any subgroup of G such that $[G, G] \subseteq A$. For any $a \in A$ and $x \in G$,

$$xax^{-1} = (xax^{-1}a^{-1})a \in A,$$

since $a \in A$ and $xax^{-1}a^{-1} = [x, a] \in [G, G] \subseteq A$. Thus, A is a normal subgroup in G.

Corollary 4.6.2. For any group G, $[G, G]$ is a normal subgroup of G.

Definition 4.6.4. For any group G, the subgroup $[G, G]$ is called the *derived subgroup* or *commutator subgroup* of G and the quotient group $G/[G, G]$ is called the *commutator quotient group* or *abelianized group*. The reason for the latter terminology is the following theorem.

Theorem 4.6.4. Let N be any normal subgroup of a group G. Then, the quotient group G/N is abelian if and only if $[G, G] \subseteq N$.

Proof: Since $ab = ba \Leftrightarrow [a, b] = e$, the identity, for any elements a and b in any group, we have

$$\begin{aligned} G/N \text{ is abelian} &\Leftrightarrow [aN, bN] = N \quad \text{for all } aN \text{ and } bN \in G/N \\ &\Leftrightarrow aN \cdot bN \cdot (aN)^{-1}(bN)^{-1} = N \quad \text{for all } a, b \in G \\ &\Leftrightarrow (aba^{-1}b^{-1})N = N \quad \text{for all } a, b, \in G \\ &\Leftrightarrow [a, b] \in N \quad \text{for all } a, b \in G \\ &\Leftrightarrow [G, G] \subseteq N. \end{aligned}$$

The following is a direct consequence of Corollary 4.6.2 and Theorem 4.6.4.

Corollary 4.6.3. For any group G, $G/[G, G]$ is an abelian group.

Theorem 4.6.4 and Corollary 4.6.3 say that the commutator subgroup $[G, G]$ is the smallest normal subgroup of G having an abelian quotient group. The transition from a group G to its commutator quotient group $G/[G, G]$ is referred to as the *abelianization* of the group G and provides a convenient procedure to produce abelian groups from nonabelian ones. Note that a group G is abelian if and only if the commutator subgroup $[G, G] = \{e\}$.

In the following, we give a procedure to find all subgroups of the quotient G/N, where N is a given normal subgroup of a group G.

Theorem 4.6.5. Let N be a normal subgroup of a group G and G/N be the quotient group. For any subgroup A of G containing N, $A/N = \{aN : a \in A\}$ is a subgroup of G/N. Further, $A \mapsto A/N$ is a one-to-one correspondence between the subgroups of G containing N and the subgroups of G/N.

Proof: For any a and $b \in A$,

$$(aN)(bN^{-1}) = (aN)(b^{-1}N) = ab^{-1}N \in A/N,$$

since A is a subgroup, and hence A/N is a subgroup of G/N. If A and B are subgroups of G containing N and $A/N = B/N$, then

$$\begin{aligned} a \in A &\Rightarrow aN \in A/N = B/N \\ &\Rightarrow aN = bN \quad \text{for some } b \in B \\ &\Rightarrow a^{-1}b \in N \\ &\Rightarrow ab^{-1} \in N \subseteq B \quad (\text{since } N \text{ is normal}) \\ &\Rightarrow a = (ab^{-1})\, b \in B \end{aligned}$$

and hence $A \subseteq B$. Similarly, $B \subseteq A$ and therefore $A = B$. Thus, $A \mapsto A/N$ is one-one. Further, if M is any subgroup of G/N, define $A = \{x \in G : xN \in M\}$. Then, A is a subgroup of G containing N and $A/N = M$.

Thus, $A \mapsto A/N$ is a one-to-one (bijective) correspondence between the subgroups of G containing N and the subgroups of G/N.

Remark 4.6.1. Note that, in the above, A/N is normal in G/N if and only if A is normal in G. Also, for any subgroups A and B containing N, $A/N \subseteq B/N$ if and if only $A \subseteq B$.

If $N = \{e\}$, then $G/N \cong G$ and, if $N = G$, then $G/N = \{N\}$, the trivial group.

EXERCISE 4(F)

1. Describe the quotient group of each of the following in the groups mentioned against them.
   (i) $\{0, 4, 8, 12\}$ in $(\mathbb{Z}_{16}, +_{16})$.
   (ii) The set E of even integers in $(\mathbb{Z}, +)$.
   (iii) $\mathbb{Z}$ in $(\mathbb{Q}, +)$.
   (iv) $\{1, -1\}$ in $(\{1, -1, i, -i\}, \cdot)$.
   (v) $\mathbb{R}$ in $(\mathbb{C}, +)$.
   (vi) $\mathbb{Q}$ in $(\mathbb{R}, +)$.
2. Let $G = <a>$ be a cyclic group of order 15 and $A = <a^3>$. Construct multiplication table representing the quotient group G/A.
3. For any group G, determine the quotient groups of the trivial normal subgroups $\{e\}$ and G.
4. Let A be a subgroup of a group G such that $x^2 \in A$ for all $x \in G$. Then prove that A is normal in G and the quotient group G/A is abelian.
5. Let $Z(G)$ be the centre of a group G. If $G/Z(G)$ is cyclic, prove that G is abelian.
6. Let N be a normal subgroup of a finite group G. Then prove the following.
   (i) $|G/N| = |G|/|N| = i_G(N)$.
   (ii) If $n = i_G(N)$, then $x^n \in N$ for all $x \in G$.
   (iii) The order of aN in G/N is a divisor of the order of a in G, for any $a \in G$. Can they be equal? Justify your answer.
7. Let N be a normal subgroup of a group G. Then prove that G is finite if and only if both N and G/N are finite.

8. A group G is said to be *finitely generated* if $G = <F>$ for some finite subset F of G. Let A be a subgroup of an abelian group such that both A and G/A are finitely generated. Then prove that G is finitely generated.

9. Let G be a group and $S = \{a^2 : a \in G\}$. Then prove that $<S>$ is a normal subgroup of G and that $G/<S>$ is an abelian group.

10. List all normal subgroups of the group $(S(X), \circ)$ of bijections of a 3-element set X onto itself and construct tables representing the quotient group of each normal subgroup $S(X)$.

11. Let $G = \{(a, b) \in \mathbb{R} \times \mathbb{R} : a \neq 0\}$ and, for any (a, b) and $(c, d) \in G$, define

$$(a, b) * (c, d) = (ac, ad + b).$$

Then prove that $(G, *)$ is a group. If $K = \{(1, b) : b \in \mathbb{R}\}$, then prove K is a normal subgroup of G.

12. Let G be a group of order $2n$, where n is odd. Prove that G contains a normal subgroup of index 2.

5

Homomorphisms of Groups

Homomorphisms play a major role in all aspects of modern algebra. One of the most difficult problems in the theory of groups is to list all finite groups having the same order. The difficulty is that we may list the same groups in different forms. The notion of isomorphism builds up an equivalence relation between groups, so that we may consider two groups belonging to a given equivalence class as the same. Just as we cannot conclude that two human beings are same, because they wear an identical set of clothes or the same person putting on a different set of clothes does not become different, groups are to be recognized as same or different on the basis information that is not readily apparent. Consider the group $(\mathbb{Z}_4, +_4)$, where $\mathbb{Z}_4 = \{0, 1, 2, 3\}$ and $+_4$ is the addition modulo 4, and the group $(G, \cdot)$, where $G = \{1, i, -1, i\}$ and '$\cdot$' is the usual multiplication of complex numbers. We have

$$\mathbb{Z}_4 = <1> = \{0, 1, 2, 3\} \quad \text{and} \quad G = \{1, i, i^2, i^3\}$$

and hence these two groups look same as we have the correspondence $f: \mathbb{Z}_4 \to G$ defined by $f(0) = 0, f(1) = i, f(2) = i^2$ and $f(3) = i^3$. This correspondence is compatible with the operations $+_4$ on $\mathbb{Z}_4$ and '$\cdot$' on G and is a bijection of $\mathbb{Z}_4$ onto G. This helps us in proving that any property of $\mathbb{Z}_4$ gives a similar property in G'. Such correspondences are called homomorphisms.

In this chapter, we focus our attention on homomorphisms of groups. A homomorphism of two groups may reveal some information about one of the groups as a deduction of known structural properties of the other. If f is a homomorphism of a group G onto another group G' and if we know all about the structure of G, then we can deduce the structure of G' also.

5.1 DEFINITION AND EXAMPLES

For any groups G and G', we are interested in the mappings f of G into G' such that the product of images of two elements in G is same as the image of the product. If * and o are the binary operations in groups G and G', respectively, then a mapping $f : G \to G'$ is called a homomorphism if the figure given below is commutative, in the

sense that, the composition of f and * is same as that of o and $f \times f$, where $f \times f : G \times G \to G' \times G'$ is defined as $(f \times f)(x, y) = (f(x), f(y))$. This is precisely expressed in the following definition.

Definition 5.1.1. Let G and G' be groups. Then, a mapping $f : G \to G'$ is called a *homomorphism of G into G′* if

$$f(ab) = f(a)f(b)$$

for all elements a and b in G.

Here, the product ab which appears on the left side of the above equation corresponds to the binary operation on G, whereas the product $f(a)f(b)$ on the right side corresponds to the binary operation on G'. As we have agreed to skip the symbol denoting the binary operation in a group and to write simple ab for $a * b$, when * is the binary operation of G and a and b are elements of G, the equation $f(ab) = f(a)f(b)$ makes sense since a and b are elements of G and $f(a)$ and $f(b)$ are those in G'. Strictly speaking, we should have written

$$f(a * b) = f(a) \circ f(b) \quad \text{for all } a, b \in G,$$

where * and o are the binary operations in the groups G and G', respectively. Examples given below make these things clear.

Example 5.1.1

1. Consider the groups $(\mathbb{Z}, +)$ and $(\mathbb{R}^+, \cdot)$ where $+$ and '·' are the usual addition and multiplication of real numbers. Let m be any positive integer and define $f: \mathbb{Z} \to \mathbb{R}^+$ by

$$f(a) = m^a \quad \text{for all } a \in \mathbb{Z}.$$

   Then, for any a and $b \in \mathbb{Z}$, we have

$$f(a + b) = m^{a+b} = m^a \cdot m^b = f(a) \cdot f(b)$$

   and hence f is a homomorphism of $(\mathbb{Z}, +)$ into $(\mathbb{R}^+, \cdot)$.

2. Consider the groups $(\mathbb{R}^+, \cdot)$ and $(\mathbb{R}, +)$ and let m be any positive integer greater than 1. Define

$$f: \mathbb{R}^+ \to \mathbb{R} \text{ by } f(a) = \log_m a \quad \text{for all } a \in \mathbb{R}^+.$$

   Then, for any a and $b \in \mathbb{R}^+$,

$$f(a \cdot b) = \log_m (ab) = \log_m a + \log_m b = f(a) + f(b).$$

   Therefore, f is a homomorphism of $(\mathbb{R}^+, \cdot)$ into $(\mathbb{R}, +)$.

3. For any group G, define $f: G \to G$ by

$$f(x) = x \quad \text{for all } x \in G.$$

   Then, clearly f is a homomorphism of G into itself and is called the *identity homomorphism*.

4. Let G and G' be groups in which e and e' are the identities, respectively. Define

$$f: G \to G' \text{ by } f(x) = e' \quad \text{for all } x \in G.$$

   Then, for any a and b in G, we have

$$f(ab) = e' = e' \cdot e' = f(a) f(b)$$

   and hence f is a homomorphism of G into G' and is called the *trivial homomorphism*.

5. Let m be an arbitrary integer and define $f: \mathbb{Z} \to \mathbb{Z}$ by

$$f(a) = ma \quad \text{for all } a \in \mathbb{Z}.$$

Then, $f(a + b) = m(a + b) = ma + mb = f(a) + f(b)$ for all a and $b \in \mathbb{Z}$. Therefore, f is a homomorphism of $(\mathbb{Z}, +)$ into itself.

6. Consider the groups $(\mathbb{Z}, +)$ and $(\mathbb{Z}_2, +_2)$, where $+_2$ is the addition modulo 2. Define

$$f: \mathbb{Z} \to \mathbb{Z}_2 \text{ by } f(a) = \begin{cases} 0, & \text{if } a \text{ is even} \\ 1, & \text{if } a \text{ is odd} \end{cases}$$

for any $a \in \mathbb{Z}$. Then, since $a + b$ is even if and only if both a and b are even or both a and b are odd, we get that

$$f(a + b) = f(a) +_2 f(b) \quad \text{for all } a \text{ and } b \in G.$$

Therefore, f is a homomorphism of $(\mathbb{Z}, +)$ into $(\mathbb{Z}_2, +_2)$.

7. Consider the group $(\mathbb{R} - \{0\}, \cdot)$ of all nonzero real numbers under the usual multiplication of real numbers and the group $\text{NSM}_2(\mathbb{R})$ of all 2×2 nonsingular matrices over $\mathbb{R}$ under the usual multiplication of matrices. Define

$$f: \text{NSM}_2(\mathbb{R}) \to \mathbb{R} - \{0\} \text{ by } f\left(\begin{pmatrix} a & b \\ c & d \end{pmatrix}\right) = ad - bc$$

for any $\begin{pmatrix} a & b \\ c & d \end{pmatrix} \in \text{NSM}_2(\mathbb{R})$.

Then, for any $A = \begin{pmatrix} a & b \\ c & d \end{pmatrix}$ and $B = \begin{pmatrix} r & s \\ t & u \end{pmatrix}$ in $\text{NSM}_2(\mathbb{R})$, we have

$$\begin{aligned} f(AB) &= f\begin{pmatrix} ar+bt & as+bu \\ cr+dt & cs+du \end{pmatrix} \\ &= (ar+bt)\,(cs+du) - (as+bu)\,(cr+dt) \\ &= arcs + ardu + btcs + btdu - ascr - asdt - bucr - budt \\ &= ardu + btcs - asdt - bucr \\ &= (ad - bc)\,(ru - st) \\ &= f(A)f(B). \end{aligned}$$

Therefore, f is a homomorphism of $(\text{NSM}_2(\mathbb{R}), \cdot)$ into $(\mathbb{R} - \{0\}, \cdot)$.

8. Let N be a normal subgroup of a group G and G/N, the quotient group of G by N. Define

$$f: G \to G/N \text{ by } f(a) = aN \quad \text{for all } a \in G.$$

Then, for any a and b in G,

$$f(ab) = (ab)N = aN \cdot bN = f(a) \cdot f(b)$$

and hence f is a homomorphism of G into G/N and is called the *natural or cannonical homomorphism*.

In the following, we derive certain important elementary properties of homomorphisms.

Theorem 5.1.1. Let $f : G \to G'$ be a homomorphism of groups. Then, the following holds:

1. $f(e) = e'$, where e and e' are the identities in G and G', respectively.
2. $f(a^{-1}) = f(a)^{-1}$ for all $a \in G$.

Proof:

1. Let $f(e) = x$. Then, x is an element in G' and hence

$$xe' = x = f(e) = f(ee) = f(e)f(e) = xx.$$

Therefore, $xx = xe'$ and, by the left cancellation law, $x = e'$. Thus, $f(e) = e'$.

2. For any $a \in G$, we have

$$f(a)f(a^{-1}) = f(aa^{-1}) = f(e) = e' = f(e) = f(a^{-1}a) = f(a^{-1})f(a)$$

and hence $f(a^{-1})$ is the inverse of $f(a)$ in G'; that is,

$$f(a^{-1}) = f(a)^{-1}.$$

Theorem 5.1.2. Let $f: G \to G'$ be a homomorphism of groups.

1. For any subgroup A of G, the image $f(A)$ is a subgroup of G'.
2. For any subgroup A' of G', the inverse image $f^{-1}(A')$ is a subgroup of G.

Proof:

1. Let A be a subgroup of G. Then,

$$f(A) = \{f(a) : a \in G\} \subseteq G'.$$

First of all, since A is a subgroup of G, we have $e \in A$ and hence

$$e' = f(e) \in f(A)$$

so that $f(A)$ is a nonempty subgroup of G'. Also,
x and $y \in f(A) \Rightarrow x = f(a)$ and $y = f(b)$ with a and $b \in A$.

$$\Rightarrow xy^{-1} = f(a)f(b)^{-1} = f(a)f(b^{-1}) = f(ab^{-1})$$
$$\in f(A) \text{ (since } A \text{ is a subgroup).}$$

Therefore, $f(A)$ is a subgroup of G'.

2. Let A' be a subgroup of G'. Then,

$$f^{-1}(A') = \{a \in G : f(a) \in A'\}.$$

Since $f(e) = e' \in A'$, we get that $e \in f^{-1}(A')$ and hence $f^{-1}(A')$ is a nonempty subset of G. Now,

$$\begin{aligned} a \text{ and } b \in f^{-1}(A') &\Rightarrow f(a) \text{ and } f(b) \in A' \\ &\Rightarrow f(ab^{-1}) = f(a)f(b^{-1}) = f(a)f(b)^{-1} \in A' \\ &\qquad \text{(since } A' \text{ is a subgroup of } G') \\ &\Rightarrow ab^{-1} \in f^{-1}(A'). \end{aligned}$$

Thus, $f^{-1}(A')$ is a subgroup of G.

Theorem 5.1.3. Let $f: G \to G'$ be a homomorphism of groups.

1. If f is a surjection and A is a normal subgroup of G, then $f(A)$ is a normal subgroup of G'.
2. If A' is a normal subgroup of G', then $f^{-1}(A')$ is a normal subgroup of G.

Proof:

1. Let f be a surjection and A be a normal subgroup of G. Then, we have already proved that $f(A)$ is a subgroup of G'. Now,

$$\begin{aligned} \in f(A) \text{ and } y \in G' &\Rightarrow x = f(a) \text{ and } y = f(b) \text{ for some } a \in A \\ &\qquad \text{and } b \in G \\ &\Rightarrow y\,x\,y^{-1} = f(b)f(a)f(b)^{-1} = f(bab^{-1}) \in f(A). \end{aligned}$$

Therefore, $yf(A)y^{-1} \subseteq f(A)$ for all $y \in G'$ and hence $f(A)$ is a normal subgroup of G'.

2. Let A' be a normal subgroup of G'. By the above theorem, $f^{-1}(A')$ is a subgroup of G. Now,

$$\begin{aligned} a \in f^{-1}(A') \text{ and } x \in G &\Rightarrow f(a) \in A' \text{ and } f(x) \in G' \\ &\Rightarrow f(xax^{-1}) = f(x)f(a)f(x)^{-1} \in A' \\ &\qquad \text{(since } A' \text{ is normal in } G') \\ &\Rightarrow xax^{-1} \in f^{-1}(A'). \end{aligned}$$

Therefore, $f^{-1}(A')$ is a normal subgroup of G.

Remark 5.1.1. Note that in Theorem 5.1.3 (1), it is necessary that f is a surjection; for, consider a group G' and a subgroup G of G' such that G is not normal in G'. Let $f: G \to G'$ be the inclusion map defined $f(x) = x$ for all

$x \in G$. Then, f is a homomorphism of G into G', G is a normal subgroup of G', but $f(G)$ $(= G)$ is not normal in G'.

For any homomorphism $f: G \to G'$, the subgroup $f^{-1}(\{e'\})$ of G is of the special importance and is called the *kernel* of f. This is formally defined in the following definition.

Definition 5.1.2. Let $f: G \to G'$ be a homomorphism of groups. Then, the *kernel* of f is defined to be the set

$$\ker f = \{a \in G : f(a) = e'\},$$

where e' is the identity in the group G'.

Theorem 5.1.4. For any homomorphism $f: G \to G'$ of groups, the kernel of f is a normal subgroup of G.

Proof: By the definition of the kernel of f, we have

$$\ker f = f^{-1}(\{e'\}),$$

where e' is the identity in G'. Since $\{e'\}$ is a normal subgroup of G', it follows from Theorems 5.1.2 (2) and 5.1.3 (2) that $\ker f$ is a normal subgroup of G.

The converse of the above theorem is also true, in the sense that, any normal subgroup of a group G is the kernel of some homomorphism of G into some group G'. This is proved in the following theorem.

Theorem 5.1.5. Let N be a normal subgroup of a group G and G/N, the quotient group of G by N. Define

$$f: G \to G/N \text{ by } f(a) = aN \quad \text{for all } a \in G.$$

Then, f is a homomorphism, whose kernel is N.

Proof: For any a and b in G,

$$\begin{aligned} f(ab) &= abN \\ &= aN \cdot bN \text{ (since } N \text{ is normal in } G) \\ &= f(a)f(b) \end{aligned}$$

and hence f is a homomorphism (see Example 5.1.1 (8) also) and

$$\begin{aligned}\ker f &= \{a \in G : f(a) = \text{the identity in } G/N\} \\ &= \{a \in G : aN = N = eN\} \\ &= \{a \in G : a \in N\} = N.\end{aligned}$$

Let us compute the kernels of homomorphisms given in Example 5.1.1.

Example 5.1.2

1. $f : \mathbb{Z} \to \mathbb{R}^+$ is defined by $f(a) = m^a$ for all $a \in \mathbb{Z}$, where m is a given positive integer. Now,

$$\begin{aligned}\ker f &= \{a \in \mathbb{Z} : f(a) = \text{the identity in } \mathbb{R}^+\} \\ &= \{a \in \mathbb{Z} : m^a = 1\} \\ &= \begin{cases} \mathbb{Z}, & \text{if } m = 1 \\ \{0\}, & \text{if } m > 1 \end{cases}.\end{aligned}$$

2. $f : \mathbb{R}^+ \to \mathbb{R}$ is defined by $f(a) = \log_m a$.

$$\begin{aligned}\ker f &= \{a \in \mathbb{R}^+ : f(a) = \text{the identity in } (\mathbb{R}, +)\} \\ &= \{a \in \mathbb{R}^+ : \log_m a = 0\} \\ &= \{1\}.\end{aligned}$$

3. For any group G, define $f(x) = x$ for all $x \in G$. Then,

$$\ker f = \{e\}.$$

4. $f : G \to G'$ is defined by $f(x) = e'$ for all $x \in G$.
   Therefore, $\ker f = G$.
5. $f : \mathbb{Z} \to \mathbb{Z}$ is defined by $f(a) = ma$ for all $a \in \mathbb{Z}$, where m is a given integer. Then,

$$\begin{aligned}\ker f &= \{a \in \mathbb{Z} : f(a) = 0, \text{ the identity in } (\mathbb{Z}, +)\} \\ &= \{a \in \mathbb{Z} \mid ma = 0\} \\ &= \begin{cases} \mathbb{Z}, & \text{if } m = 0 \\ \{0\}, & \text{if } m \neq 0 \end{cases}.\end{aligned}$$

6. $f : \mathbb{Z} \to \mathbb{Z}_2$ is defined by $f(a) = \begin{cases} 0, & \text{if } a \text{ is even} \\ 1, & \text{if } a \text{ is odd} \end{cases}$

$$\begin{aligned}\ker f &= \{a \in \mathbb{Z} : f(a) = 0, \text{ the identity in } (\mathbb{Z}_2, +_2)\} \\ &= 2\mathbb{Z}, \text{ the set of all even integers.}\end{aligned}$$

7. $f: NSM_2(\mathbb{R}) \to \mathbb{R} - \{0\}$ is defined by

$$f\begin{pmatrix} a & b \\ c & d \end{pmatrix} = ad - bc.$$

$$\ker f = \left\{\begin{pmatrix} a & b \\ c & d \end{pmatrix} \in NSM_2(\mathbb{R}) : ad - bc = 1\right\}.$$

8. The kernel of the natural map $f: G \to G/N$ defined by $f(a) = aN$, is precisely N.

In the following, we define various types of homomorphisms.

Definition 5.1.3. Let $f: G \to G'$ be a homomorphism of groups. Then, f is called

1. a *monomorphism* if f is an injection.
2. an *epimorphism* if f is a surjection.
3. an *isomorphism* if f is a bijection.
4. an *endomorphism* if $G = G'$.
5. an *automorphism* if f is a bijective endomorphism.

Example 5.1.3

1. If $m > 1$, then the map $f: \mathbb{Z} \to \mathbb{R}^+$, defined by $f(a) = m^a$, is a monomorphism of $(\mathbb{Z}, +)$ into $(\mathbb{R}^+, \cdot)$.
2. The homomorphism given in (2), (3) and (5) (if $m \neq 0$) of Example 5.1.2 are monomorphisms, while those given in (2), (3), (6) and (8) are epimorphisms.

Note that a homomorphism $f: G \to G'$ is an epimorphism if and only if $f(G) = G'$. In the following, we give a characterization of monomorphisms in terms of their kernels.

Theorem 5.1.6. Let $f: G \to G'$ be a homomorphism of groups. Then, f is a monomorphism if and only if the kernel of f is trivial, that is, $\ker f = \{e\}$.

Proof: Suppose that f is a monomorphism. Then, f is an injection and therefore

$$a \in \ker f \Rightarrow f(a) = e' = f(e) \Rightarrow a = e$$

and hence $\ker f$ contains e alone. Thus, $\ker f = \{e\}$. Conversely, suppose that f is not a monomorphism, then f is not an injection and hence there exists $a \neq b \in G$ such that $f(a) = f(b)$. Now, $ab^{-1} \neq e$ and

$$f(ab^{-1}) = f(a)f(b^{-1}) = f(a)f(b)^{-1} = f(b)f(b)^{-1} = e$$

and hence $e \neq ab^{-1} \in \ker f$, so that $\ker f \neq \{e\}$.
Recall that a mapping $f : A \to B$ is a bijection if and only if there exists a unique mapping $g : B \to A$ such that

$$f \circ g = I_B \quad \text{and} \quad g \circ f = I_A;$$

that is, $f(g(b)) = b$ for all $b \in B$ and $g(f(a)) = a$ for all $a \in A$.
This unique g is called the inverse of f and is denoted by f^{-1}. Also, in this case, for any $a \in A$ and $b \in B$, we have

$$f(a) = b \Leftrightarrow a = f^{-1}(b).$$

In the following, we prove that the inverse of a bijective homomorphism is again a homomorphism.

Theorem 5.1.7. Let $f : G \to G'$ be an isomorphism of groups. Then, $f^{-1} : G' \to G$ is also an isomorphism.

Proof: We are given that $f : G \to G'$ is a bijective homomorphism. Then, clearly $f^{-1} : G' \to G$ is a bijection. Now, for any x and $y \in G'$,

$$f(f^{-1}(xy)) = xy = f(f^{-1}(x)) \cdot f(f^{-1}(y)) = f(f^{-1}(x)f^{-1}(y))$$

and since f is a bijection, we have

$$f^{-1}(xy) = f^{-1}(x)f^{-1}(y) \quad \text{for all } x \text{ and } y \in G'.$$

Thus, f^{-1} is a homomorphism and a bijection and hence an isomorphism.

Definition 5.1.4. Two groups G and G' are said to be isomorphic if there is an isomorphism $f : G \to G'$ and, in this case, we write $G \cong G'$.

Theorem 5.1.8. The following holds for any groups G, G' and G'':

1. $G \cong G$
2. $G \cong G' \Rightarrow G' \cong G$
3. $G \cong G'$ and $G' \cong G'' \Rightarrow G \cong G''$

Proof:

1. follows from the fact that the identity map I_G on G is an isomorphism of G onto itself.
2. is consequence of Theorem 5.1.7 and
3. follows from the fact that the composition of two isomorphisms is again an isomorphism.

Example 5.1.4. Consider the groups $(\mathbb{Z}_4, +_4)$ and $(G, \cdot)$, where $G = \{1, -1, i, -i\}$ and '·' is the multiplication of complex numbers. Define $f: \mathbb{Z}_4 \to G$ by

$$f(0) = 1, f(1) = i, f(2) = i^2 = -1 \quad \text{and} \quad f(3) = i^3 = -i.$$

Then, f is an isomorphism and hence $\mathbb{Z}_4 \cong G$.

Worked Exercise 5.1.1. If $f: G \to G'$ and $g : G' \to G''$ are homomorphisms of groups, prove that $g \circ f: G \to G''$ is a homomorphism.

Answer: For any a and $b \in G$, we have

$$\begin{aligned}(g \circ f)(ab) &= g(f(ab)) \\ &= g(f(a)f(b)) \\ &= g(f(a))g(f(b)) \\ &= (g \circ f)(a)\cdot(g \circ f)(b)\end{aligned}$$

Therefore, $g \circ f$ is a homomorphism.

Worked Exercise 5.1.2. Let $f: G \to G'$ be a homomorphism of groups and a be an element of finite order in G. Prove that the order of $f(a)$ is finite in G' and that $O(f(a))$ divides $O(a)$.

Answer: Let $a \in G$ be of finite order and $O(a) = n$. Then, $a^n = e$ and $(f(a))^n = f(a^n) = f(e) = e'$.
Therefore, $f(a)$ is of finite order and, by Theorem 4.2.4 (3), $O(f(a))$ divides $O(a)$.

Worked Exercise 5.1.3. Let $f: G \to G'$ be a homomorphism of groups and $K = \ker f$. Describe the cosets of K in terms of f.

Answer: We know that $K = \ker f = \{a \in G : f(a) = e'\}$ and that K is a normal subgroup of G and hence every left coset of K is a right coset of K in G. For any $a \in G$, consider

$$\begin{aligned} aK &= \{ax : x \in K\} \\ &= \{ax : f(x) = e'\} \\ &= \{y \in G : f(y) = f(a)\}, \end{aligned}$$

since $f(ax) = f(a)f(x) = f(a)e' = f(a)$ and, if $y \in G$ is such that $f(y) = f(a)$, then $y = ax$, where $x = a^{-1}y \in \ker f = K$. Thus, $aK = f^{-1}\{f(a)\}$ for any $a \in G$.

Worked Exercise 5.1.4. Let n be a positive integer and $\mathbb{Z}_n = \{0, 1, 2, \ldots, n-1\}$. Define $f : \mathbb{Z} \to \mathbb{Z}_n$ by $f(a) = r$, where r is the remainder obtained by dividing a with n. Then, prove that f is an epimorphism of $(\mathbb{Z}, +)$ onto $(\mathbb{Z}_n, +_n)$.

Answer: Note that, for any $a \in \mathbb{Z}$, $f(a) = r$, where

$$a = qn + r,\ q \text{ and } r \in \mathbb{Z} \text{ and } 0 \leq r < n$$

and hence $f(a) \in \mathbb{Z}_n$. Therefore, $f : \mathbb{Z} \to \mathbb{Z}_n$ is a mapping. For any a and $b \in \mathbb{Z}$, let

$$f(a) = r \quad \text{and} \quad f(b) = s.$$

Then, $a = qn + r$ and $b = q'n + s$, where q and q' are integers.
Now,

$$\begin{aligned} a+b &= (q+q')n+(r+s) \\ &= \begin{cases} (q+q')n+(r+s), & \text{if } r+s<n \\ (q+q'+1)n+(r+s-n), & \text{if } r+s \geq n \end{cases} \\ &= tn + (r +_n s) \quad \text{and} \quad 0 \leq r +_n s < n \end{aligned}$$

and hence $f(a + b) = r +_n s = f(a) +_n f(b)$.
Thus, f is a homomorphism. Also, for any $0 \leq a < n$, $f(a) = a$ and hence f is a surjection also. Thus, f is an epimorphism of $(\mathbb{Z}, +)$ onto $(\mathbb{Z}_n, +_n)$.

Worked Exercise 5.1.5. Let G be an abelian group of order m and let n be any positive integer relatively prime to m. Define $f : G \to G$ by $f(a) = a^n$ for all $a \in G$. Then, prove that f is an automorphism of G.

Answer: Since G is an abelian group, we have

$$f(ab) = (ab)^n = a^n b^n = f(a)f(b)$$

for all a and $b \in G$ and hence f is a homomorphism of G into G. For any $a \in G$, $O(a)$ divides $|G| = m$ and hence

$$a^m = e \quad \text{for all } a \in G.$$

Since n is relatively prime to m, there exist integers r and s such that $rm + sn = 1$. Now, for any a and $b \in G$,

$$\begin{aligned} f(a) = f(b) &\Rightarrow a^n = b^n \quad \text{and} \quad a^m = e = b^m \\ &\Rightarrow a = a^{rm+sn} = (a^m)^r(a^n)^s \\ &= (b^m)^r(b^n)^s = b^{rm+sn} = b \end{aligned}$$

and hence f is an injection. Therefore,

$$m = |G| = |f(G)| \leq |G| = m$$

and hence $|f(G)| = |G|$ so that $f(G) = G$ (since G is finite and $f(G) \subseteq G$). Therefore, f is a surjection also. Thus, f is an isomorphism of G onto G; that is, f is an automorphism.

EXERCISE 5(A)

1. Determine which of the following are homomorphisms between the given groups:
   - (i) Consider the group $(\mathbb{Z}, +)$ and define $f: \mathbb{Z} \to \mathbb{Z}$ by $f(a) = 2^a$ for all $a \in \mathbb{Z}$.
   - (ii) Consider the groups $(\mathbb{Z}_6, +_6)$ and $(\mathbb{Z}_8, +_8)$ and define $f: \mathbb{Z}_6 \to \mathbb{Z}_8$ by $f(a) = a$ for all $a \in \mathbb{Z}_6$.
   - (iii) Consider the groups $(\mathbb{Z}_6, +_6)$ and $(\mathbb{Z}, +)$ and define $f: \mathbb{Z}_6 \to \mathbb{Z}$ by $f(a) = a$ for all $a \in \mathbb{Z}_6$.
   - (iv) Define $f: \mathbb{Z}_8 \to \mathbb{Z}_2$ by $f(a) = \begin{cases} 0, & \text{if } a \text{ is even} \\ 1, & \text{if } a \text{ is odd} \end{cases}$.
   - (v) Define $f: \mathbb{Z}_{15} \to \mathbb{Z}_2$ as in (iv) above.
   - (vi) Consider the group $(\mathbb{R}, +)$ and define $f: \mathbb{R} \to \mathbb{R}$ by $f(x) = \cos x$ for all $x \in \mathbb{R}$.
   - (vii) Let Y be a nonempty subset of a set X and consider the group $P(X)$ and $P(Y)$ under the symmetric difference of sets. Define $f: P(X) \to P(Y)$ by $f(A) = A \cap Y$ for any $A \in P(X)$.
   - (viii) Let X and Y be as in (vii) above. Define $f: P(X) \to P(Y)$ by $f(A) = Y - A$ for any $A \in P(X)$.
   - (ix) Let G be the group of all real valued continuous functions of the interval $[0, 1]$ under point-wise addition. Define $f: G \to \mathbb{R}$ by $f(\alpha) = \int_0^1 \alpha(x)\,dx$ for all $\alpha \in G$.
   - (x) Let X be any nonempty set and $\mathbb{R}^X$ be the set of all mappings of X into $\mathbb{R}$. Consider the group $(\mathbb{R}^X, +)$ where $+$ is the point-wise addition. For any $x_0 \in X$, define $f: \mathbb{R}^X \to \mathbb{R}$ by $f(\alpha) = \alpha(x_0)$ for all $\alpha \in \mathbb{R}^X$.

2. Let G and G' be groups and $f: G \to G'$ be a mapping. Prove that f is a homomorphism if and only if

$$f(ab^{-1}) = f(a)f(b)^{-1} \quad \text{for all } a \quad \text{and} \quad b \in G.$$

3. Let $\mathbb{C} - \{0\}$ and $\mathbb{R} - \{0\}$ be the groups of nonzero complex numbers and nonzero real numbers, respectively, under the usual multiplications. Prove that the map

$f: \mathbb{C} - \{0\} \to \mathbb{R} - \{0\}$ defined by $f(z) = |z|$ is a homomorphism.

4. Let G be a group and define $f: G \to G$ by $f(a) = a^{-1}$ for any $a \in G$. Prove that f is an endomorphism if and only if G is an abelian group.

5. Determine the Kernels of the homomorphisms (if they are) given in Exercise 1 above.

6. Determine all the homomorphisms of $(\mathbb{Z}, +)$ into $(\mathbb{Z}_2, +_2)$.

7. Determine all the endomorphisms of the group $(\mathbb{Z}, +)$ into itself.

8. Prove that every nontrivial endomorphism of $(\mathbb{Z}, +)$ into itself is a monomorphism.

9. Prove that there is no epimorphism of $(\mathbb{Z}, +)$ onto itself, except the identity map.

10. Consider the groups $(\mathbb{Z}, +)$ and $(\mathbb{R}, +)$ and, for any real number a, define $f_a : \mathbb{Z} \to \mathbb{R}$ by $f_a(x) = ax$ for all $x \in \mathbb{Z}$. Prove that a mapping $f: \mathbb{Z} \to \mathbb{R}$ is a homomorphism if and only if $f = f_a$ for some $a \in \mathbb{R}$.

11. Let G and G' be finite groups and $f: G \to G'$ be a homomorphism. Prove that the index of the Kernel of f in G is a divisor of $|f(G)|$.

12. Let $f: G \to G'$ be a homomorphism of groups and $a \in G$. If $O(a)$ is finite, then prove that $O(f(a))$ is also finite and is a divisor of $O(a)$. Give an example where $O(f(a))$ is a proper divisor of $O(a)$.

13. Let f and $g : G \to G'$ be homomorphisms of groups and $A = \{a \in G : f(a) = g(a)\}$. Then, prove that A is a subgroup of G.

14. Let $f: G \to G'$ be a homomorphism and G be a finite group of prime order. Then, prove that f is either trivial or a monomorphism.

15. Let $f: G \to G'$ be a homomorphism of groups and $[G, G]$ be the commutator subgroup of G (see 4, 6, 12). Then, prove that $f(G)$ is an abelian group if and only if $[G, G] \subseteq \ker f$.

16. Let G be a group and a and $b \in G$. Consider the group $\mathbb{Z} \times \mathbb{Z}$ under coordinatewise addition and define $f: \mathbb{Z} \times \mathbb{Z} \to G$ by

$$f(m, n) = a^m b^n \quad \text{for any } (m, n) \in \mathbb{Z} \times \mathbb{Z}.$$

Obtain a necessary and sufficient condition, in terms of a and b, for f to be a homomorphism.

17. Let a be an element in a group G and $n \in \mathbb{Z}^+$. Define

$$f: \mathbb{Z}_n \to G \text{ by } f(i) = a^i \quad \text{for } 0 \le i < n.$$

Obtain a necessary and sufficient condition in terms of a and n for f to be a homomorphism.

18. For any element a in a group G, define $f: \mathbb{Z} \to G$ by $f(n) = a^n$ for all $n \in \mathbb{Z}$. Prove that f is a homomorphism and determine the kernel of f.

19. Let $n \in \mathbb{Z}^+$ and G be the group of all n^{th} roots of unity under the usual multiplication of complex numbers. Prove that $G \cong (\mathbb{Z}_n, +_n)$.

20. Let Q_8 be the quaternion group with eight elements. Prove that there is a unique homomorphism $f: G \to \mathbb{Z}_2$ such that $f(i) = 0$ and $f(j) = 1$.

21. Let G and G' be finite groups of the same order and $f: G \to G'$ be a homomorphism. Then, prove that the following are equivalent to each other:
    (i) f is a monomorphism.
    (ii) f is an epimorphism.
    (iii) f is an isomorphism.

22. Let $\mathbb{R} - \{0\}$ be the group of nonzero real numbers under multiplication and $G = \{1, -1\}$. Define

$$f: \mathbb{R} \to G \text{ by } f(a) = \begin{cases} 1, & \text{if } a > 0 \\ -1, & \text{if } a < 0 \end{cases}.$$

Then, prove that f is an epimorphism.

23. Prove that there is no epimorphism of $(\mathbb{Z}, +)$ onto $(\mathbb{R}, +)$.

24. Exhibit a monomorphism of $\mathbb{Z}_8$ into $\mathbb{Z}_{24}$.

25. For any positive integers m and n, obtain a necessary and sufficient condition for having a monomorphism of $\mathbb{Z}_m$ into $\mathbb{Z}_n$.

26. List all the isomorphisms of $(\mathbb{Z}, +)$ onto $(\mathbb{Z}, +)$.

27. How many homomorphisms are there from $\mathbb{Z}$, into $\mathbb{Z}_{15}$?

28. Let G and H be any groups and H be an abelain group. Let Hom(G, H) be the set of all homomorphisms of G into H. Prove that Hom(G, H) is a group under the point-wise operation.

29. Prove that the composition of any two monomorphisms (epimorphisms) is again a monomorphism (epimorphism).

30. Let $f: G \to G'$ and $g: G' \to G''$ be homomorphisms, such that $g \circ f$ is a monomorphism. Then, prove that f is a monomorphism. If $g \circ f$ is an epimorphism, then prove that g is an epimorphism.

31. Let G be a group and $f: G \to G$ be defined by $f(x) = x^3$ for all $x \in G$. If f is monomorphism, then prove that G is abelian.

5.2 FUNDAMENTAL THEOREM OF HOMOMORPHISMS

For any homomorphism f of a group G into a group G', we know that $f(G)$ is a subgroup of G' and hence $f(G)$ is a group by itself. $f(G)$ is called a homomorphic image of G. For example, if N is any normal subgroup of a group G and G/N is the corresponding quotient group, then G/N is a homomorphic image of G, since we have the natural map $f: G \to G/N$ which is a homomorphism and $f(G) = G/N$. In the following, we prove a converse of the above; that is, any homomorphic image of a group G is isomorphic to a quotient group of G.

Theorem 5.2.1 (Fundamental Theorem of Homomorphisms). Let $f: G \to G'$ be a homomorphism of groups. Then,

$$G/\ker f \cong f(G)$$

and, in particular, if f is an epimorphism, then

$$G/\ker f \cong G'.$$

Proof: For simplicity, let $K = \ker f = \{a \in G : f(a) = e'\}$. We know that K is a normal subgroup of G and hence we have the quotient group whose elements are the cosets of K in G. Also, we know that $f(G)$ is a subgroup of G' and hence $f(G)$ is a group on its own.

$$\text{Define } g : G/K \to f(G) \quad \text{by} \quad g(aK) = f(a)$$

for any $aK \in G/K$. There is an apparent ambiguity in the definition of g. This looks like depending on a. Actually, this does not depend upon the representative a of the coset aK. For, consider

$$\begin{aligned} aK = bK &\Rightarrow a^{-1}b \in K = \ker f \\ &\Rightarrow f(a)^{-1}f(b) = f(a^{-1})f(b) = f(a^{-1}b) = e' \\ &\Rightarrow f(a) = f(b) \end{aligned}$$

This clears the ambiguity and it follows that g is well defined. For any aK and $bK \in G/K$, we have

$$\begin{aligned} g(aK \cdot bK) &= g(abK) \\ &= f(ab) \\ &= f(a)f(b) \\ &= g(aK) \cdot g(bK) \end{aligned}$$

and therefore g is a homomorphism. Since any element of $f(G)$ is of the form $f(a) = g(aK)$ for some $a \in G$ and $aK \in G/K$, g is a surjection. Also, for any aK and $bK \in G/K$,

$$\begin{aligned} g(aK) = g(bK) &\Rightarrow f(a) = f(b) \\ &\Rightarrow f(a^{-1}b) = f(a)^{-1}f(b) = e' \\ &\Rightarrow a^{-1}b \in \ker f = K \\ &\Rightarrow aK = bK. \end{aligned}$$

Therefore, g is an injection also. Thus, $g : G/K \to f(G)$ is an isomorphism and hence $G/K \cong f(G)$. If f is an epimorphism, then $f(G) = G'$ and hence $G/K \cong G'$.

For any sets A, B and C, if $\alpha : A \to B$ is a bijection and B is a subset of C, then α can be considered as an injection of A into C. On the other hand, if $\beta : A \to C$ is an injection, then β can be considered as a bijection of A onto $\beta(A)$. Now, the following is an important consequence of the Fundamental Theorem of Homomorphisms.

Theorem 5.2.2 (Factorization Theorem). Any homomorphism of groups can be expressed as a composition of an epimorphism and a monomorphism.

Proof: Let $f: G \to G'$ be a homomorphism of groups and $K = \ker f$. Then, by the Fundamental Theorem of Homomorphisms (Theorem 5.2.1), there is an isomorphism $g : G/K \to f(G)$ such that $g(aK) = f(a)$ for all $a \in G$. Now, since $f(G)$ is a subgroup of G', g can be considered as a monomorphism of G/K into G'. Also, let $h : G \to G/K$ be the natural homomorphism defined by $h(a) = aK$ for all $a \in G$. Then, clearly h is an epimorphism and we have,

$$G \xrightarrow{h} G/K \xrightarrow{g} G'$$

for any $a \in G$, $(g \circ h)(a) = g(h(a)) = g(aK) = f(a)$ are hence $f = g \circ h$. Therefore, f is the composition of the epimorphism h and the monomorphism g.

Example 5.2.1. Let n be a positive integer and $(\mathbb{Z}_n, +_n)$ be the group of integers modulo n. As in Worked Exercise 5.1.4, define $f : \mathbb{Z} \to \mathbb{Z}_n$ by $f(a) = r$, where r is the remainder obtained by dividing a with n; that is, if q and r are integers such that $a = qn + r$ and $0 \le r < n$, $f(a) = r$.

In Worked Exercise 5.1.4, we have proved that f is an epimorphism. By the Fundamental Theorem of Homomorphisms, $\mathbb{Z}/\ker f \cong \mathbb{Z}_n$, we have

$$\begin{aligned}\ker f &= \{a \in \mathbb{Z} : f(a) = 0, \text{ the identity in } \mathbb{Z}_n\} \\ &= \{a \in \mathbb{Z} : a = nq \text{ for some } q \in \mathbb{Z}\} \\ &= n\mathbb{Z} \\ &= <n>, \text{ the subgroup of } \mathbb{Z} \text{ generated by } n.\end{aligned}$$

Theorem 5.2.3. Let G, G_1 and G_2 be groups. Then, prove that $G \cong G_1 \times G_2$ if and only if there exist normal subgroups N_1 and N_2 of G such that $N_1N_2 = G$, $N_1 \cap N_2 = \{e\}$, $G_1 \cong G/N_1$ and $G_2 \cong G/N_2$.

Proof: Let e, e_1 and e_2 be identities in G, G_1 and G_2, respectively. Suppose that $G \cong G_1 \times G_2$ and let $f : G \to G_1 \times G_2$ be an isomorphism. Put $A_1 = \{e_1\} \times G_2$ and $A_2 = G_1 \times \{e_2\}$. Then, A_1 and A_2 are normal subgroups of $G_1 \times G_2$. Now, put

$$N_1 = f^{-1}(A_1) \quad \text{and} \quad N_2 = f^{-1}(A_2).$$

Then, N_1 and N_2 are normal subgroups of G. For any $x \in G$, we have $f(x) = (a_1, a_2) \in G_1 \times G_2$.Choose x_1 and x_2 in G such that

$$f(x_1) = (e_1, a_2) \quad \text{and} \quad f(x_2) = (a_1, e_2).$$

Then, $x_1 \in f^{-1}(A_1)$, $x_2 \in f^{-1}(A_2)$ and

$$f(x) = (e_1, a_2)\,(a_1, e_2) = f(x_1)\,f(x_2) = f(x_1\,x_2)$$

and hence $x = x_1x_2 \in N_1N_2$. Therefore, $N_1N_2 = G$, Also,

$$\begin{aligned}x \in N_1 \cap N_2 &\Rightarrow x \in f^{-1}(A_1) \quad \text{and} \quad x \in f^{-1}(A_2) \\ &\Rightarrow f(x) \in A_1 = \{e_1\} \times G_2 \quad \text{and} \quad f(x) \in A_2 = G_1 \times \{e_2\} \\ &\Rightarrow f(x) \in (\{e_1\} \times G_2) \cap (G_1 \times \{e_2\}) = \{e_1\} \times \{e_2\} \\ &\Rightarrow f(x) = (e_1, e_2) = f(e) \text{ (since } f \text{ is a homomorphism)} \\ &\Rightarrow x = e.\end{aligned}$$

Therefore, $N_1 \cap N_2 = \{e\}$. Next define $f_1 : G \to G_1$ and $f_2 : G \to G_2$ by

$$f_1(x) = a_1 \quad \text{and} \quad f_2(x) = a_2 \quad \text{if } f(x) = (a_1, a_2).$$

Then, f_1 and f_2 are epimorphisms and $\ker f_1 = N_1$ and $\ker f_2 = N_2$; For

$$\begin{aligned}& x \in N_1 \Leftrightarrow f(x) \in A_1 \Leftrightarrow f(x) = (e_1, a_2) \Leftrightarrow f_1(x) = e_1 \\ \text{and} \quad & x \in N_2 \Leftrightarrow f(x) \in A_2 \Leftrightarrow f(x) = (a_1, e_2) \Leftrightarrow f_2(x) = e_2.\end{aligned}$$

Therefore, $G/N_1 \cong G_1$ and $G/N_2 \cong G_2$.

Conversely suppose that N_1 and N_2 are normal subgroups of G such that $N_1 N_2 = G$, $N_1 \cap N_2 = \{e\}$, $G/N_1 \cong G_1$ and $G/N_2 \cong G_2$. Let $\alpha_1 : G/N_1 \to G_1$ and $\alpha_2 : G/N_2 \to G_2$ be isomorphisms and $\beta_1 : G \to G/N_1$ and $\beta_2 : G \to G/N_2$ be nature homomorphisms. Now define

$$\theta : G \to G_1 \times G_2 \quad \text{by} \quad \theta(x) = ((\alpha_1 \circ \beta_1)(x), (\alpha_2 \circ \beta_2)(x))$$

for any $x \in G$ since $\alpha_1, \beta_1, \alpha_2, \beta_2$ are all homomorphisms, θ is also a homomorphism. Also, for any $x \in G$,

$$\begin{aligned} \theta(x) = (e_1, e_2) &\Leftrightarrow \alpha_1(\beta_1(x)) = e_1 \quad \text{and} \quad \alpha_2(\beta_2(x)) = e_2 \\ &\Leftrightarrow \beta_1(x) = N_1 \text{ and } \beta_2(x) = N_2 \\ &\quad \text{(since } \alpha_1 \text{ and } \alpha_2 \text{ are isomorphism)} \\ &\Leftrightarrow x \in N_1 \cap N_2 = \{e\}. \end{aligned}$$

Therefore, $\ker \theta = \{e\}$ and hence θ is a monomorphism. For any $(x_1, x_2) \in G_1 \times G_2$, choose $a_1, a_2 \in G$ such that $\alpha_1(a_1N_1) = x_1$ and $\alpha_2(a_2N_2) = x_2$. Since N_1 and N_2 are normal subgroups and $N_1 \cap N_2 = \{e\}$, we get that $ab = ba$ for all $a \in N_1$ and $b \in N_2$ (for, consider $aba^{-1}b^{-1} \in N_1 \cap N_2 = \{e\}$).

Now since a_1 and $a_2 \in G = N_1N_2$, we get that

$$a_1 = r_1r_2 \text{ and } a_2 = s_1s_2 \text{ for some } r_1, s_1 \in N_1 \text{ and } r_2, s_2 \in N_2.$$

Put $x = r_2s_1$. Then,

$$a_1^{-1}x = (r_2^{-1}r_1^{-1}r_2)s_1 \in N_1 \ \ (\text{since } r_2^{-1}r_1^{-1}r_2 \in N_1 \text{ and } s_1 \in N_1)$$

and $$a_2^{-1}x = s_2^{-1}s_1^{-1}(r_2s_1) = s_2^{-1}s_1^{-1}(s_1r_2) = s_2^{-1}r_2 \in N_2$$

and hence $a_1N_1 = xN_1$ and $a_2N_2 = xN_2$. Now,

$$\begin{aligned} \theta(x) &= ((\alpha_1(\beta_1(x))), \alpha_2(\beta_2(x))) \\ &= (\alpha_1(xN_1), \alpha_2(xN_2)) \\ &= (\alpha_1(a_1N_1), \alpha_2(a_2N_2)) = (x_1, x_2). \end{aligned}$$

Thus, θ is a surjection also. Therefore, θ is an isomorphism and $G \cong G_1 \times G_2$.

Theorem 5.2.4. Let N and M be normal subgroups of a group G such that $NM = G$. Then,

$$G/N \cap M \cong G/N \times G/M.$$

Proof: Define $f: G \to G/N \times G/M$ by $f(a) = (aN, aM)$ for all $a \in G$. For any a and $b \in G$, we have

$$\begin{aligned} f(ab) &= (abN, abM) \\ &= (aN \cdot bN, aM \cdot bM) \\ &= (aN, aM) \cdot (bN, bM) \\ &= f(a) \cdot f(b) \end{aligned}$$

and hence f is a homomorphism. We shall prove that f is a surjection also. Let $(xN, yM) \in G/N \times G/M$, where x and $y \in G$. Since $NM = G$, x and $y \in NM$ and hence

$$x = rs, \quad \text{for some } r \in N \quad \text{and} \quad s \in M$$

and

$$y = tu, \quad \text{for some } t \in N \quad \text{and} \quad u \in M.$$

Now, put $a = st$. Then,

$$a^{-1}x = (st)^{-1}(rs) = t^{-1}(s^{-1}rs) \in N \text{ (since } t, r \in N)$$

and hence

$$aN = xN, \text{ Also,}$$

$$a^{-1}y = (st)^{-1}(tu) = (t^{-1}s^{-1}t)\, u \in M \text{ (since } s, u \in M)$$

and hence $aM = yM$. Therefore, $f(a) = (aN, aM) = (xN, yM)$. Thus, f is an epimorphism. By the Fundamental Theorem of Homomorphisms,

$$G/\ker f \cong G/N \times G/M.$$

Now,

$$\begin{aligned} \ker f &= \{a \in G : f(a) = \text{the identity in } G/N \times G/M\} \\ &= \{a \in G : (aN, aM) = (N, M)\} \\ &= \{a \in G : aN = N \text{ and } aM = M\} \\ &= N \cap M. \end{aligned}$$

Thus, $G/N \cap M \cong G/N \times G/M$.

Worked Exercise 5.2.1. Let m and n be relatively prime positive integers. Then, prove that $\mathbb{Z}_{mn} \cong \mathbb{Z}_m \times \mathbb{Z}_n$.

Answer: Consider the group $(\mathbb{Z}, +)$ of integers and let $N = n\mathbb{Z}$ and $M = m\mathbb{Z}$. Since $(n, m) = 1$,

$$1 = an + bm \quad \text{for some integers } a \text{ and } b$$

and hence $1 \in N + M$ so that $N + M = \mathbb{Z}$. By Theorem 5.2.4 (where we have written NM, since the binary operation in G is taken as $\cdot$),

$$\mathbb{Z}/N \cap M \cong \mathbb{Z}/N \times \mathbb{Z}/M.$$

Now, $N \cap M = k\mathbb{Z}$, where k is the least common multiple of n and m. Since $(n, m) = 1$, $k = mn$. Therefore,

$$\begin{aligned} \mathbb{Z}_{mn} &\cong \mathbb{Z}/mn\mathbb{Z} \text{ (by Example 5.2.1)} \\ &= \mathbb{Z}/N \cap M \\ &\cong \mathbb{Z}/N \times \mathbb{Z}/M \\ &= \mathbb{Z}/n\mathbb{Z} \times \mathbb{Z}/m\mathbb{Z} \cong \mathbb{Z}/N \times \mathbb{Z}/M. \end{aligned}$$

In the following, we shall classify the cyclic groups and prove that $(\mathbb{Z}, +)$ and $(\mathbb{Z}_n, +_n)$ are the only (up to isomorphism) cyclic groups.

Theorem 5.2.5. Let G be a cyclic group. If G is infinite, them $G \cong \mathbb{Z}$. If G is finite, them $G \cong \mathbb{Z}_n$ where n is the order of G.

Proof: Since G is cyclic, we can choose an element a in G such that

$$G = \langle a \rangle = \{a^n : n \in \mathbb{Z}\}.$$

Consider the group $(\mathbb{Z}, +)$ of integers and define

$$f : \mathbb{Z} \to G \text{ by } f(m) = a^m \quad \text{for any } m \in \mathbb{Z}.$$

For any m and $n \in \mathbb{Z}$,

$$f(m + n) = a^{m+n} = a^m \cdot a^n = f(m) \cdot f(n)$$

and hence f is a homomorphism. Since any element of G is of the form $a^m = f(m)$ for some $m \in \mathbb{Z}$, f is a surjection. Therefore, f is an epimorphism. By the Fundamental Theorem of Homomorphisms,

$$\mathbb{Z}/\text{Ker}\, f \cong G.$$

Since $\ker f$ is a subgroup of $(\mathbb{Z}, +)$, $\ker f = n\mathbb{Z}$ for some nonnegative integer n (by Worked Exercise 4.1.1),

$$n = 0 \Leftrightarrow \text{Ker}\, f = \{0\} \Leftrightarrow f \text{ is a monomorphism.}$$

Thus, by Theorem 4.2.7, C is infinite if and only if f is an isomorphism and hence $\mathbb{Z} \cong G$. Also,

$$G \text{ is finite} \Leftrightarrow \mathbb{Z}/\ker f \text{ is finite}$$
$$\Leftrightarrow \ker f = n\mathbb{Z}, n > 0.$$

In this case, $|G| = |\mathbb{Z}/n\mathbb{Z}| = |\mathbb{Z}_n| = n$.
Thus, if G is finite, then $G \cong \mathbb{Z}_n$, where n is the order of G.

Worked Exercise 5.2.2. For any positive integers n and m, prove that the following are equivalent to each other.

1. n and m are relatively prime.
2. $\mathbb{Z}_n \times \mathbb{Z}_m$ is cyclic.
3. $\mathbb{Z}_n \times \mathbb{Z}_m \cong \mathbb{Z}_{nm}$

Answer: (1) $\Rightarrow$ (3) is proved in Worked Exercise 5.2.1 and (3) $\Rightarrow$ (2) is trivial, since $\mathbb{Z}_{nm}$ is cyclic and, for any group G and G' such that $G \cong G'$, G is cyclic if and only if G' is cyclic. We are left with only (2) $\Rightarrow$ (1). Assume that $\mathbb{Z}_m \times \mathbb{Z}_n$ is cyclic. Since the order of $\mathbb{Z}_m \times \mathbb{Z}_n$ is mn, it follows from Theorem 5.2.5 that

$$\mathbb{Z}_m \times \mathbb{Z}_n \cong \mathbb{Z}_{mn}.$$

There must be an element (a, b) in $\mathbb{Z}_m \times \mathbb{Z}_n$ such that $O(a, b) = mn$, $0 \le a < m$ and $0 \le b < n$. Let k be the l.c.m. of m and n. Then,

$$k = ms \text{ and } k = nt \text{ for some } s \text{ and } t \in \mathbb{Z}^+.$$

$$\begin{aligned}\text{Now, } k(a, b) &= (ka, kb)\\ &= (msa, ntb)\\ &= (0, 0) \text{ since } a \in \mathbb{Z}_m \text{ and } b \in \mathbb{Z}_n.\end{aligned}$$

Therefore, $O(a, b)$ divides k and hence mn divides k.

$$k = mn = kd$$

where d is the g.c.d. of m and n. Therefore, $d = 1$; that is, m and n are relatively prime. ◀

EXERCISE 5(B)

1. If N and M are normal subgroups of a group G, prove that $G/N \cap M$ is isomorphic to a subgroup of $G/N \times G/M$.

2. For any positive integers m and n whose least common multiple is k, prove that $\mathbb{Z}_k$ is isomorphic to a subgroup of $\mathbb{Z}_n \times \mathbb{Z}_m$.

3. Suppose that G is a finite group and $\mathbb{Z}_{10}$ is a homomorphic image of G. Then, what can be said about the order of G?

4. Prove that $\mathbb{Z}_{27}$ is not a homomorphic image of $\mathbb{Z}_{72}$.

5. For any finite group G, prove that there exists a prime number p such that $\mathbb{Z}_p$ is not a homomorphic image of G.

6. Prove that the order of any homomorphic image of a finite group G must be a divisor of the order of G.

7. In Theorem 5.2.2, we have proved that any homomorphism can be expressed as a composition of an epimorphism and a monomorphism. Discuss the uniqueness of this expression.

8. Let $f: G \to G'$ and $g : G' \to G''$ be homomorphism of groups. If $g \circ f$ is a monomorphism, prove that f is a monomorphism. If $g \circ f$ is an epimorphism, prove that g is an epimorphism

9. Express $\mathbb{Z}_9$ as a homomorphic image of $\mathbb{Z}_{27}$.

10. Prove that $\mathbb{Z}_9$ is isomorphic to a subgroup of $\mathbb{Z}_{27}$.

5.3 ISOMORPHISM THEOREMS

The Fundamental Theorem of Homomorphisms is also called the *First Isomorphism Theorem*. In this section, we present two more isomorphism theorems. If N is a normal subgroup of a group G and if we are required to prove that the quotient group G/N is isomorphic to another group G', then Fundamental Theorem of Homomorphisms provides a technique. We simply exhibit an epimorphism of G onto G' whose kernel is the given normal subgroup N. We shall use this technique in proving the following two theorems.

Theorem 5.3.1 (Second Isomorphism Theorem). Let M and N be subgroups of a group G and N normal in G. Then, $M \cap N$ is a normal subgroup of M and

$$M/M \cap N \cong MN/N.$$

Proof: Since N is given to be a normal subgroup of G, $Na = aN$ for all $a \in G$ and, in particular $Na = aN$ for all $a \in M$ so that $NM = MN$. Therefore, MN is a subgroup of G and $N \subseteq MN$. Also, the normality of N in G implies the normality of N in MN and hence the quotient group MN/N is defined. Also, clearly $M \cap N$ is a subgroup of M. For any $a \in M \cap N$ and $x \in M$, we have

$$xax^{-1} \in N \text{ (since } a \in N \text{ and } N \text{ is normal in } G)$$

and $xax^{-1} \in M$ (since $a \in M$ and $x \in M$)

and therefore $xax^{-1} \in M \cap N$. Thus, $M \cap N$ is a normal subgroup of M. Now, define

$$f: M \to MN/N \quad \text{by} \quad f(m) = mN \quad \text{for all } m \in M.$$

Observe that, since $M \subseteq MN$, $mN \in MN/N$ for all $m \in M$ and therefore f is well defined. For any a and $b \in M$,

$$f(ab) = (ab)N = aN \cdot bN = f(a)f(b)$$

and hence f is a homomorphism. If $xN \in MN/N$, then

$$xN = (mn)N = m(nN) = mN = f(m)$$

for some $m \in M$ and $n \in N$. Therefore, f is an epimorphism.
By the Fundamental Theorem of Homomorphisms,

$$M/\ker f \cong MN/N.$$

Now, $\text{Ker} f = \{m \in M : f(m) = \text{the identity in } MN/N\}$

$$= \{m \in M : mN = N\}$$
$$= \{m \in M : m \in N\} = M \cap N$$

and therefore $M/M \cap N \cong MN/N$. ◀

Example 5.3.1. Consider the group $(\mathbb{Z}, +)$ of integers and let $M = \langle 3 \rangle = 3\mathbb{Z}$ and $N = \langle 5 \rangle = 5\mathbb{Z}$. Since $+$ is commutative on $\mathbb{Z}$, M and N are normal subgroups of $\mathbb{Z}$. Also,

$$M \cap N = 15\mathbb{Z} \quad \text{and} \quad M + N = \mathbb{Z}$$

and hence, by the above theorem, we have

$$3\mathbb{Z}/15\mathbb{Z} \cong \mathbb{Z}/5\mathbb{Z}.$$

Recall that $\mathbb{Z}/5\mathbb{Z} \cong \mathbb{Z}_5$, the group of integers modulo 5.

A closer examination of the cosets and operation in the quotient group $3\mathbb{Z}/15\mathbb{Z}$ reveals this to be none other than the group $(\{0, 3, 6, 9, 12\}, +_{15})$. The above isomorphism is a disguised version of the isomorphism

$$(\{0, 3, 6, 9, 12\}, +_{15}) \cong (\mathbb{Z}_5, +_5).$$

Theorem 5.3.2. Let $f : G \to G'$ be an epimorphism of groups and N be a normal subgroup of G such that $\ker f \subseteq N$. Then,

$$G/N \cong G'/f(N).$$

Proof: Since N is a normal subgroup of G and f is an epimorphism, it follows that $f(N)$ is a normal subgroup of G' (see Theorem 5.1.3) and hence the quotient group $G'/f(N)$ is defined. Now, define

$$g : G \to G'/f(N) \text{ by } g(a) = f(a)f(N)$$

for any $a \in G$. For any x and $y \in G$,

$$\begin{aligned} g(xy) &= f(xy)f(N) \\ &= f(x)f(y)f(N) \\ &= f(x)f(N)\cdot f(y)f(N) \text{ (since } f(N) \text{ is normal in } G') \\ &= g(x)g(y) \end{aligned}$$

and therefore g is a homomorphism. Also, for any $z \in G'$, there exists $a \in G$ such that $f(a) = z$ (since f is an epimorphism) and therefore

$$g(a) = f(a)f(N) = z\,f(N).$$

This implies that g is an epimorphism. Further, for any $a \in G$,

$$\begin{aligned} a \in \ker g &\Leftrightarrow g(a) = \text{the identity in } G'/f(N) \\ &\Leftrightarrow f(a)\,f(N) = f(N) \\ &\Leftrightarrow f(a) \in f(N) \\ &\Leftrightarrow f(a) = f(x) \quad \text{for some } x \in N \\ &\Leftrightarrow f(x^{-1}a) = f(x)^{-1}f(a) = e', x \in N \\ &\Leftrightarrow x^{-1}a \in \ker f \subseteq N \quad \text{and} \quad x \in N \\ &\Leftrightarrow a = x(x^{-1}a) \in N \end{aligned}$$

and therefore $\ker g = N$. Thus, by the Fundamental Theorem of Homomorphism,

$$G/N = G/\ker g \cong G'/f(N).$$

While applying the above theorem, one frequently starts with a normal subgroup of G' and use its inverse images, rather than starting with a normal subgroup of G containing the kernel. In this context, recall the one-to-one correspondence between normal subgroups of G containing the kernel and the normal subgroups of G'. In view of this, the above theorem can be rephrased as follows.

Corollary 5.3.1. Let $f : G \to G'$ be an epimorphism of groups and M be a normal subgroup in G'. Then,

$$G/f^{-1}(M) \cong G'/M.$$

Proof: Put $N = f^{-1}(M)$. Since M is a normal subgroup of G', N is a normal subgroup of G. Also,

$$\ker f = f^{-1}(\{e'\}) \subseteq f^{-1}(M) = N.$$

Further, since f is a surjection,

$$f(N) = f(f^{-1}(M)) = M.$$

Thus, by Theorem 5.3.2,

$$G/f^{-1}(M) = G/N \cong G'/f(N) = G'/M.$$

The following is another special case which is of interest on its own and is popularly called the Third Isomorphism Theorem. The reader is cautioned that there seems to be no universally accepted agreement on the numbering of these three Isomorphism theorems. However, the Fundamental Theorem of Homomorphisms deserves to be called as the First Isomorphism Theorem, since the other two are proved using this.

Theorem 5.3.3 (Third Isomorphism Theorem). Let M and N be normal subgroups of a group G such that $M \subseteq N$. The N/M is a normal subgroup of G/M and $(G/M)/(N/M) \cong G/N$.

Proof: By Remarks 4.6.1, N/M is a normal subgroup of G/M. Define f : $G/M \to G/N$ by $f(aM) = aN$ for any $a \in G$. First observe that f is well defined; for, if a and $b \in G$, then

$$aM = bM \Rightarrow a^{-1}b \in M \subseteq N \Rightarrow a^{-1}b \in N \Rightarrow aN = bN.$$

Clearly f is an epimorphism and its kernel is given by

$$\begin{aligned}\ker f &= \{aM \in G/M : a \in G \text{ and } f(aM) = \text{the identity in } G/N\}\\ &= \{aM \in G/M : a \in G \text{ and } aN = N\}\\ &= \{aM \in G/M : a \in N\} = N/M.\end{aligned}$$

Thus, by the Fundamental Theorem of Homomorphisms,

$$\frac{G/M}{N/M} = \frac{G/M}{\ker f} \cong G/N$$

Therefore,

$$\frac{(G/M)}{(N/M)} \cong \frac{G}{N}.$$

Worked Exercise 5.3.1. Let M and N be subgroups of a group G such that N is normal in G and $MN = G$. Then, prove that

$$G/N \cong M \quad \text{if and only if } M \cap N = \{e\}.$$

Answer: By the Second Isomorphism Theorem 5 3.1,

$$G/N = MN/N \cong M/M \cap N.$$

Therefore, $G/N \cong M \Leftrightarrow M/M \cap N \cong M$

$$\Leftrightarrow M \cap N = \{e\}.$$

EXERCISE 5(C)

1. Let G, G_1 and G_2 be groups and $f_1 : G \to G_1$ and $f_2 : G \to G_2$ be epimorphisms such that $\ker f_1 \subseteq \ker f_2$. Then, prove that there exists a unique homomorphism $f: G_1 \to G_2$ such that

$$f \circ f_1 = f_2$$

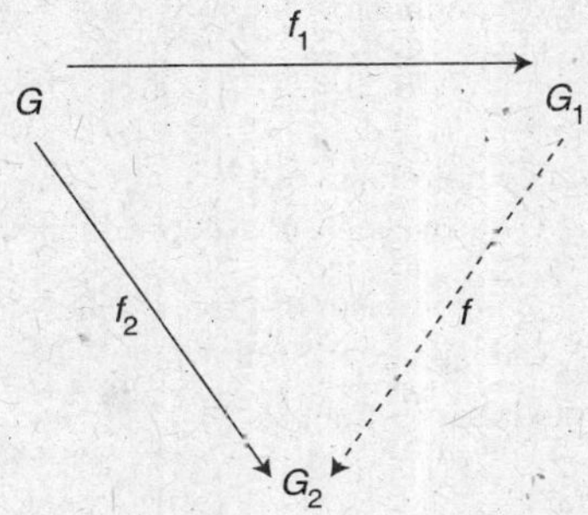

2. From the above exercise, deduce the Factorization Theorem 5.2.2.

3. Let G be the group of nonzero real numbers under the usual multiplication and $N = \{1, -1\}$. Then, prove that N is a normal subgroup of G and the quotient group G/N is isomorphic to the group of positive real numbers under multiplication.

4. Let $f: G \to G'$ be an epimorphism of groups. Then, prove that $A \to f^{-1}(A)$ is a one-to-one correspondence between the (normal) subgroups of G' and the (normal) subgroups of G containing $\ker f$.

5. Let G_8 be the group of symmetries of a square (see Example 3.2.8) and define $f: G_8 \to \mathbb{Z}_2 \times \mathbb{Z}_2$ by

$$f(e) = f(r_2) = (0, 0), f(r_1) = f(r_2) = (1, 0),$$

$$f(h) = f(0) = f(0, 1) \quad \text{and} \quad f(d_1) = f(d_2) = (1, 1).$$

Prove that f is an epimorphism and deduce that

$$G_8/Z(G_8) \cong \mathbb{Z}_2 \times \mathbb{Z}_2, \quad \text{where } Z(G_8) \text{ is the centre of } G_8.$$

6. Find all (up to isomorphism) homomorphic images of the group G_8 of symmetries of a square and exhibit the correspondence between the subgroups of G_8 containing $Z(G_8)$ and the subgroups of $G_8/Z(G_8)$.

7. For any $(a, b) \in \mathbb{R} \times \mathbb{R}$ with $a \neq 0$, define $T_{ab} : \mathbb{R} \to \mathbb{R}$ by $T_{ab}(x) = ax + b$ for all $x \in \mathbb{R}$ and let

$$G = \{T_{ab} : (a,b) \in \mathbb{R} \times \mathbb{R} \quad \text{and} \quad a \neq 0\}$$

and

$$N = \{T_{1b} : b \in \mathbb{R}\}.$$

Prove that G is a group under the composition of mappings and N is a normal subgroup of G. Further, prove that the quotient group G/N is isomorphic to the group of nonzero real numbers under the multiplication.

8. Prove the following for any epimorphism $f: G \to G'$ of groups:
   (i) For any subgroups A and B with $\ker f \subseteq A \cap B$, $f(A \cap B) = f(A) \cap f(B)$.
   (ii) For any subgroup A of G, $A \ker f = f^{-1}(f(A))$
   (iii) If $[G, G]$ is the commutator subgroup of G and $\ker f \subseteq [G, G]$, then

$$G/[G,G] \cong G'/[G',G']$$

9. Let $f: G \to \mathbb{Z}_8$ be an epimorphism of a group G onto the group $\mathbb{Z}_8$ of integers modulo 8. Prove that G has normal subgroups of index 2 and 4.

10. Let $f: G \to G'$ be an epimorphism of groups and A and A' be subgroups of G and G', respectively. Then, prove the following:
   (i) If A is of finite index in G and $\ker f \subseteq A$, then

$$i_G(A) = i_{G'}(f(A))$$

(ii) If A' is of finite index in G', then

$$i_G(f^{-1}(A')) = i_{G'}(A').$$

11. Prove that any group of order 4 is isomorphic to either $\mathbb{Z}_4$ or $\mathbb{Z}_2 \times \mathbb{Z}_2$ and hence is abelian.

12. Prove that any group of prime order p is isomorphic to $\mathbb{Z}_p$ and hence is cyclic.

13. Let $f: G \to A$ and $g: G \to B$ be homomorphisms of groups such that g is an epimorphism and $\ker g \subseteq \ker f$. Then, prove that there is a unique homomorphism $\overline{f}: B \to A$ such that $\overline{f} \circ g = f$.

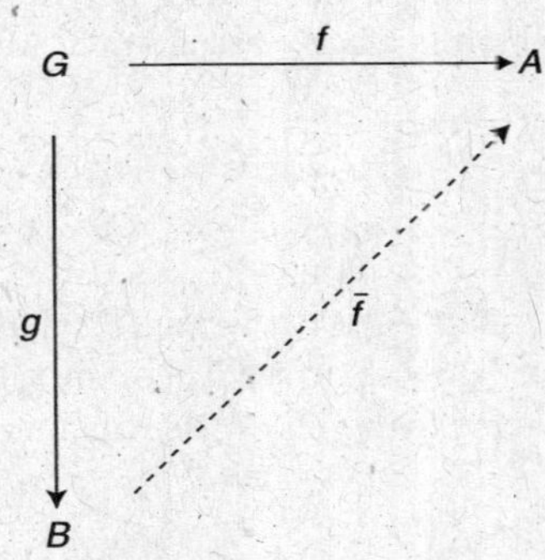

14. Prove the following in the above Exercise 13.
    (i) $\overline{f}$ is an epimorphism if and only if f is so.
    (ii) $\overline{f}$ is a monomorphism if and only if $\ker f = \ker g$.

15. Let A, N_1 and N_2 be subgroups of a group G such that N_1 and N_2 are normal in G and $A \cap N_1 = A \cap N_2$. Then, prove that $AN_1/N_2 \cong AN_2/N_1$.

16. Let A_1, A_2 and N be subgroups of a group G such that N is normal in G and $A_1N = A_2N$. Then, prove that $A_1 / A_1 \cap N \cong A_2 / A_2 \cap N$

5.4 AUTOMORPHISMS

Let us recall that a homomorphism of a group G into itself is called an *endomorphism* of G and a bijective endomorphism of G is called an *automorphism* of G. Among the endomorphisms of a group G, the automorphisms of G need special attention, for the reason that they form a group on their own under the composition of mappings and that the structure of this group reveals that of the group G itself. Even though the following is a repetition, we prefer to give an independent status for convenience and for its importance.

Definition 5.4.1. For any group G, a bijective homomorphism of G onto itself is called an *automorphism* of G. The set of all automorphisms of G will be denoted by Aut(G).

Theorem 5.4.1. For any group G, the set Aut(G) of all automorphisms of G forms a group under the composition of mappings.

Proof: Let G be a group, since the identity mapping $I_G : G \to G$, defined by $I_G(x) = x$ for all $x \in G$, is an automorphism of G, we have $I_G \in \text{Aut}(G)$ and hence Aut(G) is a nonempty set. If f and g are automorphisms of G, then the composition $f \circ g$ is also an automorphism of G. Therefore, o is a binary operation on Aut(G), which is clearly associative. Also, for any $f \in \text{Aut}(G)$, we have

$$f \circ I_G = f = I_G \circ f.$$

Therefore, the identity map I_G is the identity element in the semigroup (Aut(G), o). Further, we know that (see Theorem 5.1.7), if f is an automorphism of G, then the inverse mapping f^{-1} also an automorphism of G and

$$f \circ f^{-1} = I_G = f^{-1} \circ f.$$

Thus, Aut(G) is a group under the composition of mappings.

Example 5.4.1

1. If G is an abelian group, then the map $f : G \to G$, defined by $f(x) = x^{-1}$ for any $x \in G$, is an automorphism of G.
2. Consider the group $(\mathbb{Z}_{12}, +_{12})$ of integers modulo 12 and define $f : \mathbb{Z}_{12} \to \mathbb{Z}_{12}$ by

$$f(a) = 5a \; (= a +_{12} a +_{12} a +_{12} a +_{12} a).$$

Note that $f(a) = r$, where $5a = 12q + r$, where q and r are integers and $0 \leq r < 12$.

Then, f is clearly an endomorphism of $\mathbb{Z}_{12}$

Also,

$$\begin{aligned} \ker f &= \{a \in \mathbb{Z}_{12};\, f(a) = 0\} \\ &= \{a \in \mathbb{Z}_{12};\, 5a = 12q,\, q \in \mathbb{Z}\} \\ &= \{0\} \text{ (since } (5, 12) = 1) \end{aligned}$$

and therefore f is a monomorphism of $\mathbb{Z}_{12}$ into $\mathbb{Z}_{18}$. Since $\mathbb{Z}_{12}$ is a finite set, it follows that f is an surjection also. Thus, f is an automorphism of $\mathbb{Z}_{12}$.

In the following, we discuss about a special subgroup of Aut(G) consisting of certain special class of automorphisms which are important in the case of an abelian group.

Theorem 5.4.2. Let G be a group and $a \in G$. Define $T_a : G \to G$ by

$$T_a(x) = a\,x\,a^{-1} \quad \text{for all } x \in G.$$

Then, T_a is an automorphism of G and hence $T_a \in \text{Aut}(G)$.

Proof: For any x and $y \in G$,

$$T_a(xy) = a(xy)a^{-1} = (axa^{-1})(aya^{-1}) = T(x)T(y)$$

and hence T_a is a homomorphism. Also, for any $y \in G$, we have

$$a^{-1}ya \in G \quad \text{and} \quad T_a(a^{-1}ya) = a(a^{-1}ya)a^{-1} = y$$

and therefore T_a is a surjection. From the cancellation laws, we have

$$T_a(x) = T_a(y) \Rightarrow axa^{-1} = aya^{-1} \Rightarrow x = y$$

and hence T_a is an injection also. Thus, T_a is an automorphism of G.

Definition 5.4.2. For any element a in a group G, the automorphism T_a defined above is called the *inner automorphism of G corresponding to a.* The set of all inner automorphisms of G will be denoted by $I(G)$; that is,

$$I(G) = \{T_a : a \in G\}.$$

Theorem 5.4.3. Let G be any group and define $\alpha : G \to \text{Aut}(G)$ by

$$\alpha(a) = T_a \quad \text{for all } a \in G.$$

Then, α is a homomorphism and $I(G)$ is a subgroup of Aut(G). Also, $G/Z(G) \cong I(G)$, where $Z(G)$ is the centre of G, defined by

$$Z(G) = \{a \in G : ax = xa \quad \text{for all } x \in G\}.$$

Proof: For any a and $b \in G$, we have

$$T_{ab}(x) = (ab)x(ab)^{-1} = a(bxb^{-1})a^{-1} = T_a(T_b(x)) = (T_a \circ T_b)(x)$$

for any $x \in G$ and hence $T_{ab} = \text{T}_a \circ \text{T}_{b.}$ Therefore,

$$\alpha(ab) = \text{T}_{ab} = \text{T}_a \circ \text{T}_b = \alpha(a) \circ \alpha(b)$$

for all a and $b \in G$. Therefore, α is a homomorphism of G into Aut(G) and hence $\alpha(G) = I(G)$ is a subgroup of G. Now, let us compute the kernel of α.

$$\begin{aligned} a \in \ker \alpha &\Leftrightarrow \alpha(a) = \text{the identity in Aut}(G) \\ &\Leftrightarrow T_a = I_G \\ &\Leftrightarrow T_a(x) = I_G(x) \quad \text{for all } x \in G \\ &\Leftrightarrow a\,x\,a^{-1} = x \quad \text{for all } x \in G \\ &\Leftrightarrow ax = xa \quad \text{for all } x \in G \\ &\Leftrightarrow a \in Z(G), \text{ the centre of } G. \end{aligned}$$

Thus, by the Fundamental Theorem of Homomorphisms,

$$G/Z(G) = G/\ker\alpha \cong \alpha(G) = I(G).$$

It can be easily seen that a group G is abelian if and only if $T_a(x) = x$ for all a and $x \in G$ (that is, $T_a = I_G$ for all $a \in G$ and the group $I(G)$ is trial). If G is a nonabelian group, then there exists an automorphism $T_a \neq I_G$. In the following, we prove that a group G has a nonidentity automorphism if and only if G has atleast 3 elements.

Theorem 5.4.4. Let G be a group. Then,

$$|\text{Aut}(G)| > 1 \quad \text{if and only if } |G| > 2.$$

Proof: Suppose that $|\text{Aut}(G)| > 1$. Then, the group Aut(G) has an element other than its identity. That is, there exists an automorphism f of G such that $f \neq I_G$, the identity map. Now, we can choose an element $a \in G$ such that $f(a) \neq I_G(a) = a$. Since $f(e) = e$, it follows that $a \neq e$ and, since f is injective, $f(a) \neq f(e) = e$. Therefore, e, a and $f(a)$ are three distinct elements in G and hence $|G| \geqslant 3 > 2$.

Conversely suppose that $|G| > 2$. If G is not abelian, then $ax \neq xa$ for some a and $x \in G$ and hence

$$T_a(x) = axa^{-1} \neq x$$

so that $T_a \neq I_G \in \text{Aut}(G)$, where T_a is the inner automorphism of G corresponding to a and hence $|\text{Aut}(G)| > 1$. Therefore, we can assume that G is abelian. Then, the map $g : G \to G$, defined by $g(x) = x^{-1}$ for all $x \in G$, is an automorphism of G. If $g \neq I_G$, then $|\text{Aut}(G)| > 1$.Therefore, we can assume that $g = I_G$; that is, $g(x) = x$ or $x^{-1} = x$ or $x^2 = e$ for all $x \in G$. since $|G| > 2$, G has atleast three distinct elements; that is, G has atleast two distinct elements other than the identity. Let a and $b \in G$ such that $a \neq e$, $b \neq e$ and $a \neq b$. Put $A = \{e, a, b, ab\}$. Then,

$$(ab)a = ba^2 = be = b \quad \text{and} \quad (ab)b = ab^2 = ae = a$$

and hence A is a subgroup of G. Define $f: A \to A$ by

$$f(e) = e, f(a) = b, f(b) = a \quad \text{and} \quad f(ab) = ab.$$

Then, it can be easily verified that f is an automorphism of A. We shall extend f to an automorphism of the whole of G. Consider the set

$$\mathscr{C} = \{(B, g) : B \text{ is a subgroup of } G, A \subseteq B, g \in \text{Aut}(B) \text{ and } g/A = f\}.$$

Note that $(A, f) \in \mathscr{C}$ and hence $\mathscr{C}$ is a nonempty set. For any (B, g) and $(C, h) \in G$, define

$$(B, g) \leq (C, h) \quad \text{if and only if } B \subseteq C \quad \text{and} \quad h/B = g.$$

Then, $\leq$ is a partial order on $\mathscr{C}$. If $\{(B_i, g_i)\}_{i\in I}$ is a chain in the partially ordered set $(\mathscr{C}, \leq)$ and $B = \cup_{i\in I} B_i$, then B is a subgroup of G and, if we define $g : B \to B$ such that $g/B_i = g_i$, then (B, g) is a member in $\mathscr{C}$ and is an upper bound of $\{(B_i, g_i)\}_{i\in I}$ in $\mathscr{C}$. Thus, $(\mathscr{C}, \leq)$ satisfies the hypothesis of the Zorn's lemma which guarantees the existence of a maximal member, say (M, α), in $(\mathscr{C}, \leq)$. Since $A \subseteq M$ and $\alpha/A = f$, we have $\alpha(a) = f(a) = b \neq a$ and hence $\alpha \neq Id_M$ and therefore it is enough it we prove that $M = G$.

Otherwise, suppose that there is an element $s \in G$ such that $s \notin M$. Put $B = M \cup Ms$ and define $g : B \to B$ by $g(m) = \alpha(m)$ and $g(ms) = \alpha(m)s$ for all $m \in M$. Then, B is a subgroup of G (note that $B = MS$, where S is the subgroup $\{e, s\}$), $g \in \text{Aut}(B)$ and $g/M= \alpha$ so that $(M, \alpha) < (B, g)$, which is a contradiction to the maximality of (M, α) in $\mathscr{C}$. Thus, $M = G$, $\alpha \in \text{Aut}(\mathscr{C})$ and $\alpha \neq I_G$ so that $|\text{Aut}(\mathscr{C})| > 1$.

In the following, we completely determine all the automorphisms of a cyclic group. In fact, each automorphism of a cyclic group corresponds to a generator of the group and vice versa.

Theorem 5.4.5. Let f be an automorphism of a cyclic group G.

1. For any $a \in G$, $G = <a>$ if and only if $G = <f(a)>$.
2. If f and $g \in \text{Aut}(G)$ and a is a generator of G such that $f(a) = g(a)$, then $f = g$.

Proof: First recall that f^{-1} is also an automorphism of G.

1. Suppose that $a \in G$ such that
$$G = <a> = \{a^n : n \in \mathbb{Z}\}.$$
Then, since $f(a) \in G$, we have $<f(a)> \subseteq G$. On the other hand,
$$\begin{aligned} x \in G &\Rightarrow f^{-1}(x) \in G = <a> \\ &\Rightarrow f^{-1}(x) = a^n \quad \text{for some } n \in \mathbb{Z} \\ &\Rightarrow x = f(f^{-1}(x)) = f(a^n) = f(a)^n \in <f(a)> \end{aligned}$$
and hence $G = <f(a)>$. The converse follows from the fact that f^{-1} is an automorphism of G.
2. Let f and $g \in \text{Aut}(G)$ and $G = <a>$ such that $f(a) = g(a)$. Then, for any $x \in G$, $x = a^n$ for some $n \in \mathbb{Z}$ and
$$f(x) = f(a^n) = f(a)^n = g(a)^n = g(a)^n = g(x)$$
Therefore, $f = g$.

Corollary 5.4.1. For any cyclic group G, the number of automorphisms of G is finite and is precisely equal to the number of generators of G.

Proof: Let $G = <a>$ be a cyclic group. Then, by the above theorem $f \mapsto f(a)$ is an injection of Aut(G) into the set gen(G) of generators of G. Further, if b is any generator of G, then we can define automorphism f such that $f(a) = b$ (that is, $f(a^n) = b^n$ for any $n \in \mathbb{Z}$). Thus, $f \mapsto f(a)$ is a bijection of Aut(G) onto gen(G). From Theorems 4.2.7 and 4.2.8, gen(G) is a finite set and hence so is Aut(G) and $|\text{Aut}(G)| = |\text{gen}(G)|$.

Corollary 5.4.2

1. For any infinite cyclic group G, $|\text{Aut}(G)| = 2$.
2. For any finite cyclic group G of order n, $|\text{Aut}(G)| = \phi(n)$, where ϕ is the Euler-Totient function.

Proof: These follow from Corollary 5.4.1, Theorems 4.2.7 and 4.2.8.

Worked Exercise 5.4.1. For any groups G and H, if $G \cong H$, prove that $\text{Aut}(G) \cong \text{Aut}(H)$.

Answer: Suppose that $G \cong H$ and $\alpha; G \to H$ is an isomorphism. Define

$$\theta : \text{Aut}(G) \to \text{Aut}(H) \text{ by } \theta(f) = \alpha \circ f \circ \alpha^{-1}$$

for any $f \in \text{Aut}(G)$, since f, α and α^{-1} are isomorphisms, so is $\alpha \circ f \circ \alpha^{-1} : H \to H$. Therefore, $\alpha \circ f \circ \alpha^{-1}$ is an automorphism of H. For any f and $g \in \text{Aut}(G)$,

$$\begin{aligned}\theta(f \circ g) &= \alpha \circ (f \circ g) \circ \alpha^{-1} \\ &= (\alpha \circ f \circ \alpha^{-1}) \circ (\alpha \circ g \circ \alpha^{-1}) \\ &= \theta(f) \circ \theta(g).\end{aligned}$$

Therefore, θ is homomorphism. Further,

$$\begin{aligned}\theta(f) = \theta(g) &\Rightarrow \alpha \circ f \circ \alpha^{-1} = \alpha \circ g \circ \alpha^{-1} \\ &\Rightarrow f = (\alpha^{-1} \circ \alpha) \circ f \circ (\alpha^{-1} \circ \alpha) = \alpha^{-1} \circ (\alpha \circ f \circ \alpha^{-1}) \circ \alpha \\ &= \alpha^{-1} \circ (\alpha \circ g \circ \alpha^{-1}) \circ \alpha = (\alpha^{-1} \circ \alpha) \circ g \circ (\alpha^{-1} \circ \alpha) \\ &= g\end{aligned}$$

Therefore, θ is an injection. Also,

$$\begin{aligned} h \in \text{Aut}(H) \Rightarrow \alpha^{-1} \circ h \circ \alpha \in \text{Aut}(G) \text{ and } \theta(\alpha^{-1} \circ h \circ \alpha) = \\ \alpha \circ (\alpha^{-1} \circ h \circ \alpha) \circ \alpha^{-1} = h.\end{aligned}$$

Therefore, θ is a surjection. This θ is a bijective homomorphism and hence an isomorphism of $\text{Aut}(G)$ onto $\text{Aut}(H)$.

Worked Exercise 5.4.2. List all the automorphisms of the group $(\mathbb{Z}_n, +_n)$ for any positive integer n and, in particular, of the group $(\mathbb{Z}_{12}, +_{12})$.

Answer: Recall that, for any $1 \le r < n$, r is a generator of $\mathbb{Z}_n$ if and only if r is relatively prime to n. Therefore, by Corollary 5.4.2 (2), these are exactly $\phi(n)$ automorphisms of $\mathbb{Z}_{12}$ and these are given by

$$f_r : \mathbb{Z}_{12} \to \mathbb{Z}_{12}, \quad f_r(m) = mr (\text{mod } 12)$$

for each $1 \le r < n$, such that $(r, n) = 1$.

Consider $\mathbb{Z}_{12}$. We have $\phi(n) = 4$ since 1, 5, 7 and 11 are the only integers r such that $1 \leq r < 12$ and $(r, 12) = 1$. Therefore, there are exactly four automorphisms of $\mathbb{Z}_{12}$ and are given below.

$f_1 = I$, the identity map of $\mathbb{Z}_{12}$,
$f_5 : \mathbb{Z}_{12} \to \mathbb{Z}_{12}$ defined by $f_5(m) = 5m \pmod{12}$,
$f_7 : \mathbb{Z}_{12} \to \mathbb{Z}_{12}$ defined by $f_7(m) = 7m \pmod{12}$,
$f_{11} : \mathbb{Z}_{12} \to \mathbb{Z}_{12}$ defined by $f_{11}(m) = 11m \pmod{12}$.

The following table gives a complete description of all the four automorphisms of $\mathbb{Z}_{12}$.

	0	$\underline{1}$	2	3	4	$\underline{5}$	6	$\underline{7}$	8	9	10	$\underline{11}$
f_1	0	1	2	3	4	5	6	7	8	9	10	11
f_5	0	5	10	3	8	1	6	11	4	9	2	7
f_7	0	7	2	9	4	11	6	1	8	3	10	5
f_{11}	0	11	10	9	8	7	6	5	4	3	2	1

EXERCISE 5(D)

1. For any endomorphism f of a finite group G, prove that the following are equivalent to each other:
   (i) f is an epimorphism.
   (ii) f is an automorphism of G.
   (iii) f is a monomorphism.
2. Give an example of an infinite group G and of an isomorphism of G onto a proper subgroup of G.
3. Let $G = \{e, a, b, ab\}$ be a group of order 4 in which $a^2 = e = b^2$ and $ab = ba$. Then, determine Aut(G).
4. Let A be a subgroup of a group G, such that $f(A) \subseteq A$ for all $f \in \text{Aut}(G)$. Then, prove that A is a normal subgroup in G.
5. Let G be a group and $f \in \text{Aut}(G)$. Prove that the set $\{a \in G : f(a) = a\}$ is a subgroup of G.
6. For any group G, prove that $\{a \in G : f(a) = a \text{ for all } f \in \text{Aut}(G)\}$ is a normal subgroup of G.
7. Let G be a finite group and $f \in \text{Aut}(G)$, such that, for any $x \in G$,

$$f(x) = x \quad \text{if and only if } x = e.$$

Then, prove that any element x of G can be expressed as

$$x = y^{-1}f(y) \quad \text{for some } y \in G.$$

8. If $f^2 = I_G$ for the f in Exercise 7 above, then prove that G is an abelian group.

9. Let G be a finite group and $f \in \text{Aut}(G)$ such that f sends more than three-quarters of the elements of G onto their inverses. Then, prove that $f(a) = a^{-1}$ for all $a \in G$ and that G is abelian.

10. Let C be the commutator subgroup of a group G and $f \in \text{Aut}(G)$. Prove that $f(C) \subseteq C$.

11. If a group G has a nonidentity automorphism, then prove that G has atleast three elements.

12. Prove that any homomorphic image of a cyclic group is cyclic.

13. Prove that any homomorphic image of an abelian group is abelian.

14. Determine $\text{Aut}(\mathbb{Z}_p)$ for any prime number p.

15. Find all the automorphisms of $(\mathbb{Z}, +)$ and $(\mathbb{Z}_n, +_n)$ for any $n \in \mathbb{Z}^+$.

16. Let G be a finite cyclic group of order n and define $f_m : G \to G$ by $f_m(a) = a^m$ for all $a \in G$ and $m \in \mathbb{Z}^+$. Prove that f_m is an automorphism of G if and only if m is relatively prime with n.

17. Let G be a finite group of order $n > 2$. If $a^2 \neq e$ for some $a \in G$, then prove that G has a nonidentity automorphism.

18. If G is a noncyclic finite abelian group, prove that $\text{Aut}(G)$ is not abelian.

19. Let G be a finite group, such that $|\text{Aut}(G)| = p$, where p is a prime number. Then, prove that $|G| \leq 3$.

20. Prove that $\text{Aut}(\mathbb{Z}_2 \times \mathbb{Z}_3) \cong \text{Aut}(\mathbb{Z}_2) \times \text{Aut}(\mathbb{Z}_3)$.

6

Permutation Groups

For any nonempty set X, the set $M(X)$ of all mappings of X into itself forms a monoid under the composition of mappings in which the invertible elements (elements possessing inverses) are precisely the bijections of X onto itself. In fact, we have observed that a mapping $f : X \to X$ has left (right) inverse in $M(X)$ if and only if f is an injection (respectively, surjection) and, as such, the set of all bijections of X onto itself forms a group under the composition of mappings. Before the advent of the abstract form of a group, mathematicians were only interested in the group structure of certain sets of bijections, which were also known as permutations, when the set X is finite. In this chapter, we discuss thoroughly the structure of this type of groups.

6.1 CAYLEY'S THEOREM

Before the formation of the present day abstract concept of a group, most of the groups were in the form of a set of transformations of a particular mathematical structure, like the group of symmetries of square or of an equilateral triangle. Most finite groups appeared as groups of bijections of an n-element set onto itself for some positive integer n. The English Mathematician Cayley first noted that any abstract group can be viewed as a subgroup of the group $S(X)$ of bijections of X onto itself, for a suitable set X. In this section, we shall prove this theorem of Cayley and some of its consequences. First, we have the following definition.

Definition 6.1.1. Let X be a nonempty set. Any bijection of X onto itself is called a *permutation on X*. The set $S(X)$ of all permutations on X forms a group under the composition of mappings. Any subgroup of X is called *a group of permutations on X.*

Theorem 6.1.1. Any group is isomorphic to a group of permutations on a suitable set.

Proof: Let G be a group. For any $a \in G$, define

$$f_a : G \to G \text{ by } f_a(x) = ax \quad \text{for all } x \in G.$$

Then, for any $y \in G$, $a^{-1}y \in G$ and $f_a(a^{-1}y) = a(a^{-1}y) = y$ and therefore f_a is a surjection. Also, by the left cancellation law in the group G, f_a is an injection. Therefore, f_a is a permutation on the set G. Now, define

$$G' = \{f_a : a \in G\}.$$

We shall prove that G' is a group of permutations on the set G and that $G \cong G'$. For any a and $b \in G$, we have

$$f_{ab}(x) = (ab)x = a(bx) = f_a(f_b(x)) = (f_a \cdot f_b)(x)$$

for all $x \in G$ and hence $f_{ab} = f_a \cdot f_b$. Also, if e is the identity in the group G,

$$f_e(x) = ex = x \quad \text{for all } x \in G$$

and hence $f_e = I_G$, the identity map on G. In particular, for any $a \in G$,

$$f_a \cdot f_{a^{-1}} = f_{aa^{-1}} = f_e = I_G = f_e = f_{a^{-1}a} = f_{a^{-1}} \cdot f_a$$

and hence $f_a^{-1} = f_{a^{-1}} \in G$. Therefore, G' is a subgroup of the group $(S(G), \text{o})$; that is, G' is a group of permutations on the set G. Now, define

$$\theta : G \to G' \text{ by } \theta(a) = f_a \quad \text{for all } a \in G.$$

Then, $\theta(ab) = f_{ab} = f_a \cdot f_b$ and hence θ is a homomorphism. For any a and $b \in G$,

$$f_a = f_b \Rightarrow f_a(e) = f_b(e) \Rightarrow a = ae = be = b$$

and therefore θ is an injection. Clearly θ is a surjection. Thus, θ is an isomorphism of G onto G' and therefore $G \cong G'$. ◀

Cayley's theorem enables us to view any abstract group as a more concrete object, namely as a group of mappings and the binary operation on G as the composition of mappings. However, this has its own disadvantages. For example, if G is a finite group of order n, then $S(G)$ is a group of order $n!$ which is far bigger than n. Now, a natural question is that can we find a smaller set X such that G is isomorphic to a subgroup of $S(X)$. The following theorem is a step ahead in this direction.

Theorem 6.1.2. Let G be a group and H be a subgroup of G. Let X be the set of all left cosets of H in G. Then, there exists a homomorphism $\theta : G \to S(X)$ satisfying the following:

1. $\ker \theta$ is the largest normal subgroup of G contained in H.
2. θ is a monomorphism if and only if H contains no nontrivial normal subgroup of G.

Proof: We have $X = \{aH : a \in G\}$. For any $x \in G$, define

$$g_x : X \to X \text{ by } g_x(aH) = (xa)H$$

for all $aH \in X$, $a \in G$. Note that, for any a and $b \in G$,

$$\begin{aligned} aH = bH &\Rightarrow a^{-1}b \in H \\ &\Rightarrow (xa)^{-1}(xb) = a^{-1}x^{-1}xb = a^{-1}b \in H \\ &\Rightarrow (xa)H = (xb)H \end{aligned}$$

and hence g_x is well-defined. Also, for any a and $b \in G$,

$$\begin{aligned} g_x(aH) = g_x(bH) &\Rightarrow (xa)H = (xb)H \\ &\Rightarrow (xa)^{-1}(xb) \in H \\ &\Rightarrow a^{-1}x^{-1}xb \in H \\ &\Rightarrow a^{-1}b \in H \Rightarrow aH = bH \end{aligned}$$

and therefore g_x is an injection. Further, if $bH \in X$, then $(x^{-1}b)H \in X$ and

$$g_x((x^{-1}b)H) = (x(x^{-1}b))H = bH$$

and hence g_x is a surjection also. Thus, g_x is a permutation on X; that is, $g_x \in S(X)$ for any $x \in G$. Now, define

$$\theta : G \to S(X) \text{ by } \theta(x) = g_x \quad \text{for all } x \in G.$$

For any x and $y \in G$ and $aH \in X$ $(a \in G)$, we have

$$\theta(xy)(aH) = g_{xy}(aH) = ((xy)a)H = g_x(g_y(aH)) = (\theta(x) \cdot \theta(y))(aH)$$

and hence $\theta(xy) = \theta(x) \cdot \theta(y)$. Therefore, θ is a homomorphism of G into $S(X)$.

1. Clearly $\ker\theta$ is a normal subgroup of G and

$$\begin{aligned}\ker\theta &= \{x \in G : \theta(x) = \text{the identity in } S(X)\}\\ &= \{x \in G : g_x = I_X\}\\ &= \{x \in G : g_x(aH) = aH \quad \text{for all } a \in G\}\\ &= \{x \in G : xaH = aH \quad \text{for all } a \in G\}\\ &= \{x \in G : a^{-1}xa \in H \quad \text{for all } a \in G\}\\ &= \{x \in G : x \in aHa^{-1} \quad \text{for all } a \in G\}\\ &= \bigcap_{a \in G} aHa^{-1}\end{aligned}$$

Therefore, $\ker\theta \subseteq aHa^{-1}$ for all $a \in G$ and, in particular,

$$\ker\theta \subseteq eHe^{-1} = H.$$

If N is any normal subgroup of G contained in H, then

$$a^{-1}Na \subseteq N \subseteq H$$

and hence $N \subseteq aHa^{-1}$ for all $a \in G$, so that

$$N \subseteq \bigcap_{a \in G} aHa^{-1} = \ker\theta.$$

Thus, $\ker\theta$ is the largest normal subgroup of G contained in H.

2. θ is a monomorphism $\Leftrightarrow \ker\theta = \{e\}$

$\Leftrightarrow \{e\}$ is the largest normal subgroup of G contained in H.

$\Leftrightarrow H$ contains no nontrivial normal subgroup of G. ◀

Note that the Cayley's Theorem 6.1.1 can be deduced from the above theorem by taking $H = \{e\}$. The above theorem is an important tool in determining the existence of normal subgroups of a group G contained in a given subgroup of G.

Theorem 6.1.3. Let H be a subgroup of a finite group G such that $|G|$ is not a divisor of $i_G(H)!$ Then, H contains a nontrivial normal subgroup of G.

Proof: Let X be the set of all left cosets of H in G. Then, $|X| = i_G(H)$. By the above theorem, there exists a homomorphism $\theta : G \to S(X)$ such that $\ker\theta$ is the largest normal subgroup of G contained in H. Then, $\theta(G)$ is a subgroup of $S(X)$ and hence, by the Lagrange's theorem, $|\theta(G)|$ is a divisor of $|S(X)| = |X|!$ If $\ker\theta = \{\bar{e}\}$, then θ is a monomorphism and

$$|G| = |\theta(G)|, \quad \text{which is a divisor of } |S(X)| = i_G(H)!,$$

a contradiction to the hypothesis. Therefore, $\ker\theta$ is a nontrivial normal subgroup of G contained in H. ◀

Corollary 6.1.1. Let H be a normal subgroup of a finite group G such that $i_G(H)! < |G|$. Then, H contains a nontrivial normal subgroup of G.

Worked Exercise 6.1.1. Let A be a subgroup of order 9 in a group of order 36. Prove that A contains a normal subgroup of G whose order is 3 or 9.

Answer: $i_G(A)! = \dfrac{|G|}{|A|}! = \dfrac{36}{9}! = 4! < 36 = |G|$

and hence, by Corollary 6.1.1, A contains a nontrivial normal subgroup N of G. Since N is a subgroup of A, we get by the Lagrange's theorem that $|N|$ divides $|A| = 9$. Also, since N is nontrivial, $|N| = 3$ or 9.

Worked Exercise 6.1.2. Let G be a group of order 187. Prove that any subgroup of order 17 in G must be normal.

Answer: Let A be a subgroup of order 17 in G. Then,

$$i_G(A) = \frac{|G|}{17} = \frac{187}{17} = 11.$$

Since 17 is a prime and $17 > 11$, we get that 17 does not divide $11! = i_G(A)!$ and hence $|G|$ does not divide $i_G(A)!$ By Theorem 6.1.3, A contains a nontrivial normal subgroup of G. Let N be a nontrivial normal subgroup of G contained in A. Then, $1 < |N|$ and, by Lagrange's theorem, $|N|$ is a divisor of $|A| = 17$. Since 17 is a prime, $|N| = 17 = |A|$ and hence $N = A$. Thus, A is a normal subgroup of G.

Worked Exercise 6.1.3. Let G be a finite group of order n and p be a prime number such that $p > \frac{n}{p}$. Then prove that any subgroup of order p in G is normal in G.

Answer: Let A be a subgroup of order p in G. Then, p divides n and $i_G(A) = \frac{n}{p}$. Since $p > \frac{n}{p}$ and p is a prime, it follows that p does not divide $i_G(A)!$

By Theorem 6.1.3, A contains a nontrivial normal subgroup N of G. By the Lagrange's theorem, $|N|$ divides $|A| = p$. Since $|N| > 1$ and p is a prime, $|N| = p = |A|$ and, since $N \subseteq A$, $N = A$. Thus, A is a normal subgroup of G.

EXERCISE 6(A)

1. Which of the following mappings $f: \mathbb{R} \to \mathbb{R}$ are permutations on $\mathbb{R}$?
   (i) $f(x) = x^3 - 2$
   (ii) $f(x) = x^2 - 2$
   (iii) $f(x) = 3x + 2$
   (iv) $f(x) = 2x - 3$
   (v) $f(x) = x^3 + 6x^2 + 12x + 8$
   (vi) $f(x) = |x| - 2$
   (vii) $f(x) = \sin x$
   (viii) $f(x) = \log |x|$
   (ix) $f(x) = \begin{cases} e^x & \text{if } x \geq 0 \\ -e^{-x} & \text{if } x < 0 \end{cases}$
2. State whether the following are true or false and substantiate your answers:
   (i) Every surjection of $\mathbb{Z}_n$ onto itself is a permutation, for any $n \in \mathbb{Z}^+$.
   (ii) Every injection of $\mathbb{Z}$ into $\mathbb{Z}$ is a permutation.
   (iii) For any $n \in \mathbb{Z}^+$, every injection of $\mathbb{Z}_n$ into $\mathbb{Z}_n$ is a permutation.
   (iv) For any finite set X, every surjection of $S(X)$ into $S(X)$ is a permutation.
   (v) For any finite set X, every injection of $\mathbb{P}(X)$ into itself is a permutation, where $\mathbb{P}(X)$ is the power set of X.
   (vi) Any group G is isomorphic with a subgroup of $(S(G), \circ)$.
3. If X is a finite set with $|X| = n$, prove that $S(X)$ is a finite group of order $n!$
4. For any set X, prove that X is finite if and only if $S(X)$ is finite.
5. Let G be a group and define $g_a : G \to G$ by $g_a(x) = xa$ for any a and $x \in G$. Then prove that g_a is a permutation on G. Can we replace f_a with g_a in the proof of the Cayley's theorem?
6. Let A be a subgroup of a finite group G and $i_G(A) = m$. If A does not contain any nontrivial normal subgroup of G, then prove that $|G|$ divides $m!$
7. Let G be a group of order 396. Prove that any group of order 11 in G is normal in G.
8. Let G be a finite group of order n and p be a prime number such that p^2 does not divide n. Prove that any subgroup of order p in G is normal in G.

9. For any elements a and b in a set X, prove that there is a permutation f on X such that $f(a) = b$ and $f(b) = a$.

10. Construct a permutation f on $\mathbb{R}$ such that $f(n) = n + 1$ for all integers n.

6.2 THE SYMMETRIC GROUP S_n

Recall that a bijection of a set X onto itself is also called a permutation on X and the permutations on X form a group under the composition of mappings which is denoted by $S(X)$. It is well-known that a set X is said to be *equipotent or bijective* with another set Y if there is a bijection of X onto Y and that, in this case, we write $X \simeq Y$. The following can be easily proved by using induction on the number of elements of X.

Theorem 6.2.1. If X is a finite set and $|X| = n$, then $|S(X)| = n!$

Theorem 6.2.2. For any nonempty finite sets X and Y,

$$X \simeq Y \quad \text{if and only if } S(X) \cong S(Y).$$

Proof: Suppose that $X \simeq Y$. Then, there exists a bijection $\alpha : X \to Y$. Now, define

$$\theta : S(X) \to S(Y) \quad \text{by} \quad \theta(f) = \alpha \circ f \circ \alpha^{-1}$$

$$Y \xrightarrow{\alpha^{-1}} X \xrightarrow{f} X \xrightarrow{\alpha} Y$$

for all $f \in S(X)$. For any f and $g \in S(X)$,

$$\begin{aligned}\theta(f \circ g) &= \alpha \circ (f \circ g) \circ \alpha^{-1} \\ &= (\alpha \circ f \circ \alpha^{-1}) \circ (\alpha \circ g \circ \alpha^{-1}) = \theta(f) \circ \theta(g)\end{aligned}$$

and therefore θ is a homomorphism of groups. Also, for any f and $g \in S(X)$,

$$\begin{aligned}\theta(f) = \theta(g) &\Rightarrow \alpha \circ f \circ \alpha^{-1} = \alpha \circ g \circ \alpha^{-1} \\ &\Rightarrow \alpha^{-1} \circ (\alpha \circ f \circ \alpha^{-1}) \circ \alpha = \alpha^{-1} \circ (\alpha \circ g \circ \alpha^{-1}) \circ \alpha \\ &\Rightarrow f = g.\end{aligned}$$

Therefore, θ is an injection. Further, for any $h \in S(Y)$, $\alpha^{-1} \circ h \circ \alpha \in S(X)$ and

$$\theta(\alpha^{-1} \circ h \circ \alpha) = \alpha \circ (\alpha^{-1} \circ h \circ \alpha) \circ \alpha^{-1} = h.$$

Therefore, θ is a surjection also. Thus, θ is an isomorphism of $S(X)$ onto $S(Y)$ and hence $S(X) \cong S(Y)$. Conversely suppose that $S(X) \cong S(Y)$. Then, $S(X) \simeq S(Y)$ and hence

$$2^n = |S(X)| = |S(Y)| = 2^m, \quad \text{where } |X| = n \text{ and } |Y| = m.$$

This implies that $n = m$ and hence $|X| = |Y|$, so that $X \simeq Y$. ◀

The above theorem says that two finite sets X and Y are equipotent if and only if the groups $(S(X), \circ)$ and $(S(Y), \circ)$ are isomorphic. In particular, if X is a set with n elements, then $S(X) \cong S(I_n)$, where $I_n = \{1, 2, \ldots, n\}$. For this reason, it is enough if we study the permutations on the set $\{1, 2, \ldots, n\}$. We begin this with the following formal definition.

Definition 6.2.1. For only positive integer n, the set I_n is defined by

$$I_n = \{1, 2, \ldots, n\}$$

and the group $(S(I_n), \circ)$ of permutations on I_n is denoted by S_n and is called the *symmetric group of degree n*. Any permutation f on I_n is usually denoted by an array

$$f = \begin{pmatrix} 1 & 2 & 3 & \ldots & n \\ f(1) & f(2) & f(3) & \ldots & f(n) \end{pmatrix}$$

which symbolizes that each $1 \leq i \leq n$ is mapped onto $f(i)$, the integer that is written just below i in the array. As usual, let e denote the identity in the group S_n. Note that e is the identity mapping on I_n.

Example 6.2.1

1. Consider $I_6 = \{1, 2, 3, 4, 5, 6\}$ and define $f: I_6 \to I_6$ by $f(1) = 3, f(2) = 5, f(3) = 1, f(4) = 2, f(5) = 4$ and $f(6) = 6$. Then, f is denoted by

$$f = \begin{pmatrix} 1 & 2 & 3 & 4 & 5 & 6 \\ 3 & 5 & 1 & 2 & 4 & 6 \end{pmatrix}.$$

2. Let $f \in S_9$ be given by

$$f = \begin{pmatrix} 1 & 2 & 3 & 4 & 5 & 6 & 7 & 8 & 9 \\ 4 & 6 & 8 & 5 & 2 & 3 & 7 & 9 & 1 \end{pmatrix}.$$

   Then, f is a permutation on $\{1, 2, 3, 4, 5, 6, 7, 8, 9\}$ defined by $f(1) = 4, f(2) = 6, f(3) = 8, f(4) = 5, f(5) = 2, f(6) = 3, f(7) = 7, f(8) = 9$ and $f(9) = 1$.

Recall that the order of the symmetric group of degree n is $n!$

Theorem 6.2.3. Let n and m be positive integers. S_m is isomorphic to a subgroup of S_n if and only if $m \leq n$.

Proof: Suppose that $m \leq n$. For any $f \in S_m$, define $\theta(f) \in S_n$ as given below.

$$\theta(f)(i) = \begin{cases} f(i), & \text{if } 1 \leq i \leq m \\ i, & \text{if } m < i \leq n \end{cases}.$$

Then, it can be easily checked that θ is a monomorphism of S_m into S_n and hence $S_m \cong \theta(S_m)$, which is a subgroup of S_n. Conversely suppose that S_m is isomorphic to a subgroup of S_n. Then, $|S_m|$ divides $|S_n|$ so that $m!$ divides $n!$ which happens only when $m \leq n$. ◀

In view of the above theorem, a permutation f in S_n can be identified with a permutation in S_m for any $m \geq n$ with the understanding that $f(i) = i$ for all $n < i \leq m$.

Worked Exercise 6.2.1. Construct a table representing the symmetry group of degree 3.

Answer: There are 3! (=6) elements in the group S_3, which are given below (recall Example 3.4.3).

$$e = \begin{pmatrix} 1 & 2 & 3 \\ 1 & 2 & 3 \end{pmatrix},\ a = \begin{pmatrix} 1 & 2 & 3 \\ 2 & 3 & 1 \end{pmatrix},\ b = \begin{pmatrix} 1 & 2 & 3 \\ 3 & 1 & 2 \end{pmatrix},$$

$$c = \begin{pmatrix} 1 & 2 & 3 \\ 1 & 3 & 2 \end{pmatrix},\ d = \begin{pmatrix} 1 & 2 & 3 \\ 3 & 2 & 1 \end{pmatrix} \text{ and } s = \begin{pmatrix} 1 & 2 & 3 \\ 2 & 1 & 3 \end{pmatrix}.$$

The table representing S_3 is given below.

o	*e*	*a*	*b*	*c*	*d*	*s*
e	*e*	*a*	*b*	*c*	*d*	*s*
a	*a*	*b*	*e*	*s*	*c*	*d*
b	*b*	*e*	*a*	*d*	*s*	*c*
c	*c*	*d*	*s*	*e*	*a*	*b*
d	*d*	*s*	*c*	*b*	*e*	*a*
s	*s*	*c*	*d*	*a*	*b*	*e*

EXERCISE 6(B)

1. Consider the following elements in S_8 and compute the expressions in (i) to (viii) given below.

$$a = \begin{pmatrix} 1 & 2 & 3 & 4 & 5 & 6 & 7 & 8 \\ 3 & 7 & 4 & 6 & 8 & 5 & 1 & 2 \end{pmatrix}$$

$$b = \begin{pmatrix} 1 & 2 & 3 & 4 & 5 & 6 & 7 & 8 \\ 4 & 5 & 6 & 7 & 8 & 1 & 2 & 3 \end{pmatrix}$$

and

$$c = \begin{pmatrix} 1 & 2 & 3 & 4 & 5 & 6 & 7 & 8 \\ 8 & 7 & 6 & 5 & 1 & 2 & 3 & 4 \end{pmatrix}$$

(i) a^2b
(ii) ab^2
(iii) abc
(iv) ab^2c
(v) a^2bc
(vi) abc^2
(vii) b^2ca
(viii) c^3a

2. For any positive integer n, prove that the order of any element in S_n is finite and is a divisor of $n!$

3. Find the orders of a, b and c given in the Exercise 1 above.

4. Determine all the elements in the cyclic subgroups $<a>$, $<b>$ and $<c>$, where a, b and c are as given above.

5. Prove that S_n is abelian if and only if $n \leq 2$.

6. Can S_8 be a homomorphic image of S_{12}?

7. Can S_8 be isomorphic to a subgroup of S_{12}?

8. Compute aba^{-1}, bcb^{-1} and cac^{-1} for the elements a, b and c given in Exercise 1.

9. Determine the orders of all the elements in S_3.

10. List all the elements of S_4.

6.3 CYCLES

Consider the permutation f in S_5 given by $f(1) = 3, f(2) = 1, f(3) = 5, f(4) = 2$ and $f(5) = 4$. In the array form, f can be expressed as

$$f=\begin{pmatrix}1 & 2 & 3 & 4 & 5\\ 3 & 1 & 5 & 2 & 4\end{pmatrix}.$$

Instead of this, we can write

$$f=\begin{pmatrix}1 & & 3 & & 5 & & 4 & & 2\\ \downarrow & \nearrow & \downarrow & \nearrow & \downarrow & \nearrow & \downarrow & \nearrow & \downarrow\\ 3 & & 5 & & 4 & & 2 & & 1\end{pmatrix}$$

suggesting that $1 \to 3 \to 5 \to 4 \to 2 \to 1$; that is, f maps 1 to 3, 3 to 5, 5 to 4, 4 to 2 and 2 to 1 back. Here, the action of f on the elements of I_5 is cyclic.

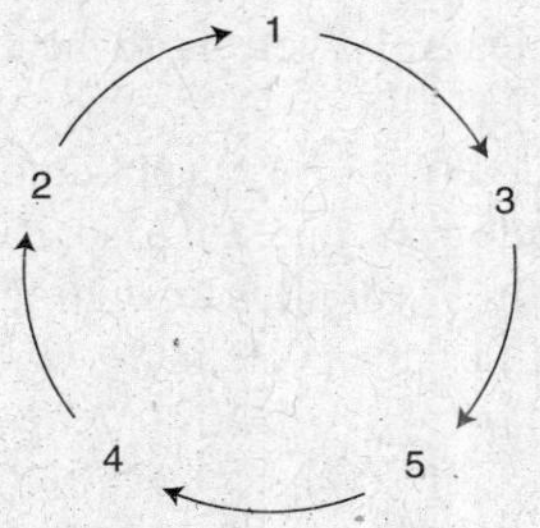

In this case, we can as well denote f by (1 3 5 4 2) hinting each element, except the last, is mapped onto the next and the last element to the first, completing the cycle. Permutations like this are called cycles, which play a vital role in the study of the permutations for the simple reason that any permutation can be expressed as a product (composition) of cycles. Before going to prove this fundamental theorem, we first have the following definition.

Definition 6.3.1. Let n be a positive integer and $i_1, i_2, \ldots, i_r$ be distinct elements in the set $I_n = \{1, 2, \ldots, n\}$. Define $f : I_n \to I_n$ by

$$f(i)=\begin{cases} i_{j+1}, & \text{if } i=i_j,\ 1\le j<r\\ i_1, & \text{if } i=i_r\\ i, & \text{if } i\ne i_j,\ 1\le j\le r\end{cases}.$$

That is, $f(i_1) = i_2, f(i_2) = i_3, \ldots, f(i_{r-1}) = i_r, f(i_r) = i_1$, and $f(i) = i$ for all $i \notin \{i_1, i_2, \ldots, i_r\}$.
The action of f is cyclic on the set $\{i_1, i_2, \ldots, i_r\}$ and f is identity on the complement of $\{i_1, i_2, \ldots, i_r\}$.

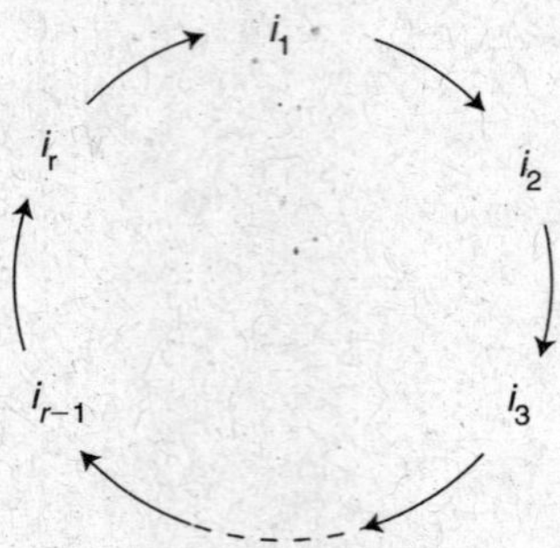

For this reason, f is called *a cycle of length r* or simply an *r-cycle* and is denoted by $(i_1, i_2, \ldots, i_r)$ which is called the *cyclic representation* or *the cyclic form* of the r-cycle. A 1-cycle is to be interpreted as the identity permutation.

The cyclic representation $(i_1\ i_2 \ldots i_r)$ is not unique. For example, $(i_2\ i_3 \ldots i_r\ i_1)$ represents the same cycle as $(i_1\ i_2 \ldots i_r)$. Also, if $a = (i_1\ i_2 \ldots i_r)$ is an r-cycle in S_n, then a is an r-cycle in S_m for all $m \geq n$. In fact, if m is the maximum of $i_1, i_2, \ldots, i_r$, then $(i_1\ i_2 \ldots i_r)$ is an r-cycle in S_m.

Example 6.3.1

1. $a = (2\ 5\ 3\ 4\ 6)$ is a 6-cycle in S_6 and hence in S_m for any $m \geq 6$. a is defined by

$$a(2) = 5, a(5) = 3, a(3) = 4, a(4) = 6, a(6) = 2$$
$$\text{and} \quad a(i) = i \quad \text{for all } i \notin \{2, 3, 4, 5, 6\}.$$

2. Let $a = \begin{pmatrix} 1 & 2 & 3 & 4 & 5 & 6 & 7 & 8 \\ 3 & 6 & 5 & 7 & 2 & 4 & 8 & 1 \end{pmatrix}$. Let us express this in cyclic form.
We have $a(1) = 3, a(3) = 5, a(5) = 2, a(2) = 6, a(6) = 4, a(4) = 7, a(7) = 8$ and $a(8) = 1$. Therefore,

$$a = (1\ 3\ 5\ 2\ 6\ 4\ 7\ 8)$$

Note that $a(1) = 3, a^2(1) = 5, a^3(1) = 2, a^4(1) = 6, a^5(1) = 4, a^6(1) = 7, a^7(1) = 8$ and $a^8(1) = 1$.

The example given in (2) above can be extended to any r-cycle as given in the following.

Theorem 6.3.1. Let a be a cycle in S_n and $a(i) \neq (i)$ for some $i \in I_n$. Then, $a = (i\ c(i)\ c^2(i)\ \dots\ c^{r-1}(i))$ for some $r > 1$.

Proof: Let $a = (i_1\ i_2\ \dots\ i_r)$. Since $a(i) \neq i$, we get that $i = i_k$ for some $1 \leq k \leq r$. Then,

$$\begin{aligned} a(i) &= a(i_k) = i_{k-1} \\ a^2(i) &= a(i_{k+1}) = i_{k+2} \\ a^{r-k}(i) &= i_{r+r-k} = i_r \\ a^{r-k+1}(i) &= a(i_r) = i_1 \\ &\vdots \\ a^{r-1}(i) &= i_{k-1} \\ \text{and} \qquad a^r(i) &= i \end{aligned}$$

Therefore,
$$\begin{aligned} a &= (i_k\ i_{k+1}\ \dots\ i_r\ i_1\ \dots\ i_{k-1}) \\ &= (i\ a(i)\ a^2(i)\ \dots\ a^{r-1}(i)). \end{aligned}$$ ◀

Theorem 6.3.2. Let a be an r-cycle in S_n. Then,

1. $O(a) = r$ (that is, length of a is same as the order of a)
2. $a^{-1} = a^{r-1} = (i_r\ i_{r-1}\ \dots\ i_2\ i_1)$, where $a = (i_1\ i_2\ \dots\ i_r)$
3. For any positive integer m, $a^m = e$ if and only if r divides m.

Proof: Let $a = (i_1\ i_2\ \dots\ i_r)$. Then,

$$a = (i_k\ i_{k+1}\ \dots\ i_r\ i_1\ i_2\ \dots\ i_{k-1}) \quad \text{for any } 1 \leq k \leq r.$$

From the above theorem, it follows that $a^r(i_k) = i_k$ for all $1 \leq k \leq n$ and hence a^r coincides with the identity permutation. Therefore, $a^r = e$. Also, for any $1 \leq k < r$,

$$a^k(i_1) = i_{k+1} \neq i_1$$

and hence $a^k \neq e$ for all $1 \leq k < r$. Thus, r is the smallest positive integer such that $a^r = e$ and hence the order of a is r, which gives (1).

(2) Since $a^r = e$, we get that $a^{-1} = a^{r-1}$. By the above theorem, we get that

$$a^{r-1}(i_r) = i_{r-1}$$

$$a^{r-1}(i_{r-1}) = i_{r-2}, \text{ etc.}$$

$$\text{and hence} \quad a^{r-1} = (i_r\, i_{r-1} \ldots i_2\, i_1).$$

(3) This is clear, since $O(a) = r$. ◀

Note: If $a = (i_1\, i_2 \ldots i_r)$ is an r-cycle, then $a^k(i_j) = i_s$, where $s \equiv j + k \pmod r$.

Definition 6.3.2. A 2-cycle in S_n is called a *transposition*.

Note that any transposition a is a cycle of the form $(i\, j)$ where $i \neq j \in I_n$ and that a interchanges the positions of i and j and fix all the other elements of I_n. That is,

$$a(i) = j, a(j) = i \quad \text{and} \quad a(k) = k \quad \text{for all } k \in I_n - \{i, j\}.$$

Theorem 6.3.3. Any r-cycle in S_n is a product of $r - 1$ transpositions. In fact, if $a = (i_1\, i_2 \ldots i_r)$ is an r-cycle, then

$$a = (i_1\, i_r) \circ (i_1\, i_{r-1}) \circ \ldots \circ (i_1\, i_2).$$

Proof: Let $a = (i_1\, i_2 \ldots i_r)$. If $k \in I_n - \{i_1, i_2, \ldots, i_r\}$, then $a(k) = k = (i_1\, i_j)(k)$ for all $2 \leq j \leq r$ and hence

$$a(k) = ((i_1, i_r) \circ (i_1\, i_{r-1}) \circ \ldots \circ (i_1\, i_2))(k).$$

On the other hand, for $1 \leq j < r$

$$\begin{aligned} a(i_j) = i_{j+1} &= ((i_1\, i_r) \circ (i_1\, i_{r-1}) \circ \ldots \circ (i_1\, i_2))(i_j) \\ ((i_1\, i_r) \circ (i_1\, i_{r-1}) \circ \ldots \circ (i_1\, i_2))(i_j) &= ((i_1\, i_r) \circ (i_1\, i_{r-1}) \circ \ldots \circ (i_1\, i_j))\,(i_j) \\ &= (i_1\, i_r) \circ \ldots \circ (i_1\, i_{j+1})(i_1) \\ &= (i_1\, i_r) \circ \ldots \circ (i_1\, i_{j+2})(i_{j+1}) \\ &= i_{j+1} = a(i_j) \end{aligned}$$

and $(i_1\, i_r) \circ (i_1\, i_{r-1}) \circ \ldots \circ (i_1\, i_2)(i_r) = (i_1 i_r)(i_r) = i_1 = a(i_r)$.

Thus, $a = (i_1\, i_r) \circ (i_1\, i_{r-1}) \circ \ldots \circ (i_1\, i_2)$. ◀

Example 6.3.2. Let $a = (2\ 5\ 3\ 6\ 1\ 4)$ be a 6-cycle in S_6. Then,

$$a = (2\ 4) \circ (2\ 1) \circ (2\ 6) \circ (2\ 3) \circ (2\ 5).$$

Definition 6.3.3. For any permutation f in S_n, the *support* of f is defined as the set $\{i \in S_n : f(i) \neq i\}$ of all elements in I_n which are not fixed by f. The support of f will be denoted by supp(f).

Note that if a is the r-cycle $(i_1\ i_2\ ..\ i_r)$, then the support of a is precisely the set $\{i_1, i_2, \ldots, i_r\}$.

Examples 6.3.3

1. Let $f = \begin{pmatrix} 1 & 2 & 3 & 4 & 5 & 6 & 7 & 8 & 9 \\ 6 & 5 & 3 & 7 & 4 & 1 & 2 & 8 & 9 \end{pmatrix}$.

   Then, supp(f) = $\{1, 2, 4, 5, 6, 7\}$.

2. Let $f = \begin{pmatrix} 1 & 2 & 3 & 4 & 5 \\ 4 & 2 & 3 & 5 & 1 \end{pmatrix}$ and $g = \begin{pmatrix} 1 & 2 & 3 & 4 & 5 & 6 & 7 \\ 5 & 6 & 7 & 3 & 4 & 2 & 1 \end{pmatrix}$.

   Then, supp(f) = $\{1, 4, 5\}$ and supp(g) = $\{1, 2. 3, 4, 5, 6, 7\}$.

   Also $f \circ g = \begin{pmatrix} 1 & 2 & 3 & 4 & 5 & 6 & 7 \\ 1 & 6 & 7 & 3 & 5 & 2 & 4 \end{pmatrix}$

   and $g \circ f = \begin{pmatrix} 1 & 2 & 3 & 4 & 5 & 6 & 7 \\ 3 & 6 & 7 & 4 & 5 & 2 & 1 \end{pmatrix}$

   and therefore supp($f \circ g$) = $\{2, 3, 4, 6, 7\}$
   and supp($g \circ f$) = $\{1, 2, 3, 6, 7\}$.

Definition 6.3.4. Two permutations f and g in S_n are said to be *disjoint* if their supports are disjoint sets.

Permutations $f_1, f_2, \ldots, f_r$ are said to be disjoint if they are pair-wise disjoint; equivalently, if $f_k(i) \neq i$ for some k, then $f_j(i) = (i)$ for all $j \neq k$.

Theorem 6.3.4. Any two disjoint permutations in S_n commute; that is, $f \circ g = g \circ f$ if f and g are disjoint.

Proof: Let f and g be disjoint permutations in S_n. Suppose that supp(f) = A and supp(g) = B. Then, $A \cap B = \varnothing$ and hence, for any $1 \leq i \leq n$, $i \notin A$ or $i \notin B$; that is, $f(i) = i$ or $g(i) = i$.
If $i \notin A \cup B$, then $f(i) = i = g(i)$ and hence

$$(f \circ g)(i) = f(g(i)) = f(i) = i = (g \circ f)(i).$$

If $i \in A$, then $f(i) \neq i$ and $g(i) = i$ and hence

$$(f \circ g)(i) = f(g(i)) = f(i) = g(f(i)) = (g \circ f)(i)$$

(since $f(i) \neq i$ and f is an injection, $f(f(i)) \neq f(i)$ and hence $f(i) \in A$ so that $f(i) \notin B$ and $g(f(i)) = f(i)$). Similarly, if $i \in B$, then $g(i) \neq i$ and $f(i) = i$ and hence $(f \circ g)(i) = (g \circ f)(i)$. Thus, $f \circ g = g \circ f$. ◀

In the following, we prove a fundamental theorem on permutations which will be an important tool in the study of permutations.

Theorem 6.3.5. Any nonidentity permutation f in S_n can be expressed as

$$f = a_1 \circ a_2 \circ \ldots \circ a_s$$

where $a_1, a_2, \ldots, a_s$ are pair-wise disjoint cycles each of length atleast two. This expression of f is unique except for the order of occurrences of the cycles a_i.

Proof: Let $e \neq f \in S_n$ and $A = \text{supp}(f)$. For any i and $j \in A$, define $i \sim j$ if and only if $f^r(i) = j$ for some $r \in \mathbb{Z}$. Note that $f^0(i) = e(i) = i$ and that $f^r(i) = j$ if and only if $f^{-r}(j) = i$. These imply that $\sim$ is an equivalence relation on $A = \text{supp}(f)$. Let $\tilde{i}$ be the equivalence class of $\sim$ containing $\sim$. Then, we shall prove the following for any $i \in A$.

1. $\tilde{i} = \{f^r(i) : r \in \mathbb{Z}\}$.
2. There exist $r \in \mathbb{Z}^+$ such that
   $\tilde{i} = \{i, f(i), f^2(i), \ldots, f^{r-1}(i)\}$,
   $f^r(i) = i$ and $f^s(i) \neq f^t(i)$ for all $0 \leq s \neq t < r$.
3. The restriction of f to $\tilde{i}$ is an r-cycle, $r > 1$.

1. Follows from the definition of $\sim$.
2. First note that $j \in \tilde{i} \Rightarrow f(j) \in \tilde{i}$. Since I_n and hence supp(f) is finite, $\tilde{i}$ is also a finite set, therefore, there exists least positive integer r such that i, $f(i), f^2(i), \ldots, f^{r-1}(i)$ are all distinct and $f^r(i) = i$. Thus, we have (2).
3. The restriction of f to $\tilde{i}$ is the r-cycle

$$(i\, f(i)\, f^2(i) \ldots f^{r-1}(i)).$$

Since $f^r(i) = i$ and $i \in \text{supp}(f)$, it follows that $r \neq 1$ and hence $r > 1$.

Now, since supp(f) is finite, the number of equivalence classes of $\sim$ in supp(f) is finite. Let

$$\tilde{i}_1, \tilde{i}_2, \ldots, \tilde{i}_s$$

be all the distinct equivalence classes of ~ in supp(f). These are pair-wise disjoint and their union is the support of f. Corresponding to each $\tilde{i}_j$, $1 \le j \le s$, define $a_j : I_n \to I_n$ by

$$a_j(k) = \begin{cases} f(k), & \text{if } k \in \tilde{i}_j \\ k, & \text{if } k \notin \tilde{i}_j \end{cases}.$$

The restrictions of a_j and f to $\tilde{i}_j$ are equal and hence, by (3) above, a_j is a cycle of length atleast two and clearly

$$f(k) = (a_1 \circ a_2 \circ \ldots \circ a_s)(k) \quad \text{for all } k \in I_n.$$

Thus, $\qquad f = a_1 \circ a_2 \circ \ldots \circ a_s$.
Also, since distinct equivalence classes are disjoint, $a_1, a_2, \ldots, a_s$ are pair-wise disjoint cycles, each of length atleast two.

The uniqueness of this representation of f is a direct consequence of the facts that a_j and f coincide on $\tilde{i}_j$ and a_j is the identity outside $\tilde{i}_j$. ◀

Corollary 6.3.1. For any integer $n > 1$, every permutation in S_n is a product of transpositions.

Proof: This is a consequence of the above theorem and the fact that any cycle is a product of transpositions (see Theorem 6.3.3). Note that the identity permutation e can be expressed as

$$e = (i\ j) \circ (i\ j)$$

where $(i\ j)$ is any transposition in S_r (since $n > 1$, there is a transposition in S_n). ◀

Note 6.3.1. An algorithm is given below for expressing a given permutation as a product of disjoint cycles.
Let f be a permutation in S_n and

$$A = \text{supp}(f) = \{i \in I_n : f(i) \ne i\}.$$

Choose $i_1 \in A$ and consider

$$i_1, f(i_1), f^2(i_1), f^3(i_1), \ldots$$

There should exist least $r_1 > 1$ (since all these are elements in the finite set A) at which $f^{r_1}(i_1) = i_1$.

Now, consider the r_1-cycle

$$a_1 = (i_1 f(i_1) f^2(i_1) \ldots f^{r_1-1}(i_1))$$

Clearly $r > 1$ and $f^{r_1}(i_1) = i_1$. Next choose $i_2 \in A - \{i_1, f(i_1), \ldots, f^{r_1-1}(i_1)\}$ and consider

$$i_2, f(i_2), f^2(i_2), f^3(i_2), \ldots$$

By the same argument given by, there exists $r_2 > 1$ such that $f^{r_2}(i_2) = i_2$ and

$$i_2, f(i_2), f^2(i_2), \ldots, f^{r_2-1}(i_2)$$

are all distinct. Now, consider the r_2-cycle

$$a_2 = (i_2 f(i_2) f^2(i_2) \ldots f^{r_2-1}(i_2)).$$

Next choose $i_3 \in A - \{i_1, f(i_1), \ldots, f^{r_1-1}(i_1), i_2, f(i_2), \ldots, f^{r_2-1}(i_2)\}$ and continue the above process. This process terminates when all the elements of the support of f exhaust. Then, we get disjoint cycles $a_1, a_2, \ldots, a_s$ such that $f = a_1 \circ a_2 \circ \ldots \circ a_s$. Since a_i's are pair-wise disjoint, they commute with each other and hence

$$f = a_{\sigma(1)} \circ a_{\sigma(2)} \circ \ldots \circ a_{\sigma(s)}$$

for any permutation σ on $\{1, 2, \ldots, s\}$.

Worked Exercise 6.3.1. Express $f = \begin{pmatrix} 1 & 2 & 3 & 4 & 5 & 6 & 7 & 8 & 9 \\ 3 & 6 & 5 & 4 & 1 & 2 & 8 & 9 & 7 \end{pmatrix}$ as a product of disjoint cycles and as a product of transpositions.

Answer: We have supp(f) = $\{1, 2, 3, 5, 6, 7, 8, 9\}$.
Now $1 \in$ supp(f) and consider

$$1, f(1) = 3, f^2(1) = 5, f^3(1) = 1.$$

(1 3 5) is the cycle $(1\, f(1)\, f^2(1))$.

Next choose $2 \in$ supp(f) $- \{1, 3, 5\}$ and consider

$$2, f(2) = 6, f^2(2) = 2.$$

(2 6) is the cycle $(2\, f(2))$.
Next choose $7 \in$ supp(f) $- \{1, 3, 5, 2, 6\}$ and consider

$$7, f(7) = 8, f^2(7) = 9, f^3(7) = 7.$$

(7 8 9) is the cycle $(7\ f(7)\ f^2(7))$. We have exhausted all the elements of supp(f) and hence

$$f = (1\ 3\ 5) \circ (2\ 6) \circ (7\ 8\ 9).$$

By using Theorem 6.3.3, we have

$$\begin{aligned} (1\ 3\ 5) &= (1\ 5) \circ (1\ 3) \\ \text{and} \quad (7\ 8\ 9) &= (7\ 9) \circ (7\ 8) \\ \text{and hence} \quad f &= (1\ 5) \circ (1\ 3) \circ (2\ 6) \circ (7\ 9) \circ (7\ 8). \end{aligned}$$

Worked Exercise 6.3.2. Let $i_1, i_2, \ldots, i_r$ be given distinct elements in I_n. How many distinct r-cycles can be formed using all the $i_1, i_2, \ldots, i_r$.

Answer: Any permutation σ on $\{1, 2, \ldots, r\}$ gives us an r-cycle $(i_{\sigma(1)}\ i_{\sigma(2)} \ldots i_{\sigma(r)})$ and every r-cycle formed using all the $i_1, i_2, \ldots, i_r$ must be of this form. But, we have discount repetitions, since if $(i_{\sigma(1)}\ i_{\sigma(2)} \ldots i_{\sigma(r)})$ is an r-cycle, then $(i_{\sigma(k)}\ i_{\sigma(k+1)} \ldots i_{\sigma(r)}\ i_{\sigma(1)}\ i_{\sigma(2)} \ldots i_{\sigma(k-1)})$ is the same r-cycle for any $1 \leq k \leq r$. Therefore, there are exactly $\frac{r!}{r} (= (r-1)!)$ distinct r-cycles formed by using all the $i_1, i_2, \ldots, i_r$.

Expressing a permutation as a product of disjoint cycles is an important tool in determining the order of that permutation. The following result gives us a formula for the order of a permutation.

Theorem 6.3.6. Let $f_1, f_2, \ldots, f_r$ be pair-wise disjoint permutations in S_n and $f = f_1 \circ f_2 \circ \ldots \circ f_s$. Then,

$$O(f) = \text{l.c.m. of } \{O(f_1), O(f_2), \ldots, O(f_s)\}.$$

Proof: Let $O(f_i) = r_i$, $r = $ l.c.m. of $\{r_1, r_2, \ldots, r_s\}$ and $r = r_i t_i$, $t_i \in \mathbb{Z}^+$. Since f_i's are pair-wise disjoint, we get that

$$f_i \circ f_j = f_j \circ f_i \quad \text{for all } 1 \leq i, j \leq s.$$

Now,

$$\begin{aligned} f^r &= (f_1 \circ f_2 \circ \ldots \circ f_s)^r \\ &= f_1^r \circ f_2^r \circ \ldots \circ f_s^r \ (\text{since } f_i \circ f_j = f_j \circ f_i) \\ &= f_1^{r_1 t_1} \circ f_2^{r_2 t_2} \circ \cdots \circ f_s^{r_s t_s} \\ &= e \ (\text{since } O(f_i) = r_i,\ f_i^{r_i} = e). \end{aligned}$$

On the other hand, for any positive integer t,

$$
\begin{aligned}
f^t = e &\Rightarrow f_1^t \circ f_2^t \circ \ldots \circ f_s^t = e \\
&\Rightarrow f_1^t = f_2^t = \cdots = f_s^t = e \quad \text{(since } f_i\text{'s are disjoint)} \\
&\Rightarrow \mathrm{O}(f_i) \text{ divides } t \text{ for all } 1 \le i \le s \\
&\Rightarrow r_i \text{ divides } t \text{ for all } 1 \le i \le s \\
&\Rightarrow r \text{ divides } t \text{ (since } r = \text{l.c.m. of } \{r_1, \ldots, r_s\}).
\end{aligned}
$$

Thus, $\mathrm{O}(f) = r = \text{l.c.m. of } \{\mathrm{O}(f_1), \mathrm{O}(f_2), \ldots, \mathrm{O}(f_s)\}$.

Corollary 6.3.2. Let $f = a_1 \circ a_2 \circ \ldots \circ a_s$, where $a_1, a_2, \ldots, a_s$ are pair-wise disjoint cycles of length $r_1, r_2, \ldots, r_s$, respectively. Then,

$$\mathrm{O}(f) = \text{l.c.m. of } \{r_1, r_2, \ldots, r_s\}.$$

Proof: Since a_i is a cycle of length r_i, we have $\mathrm{O}(a_i) = r_i$ and hence

$$
\begin{aligned}
\mathrm{O}(f) &= \text{l.c.m. of } \{\mathrm{O}(a_1), \mathrm{O}(a_2), \ldots, \mathrm{O}(a_s)\} \\
&= \text{l.c.m. of } \{r_1, r_2, \ldots, r_s\}.
\end{aligned}
$$ ◀

EXERCISE 6(C)

1. State whether each of the following is true or false:
   - (i) Every cycle is a transposition.
   - (ii) Every transposition is a cycle.
   - (iii) Every cycle is a permutation.
   - (iv) Every permutation is a cycle.
   - (v) Every transposition is a permutation.
   - (vi) Every permutation is a product of disjoint transpositions.
   - (vii) $f^2 = e$ for any transposition f.
   - (viii) $f^2 = e$ implies that f is a transposition.
   - (ix) For any f and g in S_n, $f \circ g = e \Rightarrow f = e = g$
   - (x) $\mathrm{supp}(f) = \mathrm{supp}(g)$ if and only if $f = g$.
2. Which of the following are cycles? If they are cycles, then express them in cyclic representation.
   - (i) $\begin{pmatrix} 1 & 2 & 3 & 4 & 5 & 6 & 7 & 8 \\ 4 & 6 & 8 & 2 & 3 & 5 & 1 & 7 \end{pmatrix}$

(ii) $\begin{pmatrix} 1 & 2 & 3 & 4 & 5 & 6 \\ 2 & 3 & 4 & 1 & 6 & 5 \end{pmatrix}$

(iii) $\begin{pmatrix} 1 & 2 & 3 & 4 & 5 & 6 & 7 & 8 \\ 2 & 3 & 4 & 5 & 6 & 7 & 8 & 1 \end{pmatrix}$

(iv) $\begin{pmatrix} 1 & 2 & 3 & 4 & 5 & 6 & 7 & 8 & 9 \\ 8 & 6 & 7 & 9 & 5 & 1 & 2 & 3 & 4 \end{pmatrix}$

3. Express the following as products of disjoint cycles, each of length atleast two. Also express each of following as a product of transpositions.

(i) $\begin{pmatrix} 1 & 2 & 3 & 4 & 5 & 6 & 7 & 8 & 9 \\ 4 & 3 & 2 & 1 & 5 & 8 & 6 & 7 & 9 \end{pmatrix}$

(ii) $\begin{pmatrix} 1 & 2 & 3 & 4 & 5 & 6 \\ 3 & 2 & 4 & 1 & 6 & 5 \end{pmatrix} \text{o} \begin{pmatrix} 1 & 2 & 3 & 4 & 5 & 6 \\ 5 & 1 & 4 & 6 & 2 & 3 \end{pmatrix}$

(iii) $\begin{pmatrix} 1 & 2 & 3 & 4 & 5 & 6 & 7 & 8 \\ 2 & 3 & 1 & 4 & 6 & 5 & 7 & 8 \end{pmatrix} \text{o} \begin{pmatrix} 1 & 2 & 3 & 4 & 5 & 6 \\ 5 & 6 & 2 & 3 & 4 & 1 \end{pmatrix}$

(iv) $\begin{pmatrix} 1 & 2 & 3 \\ 2 & 3 & 1 \end{pmatrix} \text{o} \begin{pmatrix} 1 & 2 & 3 & 4 & 5 & 6 \\ 3 & 1 & 2 & 5 & 6 & 4 \end{pmatrix}$

(v) $(1\ 3\ 7\ 4\ 6) \text{o} (2\ 3\ 5\ 6\ 4) \text{o} (8\ 7\ 6\ 2\ 4\ 3\ 5)$

(vi) $(1\ 2\ 3\ 4) \text{o} (2\ 3\ 4\ 5) \text{o} (3\ 4\ 5\ 6)$

4. Determine the order of each of the following permutations.

(i) $\begin{pmatrix} 1 & 2 & 3 & 4 & 5 & 6 & 7 & 8 & 9 \\ 2 & 4 & 6 & 1 & 7 & 3 & 8 & 9 & 5 \end{pmatrix}$

(ii) $\begin{pmatrix} 1 & 2 & 3 & 4 & 5 & 6 \\ 4 & 3 & 2 & 1 & 6 & 5 \end{pmatrix} \text{o} \begin{pmatrix} 1 & 2 & 3 & 4 & 5 \\ 2 & 3 & 4 & 1 & 5 \end{pmatrix}$

(iii) $(5\ 4\ 3\ 2) \text{o} (1\ 2\ 3\ 4\ 5\ 6) \text{o} (2\ 4\ 6\ 1\ 3\ 5)$

(iv) $(8\ 7\ 6\ 9\ 3\ 4) \text{o} (4\ 3\ 9\ 6\ 7\ 8) \text{o} (3\ 4\ 5\ 6\ 7\ 8\ 9)$

5. For any permutations f and g in S_n, prove that $\text{supp}(f \text{ o } g) \subseteq \text{supp}(f) \cup \text{supp}(g)$.

6. If f and g are disjoint permutations in S_n, prove that

$$\text{supp}(f \text{ o } g) = \text{supp}(f) \cup \text{supp}(g)$$

and
$$\text{supp}(f) \cap \text{supp}(g) = \varnothing.$$

7. If $f = f_1 \text{ o } f_2 \text{ o } \ldots \text{ o } f_r$ and f_i's are pair-wise disjoint permutations in S_n, prove that $\text{supp}(f_1), \ldots, \text{supp}(f_r)$ form a partition of $\text{supp}(f)$.

8. For $f \in S_n$ and for $m \in \mathbb{Z}$, prove that $\text{supp}(f^m) \subseteq \text{supp}(f)$.

9. For any disjoint permutations f and g, prove that $f \text{ o } g = e$ if and only if $f = e = g$.

10. For any permutation f in S_n, prove that supp(f) $= \varnothing$ if and only if $f = e$.

11. Let $f \in S_n$. Prove that f is a transposition if and only if supp(f) is a 2-element set.

12. Prove that f is a 3-cycle if and only if supp(f) is a 3-element set. Can this be extended for any r-cycle?

13. Let $e \neq f \in S_n$. Prove that $f^2 = e$ if and only if f is a product of disjoint transpositions.

14. For any f and $g \in S_n$, prove that $O(f) = O(g f g^{-1})$.

15. For any $1 \leq r \leq n$, prove that there is an element in S_n whose order is r.

16. If a is an r-cycle and r is odd, prove that a^2 is also a cycle.

17. If r is even in Exercise 1b, then can a^2 be a cycle? Substantiate your answer.

18. If a is an r-cycle and $1 \leq s < r$ such that r and s are relatively prime, then prove that a^s is also an r-cycle.

19. For any r-cycle a, prove that $f \circ a \circ f^{-1}$ is also an r-cycle for any permutation f.

20. If a and b are disjoint cycles, prove that $f \circ a \circ f^{-1}$ and $f \circ b \circ f^{-1}$ are also disjoint cycles.

21. If $f = a_1 \circ a_2 \circ \ldots \circ a_s$ is a representation of a permutation f in S_n as a product of disjoint cycles and g is any permutation in f, then prove that

$$g \circ f \circ g^{-1} = (g \circ a_1 \circ g^{-1}) \circ (g \circ a_2 \circ g^{-1}) \circ \ldots \circ (g \circ a_s \circ g^{-1})$$

is a representation of $g \circ g \circ g^{-1}$ as a product of disjoint cycles.

22. For any positive integer n, a *partition* of n is defined to be a finite sequence $r_1, r_2, \ldots, r_s$ of positive integers such that $r_1 \leq r_2 \leq \cdots \leq r_s$ and $r_1 + r_2 + \cdots + r_s = n$. List all the partitions of 4 and 5.

23. For any permutation f in S_n, let $|f|$ denote the number of elements in the support of f. Let $f = a_1 \circ a_2 \circ \ldots \circ a_s$ where $a_1, a_2, \ldots, a_s$ are disjoint cycles, each of length greater than 1, such that $|a_1| \leq |a_2| \leq \cdots \leq |a_s|$. Then prove that $|a_1|, |a_2|, \ldots, |a_s|$ is a partition of $|f|$.

24. Prove that any permutation in S_n determines a partition of n such that f and g determine the same partition of n if and only if $g = h \circ f \circ h^{-1}$ for some $h \in S_n$.

25. Prove that S_n is generated by the $n - 1$ transpositions (1 2), (1 3), (1 4), …, (1 n).

26. Prove that S_n is generated by (1 2) and (1 2 3 4 … n).

27. Prove that S_n is generated by the transpositions (1 2), (2 3), (3 4), …, (n − 1 n).

28. If $f = (i_1\ i_2 \ldots i_r)$ is an r-cycle in S_n, then prove that $g f g^{-1} = (g(i_1)\ g(i_2) \ldots g(i_n))$ for all $g \in S_n$.

29. For any f and $g \in S_n$, prove that f is an r-cycle if and only if $g f g^{-1}$ is an r-cycle.

30. Prove that any group of order 6 is isomorphism to either $\mathbb{Z}_6$ or S_3.

6.4 ALTERNATING GROUP A_n AND DIHEDRAL GROUP D_n

We have proved in the previous section that every permutation can be expressed as a product of transpositions. This expression is not unique unless the transpositions involved are disjoint. For example, for any distinct i, j, k and l in I_n,

$$(i\,j) \circ (k\,l) \circ (j\,i) = (k\,l)$$
$$(i\,j) \circ (j\,i) = e = (k\,l) \circ (l\,k).$$

Even though an expression of a permutation as a product of transpositions is not unique, the number of transpositions involved in any expression of a given permutation is always even or always odd. That is, if a permutation f can be expressed as a product of even number of transpositions, then any expression of f as a product of transpositions contains even number of transpositions. In order to prove this, we first have the following definition.

Definition 6.4.1. Let f be a nonidentity permutation in S_n and

$$f = a_1 \circ a_2 \circ \ldots \circ a_s$$

where $a_1, a_2, \ldots, a_s$ are pair-wise disjoint cycles. Then, the *Cauchy index* of f is defined as

$$\mathrm{O}(a_1) + \mathrm{O}(a_2) + \cdots + \mathrm{O}(a_s) - s$$

and is denoted by CI(f). Also, for the identity permutation e, we define CI(e) to be 0.

Since any $f \neq e$ can be uniquely, except for the order of occurrences of the cycles, expressed as a product of disjoint cycles, the Cauchy index of f is well-defined and CI(f) is always a positive integer for any $f \neq e$. Note that,

$$\mathrm{CI}(f) = \sum_{i=1}^{s} \mathrm{O}(a_i) - s = |f| - s$$

where $e \neq f = a_1 \circ a_2 \circ \ldots \circ a_s$ is an expression of f as a product of disjoint cycles a_i's and $|f|$ is the number of elements in the support of f. Since each a_i is a cycle of length atleast two, $\mathrm{O}(a_i) > 1$ for all $1 \leq i \leq s$ and hence $|f| > s$, so that CI(f) > 0.

Examples 6.4.1

1. Let $f = \begin{pmatrix} 1 & 2 & 3 & 4 & 5 & 6 & 7 & 8 & 9 \\ 4 & 1 & 3 & 2 & 7 & 9 & 8 & 5 & 6 \end{pmatrix}$

   Then, $f = (1\ 4\ 2) \circ (5\ 7\ 8) \circ (6\ 9) = a_1 \circ a_2 \circ a_3$

   Therefore, $\mathrm{CI}(f) = \mathrm{O}(a_1) + \mathrm{O}(a_2) + \mathrm{O}(a_3) - 3$
   $= 3 + 3 + 2 - 3 = 5$

2. Let $f = (1\ 2) \circ (2\ 5) \circ (4\ 5) \circ (3\ 7) \circ (9\ 3) \circ (6\ 9) \circ (8\ 9)$. Then, we should express f as a product of disjoint cycles, to find the Cauchy index of f. We have

$$f = (1\ 2\ 5\ 4) \circ (3\ 9\ 8\ 6\ 7) = a_1 \circ a_2$$

   Therefore, $\mathrm{CI}(f) = \mathrm{O}(a_1) + \mathrm{O}(a_2) - 2$
   $= 4 + 5 - 2 = 7$

Theorem 6.4.1. If $f \in S_n$ is a product of s transpositions, then

$$\mathrm{CI}(f) + s \text{ is even.}$$

Proof: Let $f \in S_n$ and $f = a_1 \circ a_2 \circ \ldots \circ a_t$, where $a_1, a_2, \ldots, a_t$ are pair-wise disjoint cycles of orders $r_1, r_2, \ldots, r_t$, respectively. Then, we have

$$\mathrm{CI}(f) = r_1 + r_2 + \cdots + r_t - t.$$

We shall prove that

$$\mathrm{CI}((i\,j) \circ f) = \mathrm{CI}(f) \pm 1$$

for any transposition (i, j). To find $\mathrm{CI}((i, j) \circ f)$, we should first express $(i\,j)$ of as a product of disjoint cycles. We do this in the following cases:

Case (1): Suppose that $(i\,j)$ is disjoint with f. In this case, $(i\,j)$ is disjoint with each a_i and

$$(i\,j) \circ f = (i\,j) \circ a_1 \circ a_2 \circ \ldots \circ a_t$$

which is a product of $t + 1$ number of pair-wise disjoint cycles and therefore

$$\begin{aligned} \mathrm{CI}((i\,j) \circ f) &= 2 + r_1 + r_2 + \cdots + r_t - (t + 1) \\ &= (r_1 + r_2 + \cdots + r_t - t) + 1 \\ &= \mathrm{CI}(f) + 1 \end{aligned}$$

Case (2): Suppose that $i \in \mathrm{supp}(f)$ and $j \notin \mathrm{supp}(f)$. Then, $f(i) \neq i$ and $f(j) = j$. We can assume, without loss of generality, that $i \in \mathrm{supp}(a_1)$ and $i \notin \mathrm{supp}(a_k)$ for all $1 < k \leq t$ and that

$$a_1 = (i\ k_1\ k_2 \ldots k_r).$$

Then, $r_1 = O(a_1) = r + 1$ and

$$(i\,j) \circ f = (i\,j) \circ (i\,k_1\,k_2 \ldots k_r) \circ a_2 \circ a_3 \circ \ldots \circ a_t$$
$$= (i\,k_1\,k_2 \ldots k_r\,j) \circ a_2 \circ a_3 \circ \ldots \circ a_t$$

which is a product of t number of disjoint cycles and therefore

$$\begin{aligned} \text{CI}((i\,j) \circ f) &= (r + 2) + r_2 + \cdots + r_t - t \\ &= (r_1 + 1) + r_2 + \cdots + r_t - t \\ &= (r_1 + r_2 + \cdots + r_t - t) + 1 \\ &= \text{CI}(f) + 1 \end{aligned}$$

Case (3): Suppose that both i and j belong to supp(f).

(i) Suppose that both i and j involve in the same cycle, say a_1.
Then, $a_1 = (i\,k_1 \ldots k_r\,j\,m_1 \ldots m_u)$
and $r_1 = O(a_1) = r + u + 2$.
Now, $(i\,j) \circ f = (i\,j) \circ (i\,k_1 \ldots k_r\,j\,m_1 \ldots m_u) \circ a_2 \circ \ldots \circ a_t$
$= (i\,k_1 \ldots k_r) \circ (j\,m_1 \ldots m_u) \circ a_2 \circ \ldots \circ a_t$
which is a product of $t + 1$ number of disjoint cycles and therefore
$$\begin{aligned} \text{CI}((i\,j) \circ f) &= (r + 1) + (u - 1) + r_2 + \cdots + r_t - (t + 1) \\ &= (r + u + 2) + r_2 + \cdots + r_t - (t + 1) \\ &= (r_1 + r_2 + \cdots + r_t - t) - 1 \\ &= \text{CI}(f) - 1. \end{aligned}$$

(ii) Suppose that i and j involve in distinct cycles of f, say

$$a_1 = (i\,k_1 \ldots k_r) \quad \text{and} \quad a_2 = (j\,m_1 \ldots m_u).$$

Then, $r_1 = O(a_1) = r + 1$ and $r_2 = O(a_2) = u + 1$
Now,

$$\begin{aligned} (i\,j) \circ f &= (i\,j) \circ a_1 \circ a_2 \circ \ldots \circ a_t \\ &= (i\,j) \circ (i\,k_1 \ldots k_r) \circ (j\,m_1 \ldots m_u) \circ a_3 \circ \ldots \circ a_t \\ &= (i\,k_1 \ldots k_r\,j\,m_1 \ldots m_u) \circ a_3 \circ \ldots \circ a_t \end{aligned}$$

which is a product $t - 1$ number of disjoint cycles.
Therefore, $\text{CI}((i\,j) \circ f) = (r + u + 2) + r_3 + \cdots + r_t - (t - 1)$
$= (r_1 + r_2 + r_3 + \cdots + r_t - t) + 1$
$= \text{CI}(f) + 1.$

Thus, in any case, $\text{CI}((i\,j) \circ f) = \text{CI}(f) \pm 1$ for any transposition $(i\,j)$.

Now, let f be a product of s transpositions; that is, $f = c_1 \circ c_2 \circ \ldots \circ c_s$, where $c_1, c_2, \ldots, c_s$ are transpositions. Then consider

$$\begin{aligned} 0 = \mathrm{CI}(e) &= \mathrm{CI}(f^{-1} \circ f) \\ &= \mathrm{CI}((c_1 \circ c_2 \circ \ldots \circ c_s)^{-1} \circ f) \\ &= \mathrm{CI}(c_s \circ c_{s-1} \circ \ldots \circ c_2 \circ c_1 \circ f) \\ &= \mathrm{CI}(c_{s-1} \circ \ldots \circ c_2 \circ c_1 \circ f) \pm 1 \\ &= \ldots \\ &= \mathrm{CI}(f) \pm 1 \pm \cdots \pm 1 \\ &= \mathrm{CI}(f) + p - q \end{aligned}$$

where p and q are nonnegative integers such that $p + q = s$. Therefore,

$$\mathrm{CI}(f) + p = q$$
$$\text{and} \quad \mathrm{CI}(f) + s = \mathrm{CI}(f) + p + q = 2q.$$

Since $2q$ is always even, it follows that $\mathrm{CI}(f) + s$ is even. ◀

Corollary 6.4.1. Let $a_1, a_2, \ldots, a_r$ and $b_1, b_2, \ldots, b_s$ be transpositions such that

$$a_1 \circ a_2 \circ \ldots \circ a_r = b_1 \circ b_2 \circ \ldots \circ b_s.$$

Then, $r + s$ is even and hence either both r and s are even or both r and s are odd.

Proof: Let $f = a_1 \circ a_2 \circ \ldots \circ a_r = b_1 \circ b_2 \circ \ldots \circ b_s$.
Then, by the above theorem, $\mathrm{CI}(f) + r$ and $\mathrm{CI}(f) + s$ are even and hence

$$\mathrm{CI}(f) + r + \mathrm{CI}(f) + s \text{ is even.}$$

So that $r + s$ is even. The later assertion follows from the fact that $r + s$ is odd if and only if one of r and s is odd and the other is even. ◀

Corollary 6.4.2. If a permutation can be expressed as a product of even number of transpositions, then it cannot be expressed as a product of odd number of transpositions.

Definition 6.4.2. A permutation in S_n is called an *even permutation* if it can be expressed as a product of even number of transpositions. A permutation which is not even is called an *odd permutation*.

Examples 6.4.2

1. The permutation $f = \begin{pmatrix} 1 & 2 & 3 & 4 & 5 & 6 & 7 & 8 & 9 \\ 4 & 6 & 5 & 7 & 2 & 3 & 1 & 9 & 8 \end{pmatrix}$ can be expressed as

$$\begin{aligned} f &= (1\ 4\ 7) \circ (2\ 6\ 3\ 5) \circ (8\ 9) \\ &= (1\ 7) \circ (1\ 4) \circ (2\ 5) \circ (2\ 3) \circ (2\ 6) \circ (8\ 9) \end{aligned}$$

   which is a product of six transpositions. Therefore, f is an even permutation.
2. Any r-cycle can be expressed as a product of $r - 1$ transpositions and therefore an r-cycle is an even permutation if and only if r is odd.
3. Any three cycle $(i\ j\ k) = (i\ k) \circ (i\ j)$ is an even permutation.
4. Any transposition is an odd permutation.
5. The identity e in S_n is an even permutation if $n > 1$.

Theorem 6.4.2. For any integer $n > 1$, the set of all even permutations in S_n is a normal subgroup of S_n and is of index 2 in S_n.

Proof: Let $n > 1$ and

$$A_n = \text{the set of all even permutations in } S_n.$$

Consider the group $G = \{1, -1\}$ under the usual multiplication of real numbers. Define

$$\theta : S_n \to G \quad \text{by} \quad \theta(f) = \begin{cases} 1, & \text{if } f \text{ is even} \\ -1, & \text{if } f \text{ is odd} \end{cases}.$$

Since $n > 1$, we have $(1, 2) \in S_n$ and $(1, 2)$ is an odd permutation and hence $\theta((1, 2)) = -1$. Also the identity

$$e = (1\ 2) \circ (1\ 2)$$

and hence e is an even permutation so that $\theta(e) = 1$. Therefore, θ is a surjection. For any f and g in S_n, note that $f \circ g$ is even if and only if either both f and g are even or both f and g are odd. Also, for any a and b in G, the product $ab = 1$ if and only if either $a = 1 = b$ or $a = -1 = b$. From these, it follows that θ is a homomorphism and hence an epimorphism. Also,

$$\ker \theta = \{f \in S_n : \theta(f) = 1\} = A_n.$$

The kernel of any homomorphism is a normal subgroup of the domain group and hence A_n is normal subgroup of S_n. Further, by the Fundamental Theorem of Homomorphisms,

$$S_n / A_n \cong G.$$

Therefore, we have

$$i_{S_n}(A_n) = \frac{|S_n|}{|A_n|} = |S_n / A_n| = |G| = 2.$$

Thus, the index of A_n in S_n is 2.

Corollary 6.4.3. For any $n > 1$, $|A_n| = \frac{|S_n|}{2} = \frac{n!}{2}$. ◀

Definition 6.4.3. For any $n > 1$, the group A_n of all even permutations in S_n is called the *alternating group of degree n.*

Worked Exercise 6.4.1. Construct a table representing the alternating group A_3 of degree 3.

Answer: First note that $|A_3| = \frac{3!}{2} = 3$.
The identity e and the 3-cycles $a = (1\ 2\ 3)$ and $b = (1\ 3\ 2)$ are the only even permutations in S_3.
Therefore, $A_3 = \{e, a, b\}$. Note that

$$a^2 = b,\ a^3 = e,\ b^2 = a \quad \text{and} \quad b^3 = e$$

o	e	a	b
e	e	a	b
a	a	b	e
b	b	e	a

Worked Exercise 6.4.2. List all the elements of A_4 and construct a table representing the group A_4.

Answer: First note that $|A_4| = \frac{4!}{2} = 12$.
There are eight 3-cycles, each of which is an even permutation. These are

$$a_1 = (1\ 2\ 3),\ a_2 = (1\ 3\ 2)$$
$$b_1 = (2\ 3\ 4),\ b_2 = (2\ 4\ 3)$$
$$c_1 = (1\ 2\ 4),\ c_2 = (1\ 4\ 2)$$
$$d_1 = (1\ 3\ 4),\ d_2 = (1\ 4\ 3)$$

Also, $p = (1\ 2) \circ (3\ 4)$

$q = (1\ 3) \circ (2\ 4)$

and $r = (1\ 4) \circ (2\ 3)$

are also even permutations. There are only 12 even permutations and therefore the above together with the identity form A_4. That is,

$$A_4 = \{e, a_1, a_2, b_1, b_2\ c_1, c_2\ d_1, d_2\ p, q, r\}.$$

Here, we have

$$\begin{aligned} a_1^2 &= a_2, \quad a_2^2 = a_1, \quad a_1^3 = e = a_2^3 \\ b_1^2 &= b_2, \quad b_2^2 = b_1, \quad b_1^3 = e = b_2^3 \\ c_1^2 &= c_2, \quad c_2^2 = c_1, \quad c_1^3 = e = c_2^3 \\ d_1^2 &= d_2, \quad d_2^2 = d_1, \quad d_1^3 = e = d_2^3 \end{aligned}$$

$$p^2 = q^2 = r^2 = e$$

The following table represents the group A_4.

o	e	a_1	a_2	b_1	b_2	c_1	c_2	d_1	d_2	p	q	r
e	e	a_1	a_2	b_1	b_2	c_1	c_2	d_1	d_2	p	q	r
a_1	a_1	a_2	e	p	c_1	q	d_2	b_1	r	d_1	b_2	c_2
a_2	a_2	e	a_1	d_1	q	b_2	r	p	c_2	b_1	c_1	d_2
b_1	b_1	q	c_2	b_2	e	d_1	p	r	a_1	a_2	d_2	c_1
b_2	b_2	d_2	p	e	b_1	r	a_2	c_1	q	c_2	a_1	d_1
c_1	c_1	r	d_1	a_1	p	c_2	e	q	b_2	d_2	a_2	b_1
c_2	c_2	b_1	q	r	d_2	e	c_1	a_2	p	b_2	d_1	a_1
d_1	d_1	c_1	r	q	a_2	p	b_1	d_2	e	a_1	c_2	b_2
d_2	d_2	p	b_2	c_2	r	a_1	q	e	d_1	c_1	b_1	a_2
p	p	b_2	d_2	c_1	a_1	b_1	d_1	c_2	a_2	e	r	q
q	q	c_2	b_1	a_2	d_1	d_2	a_1	b_2	c_1	r	e	p
r	r	d_1	c_1	d_2	c_2	a_2	b_2	a_1	b_1	q	p	e

Alternating group of degree 4

We know that any 3-cycle is an even permutation. In fact, every even permutation is a product of 3-cycles and hence the 3-cycles generate the group A_n. This is proved in the following theorem.

Theorem 6.4.3. Let $n > 2$ and $i \neq j \in I_n = \{1, 2, ..., n\}$. Then, the alternating group A_n is generated by the 3-cycles of the form $(i\ j\ k)$, where $k \in I_n - \{i, j\}$.

Proof: Let $S = \{(i\ j\ k) : k \in I_n - \{i, j\}\}$ Clearly $S \subseteq A_n$. Any even permutation must be a product of terms of the form

$$(a\ b) \circ (c\ d) \quad \text{or} \quad (a\ b) \circ (a\ c)$$

where a, b, c and d are distinct elements of I_n. Since

$$\begin{aligned} & (a\ b) \circ (c\ d) = (a\ c\ b) \circ (a\ c\ d) \\ \text{and} \quad & (a\ b) \circ (a\ c) = (a\ c\ b), \end{aligned}$$

it follows that A_n is generated by the set of all 3-cycles in S_n. Next, we prove that any 3-cycle can be expressed as product of 3-cycles in S. Any 3-cycle is of the form $(i\ j\ a)$ or $(i\ a\ j)$ or $(i\ a\ b)$ or $(j\ a\ b)$ or $(a\ b\ c)$, where a, b and c are distinct elements $I_n - \{i, j\}$. Now, we have

$$\begin{aligned} (i\ a\ j) &= (i\ j\ a) \circ (i\ j\ a) \\ (i\ a\ b) &= (i\ j\ b) \circ (i\ j\ a) \circ (i\ j\ a) \\ (j\ a\ b) &= (i\ j\ b) \circ (i\ j\ b) \circ (i\ j\ a) \\ \text{and} \quad (a\ b\ c) &= (i\ j\ a)^2 \circ (i\ j\ c) \circ (i\ j\ b)^2 \circ (i\ j\ a) \end{aligned}$$

Thus, every 3-cycle is a product of members of S and hence A_n is generated by S. ◀

Corollary 6.4.4. For any $n \geq 2$, A_n is the smallest subgroup of S_n containing all the 3-cycles in S_n.

Theorem 6.4.4. Let $n > 2$ and N be a normal subgroup of A_n. If N contains a 3-cycle, then $N = A_n$.

Proof: Suppose that N contains a 3-cycle $(i\ j\ k)$, where i, j and k are some distinct elements in I_n.
For any $a \in I_n - \{i, j, k\}$, we have

$$\begin{aligned} (i\ j\ a) &= (i\ j) \circ (k\ a) \circ (i\ j\ k)^2 \circ (k\ a) \circ (i\ j) \\ &= (i\ j) \circ (k\ a) \circ (i\ j\ k)^2 \circ ((i\ j) \circ (k\ a))^{-1} \in fNf^{-1}, \end{aligned}$$

where $f = (i\ j) \circ (k\ a) \in A_n$.

Since N is normal in A_n, it follows that $(i\ j\ a) \in N$. By Theorem 6.4.3, $N = A_n$.

In the following, we prove an important property of the alternating groups; namely, all alternating groups, except A_4, are simple in the following sense.

Definition 6.4.4. A nontrivial group G is said to be *simple* if G does not contain any nontrivial proper normal subgroups; that is, $\{e\}$ and G are the only normal subgroups of G.

Example 6.4.3

1. Any finite group of prime order is simple; for if $|G| = p$, where p is a prime number and H is a subgroup of G, then, by the Lagrange's theorem, $|H|$ divides $|G|$ and hence $|H| = 1$ or p which implies that $H = \{e\}$ or $H = G$.
2. A_4 is a not a simple group, since

$$N = \{e, p\, q, r\}$$

is a normal subgroup of A_4 (see Worked Exercise 6.4.2)
3. In the following, we prove that A_n is simple for any $n \neq 4$.

Theorem 6.4.5. Let $n > 2$. Then, the alternative group A_n is simple if and only if $n \neq 4$.

Proof: Since A_3 is a group of order $3\left(= \frac{3!}{2}\right)$, A_3 is simple (see Example 6.4.3 (1)). Also A_4 is not simple by Example 6.4.3 (2). Now, let $n > 4$. We shall prove that A_n is simple. Let N be a normal subgroup of A_n and $N \neq \{e\}$. By Theorem 6.4.4, it is enough if we can prove that N contains a 3-cycle. We prove this by distinguishing the following cases.

Case (1): Suppose that N contains an element f such that f is the product of disjoint cycles, atleast one of which is of length $r \geq 4$. Then,

$$f = (i_1\, i_2 \ldots i_r) \circ \beta$$

where $r \geq 4$ and $(i_1\, i_2 \ldots i_r)$ and β are disjoint. Now, put $\alpha = (i_1\, i_2\, i_3)$. Then, $\alpha \in A_n$ and, by the normality of N in A_n, $\alpha \circ f \circ \alpha^{-1} \in N$. Now, we have

$$\begin{aligned}(i_1\, i_2\, i_r) &= \beta^{-1} \circ (i_1\, i_r\, i_{r-1} \cdots i_2) \circ \alpha \circ (i_1\, i_2 \cdots i_r) \circ \beta \circ (i_1\, i_3\, i_2)\\ &= f^{-1} \circ (\alpha \circ f \circ \alpha^{-1}) \in N\end{aligned}$$

Thus, N contains a 3-cycle.

Case (2): Suppose that N contains an element f such that f is the product of disjoint cycles, atleast two of which are 3-cycles. Then,

$$f = (i_1\ i_2\ i_3) \circ (i_4\ i_5\ i_6) \circ \beta$$

where $(i_1\ i_2\ i_3)$, $(i_4\ i_5\ i_6)$ and β are disjoint.
Put $\alpha = (i_1\ i_2\ i_4)$. Then, $\alpha \in A_n$ and, by the normality of N in A_n, $\alpha \circ f \circ \alpha^{-1} \in N$.
Now, consider

$$\begin{aligned}(i_1\ i_4\ i_2\ i_6\ i_3) &= \beta^{-1} \circ (i_4\ i_6\ i_5) \circ (i_1\ i_3\ i_2) \circ (i_1\ i_2\ i_4) \circ (i_1\ i_2\ i_3) \\ &\quad \circ (i_4\ i_5\ i_6) \circ \beta \circ (i_1\ i_4\ i_2) \\ &= f^{-1} \circ (\alpha \circ f \circ \alpha^{-1}) \in N\end{aligned}$$

Therefore, N has a 5-cycle and hence, by Case (1), N contains a 3-cycle.

Case (3): Suppose that N contains an element f which is the disjoint product of one 3-cycle and some 2-cycles. Then,

$$f = (i_1\ i_2\ i_3) \circ \beta$$

where $(i_1\ i_2\ i_3)$ and β are disjoint and β is a product of disjoint 2-cycles. Now we have

$$\begin{aligned}(i_1\ i_3\ i_2) &= (i_1\ i_2\ i_3)^2 \\ &= (i_1\ i_2\ i_3)^2 \circ \beta^2 \text{ (since } \beta^2 = e) \\ &= (i_1\ i_2\ i_3) \circ \beta \circ (i_1\ i_2\ i_3) \circ \beta \\ &= f^2 \in N.\end{aligned}$$

Thus, N contains a 3-cycle.

Case (4): Suppose that every element of N is the product of (an even number of) disjoint 2-cycles.
Let $f \in N$ such that $f = (i_1\ i_2) \circ (i_3\ i_4)$
where (i_1, i_2), (i_3, i_4) and N are disjoint. Put $\alpha = (i_1\ i_2\ i_3)$
Then, $\alpha \in A_n$ and therefore $f^{-1} \circ (\alpha \circ f \circ \alpha^{-1}) \in N$. But

$$\begin{aligned}f^{-1} \circ (\alpha \circ f \circ \alpha^{-1}) &= \beta^{-1} \circ (i_3 \cdot i_4) \circ (i_1 \cdot i_2) \circ (i_1 \cdot i_2 \cdot i_3) \circ (i_1 \cdot i_2) \circ \\ &\quad (i_3\ i_4) \circ \beta \circ (i_1\ i_3\ i_2) \\ &= (i_1\ i_3) \circ (i_2\ i_4)\end{aligned}$$

Now, put $\gamma = (i_1\ i_3) \circ (i_2\ i_4)$. We have $\gamma \in N$. since $n \geq 5$, we can choose $j \in I_n - \{i_1, i_2, i_3, i_4\}$. Put $\theta = (i_1\ i_3\ j)$.

Then, $\theta \in A_n$ and $\gamma \circ (\theta \circ \gamma \circ \theta^{-1}) \in N$. Also.

$$\gamma \circ (\theta \circ \gamma \circ \theta^{-1}) = (i_1 i_3) \circ (i_2 i_4) \circ (i_1 i_3 j) \circ (i_1 i_3) \circ (i_2 i_4) \circ (i_1 j i_3)$$
$$= (i_1 i_3 j).$$

Therefore, $(i_1\ i_3\ j) \in N$. Thus, N contains a 3-cycle. Thus, in any case, N contains a 3-cycle and, by Theorem 6.4.4, $N = A_n$. Thus, A_n is simple for all $n \geq 5$. ◀

Theorem 6.4.6. For any $n > 1$, the alternating group A_n is the only subgroup of index 2 in the symmetric group S_n.

Proof: Let H be a subgroup of index 2 in S_n. Then, there is only one left coset of H other than H. Then, $fH = S_n - H$ and hence $fH = gH$ for any f and $g \in S_n - H$. In particular, $fH = f^{-1}H$ for any $f \in S_n - H$ (note that $f \in H$ if and only if $f^{-1} \in H$) and hence $f^2 \in H$ for all $f \in S_n$.
If $\alpha = (i\ j\ k)$ is any 3-cycle in S_n, then $\alpha^3 = e$ and hence

$$\alpha^{-1} = \alpha^2 \in H \quad \text{and therefore } \alpha \in H.$$

Therefore, H contains all 3-cycles. By Corollary 6.4.4, $A_n \subseteq H$. But

$$\frac{|S_n|}{|H|} = i_{S_n}(H) = 2 = i_{S_n}(A_n) = \frac{|S_n|}{|A_n|}$$

and hence $|H| = |A_n|$. Thus, $H = A_n$. ◀

In the following, we shall exhibit another special subgroup of the symmetric group S_n and prove a characterization theorem for it.

Definition 6.4.5. Let $n \geq 3$ and θ and $\phi \in S_n$ be defined as follows:

$$\theta = \text{the } n\text{-cycle } (1\ 2\ 3\ \ldots\ n)$$

$$\text{and} \quad \phi = \begin{pmatrix} 1 & 2 & 3 & 4 & \ldots & n \\ 1 & n & n-1 & n-2 & \ldots & 2 \end{pmatrix}$$

Note that $\phi = \prod_{2 \leq i \leq n+2-i} (i \quad n+2-i)$, the product of the transpositions $(i \quad n+2-i)$. Let D_n be the subgroup of S_n generated by $\{\theta, \phi\}$. D_n is called the *dihedral group of degree n*.
The dihedral group D_n is completely characterized by certain properties of its generators θ and ϕ.

Theorem 6.4.7. Let $n \geq 3$. Then, a group G is isomorphic to the dihedral group D_n of degree n if and only if G is generated by two elements a and b satisfying the following.

1. $O(a) = n$
2. $O(b) = 2$
3. $aba = b$

Proof: Let G be a group. Suppose that $G \cong D_n$ and $f : G \to D_n$ is an isomorphism. Recall that D_n is generated by θ and ϕ, where

$$\theta = (1 \quad 2 \quad 3 \quad \ldots \quad n)$$

$$\text{and} \quad \phi = \begin{pmatrix} 1 & 2 & 3 & 4 & \cdots & n \\ 1 & n & n-1 & n-2 & \cdots & 2 \end{pmatrix}.$$

Being an n-cycle, θ is of order n. Also,

$$\phi = \prod_{2 \leq i \leq n+2-i} (i, \quad n+2-i) = (2 \quad n) \circ (3 \quad n-1) \circ (4 \quad n-2) \circ \cdots$$

which is a product of disjoint transpositions. Since the order of any transposition is 2, we get that $O(\phi) = 2$. Also, it can be easily verified that

$$\theta \circ \phi \circ \theta = \phi.$$

If we choose a and $b \in G$ such that $f(a) = \theta$ and $f(b) = \phi$, then we get that G is generated by a and b, $O(a) = n$, $O(b) = 2$ and $aba = b$.
Conversely suppose that G is generated by two elements a and b, such that $O(a) = n$, $O(b) = 2$ and $aba = b$ (or $ab = ba^{-1}$ or $ba = a^{-1}b$). Then, for any integers j and k,

$$ba^k = a^{-1}ba^{k-1} = a^{-2}ba^{k-2} = \cdots = a^{-k}b$$

$$\text{and} \quad b^j a^k = b^{j-1}a^{-k}b = b^{j-2}a^k b^2 = \cdots = a^{(-1)^j k} b^j.$$

From these relations and from the hypothesis that G is generated by a and b, it follows that every element of G can be expressed as $a^k b^j$ for some integers k and j. Since $O(a) = n$,

$$\{a^i : i \in \mathbb{Z}\} = \langle a \rangle = \{e, a, a^2, \ldots, a^{n-1}\}.$$

Also, since $O(b) = 2$, we have $b = b^{-}$ and hence

$$b^j = \begin{cases} e, & \text{if } j \text{ is even} \\ b, & \text{if } j \text{ is odd} \end{cases}.$$

Thus, $G = \{a^k b^j : 0 \leq k < n \text{ and } j = 0 \text{ or } 1\}$.
Further, suppose that $a^k b^j = a^r b^s$, where $0 \leq k, r < n$ and $j, s \in \{0, 1\}$. Then,

$$a^{k-r} = b^{s-j}.$$

If $s \neq j$, then $s - j = 1$ or -1, so that $a^{k-r} = b$ and

$$a^{k-r+2} = a(a^{k-r})a = aba = b = a^{k-r}$$

and hence $a^2 = e$, which is a contradiction (since $O(a) = n > 2$). Therefore, $s = j$ and $a^{k-r} = e$ so that $k = r$.
Thus, every element of G can be uniquely expressed as $a^k b^j$ where $0 \leq k < n$ and $j = 0$ or 1. Now, define

$$f: G \to D_n \quad \text{by} \quad f(a^k b^j) = \theta^k \phi^j$$

for all $0 \leq k < n$ and $j = 0$ or 1. Using the fact that θ and ϕ satisfy the same conditions (1), (2) and (3) in D_n as a and b in G, it can easily checked that f is an isomorphism. Thus, $G > D_n$. ◀

Corollary 6.4.5. The order of the dihedral group D_n of degree n is $2n$.

Proof: This is an immediate consequence of the fact that any element of D_n can be uniquely expressed as $\theta^k \phi^j$, where $0 \leq k < n$ and $j = 0$ or 1.

Recall from Example 3.2.8 that the group of symmetries of a square (a regular 4-gon) is of order 8. In fact, we prove in the following theorem that the group of symmetries of a square is isomorphic to the dihedral group D_4 of degree 4.

Theorem 6.4.8. Let $n \geq 3$ and D'_n be the group of all symmetries of a regular n-gon (a polygon of n equal sides). Then,

$$D_n \cong D'_n.$$

Proof: Consider a regular n-gon. Without loss of generality, we can assume that one of the vertices of the n-gon lies on the X-axis. First observe that D'_n is generated by the rotation a and the reflection b, where a and b are analytically given by

$$a(x, y) = \left(x\cos\frac{2\pi}{n} - y\sin\frac{2\pi}{n},\ x\sin\frac{2\pi}{n} + y\cos\frac{2\pi}{n}\right)$$

and $\quad b(x, y) = (x, -y)$.

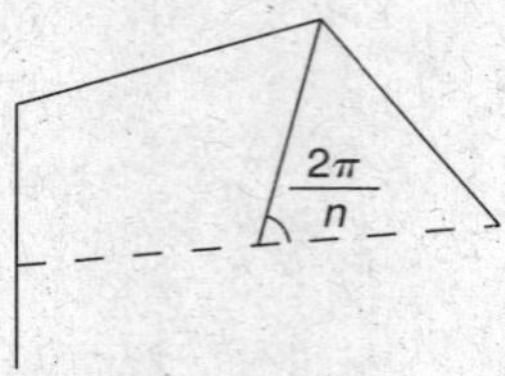

It is clear that $O(a) = n$, $O(b) = 2$ and $aba = b$. Therefore, by Theorem 6.4.7, $D'_n \cong D_n$.

Corollary 6.4.6. The group of symmetries of a square is isomorphic to the dihedral group D_4 of degree 4 and the group of symmetries of an equilateral triangle (see Worked Exercise 3.3.7) is isomorphic to D_3.

Worked Exercise 6.4.3. Prove that the dihedral group D_n, $n \geq 3$, is not simple.

Answer: D_n is generated by two elements θ and ϕ, where $O(\theta)^n$, $O(\phi) = 2$ and $\theta \circ \phi \circ \theta = \phi$. Consider the subgroup A generated by θ in D_n. Then, $|A| = O(\theta) = n$. Since $|D_n| = 2n$, the index of A in D_n is 2 and hence A is normal in D_n, $A \neq \{e\}$ and $A \neq D_n$. Thus, D_n has a nontrivial proper normal subgroup and hence D_n is not simple.

EXERCISE 6(D)

1. State whether the following are true or false and substantiate your answers.
   - (i) The Cauchy index of a permutation f is equal to the number of elements in the support of f.
   - (ii) The order of A_5 is even.
   - (iii) A_3 is an abelian group.
   - (iv) A_4 is an abelian group.
   - (v) $|A_5| = 120$
   - (vi) A_2 is trivial.

(vii) The Cauchy index of any permutation is even.

(viii) The set of all odd permutations is a subgroup of S_n.

(ix) $CI(f \circ g) = CI(f) + CI(g)$ for all f and $g \in S_n$.

(x) For any $n > 2$, A_n has a subgroup of order 3.

2. Determine the Cauchy index of each of the following permutation.

(i) $\begin{pmatrix} 1 & 2 & 3 & 4 & 5 & 6 & 7 & 8 & 9 \\ 3 & 4 & 2 & 1 & 5 & 7 & 8 & 9 & 6 \end{pmatrix}$

(ii) $\begin{pmatrix} 1 & 2 & 3 & 4 & 5 & 6 & 7 & 8 & 9 \\ 3 & 4 & 6 & 7 & 1 & 9 & 8 & 2 & 5 \end{pmatrix}$

(iii) $(3\,5\,2\,6\,4) \circ (4\,5\,6\,7\,3) \circ (7\,2\,8\,3\,9)$

(iv) $(7\,8\,3\,4\,9\,6) \circ (5\,6\,7\,3) \circ (4\,2\,6\,8)$

3. Which of the following are even permutations?

(i) $\begin{pmatrix} 1 & 2 & 3 & 4 & 5 & 6 & 7 & 8 \\ 7 & 8 & 6 & 5 & 3 & 4 & 2 & 1 \end{pmatrix}$

(ii) $\begin{pmatrix} 1 & 2 & 3 & 4 & 5 & 6 & 7 & 8 & 9 \\ 9 & 8 & 7 & 6 & 5 & 4 & 3 & 2 & 1 \end{pmatrix}$

(iii) $(2\,4\,8\,6) \circ (4\,3\,2\,6\,1) \circ (8\,5\,3)$

(iv) $(8\,4\,6\,5\,3) \circ (7\,8\,9\,4\,5\,6) \circ (1\,2\,3\,4\,5\,6)$

4. Prove that the Cauchy index of any permutation f is equal to that of f^{-1}.

5. Prove that $CI(f)$ is a nonnegative integer for any f in S_n.

6. For any permutations f and g in S_n, prove that that $CI(f) = CI(g \circ f \circ g^{-1})$.

7. For any $f \in S_n$, prove that $CI(f) = 0 \Leftrightarrow f$ is the identity.

8. Prove that $CI(f) = 1$ if and only if f is a transposition.

9. Determine all the permutations in S_4 whose Cauchy index is two.

10. Determine all the elements in S_6 whose Cauchy index is one.

11. Find all the positive integers n for which A_n is a cyclic group.

12. Prove that A_n is abelian if and only if $n < 4$.

13. List all odd permutations in S_3 and S_4.

14. For any $n > 1$, prove that the number of odd permutations in S_n is equal to that of even permutations in S_n.

15. How many odd permutations are there in S_8?

16. Show that A_8 contains an element of order 15.

17. What is the maximum order of any element in A_{10}?

18. Prove that f^2 is even for all permutations f in S_n.

19. Find all maximal subgroups M of S_n such that $f^2 \in M$ for all $f \in S_n$.

20. Prove that there is no proper subgroup of S_n containing A_n properly.

21. Let H be a subgroup of S_n containing an odd permutation. Then prove that exactly half of the number of elements in H are even.

22. Prove that the order of any subgroup of S_n containing an odd permutation is even.

23. If f is an odd permutation in S_n, then prove that fA_n is the set of all odd permutations in S_n.

24. Prove that any element of order 5 in S_6 is a 5-cycle.

25. Prove that $f \circ g \circ f^{-1} \circ g^{-1}$ is even for all f and $g \in S_n$.

26. Prove that there is no subgroup of order 6 in A_4. What does this say about the Lagrange's theorem.

27. How many elements of order 5 are there in S_7?

28. Determine the centralizers of (2 4 1) and (1 2) o (3 4) in A_4.

29 If $\alpha \in S_7$ such that $\alpha^4 = (2\ 1\ 4\ 3\ 5\ 6\ 7)$, then what is α?

30. Prove that (1 2 3 4) is not a product of 3-cycles in S_n for all $n \geq 4$.

31. If $f = (1\ 2\ 3) \circ (1\ 4\ 5)$, then express f^{99} as a product of disjoint cycles.

32. If $f = (1\ 3\ 5\ 7\ 9\ 8\ 6) \circ (2\ 4\ 10)$, what is the smallest positive integer n for which $f^n = f^{-5}$?

33. Prove that $\{f \in S_6 : f(3) = 3 \text{ and } f(5) = 5\}$ is a subgroup of S_6. What is its order?

34. How many elements of order 7 are there in A_8?

35. If $f = (9\ 7\ 5\ 3\ 1) \circ (6\ 4\ 2) \circ (8\ 10)$ and f^m is a 5-cycle, then what can we say about m?

36. If f is a 10-cycle, then find all the integers m between 2 and 10 for which f^m is also a 10-cycle.

37. For any f and g in S_n, prove that g is even if and only if $f \circ g \circ f^{-1}$ is even.

38. Prove that the set of all odd permutations in S_n is a coset of A_n in S_n.

39. Prove that the centre of the group S_n is trivial for any $n \leq 3$.

40. How many 3-cycles are there in S_5?

41. For any $1 < r \leq n$, derive a formula for the number of r-cycles in S_n.

42. How many 4-cycles are there in A_8.

43. For any $1 < r \leq n$, derive a formula for the number of r-cycles in A_n.

44. Let $f \in S_n$, such that the order of f in the group S_n is odd. Prove that f is an even permutation.

45. Let $n \geq 3$. Let G be the multiplicative group of matrices over complex numbers generated by

$$A = \begin{pmatrix} e^{2\pi i/n} & 0 \\ 0 & e^{2\pi i/n} \end{pmatrix} \quad \text{and} \quad B = \begin{pmatrix} 0 & 1 \\ 1 & 0 \end{pmatrix}.$$

Prove that G is isomorphic with the dihedral group D_n of degree n.

46. If a is the generator of order n in D_n, prove that $<a>$ is normal in D_n and $D_n/<a> \cong \mathbb{Z}_2$.

47. List all the normal subgroups of D_n.

48. For any $n > 1$, prove that the alternating group A_n is the only subgroup of index 2 in S_n.

49. Determine all the subgroups of A_4.

50. Prove that $Z(D_n)$, the centre of the dihedral group of degree n, is of the order 1 or 2 according as n is odd or even.

7

Group Actions on Sets

Before the concept of an abstract group took its present shape, the theory of groups dealt only with permutation groups. Abstract groups were introduced much later in order to focus attention on those properties of permutation groups that concern the resultant composition (the binary operation in the permutation groups) only and do not refer to the set on which the permutations act. However, we have seen that any group can be identified (isomorphic) with a group of permutations on some set. Switching back from the abstract point of view to the concrete case of a permutation group is often useful in the abstract theory. The use of permutation groups provides certain counting techniques which play an important role in the theory of finite groups. In this chapter, we provide a procedure for passing from the abstract point of view to the concrete case of permutations, by introducing the concept of 'a group acting on a set' and develop certain counting techniques in finite groups.

7.1 ACTION OF A GROUP ON A SET

Cayley's theorem states that any group is isomorphic with a subgroup of the group $S(X)$ of permutations on some set X (bijections of X onto itself). Suppose that G is a subgroup of $S(X)$. Then, for any $a \in G$ and $x \in X$, there corresponds an element $a(x)$ of X and this correspondence satisfies the following properties.

(i) $e(x) = x$
(ii) $(a \cdot b)(x) = a(b(x))$

for all $x \in X$ and a and $b \in G$, where e is the identity in G. This is abstracted in the following definition.

Definition 7.1.1. Let G be a group and X be any nonempty set. A mapping $\theta : G \times X \to X$ is called an *action of G on X* if it satisfies the following properties.

1. $\theta(e, x) = x$ for all $x \in X$, where e is the identity in the group G.
2. $\theta(ab, x) = \theta(a, \theta(b, x))$ for all $x \in X$ and $a, b \in G$.

We say that *G acts on X* if there is an action of G on X.

Example 7.1.1

1. Let X be a nonempty set and G be a subgroup of the group $S(X)$ of permutations on X (bijections of X onto itself). Define

$$\theta : G \times X \to X \quad \text{by } \theta(a, x) = a(x)$$

for any $a \in G$ and $x \in X$. Then, it can be easily verified that θ is an action of G on X. Note that the identity in G is the identity mapping and the binary operation in the group G is the composition of mappings. This action is called the *natural action* of G.

2. Let G be a group and X be the set G itself. Define $\theta : G \times X \to X$ by $\theta(a, x) = a \cdot x$, where $a \cdot x$ is the product of a and x in the group G. Then clearly θ is an action of G on itself and is called the *action of G on itself by left translation*. The action of an element a in G on an element x in X $(= G)$ is simply multiplying x by a on the left.

3. Again let G be a group and X be the set G. Define $\theta : G \times X \to X$ by $\theta(a, x) = xa^{-1}$. Then,

$$\theta(e, x) = xe^{-1} = xe = x$$
$$\text{and} \quad \theta(ab, x) = x(ab)^{-1} = x(b^{-1}a^{-1}) = (xb^{-1})a^{-1} = \theta(a, \theta(b, x))$$

for all $x \in X$ and $a, b \in G$. Therefore, θ is an action of G on itself and is called the *action by right translation.*

4. By clubbing the above two actions of a group G on itself, we get another important action of G. Let G be a group and X be the set G itself. Define

$$\theta : G \times X \to X \quad \text{by } \theta(a, x) = axa^{-1}$$

for all $a \in G$ and $x \in X (= G)$. axa^{-1} is called the *conjugate of x corresponding to a*. For any a and $b \in G$ and $x \in X(= G)$, we have

$$\theta(e, x) = exe^{-1} = x$$
and $$\theta(ab, x) = (ab)x(ab)^{-1} = a(bxb^{-1})a^{-1} = \theta(a, \theta(b, x)).$$

Therefore, θ is an action of G on itself and is called the *action by conjugation*.

5. Let H be a subgroup of a group G and X be the set of all left cosets of H in G. Define

$$\theta : G \times X \to X \quad \text{by } \theta(a, xH) = axH$$

for all $a \in G$ and $xH \in X$ with $x \in G$. First note that θ is well defined; for

$$\begin{aligned} xH = yH &\Rightarrow x^{-1}y \in H \\ &\Rightarrow (ax)^{-1}(ay) = x^{-1}a^{-1}ay = x^{-1}y \in H \\ &\Rightarrow axH = ayH \end{aligned}$$

Then, θ is an action of G on the set of left cosets of H in G.

6. Let G be a group and X be the set of all subgroups of G. Define $\theta : G \times X \to X$ by

$$\theta(a, H) = aHa^{-1} \quad \text{for any } a \in G \quad \text{and} \quad H \in X.$$

Then, θ is an action of G on the set of all subgroups of G.

As in the case of the binary operation in a group, we simply write ax for $\theta(a, x)$, where θ is an action of G on X, $a \in G$ and $x \in X$. This is only for simplicity and convenience. The defining conditions for an action can be rewritten as

1. $ex = x$ for all $x \in X$
2. $(ab)x = a(bx)$ for all $a, b \in G$ and $x \in X$.

The condition (2) is not the associative law, for we are not dealing with a binary operation on a set. a and b are elements of the group G and x is an element of X. There should not be any confusion with this notation. One should understand as per the context.

There may be several actions of the same group on the same set, as given in the examples (2), (3) and (4) above. In the following, we establish an interrelation between the actions of a given group G on a given set X and the homomorphisms of G into the group $S(X)$ of permutations on X.

Theorem 7.1.1. Let θ be an action of a group G on a nonempty set X. For each $a \in G$, define

$$f_\theta(a) : X \to X \quad \text{by } f_\theta(a)(x) = \theta(a, x)$$

for any $x \in X$. Then, $f_\theta(a)$ is a permutation on X and f_θ defines a homomorphism of G into $S(X)$. Further, $\theta \mapsto f_\theta$ is a bijection of the set of all actions of G on X onto the set Hom$(G, S(X))$ of homomorphisms of G into $S(X)$.

Proof: First note that, for any $a \in G$ and $x, y \in X$,

$$\begin{aligned} f_\theta(a)(x) = f_\theta(a)(y) &\Rightarrow \theta(a, x) = \theta(a, y) \\ &\Rightarrow \theta(a^{-1}, \theta(a, x)) = \theta(a^{-1}, \theta(a, y)) \\ &\Rightarrow \theta(a^{-1}a, x) = \theta(a^{-1}a, y) \\ &\Rightarrow \theta(e, x) = \theta(e, y) \\ &\Rightarrow x = y. \end{aligned}$$

Therefore, $f_\theta(a) : X \to X$ is an injection. Also, for any $y \in X$, we have $\theta(a^{-1}, y) \in X$ and

$$f_\theta(a)(\theta(a^{-1}, y)) = \theta(a, \theta(a^{-1}, y)) = \theta(aa^{-1}, y) = \theta(e, y) = y$$

and hence f_θ is a surjection also. Therefore, $f_\theta(a)$ is a permutation on X; that is, $f_\theta(a) \in S(X)$ and hence f_θ can be considered as a mapping of G into $S(X)$. Also, for any a and $b \in G$, we have

$$f_\theta(ab)(x) = \theta(ab, x) = \theta(a, \theta(b, x)) = (f_\theta(a) \cdot f_\theta(b))(x)$$

for all $x \in X$ and therefore $f_\theta(a, b) = f_\theta(a) \cdot f_\theta(b)$. This means that f_θ is a homomorphism of G into $S(X)$. If θ_1 and θ_2 are two actions of G on X, such that $f_{\theta_1} = f_{\theta_2}$ then for any $(a, x) \in G \times X$,

$$\theta_1(a,x) = f_{\theta_1}(a)(x) = f_{\theta_2}(a)(x) = \theta_2(a,x)$$

and hence $\theta_1 = \theta_2$. Thus, $\theta \mapsto f_\theta$ is an injection.
Next, let $\alpha \in$ Hom$(G, S(X))$ and define $\theta : G \times X \to X$ by

$$\theta(a, x) = \alpha(a)(x) \quad \text{for all } a \in G \text{ and } x \in X.$$

Then, it can be verified that θ is an action of G on X and $f_\theta = \alpha$. Thus, $\theta \mapsto f_\theta$ is a bijection of the set of actions of G on X onto the set Hom$(G, S(X))$ of homomorphisms of G into $S(X)$. ◀

Definition 7.1.2. For any action θ of a group G on a set X, the homomorphism $f_\theta : G \to S(X)$ defined above is called the *homomorphism associated with the action* θ. If f_θ is an injection, then θ is called an *effective action* of G on X and, in this case, we say that *G acts on X effectively*. The kernel of f_θ is called the kernel of the action θ.

$$\begin{aligned}\text{That is, } \ker\theta &= \{a \in G : f_\theta(a) = \text{the identity in } S(X)\}\\ &= \{a \in G : f_\theta(a)(x) = x \text{ for all } x \in X\}\\ &= \{a \in G : \theta(a, x) = x \text{ for all } x \in X\}\end{aligned}$$

Therefore, an action θ is effective if and only if $\ker\theta = \{e\}$; that is, e is the only element a in G such that $\theta(a, x) = x$ for all $x \in X$. For example, if θ is the action of G on itself by left translation (see Example 7.1.1 (2)), then θ is effective. On the other hand, if θ is the action of G on itself by conjugation (that is, $\theta(a, x) = axa^{-1}$), then θ is effective if and only if the centre $Z(G)$ of G is trivial, since

$$\begin{aligned}Z(G) &= \{a \in G : ax = xa \text{ for all } x \in G\}\\ &= \{a \in G : axa^{-1} = x \text{ for all } x \in G\}\\ &= \{a \in G : \theta(a, x) = x \text{ for all } x \in G\}\\ &= \ker\theta.\end{aligned}$$

Since a group G is abelian (commutative) if and only if $Z(G) = G$, we can consider the effectiveness of the conjugacy action as a measure of the commutativity of the group.

Worked Exercise 7.1.1. Let X be the set of all complex number with unit modulus and G be the additive group of real numbers. Define $\theta : G \times X \to X$ by $\theta(a, x) = e^{ia}x$ for any $a \in G$ and $x \in X$. Then prove that θ is an action of G on X. Is θ effective?

Answer: Note the 0 is the identity in $G\ (= (\mathbb{R}, +))$ and that the usual addition $+$ is the binary operation on G.

$$\theta(0, x) = e^{i0}x = 1x = x \quad \text{for all } x \in X.$$

Also, for any a and $b \in G$ and $x \in X$,

$$\theta(a+b, x) = e^{i(a+b)}x = e^{ia}(e^{ib}x) = \theta(a, \theta(b, x)).$$

Therefore, θ is an action of G on X. Note that,

$$\begin{aligned}\ker\theta &= \{a \in G : \theta(a, x) = x \text{ for all } x \in X\}\\ &= \{a \in G : e^{ia}x = x \text{ for all } x \in X\}\\ &= \{a \in G : e^{ia} = 1\}\\ &= \{2n\pi : n \in \mathbb{Z}\}\end{aligned}$$

Therefore, θ is not effective.

Worked Exercise 7.1.2. Let θ be an action of a group on a set X. For any $A \subseteq X$, define

$$\theta'(a, A) = \{\theta(a, x) : x \in A\}.$$

Then prove that θ' is an action of G on the power set of X and that $\ker\theta = \ker\theta'$.

Answer: For any $A \subseteq X$,

$$\begin{aligned}\theta'(e, A) &= \{\theta(e, x) : x \in A\} = A\\ \text{and}\quad \theta'(a, \theta'(b, A)) &= \{\theta(a, y) : y \in \theta'(b, A)\}\\ &= \{\theta(a, \theta(b, x)) : x \in A\}\\ &= \{\theta(ab, x) : x \in A\}\\ &= \theta'(ab, A)\end{aligned}$$

Therefore, θ' is an action of G on $\mathbb{P}(X)$, the power set of X. For any $a \in G$,

$$\begin{aligned}a \in \ker\theta &\Leftrightarrow \theta(a, x) = x \quad \text{for all } x \in X\\ &\Leftrightarrow \theta'(a, A) = A \quad \text{for all } A \subseteq X\\ &\qquad\qquad (\text{note that } \theta'(a, \{x\}) = \{\theta(a, x)\})\\ &\Leftrightarrow a \in \ker\theta'.\end{aligned}$$

Therefore, $\ker\theta = \ker\theta'$; in particular, θ' is effective if and only if θ is effective.

EXERCISE 7(A)

1. Let G be a group and define $\theta : G \times G \to G$ by $\theta(a, x) = xa$. Then prove that θ is an action of G on itself if and only if G is abelian.
2. Let $G = S_5$ and $X = \{x_1, x_2, x_3, x_4, x_5\}$ and define $\theta : G \times X \to X$ by $\theta(f, x_i) = x_{f(i)}$ for any $f \in S_5$ and $1 \leq i \leq 5$. Then prove that θ is an action of G on X. Is it effective?

3. Let X be the set of vertices 1, 2, 3, 4 of a square and $G = D_4$, the dihedral group of degree 4. Define $\theta : D_4 \times X \to X$ by $\theta(f, i) = f(i)$. Then prove that θ is an action of D_4 on X.

4. Let H be a subgroup of a group G and X be the set of left cosets of H in G. Define $\theta : G \times X \to X$ by

$$\theta(a, xH) = axa^{-1}H \quad \text{for } a \in G \quad \text{and} \quad xH \in X, x \in G.$$

Then prove that θ is an action of G on X. Is it effective? What is the Kernel of θ?

5. Let H be a subgroup of finite index n in a group G. Then prove that there is a homomorphism $f: G \to S_n$ whose kernel is $\bigcap_{a \in G} aHa^{-1}$.

6. If H is a normal subgroup of finite index n in a group G, then prove that G is isomorphic to a subgroup of S_n.

7. If G is a simple group and H is a proper subgroup of index n in G, then prove that G is isomorphic to a subgroup of S_n.

8. Let G be the group of symmetries of a cube. Prove that there are nontrivial actions of G on each of the set of edges, the set of faces, the set of vertices and the set of diagonals of the cube.

9. Let θ be an action of a group on a set X and define $\theta' : G \times (X \times X) \to (X \times X)$ by

$$\theta^2(a, (x, y)) = (\theta(a, x), \theta(a, y))$$

for any $a \in G$ and x and $y \in X$. Then prove that θ^2 is an action of G on $X \times X$ and that θ^2 is effective if and only if θ is effective.

10. Let G be a group and X_1 and X_2 be nonempty sets, let θ_1 and θ_2 be actions of G on X_1 and X_2, respectively. Define $(\theta_1 \times \theta_2) : G \times (X_1 \times X_2) \to X_1 \times X_2$ by

$$(\theta_1 \times \theta_2)(a, (x_1, x_2)) = (\theta_1(a, x_1), \theta_2(a, x_2))$$

for any $a \in G$ and $(x_1, x_2) \in X_1 \times X_2$ Prove that $\theta_1 \times \theta_2$ is an action of G on $X_1 \times X_2$ and that

$$\ker(\theta_1 \times \theta_2) = \ker \theta_1 \cap \ker \theta_2.$$

11. Let a group G act on itself by left translation. Prove that the action is effective and deduce the Cayley's theorem.

12. Let H be a subgroup of a group G and X be the set of right cosets of H in G. Prove that $(a, Hx) \mapsto Hxa^{-1}$ is an action of G on X which is effective if and only if H contains no nontrivial normal subgroup of G.

13. Let H be a proper subgroup of a finite group G. Then prove that $G \neq \bigcup_{a \in G} aHa^{-1}$.

14. Let p be the smallest prime dividing the order of a finite group G. Then prove that any subgroup of index p in G is normal in G.

15. Derive from the above that any subgroup of index 2 in a group is normal.

16. Prove that any subgroup of order 539 in a group of order 2695 is normal.

17. Let p and q be distinct primes such that $p < q$. Prove that any subgroup of order q in a group of order pq is normal.

18. Let G be a group of odd order. Then prove that any subgroup of index 3 is normal in G.

7.2 ORBITS AND STABILIZERS

In this section, we introduce two important concepts regarding the actions of groups on sets. When θ is an arbitrary action of a group G on a set X, we simply write, as agreed earlier, ax for $\theta(a, x)$ for any $a \in G$ and $x \in X$, when there is no ambiguity about the action θ. Recall that a group is a pair $(G, *)$ where G is a nonempty set and $*$ is a binary operation on G satisfying certain conditions. However, we use to simply say that G is a group without specifically mentioning about the binary operation and further we used to write simply ab for $a * b$, where a and $b \in G$. Likewise, when there is an action of a group G on a set X, we simply say that G acts on X without specifically mentioning the action of G on X. It is understood that is an action $(a, x) \mapsto ax$ sending any pair (a, x) in $G \times X$ onto the element ax.

Definition 7.2.1. Let a group G act on a set X and $x \in X$. The *orbit of* x is defined to be the set

$$O(x) = \{ax : a \in G\}.$$

Before going to certain examples, we prove an important elementary property of orbits in the following theorem.

Theorem 7.2.1. Let a group G act on a set X. Then, the orbits of elements of X form a partition of X. That is, any two orbits are either equal or disjoint subsets of X and the union of all orbits is equal to X.

Proof: For any $x \in X$, we have $O(x) = \{ax : a \in G\}$. Clearly $O(x)$ is a subset of X for each $x \in X$. Also, since

$$x = ex \in O(x)$$

We have $X = \bigcup_{x \in X} O(x)$. Now, let x and $y \in X$ such that the orbits $O(x)$ and $O(y)$ are not disjoint. Then, choose $z \in O(x) \cap O(y)$. Thus,

$$z = ax = by \text{ for some } a \text{ and } b \in G$$

and hence $x = (a^{-1}b)y$ and $y = (b^{-1}a)x$ so that $O(x) \subseteq O(y)$ and $O(y) \subseteq O(x)$ and therefore $O(x) = O(y)$. Then,

$$O(x) \cap O(y) = \varnothing \quad \text{or} \quad O(x) = O(y).$$ ◀

Definition 7.2.2. Let a group G act on a set X and $x \in X$. The *stabilizer of x* is defined to be the set

$$\text{St}(x) = \{a \in G : ax = x\}.$$

Theorem 7.2.2. Let a group G act on a set X and $x \in X$. Then, the stabilizer $St(x)$ is a subgroup of G and there is a bijection of the orbit $O(x)$ onto the set of left cosets of the stabilizer $\text{St}(x)$ in G.

Proof: Since $ex = x$, where e is the identity in the group G,

$$e \in \text{St}(x) = \{a \in G : ax = x\}$$

and hence $\text{St}(x)$ is a nonempty subset of G. Also,

$$\begin{aligned} a \text{ and } b \in \text{St}(x) &\Rightarrow ax = x = bx \\ &\Rightarrow (ab)x = a(bx) = ax = x \\ &\Rightarrow ab \in \text{St}(x) \end{aligned}$$

$$\begin{aligned} \text{and} \quad a \in \text{St}(x) &\Rightarrow ax = x \\ &\Rightarrow a^{-1}x = a^{-1}(ax) = (a^{-1}a)x = ex = x \\ &\Rightarrow a^{-1} \in \text{St}(x). \end{aligned}$$

Therefore, the stabilizer $\text{St}(x)$ is a subgroup of G. Next, let A be the set of all left cosets of $\text{St}(x)$ in G. That is,

$$A = \{a\text{St}(x) : a \in G\}.$$

Define $\alpha : O(x) \to A$ by $\alpha(ax) = a\text{St}(x)$.
For any a and $b \in G$, we have

$$\begin{aligned} ax = bx &\Leftrightarrow a^{-1}bx = x \\ &\Leftrightarrow a^{-1}b \in \text{St}(x) \\ &\Leftrightarrow a\text{St}(x) = b\text{St}(x). \end{aligned}$$

Therefore, α is well defined and is an injection. Clearly α is a surjection also. Thus, α is a bijection of $O(x)$ onto the set of left cosets of $\text{St}(x)$ in G. ◀

Corollary 7.2.1. Let G be a group acting on a finite set X and $x \in X$. Then, the number of elements in the orbit of x is equal to the index of the stabilizer of x in G. That is, $|\mathrm{O}(x)| = i_G(\mathrm{St}(x)) = \dfrac{|G|}{|\mathrm{st}(x)|}$.

Now, we shall take up various examples of actions of groups on sets and determine the orbits and stabilizers of arbitrary elements of X.

Example 7.2.1

1. Let X be any nonempty set and G be any group. Define $\theta : G \times X \to X$ by

$$\theta(a, x) = x \quad \text{for all } a \in G \quad \text{and} \quad x \in X.$$

Then clearly θ is an action of G on X and is called the *trivial action*. Here, for any $x \in X$,

$$\begin{aligned} &\text{the orbit } \mathrm{O}(x) = \{x\} \\ \text{and the}\ &\text{stabilizer } \mathrm{St}(x) = G. \end{aligned}$$

2. Consider the action of any group G on itself by left translation (see Example 7.1.1 (2)). Here, $X = G$ and, for any $x \in X (= G)$,

$$\text{the orbit } \mathrm{O}(x) = \{ax : a \in G\} = X = G$$

since any $b \in G$ can be written as $b = (bx^{-1})x$, $bx^{-1} \in G$.
Also, the stabilizer $\mathrm{St}(x) = \{a \in G : ax = x\} = \{e\}$.

3. Consider the action of a group on itself by conjugation (see Example 7.1.1 (4)), where $\theta : G \times G \to G$ is defined by $\theta(a, x) = axa^{-1}$, for all a and $x \in G$. Here, for any $x \in X$,

$$\begin{aligned} \text{the orbit } \mathrm{O}(x) &= \{\theta(a, x) : a \in G\} \\ &= \{axa^{-1} : a \in G\} \\ \text{and the stabilizer } \mathrm{St}(x) &= \{a \in G : \theta(a, x) = x\} \\ &= \{a \in G : axa^{-1} = x\} \\ &= \{a \in G : ax = xa\} \end{aligned}$$

The orbit $\mathrm{O}(x)$ is called the *conjugacy class of x in G* and is usually denoted by $C(x)$. Also, the stabilizer $\mathrm{St}(x)$ is called the *centralizer of x in G* and is usually denoted by $\mathrm{Cent}_G(x)$. That is,

$$\begin{aligned} C(x) &= \{axa^{-1} : a \in G\} \\ \text{and} \quad \mathrm{Cent}_G(x) &= \{a \in G : ax = xa\}. \end{aligned}$$

4. Let H be a subgroup of a group G and X be the set of left cosets of H in G. Define $\theta : G \times X \to X$ by

$$\theta\,(a, xH) = (axa^{-1})H$$

for any $a \in G$ and $xH \in X$, $x \in G$. Then, θ is an action of G on X. The orbit of any xH, $x \in G$, is

$$\begin{aligned} \mathrm{O}(xH) &= \{\theta(a, xH) : a \in G\} \\ &= \{axa^{-1}H : a \in G\} \end{aligned}$$

and the stabilizer of xH is

$$\begin{aligned} \mathrm{St}(xH) &= \{a \in G : \theta(a, xH) = xH\} \\ &= \{a \in G : axa^{-1}H = xH\}. \end{aligned}$$

Definition 7.2.3. An action of a group G on a set X is said to *transitive* if there is only one orbit in X; that is, $\mathrm{O}(x) = X$ for all $x \in X$ or, equivalently $\mathrm{O}(x) = \mathrm{O}(y)$ for all x and $y \in X$. In this case, we say that G *acts transitively* on X.

Clearly an action is transitive if and only if, for any x and $y \in X$, there is an element a in the group G such that $ax = y$; that is, any element of X can be transformed to any other element of X by means of the action of an element of the group.

Example 7.2.2

1. The action of a group G on itself by left (right) translation (see Example 7.1.1 (2) and (3)) is transitive, since, for any x and $y \in G$, $yx^{-1} \in G$ and $y^{-1}x \in G$ and

$$(yx^{-1})x = y \quad \text{and} \quad x(y^{-1}x)^{-1} = y.$$

2. The action of a group G on itself by conjugation (see Example 7.1.1 (4)) is not transitive, in general. This is transitive if and only if there is only one conjugacy class or, equivalently, any two elements of the group or conjugates to each other.

A group may act on the same set (or on two different sets) differently. In the following, we define equivalence of two such actions in a natural way.

Definition 7.2.4. Let a group G act on two sets X and X'. These two actions are said to *equivalent* if there is a bijection $\alpha : X \to X'$ such that

$$\alpha(ax) = a\alpha(x)$$

for all $a \in G$ and $x \in X$; that is, if $\theta : G \times X \to X$ and $\theta' : G \times X' \to X'$ are actions of G on X and X', respectively, then θ and θ' are said to be equivalent if

$$\alpha \cdot f_\theta(a) = f_{\theta'}(a) \cdot \alpha$$

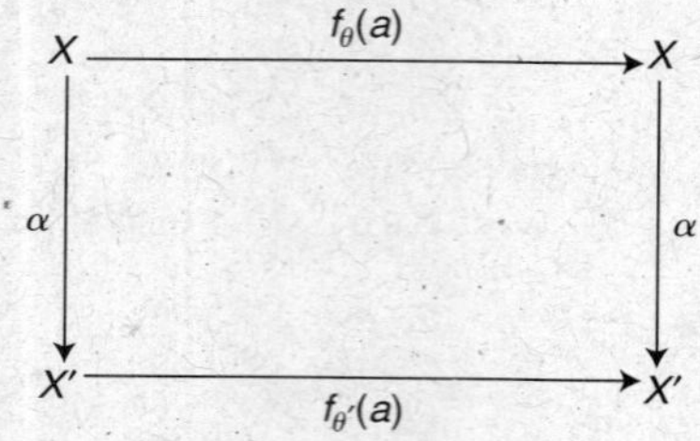

for all $a \in G$, where $f_\theta(a)$ and $f_{\theta'}(a)$ are the permutations on X and X' corresponding to the actions θ and θ', respectively (see Theorem 7.1.1). In other words,

$$\alpha(\theta(a, x)) = \theta'(a, \alpha(x))$$

for all $a \in G$ and $x \in X$.

Example 7.2.3. Consider the actions of a group G on itself by left translation and right translation defined by

$$\theta(a, x) = ax \quad \text{and} \quad \theta'(a, x) = xa^{-1}$$

(see Example 7.1.1 (2) and (3)). Define $\alpha : G \to G$ by $\alpha(x) = x^{-1}$ for any $x \in G$. Then, α is a bijection and, for any a and $x \in G$,

$$\alpha(\theta(a, x)) = \alpha(ax) = (ax)^{-1} = x^{-1}a^{-1} = \theta'(a, \alpha(x)).$$

Therefore, the actions θ and θ' are equivalent.

In the following, we obtain an internal characterization of transitive actions. This is an extension of Theorem 7.2.2 which gives us that the orbit of any element x in X is bijective with the set of left cosets of the stabilizer of x in G.

Theorem 7.2.3. Let a group G act transitively on a set X, $x \in X$ and $H = \mathrm{St}(x)$. Then, the action of G on X is equivalent to the action of G on the set of left cosets of H in G by left translation.

Proof: Let θ be the given transitive action on X and Y be the set of left cosets of H in G. Let θ' be the action of G on Y by left translation; that is, $\theta' : G \times Y \to Y$ is defined by

$$\theta'(a, bH) = abH \quad \text{for any } a \in G \quad \text{and} \quad bH \in Y.$$

Since the action θ is transitive, we get that the orbit $O(x) = X$. Let $\alpha : X \to Y$ be the bijection given in the proof of Theorem 7.2.2; that is,

$$\alpha(ax) = aH \quad \text{for any } a \in G.$$

Now, for any $a \in G$ and $y \in X$, choose $b \in G$ such that $y = bx$ (since X is the orbit of x) and we have

$$\begin{aligned}\alpha(\theta(a, y)) &= \alpha(ay) = \alpha(abx) \\ &= abH = \theta'(a, bH) \\ &= \theta'(a, \alpha(bx)) = \theta'(a, \alpha(y)).\end{aligned}$$

Thus, the actions θ and θ' are equivalent. ◀

Among the transitive actions, there is a special class, namely primitive actions, which deserves an emphasis. In the following, we define primitive actions and characterize these in terms of the stabilizers.

Definition 7.2.5. Let θ be an action of a group G on a set X. An equivalence relation ψ on X is said to be *compatible with the action* θ if, for any x and $y \in X$,

$$(x, y) \in \psi \Rightarrow (\theta(a, x), \theta(a, y)) \in \psi \quad \text{for all } a \in G.$$

Clearly the whole of $X \times X$ and the diagonal $\Delta_X = \{(x, x) : x \in X\}$ are equivalence relation X which are compatible with every action of G on X. These two equivalence relations are called *trivial relations*.

Definition 7.2.6. An action of a group G on a set X is called *primitive* if $X \times X$ and Δ_X are the only equivalence relations on X which are compatible with the action. An action which is not primitive is called *imprimitive*.

Theorem 7.2.4. Let θ and θ' be equivalent actions of G on X and X', respectively. Then, θ is primitive if and only if so is θ'.

Proof: Since θ and θ' are equivalent, there exists a bijection $\alpha : X \to X'$ such that

$$\alpha(\theta(a, x)) = \theta'(a, \alpha(x))$$

for any $a \in G$ and $x \in X$. For any equivalence relation ψ on X, let

$$\alpha(\psi) = \{(\alpha(x), \alpha(y)\text{: } (x, y) \in \psi\}.$$

Then, $\alpha(\psi)$ is an equivalence relation on X' and $\psi \mapsto \alpha(\psi)$ is a one-to-one correspondence between the equivalence relations on X and those on X'. Now, theorem follows from the fact that ψ is compatible with the action θ on X if and only if $\alpha(\psi)$ is compatible with the action θ' on X'. ◀

Theorem 7.2.5. Let θ be an action of a group G on a set X. Then, θ is imprimitive if and only if there exists a proper subset Y of X with $|Y| > 1$ such that, for any $a \in G$, either $\theta(a, Y) = Y$ or $\theta(a, Y) \cap Y = \emptyset$, where $\theta(a, Y) = \{\theta(a, y) : y \in Y\}$.

Proof: Suppose that θ is imprimitive. Then, there exists an equivalence relation ψ on X which is compatible with θ such that $\psi \neq X \times X$ and $\psi \neq \Delta_X$. Choose $x \neq y \in X$ such that $(x, y) \in \psi$. Put $Y =$ the equivalence class of ψ containing x. That is, $Y = \psi(x) = \{z \in X : (x, z) \in \psi\}$. Since $x \neq y \in Y, |Y| > 1$. Also, since $\psi \neq X \times X$, Y is a proper subset of X. Now, let $a \in G$ and $\theta(a, Y) \cap Y \neq \emptyset$. Then, choose $z \in Y$ such that $\theta(a, z) \in Y = \psi(x)$. Since $(x, \theta(a, z)) \in \psi$ and ψ is compatible with θ, we get that

$$(\theta(a, x), \theta(a, \theta(a, z))) \in \psi$$

and hence $(\theta(a, x), \theta(a, z)) \in \psi$ so that $(x, \theta(a, x)) \in \psi$. Now, it can be easily verified that $\theta(a, Y) = Y$. Conversely suppose that there is a proper subset Y of X such that $|Y| > 1$ and, for any $a \in G$, either $\theta(a, Y) = Y$ or $\theta(a, Y) \cap Y = \emptyset$. From this it follows that, for any a and $b \in G$,

$$\text{either} \quad \theta(a, Y) = \theta(b, Y) \quad \text{or} \quad \theta(a, Y) \cap \theta(b, Y) = \emptyset.$$

Put $Z = X - (\bigcup_{a \in G} \theta(a, Y))$. Then, $\mathscr{C} = \{\theta(a, Y) : a \in G\} \cup \{Z\}$ is a partition of X and the corresponding equivalence relation ψ on X is compatible with the action θ. Recall that

$$\psi = \{(x, y) \in X \times X : \text{both } x \text{ and } y \text{ belong to the same set in } \mathscr{C}\}.$$

Since $Y = \theta(e, Y)$ is an equivalence class and $Y \neq X$, it follows that $\psi \neq X \times X$. Also, since $|Y| > 1$, $\psi \neq \Delta_X$. Thus, θ is imprimitive. ◀

Theorem 7.2.6. Let θ be an action of a group on a set X and define

$$\psi_\theta = \{(x, y) \in X \times X : \theta(a, x) = y \quad \text{for some } a \in G\}.$$

Then, ψ_θ is an equivalence relation on X, that is, compatible with the action θ.

Proof: Note that $(x, y) \in \psi_\theta \Leftrightarrow y \in O(x)$, the orbit of x

$$\Leftrightarrow x \in O(y), \quad \text{the orbit of } y$$

and hence ψ_θ is precisely the equivalence relation corresponding to the partition of X consisting the orbits of elements of X. Now, for any $a \in G$,

$$\begin{aligned}(x, y) \in \psi_\theta &\Leftrightarrow \theta(b, x) = y \quad \text{for some } b \in G \\ &\Leftrightarrow \theta(b, \theta(a, x)) = \theta(ab, x) = \theta(a, \partial(b, x)) = \theta(a, y) \\ &\Leftrightarrow (\theta\,(a, x), \theta(a, y)) \in \psi_\theta.\end{aligned}$$

Therefore, ψ_θ is compatible with the action θ. ◀

Corollary 7.2.2. If an action θ of G on X is primitive, then either θ is trivial (that is, $\theta(a, x) = x$ for all $a \in G$ and $x \in X$) or θ is transitive.

Proof: If θ is primitive, then $\psi_\theta = \Delta_X$ or $X \times X$ and hence all orbits or singleton sets or there is only one orbit which mean that either θ is trivial or transitive.

In particular, a nontrivial primitive action must be necessarily transitive and hence the class of nontrivial primitive actions of a group on a set X is a subclass of the transitive actions of G on X. But in general a transitive action need not be primitive. For consider the following example.

Example 7.2.4. Let a group G act on itself by left translation. (see Example 7.1.1 (2)) and H be a nontrivial proper subgroup of G. The action θ of G on itself is defined by $\theta(a, x) = ax$, the product of a and x in G. For any x and $y \in G$, we have $y = (yx^{-1})x = \theta(yx^{-1}, x)$ and $yx^{-1} \in G$ and therefore the action θ is transitive. Define a relation ψ on G by

$$(x, y) \in \psi \quad \text{if and only if } x^{-1}y \in H.$$

It can be easily seen that ψ is an equivalence relation on G. Also for any $a \in G$,

$$\begin{aligned}(x, y) \in \psi &\Rightarrow x^{-1}y \in H \\ &\Rightarrow (ax)^{-1}(ay) = x^{-1}a^{-1}ay = x^{-1}y \in G \\ &\Rightarrow (ax, ay) \in \psi\end{aligned}$$

and hence ψ is compatible with the action θ. Also since $H \neq \{e\}$, $\psi \neq \triangle_H$ and since $H \neq G$, $\psi \neq G \times G$. Therefore, the action θ is imprimitive.

Note that in the above example, any equivalence class of ψ is simply a left coset of H in G. If G has no nontrivial proper subgroups, then the above action is primitive and vice versa. In other words, the action of G on itself by left translation is primitive if and only if $\{e\}$ is a maximal subgroup of G.

This is generalised in the following theorem. First, let us call a proper subgroup K of a group G *maximal* if there is no proper subgroup of G containing K properly; that is, for any subgroup H of G, $K \subseteq H \subseteq G$ implies that either $H = K$ or $H = G$.

Theorem 7.2.7. Let a group G act transitively on a set X with $|X| \geq 2$. Then, the action is primitive if and only if the stabilizer of any $x \in X$ is a maximal subgroup of G.

Proof: For any $x \in X$, let $S = \text{St}(x)$, the stabilizer of x in G. By Theorem 7.2.3, the given action G on X is equivalent to the action θ of G on the set G/S of left cosets of S in G by left translation.
Here, $\theta : G \times G/S \to G/S$ is defined by $\theta(a, xS) = axS$ for any $a \in G$ and $xS \in G/S$, $x \in G$, where $G/S = \{xS : x \in G\}$ (note that G/S is not the quotient group, unless S is a normal subgroup of G). Therefore, by Theorem 7.2.4, the given action of G on X is primitive if and only if θ is primitive.
First suppose that $S = \text{St}(x)$ is not a maximal subgroup of G for some $x \in X$. Then, choose a subgroup H of G such that $S \subsetneqq H \subsetneqq G$.
Let $Y = \{xS : x \in H\}$.
Since $H \subsetneqq G$, Y is a proper subset of G/S. Also, since $S \subsetneqq H$, Y has atleast two elements. Further, for any $a \in G$,

$$\begin{aligned}
\theta(a, Y) \cap Y \neq \varnothing &\Rightarrow xS = \theta(a, yS) \quad \text{for some } x, y \in H \\
&\Rightarrow xS = ayS, \quad \text{for some } x, y \in H \\
&\Rightarrow x^{-1}ay \in S \subseteq H, x, y \in H \\
&\Rightarrow a \in xHy^{-1} = H \\
&\Rightarrow \theta(a, Y) = \{axS : x \in H\} = \{yS : y \in H\} = Y.
\end{aligned}$$

Thus, by Theorem 7.2.5, θ is imprimitive and hence the given action of G on X is imprimitive. Conversely, suppose that the given action of G on X is imprimitive. Then, again by Theorem 7.2.5, there exists a proper subset Y of X with $|Y| > 1$ such that, for any $a \in G$,

$$\text{either } aY = Y \quad \text{or} \quad aY \cap Y = \varnothing.$$

Choose $x \in Y$. We shall prove that the stabilizer St(x) is not a maximal subgroup of G.

$$\text{Put } H = \{a \in G : aY = Y\}.$$

Clearly H is a subgroup of G. Also,

$$\begin{aligned} a \in \text{St}(x) &\Rightarrow ax = x \\ &\Rightarrow aY \cap Y \neq \varnothing \\ &\Rightarrow aY = Y \\ &\Rightarrow a \in H. \end{aligned}$$

Therefore, St(x) $\subseteq H \subseteq G$. Since $|Y| > 1$, we can choose $y \in Y$ such that $y \neq x$. Since the action of G is transitive, there exists $a \in G$ such that $ax = y \neq x$. Now, $a \notin$ St(x) and, since $ax = y \in aY \cap Y$, it follows that $aY = Y$ and hence $a \in H$. Therefore, St(x) $\subsetneqq H$. Further, since Y is a proper subset of X, we can choose $z \in X$ such that $z \notin Y$. Again by the transitivity of the action, there exists $a \in G$ such that $ax = z \notin Y$, so that $aY \neq Y$ and hence $a \notin H$. Therefore, $H \subsetneqq G$. Now, since

$$\text{St}(x) \subsetneqq H \subsetneqq G,$$

if follows that St(x) is not a maximal subgroup of G. ◀

Worked Exercise 7.2.1. Let G be a finite group of prime order. Suppose that G acts on a set X and $x \in X$ such that $ax = x$ for some $a \neq e$ in G. Then prove that $bx = x$ for all $b \in G$.

Answer: Consider the stabilizer St(x), which is a subgroup of G. Since the order of G is prime, by the Lagrange's theorem $|\text{St}(x)| = 1$ or $|G|$ and hence St(x) $= \{e\}$ or G. If there is $a \neq e$, such that $ax = x$, then St(x) $\neq \{e\}$ and hence St(x) $= G$, so that $bx = x$ for all $b \in G$.

EXERCISE 7(B)

1. Let H be a subgroup of a group G and

$$N = \bigcap_{x \in G} xHx^{-1}.$$

Then prove that N is largest normal subgroup of G contained in H.

2. Prove that the N given in (1) above is kernel of the action of G on the set of left cosets of H in G by left translation.

3. Prove that a subgroup H of a group contains no nontrivial normal subgroups of G if and only if the action of G on the set of cosets of H in G by left translation is effective.

4. Let H be a subgroup of a group G and X be the set of right cosets of H in G. Then prove that

$$(a, Hx) \mapsto Hxa^{-1}$$

is an action of G on X whose kernel is the largest normal subgroup of G contained in H.

5. Let a group G act on a set X and Y be a subset of X. Let $\text{St}(Y) = \{a \in G : ay = y$ for all $y \in Y\}$. Prove that $\text{St}(Y)$ is a subgroup of G.

6. Let G be the group $(\mathbb{R}, +)$ and $X = \mathbb{R}^2$ be the two-dimensional Euclidean space. For any $a \in G$, let r_a be the rotation of the plane about the origin through a radius. Prove that $a \mapsto r_a$ is a homomorphism of G into $S(X)$ and hence yields an action of G on X. Is this action effective? Give a geometrical description of the orbit of a point P in $\mathbb{R}^2$. What is the stabilizer of a point P?

7. Let a group G act on a set X. Prove that the action of G on X is effective if and only if, for any a and $b \in G$,

$$ax = bx \quad \text{for all } x \in X \Rightarrow a = b.$$

8. Let a group G act on a set X primitively and N be a nontrivial normal subgroup of G. Then prove that the induced action of N on X is transitive.

9. Let two groups G_1 and G_2 act on sets X_1 and X_2, respectively and $X_1 \cap X_2 = \varnothing$. Define an action of $G_1 \times G_2$ on $X_1 \cup X_2$ by

$$(a_1, a_2)x_1 = a_1x_1 \quad \text{and} \quad (a_1, a_2)x_2 = a_2x_2$$

for any $(a_1, a_2) \in G_1 \times G_2$, $x_1 \in X_1$ and $x_2 \in X_2$. Prove that this is an action, which is not transitive.

10. An action of a group G on a set X is called doubly transitive if, for any x_1, x_2, y_1 and $y_2 \in X$, there exists $a \in G$ such that $ax_1 = y_1$ and $ax_2 = y_2$. Prove that any doubly transitive action is primitive. Is the converse true?

7.3 CERTAIN COUNTING TECHNIQUES

We have observed in Theorem 7.2.1 that the orbits form a partition of X, when a group G acts on X and that, in Theorem 7.2.2, we have proved that the number of elements in any finite orbit is precisely the index of the stabilizer of the corresponding element. We consolidate these ideas in proving the following theorem.

Theorem 7.3.1. Let G be group acting on a finite set X and $O(x_1)$, $O(x_2)$, …, $O(x_n)$ be all the distinct orbits in X. Then, the number of elements in X can be obtained by the formula

$$|X| = \sum_{i=1}^{n} |O(x_i)| = |G| \left(\sum_{i=1}^{n} \frac{1}{|\mathrm{St}(x_i)|} \right).$$

Proof: Since the orbits form a partition of X, we have

$$\bigcup_{i=1}^{n} O(x_i) = X \quad \text{and} \quad O(x_i) \cap O(x_j) = \varnothing \quad \text{for} \quad i \neq j.$$

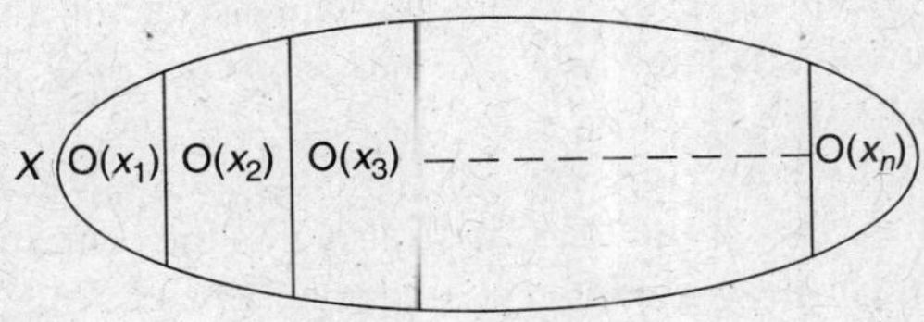

Therefore, the total number of elements in X is equal to the sum of the numbers of elements in the orbits. That is,

$$\begin{aligned}
|X| &= |O(x_1)| + |O(x_2)| + \cdots + |O(x_n)| \\
&= i_G(\mathrm{St}(x_1)) + \cdots + i_G(\mathrm{St}(x_n)) \text{ (by Theorem 7.2.2)} \\
&= \frac{|G|}{|S(x_1)|} + \cdots + \frac{|G}{|S(x_n)|} \\
&= |G| \left(\sum_{i=1}^{n} \frac{1}{|\mathrm{St}(x_i)|} \right).
\end{aligned}$$

◀

We shall apply the above formula to a special action of a group on itself and derive an important formula for the order of a finite group, namely, the class equation. First let us recall that $(a, x) \mapsto axa^{-1}$ is an action of a group G on itself and is called the action of G on itself by conjugation. Here, for any $x \in G$, the orbit of x is simply the conjugate class $C(x)$ of x in G; that is,

$$\mathrm{O}(x) = \{axa^{-1} : a \in G\} = C(x).$$

and the stabilizer of x is given by

$$\begin{aligned}\mathrm{St}(x) &= \{a \in G : axa^{-1} = x\}\\ &= \{a \in G : ax = xa\}\end{aligned}$$

which is known as the *centralizer* of x in G and is denoted by $\mathrm{Cent}_G(x)$. By Corollary 7.2.1, we have $|C(x)| = i_G(\mathrm{Cent}_G(x))$; that is, the number of elements in the conjugate class of x is equal to the index of the centralizer of x in G.

Theorem 7.3.2 (The class equation). Let G be a finite group. Then,

$$|G| = \sum_{i=1}^{n} i_G(\mathrm{Cent}_G(x_i)) + |Z(G)|$$

where $Z(G)$ is the centre of G and $x_1, \ldots, x_n$ are elements of G such that $C(x_1)$, $C(x_2), \ldots, C(x_n)$ are all the distinct conjugacy classes, each with more than one element. This equation is known as the *class equations of G.*

Proof: Consider the action of G on itself by conjugation. Then, the orbit of x is the conjugacy class of x and the stabilizer of x is the centralizer of x. Since the orbits form a partition of G, the conjugacy classes form a partition of G. Therefore,

$$|G| = \sum_{i=1}^{m} |C(y_i)| = \sum_{i=1}^{m} i_G(\mathrm{Cent}_G(y_i)).$$

Now, we shall distinguish two types of conjugacy class, namely classes each with only one element and classes each with more than one element. Note that, for any $x \in G$, the conjugacy class of x is

$$C(x) = \{axa^{-1} : a \in G\}.$$

Therefore, $|C(x)| = 1 \Leftrightarrow C(x) = \{x\}$

$$\begin{aligned}&\Leftrightarrow axa^{-1} = x \quad \text{for all } a \in G\\ &\Leftrightarrow ax = xa \quad \text{for all } a \in G\\ &\Leftrightarrow x \in Z(G), \quad \text{the centre of } G.\end{aligned}$$

Therefore, each element of $Z(G)$ contributes a singleton conjugacy class and vice versa. If $C(x_1)$, $C(x_2)$, …, $C(x_n)$ are all the distinct conjugacy classes each with more than one element, then

$$\begin{aligned}|G| &= |Z(G)| + \sum_{i=1}^{n} |C(x_i)| \\ &= |Z(G)| + \sum_{i=1}^{n} i_G(\text{Cent}_G(x_i)) \\ &= |Z(G)| + |G| \left(\sum_{i=1}^{n} \frac{1}{|\text{Cent}(x_i)|} \right).\end{aligned}$$ ◀

For any subset A of a group G and for any element x in G, the set

$$xAx^{-1} = \{xax^{-1} : a \in A\}$$

is called the conjugate of A corresponding to x. The map $(x, A) \mapsto xAx^{-1}$ is an action of G on the power set of G, with respect to which the orbit of A is

$$C(A) = \{xAx^{-1} : x \in G\}$$

and the stabilizer of A is

$$\begin{aligned}\text{Cent}_G(A) &= \{x \in G : xAx^{-1} = A\} \\ &= \{x \in G : xA = Ax\}.\end{aligned}$$

$C(A)$ and $\text{Cent}_G(A)$ are respectively called the conjugacy class of A in $\mathbb{P}(G)$ and the centralizer or normalizer of A in G. Clearly, for any subset A of a finite group G,

$$|C(A)| = i_G(\text{Cent}_G(A)) = \frac{|G|}{|\text{Cent}_G(A)|}.$$

In the following, we give certain important applications of the class equation of a finite group G proved in Theorem 7.3.2. The following is a simple consequence of the discussion made above.

Theorem 7.3.3. Let K be a subgroup of a finite group G. Then, the number of subgroups of G conjugate to K is equal to the index of the normalizer of K in G.

Theorem 7.3.4. Let p be a prime number and n be a positive integer. Let G be a group of order p^n. Suppose that G acts on a finite set X and

$$X_0 = \{x \in X : ax = x \text{ for all } a \in G\}.$$

Then, $|X| \equiv |X_0| \pmod p$.

Proof: We have to prove that $|X| - |X_0|$ is a multiple of p. Observe that the orbit of an element $x \in X$ is a singleton set if and only if $x \in X_0$. Therefore, there are exactly $|X_0|$ number of singleton orbits in X. Let $O(x_1)$, $O(x_2)$, …, $O(x_n)$ be all the distinct orbits each with more than one element. Then, by Theorem 7.3.1, we have

$$|X| = |X_0| + \sum_{i=1}^{n} |O(X_i)| = |X_0| + \sum_{i=1}^{n} i_G(\mathrm{St}(x_i)).$$

Note that the stabilizer $\mathrm{St}(x_i)$ is a subgroup of G and $|G| = p^n$. By the Lagrange's theorem, $|\mathrm{St}(x_i)|$ is a divisor of p^n and

$$\frac{|G|}{|\mathrm{St}(x_i)|} = i_G(\mathrm{St}(x_i)) = |O(x_i)| > 1$$

for each $1 \leq i \leq n$. Therefore, for each $1 \leq i \leq n$,

$$i_G(\mathrm{St}(x_i)) = p^{n_i} \text{ for some } n_i > 0.$$

Therefore, $|X| = |X_0| + \sum_{i=1}^{n} p^{n_i} = |X_0| + p(\sum_{i=1}^{n} p^{n_i - 1})$
and thus $|X| - |X_0|$ is a multiple of p, so that

$$|X| \equiv |X_0| \pmod p.$$ ◀

Theorem 7.3.5. Let G be a group of order p^n, where p is a prime and n be a positive integer. Then, the centre of G is nontrivial; that is, $|Z(G)| > 1$.

Proof: Consider the action of G on itself by conjugation. Then, by Theorem 7.3.4,

$$|G| \equiv |G_0| \pmod p,$$

where $G_0 = \{x \in G : axa^{-1} = x \text{ for all } a \in G\} = Z(G)$. Therefore, $p^n = |G| \equiv |Z(G)| \pmod p$, which implies that $|Z(G)|$ is a multiple of p. Since $e \in Z(G)$, $|Z(G)| > 0$ and hence $|Z(G)| > 1$, so that $Z(G) \neq \{e\}$. ◀

Theorem 7.3.6. Let p be a prime number. Then, any group of order p^2 is abelian.

Proof: Let G be a group of order p^2 and $Z(G)$ its centre. By Theorem 7.3.5, $|Z(G)| > 1$ and, by the Lagrange's theorem $|Z(G)|$ is a divisor of $|G| = p^2$. Therefore, $|Z(G)| = p$ or p^2. Suppose that $|Z(G)| = p$. Then, $Z(G)| \subsetneqq G$ and hence we can choose $a \in G$ such that $a \notin Z(G)$. Then, there exists $x \in G$ such that $ax \neq xa$. Consider the centralizer of x

$$\text{Cent}_G(x) = \{y \in G : xy = yx\}.$$

$\text{Cent}_G(x)$ is a subgroup of G containing $Z(G)$ properly (since $x \in \text{Cent}_G(x)$ and $x \notin Z(G)$). Then follows that $|\text{Cent}_G(x)|$ is a divisor of p^2 and $|\text{Cent}_G(x)| > |Z(G)| = p$. Therefore, $|\text{Cent}_G(x)| = p^2 = |G|$ and hence $\text{Cent}_G(x) = G$ which is a contradiction, since $a \notin \text{Cent}_G(x)$. Therefore, being $|Z(G)| = p$ is impossible and hence $|Z(G)| = p^2$, so that $Z(G) = G$ and hence G is abelian. ◀

In the following, we prove a theorem of Burnside which has applications in combinatories. When a group G act on a set X, then, for any $a \in G$, define

$$X_a = \{x \in X : ax = x\}.$$

That is, X_a is the set of all elements of X which are fixed by the action of a. Note that $X_e = X$, where e is the identity in group G.

Theorem 7.3.7 (Burnside's theorem). Let a finite group G act on a finite set X and n be the number of orbits in X. Then,

$$n = \frac{1}{|G|}\sum_{a \in G} |X_a|.$$

Proof: Consider the set

$$A = \{(a, x) \in G \times X : ax = x\}.$$

Note that, for any fixed element x in X, the number of pairs (a, x) in A is precisely equal to the order of the stabilizer $\text{St}(x)$. Also, for any fixed element a in G, the number of pairs (a, x) in A is exactly equal to $|X_a|$, where $X_a = \{x \in X : ax = x\}$. Therefore, we have

$$\sum_{x \in X} |\text{St}(x)| = |A| = \sum_{a \in G} |X_a| \cdot$$

By Corollary 7.2.1,

$$|\mathrm{St}(x)| = \frac{|G|}{|\mathrm{O}(x)|},$$

where O(x) is the orbit of x. Note that O(x) = O(y) for all $y \in$ O(x). Since X is finite, the number of orbits in X is finite. Let O(x_1), O(x_2), …, O(x_n) be all the distinct orbits in X. Then,

$$\begin{aligned}\sum_{x\in X}|\mathrm{St}(x)| &= \sum_{x\in X}\frac{|G|}{|\mathrm{O}(x)|}\\ &= |G|\sum_{i=1}^{n}\sum_{x\in \mathrm{O}(x_i)}\frac{1}{|\mathrm{O}(x)|}\\ &= |G|\sum_{i=1}^{n}\sum_{x\in \mathrm{O}(x_i)}\frac{1}{|\mathrm{O}(x_i)|}\\ &= |G|\sum_{i=1}^{n}|\mathrm{O}(x_i)|\frac{1}{|\mathrm{O}(x_i)|}\\ &= |G|\cdot n\end{aligned}$$

Thus, $n = \frac{1}{|G|}\sum_{x\in X}|\mathrm{St}(x)| = \frac{1}{|G|}\sum_{a\in G}|X_a|$. ◀

Worked Exercise 7.3.1. Let p be a prime number and n be a positive integer. Find the number of different necklaces formed by p beads, where the beads can have any of n different colours.

Answer: Let ($\mathbb{Z}_p$, +) be the additive group of integers modulo p and X be the set of all possible necklaces. Since there are p beads in each of the necklaces and each bead can have any of n different colours, $|X| = n^p$.

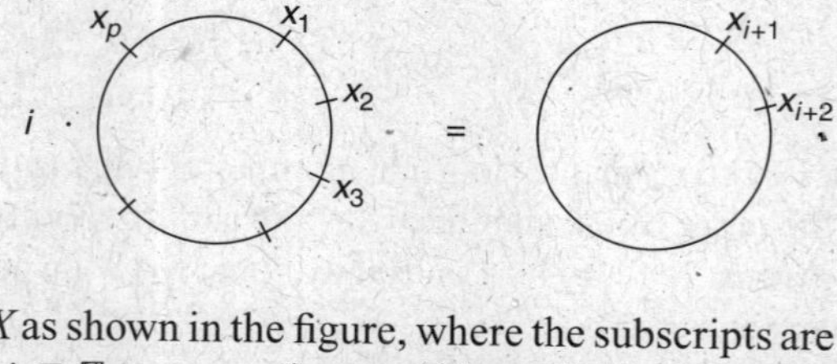

Let $\mathbb{Z}_p$ act on X as shown in the figure, where the subscripts are modulo p. The action of any $i \in \mathbb{Z}_p$ on any given necklace yields the same necklace; only the beads are permuted cyclically. Therefore, the number of orbits in X is same as the number of different necklaces, which can be computed by using

Theorem 7.3.7. First, let us compute $X_i = \{x \in X : ix = x\}$, for any $i \in \mathbb{Z}_p$. Clearly, $X_0 = X$ and hence $|X_0| = n^p$. For any $0 \neq i \in \mathbb{Z}_p$,

$$\begin{aligned} X_i &= \{x \in X : ix = x\}, \\ &= \{x \in X : (i +_p j)x = x \text{ for all } 0 \le j < p\} \\ &= \{x \in X : jx = x \text{ for all } j \in \mathbb{Z}_p\}, \end{aligned}$$

since $\mathbb{Z}_p$ is a cyclic group of prime order and hence any nonidentity element generates $\mathbb{Z}_p$. Therefore, for and $0 \neq i \in \mathbb{Z}_p$, X_i consists those necklaces which are unchanged by permutation and hence X_i consists of those necklaces in which all the beads are of same colour. Since we are given with n different colours, it follows that $|X_i| = n$ for all $0 \neq i \in \mathbb{Z}_p$. Thus, by the Burnside's theorem, the number of different necklaces (the number of orbits in X) is

$$\begin{aligned} \frac{1}{|\mathbb{Z}_p|}\sum_{i\in\mathbb{Z}_p}|X_i| &= \frac{1}{p}\sum_{i=0}^{p-1}|X_i| \\ &= \frac{1}{p}(n^p + (n + \cdots + n)) \\ &= \frac{1}{p}n^p + (p-1)n \\ &= \frac{n}{p}(n^{p-1} + p - 1) \end{aligned}$$

Worked Exercise 7.3.2. Let G be a group and $a \in G$ such that $O(a) > 1$. Suppose that G has exactly two conjugacy classes. Then prove that $|G| = 2$.

Answer: Let $O(a) = n > 1$. Then, $a \neq e$. Since G has exactly two conjugacy classes and $\{e\}$ is a singleton conjugacy class, it follows that $\{e\}$ and $C(a)$ are the only conjugacy classes. Therefore,

$$G - \{e\} = C(a) = \{xax^{-1} : x \in G\}.$$

Since $O(a) = O(xax^{-1})$, it follows that $O(b) = n$ for all $b \neq e$ and, in particular $b^n = e$ for all $b \neq e$. Now, we prove that n is a prime. Since $n > 1$, we can choose a prime p dividing n. Then, $O(a^p) = \frac{O(a)}{p} = \frac{n}{p} \neq n$ and hence $a^p = e$, so that $p = O(a) = n$. Thus, n is a prime. Next, consider a^2. If $a^2 \neq e$, then $a^2 \in C(a)$ and hence $a^2 = xax^{-1}$ for some $x \in G$, so that $(a^2)^m = x^m a x^{-m}$ for all $m > 0$ and

$$e = a^{2n} = (xax^{-1})^n = x^n a x^{-1} = eae^{-1} = a,$$

which is a contradiction. Therefore, $a^2 = e$ and $O(a) = 2$. Thus, $b^2 = e$ for all $b \in G$ (since $O(b) = O(a)$ for all $b \neq e$). Therefore, G is abelian and hence $C(a) = \{a\}$. Thus, $G = \{e, a\}$ and $|G| = 2$.

Worked Exercise 7.3.3. Let p be a prime number and n be a positive integer. Let G be a group of order p^n and N be a nontrivial normal subgroup of G. Prove that $Z(G) \cap N$ is nontrivial, whole $Z(G)$ is the centre of G.

Answer: Since N is a nontrivial subgroup of G, $|N| > 1$ and $|N|$ is a divisor of $|G| = p^n$. Therefore, $|N| = p^m$ for some $m > 0$. Also, since N is normal in G, $axa^{-1} \in N$ for all $x \in N$ and $a \in G$. Therefore, G acts on N by conjugation (the action is $(a, x) \to axa^{-1}$). By Theorem 7.3.4,

$$|N| \equiv |N_0| \pmod{p}$$

where
$$\begin{aligned} N_0 &= \{x \in N : axa^{-1} = x \text{ for all } a \in G\} \\ &= \{x \in N : ax = xa \text{ for all } a \in G\} \\ &= N \cap Z(G) \end{aligned}$$

Since p divides both $|N|$ and $|N| - |N_0|$, it follows that $|N_0|$ is a multiple of p. Also, since $e \in Z(G) \cap N = N_0$, $|N_0| > 0$ and hence $|N_0| \geq p > 1$. Thus, $Z(G) \cap N$ is nontrivial.

EXERCISE 7(C)

1. Determine all the distinct conjugacy classes in each of the following and verify that the number of elements in each conjugacy class is a divisor of the order of the group
   - (i) The symmetric group S_3 of degree three.
   - (ii) The alternating group A_4 of degree four.
   - (iii) The symmetric group S_4 of degree four.
   - (iv) The dihedral group D_4 of degree four.
2. Find the number of different (distinguishable) dice that can be made by marking the faces of a cube using one to six dots.
3. How many different tetrahedral dice can be made by marking the faces of a regular tetrahedron using one to four dots?
4. How many different ways can seven people be seated at a round table, where there is no distinguishable leader to the table?
5. Find the number of different ways the edges of an equilateral triangle can be painted if four different colours of paint are available, assuming only one colour is used on each edge, and the same colour may be used on different edges.

6. Repeat Exercise 5 above with the assumption that a different colour is used on each edge.

7. For any proper subgroup H of a finite group G, prove that $G \neq \bigcup_{a \in G} aHa^{-1}$.

8. Let H be a proper subgroup of finite index in a group G. Prove that H contains a normal subgroup N that is of finite index.

9. Let G be a group such that any proper subgroup is contained in a maximal subgroup of finite index in G and that any two maximal subgroups of G are conjugate to each other in G. Then prove that G is cyclic.

10. Let N be a normal subgroup of a finite group G such that the order and the index of N are relatively prime. If a is an element of G, such that $O(a)$ divides $|N|$, then prove that $a \in N$.

11. Let G be a group and $H = \{a \in G : C(a) \text{ is finite}\}$, where $C(a)$ is the conjugacy class of a. Then prove that H is a subgroup of G.

12. Let N be a normal subgroup of order 3 in a group G such that $N \not\subseteq Z(G)$. Then prove that G has a subgroup of index 2.

13. Find the number of different necklaces that can be formed with five beads and two colours.

14. Determine the number of different necklaces that can be formed with six beads and two colours.

15. Find the number of neckties having n strips (of equal width) of K distinct colours.

16. Let S be a subset of a group G and

$$C(S) = \{aSa^{-1} : a \in G\}$$
$$\text{and} \quad N(S) = \{a \in G : aSa^{-1} = S\}.$$

Prove that $N(S)$ is a subgroup of G and that $C(S)$ is bijective with the set of left cosets of $N(S)$ in G. $N(S)$ is called the *normalizer of S in G*.

17. Let G be a group of order p^n, where p is a prime and $n \in \mathbb{Z}^+$. If A is a proper subgroup of G, then prove that A is properly contained in the normalizer of A in G.

18. If G is a group of order p^n (p is a prime and $n \in \mathbb{Z}^+$) and A is a subgroup of order p^{n-1} in G, then prove that A is normal in G.

19. Prove that any subgroup of order 343 in a group of order 2401 is normal.

20. Find the number of different necklaces formed by 11 beads, where each bead can have any of the five given different colours.

7.4 CAUCHY AND SYLOW THEOREMS

The Lagrange's theorem states that, for any finite group G, if d is the order of a subgroup of G, then d is a divisor of the order of G. The converse of this is not true. That is, if d is a divisor of the order of G, we may not find a subgroup of order d in G. For, consider the alternating group A_4 of degree 4 whose order is 12. Even though 6 is a divisor of the order of A_4, there is no subgroup of order 6 in A_4. However, in certain special cases, the converse of the Lagrange's theorem is true. In particular, when the divisor d is a prime or a power of a prime, then there always exists a subgroup of order d. We prove these and certain important consequences of these in this section.

Recall that any group of prime order is cyclic and hence, for any group G and for any prime number p, the existence of a subgroup of order p in G is equivalent to the existence of an element of order p in G.

Theorem 7.4.1 (Cauchy's Theorem). Let G be a finite group and p be a prime number such that p divides the order of G. Then, G has an element of order p.

Proof: Consider the set

$$X = \{(x_1, x_2, \ldots, x_p) : x_i \in G \quad \text{and} \quad x_1, x_2, \ldots, x_p = e\}$$

Then, $|X| = |G|^{p-1}$, since, for any $(x_1, x_2, \ldots, x_{p-1}) \in G^{p-1}$, $(x_1, x_2, \ldots, x_{p-1}, x_p) \in X$, where $x_p = (x_1, x_2, \ldots, x_{p-1})^{-1}$ and vice versa. Since p divides $|G|$ and $p - 1 > 0$, it follows that p divides $|X|$.

Consider the group $\mathbb{Z}_p$ of integers modulo p. We shall define an action of $\mathbb{Z}_p$ on X as follows: for any $x = (x_1, x_2, \ldots, x_p) \in X$ and $i \in \mathbb{Z}_p = \{0, 1, 2, \ldots, p-1\}$, define

$$i \cdot (x_1, x_2, \ldots, x_p) = (x_{i+1}, x_{i+2}, \ldots, x_p, x_1, x_2, \ldots, x_i).$$

Since $(x_1\, x_2 \ldots x_i)\,(x_{i+1}\, x_{i+2} \cdots x_p) = e$, it follows that

$$(x_{i+1} \cdots x_p) \cdot (x_1 \cdots x_p) = e$$

and hence the above defines a mapping of $\mathbb{Z}_p \times X$ into X. It can be easily verified that this is an action of $\mathbb{Z}_p$ on X. Therefore, by Theorem 7.3.4,

$$|X| \equiv |X_0| (\text{mod}\, p),$$

$$\begin{aligned} \text{where } X_0 &= \{x \in X : ix = x \text{ for all } i \in \mathbb{Z}_p\} \\ &= \{(x_1, x_2, \ldots, x_p) \in X : 1 \cdot (x_1, x_2, \ldots, x_p) = (x_1, x_2, \ldots, x_p)\} \\ &= \{(x_1, x_2, \ldots, x_p) \in X : (x_2, x_3, \ldots, x_p, x_1) = (x_1, x_2, \ldots, x_p)\} \\ &= \{(x_1, x_2, \ldots, x_p) \in X : x_1 = x_2 = \cdots = x_p\} \\ &= \{(x, x, \ldots, x) : x \in G \text{ and } x^p = e\}. \end{aligned}$$

Since p divides $|X|$ and $|X| - |X_0|$, it follows that p divides $|X_0|$. Also, since $(e, e, \ldots, e) \in X_0$, we get that $|X_0| > 0$. Therefore, $|X_0| = p^n$ for some $n \in \mathbb{Z}^+$ and, in particular,

$$|X_0| \geq p > 1.$$

Thus, there exists $(x, x, \ldots, x) \in X_0$ other then $(e, e, \ldots, e)$ and hence there exists $x \neq e$ in G such that $x^p = e$. Then, $O(x) > 1$ and $O(x)$ is a divisor of p. Since p is prime, it follows that $O(x) = p$. Thus, x is an element of order p in G. ◀

Corollary 7.4.1. Let G be a finite group and p be a prime divisor of the order of G. Then, G has a subgroup of order p.

Proof: If $x \in G$ such that $O(x) = p$, then

$$<x> = \{e, x, x^2, \ldots, x^{p-1}\}$$

is a subgroup of order p in G.

Definition 7.4.1. Let p be any prime number. A group G is called a *p-group* if the order of every element of G is a power of p; that is, for any $a \in G$, $O(a) = p^n$ for some integer $n \geq 0$.

Theorem 7.4.2. A finite group G is a p-group (where p is a prime) if and only if $|G| = p^m$ for some nonnegative integer m.

Proof: Let G be a finite group and p be a prime number. Suppose that $|G| = p^m$, $0 \leq m \in \mathbb{Z}$. For any $a \in G$, $O(a)$ divides $|G| = p^m$ and hence $O(a) = p^m$ for some $0 \leq n \leq m$. Therefore, G is a p-group. Conversely suppose that $|G| \neq p^m$ for any $m \geq 0$. Then, there exists a prime $q \neq p$ such that q divides $|G|$. By the Cauchy's theorem, there exists an element a of order q in G and therefore G is not a p-group. ◀

In the proof of the above theorem, the finiteness of G is necessary. For any prime p, there are infinite p-groups. Consider the following example.

Example 7.4.1. Let p be any prime number and

$$H = \left\{ \frac{a}{p^n} : a \text{ and } n \in \mathbb{Z} \text{ and } n \geq 0 \right\}.$$

Then, $(H, +)$ is a group, where $+$ is the usual addition of rational numbers. Any integer a can be expressed as $a/p^0 \in H$. Therefore, $\mathbb{Z}$ is a subgroup of H. Now, let

$$G = \text{the quotient group } H/\mathbb{Z}.$$

Then, for any positive integers n and m,

$$\frac{1}{p^n} - \frac{1}{p^m} \notin \mathbb{Z} \quad \text{and hence } \frac{1}{p^n} + \mathbb{Z} \neq \frac{1}{p^m} + \mathbb{Z}.$$

Therefore, G is an infinite group. Let x be any element of G. Then,

$$x = \frac{a}{p^n} + \mathbb{Z} \quad \text{for some } a \text{ and } n \in \mathbb{Z},\ n \geq 0.$$

Since $p^n x = p^n((a/p^n) + \mathbb{Z}) = a + \mathbb{Z} = \mathbb{Z}$, it follows that $O(x)$ is a divisor of p^n and hence $O(x) = p^m$ for some $0 \leq m \leq n$. Thus, G is a p-group.

Definition 7.4.2. A subgroup H of a group G is called a *p-subgroup of G* if it is a p-group; that is, every element of H is of order p^n for some $n \geq 0$.

Clearly the trivial subgroup $\{e\}$ is a p-subgroup of any group G, for any prime p. Also, every subgroup of a p-group is a p-subgroup. In the following, we prove an important result on p-subgroups which plays a crucial role in the proofs of Sylow theorems. For any subgroup H of a group G, the set

$$N_G(H) = \{a \in G : aHa^{-1} = H\}$$

is called the *normalizer of H in G*. Recall that $N_G(H)$ is precisely the stabilizer of H when G acts on the power set $\mathbb{P}(G)$ by conjugation. Also, clearly $N_G(H)$ is the largest subgroup of G containing H as normal subgroup; that is, for any subgroup A of G containing H,

$$H \text{ is normal in } A \Leftrightarrow A \subseteq N_G(H).$$

Theorem 7.4.3. Let G be a finite group and p be a prime number. Let H be a p-subgroup of G. Then,

$$i_G(H) \equiv i_{N_G(H)}(H) (\text{mod } p),$$

where $i_G(H)$ is the index of H in G.

Proof: Let X be the set of all left cosets of H in G. Then, H acts on X by left translation; that is, $(h, xH) \mapsto hxH$ is an action of H on X. Then, by Theorem 7.3.4,

$$i_G(H) = |X| \equiv |X_0| (\text{mod } p)$$

where

$$\begin{aligned} X_0 &= \{xH \in X : hxH = xH \text{ for all } h \in H\} \\ &= \{xH \in X : x^{-1}hx \in H \text{ for all } h \in H\} \\ &= \{xH : x \in G \text{ and } x^{-1}Hx \subseteq H\} \\ &= \{xH : x \in G \text{ and } xHx^{-1} = H\} \\ &= \{xH : x \in N_G(H)\} \end{aligned}$$

Therefore, $|X_0| = i_{N_G(H)}(H)$, the index of H in $N_G(H)$. Thus, $i_G(H) \equiv i_{N_G(H)}(H)$. (mod p). ◀

Corollary 7.4.2. Let H be a p-subgroup of a finite group G such that p divides $i_G(H)$. Then, $N_G(H) \neq H$.

Proof: By the above theorem,

$$i_G(H) \equiv i_{N_G(H)}(H) (\text{mod } p).$$

Therefore, p divides $i_G(H) - i_{N_G(H)}(H)$. By the hypothesis, p divides $i_G(H)$ also. Therefore, p divides $i_{N_G(H)}(H)$ also, and hence,

$$\frac{N_G(H)}{|H|} = i_{N_G(H)}(H) = ps \quad \text{for some } s \in \mathbb{Z}^+.$$

which implies that $|H| < |N_G(H)|$ so that H is a proper subgroup of $N_G(H)$.

When p^n divides the order of a finite group G, the following theorem guarantees the existence of a subgroup of order p^n in G.

Theorem 7.4.4 (Sylow Theorem – I). Let G be a finite group, p be a prime number and n be a nonnegative integer such that p^n divides the order of the group G. Then, G has a subgroup of order p^n.

Proof: We shall prove the theorem by induction on n. The theorem is trivial for $n = 0$ and, for $n = 1$, the theorem is a consequence of Corollary 7.4.1. Now, let $n > 1$ and p^n divides $|G|$. Then, p^{n-1} divides $|G|$ and hence, by induction; there exists a subgroup H of order p^{n-1} in G. By Theorem 7.4.3,

$$i_G(H) \equiv i_{N_G(H)}(H) \pmod p.$$

Since p^n divides $|G| = |H| \cdot i_G(H) = p^{n-1} \cdot i_G(H)$, it follows that p divides $i_G(H)$ and therefore p divides $i_{N_G(H)}(H)$. Recall that $N_G(H) = \{a \in G : aHa^{-1} = H\}$ and hence H is a normal subgroup of $N_G(H)$ and therefore we have the quotient group $N_G(H)/H$ whose order is $i_{N_G(H)}(H)$. Now, $N_G(H)/H$ is a group whose order is divisible by p. Therefore, by Corollary 7.4.1, $N_G(H)/H$ has a subgroup K of order p. Then, $K = A/H$ where A is a subgroup of $N_G(H)$ containing H. Now,

$$|A| = |A/H| \cdot |H| = |K|\,|H| = p \cdot p^{n-1} = p^n$$

and, since A is a subgroup of $N_G(H)$ which is a subgroup of G, it follows that A is a subgroup of order p^n in G. ◀

Corollary 7.4.3. Let H be a subgroup of order p^{n-1} in a finite group G, where p is a prime number and n is a positive integer. If p^n divides $|G|$, then there exists a subgroup A of order p^n such that H is a normal subgroup of A.

Proof: In the proof of the above theorem, we have $H \subseteq A \subseteq N_G(H)$ and hence H is a normal subgroup of A (since $a \in A \Rightarrow a \in N_G(H) \Rightarrow aHa^{-1} = H$).

Corollary 7.4.4. Let G be a finite group and p be a prime number. Then, every p-subgroup of G is contained in a maximal p-subgroup.

Proof: Follows from the fact that a subgroup H of G is a p-subgroup if and only if the order of H is a power of p and from the above corollary.

Definition 7.4.3. For any prime number p, a maximal p-subgroup of a finite group G is called a *Sylow p-subgroup* of G.

Note that the order of a Sylow p-subgroup of G must be the largest power of p dividing the order of G. In fact, a subgroup H of G is a Sylow p-subgroup of G if and only if $|H| = p^n$, where p^n divides $|G|$ and p^{n+1} does not divide $|G|$.

By the Sylow Theorem – I (7.4.4), for any prime p and a finite group G, maximal p-subgroups exist in G. If p does not divide $|G|$, then $p^0(=1)$ is the largest power of p dividing $|G|$ and hence $\{e\}$ is the only Sylow p-subgroup of G.

Example 7.4.2. Consider the alternating group A_4 of degree 4. The order of A_4 is $12 = 2^2 \cdot 3^1$. For any prime p other than 2 and 3, $\{e\}$ is the Sylow p-subgroup of A_4. Also, any subgroup of order 4 is a Sylow 2-subgroup and any subgroup of order 3 is a Sylow 3-subgroup. Clearly any Sylow 3-subgroup is a cyclic subgroup generated by a 3-cycle in S_4. Further, any subgroup of order 4 in A_4 is not cyclic, since a 4-cycle is not an even permutation. It can be easily checked that any Sylow 2-subgroup of A_4 must be necessarily of the form $\{e, \alpha, \beta, \alpha\beta\}$ where each of α and β is a product of two disjoint transpositions, and that there are three Sylow 2-subgroups of A_4.

Note that, for any prime p and for any finite group G, a subgroup H of G is a Sylow p-subgroup if and only if aHa^{-1} is also a Sylow p-subgroup for every $a \in G$, since H and aHa^{-1} are of same order. Therefore, if H is a Sylow p-subgroup of G, then any conjugate of H in G is also a Sylow p-subgroup and, in the following, we prove that $\{aHa^{-1} : a \in G\}$ is the complete list of Sylow p-subgroups of G.

Theorem 7.4.5 (Sylow Theorem – II). Let p be a prime number and G be a finite group. If S is a Sylow p-subgroup of G and H is a p-subgroup of G, then $H \subseteq aSa^{-1}$ for some $a \in G$. In particular, any two Sylow p-subgroups of G are conjugate to each other.

Proof: Let S be a Sylow p-subgroup and H be any p-subgroup of G and let X be the set of all left cosets of S in G. Since $|S|$ is the largest power of p dividing $|G|$ and $|X| = i_G(S) = |G|/|S|$, it follows that p does not divide $|X|$. Now, H acts on X by left translation; that is, $(h, xS) \mapsto hxS$ is an action of H on X. Then, by Theorem 7.3.4,

$$i_G(S) = |X| \equiv |X_0| (\text{mod}\, p),$$

where

$$\begin{aligned} X_0 &= \{xS \in X \text{ and } hxS = xS \text{ for all } h \in H\} \\ &= \{xS : x \in G \text{ and } x^{-1}hx \in S \text{ for all } h \in H\} \\ &= \{xS : x \in G \text{ and } H \subseteq xSx^{-1}\}. \end{aligned}$$

Since p divides $|X| - |X_0|$ and p does not divide $|X|$, it follows that p does not divide $|X_0|$ and hence $|X_0| \neq 0$; that is, X_0 is a nonempty set. Thus, there exists $x \in G$ such that $H \subseteq xSx^{-1}$. In particular, if H is also a Sylow p-subgroup of G, then $|H| = |S|$ and $H \subseteq xSx^{-1}$ and hence $H = xSx^{-1}$ for some $x \in G$. ◀

Corollary 7.4.5. Let S be a Sylow p-subgroup of a finite group G. Then, S is normal in G if and only if S is the only Sylow p-subgroup of G.

Corollary 7.4.6. Let G be a finite abelian group. Then, for each prime number p, G has a unique Sylow p-subgroup.

Sylow Theorem − II describes all Sylow p-subgroups in terms of a given Sylow p-subgroup. However, it does not give the exact number of Sylow p-subgroups. In the following, we derive certain formulae to find the exact number of Sylow p-subgroups.

Theorem 7.4.6 (Sylow Theorem − III). Let G be a finite group. For any prime p, let n_p be the number of Sylow p-subgroups of G. Then,

1. n_p divides $|G|/p^n$, where p^n is the largest power of p dividing $|G|$,
2. n_p divides $|G|$ and
3. $n_p = mp + 1$ for some nonnegative integer m.

Proof:

1. Let X be the set of all subgroups of G. Then, G acts on X by conjugation. Let P be a Sylow p-subgroup of G. By the Sylow Theorem − II (7.4.5), the orbit of P in X is precisely the set of all Sylow p-subgroups of G. We know that

$$n_p = |\mathrm{O}(P)| = i(N_G(P)),$$

where $N_G(P) = \{a \in G : aPa^{-1} = P\}$. Let p^n be the largest power of p dividing $|G|$. Then, $|P| = p^n$ and, by Worked Exercise 4.4.1 (1),

$$i_{N_{G(P)}}(P) \cdot n_p = i_{N_G(P)}(P) \cdot i_G(N_G(P)) = i_G(P) = \frac{|G|}{p^n}$$

and hence n_p is a divisor of $|G|/p^n$.
2. This a single consequence of (1).
3. Let Y be the set of all Sylow p-subgroups of G and $S \in Y$. Let S act on Y by conjugation. Then, by Theorem 7.3.4,

$$|Y| \equiv |Y_0| (\mathrm{mod}\; p),$$

where $Y_0 = \{P \in Y : aPa^{-1} = P \text{ for all } a \in S\}$

$= \{P \in Y : S \subseteq N_G(P)\}$,

where $N_G(P) = \{a \in G : aPa^{-1} = P\}$. We shall prove that Y_0 is a singleton set. Consider $P \in Y_0$. Then,

$$aPa^{-1} = P \text{ and hence } aP = Pa \text{ for all } a \in S$$

so that $SP = PS$ which implies that SP is a subgroup of G. Also, since S and $P \subseteq N_G(P)$, it follows that $S \subseteq SP \subseteq N_G(P)$. Since S and P are p-subgroups of G. SP is also a p-subgroup of G. By the maximality of S, it follows that $S = SP$ and hence $P \subseteq S$. Since P and S are subgroups of the same order (because both of these are Sylow p-subgroups), we get that $P = S$. Thus, $Y_0 = \{S\}$ and $|Y_0| = 1$ and therefore

$$n_p = |Y| \equiv 1 (\text{mod } p).$$

That is, $n_p - 1$ is divisible by p or $n_p = mp + 1$ for some nonnegative integer m. ◀

In the following theorem, we prove that the converse of the Lagrange's theorem holds good for finite abelian groups.

Theorem 7.4.7. Let G be a finite abelian group and d be a positive divisor of $|G|$. Then, G has a subgroup of order d.

Proof: The theorem is trivial if $|G| = 1$. Therefore, we can assume that $|G| > 1$. Let us suppose that

$$|G| = p_1^{r_1} p_2^{r_2} \cdots p_k^{r_k},$$

where $p_1, p_2, \ldots, p_k$ are distinct primes and $r_1, r_2, \ldots, r_k$ are positive integers. Then, since d is a divisor of $|G|$,

$$d = p_1^{s_1} p_2^{s_2} \cdots p_k^{s_k},$$

where $s_1, s_2, \ldots, s_k$ are integers such that $0 \leq s_i \leq r_i$ for all $1 \leq i \leq k$. Now, for each i, $p_i^{s_i}$ is a divisor of d and d is a divisor of $|G|$ and hence $p_i^{s_i}$ is a divisor of $|G|$. Therefore, by Sylow Theorem – I (7.4.4) there exists a subgroup A_i of G such that $|A_i| = p_i^{s_i}$. For any $i \neq j$, $A_i \cap A_j$ is a subgroup of A_i as well as of A_j and hence, by Lagrange's theorem $|A_i \cap A_j|$ is a common divisor of $|A_i|$ $(= p_i^{s_i})$ and $|A_j|$ $(= p_j^{s_j})$. Since p_i and p_j are distinct primes, $p_i^{s_i}$ and $p_j^{s_j}$ are relatively prime and hence $|A_i \cap A_j| = 1$ for all $i \neq j$. Also, since G is an abelian group, A_iA_j is a subgroup of G and

$$|A_iA_j| = \frac{|A_i||A_j|}{|A_i \cap A_j|} = p_i^{s_i} p_j^{s_j}.$$

This argument can be extended inductively to prove that

$$|A_1A_2 \cdots A_i| = |A_1||A_2| \cdots |A_i| = p_1^{s_1} p_2^{s_2} \cdots p_i^{s_i}$$

for any $1 \leq i \leq k$. Now, put $A = A_1 A_2 \ldots A_k$. Then, A is a subgroup of order $p_1^{s_1} p_2^{s_2} \cdots p_k^{s_k} = d$ in G. ◀

Worked Exercise 7.4.1. Prove that any group of order 15 is cyclic.

Answer: Let G be a group of order 15. Let n_3 and n_5 be the number of Sylow 3-subgroups and Sylow 5-subgroups, respectively. Then, by Sylow Theorem − III (7.4.6),

$$n_3 \text{ divides } \frac{|G|}{3} = 5 \quad \text{and} \quad n_3 = 3m + 1, m \geq 0$$

$$n_5 \text{ divides } \frac{|G|}{5} = 3 \quad \text{and} \quad n_5 = 5s + 1, s \geq 0.$$

These imply that $n_3 = 1 = n_5$. Therefore, there is a unique subgroup A of order 3 in G (note that 3^1 is the largest power of 3 dividing $|G|$ and 5^1 is the largest power of 5 dividing $|G|$) and hence A is normal. Similarly, there is a normal subgroup B of order 5 in G. Since $|A|$ and $|B|$ are relatively prime, we get that $A \cap B = \{e\}$. From this, we have

$$|AB| = \frac{|A||B|}{|A \cap B|} = |A||B| = 3 \cdot 5 = 15 = |G|$$

and hence $AB = G$. Any element of G can be uniquely expressed as ab with $a \in A$ and $b \in B$ ($a_1 b_1 = a_2 b_2 \Rightarrow a_2^{-1} a_1 = b_2 b_1^{-1} \in A \cap B = \{e\} \Rightarrow a_2^{-1} a_1 = e = b_2 b_1^{-1} \Rightarrow a_1 = a_2$ and $b_1 = b_2$). Also, for any $a \in A$ and $b \in B$, $a(ba^{-1}b^{-1}) = (aba^{-1})b^{-1} \in A \cap B = \{e\}$ and hence $aba^{-1}b^{-1} = e$ or $ab = ba$. From these, it can be verified that $(a, b) \mapsto ab$ is an isomorphism of $A \times B$ onto G. Further, $A \cong \mathbb{Z}_3$ and $B \cong \mathbb{Z}_5$

$$\therefore G \cong A \times B \cong Z_3 \times Z_5 \cong Z_{15}.$$

Thus, G is cyclic.

The above result is extended to any groups of order pq, where p and q are primes, $p > q$ and q does not divide $p - 1$. In the above result, we have $15 = 5 \cdot 3$ and 3 does not divide $5 - 1$.

Worked Exercise 7.4.2. Let G be a group of order pq, where p and q are distinct primes, $p > q$ and q does not divide $p - 1$. Then prove that G is cyclic.

Answer: Let n_p and n_q be the number of Sylow p-subgroups and the number of Sylow q-subgroups, respectively. Then, we have

$$n_p = mp + 1, m \geq 0, n_q = sq+1, s \geq 0,$$

n_p divides q and n_q divides p. Since $p > q$ and $n_p = mp + 1, m \geq 0$, it follows that $n_p = 1$ and hence there exists unique Sylow p-subgroup A in G and this A must be normal subgroup of order p. Also, since n_q divides p and p is a prime, $n_q = 1$ or p; but $sq + 1 = n_q \neq p$ (otherwise $sq+1 = p$ and q divides $p - 1$, which is a contradiction to the hypothesis). Therefore, $n_q = 1$ and hence G has a unique Sylow q-subgroup B, which becomes a normal subgroup of order q in G. As in 3.4.19, we get that $A \cap B = \{e\}$ and $AB = G$. Therefore, as in the above exercise,

$$G \cong A \times B \cong \mathbb{Z}_p \times \mathbb{Z}_q \cong \mathbb{Z}_{pq}.$$

Thus, G is cyclic.

Worked Exercise 7.4.3. Prove that there are no simple groups of order 63.

Answer: Let G be a group of order $63 = 3^2 \cdot 7$. Let n_7 be the number of Sylow 7-subgroups in G. Then, $n_7 = 7m + 1, m \geq 0$ and n_7 divides 63. From these two, we can infer that $m = 0$ and $n_7 = 1$. Therefore, there is a unique Sylow 7-subgroup H of G. Then, H is a normal subgroup of order 7 in G and hence $H \neq \{e\}$ and $H \neq G$. Thus, G is not simple.

Worked Exercise 7.4.4. Let S be a Sylow p-subgroup of a finite group G. Then prove that $N_G(N_G(S)) = N_G(S)$, where $N_G(S)$ is the normalizer of S in G.

Answer: We have $N_G(S) = \{a \in G : aSa^{-1} = S\}$ and

$$N_G(N_G(S)) = \{a \in G : aN_G(S)a^{-1} = N_G(S)\}.$$

For simplicity, let $N = N_G(S)$.
First note that every conjugate of S is a Sylow p-subgroup of G. Also, if H is a subgroup of G such that $aSa^{-1} \subseteq H$, then aSa^{-1} is a Sylow p-subgroup of H. Clearly S is a Sylow p-subgroup of $N_G(S) = N$. Further, if T is any Sylow p-subgroup of N, then $T = aSa^{-1}$ for some $a \in N$ and hence $T = S$. Therefore, S is the only Sylow p-subgroup of N. Now, consider

$$\begin{aligned} a \in N_G(N) &\Rightarrow aNa^{-1} = N \\ &\Rightarrow aSa^{-1} \subseteq aNa^{-1} = N \\ &\Rightarrow aSa^{-1} = S \\ &\Rightarrow a \in N_G(S) = N. \end{aligned}$$

Therefore, $N_G(N) \subseteq N$. Since N is always contained in $N_G(N)$, it follows that $N_G(N) = N$.

EXERCISE 7(D)

1. State whether each of the following is true or false and substantiate your answer.
   (i) For any prime p and for any finite group G, there is a Sylow p-subgroup of G.
   (ii) The order of a Sylow 3-subgroup of a group of order 108 is 27.
   (iii) Any Sylow 3-subgroup of a group of order 54 is normal.
   (iv) There exists a subgroup of order 16 in a group of order 216.
   (v) Any group of order 159 is simple.
   (vi) Any group of order 159 is cyclic.
   (vii) A group of prime power order has no Sylow p-subgroups.
   (viii) Every p-subgroup of a finite group is a Sylow p-subgroup.
   (ix) Any group of order 121 is abelian.
   (x) Any group of order 8 is abelian.
2. Determine all the Sylow p-subgroups of the following groups for all the primes p.
   (i) $\mathbb{Z}_{24}$, the group of integers modulo 24.
   (ii) S_3, the symmetric group of degree 3.
   (iii) S_4, the symmetric group of degree 4.
   (iv) A_4, the alternating group of degree 4.
   (v) $\mathbb{Z}_3 \times \mathbb{Z}_3$.
3. Prove that any group of order 45 has a normal subgroup of order 9.
4. Prove that there is no simple group of order 56.
5. Show that a normal p-subgroup of a finite group is contained in every Sylow p-subgroup.
6. For any fixed prime p, prove that the intersection of all Sylow p-subgroups of a group G is a normal subgroup of G.
7. Prove that there are no simple groups of order 255.

8. If p is a prime and r and n are positive integers such that $n < p$, then prove that there are no simple groups of order $p^r \cdot n$.

9. Let G be a group of order p^n, where p is a prime and $n \in \mathbb{Z}^+$. Prove that there are normal subgroups A_i for $0 \leq i \leq n$ such that $|A_i| = p^i$ and $A_i \subset A_{i+1}$ for all $0 \leq i < n$.

10. Deduce from above that there are no simple groups of order p^n, for any prime p and $n \geq 2$.

11. Prove that no group of order 30 or 36 or 48 is simple.

12. Show that any group of order 225 is cyclic.

13. Prove that there is exactly one, up to isomorphism, group of order 323.

14. Prove that any group of order 899 or 961 is cyclic.

15. Prove that no group of order 160 is simple.

16. Prove the following in the symmetric group S_r of degree n.
    (i) If $a = (i_1\, i_2 \ldots i_r)$ is an r-cycle, then $f \cdot a \cdot f^{-1} = (f(i_1)\, f(i_2) \ldots f(i_r))$, which is again an r-cycle.
    (ii) Any two cycles of same length are conjugate to each other.
    (iii) Two permutations f and g in S_n are conjugates to each other if and only if $f = a_1 \cdot a_2 \cdot \cdots \cdot a_k$ and $g = b_1 \cdot b_2 \cdot \cdots \cdot b_k$, where a_i's are disjoint cycles and b_i's are disjoint cycles such that $O(a_i) = O(b_i)$ for all $1 \leq i \leq k$.
    (iv) A finite sequence $0 < r_1 \leq r_2 \leq \cdots \leq r_k$ of positive integers is said to be a *partition* of n if $r_1 + r_2 + \cdots + r_k = n$. Then, the number of conjugate classes in S_n is equal to the number of partitions of n.

17. Determine all the conjugacy classes in S_4 and write down the class equation of S_4.

18. Prove that the centre of S_n is trivial for any $n > 2$.

19. Let G be a group of order 341. Prove that any subgroup of order 31 is normal in G.

20. Let p be a prime and N be a normal subgroup of a group G. Prove that G is a p-group if and only if both N and G/N are p-groups.

21. Let N be a normal subgroup of order p in a p-group G, where p is prime. Then prove that N is contained in the centre of G.

22. Let p be a prime and $n \in \mathbb{Z}^+$ such that $p > n$. Prove that any subgroup of order p in a group G of order p^n is normal in G.

23. If a group G contains a proper subgroup of finite index, then prove that G contains a proper normal subgroup of finite index.

24. Give an example of an infinite 7-group.

25. Let G be an infinite p-group. Then prove that either G has a subgroup of order p^n for each positive integer n or there exists a positive integer m such that every finite subgroup of G is of order $\leq p^m$.

26. Let f be an endomorphism of a finite group G. If S is a normal Sylow p-subgroup of G, then prove that $f(a) \in S$ for all $a \in S$.

27. Let p and q be distinct primes and $p > q$. Prove that any group of order $p^n q$ contains a unique normal subgroup of index q.

28. Prove that any finite abelian group of square-free order is cyclic (an integer is said to be square-free if it is not divisible by m^2 for any integer $m > 1$).

29. Let G be a finite abelian group and p be a prime number. Let

$$S_p = \{a \in G : \mathrm{O}(a) = p^r \text{ for some } r \geq 0\}.$$

Then prove that S_p is the unique Sylow p subgroup of G.

30. Use the above to find all Sylow p-subgroups of $\mathbb{Z}_{30}$, for each prime p.

8

Structure Theory of Groups

It is well known that any cyclic group is abelian and the product of any class of abelian groups is abelian. In this chapter, we prove the celebrated theorem known as the Fundamental Theorem of finitely generated abelian groups which states that any finitely generated abelian group is a product of finite number of cyclic groups. This amounts to saying that the cyclic groups are like 'building blocks' for the finite or finitely generated abelian groups. Since any cyclic group is isomorphic to the group $\mathbb{Z}$ of integers or the group $\mathbb{Z}_n$ of integers modulo n for some positive integers, the Fundamental Theorem implies that any finitely generated abelian group is isomorphic to the product of a finite number of copies of $\mathbb{Z}$ and $\mathbb{Z}_n$'s. This facilitates to a great extent the study of finitely generated abelian groups and, in particular, finite abelian groups. In fact, we derive a precise formula for the number of abelian groups of a given order n in terms of the partitions of n.

8.1 DIRECT PRODUCTS

It is well known that, for any groups $G_1, G_2, \ldots, G_n$, the Cartesian product $G = G_1 \times G_2 \times \cdots \times G_n$ can be made into a group by defining

$$(a_1, a_2, \ldots, a_n), (b_1, b_2, \ldots, b_n) = (a_1b_1, a_2b_2, \ldots, a_nb_n)$$

for any $(a_1, a_2, \ldots, a_n)$ and $(b_1, b_2, \ldots, b_n) \in G$. In this group, $(e_1, e_2, \ldots, e_n)$ is the identity, where e_i is the identity in G_i and, for any $(a_1, a_2, \ldots, a_n) \in G$, $(a_1^{-1}, a_2^{-1}, \ldots, a_n^{-1})$ is the inverse of $(a_1, a_2, \ldots, a_n)$, where a_i^{-1} is the inverse of a_i in G_i. This group G is called the *direct product*, or simply, the *product* of $G_1, G_2, \ldots, G_n$ and is denoted by $\prod_{i=1}^{n} G_i$ or $G_1 \times G_2 \times \cdots \times G_n$. If a group H is isomorphic to $G_1 \times G_2 \times \cdots \times G_n$, then H is said to be decomposed into product of groups $G_1 \times G_2 \times \cdots \times G_n$. In this section, we obtain equivalent conditions for the decompositions of a group G into products of groups in terms of normal subgroups of G and the corresponding quotient groups.

If A and B are normal subgroups of a group G such that $AB = G$ and $A \cap B = \{e\}$, then we have proved (see 7. …) that the map $(a, b) \mapsto ab$ is an isomorphism of $A \times B$ onto G. This is extended further in the following theorem.

Theorem 8.1.1. Let $G, G_1, G_2, \ldots, G_n$ be groups. Then, $G \cong G_1 \times G_2 \times \cdots \times G_n$ if and only if there exist normal subgroups $N_1, N_2, \ldots, N_n$ of G satisfying the following:

1. $N_1 N_2 \ldots N_n = G$.
2. For each $1 \le i \le n$, $N_i \cap (N_1 \ldots N_{i-1} N_{i+1} \ldots N_n) = \{e\}$.
3. $N_i \cong G_i$ for each $1 \le i \le n$.

Proof: Let $H = G_1 \times G_2 \times \cdots \times G_n$. For each $1 \le i \le n$, let

$$M_i = \{(a_1, a_2, \ldots, a_n) \in H : a_j = e_j \text{ for all } j \ne i\}.$$

Then, it can be easily verified that each M_i is a subgroup of H. For any $a = (a_1, \ldots, a_n) \in M_i$ and $x = (x_1, \ldots, x_n) \in H$, we have,

$$(xax^{-1})_j = x_j a_j x_j^{-1} = x_j e_j x_j^{-1} = e_j \quad \text{for all } j \ne i$$

and hence $xax^{-1} \in M_i$. Therefore, M_i is a normal subgroup of H. Further, any $x = (x_1, x_2, \ldots, x_n) \in H$ can be expressed as

$$x = (x_1, e_2, \ldots, e_n)\cdot(e_1, x_2, e_3, \ldots, e_n) \ldots (e_1, e_2, \ldots, e_{n-1}, x_n)$$

and therefore $M_1 M_2 \cdots M_n = H$. Also, for any $1 \le i \le n$, if $x \in M_i \cap (M_1 \cdots M_{i-1} M_{i+1} \cdots M_n)$, then $x_j = e_j$ for all $j \ne i$, and $x = x_i \cdots x_{i-1} x_{i+1} \cdots x_n$ for some $x_j \in M_j, j \ne i$, so that $x_i = (x_1)_i \cdots (x_{i-1})_i (x_{i+1})_i \cdots (x_n)_i = e_i \cdots e_i e_i \cdots e_i = e_i$ and therefore $x = (e_1, e_2, \ldots, e_n) = e$, the identity in H. Thus,

$$H_i \cap (H_1 \cdots H_{i-1} H_{i+1} \cdots H_n) = \{e\} \quad \text{for each } 1 \le i \le n.$$

Also, $x \mapsto x_i$ is an isomorphism of M_i onto G_i and hence $M_i \cong G_i$.
Now suppose that $G \cong G_1 \times G_2 \times \cdots \times G_n$ and f is an isomorphism of G onto $G_1 \times G_2 \times \cdots \times G_n$. By taking $N_i = f^{-1}(M_i)$ for each $1 \le i \le n$, it follows that $N_1, N_2, \ldots, N_n$ satisfy all the three required conditions.
Conversely, suppose that $N_1, N_2, \ldots, N_n$ be normal subgroups of G such that

1. $N_1 N_2 \cdots N_n = G$.
2. $N_i \cap (N_1 \cdots N_{i-1} N_{i+1} \cdots N_n) = \{e\}$ for each $1 \le i \le n$.
3. $N_i \cong G_i$ for each $1 \le i \le n$.

Define $g : N_1 \times N_2 \times \cdots \times N_n \to G$ by

$$g(a_1, a_2, \ldots, a_n) = a_1 a_2 \cdots a_n.$$

We shall prove that g is an isomorphism so that

$$G \cong N_1 \times N_2 \times \cdots \times N_n \cong G_1 \times G_2 \times \cdots \times G_n.$$

First observe that, for any $i \neq j$ and $a \in N_i$ and $b \in N_j$, we have

$$(ab)(a^{-1}b^{-1}) = (aba^{-1})b^{-1} = a(ba^{-1}b^{-1}) \in N_i \cap N_j = \{e\},$$

since N_i and N_j are normal subgroups of G and

$$N_i \cap N_j \subseteq N_i \cap (N_1 \cdots N_{i-1} N_{i+1} \cdots N_n) = \{e\}.$$

Therefore, $ab(ba)^{-1} = ab(a^{-1}b^{-1}) = e$ so that $ab = ba$. Now, for any $(a_1, \ldots, a_n)$ and $(b_1, \ldots, b_n) \in N_1 \times \cdots \times N_n$,

$$\begin{aligned} g((a_1, \ldots, a_n)(b_1, \ldots, b_n)) &= g(a_1 b_1, \ldots, a_n b_n) \\ &= (a_1 b_1)(a_2 b_2) \cdots (a_n b_n) \\ &= a_1 (b_1 a_2)(b_2 a_3) \cdots (b_{n-1} a_n) b_n \\ &= a_1 a_2 \cdots a_n b_1 b_2 \cdots b_n \\ &= g(a_1, \ldots, a_n) g(b_1, \ldots, b_n). \end{aligned}$$

Thus, g is a homomorphism of G into $N_1 \times \cdots \times N_n$. For any $(a_1, a_2, \ldots, a_n) \in N_1 \times \cdots \times N_n$, we have

$$g(a_1, \ldots, a_n) = e \Rightarrow a_1 a_2 \ldots a_n = e$$

$$\Rightarrow a_2 \cdots a_n = a_1^{-1} \in N_1 \cap (N_2 \cdots N_n) = \{e\}$$
$$\Rightarrow a_1 = e = a_2 \cdots a_n$$
$$\Rightarrow a_1 = e \quad \text{and} \quad a_2^{-1} = a_3 \cdots a_n \in N_2 \cap (N_3 \cdots N_n) = \{e\}$$
$$\Rightarrow a_1 = e = a_2 = \cdots = a_n$$
$$\Rightarrow (a_1, \ldots, a_n) = (e_1, \ldots, e_n)$$

and hence $\ker g = \{e\}$ so that g is an injection. Since $G = N_1N_2 \cdots N_n$, it follows that g is a surjection also. Thus, g is an isomorphism of $N_1 \times \cdots \times N_n$ onto G. Therefore,

$$G_1 \times G_2 \times \cdots \times G_n \cong N_1 \times N_2 \times \cdots \times N_n \cong G. \qquad ◀$$

Corollary 8.1.1. Let G, G_1 and G_2 be groups. Then, $G \cong G_1 \times G_2$ if and only if there exist normal subgroups N_1 and N_2 of G such that $G = N_1N_2$, $N_1 \cap N_2 = \{e\}$, $G_1 \cong N_1$ and $G_2 \cong N_2$.

Corollary 8.1.2. Let $N_1, N_2, \ldots, N_n$ be subgroups of a group G. Then, the mapping $(a_1, a_2, \ldots, a_n) \mapsto a_1a_2 \cdots a_n$ is an isomorphism of $N_1 \times N_2 \times \cdots \times N_n$ onto G if and only if the following are satisfied.

1. Each N_i is a normal subgroup of G.
2. $N_1N_2 \cdots N_n = G$.
3. $N_i \cap (N_1 \cdots N_{i-1} N_{i+1} \cdots N_n) = \{e\}$ for each $1 \leq i \leq n$.

We obtain another characterization of a decomposition of a group G in terms of its quotient groups. First recall that quotient groups of G are precisely (up to isomorphism) the homomorphic images of G. If N is a normal subgroup of a group G, then the natural map $x \mapsto xN$ is an epimorphism of G onto the quotient group G/N. If N_1 and N_2 are two normal subgroups, then clearly the map $x \mapsto (xN_1, xN_2)$ is a homomorphism of G into $G/N_1 \times G/N_2$. We obtain a necessary and sufficient condition for this map to be a surjection in the following theorem.

Theorem 8.1.2. Let N_1 and N_2 be normal subgroups of a group G. Define

$$f: G \to G/N_1 \times G/N_2 \quad \text{by} \quad f(x) = (xN_1, xN_2)$$

for any $x \in G$. Then, f is an epimorphism if and only if $N_1N_2 = G$.

Proof: Clearly f is always a homomorphism. Suppose that $N_1N_2 = G$. Let $(x_1N_1, x_2N_2) \in G/N_1 \times G/N_2$, where x_1 and $x_2 \in G = N_1N_2$. Then, we can write

$$x_1 = a_1a_2 \quad \text{and} \quad x_2 = b_1b_2,$$

where a_1 and $b_1 \in N_1$ and a_2 and $b_2 \in N_2$. Now, put $x = b_1a_2$. Then, we have

$$x^{-1}x_1 = (b_1a_2)^{-1}a_1a_2 = a_2^{-1}(b_1^{-1}a_1)\,a_2 \in N_1,$$

since $b_1^{-1}a_1 \in N_1$ and N_1 is normal in G. Therefore, $xN_1 = x_1N_1$. Also,

$$x^{-1}x_2 = (b_1a_2)^{-1}x_2 = a_2^{-1}b_1^{-1}(b_1b_2) = a_2^{-1}b_2 \in N_2$$

and hence $xN_2 = x_2N_2$. Therefore,

$$f(x) = (xN_1, xN_2) = (x_1N_1, x_2N_2).$$

Thus, f is a surjection and hence an epimorphism. Conversely suppose f is an epimorphism. Let $x \in G$. Consider the element $(N_1, xN_2) \in G/N_1 \times G/N_2$. Since f is a surjection, there exists $a \in G$ such that

$$(aN_1, aN_2) = f(a) = (N_1, xN_2).$$

Then, $aN_1 = N_1$ and $aN_2 = xN_2$ and hence $a \in N_1$ and $a^{-1}x \in N_2$ so that $x = a(a^{-1}x) \in N_1N_2$. Thus, $N_1N_2 = G$. ◀

Corollary 8.1.3. Let N_1 and N_2 be normal subgroups of a group G such that $N_1N_2 = G$. Then,

$$G/N_1 \cap N_2 \cong G/N_1 \times G/N_2$$

and in particular, when $N_1 \cap N_2 = \{e\}$, $G \cong G/N_1 \times G/N_2$.

Proof: In the above theorem, we have an epimorphism $f: G \to G/N_1 \times G/N_2$ whose Kernel is given by

$$\begin{aligned}\ker f &= \{x \in G : f(x) = (eN_1, eN_2)\\ &= \{x \in G : xN_1 = N_1 \text{ and } xN_2 = N_2\}\\ &= N_1 \cap N_2\end{aligned}$$

and therefore, by the fundamental theorem of homomorphisms

$$G/N_1 \cap N_2 \cong G/N_1 \times G/N_2.$$

These are generalised in the following theorem. ◀

Theorem 8.1.3. Let $N_1, N_2, \ldots, N_n$ be normal subgroups of a group G and $f: G \to G/N_1 \times G/N_2 \times \cdots \times G/N_n$ be defined by

$$f(x) = (xN_1, xN_2, \ldots, xN_n).$$

Then, f is an epimorphism if and only if, for each $1 \le i \le n$,

$$N_i \cdot \left(\bigcap_{j \ne i} N_j\right) = G.$$

Proof: Clearly f is a homomorphism. Suppose that f is epimorphism. Fix $1 \le i \le n$. For any $x \in G$, consider the element

$$(xN_1, \ldots, xN_{i-1}, N_i, xN_{i+1}, \ldots, xN_n) \in G/N_1 \times \cdots \times G/N_n.$$

Since f is a surjection, there exists $a \in G$ such that

$$\begin{aligned}(xN_1, \ldots, xN_{i-1}, N_i, xN_{i+1}, \ldots, N_n) &= f(a)\\ &= (aN_1, aN_2, \ldots, aN_n).\end{aligned}$$

Therefore, $xN_j = aN_j$ for all $j \ne i$ and $N_i = aN_i$ and hence $a \in N_i$ and $a^{-1}x \in N_j$ for all $j \ne i$, so that

$$x = a(a^{-1}x) \in N_i . \left(\bigcap_{j \ne i} N_j\right).$$

Thus, $G = N_i \cdot (\bigcap_{j \ne i} N_j)$.
Conversely suppose that $N_i \cdot (\bigcap_{j \ne i} N_j) = G$ for each $1 \le i \le n$. Then, $N_i N_j = G$ for all $i \ne j$ and

$$\left(\bigcap_{i=1}^{m-1} N_i\right) \cdot N_m = G \text{ for all } 1 < m \le n$$

(note that, for any subgroups A and B of a group G, $AB = G$ if and only if $BA = G$). We shall use induction on n to prove that f is a surjection. If $n = 1$, it is trivial. If $n = 2$, Theorem 8.1.2 gives the result. Let $n > 2$ and assume the result for $n - 1$. Let

$$(x_1N_1, x_2N_2, \ldots, x_nN_n) \in G/N_1 \times G/N_2 \times \cdots \times G/N_n,$$

where $x_1, x_2, \ldots, x_n \in G$. Then, there exists $y \in G$ such that

$$yN_i = x_iN_i \quad \text{for all } 1 \leq i \leq n - 1$$

and hence $y^{-1}x_i \in N_i$ for all $1 \leq i \leq n - 1$.
Now, put $M = N_1 \cap N_2 \cap \cdots \cap N_{n-1}$. Then, $MN_n = G$ and hence, by Theorem 8.1.2, $a \mapsto (aM, aN_n)$ is an epimorphism of G onto $G/M \times G/N_n$. We have $(yM, x_nN_n) \in G/M \times G/N_n$. Therefore, there exists $x \in G$ such that

$$xM = yM \quad \text{and} \quad xN_n = x_nN_n$$

so that $x^{-1}y \in M$ and $x^{-1}x_n \in N_n$. For any $1 \leq i \leq n - 1$, we have

$$x^{-1}x_i = (x^{-1}y)(y^{-1}x_i) \in N_i \quad \text{for all } 1 \leq i \leq n - 1$$

and hence $xN_i = x_iN_i$ for all $1 \leq i \leq n$, so that

$$f(x) = (xN_1, \ldots, xN_n) = (x_1N_1, \ldots, x_nN_n).$$

Thus, f is a surjection and hence an epimorphism. ◀

Corollary 8.1.4. Let $N_1, N_2, \ldots, N_n$ be normal subgroups of a group G and define

$$f : \frac{G}{\bigcap_{i=1}^{n} N_i} \rightarrow \frac{G}{N_1} \times \cdots \times \frac{G}{N_n}$$

by

$$f\left(x\left(\bigcap_{i=1}^{n} N_i\right)\right) = (xN_1, \cdots, xN_n).$$

Then, f is an isomorphism if and only if, for each $1 \leq i \leq n$,

$$N_i \cdot \left(\bigcap_{j \neq i} N_j\right) = G.$$

Corollary 8.1.5. Let $G, G_1, G_2, \ldots, G_n$ be groups. Then, $G \cong G_1 \times G_2 \times \cdots \times G_n$ if and only if there exist normal subgroups $N_1, N_2, \ldots, N_n$ of G satisfying the following:

1. $N_1 \cap N_2 \cap \cdots \cap N_n = \{e\}$.
2. $N_i \cdot (\bigcap_{j \neq i} N_j) = G$, for all $1 \leq i \leq n$.
3. $G/N_i \cong G_i$, for all $1 \leq i \leq n$.

Proof: Let $H = G_1 \times G_2 \times \cdots \times G_n$ and define

$$p_i : H \to G_i \text{ by } p_i(x_1, x_2, \ldots, x_n) = x_i.$$

Then, it can be easily verified that p_i is an epimorphism for each $1 \leq i \leq n$ and hence $H/\ker p_i \cong G_i$. Put $M_i = \ker p_i$. Then, M_i is a normal subgroup of H for each $1 \leq i \leq n$ and $H/M_i \cong G_i$. Also, $M_1 \cap M_2 \cap \cdots \cap M_n = \{e\}$, where $e = (e_1, e_2, \ldots, e_n)$ and e_i is the identity in G_i. For any $1 \leq i \leq n$ and $x = (x_1, x_2, \ldots, x_n) \in H$, we can write

$$x = (x_1, \ldots, x_{i-1}, e_i, x_{i+1}, \ldots, x_n) \cdot (e_1, \ldots, e_{i-1}, x_i, e_{i+1}, \ldots, e_n)$$

which is an element in $M_i \cdot (\bigcap_{j \neq i} M_j) = H$. If $f : G \to H$ is an isomorphism and $N_i = f^{-1}(M_i)$ for $1 \leq i \leq n$, then $N_1, N_2, \ldots, N_n$ satisfy all the required properties. Converse follows from Theorem 8.1.3 and Corollary 8.1.4. ◀

Worked Exercise 8.1.1. Let $A_1, A_2, \ldots, A_n$ be subgroups of a group G and $G = A_1 A_2 \cdots A_n$.
Define $f : A_1 \times A_2 \times \cdots \times A_n \to G$ by

$$f(a_1, a_2, \ldots, a_n) = a_1 a_2 \ldots a_n.$$

Then prove that f is an isomorphism if and only if each A_i is a normal subgroup of G and any element $a \in G$ can be uniquely expressed as $a = a_1 a_2 \cdots a_n$ for some $a_i \in A_i$, $1 \leq i \leq n$.

Answer: Suppose that f is an isomorphism. For each $1 \leq i \leq n$, let

$$B_i = \{(a_1, \ldots, a_n) \in A_1 \times \cdots \times A_n : a_j = e_j \text{ for all } j \neq i\}.$$

Then, as in Theorem 8.1.1, we can verify that each B_i is a normal subgroup of $A_1 \times \cdots \times A_n$ and f/B_i is an isomorphism of B_i onto A_i. Therefore, A_i is a normal subgroup of G for each $1 \leq i \leq n$. Since f is an isomorphism and, in particular, a bijection, any element of G can be uniquely expressed as a product $a_1 a_2 \cdots a_n$, where $a_i \in A_i$.

Conversely, suppose that each A_i is a normal subgroup of G and any element of G can be uniquely expressed as $a_1 a_2 \cdots a_n$, $a_i \in A_i$. Then, clearly f is a bijection. To prove that f is a homomorphism, first observe that $A_i \cap A_j = \{e\}$ for any $i \neq j$; for, if $a \in A_i \cap A_j$ and $i \neq j$, then

$$f(e, \ldots, e, \underset{i^{\text{th}}}{a}, e, \ldots, e) = f(e, \ldots, e, \underset{j^{\text{th}}}{a}, e, \ldots, e)$$

and hence $a = e$. From this we get that, for any $i \neq j$, $a_i \in A_i$ and $a_j \in A_j$,

$$a_i(a_j a_i^{-1} a_j^{-1}) = (a_i a_j a_i^{-1})\, a_j^{-1} \in A_i \cap A_j = \{e\}$$

and therefore $a_i a_j = a_j a_i$. From this, it follows that f is a homomorphism. Thus, f is an isomorphism.

Worked Exercise 8.1.2. Let G be a finite nontrivial group such that $a^2 = e$ for all $a \in G$. Then prove that $G \cong C_1 \times C_2 \times \cdots \times C_n$, where $n > 0$ and each C_i is a cyclic group of order 2 and deduce that $|G| = 2^n$.

Answer: Since $a^2 = e$, we have $a = a^{-1}$ for all $a \in G$ and hence $ab = (ab)^{-1} = b^{-1}a^{-1} = ba$ for all a and $b \in G$. Therefore, G is an abelian group. Since G is nontrivial, choose $a_1 \neq e$ in G and let $C_1 = \{e, a_1\}$. Then, C_1 is a normal subgroup of G and C_1 is a cyclic group of order 2.

If $G = C_1$, we are through. Otherwise, choose $a_2 \in G$ such that $a_2 \notin C_1$ and let $C_2 = \{e, a_2\}$. Then, $C_1 \cap C_2 = \{e\}$ and $C_1C_2 \cong C_1 \times C_2$. Also $C_1 \subsetneqq C_1C_2$. Again, if $C_1C_2 = G$, we are through. Otherwise, choose $a_3 \in G$ such that $a_3 \notin C_1C_2$ and continue the process to get cyclic subgroups $C_1, C_2, C_3, \ldots$, each of order 2 and

$$C_1 \subsetneqq C_1C_2 \subsetneqq C_1C_2C_3 \subsetneqq \cdots$$

Since G is finite, the process should terminate at a finite stage and then

$$G \cong C_1 \times C_2 \times \cdots \times C_n.$$

Worked Exercise 8.1.3. Prove that any group of order p^2 where p is a prime, is isomorphic to either $\mathbb{Z}_{p^2}$ or $\mathbb{Z}_p \times \mathbb{Z}_p$.

Answer: Let G be a group of order p^2, where p is a prime. Then, by Theorem 7.3.6, G is abelian and hence every subgroup of G is normal. If G is cyclic, then by Theorem 7.3.6, $G \cong \mathbb{Z}_{p^2}$. Suppose that G is not cyclic, then

$O(a) \neq |G| = p^2$ for all $a \in G$. Choose $a \neq e$ in G and let $A = \langle a \rangle$. Then, $O(a)$ is a divisor of p^2 and therefore $O(a) = p$. Since $A \neq G$, there exists $b \in G$ such that $b \notin A$. Let $B = \langle b \rangle$. Then, $|A \cap B| = 1$ or p. If $|A \cap B| = p$, then $A \cap B = A$ and hence $A \subseteq B$ so that $A = B$, which is not true. Therefore, $|A \cap B| = 1$ and hence $A \cap B = \{e\}$. Also

$$|AB| = \frac{|A||B|}{|A \cap B|} = \frac{|A||B|}{1} = p^2 = |G|$$

and hence $AB = G$. Therefore, by Corollary 8.1.1,

$$G \cong A \times B \cong Z_p \times Z_p,$$

since any group of prime order is cyclic.

Worked Exercise 8.1.4. Let G be a cyclic group of order $p_1^{r_1} p_2^{r_2} \ldots p_n^{r_n}$, where $p_1, p_2, \ldots, p_n$ are distinct primes and $r_1, r_2, \ldots, r_n$ are positive integers. Prove that G is isomorphic to a product of cyclic groups $A_1, A_2, \ldots, A_n$ where $|A_i| = p_i^{r_i}$.

Answer: Since G is given to be cyclic, there exists $a \in G$ such that $G = \langle a \rangle$ and $O(a) = p_1^{r_1} p_2^{r_2} \ldots p_n^{r_n}$. For each $1 \leq i \leq n$, let $m_i = \prod_{j \neq i} p_j^{r_j}$ and $b_i = a^{m_i}$. Then, $O(b_i) = \dfrac{O(a)}{m_i} = p_i^{r_i}$. Let $A_i = \langle b_i \rangle$. Then, A_i is a cyclic subgroup of order $p_i^{r_i}$. Since $p_1, p_2, \ldots, p_n$ are distinct primes, $(p_i^{r_i}, m_i) = 1$ for each $1 \leq i \leq n$. From this, it follows that

$$A_i \cap (A_1 \ldots A_{i-1} A_{i+1} \ldots A_n) = \{e\}$$

for each $1 \leq i \leq n$. Also, $A_1 A_2 \cdots A_n = G$. Note that G is abelian (being cyclic) and hence A_i's are normal subgroups of G. Therefore,

$$G \cong A_1 \times A_2 \times \cdots \times A_n.$$

EXERCISE 8(A)

1. State whether each of the following is true or false and substantiate your answer.
   (i) Any group of order 25 is cyclic.
   (ii) If G is a group of order 9 and G is not cyclic, Then, $G \cong \mathbb{Z}_3 \times \mathbb{Z}_3$.
   (iii) Any group of order 121 is abelian.
   (iv) Any group of order 8 is abelian.

(v) Any cyclic group of order 180 can be decomposed as a product of non-trivial groups.

(vi) $\mathbb{Z}_{36} \cong \mathbb{Z}_9 \times \mathbb{Z}_4$.

(vii) If G is a group of order 36, than, $G \cong \mathbb{Z}_9 \times \mathbb{Z}_4$.

(viii) $\mathbb{Z}/12\mathbb{Z} \cong \mathbb{Z}/4\mathbb{Z} \times \mathbb{Z}/3\mathbb{Z}$.

2. Prove that $\mathbb{Z}_4$ cannot be decomposed as a product of groups of order 2.

3. Show that $\mathbb{Z}_8$ cannot be decomposed as a product of two nontrivial subgroups.

4. Let $A_1, A_2, \ldots, A_n$ be subgroups of a group G such that $A_1 A_2 \cdots A_n = G$. Prove that the map $f : A_1 \times \cdots \times A_n \to G$, defined by $f(a_1, \ldots, a_n) = a_1 a_2 \cdots a_n$, is an isomorphism if and only if each A_i is normal in G and, for any $a_i \in A_i$,

$$a_1 a_2 \cdots a_n = e \Rightarrow a_1 = a_2 = \cdots = a_n = e.$$

5. Let G be a group of order pq, where p and q are distinct primes. If A and B are normal subgroups of orders p and q, respectively, then prove that G is cyclic.

6. Prove that $\mathbb{Z}_{10}$ is isomorphic to the product of the subgroups $A = \{0, 2, 4, 6, 8\}$ and $B = \{0, 5\}$.

7. Let G be a cyclic group of order mn where m and n are relatively prime positive integers. Prove that there exist subgroups A and B of orders m and n, respectively such that $G \cong A \times B$.

8. Is $\mathbb{Z} \times \mathbb{Z}$ cyclic?

9. Let G be a finite abelian group of order $p_1^{r_1} p_2^{r_2} \ldots p_n^{r_n}$, where $p_1, p_2, \ldots, p_n$ are distinct primes and $r_1, r_2, \ldots, r_n$ are positive integers. For each $1 \leq i \leq n$, let

$$S_i = \{a \in G : O(a) = p^s \text{ for some } 0 \leq s \in \mathbb{Z}\}.$$

Prove that each S_i is a subgroup of G and G is isomorphic to the product $S_1 \times S_2 \times \cdots \times S_n$.

10. Let $m = p_1^{r_1} p_2^{r_2} \ldots p_n^{r_n}$, where p_i's are distinct primes r_i's are positive integers. Prove that

$$\mathbb{Z}_m \cong \mathbb{Z}_{p_1^{r_1}} \times \mathbb{Z}_{p_2^{r_2}} \times \cdots \times \mathbb{Z}_{p_n^{r_n}}.$$

11. A nontrivial group is said to be indecomposable, if it is not isomorphic to the product of two nontrivial groups. Prove that $\mathbb{Z}_p$ is indecomposable.

12. Prove that the group of symmetries of a square, the group $(\mathbb{Z}, +)$ and the group $(\mathbb{Q}, +)$ are all indecomposable.

13. Prove that the group $(\mathbb{Z}_n, +_n)$ is indecomposable if and only if $n = p^r$ for some prime p and $r > 0$.

14. Let A and B be normal subgroups of a finite group G such that $|A| \cdot |B| = |G|$ and $|A|$ and $|B|$ are relatively prime. Then prove that $G \cong A \times B$.

15. Let A and B be subgroups of a group G such that $G \cong A \times B$. For any normal subgroup N of G, prove that either N is contained in the centre of G or N has nontrivial intersection with A or B.

16. Prove that the symmetric group S_3 is indecomposable.

17. Let $f: G \to G'$ be an epimorphism of groups and N be a normal subgroup of G. If the restriction of f to N is an isomorphism of N onto G', prove that $G \cong N \times (\ker f)$.

18. Let $G_1, G_2, \ldots, G_n$ be groups and $G = G_1 \times G_2 \times \cdots \times G_n$. For any group H, prove that a mapping $f: H \to G$ is a homomorphism if and only if $p_i \cdot f: H \to G_i$ is a homomorphism for each $1 \le i \le n$, where $p_i : G \to G_i$ is the ith projection.

19. Let A and B be any normal subgroups of a group G. If the natural map $f: G \to G/_A$ (defined be $f(x) = xA$) induces an isomorphism of B onto G/A, then prove that $G \cong A \times B$.

20. Prove the following for any groups G_1, G_2 and G_3:
    - (i) $G_1 \times G_2 \cong G_2 \times G_1$
    - (ii) $(G_1 \times G_2) \times G_3 \cong G_1 \times (G_2 \times G_3)$
    - (iii) $G_1 \cong G_2 \Rightarrow G_1 \times G_3 \cong G_2 \times G_3$
    - (iv) $G_1 \times G_2$ is abelian $\Leftrightarrow$ G_1 and G_2 are abelian.
    - (v) $G_1 \times G_2$ is cyclic $\Rightarrow$ G_1 and G_2 are cyclic.
    - (vi) The converse of (v) is not true.

8.2 FINITELY GENERATED ABELIAN GROUPS

The study of finitely generated abelian groups and, in particular, finite abelian groups is one of the richest and deepest branches of the whole of group theory. No other general class of groups is the structure as completely known or as easily described. The overall strategy in the study of a structure theory of any algebraic system is to express, in some sense, a complicated algebraic system in terms of those which are better behaved and whose structure is well known. For example, the structure of a cyclic group is well known and in fact, we have proved earlier that any cyclic group is isomorphic to the additive group $\mathbb{Z}$ of integers or to the group $\mathbb{Z}_n$ of integers modulo n according as the group is infinite or finite of order n, respectively. Here, we prove a fundamental theorem which states that a group is a finitely generated abelian group if and only if it is isomorphic to the product of a finite number of cyclic groups.

We first consider the case of finite abelian groups. Let us recall that a group G is called a p-group, where p is a given prime number, if the order of any element of G is a power of p and that a finite group is a p-group if and only if it is of order p^n for some nonnegative integer n. The following is a central topic in the structure theory of finite abelian groups.

Theorem 8.2.1. Any finite abelian group is isomorphic to a product of p-groups.

Proof: The trivial group $\{e\}$ is a p-group, for any prime p (since its order is $1 = p^0$). Let G be a nontrivial abelian group of order n and $n > 1$. We can write

$$n = p_1^{r_1} p_2^{r_2} \dots p_k^{r_k},$$

where $p_1, p_2, \dots, p_k$ are distinct primes and $r_1, r_2, \dots, r_k$ are positive integers. From Sylow Theorem I (7.4.4), there exist subgroups $A_1, A_2, \dots, A_k$ of G such that $|A_i| = p_i^{r_i}$ for $1 \leq i \leq k$. Since the group G is abelian, each A_i is a normal subgroup of G. Also, since p_i's are distinct primes, it can be easily verified that $A_i \cap A_j = \{e\}$ (for, the order of $A_i \cap A_j$ is a common divisor of $|A_i|$ and $|A_j|$) for $i \neq j$.

Further the order of $A_i \cap (A_1 \cdots A_{i-1} A_{i+1} \cdots A_k)$ is a common divisor of $|A_i| = (= p_i^{r_i})$ and $|A_1 \cdots A_{i-1} A_{i+1} \cdots A_k| (= \prod_{j \neq i} p_j^{s_j})$ and, since $p_i^{r_i}$ and $\prod_{j \neq i} p_j^{s_j}$ are relatively prime, it follows that $|A_i \cap (A_1 \cdots A_{i-1} A_{i+1} \cdots A_k| = 1$ and hence

$$A_i \cap (A_1 \cdots A_{i-1} A_{i+1} \cdots A_k) = \{e\}.$$

Also, it can be proved that $A_1 A_2 \dots A_k = G$.
Therefore, by Theorem 8.1.1,

$$G \cong A_1 \times A_2 \times \cdots \times A_k$$

and each A_i is a p_i-group (actually A_i is the unique Sylow p_i-subgroup of G).

Definition 8.2.1. Let G be a finite abelian group. For any prime p, let

$$G_p = \{a \in G : \mathrm{O}(a) = p^r \text{ for some } 0 \leq r \in \mathbb{Z}\}.$$

It can be easily verified that G_p is a p-subgroup of G. In fact, G_p is the unique Sylow p-subgroup of G. The following is a simple consequence of the proof of the above theorem.

Corollary 8.2.1. Let G be a finite abelian group. Then,

$$G \cong G_{p_1} \times G_{p_2} \times \cdots \times G_{p_k},$$

where $p_1, p_2, \ldots, p_k$ are all the distinct primes dividing the order of G.

The two results proved above reduce the study of arbitrary finite abelian groups to the study of finite abelian p-groups. The basic result on p-groups from which the whole structure theory can be pinned down is proved in the following theorem.

Theorem 8.2.2. Let p be an arbitrarily fixed prime number. Then, any finite abelian p-group is isomorphic to a product of a finite number of cyclic p-groups.

Proof: Let G be a finite abelian p-group. Then, $|G| = p^n$ for some nonnegative integer n. We shall use induction on n. If $n = 0$, there is nothing to prove, since G becomes trivial. If $n = 1$, then G is a group of order p, which is a prime, and hence G is itself a cyclic p-group. Now, let $n > 1$ and suppose that the theorem holds good for all groups of order p^m with $m < n$.
Since G is a p-group, the order of any element of G is a power of p. Let a be an element of maximal order in G and $O(a) = p^k$, where $k \leq n$. Put $H = \langle a \rangle$, the cyclic subgroup of G generated by a. If $k = n$, then $H = G$ and hence G is itself a cyclic p-group. Suppose that $k < n$ and consider the quotient group G/H whose order is p^{n-k} and $n - k < n$. By the induction hypothesis,

$$G/H \cong G_1 \times G_2 \times \cdots \times G_r,$$

where G_i, $1 \leq i \leq r$, is a cyclic p-group. By Theorem 8.1.1, there exist subgroups $A_1, A_2, \ldots, A_r$ of G/H such that $A_i \cong G_i$ for $1 \leq i \leq r$, $A_1 A_2 \cdots A_r = G/H$ and

$$A_i \cap (A_1 \cdots A_{i-1} A_{i+1} \cdots A_r) = \{H\}.$$

For any $1 \leq i \leq n$ (note that H is the identity in G/H).
Therefore, there exist subgroups $H_1, H_2, \ldots, H_r$ of G such $A_i = H_i/H$ for $1 \leq i \leq r$. Now, these subgroups H_i's satisfy the following:

1. $H \subseteq H_i$ for all $1 \leq i \leq r$
2. $G_i \cong H_i/H$ for all $1 \leq i \leq r$
3. $G/H = H_1/H \cdot H_2/H \cdot \cdots \cdot H_r/H$
4. $H_i \cap (H_1 \cdots H_{i-1} H_{i+1} \cdots H_n) = H$ for all $1 \leq i \leq n$.

Since G_i and hence H_i/H is cyclic, we get a coset b_iH generating H_i/H. Let $|H_i/H| = p^{j_i}$ for each $1 \leq i \leq r$. Then, the order of b_iH in G/H is equal to p^{j_i}.
In the next step, we produce a representative c_i of b_iH such that the order of c_i in G is equal to the order of b_iH in G/H. For convenience, let us write temporarily b for b_i and j for j_i. Since $(bH)^{p^j} = H$, we have

$$b^{p^j} \in H = <a>$$

and hence $b^{p^j} = a^s$ for some s. Since the order of a is maximal and $O(a) = p^k$, we have $O(b) \leq p^k$ and therefore $b^{p^k} = e$ which implies that

$$a^{sp^{k-j}} = (a^s)^{p^{k-j}} = (b^{p^j})^{p^{k-j}} = b^{p^k} = e.$$

Therefore, $O(a)$ divides sp^{k-j}; that is, p^k divides sp^{k-j} and hence p^j divides s. Let $s = tp^j$ and $c = ba^{-t}$. Then, $b^{-1}c = a^{-t} \in <a> = H$ and

$$c^{p^j} = (ba^{-t})^{p^j} = b^{p^j} a^{-tp^j} = a^s a^{-s} = e$$

which implies that $O(c) = p^j$.
Restoring the index i, we have the following:
For each $1 \leq i \leq r$, there is an element $c_i \in G$ and an integer ν_i such that

$$c_iH = b_iH \quad \text{and} \quad c_i^{\nu_i} = e.$$

Let K be the subgroup of G generated by $c_1, c_2, \ldots, c_r$. We shall prove that $HK = G$ and $H \cap K = \{e\}$, so that $G \cong H \times K$.
Consider an arbitrary element $x \in G$. Then, $xH \in G/H$ and hence

$$\begin{aligned} xH &= (b_1H)^{n_1} (b_2H)^{n_2} \cdots (b_rH)^{n_r} \text{ for some integers } n_i \\ &= (c_1H)^{n_1} (c_2H)^{n_2} \cdots (c_rH)^{n_r} \\ &= (c_1^{n_1} c_2^{n_2} \cdots c_r^{n_r})H. \end{aligned}$$

Therefore, $x = (c_1^{n_1} c_2^{n_2} \cdots c_r^{n_r})y$ for some $y \in H$. Thus, $x \in KH = HK$.
Thus, $HK = G$.
To prove that $H \cap K = \{e\}$, consider an element x in $H \cap K$. Since $x \in K$, there exist integers m_i such that

$$x = c_1^{m_1} c_2^{m_2} \cdots c_r^{m_r}.$$

Now,

$$H = xH = (c_1H)^{m_1}(c_2H)^{m_2}\cdots(c_rH)^{m_r}$$
$$= (b_1H)^{m_1}(b_2H)^{m_2}\cdots(b_rH)^{m_r}.$$

From the property (4) of H_i's above, it follows that

$$(b_iH)^{m_i} = H \quad \text{for all } 1 \le i \le r.$$

Since the order of b_iH in G/H is equal to the order of c_i (which is equal to p^{j_i}), we get that p^{j_i} divides m_i and hence $(c_i)^{m_i} = e$ for all $1 \le i \le r$, so that $x = e$. Thus, $H \cap K = \{e\}$. Again by Theorem 8.1.1, we have

$$G \cong H \times K.$$

From the induction hypothesis, there exist cyclic p-groups $K_1, K_2, \ldots, K_w$ such that

$$K \cong K_1 \times \cdots \times K_w.$$

Now, H is also a cyclic p-group and

$$G \cong H \times K \cong H \times K_1 \times \cdots \times K_w.$$

This completes the proof. ◀

Theorems 8.2.1 and 8.2.2 together yield the following fundamental structure theorem.

Theorem 8.2.3 (Fundamental structure theorem for finite abelian groups). Any finite abelian group is isomorphic to a product of cyclic p-groups.

Since any cyclic group of order n is isomorphic to the group $\mathbb{Z}_n$ of integers modulo n, we have the following corollary.

Corollary 8.2.2. Let G be any nontrivial finite abelian group. Then,

$$G \cong \mathbb{Z}_{p_1^{n_1}} \times \mathbb{Z}_{p_2^{n_2}} \times \cdots \times \mathbb{Z}_{p_r^{n_r}},$$

where $p_1, p_2, \ldots, p_r$ are (not necessarily distinct) primes and $n_1, n_2, \ldots, n_r$ are positive integers.

We derive a formula to find the exact number of distinct (nonisomorphic) abelian groups of a given order n. For example,

$$\mathbb{Z}_4 \quad \text{and} \quad \mathbb{Z}_2 \times \mathbb{Z}_2$$

are the only distinct abelian groups of order 4 and

$$\mathbb{Z}_8, \mathbb{Z}_4 \times \mathbb{Z}_2 \quad \text{and} \quad \mathbb{Z}_2 \times \mathbb{Z}_2 \times \mathbb{Z}_2$$

are all the distinct abelian groups of order 8. Before going for the derivation of the formula, we collect few miscellaneous facts about abelian groups that will be used in the derivation of the formula.

For any abelian group G and for any integer m, the sets

$$G^m = \{a^m : a \in G\}$$

and $G(m) = \{a \in G : a^m = e\}$
are subgroups of G. For any prime p and a positive integer n, it can be easily verified that

$$\mathbb{Z}_{p^n}^{p^m} \cong \mathbb{Z}_{p^{n-m}} \quad \text{for any } m < n$$

$$\text{and} \quad \mathbb{Z}_{p^n}(p) \cong \mathbb{Z}_p.$$

If $G_1, G_2, \ldots, G_r$ are groups and $f: G \to G_1 \times \cdots \times G_r$ is an isomorphism, then f induces isomorphisms. For any groups $G, G_1, G_2, \ldots, G_r$, if $G \cong G_1 \times G_2 \times \cdots \times G_r$, then $G^m \cong G_1^m \times G_2^m \times \cdots \times G_r^m$

$$\text{and} \quad G(m) \cong G_1(m) \times G_2(m) \times \cdots \times G_r(m)$$

for any integer m.

Definition 8.2.2. Let n be any positive integer. A finite sequence $\{n_1, n_2, \ldots, n_r\}$ of positive integers is said to be a *partition* of n if

$$n_1 \leq n_2 \leq \cdots \leq n_r \quad \text{and} \quad n_1 + n_2 + \cdots + n_r = n.$$

Let $P(n)$ denote the number of partitions of n. For example, $P(1) = 1$,
$P(2) = 2$, since $\{1, 1\}$ and $\{2\}$ are the only partitions of 2.
$P(3) = 3$, since $\{1, 1, 1\}$, $\{1, 2\}$ and $\{3\}$ are the only partitions of 3.
$P(4) = 5$ and $P(5) = 7$.

Theorem 8.2.4. Let p be a prime and n be a positive integer. Then, there are exactly $P(n)$ number of distinct (nonisomorphic) abelian groups of order p^n, where $P(n)$ is the number of partitions of n.

Proof: For any partition $\{n_1, n_2, \ldots, n_r\}$ of n, consider the product

$$\mathbb{Z}_{p^{n_1}} \times \mathbb{Z}_{p^{n_2}} \times \cdots \times \mathbb{Z}_{p^{n_r}}$$

which is an abelian group of order $p^{n_1} \cdot p^{n_2} \ldots p^{n_r} = p^{n_1+n_2+\cdots+n_r} = p^n$.
We shall prove that the correspondence

$$\{n_1,\ n_2,\ \ldots,\ n_r\} \mapsto \mathbb{Z}_{p^{n_1}} \times \mathbb{Z}_{p^{n_2}} \times \cdots \times \mathbb{Z}_{p^{n_r}}$$

is a bijection between the set of all partitions on n and the set of all distinct (nonisomorphic) abelian groups of order p^n. If G is an abelian group of order p^n, then, by Theorem 8.2.2,

$$G \cong \mathbb{Z}_{p^{n_1}} \times \mathbb{Z}_{p^{n_2}} \times \cdots \times \mathbb{Z}_{p^{n_r}}$$

(since any cyclic p-group must be isomorphic to $\mathbb{Z}_{p^m}$ for some $m > 0$) and, in this case,

$$p^n = |G| = |\mathbb{Z}_{p^{n_1}} \times \mathbb{Z}_{p^{n_2}} \times \cdots \times \mathbb{Z}_{p^{n_r}}| = p^{n_1+n_2+\cdots+n_r}$$

and hence $n = n_1 + n_2 + \cdots + n_r$. We can rearrange $\mathbb{Z}_{p^{n_i}}$'s such that $n_1 \leq n_2 \leq \cdots \leq n_r$ and therefore $\{n_1, n_2, \ldots, n_r\}$ is a partition of n. Therefore, the above correspondence is a surjection. To prove the injectivity of this correspondence, let $\{n_1, n_2, \ldots, n_r\}$ and $\{m_1, m_2, \ldots, m_s\}$ be partitions of n such that

$$G \cong \mathbb{Z}_{p^{n_1}} \times \mathbb{Z}_{p^{n_2}} \times \cdots \times \mathbb{Z}_{p^{n_r}} \cong \mathbb{Z}_{p^{m_1}} \times \mathbb{Z}_{p^{m_2}} \times \cdots \times \mathbb{Z}_{p^{m_s}}.$$

Then, we have to prove that $r = s$ and $n_i = m_i$ for all $1 \leq i \leq s$. From the discussion made before Definition 8.2.2, we have

$$\begin{aligned}
\underbrace{\mathbb{Z}_p \times \mathbb{Z}_p \times \cdots \times \mathbb{Z}_p}_{(r\text{ times})} &\cong \mathbb{Z}_{p^{n_1}}(p) \times \mathbb{Z}_{p^{n_2}}(p) \times \cdots \times \mathbb{Z}_{p^{n_r}}(p) \\
&\cong (\mathbb{Z}_{p^{n_1}} \times \mathbb{Z}_{p^{n_2}} \times \cdots \times \mathbb{Z}_{p^{n_r}})(p) \\
&\cong G(p) \\
&\cong (\mathbb{Z}_{p^{m_1}} \times \mathbb{Z}_{p^{m_2}} \times \cdots \times \mathbb{Z}_{p^{m_s}})(p) \\
&\cong \underbrace{\mathbb{Z}_p \times \mathbb{Z}_p \times \cdots \times \mathbb{Z}_p}_{(s\text{ times})}
\end{aligned}$$

and therefore $p^r = p^s$ and hence $r = s$.

Thus, we have proved that if a group G is isomorphic to a product of r number of $\mathbb{Z}_{p^{n_i}}$'s, then any expression of G as an isomorphic copy of a product of $\mathbb{Z}_{p^{m_i}}$'s has exactly r number of factors.

Now, suppose $n_i \neq m_i$ for some $1 \leq i \leq r$. Choose j, $1 \leq j \leq r$ such that $n_i = m_i$ for all $i < j$ and $n_j \neq m_j$. We may assume that $n_j < m_j$. Since $(\mathbb{Z}_{p^{n_i}})^{p^{n_j}} = \{0\}$ for $i \leq j$ (note that 0 is the identity in $\mathbb{Z}_{p^{n_i}}$), we have

$$\begin{aligned} G^{p^{n_j}} &\cong (\mathbb{Z}_{p^{n_1}} \times \mathbb{Z}_{p^{n_2}} \times \cdots \times \mathbb{Z}_{p^{n_r}})^{p^{n_j}} \\ &\cong \mathbb{Z}_{p^{n_1}}^{p^{n_j}} \times \mathbb{Z}_{p^{n_2}}^{p^{n_j}} \times \cdots \times \mathbb{Z}_{p^{n_r}}^{p^{n_j}} \\ &\cong \mathbb{Z}_{p^{n_{j+1}-n_j}} \times \mathbb{Z}_{p^{n_{j+2}-n_j}} \times \cdots \times \mathbb{Z}_{p^{n_r-n_j}} \end{aligned}$$

and $n_{j+1} - n_j \leq n_{j+2} - n_j \leq \cdots \leq n_r - n_j$. Clearly there are at most $r - j$ nonzero factors in the above decomposition of $G^{p^{n_j}}$. Similarly, since $n_i = m_i$ for $i < j$ and $n_j < m_j$ and since $G \cong \mathbb{Z}_{p_1^{m_1}} \times \cdots \times \mathbb{Z}_{p_r^{m_r}}$, we get that

$$G^{p^{n_j}} \cong \mathbb{Z}_{p^{m_j-n_j}} \times \mathbb{Z}_{p^{m_{j+1}-n_j}} \times \cdots \times \mathbb{Z}_{p^{m_r-n_j}}$$

and $1 \leq m_j - n_j \leq m_{j+1} - n_j \leq \cdots \leq m_r - n_j$. Clearly, there are atleast $r - j + 1$ nonzero factors in the second decomposition of $G^{p^{n_j}}$. Therefore, we have two decompositions of the group $G^{p^{n_j}}$ as a product of cyclic p-groups and the number of factors in the first decomposition is less than the number of factors in the second decomposition. This contradicts the conclusion obtained just after proving that $r = s$. Thus, we must have $n_i = m_i$ for all $1 \leq i \leq r$. This proves that there is a bijection between the set of partitions of n and the set of distinct (nonisomorphic) abelian groups of order p^n and hence the numbers of members in both the sets are same. ◀

Corollary 8.2.3. Let $m = p_1^{n_1} p_2^{n_2} \ldots p_r^{n_r}$, where $p_1, p_2, \ldots, p_r$ are distinct primes and $n_1, n_2, \ldots, n_r$ are positive integers. Then, the number of distinct (nonisomorphic) abelian groups of order m is equal to $p(n_1)\, p(n_2) \ldots p(n_r)$, where $p(n_i)$ is the number of partitions of n_i.

Proof: Any abelian group of order m must be isomorphic with a product

$$G_1 \times G_2 \times \cdots \times G_r,$$

where each G_i is an abelian group of order $p_i^{n_i}$, $1 \leq i \leq r$, and conversely any such product is an abelian group of order m. For each $1 \leq i \leq r$, we know (by the above theorem) that there are $p(n_i)$ number of distinct (nonisomorphic) abelian groups of order $p_i^{n_i}$ and hence the corollary. ◀

Corollary 8.2.4. Let m be any square-free positive integer (that is, m is not divisible by any perfect square greater than 1). Then, any abelian group of order m is cyclic and hence $\mathbb{Z}_m$ is the only (up to isomorphism) abelian group of order m.

Proof: Since m is square-free, we can write

$$m = p_1 p_2 \cdots p_r$$

where $p_1, p_2, \ldots, p_r$ and distinct primes. By Corollary 8.2.3, the number of distinct abelian groups of order m is $p(1)\, p(1) \cdots p(1) = 1$. We know that the group $\mathbb{Z}_n$ of integers modulo n is an abelian group of order n. Thus, $\mathbb{Z}_n$ is the only (up to isomorphism) abelian group of order n. ◀

Worked Exercise 8.2.1. Find the number of abelian groups of order 7,200.

Answer: We have to first express 7,200 as a product of primes. We have $7{,}200 = 2^5 \times 3^2 \times 5^2$ and hence, by Corollary 8.2.3, the number of distinct abelian groups of order 7,200 is $p(5)\,p(2)\,p(2)$, where $p(n)$ denotes the number of partitions of n. Note that $p(2) = 2$, since $\{1, 1\}$ and $\{2\}$ are the only partitions of 2. Coming to $p(5)$, note that

$$\{1, 1, 1, 1, 1\}, \{1, 1, 2\}, \{1, 1, 3\},$$
$$\{1, 2, 2\}, \{1,4\}, \{2, 3\} \quad \text{and} \quad \{5\}$$

are the only partitions of 5 and hence $p(5) = 7$. Thus, there are exactly $7 \times 2 \times 2$ $(= 28)$ abelian groups of order 7,200.

Worked Exercise 8.2.2. Prove that any abelian group of order 2,310 is cyclic.

Answer: Note that $2{,}310 = 2 \times 3 \times 5 \times 7 \times 11$, which is a product of distinct primes and hence square-free. By Corollary 8.2.4, the group $\mathbb{Z}_{2,310}$ of integers modulo 2,310 is the only (up to isomorphism) abelian group of order 2,310.

Worked Exercise 8.2.3. List all (up to isomorphism) abelian groups each of order 240.

Answer: First, we have to express 240 as a product of primes.

$$240 = 2^4 \cdot 3^1 \cdot 5^1$$

Therefore, there are $p(4)\,p(1)\,p(1)$ $(= 5 \times 1 \times 1 = 5)$ abelian groups of order 240. To list these, we have to write down all the partitions of 4 and 1.

Partitions of 4 are $\{1, 1, 1, 1\}$, $\{1, 1, 2\}$, $\{1, 3\}$, $\{2,2\}$ and $\{4\}$. Partitions of 1 is $\{1\}$.

Thus, the following five are all (up to isomorphism) the abelian groups of order 240.

$$\mathbb{Z}_2 \times \mathbb{Z}_2 \times \mathbb{Z}_2 \times \mathbb{Z}_2 \times \mathbb{Z}_3 \times \mathbb{Z}_5$$
$$\mathbb{Z}_2 \times \mathbb{Z}_2 \times \mathbb{Z}_{2^2} \times \mathbb{Z}_3 \times \mathbb{Z}_5$$
$$\mathbb{Z}_2 \times \mathbb{Z}_{2^3} \times \mathbb{Z}_3 \times \mathbb{Z}_5$$
$$\mathbb{Z}_{2^2} \times \mathbb{Z}_{2^2} \times \mathbb{Z}_3 \times \mathbb{Z}_5$$
$$\mathbb{Z}_{2^4} \times \mathbb{Z}_3 \times \mathbb{Z}_5$$

Note that $\mathbb{Z}_n \times \mathbb{Z}_m \cong \mathbb{Z}_{nm}$ if n and m are relatively prime and therefore $\mathbb{Z}_2 \times \mathbb{Z}_3 \times \mathbb{Z}_5 \cong \mathbb{Z}_{30}$. The above five groups are isomorphic to the following, respectively.

$$\mathbb{Z}_2 \times \mathbb{Z}_2 \times \mathbb{Z}_2 \times \mathbb{Z}_{30} \; (\cong \mathbb{Z}_2 \times \mathbb{Z}_2 \times \mathbb{Z}_6 \times \mathbb{Z}_{10})$$
$$\mathbb{Z}_2 \times \mathbb{Z}_2 \times \mathbb{Z}_{60} \; (\cong \mathbb{Z}_2 \times \mathbb{Z}_6 \times \mathbb{Z}_{20})$$
$$\mathbb{Z}_2 \times \mathbb{Z}_{120} \; (\cong \mathbb{Z}_8 \times \mathbb{Z}_{30})$$
$$\mathbb{Z}_{12} \times \mathbb{Z}_{20}$$
$$\mathbb{Z}_{240}$$

Worked Exercise 8.2.4. List all (up to isomorphism) abelian groups of order 3,375.

Answer: We have $3{,}375 = 3^3 \times 5^3$ and hence there are 9 $(= p(3) \cdot p(3) = 3 \times 3)$ abelian groups of order 3,375. The partitions of 3 are

$$\{1, 1, 1\}, \{1, 2\} \quad \text{and} \quad \{3\}.$$

Therefore, the required groups are given below.

$$\mathbb{Z}_3 \times \mathbb{Z}_3 \times \mathbb{Z}_3 \times \mathbb{Z}_5 \times \mathbb{Z}_5 \times \mathbb{Z}_5 \ (\cong \mathbb{Z}_{15} \times \mathbb{Z}_{15} \times \mathbb{Z}_{15})$$
$$\mathbb{Z}_3 \times \mathbb{Z}_3 \times \mathbb{Z}_3 \times \mathbb{Z}_5 \times \mathbb{Z}_{5^2} \ (\cong \mathbb{Z}_3 \times \mathbb{Z}_{15} \times \mathbb{Z}_{75})$$
$$\mathbb{Z}_3 \times \mathbb{Z}_3 \times \mathbb{Z}_3 \times \mathbb{Z}_{5^3} \ (\cong \mathbb{Z}_3 \times \mathbb{Z}_3 \times \mathbb{Z}_{375})$$
$$\mathbb{Z}_3 \times \mathbb{Z}_{3^2} \times \mathbb{Z}_5 \times \mathbb{Z}_5 \times \mathbb{Z}_5 \ (\cong \mathbb{Z}_{15} \times \mathbb{Z}_{45} \times \mathbb{Z}_5)$$
$$\mathbb{Z}_3 \times \mathbb{Z}_{3^2} \times \mathbb{Z}_5 \times \mathbb{Z}_{5^2} \ (\cong \mathbb{Z}_{15} \times \mathbb{Z}_{225})$$
$$\mathbb{Z}_3 \times \mathbb{Z}_{3^2} \times \mathbb{Z}_{5^3} \ (\cong \mathbb{Z}_9 \times \mathbb{Z}_{375})$$
$$\mathbb{Z}_{3^3} \times \mathbb{Z}_5 \times \mathbb{Z}_5 \times \mathbb{Z}_5 \ (\cong \mathbb{Z}_{135} \times \mathbb{Z}_5 \times \mathbb{Z}_5)$$
$$\mathbb{Z}_{3^3} \times \mathbb{Z}_5 \times \mathbb{Z}_{5^2} \ (\cong \mathbb{Z}_{135} \times \mathbb{Z}_{25})$$
$$\mathbb{Z}_{3^3} \times \mathbb{Z}_{5^3} \ (\cong \mathbb{Z}_{3,375})$$

Next, we prove a most general form of the fundamental structure theorem for finitely generated abelian groups. For convenience, we shall denote the binary operation in an abelian group by $+$ instead of $\cdot$ and, as such we write 0 for the identity element of an abelian group $(G, +)$. The inverse of a will be denoted by $-a$ instead of a^{-1} and write na instead of a^n, where a is an arbitrary element of an abelian group $(G, +)$ and n is an integer. That is, na is defined as

$$na = \begin{cases} 0 & \text{if } n = 0, \text{ the integer zero} \\ (n-1)a + a & \text{if } n > 0 \\ (-n)(-a) & \text{if } n < 0 \end{cases}$$

Definition 8.2.3. Let $(G, +)$ be an abelian group and $S \subseteq G$. If $G = \langle S \rangle$, the smallest subgroup of G containing S, then S is called a *generating set* for G. G is called *finitely generated* if there exists a finite generating set for G.

If $S = \{s_1, s_2, \ldots, s_r\}$ is a finite generating set for an abelian group $(G, +)$, then any element a of G can be expressed as

$$a = n_1 s_1 + n_2 s_2 + \cdots + n_r s_r$$

for some integers $n_1, n_2, \ldots, n_r$. In fact, for any subset S of G, we have

$$\langle S \rangle = \{n_1 s_1 + n_2 s_2 + \cdots + n_r s_r : s_i \in S \text{ and } n_i \in \mathbb{Z}\}.$$

The expression $a = n_1 s_1 + n_2 s_2 + \cdots + n_r s_r$ need not be unique. In this context, we have the following theorem.

Theorem 8.2.5. Let $(G, +)$ be an abelian group. Then, G is isomorphic to the product of a finite number of copies of $(\mathbb{Z}, +)$ if and only if there exists a finite generating set $\{s_1, s_2, \ldots, s_n\}$ such that any element a of G can be uniquely expressed as

$$a = n_1 s_1 + n_2 s_2 + \cdots + n_r s_r$$

for some integers $n_1, n_2, \ldots, n_r$.

Proof: Let $G \cong \mathbb{Z} \times \mathbb{Z} \times \cdots \times \mathbb{Z}$ (r copies of $\mathbb{Z}$) and $f: G \to \mathbb{Z} \times \mathbb{Z} \times \cdots \times \mathbb{Z}$ be an isomorphism. For each $1 \le i \le r$, let $t_i = (0, \ldots, 0, 1, 0, \ldots, 0)$, the element whose i^{th} component is 1 and all other components are zeros. Then, any element $t \in \mathbb{Z} \times \cdots \times \mathbb{Z}$ can be uniquely expressed as

$$t = (n_1, n_2, \ldots, n_r) = n_1 t_1 + n_2 t_2 + \cdots + n_r t_r.$$

If we consider $s_i = f^{-1}(t_i)$ for $1 \le i \le r$, then $s_1, s_2, \ldots, s_r$ satisfy the required property. Conversely, suppose that $s_1, s_2, \ldots, s_r$ are elements in S satisfying the given properties. Then, it can be easily verified that the map

$$(n_1, n_2, \ldots, n_r) \mapsto n_1 s_1 + n_2 s_2 + \cdots + n_r s_r$$

is an isomorphism of $\mathbb{Z} \times \mathbb{Z} \times \cdots \times \mathbb{Z}$ (r copies of $\mathbb{Z}$) onto G. ◀

The following result provides a complete characterization of finitely generated abelian groups. First, let us agree, for convenience, to write

$$\sum_{i=1}^{n} a_i \text{ for } a_1 + a_2 + \cdots + a_n,$$

where $a_1, a_2, \ldots, a_n$ are any elements of an abelian group $(G, +)$.

Theorem 8.2.6 (Fundamental structure theorem for finitely generated abelian groups). Let $(G, +)$ be a finitely generated abelian nontrivial group. Then, G is isomorphic to the product of a finite number of cyclic groups A_i; that is,

$$G \cong A_1 \times A_2 \times \cdots \times A_k,$$

where each A_i is a nontrivial cyclic group such that either all of the A_i's are infinite or for some s, $1 \le s \le k$, $A_1, A_2, \ldots, A_s$ are of orders $m_1, m_2, \ldots, m_s$ respectively with m_i divides m_{i+1} for each $1 \le i < s$ and $A_{s+1}, \ldots, A_k$ are infinite.

Proof: Since $(G, +)$ is a finitely generated abelian group, there are finite sets generating G. Let k be the least positive integer such that G has a k-element generating set. We shall use induction on k. If $k = 1$, then G is generated by a single element and hence G itself is cyclic. Suppose that $k > 1$ and assume that the theorem is true for all abelian groups generated by a set of $k - 1$ elements.

If G has a generating set $\{a_1, a_2, \ldots, a_k\}$ such that

$$n_1a_1 + \cdots + n_ra_r = 0 \Rightarrow n_1 = \cdots = n_r = 0 \qquad (*)$$

then any element of G can be uniquely expressed as $n_1a_1 + \cdots + n_ra_r$, $n_i \in \mathbb{Z}$ and hence by Theorem 8.2.5,

$$G \cong \mathbb{Z} \times \cdots \times \mathbb{Z} \ (r \text{ copies}).$$

Next, suppose that G has no generating set $\{a_1, \ldots, a_k\}$ satisfying the property (*). Then, for any generating set $\{a_1, \ldots, a_k\}$, there exists integers $n_1, \ldots, n_k$, not all zero, such that $n_1a_1 + \cdots + n_ka_k = 0$. Since

$$\sum_{i=1}^{k} n_ia_i = 0 \Rightarrow \sum_{i=1}^{k} (-n_i)a_i = -\sum_{i=1}^{k} n_ia_i = 0.$$

There is an equation of the form $n_1a_1 + \cdots + n_ka_k = 0$ with one of n_i's positive. Let T be the set of all positive integers occurring in equations of the form $n_1a_1 + \cdots + n_ka_k = 0$, where $\{a_1, \ldots, a_k\}$ is a generating set for G. The above discussion implies that T is a nonempty set of positive integers. Let m_1 be the least positive integer in T. We can assume that

$$m_1a_1 + n_2a_2 + \cdots + n_ka_k = 0 \qquad (1)$$

for some generating set $\{a_1, a_2, \ldots, a_k\}$ for G and integers $n_2, \ldots, n_k$. We shall prove that m_1 divides each n_i, $2 \le i \le k$. By the division algorithm, let us write

$$n_i = q_im_1 + r_i, 0 \le r_i < m_1, q_i \quad \text{and} \quad r_i \in \mathbb{Z}.$$

Then, from Equation (1),

$$m_1b_1 + r_2a_2 + \cdots + r_ka_k = 0 \qquad (2)$$

where $b_1 = a_1 + q_2a_2 + \cdots + q_ka_k$. If $b_1 = 0$, then $a_1 = -q_2a_2 - q_3a_3 - \cdots - q_ka_k$ and hence $\{a_2, a_3, \ldots, a_k\}$ is a $k - 1$ element generating set for G which is a contradiction to the least property of k.

Therefore, $b_1 \neq 0$. Also,

$$a_1 = b_1 - q_2a_2 - \cdots - q_ka_k$$

and therefore $\{b_1, a_2, \ldots, a_k\}$ is a k-element generating set for G and $r_2, \ldots, r_k$ are nonnegative integers less than m_1 occurring in Equation (2). By the least property of m_1, it follows that $r_2 = r_3 = \cdots = r_k = 0$ and hence $m_i = q_im_1$ for $2 \leq i \leq k$. Also, from Equation (2), we have

$$m_1b_1 = 0.$$

Put $A_1 = <b_1>$, the subgroup generated by b_1 in G. Put $H_1 = <a_2, \ldots, a_k>$, the subgroup generated by $\{a_2, \ldots, a_k\}$ in G. Since $\{b_1, a_2, \ldots, a_k\}$ is a generating set, we get that $A_1 + H_1 = G$. Also, we prove that $A_1 \cap H_1 = \{0\}$. If $x \in A_1 \cap H_1$, then, since $m_1b_1 = 0$,

$$\begin{aligned} x &= mb_1 \quad \text{for some } 0 \leq m < m_1 \\ \text{and} \quad x &= n_2a_2 + \cdots + n_ka_k \quad \text{for some } n_2, \ldots, n_k \in \mathbb{Z} \end{aligned}$$

and therefore $mb_1 + (-n_2)a_2 + \cdots + (-n_k)a_k = 0$. By the least property of m_1, we get that $m = 0$ (otherwise $m \in X$ and $m < m_1$) and hence $x = 0$. Therefore, $A_1 \cap H_1 = \{0\}$. By Theorem 8.1.1, $G \cong A_1 \times H_1$. Also, H_1 is generated by $k - 1$ elements $a_2, \ldots, a_k$ and it is not generated by a set with less than $k - 1$ elements (otherwise G would be generated by a set with less than k elements, which is a contradiction). Therefore, by the induction hypothesis,

$$H_1 \cong A_2 \times \cdots \times A_k$$

where each A_i is a cyclic group such that either all the $A_2, A_3, \ldots, A_r$ are infinite or for some s, $2 \leq s \leq r$, $A_2, \ldots, A_s$ are of orders $m_2, \ldots, m_s$ respectively with m_i divides m_{i+1} for all $2 \leq i < s$ and $A_{s+1}, \ldots, A_r$ are infinite. The proof is complete, if we can show that m_1 divides m_2 also. To do this, let $A_i = <b_i>$ and b_i be of order m_i for $2 \leq i < k$. Then, $\{b_1, b_2, \ldots, b_k\}$ is a generating set for G and

$$m_1b_1 + m_2b_2 + 0b_3 + \cdots + 0b_k = 0.$$

By an argument similar to that made after Equation (1), it follows that m_1 divides m_2. Thus, we have

$$G \cong A_1 \times H_1 \cong A_1 \times A_2 \times \cdots \times A_k$$

where $A_1, A_2, \ldots, A_k$ are cyclic groups satisfying the required properties. This completes the proof of the theorem. ◀

Since any infinite cyclic group is isomorphic to the group $(\mathbb{Z}, +)$ and any finite cyclic group of order n is isomorphic to the group $(\mathbb{Z}_n, +_n)$, the following is an immediate consequence of the above theorem.

Corollary 8.2.5. Let G be a finitely generated abelian group. Then,

$$G \cong \underbrace{\mathbb{Z} \times \cdots \times \mathbb{Z}}_{(r \text{ components})} \times \mathbb{Z}_{m_1} \times \cdots \times \mathbb{Z}_{m_s}$$

where r and s are nonnegative integers and $m_1, \ldots, m_s$ are positive integers such that m_i divides m_{i+1} for all $1 \leq i < s$.

Worked Exercise 8.2.5. State and prove the converse of Corollary 8.2.5.

Answer: If a group G is isomorphic to a product

$$\mathbb{Z} \times \cdots \times \mathbb{Z} \times \mathbb{Z}_{m_1} \times \cdots \times \mathbb{Z}_{m_s}$$

then G is a finitely generated abelian group.
In fact, we prove that the product of finite number of cyclic groups is a finitely generated abelian group.

Let $G_1, G_2, \ldots, G_n$ be cyclic groups and

$$G = G_1 \times G_2 \times \cdots \times G_n$$

since each G_i is abelian, the product G is also abelian. Let a_i be a generator for G_i and e_i be the identity in G_i. Put x_i be the element in G whose i^{th} component is a_i and other components are identities. That is,

$$x_i = (e_1, \ldots, e_{i-1}, a_i, e_{i+1}, \ldots, e_n).$$

Then, any element $g \in G$ can be written as

$$\begin{aligned} g &= (g_1, \ldots, g_n),\ g_i \in G_i \\ &= (a_1^{r_1}, \ldots, a_n^{r_n}),\ r_i \in \mathbb{Z} \\ &= x_1^{r_1}\, x_2^{r_2} \ldots x_n^{r_n} \in \langle x_1, \ldots, x_n \rangle. \end{aligned}$$

Therefore, G is generated by $\{x_1, x_2, \ldots, x_n\}$. Thus, G is a finitely generated abelian group.

Worked Exercise 8.2.6. Let G be a finitely generated abelian group. Prove that G is finite if and only if the order of every element of G is finite.

Answer: If G is finite and $a \in G$, then $O(a) \leq |G|$ (in fact, $O(a)$ is a divisor of $|G|$) and hence $O(a)$ is finite. Conversely, suppose that $O(a)$ is finite for every $a \in G$. By Corollary 8.2.5,

$$G \cong \underbrace{\mathbb{Z} \times \cdots \times \mathbb{Z}}_{(r \text{ components})} \times \mathbb{Z}_{m_1} \times \cdots \times \mathbb{Z}_{m_s}$$

where r and s are nonnegative integers and $m_1, \ldots, m_s$ are positive integers. If $r > 0$, then consider

$$a = (2, 0, \ldots, 0, 0, \ldots, 0)$$

then a is of infinite order and hence the element x in G corresponding to a is also of infinite order, which is a contradiction to our hypothesis. Thus, $r = 0$ and hence

$$G \cong \mathbb{Z}_{m_1} \times \cdots \times \mathbb{Z}_{m_s}$$

since each $\mathbb{Z}_{m_i}$ is finite, so is G.

EXERCISE 8(B)

1. State whether each of the following is True or False. Substantiate your answer.
   - (i) $\mathbb{Z}_2 \times \mathbb{Z}_2 \cong \mathbb{Z}_4$.
   - (ii) $\mathbb{Z}_3 \times \mathbb{Z}_{12}$ is cyclic.
   - (iii) $\mathbb{Z}_6 \times \mathbb{Z}_{25}$ is cyclic.
   - (iv) There is exactly one abelian group of order 105.
   - (v) There is exactly one group of order 30.
   - (vi) Any abelian group of order 165 is isomorphic to $\mathbb{Z}_{165}$.
   - (vii) The number of abelian groups of order 24 is 3.
   - (viii) The number of groups of order 24 is 3.
   - (ix) Any abelian group of order divisible by 7 contains a cyclic subgroup of order 7.
   - (x) Any abelian group of order divisible by 9 contains a cyclic subgroup of order 9.

2. Prove that the following are equivalent to each other for any positive integers m and n.
   (a) $\mathbb{Z}_m \times \mathbb{Z}_n$ is cyclic.
   (b) $\mathbb{Z}_m \times \mathbb{Z}_n \cong \mathbb{Z}_{mn}$.
   (c) m and n are relatively prime.
   (d) $\mathbb{Z} = \langle m, n \rangle$, the subgroup generated by m and n in the group $(\mathbb{Z}, +)$.

3. Prove that any abelian group of order 8 is isomorphic to one of the following.

$$\mathbb{Z}_8, \mathbb{Z}_4 \times \mathbb{Z}_2, \mathbb{Z}_2 \times \mathbb{Z}_2 \times \mathbb{Z}_2.$$

4. Let $G_1, G_2, \ldots, G_n$ be groups and $a_i \in G_i$, $1 \leq i \leq n$. Let $a = (a_1, a_2, \ldots, a_n)$. Prove that $O(a)$ is finite in the product $G_1 \times G_2 \times \cdots \times G_n$ if and only if $O(a_i)$ is finite in G_i for each $1 \leq i \leq n$ and, in this case, $O(a) = \text{l.c.m.}\ \{O(a_1), \ldots, O(a_n)\}$.

5. Prove that, for any $n > 1$, $\mathbb{Z}_{n^2}$ is not isomorphic with $\mathbb{Z}_n \times \mathbb{Z}_n$.

6. Let $\mathbb{Z}(G)$ denote the centre of any group G. For any groups $G_1, G_2, \ldots, G_n$, prove that $\mathbb{Z}(G_1 \times G_2 \times \cdots \times G_n) = \mathbb{Z}(G_1) \times \mathbb{Z}(G_2) \times \cdots \times \mathbb{Z}(G_n)$.

7. Let $[G, G]$ denote the commutator subgroup of a group G. For $G = G_1 \times \cdots \times G_n$, prove that

$$[G, G] = [G_1, G_1] \times \cdots \times [G_n, G_n].$$

8. Let N_1 and N_2 be normal subgroups of groups G_1 and G_2, respectively. Prove that $N_1 \times N_2$ is a normal subgroup of $G_1 \times G_2$ and that

$$G_1/N_1 \times G_2/N_2 \cong (G_1 \times G_2)/N_1 \times N_2.$$

9. Let N and M be normal subgroups of a group G such that $N \cap M = \{e\}$. Then prove that G is isomorphic to a subgroup of $G/M \times G/N$.

10. Prove that any cyclic p-group is finite.

11. Prove that any finite abelian p-group is generated by its elements of highest order.

12. Show that any homomorphic image of a p-group is a p-group and product of p-groups is also a p-group.

13. Let G be a finite cyclic group and p be a prime dividing the order of G. Prove that there are exactly $p - 1$ elements in G each having order p.

14. Prove that a nontrivial finite abelian group is cyclic if and only if it is isomorphic to $\mathbb{Z}_{p_1^{n_1}} \times \mathbb{Z}_{p_2^{n_2}} \times \cdots \times \mathbb{Z}_{p_r^{n_r}}$ for some distinct primes $p_1, \ldots, p_r$ and positive integers $n_1, \ldots, n_r$.

15. Determine all (up to isomorphism) abelian group of order 144, 625 and 1,94,481.

16. Prove that a cyclic group is indecomposable if and only if it is either infinite or of prime power order.

17. Describe all the positive integers n for which there is exactly one (up to isomorphism) abelian group of order n.

18. Let $\{a_1, a_2, \ldots, a_r\}$ be a generating set for an abelian group G. Prove that the following are equivalent to each other.

    (1) $(n_1, n_2, \ldots, n_r) \mapsto \sum_{i=1}^{r} n_i a_i$ is an isomorphism of $\mathbb{Z}\times\mathbb{Z}\times\cdots\times\mathbb{Z}$ (r copies) onto G.

    (2) Any element a of G can be uniquely expressed as

    $$a = n_1 a_1 + \cdots + n_r a_r, \text{ for integers } n_1, n_2, \ldots, n_r$$

    (3) For any integers $n_1, n_2, \ldots, n_r$,

    $$n_1 a_1 + n_2 a_2 + \cdots + n_r a_r = 0 \Rightarrow n_1 = n_2 = \cdots = n_r = 0$$

19. Let G be a finitely generated abelian group. Prove that G is finite if and only if G is isomorphic with a product of finitely many cyclic groups.

20. Let G be a nontrivial abelian group. Prove that G is finite if and only if

    $$G \cong \mathbb{Z}_{m_1} \times \cdots \times \mathbb{Z}_{m_r}$$

    for some positive integers $m_1, \ldots, m_r$ such that $m_1 > 1$ and m_i divides m_{i+1} for all $1 \le i < r$.

8.3 INVARIANTS OF FINITE ABELIAN GROUPS

The fundamental structure theorem for finitely generated abelian groups (Theorem 8.2.6) can be applied to finite abelian groups to associate a unique finite division sequence of positive integers with each finite abelian group. First, let us define the following.

Definition 8.3.1. A finite sequence $m_1, m_2, \ldots, m_r$ of positive integers is called a *division sequence* if $1 < m_1$ and m_i divides m_{i+1} for each $1 \le i < r$. A division sequence is denoted by

$$1 < m_1 | m_2 | \ldots | m_r.$$

In general, if a and b are integers, we write a/b to say that a divides b; that is, $ac = b$ for some integer c.

Theorem 8.3.1. Let G be a nontrivial finite abelian group. Then, there exists a unique division sequence

$$1 < m_1 | m_2 | \ldots | m_r$$

such that $G \cong Z_{m_1} \times Z_{m_2} \times \cdots \times Z_{m_r}$.

Proof: By Theorem 8.2.6, there exists nontrivial finite cyclic groups G_1, G_2, …, G_r of orders $m_1, m_2, \ldots, m_r$, respectively such that m_i divides $m_1, m_2, \ldots, m_r$ respectively such that m_i divides m_{i+1} for each $1 \le i < r$ and

$$G \cong G_1 \times G_2 \times \cdots \times G_r.$$

Since any cyclic group of order m is isomorphic to $\mathbb{Z}_m$, we get that $1 < m_1|m_2| \ldots |m_r$ is a division sequence and

$$G \cong \mathbb{Z}_{m_1} \times \mathbb{Z}_{m_2} \times \cdots \times \mathbb{Z}_{m_r}. \tag{1}$$

Suppose that $1 < n_1|n_2| \ldots |n_s$ is any division sequence such that

$$G \cong \mathbb{Z}_{n_1} \times \mathbb{Z}_{n_2} \times \cdots \times \mathbb{Z}_{n_s}. \tag{2}$$

We shall prove that $r = s$ and $m_i = n_i$ for each $1 \le i \le r$. First note that the order of any element of $\mathbb{Z}_{m_i}$ is a divisor of m_i and hence that of G is a divisor of the l.c.m. $\{m_1, m_2, \ldots, m_r\}$ which is equal to m_r (since m_i divides m_r for all $1 \le i \le r$). Therefore, the order of any element of G is a divisor of m_r. Since the element $(0, \ldots, 0, 1)$ is of order n_s in $\mathbb{Z}_{n_1} \times \cdots \times \mathbb{Z}_{n_s} \cong G$, there exists an element of order n_s in G and therefore n_s divides m_r. Similarly, by symmetry, we can prove that m_r divides n_s and hence $m_r = n_s$. From the decompositions (1) and (2), we have

$$\begin{aligned} m_{r-1}G &\cong (m_{r-1}\,\mathbb{Z}_{m_1}) \times \cdots \times (m_{r-1}\,\mathbb{Z}_{m_{r-1}}) \times (m_{r-1}\,\mathbb{Z}_{m_r}) \\ &\cong (m_{r-1}\,\mathbb{Z}_{n_1}) \times \cdots \times (m_{r-1}\,\mathbb{Z}_{n_{s-1}}) \times (m_{r-1}\,\mathbb{Z}_{n_s}), \end{aligned}$$

where mA stands for $\{ma : a \in A\}$. Since $m_i|m_{r-1}$ for all $1 \le i \le r-1$, it follows that $m_{r-1}\,\mathbb{Z}_{m_i} = \{0\}$ for each $1 \le i \le r-1$ and hence

$$|m_{r-1}G| = |m_{r-1}\,\mathbb{Z}_{m_r}| = |m_{r-1}\mathbb{Z}_{n_s}|.$$

This implies that $|m_{r-1}\mathbb{Z}_{n_j}| = 1$ for each $1 \le j \le s-1$ and, in particular $|m_{r-1}\,\mathbb{Z}_{n_{s-1}}| = 1$ so that n_{s-1} divides m_{r-1}. Similarly, by interchanging m_i's and n_j's, we get that m_{r-1} divides n_{s-1}. Therefore, $m_{r-1} = n_{s-1}$. Continuing this process, we can prove that $m_{r-i} = n_{s-i}$ for $i = 0, 1, 2, \ldots$. Since $m_1\, m_2 \ldots m_r = |G| = n_1\, n_2 \cdots n_s$, we get that $r = s$ and $m_i = n_i$ for all $1 \le i \le r$. ◀

Definition 8.3.2. If G is an abelian group and $1 < m_1|m_2| \ldots |m_r$ is a division sequence such that

$$G \cong \mathbb{Z}_{m_1} \times \mathbb{Z}_{m_2} \times \cdots \times \mathbb{Z}_{m_r}$$

then G is said to be of type $(m_1, m_2, \ldots, m_r)$ and the integers $m_1, m_2, \ldots, m_r$ are called the *invariants* of G. The following is an immediate consequence of Theorem 8.3.1.

Corollary 8.3.1. Let n be a positive integer greater than 1. Then, the number of abelian groups of order n is equal to the number of division sequences $1 < m_1|m_2| \ldots |m_r$ such that $n = m_1 m_2 \ldots m_r$.

Example 8.3.1. Let us find the invariants of the group $G = \mathbb{Z}_6 \times \mathbb{Z}_8 \times \mathbb{Z}_5$. First note that 6, 8, 5 do not form a division sequence. We have

$$\begin{aligned} G = \mathbb{Z}_6 \times \mathbb{Z}_8 \times \mathbb{Z}_5 &\cong \mathbb{Z}_2 \times \mathbb{Z}_3 \times \mathbb{Z}_8 \times \mathbb{Z}_5 \\ &\cong \mathbb{Z}_2 \times \mathbb{Z}_{120} \end{aligned}$$

and 2 divides 120. Therefore, the invariants of G are 2, 120.

Worked Exercise 8.3.1. Determine the invariants of the group

$$G = \mathbb{Z}_2 \times \mathbb{Z}_4 \times \mathbb{Z}_9 \times \mathbb{Z}_{27} \times \mathbb{Z}_{81}.$$

Answer: By inspection, we can see that

$$G \cong \mathbb{Z}_9 \times \mathbb{Z}_{54} \times \mathbb{Z}_{324}$$

and $9|54|324$ is a division sequence. Therefore, 9, 54, 324 are the invariants of G.

The following provides an algorithm to find the invariants of a given finite abelian group. Let us recall that a partition of a positive integer n is a finite sequence $\{n_1, n_2, \ldots, n_r\}$ of positive integers such that $n_1 \le n_2 \le \cdots \le n_r$ and $n_1 + n_2 + \cdots + n_r = n$. For the convenience and for the purpose of proving the following result, we relax the definition of a partition of n by including certain zeros in the beginning. Accordingly, a partition of n is a sequence of nonnegative integers $\{n_1, n_2, \ldots, n_r\}$ such that $0 \le n_1 \le n_2 \le \cdots \le n_r$ and $n_1 + n_2 + \cdots + n_r = n$.

Theorem 8.3.2. Let G be an abelian group of order $n = p_1^{m_1} p_2^{m_2} \cdots p_k^{m_k}$ where $p_1, p_2, \ldots, p_k$ are distinct primes and $m_1, m_2, \ldots, m_k$ are positive integers. For each $1 \le i \le k$, let $\{n_{i_1}, n_{i_2}, \ldots, n_{i_r}\}$ be a partition of m_i such that

$$G \cong (\mathbb{Z}_{p_1^{m_{11}}} \times \cdots \times \mathbb{Z}_{p_1^{m_{1r}}}) \times \cdots \times (\mathbb{Z}_{p_k^{n_{k1}}} \times \cdots \times \mathbb{Z}_{p_k^{n_{kr}}}).$$

For each $1 \le j \le r$, let $s_j = p_1^{n_{1j}} \ldots p_k^{n_{kj}}$. Then, $s_1, s_2, \ldots, s_r$ are the invariants of G.

Proof: Note that $0 \le n_{i_1} \le n_{i_2} \le \cdots \le n_{i_r}$ and $n_{i1} + n_{i2} + \cdots + n_{ir} = m_i$ for each $1 \le i \le k$. Let

$$G_j = \mathbb{Z}_{p_1^{n_{1j}}} \times \mathbb{Z}_{p_2^{n_{2j}}} \times \ldots \mathbb{Z}_{p_k^{n_{kj}}}, \quad \text{for } 1 \le j \le r.$$

Since $p_1, p_2, \ldots, p_k$ are distinct primes, it follows that $p_1^{n_{1j}}, p_2^{n_{2j}}, \ldots, p_k^{n_{kj}}$ are pair-wise relatively prime and hence

$$G_j \cong \mathbb{Z}_{s_j}, \quad \text{where } s_j = p_1^{n_{1j}} p_2^{n_{2j}} \ldots p_k^{n_{kj}}.$$

Now, $G \cong G_1 \times G_2 \times \cdots \times G_r \cong \mathbb{Z}_{s_1} \times \mathbb{Z}_{s_2} \times \cdots \times \mathbb{Z}_{s_r}$.
Also, since $n_{ij} \le n_{ij+1}$ for $1 \le i \le k$ and $1 \le j \le r$, we get that $p_i^{n_{ij}}$ divides $p_i^{n_{ij+1}}$ and hence s_j divides s_{j+1} for each $1 \le j < r$. Thus, $s_1, s_2, \ldots, s_r$ are the invariants of G.

Example 8.3.2. Let G be the set of all positive integers less than 100 and relatively prime to 100. Then, G is an abelian group under multiplication modulo 100. Let us find the invariants of G.
First, we shall list all elements of G and find the order G. We have $|G| = \phi(100) = 40 = 2^3 \times 5$

$$G = \{1, 3, 7, 9, 11, 13, 17, 19, 21, 23, 27, 29, 31, 33, 37, 39, 41, 43, 47, 49, 51, 53, 57, 59, 61, 63, 67, 69, 71, 73, 77, 79, 81, 83, 87, 89, 91, 93, 97, 99\}.$$

One can easily verify the $G = \langle 3 \rangle$, the cyclic subgroup generated by 3. Therefore, G is a cyclic group of order 40 and hence $G \cong \mathbb{Z}_{40}$. 40 is the only invariant of G.

EXERCISE 8(C)

1. Determine the invariants of each of the following groups.
   (i) $\mathbb{Z}_2 \times \mathbb{Z}_2 \times \mathbb{Z}_5 \times \mathbb{Z}_5$.
   (ii) $\mathbb{Z}_9 \times \mathbb{Z}_{25} \times \mathbb{Z}_7 \times \mathbb{Z}_4 \times \mathbb{Z}_{121}$.
   (iii) $\mathbb{Z}_8 \times \mathbb{Z}_7 \times \mathbb{Z}_{49} \times \mathbb{Z}_9$.
   (iv) $\mathbb{Z}_2 \times \mathbb{Z}_4 \times \mathbb{Z}_8 \times \mathbb{Z}_9 \times \mathbb{Z}_3 \times \mathbb{Z}_5 \times \mathbb{Z}_{25}$.
2. Let X be a set with 5-elements and $G = (\mathbb{P}(X), \oplus)$, where $\mathbb{P}(X)$ is the power set of X and $\oplus$ is the symmetric difference operation. Determine the invariants of G.
3. Let G be the group of all positive integers less than 47 and relatively prime to 47, under the multiplication modulo 47 Then find the invariants of G.
4. Let G be the group of all mappings of a 4-element set into the group $(\mathbb{Z}_4, +_4)$ under point-wise operation. Determine all the invariants of G.
5. Determine the invariants of each abelian group of order less than or equal to 30.

8.4 GROUPS OF SMALL ORDER

We conclude the discussion on group theory with the complete description of all groups of order less than or equal to 20. We have derived an exact formula for the number of abelian groups of a given order n and an algorithm to list all these groups, up to isomorphism. However, there is no precise formula for the number of all groups (nonabelian groups) of a given order n. In this section, we describe these groups (up to isomorphism) of order n for $n \leq 20$.

Let us first recall that any group of prime order is cyclic and hence abelian and that any group of order p^2, where p is prime, is abelian and there are only two such groups, namely $\mathbb{Z}_p$ or $\mathbb{Z}_p \times \mathbb{Z}_p$. Further, we have proved that any group of order pq, where $p < q$ are primes such that p does not divide $q - 1$, is cyclic. In the following, we prove that there is a unique nonabelian group of order pq, when p divides $q - 1$.

Theorem 8.4.1. Let p and q be primes such that $p < q$ and p divides $q - 1$ and G be a nonabelian group of order pq. Then, G is a group generated by two elements a and b satisfying the following:

1. $a^p = 1 = b^q$, $a \neq e$ and $b \neq e$.
2. $a^{-1}ba = b^r$, where $r \not\equiv 1 (\text{mod } q)$ and $r^p \equiv 1 (\text{mod } q)$.

Also, G is the unique (up to isomorphism) nonabelian group of order pq.

Proof: By Sylow Theorem III, the number of Sylow q-subgroups of G is $kq + 1$ which divides $|G| = pq$. This implies that $kq + 1 = 1$ and hence there exists exactly one Sylow q-subgroup B which is a normal subgroup of order q in G. Since q is prime, it follows that B is cyclic and hence $B = \langle b \rangle$ for some $b \in G$ such that $b^q = e \neq b$. Also, the number of Sylow p-subgroups is of the form $mp + 1$ for some $m \geq 0$ and is divisor of $|G| = pq$. This implies that $mp + 1 = 1$ or q. If $mp + 1 = 1$, then, as above, there exists a unique subgroup A of order p which becomes normal in G and, in this case $A \cap B = \{e\}$ and $AB = G$, so that $G \cong A \times B \cong \mathbb{Z}_p \times \mathbb{Z}_q \cong \mathbb{Z}_{pq}$ and hence G is cyclic and abelian, which is a contradiction to the hypothesis that G is nonabelian. Therefore, $mp + 1 = q$; that is, the number of Sylow p-subgroups is q.
Let A be a Sylow p-subgroup of G. Then, $|A| = p$ and hence $A = \langle a \rangle$ for some $a \in G$ such that $a^p = e \neq a$. Consider the subgroup $\langle a, b \rangle$ generated by a and b. Since A and B are contained in $\langle a, b \rangle$, we have $AB \subseteq \langle a, b \rangle$. Also, $A \cap B = \{e\}$ and hence

$$|AB| = \frac{|A||B|}{|A \cap B|} = \frac{p \cdot q}{1} = |G|$$

and hence $AB = G$, so that $\langle a, b \rangle = G$.
Since $B = \langle b \rangle$ is a normal subgroup of G, $a^{-1}ba = b^r$ for integer r, then $r \not\equiv 1 \pmod{q}$, (otherwise, $a^{-1}ba = b$ and hence $ba = ab$ which implies that G is abelian, a contradiction). Now, we have

$$\begin{aligned}
a^{-1}ba = b^r &\Rightarrow a^{-1}b^2a = (a^{-1}ba)(a^{-1}ba) = b^{2r} \\
&\Rightarrow a^{-1}b^r a = b^{r^2} \quad \text{(by induction)} \\
&\Rightarrow a^{-1}(a^{-1}ba)a = a^{-1}b^r a = b^{r^2} \\
&\Rightarrow a^{-2}ba^2 = b^{r^2} \\
&\Rightarrow a^{-p}ba^p = b^{r^p} \quad \text{(by induction)} \\
&\Rightarrow b = ebe = a^{-p}ba^p = b^{r^p} \\
&\Rightarrow r^p \equiv 1 \pmod{q} \text{ (since } O(b) = q).
\end{aligned}$$

Thus, G is generated by a and b, which satisfy the following:

1. $a^p = e = b^q$, $a \neq e$ and $b \neq e$.
2. $a^{-1}ba = b^r$, $r \not\equiv 1 \pmod{q}$ and $r^p \equiv 1 \pmod{q}$.

On the other hand, if G' is any nonabelian group of order pq, then G' is generated by elements a and b satisfying the above properties (1) and (2). Now, $AB = G'$, where $A = \langle a \rangle$ and $B = \langle b \rangle$ and hence

$$G' = \{a^i b^j : 0 \leq i < p \text{ and } 0 \leq j < q\}.$$

Since the solutions of $x^p \equiv 1 \pmod{q}$, $x \not\equiv 1 \pmod{q}$, are $r, r^2, \ldots, r^{p-1}$, it follows that $G \cong G'$, since replacing a by a^j as a generator of $<a>$ replaces r by r^j. This completes the proof.

Corollary 8.4.1. Let G be a group of order pq, where p and q are primes such that $p < q$. Then, either G is cyclic or G is a nonabelian group generated by two elements a and b satisfying the properties (1) and (2) above.

Now, we list all groups of order less than or equal to 20. Some of the proofs involved in the listing are left as exercises to the reader.

1. The trivial group $\{e\}$ is the only group of order 1.
2. $\mathbb{Z}_2$ is the only (up to isomorphism) group of order 2, since 2 is prime.
3. $\mathbb{Z}_3$ is the only (up to isomorphism) group of order 3.
4. Any group of order $4(= 2^2)$ is abelian and there are only two groups of order 4, one cyclic and the other noncyclic abelian group; namely $\mathbb{Z}_4$ and $\mathbb{Z}_2 \times \mathbb{Z}_2$.
5. Since 5 is prime, $\mathbb{Z}_5$ is the only group of order 5.
6. There are two groups of order 6, one is cyclic and the other is a nonabelian group (see Corollary 8.4.1). These are $\mathbb{Z}_6$ and the symmetric group S_3. Since $6 = 2 \cdot 3$, 2 and 3 are primes, there is exactly one abelian group of order 6 which is the cyclic group $\mathbb{Z}_6$. S_3 is a nonabelian group of order 6 and S_3 is generated by the elements

$$a = (1\ 2) \quad \text{and} \quad b = (1\ 2\ 3),$$
$$a^2 = e = b^3, a \neq e, b \neq e \quad \text{and}$$
$$a^{-1}ba = (1\ 2)\,(1\ 2\ 3)\,(1\ 2) = (1\ 3\ 2) = b^2 \text{ (see Theorem 8.4.1).}$$

7. Since 7 is prime, $\mathbb{Z}_7$ is the only group of order 7.
8. There are 5 groups of order 8, 3 abelian and 2 nonabelian. $\mathbb{Z}_2 \times \mathbb{Z}_2 \times \mathbb{Z}_2$, $\mathbb{Z}_4 \times \mathbb{Z}_2$ and $\mathbb{Z}_8$ are the only abelian groups of order 8.
   The quaternion group $Q_8 = \{1, -1, i, -i, j, -j, k, -k\}$ and the dihedral group D_4 are the only nonabelian groups of order 8.
9. Any group of order 9 ($= 3^2$ and 3 is prime) is abelian. There are only two groups of order 9, one is cyclic $\mathbb{Z}_9$ and the other is noncyclic abelian $\mathbb{Z}_3 \times \mathbb{Z}_3$.
10. There are only two groups of order 10, one is the cyclic group $\mathbb{Z}_{10}$ and the other is the dihedral group D_5.
11. Since 11 is prime, the cyclic group $\mathbb{Z}_{11}$ is the only group of order 11.
12. There are five groups of order 12, two abelian and three nonabelian. $\mathbb{Z}_{12}$ and $\mathbb{Z}_2 \times \mathbb{Z}_6$ are the abelian groups of order 12.

The alternating group A_4 (the group of even permutations in S_4), the dihedral group D_6 and the group G described below are the only nonabelian groups of order 12. Note that

$$D_6 \cong S_3 \times \mathbb{Z}_2 \cong D_3 \times \mathbb{Z}_2.$$

G is the group generated by two elements a and b such that $O(a) = 4$, $O(b) = 3$ and $ab = b^{-1}a$.

13. Since 13 is a prime, the cyclic group $\mathbb{Z}_{13}$ is the only group of order 13.
14. There are two groups of order 14, one is the cyclic group $\mathbb{Z}_{14}$ and the other is the dihedral group D_7.
15. Any group of order 15 is cyclic (since $15 = 3 \cdot 5$ and 3 does not divide $5 - 1$) and hence $\mathbb{Z}_{15}$ is the only group of order 15.
16. There are totally 14 groups of order 16 out of which 5 are abelian and 9 are nonabelian.

    The abelian groups of order 16 are $\mathbb{Z}_{16}$, $\mathbb{Z}_8 \times \mathbb{Z}_2$, $\mathbb{Z}_4 \times \mathbb{Z}_4$, $\mathbb{Z}_4 \times \mathbb{Z}_2 \times \mathbb{Z}_2$ and $\mathbb{Z}_2 \times \mathbb{Z}_2 \times \mathbb{Z}_2 \times \mathbb{Z}_2$.

    The nonabelian groups of order 16 are given below.

    (i) The dihedral group D_8.

    (ii) $D_4 \times \mathbb{Z}_2$, where D_4 is the dihedral group of degree 4.

    (iii) $Q_8 \times \mathbb{Z}_2$, where Q_8 is the eight element quaternion group.

    (iv) The group generated by two elements a and b such that $O(a) = 8$, $O(a) = 2$ and $ab = ba^3$.

    (v) $G = \langle a, b \rangle$, where $O(a) = 8$, $O(b) = 2$ and $ab = ba^5$.

    (vi) $G = \langle a, b \rangle$, where $O(a) = 4 = O(b)$ and $ab = ba^3$.

    (vii) $G = \langle a, b, c \rangle$, where $O(a)= 4$, $O(b) = 2 = O(c)$, $cbca^2b = 1$, $bab = a$ and $cac = a$.

    (viii) $G = \langle a, b \rangle$, where $O(a) = 8$, $a^4 = b^2$ and $aba = b$.

    (ix) $G = \langle a, b \rangle$, where $O(a) = 4 = O(b)$, $abab = e$ and $ab^3 = ba^3$.
17. 17 is a prime and hence $\mathbb{Z}_{17}$ is the only group of order 17.
18. There are five groups of order 18, two abelian and three nonabelian. $\mathbb{Z}_{18}$ and $\mathbb{Z}_3 \times \mathbb{Z}_6$ are the abelian groups and the nonabelian groups are D_9, $S_3 \times \mathbb{Z}_3$ and the group $G = \langle a, b, c \rangle$ such that $O(a) = 2$, $O(b) = 3 = O(c)$, $bc = cb$, $bab = a$ and $cac = a$.
19. 19 is a prime and hence $\mathbb{Z}_{19}$ is the only group of order 19.
20. There are five groups of order 20, two abelian and three nonabelian. $\mathbb{Z}_{20}$ and $\mathbb{Z}_2 \times \mathbb{Z}_{10}$ are the abelian groups, while the nonabelian groups are the dihedral group D_{10}, the group $G = \langle a, b \rangle$ where $O(a) = 4$, $O(b) = 5$ and $bab = a$ and the group $H = \langle x, y \rangle$ where $O(a) = 4$, $O(b) = 5$ and $ba = ab^2$.

PART III
Ring Theory

9

Rings

It is well known that there are two familiar binary operations, namely the addition $+$ and the multiplication $\cdot$ on the set $\mathbb{Z}$ of integers and that $(\mathbb{Z}, +)$ is an abelian group where as $(\mathbb{Z}, \cdot)$ is only a semigroup. We have earlier worked with algebraic systems, namely semigroups, monoids and groups, where there is only one binary operation in each. Now, in this chapter, we initiate the study of abstract algebraic systems having two binary operations as in the case of integers. Also, we have the rational number system, the real number system, the complex number system, the set of all $n \times n$ matrices, where in each of these cases we have two binary operations satisfying certain connective properties in addition to the properties satisfied by the individual operations. We introduce a common abstraction of these in the form of a ring and develop a general elementary theory of rings. A ring is basically a combination of an abelian group and a semigroup and therefore a previous knowledge of groups and semigroups will be of considerable help. Most of the important concepts in group theory have natural extensions to ring theory.

9.1 EXAMPLES AND ELEMENTARY PROPERTIES

When we have two binary operations say $*$ and o on a set X, in order to get information about the algebraic system $(X, *, o)$ more than we could obtain by studying each operation separately, these must be some relationship between the two operations. The most common requirement is that one of them be distributive over the other: the operation * is said to left distributive over o if

$$a * (b \circ c) = (a * b) \circ (a * c) \quad \text{for all } a, b \text{ and } c \in X$$

and * is said to be right distributive over o if

$$(a \circ b) * c = (a * c) \circ (b * c) \quad \text{for all } a, b \text{ and } c \in X.$$

We say that * is distributive over o if it is both left and right distributive over o.

In this section, we present a formal definition of a ring with several illustrative examples and prove certain important elementary properties of rings.

Definition 9.1.1. A triple $(R, +, \cdot)$ is called a *ring* if R is a nonempty set and $+$ and $\cdot$ are binary operations on R satisfying the following:

I. $(R, +)$ is an abelian group; that is,
   (I.1) $a + b = b + a$ for all a and $b \in R$
   (I.2) $a + (b + c) = (a + b) + c$ for all a, b and $c \in R$
   (I.3) There is a (unique) element 0 such that

$$a + 0 = a \quad \text{for all } a \in R \quad \text{and}$$

   (I.4) For each $a \in R$, there exists (unique) element $-a \in R$ such that $a + (-a) = 0$.

II. $(R, \cdot)$ is a semigroup; that is,

$$a \cdot (b \cdot c) = (a \cdot b) \cdot c \quad \text{for all } a, b \text{ and } c \in R$$

III. The operation $\cdot$ is distributive over the operation $+$; that is,

$$\begin{aligned} & a \cdot (b + c) = (a \cdot b) + (a \cdot c) \\ \text{and} \quad & (a + b) \cdot c = (a \cdot c) + (b \cdot c) \quad \text{for all } a, b \text{ and } c \in R. \end{aligned}$$

One should clearly understand that $+$ and $\cdot$ are abstract binary operations and not ordinary addition and multiplication of integers or real numbers. However, for convenience, we call the operation $+$ as *addition* and the operation $\cdot$ as *multiplication*. In the light of this terminology, it is natural then to speak of the abelian group $(R, +)$ as the *additive group* of the ring $(R, +, \cdot)$ and of $(R, \cdot)$ as the *multiplicative semigroup* of the ring $(R, +, \cdot)$.

Recall from the group theory that the element 0 in (I.3) above is unique and is called the *additive identity* or the *zero element* in R. Also, for any $a \in R$, the element $-a$ in (I.4) above is unique and is called the *additive inverse* of a.

In order to minimize the use of parentheses (brackets) in expressions involving both the operations $+$ and $\cdot$, let us stipulate that multiplication is to be performed before addition. Accordingly, the expression $a \cdot b + c$ stands for $(a \cdot b) + c$ and not for $a \cdot (b + c)$. Also, because of the generalised associative law, parentheses can also be omitted when we write sums or products of more than two elements. For example, we write

$$a_1 \cdot b_1 + a_2 \cdot b_2 + a_3 + a_4 \cdot b_4$$
$$\text{instead of} \quad ((a_1 \cdot b_1) + (a_2 \cdot b_2)) + (a_3 + (a_4 \cdot b_4)).$$

It is needless to say that $a + b$ is called the *sum* of a and b and $a \cdot b$ is called the *product* of a and b in this order. Also, as usual, we write $a - b$ for $a + (-b)$ and ab for $a \cdot b$, when there is no ambiguity about the multiplication. Note that, as in the case of groups, we simply say that 'R is a ring' instead of saying that '$(R, +, \cdot)$ is a ring' when there is no ambiguity about the operations $+$ and $\cdot$.

Further, notice that the operations $+$ and $\cdot$ in a ring R cannot be interchanged, for $(R, \cdot)$ may not be a group at all and $+$ may not be distributive over $\cdot$. In fact, except in the trivial case when $R = \{0\}$, $(R, \cdot)$ is not a group. These things will be more clear, when we consider the following examples.

Example 9.1.1

1. Let $\mathbb{Z}$ be the set of all integers, $\mathbb{Q}$ be the set of all rational numbers and $\mathbb{R}$ be the set of all real numbers. Then, $(\mathbb{Z}, +, \cdot)$, $(\mathbb{Q}, +, \cdot)$ and $(\mathbb{R}, +, \cdot)$ are all rings, where $+$ and $\cdot$ are the usual addition and multiplication of real numbers. In each of these cases, the number 0 is the zero element (that is, the additive identity).
2. Let $(G, +)$ be an abelian (commutative) group in which 0 is the identity element. Define $a \cdot b = 0$ for all a and $b \in G$. Then, $(G, +, \cdot)$ is a ring, since $\cdot$ is clearly associative and distributive over $+$, for,

$$a \cdot (b + c) = 0 = 0 + 0 = (a \cdot b) + (a \cdot c)$$
$$\text{and} \quad (a + b) \cdot c = 0 = 0 + 0 = (a \cdot c) + (b \cdot c)$$

for any a, b and $c \in G$. The multiplication here is called the *trivial multiplication in an abelian group*. Rings of this type are called *zero rings*.

3. Let $M_2(\mathbb{R})$ be the set of all 2×2 matrices over $\mathbb{R}$ (that is, with entries as real numbers). For any matrices $A = \begin{pmatrix} a_{11} & a_{12} \\ a_{21} & a_{22} \end{pmatrix}$ and $B = \begin{pmatrix} b_{11} & b_{12} \\ b_{21} & b_{22} \end{pmatrix}$ in $M_2(\mathbb{R})$, define

$$A + B = \begin{pmatrix} a_{11} + b_{11} & a_{12} + b_{12} \\ a_{21} + b_{21} & a_{22} + b_{22} \end{pmatrix}$$

and $$A \cdot B = \begin{pmatrix} a_{11}b_{11} + a_{12}b_{21} & a_{11}b_{12} + a_{12}b_{22} \\ a_{21}b_{11} + a_{22}b_{21} & a_{21}b_{12} + a_{22}b_{22} \end{pmatrix}$$

Then, $(M_2(\mathbb{R}), +, \cdot)$ is a ring and is called *the ring of* 2×2 *matrices over* $\mathbb{R}$. One should carefully check the validity of the axioms I, II and III of Definition 9.1.1 carefully. Note that the matrix in which all the entries are zero is the zero element and is called the *zero matrix*. Also, the additive inverse $-A$ of A is given by

$$-A = \begin{pmatrix} -a_{11} & -a_{12} \\ -a_{21} & -a_{22} \end{pmatrix}.$$

4. Let $\mathbb{C}$ denote the set of all complex numbers; that $\mathbb{C}$ is the set of all expressions of the form $a + ib$, where a and b are arbitrary real numbers. For any $x = a + ib$ and $y = c + id$ in $\mathbb{C}$, define

$$x = y \Leftrightarrow a = c \quad \text{and} \quad b = d$$
$$x + y = (a + b) + i(b + d)$$
$$x \cdot y = (ac - bd) + i(ad + bc),$$

where $a + b$, ac, etc. are the sums and products in the ring $(\mathbb{R}, +, \cdot)$. Then, $(\mathbb{C}, +, \cdot)$ is a ring and is called *the ring of complex numbers*. Note that $0 + i0$ is the zero element in $\mathbb{C}$, which is also denoted simply by 0. Further, the additive inverse $-x$ of $x = a + ib$ is given by

$$-x = (-a) + i(-b)$$

As usual, we denote $a + i0$ by a, $0 + ib$ by ib and $0 + i1$ by i. As per this notation, note that $i \cdot i = -1$.

5. In the above example, the operation + is defined coordinate wise. If multiplication is also defined as coordinate wise (considering a and b as first and second coordinates of $a + ib$), then $\mathbb{C}$ together with these coordinate wise addition and multiplication forms a ring.
6. The procedure in 5 above can be generalised as follows. Let $(R_1, +, \cdot)$, $(R_2, +, \cdot)$, ..., $(R_n, +, \cdot)$ be any rings and

$$R = R_1 \times R_2 \times \cdots \times R_n = \{(a_1, a_2, \ldots, a_n) : a_i \in R_i \text{ for } 1 \le i \le n\}.$$

For any $a = (a_1, a_2, \ldots, a_n)$ and $b = (b_1, b_2, \ldots, b_n)$ in R, Define

$$a + b = (a_1 + b_1, a_2 + b_2, \ldots, a_n + b_n)$$
$$\text{and} \quad a \cdot b = (a_1 \cdot b_1, a_2 \cdot b_2, \ldots, a_n \cdot b_n).$$

Then, $(R, +, \cdot)$ is a ring and is called the *product* of $R_1, R_2, \ldots, R_n$ and is denoted by $\prod_{i=1}^{n} R_i$ or, simply, $R_1 \times R_2 \times \cdots \times R_n$. Note that, $(0, 0, \ldots, 0)$ is the zero element in the product, where 0 stands for the zero element in each R_i. The additive inverse of any $a = (a_1, a_2, \ldots, a_n)$ is given by

$$-a = (-a_1, -a_2, \ldots, -a_n).$$

7. Let $(R, +, \cdot)$ be any ring and X be any nonempty set. Let R^X be the set of all mappings of X into R. For any f and $g \in R^X$, define $f + g$ and $f \cdot g : X \to R$ by

$$(f + g)(x) = f(x) + g(x)$$
$$\text{and} \quad (f \cdot g)(x) = f(x) \cdot g(x), \quad \text{for all } x \in X,$$

where the operations $+$ and $\cdot$ on the right side are those in R. Then, $(R^X, +, \cdot)$ is a ring. The operations $+$ and $\cdot$ on R^X defined above are called *point-wise addition* and *point-wise multiplication*. The constant map which maps each element of X onto the zero element in R will be the zero element in R^X and the additive inverse of f is defined by $(-f)(x) = -f(x)$ for all $x \in X$.

8. Let n be any positive integer and consider the group $(\mathbb{Z}_n, +_n)$, where

$$\mathbb{Z}_n = \{0, 1, 2, \ldots, n - 1\}$$

and $+_n$ is the addition modulo n. Recall that $+_n$ is defined on $\mathbb{Z}_n$ by

$$a +_n b = \begin{cases} a + b & \text{if } a + b < n \\ a + b - n & \text{if } a + b \geq n \end{cases},$$

for any a and $b \in \mathbb{Z}_n$. Note that $a +_n b$ is precisely the remainder obtained by dividing the usual sum $a + b$ by n. Now, extend this to the multiplication also, by defining

$$a \cdot_n b = r, \quad \text{where } 0 \leq r < n, ab = qn + r, q \quad \text{and} \quad r \in \mathbb{Z},$$

for any a and $b \in \mathbb{Z}_n$. Note that $a \cdot_n b$ is precisely the remainder obtained by dividing the usual product ab by n. This operation $\cdot_n$ is called the *multiplication modulo n*. $\mathbb{Z}_n$ is a finite set with n elements and the addition $+_n$ and multiplication $\cdot_n$ modulo n are given in the following table for $n = 6$.

$+_6$	0	1	2	3	4	5
0	0	1	2	3	4	5
1	1	2	3	4	5	0
2	2	3	4	5	0	1
3	3	4	5	0	1	2
4	4	5	0	1	2	3
5	5	0	1	2	3	4

$(\mathbb{Z}_6, +_6)$

$\cdot_6$	0	1	2	3	4	5
0	0	0	0	0	0	0
1	0	1	2	3	4	5
2	0	2	4	0	2	4
3	0	3	0	3	0	3
4	0	4	2	0	4	2
5	0	5	4	3	2	1

$(\mathbb{Z}_6, \cdot_6)$

It can be proved that $(\mathbb{Z}_n, +_n, \cdot_n)$ is a ring and is called the ring of integers modulo n.

9. Let R be a set consisting of only one element, say $R = \{a\}$. Then, the only way of defining binary operation on R is $a + a = a$ and $a \cdot a = a$ and $(R, +, \cdot)$ is a ring in which a itself is the zero element and $-a = a$. This ring is called the *trivial ring*. When we say that R is a nontrivial ring, it means that R contains atleast two elements.
10. Let X be any set and $\mathbb{P}(X)$ be the set of all subsets of X. For any A and B in $\mathbb{P}(X)$, define

$$A + B = (A - B) \cup (B - A)$$
$$\text{and} \quad A \cdot B = A \cap B.$$

Then, we shall prove that $(\mathbb{P}(X), +, \cdot)$ is a ring. Recall that the empty set $\varnothing$ is the zero element and that $(\mathbb{P}(X), +)$ is an abelian group. Clearly $\cdot$ is associative. Also, for any A, B and C in $\mathbb{P}(X)$, we have

$$\begin{aligned} A \cdot (B + C) &= A \cap ((B - C) \cup (C - B)) \\ &= (A \cap (B - C)) \cup (A \cap (C - B)) \\ &= ((A \cap B) - (A \cap C)) \cup ((A \cap C) - (A \cap B)) \\ &= (A \cap B) + (A \cap C) \\ &= (A \cdot B) + (A \cdot C) \end{aligned}$$

Since $\cap$ is a commutative operation, there is no need to verify that $\cdot$ is right distributive over $+$. Thus, $(\mathbb{P}(X), +, \cdot)$ is a ring.

In the following, we prove certain important elementary properties of rings.

Theorem 9.1.1. Let $(R, +, \cdot)$ be a ring. Then, the following holds good for any elements a, b and c in R.

1. $0a = 0 = a0$, where 0 is the zero element in R.
2. $a(-b) = -(ab) = (-a)b$
3. $(-a)(-b) = ab$
4. $a(b - c) = ab - ac$
5. $(a - b)c = ac - bc$.

Proof:

1. We have

$$0 + 0a = 0a = (0 + 0)a = 0a + 0a$$

and, by the cancellation law in the group $(R, +)$, it follows that $0 = 0a$. Also,

$$0 + a0 = a0 = a(0 + 0) = a0 + a0$$

and hence $0 = a0$.

2. Consider

$$ab + a(-b) = a(b + (-b)) = a0 = 0 \qquad \text{(by (1))}$$

and hence $a(-b)$ is the additive inverse of ab. That is, $-(ab) = a(-b)$. Similarly

$$ab + (-a)b = (a + (-a))b = 0b = 0 \qquad \text{(by (1))}$$

and therefore, $(-a)b = -(ab)$.

3. We have $(-a)(-b) = -((-a)(b)) = -(-(ab)) = ab$
4. $a(b - c) = a(b + (-c)) = ab + a(-c) = ab - ac$
5. $(a - b)c = (a + (-b))c = ac + (-b)c = ac - bc$. ◀

Definition 9.1.2. A ring $(R, +, \cdot)$ is said to be *commutative* if the multiplication $\cdot$ is a commutative operation; that is,

$$a \cdot b = b \cdot a \quad \text{for all } a \text{ and } b \in R.$$

Note that the additive operation $+$ is always commutative in any ring R and therefore, by a commutative ring R, we only mean that the multiplication is commutative.

Definition 9.1.3. A ring $(R, +, \cdot)$ is said to be a *ring with unity* or a *ring with identity* if there exists an element e in R such that

$$a \cdot e = a = e \cdot a \quad \text{for all } a \in R.$$

The element e, if exists, is unique.

Note that R always has the additive identity, namely the zero element 0. By a ring with unity, we only mean that R has multiplicative identity also. The multiplicative identity, if exists, is usually denoted by 1, with due respect to the convention that the multiplicative identity in the real number system $\mathbb{R}$ is 1, and is called the *unity* or *identity* in the ring.

Theorem 9.1.2. Let $(R, +, \cdot)$ be a ring with unity. Then, R is trivial if and only if $0 = 1$ in R (that is, the additive identity coincides with the multiplicative identity in the ring R).

Proof: If $0 = 1$, then, for any $a \in R$,

$$a = 1a = 0a = 0$$

and hence $R = \{0\}$. The converse is trivial. ◀

Example 9.1.2

1. Each of the rings $(\mathbb{Z}, +, \cdot)$, $(\mathbb{Q}, +, \cdot)$, $(\mathbb{R}, +, \cdot)$, $(\mathbb{C}, +, \cdot)$, $(\mathbb{Z}_n, +_n, \cdot_n)$ for any $n \in \mathbb{Z}^+$ and $(\mathbb{P}(X), +, \cap)$ is a commutative ring with unity. X is the unity element in $(\mathbb{P}(X), +, \cap)$, while 1 is the unity element in all these other rings.
2. Any zero ring (see Example 9.1.1 (2)) is commutative and has no unity element, unless it is trivial.
3. The ring $M_2(\mathbb{R})$ of 2×2 matrices is with unity, where $\begin{pmatrix} 1 & 0 \\ 0 & 1 \end{pmatrix}$ is the unity. However $M_2(\mathbb{R})$ is not commutative, for consider $A = \begin{pmatrix} 0 & 2 \\ 0 & 1 \end{pmatrix}$ and $B = \begin{pmatrix} 1 & 0 \\ 2 & 0 \end{pmatrix}$.

Then, $A \cdot B = \begin{pmatrix} 0 & 2 \\ 0 & 1 \end{pmatrix} \cdot \begin{pmatrix} 1 & 0 \\ 2 & 0 \end{pmatrix} = \begin{pmatrix} 4 & 0 \\ 2 & 0 \end{pmatrix}$

and $B \cdot A = \begin{pmatrix} 1 & 0 \\ 2 & 0 \end{pmatrix} \cdot \begin{pmatrix} 0 & 2 \\ 0 & 1 \end{pmatrix} = \begin{pmatrix} 0 & 2 \\ 0 & 4 \end{pmatrix}$

and hence $A \cdot B \neq B \cdot A$.

4. The ring $R = R_1 \times R_2 \times \cdots \times R_n$ given in Example 9.1.1 (6) is commutative if and only if each R_i is commutative and R has unity if and only if each R_i is so.
5. In Example 9.1.1 (7) also, R^X is commutative if and only if R is commutative. Further R^X has unity if and only if R is so.

Worked Exercise 9.1.1. Prove that $(\mathbb{Z}_n, +_n, \cdot_n)$ is a commutative ring with unity for any positive integer n, where $+_n$ and $\cdot_n$ are addition and multiplication modulo n, respectively.

Answer: Recall that $\mathbb{Z}_n = \{0, 1, 2, \ldots, n - 1\}$ and that we have already proved in group theory that $(\mathbb{Z}_n, +_n)$ is an abelian group.
Let a, b and $c \in \mathbb{Z}_n$. Suppose that

$$a \cdot_n b = r, \quad r \cdot_n c = s$$
$$b \cdot_n c = t \quad \text{and} \quad a \cdot_n t = u.$$

Then, $ab = qn + r$, $rc = q_1 n + s$

$$bc = pn + t \quad \text{and} \quad at = p_1 n + u,$$

where q, q_1, p and $p_1 \in \mathbb{Z}$ and r, s, t and $u \in \{0, 1, 2, \ldots, n - 1\}$.
Since the usual multiplication of integers is associative, we have $(ab)c = a(bc)$ and therefore

$$(qn + r)c = a(pn + t)$$
$$\therefore \quad qnc + q_1 n + s = apn + p_1 n + u$$
$$\text{i.e.,} \quad (qc + q_1)n + s = (ap + p_1)n + u.$$

By the uniqueness of the quotient and the remainder in the division algorithm, it follows that

$$qc + q_1 = ap + p_1 \quad \text{and} \quad s = u.$$

In particular, $(a \cdot_n b) \cdot_n c = r \cdot_n c = s = u = a \cdot_n t = a \cdot_n (b \cdot_n c)$.

Thus, $\cdot_n$ is associative. Also, since the usual multiplication is commutative, $\cdot_n$ is also commutative.
To prove the distributivity of $\cdot_n$, let a, b and $c \in \mathbb{Z}_n$

$$\begin{aligned} \text{and} \quad b +_n c = x,\ a \cdot_n (b +_n c) &= y \\ a \cdot_n b = z \quad \text{and} \quad a \cdot_n c &= v \\ \text{so that} \quad b + c = qn + x,\ ax &= q'n + y, \\ ab = pn + z \quad \text{and} \quad ac &= p'n + v, \end{aligned}$$

where q, q', p and $p' \in \mathbb{Z}$ and x, y, z and $v \in \{0, 1, 2, \ldots, n-1\}$.
Let $z + v = jn + t,\ 0 \leq t < n$,
so that $z +_n v = t$.
Now, by the distributivity of the usual multiplication over the usual addition in $\mathbb{Z}$, we have

$$a(b + c) = ab + ac$$

and hence $a(qn + x) = (pn + z) + (p'n + v)$.
Therefore, $aqn + q'n + y = (p + p')n + jn + t$

$$\therefore (aq + q')n + y = (p + p' + j)n + t$$

and hence $y = t$, so that

$$a \cdot_n (b +_n c) = y = t = z +_n v = (a \cdot_n b) +_n (a \cdot_n c).$$

Thus, $\cdot_n$ is distributive over $+_n$. Therefore, $(\mathbb{Z}_n, +_n, \cdot_n)$ is a commutative ring. If $n = 1$, then $\mathbb{Z}_n$ is trivial. If $n > 1$, then 1 is the unit element in $\mathbb{Z}_n$. Thus, $(\mathbb{Z}_n, +_n, \cdot_n)$ is a commutative ring with unity.

Worked Exercise 9.1.2. Prove that, for any set X, $(\mathbb{P}(X), +, \cap)$ is a commutative ring with unity.

Answer: We have already proved that $(\mathbb{P}(X), +)$ is an abelian group, where $+$ is the symmetric difference operation. For any subsets A, B and C of X, we have $(A \cap B) \cap C = A \cap (B \cap C)$ and $A \cap B = B \cap A$ and hence $(\mathbb{P}(X), \cap)$ is a semigroup. In Example 9.1.1 (10), we have proved the distributivity of $\cap$ over $+$. Thus, $(\mathbb{P}(X), +, \cap)$ is a commutative ring. Also, since

$$X \cap A = A \quad \text{for all } A \in \mathbb{P}(X),$$

X is the unity element in $\mathbb{P}(X)$. Thus, $(\mathbb{P}(X), +, \cap)$ is a commutative ring with unity.

Worked Exercise 9.1.3. Let $\mathbb{Z}[i] = \{a + ib : a \text{ and } b \in \mathbb{Z}\}$. Then, prove that $(\mathbb{Z}[i], +, \cdot)$ is a commutative ring with unity, where $+$ and $\cdot$ are the usual addition and multiplication of complex numbers, defined in Example 9.1.1 (4).

Answer: Clearly, $(\mathbb{Z}[i], +)$ is an abelian group. Let x, y and $z \in \mathbb{Z}[i]$. Then

$$x = a + ib, y = c + id \quad \text{and} \quad z = r + is$$

where a, b, c, d, r and $s \in \mathbb{Z}$. Then,

$$\begin{aligned} x \cdot (y \cdot z) &= (a + ib)\,((cr - ds) + i(cs + dr)) \\ &= (a(cr - ds) - b(cs + dr)) + i(a(cs + dr) + b(cr - ds)) \\ (x \cdot y) \cdot z &= ((ac - bd) + i(ad + bc)) \cdot (r + is) \\ &= ((ac - bd)r - (ad + bc)s) + i((ac - bd)s + (ad + bc)r) \\ &= x \cdot (y \cdot z) \end{aligned}$$

Therefore, $(\mathbb{Z}[i], \cdot)$ is a semigroup, clearly $\cdot$ is commutative.

$$\begin{aligned} \text{Also, } x \cdot (y + z) &= (a + ib) \cdot ((c + r) + i(d + s)) \\ &= (a(c + r) - b(d + s)) + i(b(c + r) + a(d + s)) \\ &= [(ac - bd) + i(ad + bc)] + [(ar - bs) + i(as + br)] \\ &= x \cdot y + x \cdot z \end{aligned}$$

Thus, $\cdot$ distributes over $+$. Also, $1 (= 1 + i0)$ is the unity in $\mathbb{Z}[i]$. Thus, $(\mathbb{Z}[i], +, \cdot)$ is a commutative ring with unity. $\mathbb{Z}[i]$ is called the *ring of Gaussian integers*.

Recall the following from group theory.

Definition 9.1.4. Let $(R, +, \cdot)$ be a ring. Then, for any $a \in R$ and $n \in \mathbb{Z}$, we define na inductively as follows.

$$na = \begin{cases} 0 & \text{if } n = 0 \\ (n-1)a + a & \text{if } n > 0. \\ (-n)(-a) & \text{if } n < 0 \end{cases}$$

That is, $0a = 0$, $1a = a$, $2a = a + a$, $3a = a + a + a$, $(-2)a = 2(-a) = (-a) + (-a)$, etc.

Definition 9.1.5. For any element a in a ring R and positive integer n, we define

$$a^n = \begin{cases} a & \text{if } n=1 \\ a^{n-1} \cdot a & \text{if } n>1 \end{cases}.$$

That is $a^2 = a \cdot a$, $a^3 = a \cdot a \cdot a$, etc. If the ring R has unity 1, then we define $a^0 = 1$. If the element a is multiplicatively invertible in R and a' is its inverse in R, then we define $a^{-n} = (a')^n$ for all $n > 0$.

EXERCISES 9(A)

1. Which of the following are rings· Substantiate your answers (here + and · are usual addition and multiplication of numbers).
   - (i) $(\mathbb{Z}^+, +, \cdot)$
   - (ii) $(6\mathbb{Z}, +, \cdot)$
   - (iii) $(E, +, \cdot)$, where E is the set of even integers.
   - (iv) $(O, +, \cdot)$, where O is the set of odd integers.
   - (v) $(\mathbb{P}(X), \cap, \cup)$, where $\mathbb{P}(X)$ is the power set of a set X.
   - (vi) $(\mathbb{P}(X), \cup, \cap)$
   - (vii) $(\mathbb{P}(X), +, \cup)$
   - (viii) $(\mathbb{Z}[\sqrt{2}], +, \cdot)$, where $\mathbb{Z}[\sqrt{2}] = \{a + b\sqrt{2} : a, b \in \mathbb{Z}\}$.
   - (ix) $(\mathbb{Q}[\sqrt{2}], +, \cdot)$, where $\mathbb{Q}[\sqrt{2}] = \{a + b\sqrt{2} : a$ and b are rational numbers$\}$.
   - (x) $(\mathbb{Z}_n, \cdot_n, +_n)$
   - (xi) $(\mathbb{Q} - \{0\}, +, \cdot)$
   - (xii) $(\mathbb{R} - \mathbb{Q}, +, \cdot)$
2. Compute the following in the given rings.
   - (i) $13 \cdot_{15} 8$ in $(\mathbb{Z}_{15}, +_{15}, \cdot_{15})$
   - (ii) $7 \cdot_{10} 9$ in $(\mathbb{Z}_{10}, +_{10}, \cdot_{10})$
   - (iii) $A \cdot A \cdot A$ in $(M_2(\mathbb{R}), +, \cdot)$ where $A = \begin{pmatrix} 1 & 1 \\ 1 & 1 \end{pmatrix}$.
   - (iv) $34 \cdot_{100} 67$ in $(\mathbb{Z}_{100}, +_{100}, \cdot_{100})$
   - (v) $8 +_{10} (9 \cdot_{10} 6) + (7 \cdot_{10} 4)$ in $(\mathbb{Z}_{10}, +_{10}, \cdot_{10})$.
   - (vi) $(A + B) \cap C$ in $(\mathbb{P}(X), +, \cap)$, where $A = \{2, 3, 4, 5\}$, $B = \{3, 5, 6, 7\}$ and $C = \{1, 2, 3\}$.
3. Prove that a ring $(R, +, \cdot)$ is commutative if and only if $(a + b)(a - b) = a^2 - b^2$ for all a and b in R.

4. Let $(R, +, \cdot)$ be a ring. Prove that the following are equivalent to each other.
   (i) $+$ distributes over $\cdot$.
   (ii) R is trivial; that is, $R = \{0\}$.
   (iii) $(R, \cdot, +)$ is a ring.
   (iv) $a + b = ab$ for all a and b in R.

5. Let a and b be two elements of a ring such that $a \cdot b = b \cdot a$. Prove the following for any positive integer n.

$$(a + b)^n = a^n + nc_1\, a^{n-1}b + nc_2\, a^{n-2}b^2 + \cdots + nc_{n-1}\, ab^{n-1} + b^n$$
$$= \sum_{r=0}^{n} nc_r a^{n-r} b^r, \quad \text{where } nc_r = \frac{n!}{r!(n-r)!}.$$

6. For any prime p and a and $b \in \mathbb{Z}_p$, prove that $(a + b)^p = a^p + b^p$.

7. Prove that the commutativity of the operation $+$ in a ring $(R, +, \cdot)$ is a consequence of the other axioms of a ring.

8. Let $(R, +, \cdot)$ be a ring with unity 1. Define new operations $\oplus$ and $\odot$ on R as follows for any a and $b \in R$.

$$a \oplus b = a + b + 1$$
$$a \odot b = a \cdot b + a + b.$$

   Prove that $(R, \oplus, \odot)$ is a ring with unity and that $(R, +, \cdot)$ is commutative if and only if $(R, \oplus, \odot)$ is commutative.

9. Let $(R, +, \cdot)$ be a ring such that $(a^2 + a)x = x(a^2 + a)$ for all a and $x \in R$. Then prove that $(R, +, \cdot)$ is a commutative ring.

10. Let $(R, +, \cdot)$ be a ring such that, for any a, b and $c \in R$,

$$ab = ca \Rightarrow a = 0 \quad \text{or} \quad b = c.$$

   Then prove that $(R, +, \cdot)$ is a commutative ring.

11. Prove that a ring $(R, +, \cdot)$ is commutative if the group $(R, +)$ is cyclic.

12. Let $(R, +, \cdot)$ be a ring and n be an integer such that $n > 1$ and $x^n = x$ for all $x \in R$. Then prove that, for any a and $b \in R$,

$$ab = 0 \Leftrightarrow ba = 0.$$

13. Prove that the set $\{0, 2, 4\}$ is a commutative ring with unity with respect to addition and multiplication modulo 6.

14. Give an example of a finite noncommutative ring.

9.2 CERTAIN SPECIAL ELEMENTS IN RINGS

In any ring, we have the additive identity 0 and, in certain rings, there is unity which is the multiplicative identity. In this section, we shall introduce certain other special elements in a ring and discuss their properties.

Definition 9.2.1. An element a in a ring $(R, +, \cdot)$ is said to be an *idempotent* if $a \cdot a = a$.

Example 9.2.1

1. The zero element 0 and the unity, if it exists, in any ring are idempotents.
2. 3 and 4 are idempotents in $(\mathbb{Z}_6, +_6, \cdot_6)$, since $3 \cdot_6 3 = 3$ and $4 \cdot_6 4 = 4$; 5 is not an idempotent, since $5 \cdot_6 5 = 1$ in $\mathbb{Z}_6$.
3. In the ring $\mathbb{Z}$ of integers, 0 and 1 are the only idempotents.

Definition 9.2.2. A ring $(R, +, \cdot)$ in which every element is an idempotent is called a *Boolean ring*.

Example 9.2.2

1. $(\mathbb{Z}_2, +_2, \cdot_2)$, the ring of integers modulo 2 is a Boolean ring.
2. For any set X, $(\mathbb{P}(X), +, \cap)$ is a Boolean ring, since $A \cap A = A$ for all $A \subseteq X$.
3. For any set X, the set $\mathbb{Z}_2^X$ of all mappings of X into $\mathbb{Z}_2$ is a Boolean ring under the point-wise operations (see Example 9.1.1 (7)), since $\mathbb{Z}_2$ is a Boolean ring.

Note that examples (2) and (3) given in Example 9.2.2 are not different. They appear to be the same in the sense given below.

Theorem 9.2.1. Let X be any set. For any $A \subseteq X$, define

$$\chi_A : X \to \mathbb{Z}_2 \quad \text{by} \quad \chi_A(x) = \begin{cases} 1 & \text{if } x \in A \\ 0 & \text{if } x \notin A \end{cases}.$$

Then, $A \mapsto \chi_A$ is a bijection of $\mathbb{P}(X)$ onto $\mathbb{Z}_2^X$ such that, for any A and B in $\mathbb{P}(X)$

$$\chi_{A+B} = \chi_A +_2 \chi_B \quad \text{and} \quad \chi_{A \cap B} = \chi_A \cdot_2 \chi_B.$$

Proof: If A and $B \in \mathbb{P}(X)$ and $a \in A - B$, then

$$\chi_A(a) = 1 \text{ and } \chi_B(a) = 0$$

and hence $\chi_A \neq \chi_B$ if $A \neq B$. Thus, $A \mapsto \chi_A$ is an injection. Further, if $f \in \mathbb{Z}_2^X$ and $A = f^{-1}(\{1\})$, then $\chi_A = f$. Thus, $A \mapsto \chi_A$ is a bijection of $\mathbb{P}(X)$ onto $\mathbb{Z}_2^X$. The other assertions follow from the definitions of the operations $+$ on $\mathbb{P}(X)$ and $\mathbb{Z}_2^X$, $\cap$ on $\mathbb{P}(X)$ and the point-wise operation $\cdot_2$ on $\mathbb{Z}_2^X$. ◀

Theorem 9.2.2. For any elements a and b in a Boolean ring $(R, +, \cdot)$,

$$a + a = 0; \quad \text{that is, } a = -a$$
$$\text{and} \quad ab = ba$$

and hence every Boolean ring is commutative.

Proof: Let $(R, +, \cdot)$ be a Boolean ring and a and $b \in R$. Then, consider

$$\begin{aligned} a + b = (a + b)^2 &= (a + b)\,(a + b) \\ &= a^2 + ab + ba + b^2 \\ &= a + ab + ba + b. \end{aligned}$$

From the cancellation laws in the group $(R, +)$, we have

$$0 = ab + ba.$$

In particular, by taking $a = b$, we have

$$0 = aa + aa = a + a$$

Also, $ab = -(ba) = (-b)a = ba$.
Thus, $(R, +, \cdot)$ is a commutative ring. ◀

Definition 9.2.3. Let $(R, +, \cdot)$ be a ring and $a \in R$. Then, a is called a *nilpotent* if $a^n = 0$ for some positive integer n.

Example 9.2.3

1. The zero element 0 in any ring R is nilpotent, since $0^1 = 0$
2. 6 is a nilpotent element in $\mathbb{Z}_8$, since
$$6^3 = (6 \cdot_8 6) \cdot_8 6 = 4 \cdot_8 6 = 0$$
3. Except 0, no element in the ring of integers is nilpotent.

Definition 9.2.4. A nonzero element a in a ring R is said to be a *zero-divisor* if there exists a nonzero element b such that $ab = 0 = ba$.

Note that, for two elements a and b in a ring R, it is quite possible that $ab = 0$ and $ba \neq 0$. For consider the example (4) in the following example.

Example 9.2.4

1. The ring $(\mathbb{Z}, +, \cdot)$ of integers has no zero divisors, since, for any integers a and b, $ab = 0$ only when $a = 0$ or $b = 0$.
2. 2 and 3 are zero divisors in $(\mathbb{Z}_6, +_6, \cdot_6)$.
3. If X is a set and A is a nonempty proper subset of X, then $A \cap (X - A) = \varnothing$, $A \neq \varnothing$ and $X - A \neq \varnothing$ and hence A is a zero divisor in $(\mathbb{P}(X), +, \cap)$.
4. Consider the ring $M_2(\mathbb{R})$ of 2×2 matrices. Let

$$A = \begin{pmatrix} 1 & 0 \\ 1 & 0 \end{pmatrix} \quad \text{and} \quad B = \begin{pmatrix} 0 & 0 \\ 1 & 1 \end{pmatrix}.$$

Then $$A \cdot B = \begin{pmatrix} 0 & 0 \\ 0 & 0 \end{pmatrix} \quad \text{and} \quad B \cdot A = \begin{pmatrix} 0 & 0 \\ 2 & 0 \end{pmatrix} \neq \begin{pmatrix} 0 & 0 \\ 0 & 0 \end{pmatrix}.$$

However, there is some other C in $M_2(\mathbb{R})$ such that $A \cdot C = 0 = C \cdot A$; for, take $C = \begin{pmatrix} 0 & 0 \\ 1 & -1 \end{pmatrix}$. Then, $AC = 0 = CA$ and therefore A is zero divisor.

Definition 9.2.5. Let $(R, +, \cdot)$ be a ring and $a \in R$. If R is a ring with unity and a has multiplicative inverse in R, then a is called a *unit* or *multiplicatively invertible* or, simply, *invertible*. That is, a is a unit in R if there exists $b \in R$ such that $ab = 1 = ba$.

Note that, in any ring $(R, +, \cdot)$, every element a has additive inverse, namely $-a$. However, an element in a ring with unity may not possess multiplicative inverse. Elements possessing multiplicative inverse need special attention. First, Let us consider the following example.

Example 9.2.5

1. In any ring, the unity (if it exists) is a unit.
2. In the ring of integers $\mathbb{Z}$, 1 and -1 are the only units and, for each of them, the multiplicative inverse is itself.
3. In the ring of rational numbers, or in the ring of real numbers, or in the ring of complex numbers, every nonzero element is a unit.
4. The zero element is a unit in a ring R if and only if R is trivial; since $0a = 0 = a0$ for all $a \in R$.

If a is a unit in a ring R, then the element $b \in R$ such that $ab = 1 = ba$ is unique and is denoted by a^{-1} and is called the inverse of a. Note that the additive inverse of a is denoted by $-a$, while the multiplicative inverse (if exists) of a is denoted by a^{-1}.

Theorem 9.2.3. Let $(R, +, \cdot)$ be a ring with unity If a and b are units in R, then so is their product ab and $(ab)^{-1} = b^{-1}a^{-1}$. Also, the set $U(R)$ of all units in R forms a group under multiplication.

Proof: If $aa^{-1} = 1 = a^{-1}a$ and $bb^{-1} = 1 = b^{-1}a$, then $(ab)(b^{-1}a^{-1}) = a(bb^{-1})a^{-1} = a_1a^{-1} = 1$ and $(b^{-1}a^{-1})(ab) = b^{-1}(a^{-1}a)b = b_1{}^{-1}b = 1$ and hence ab is a unit and $(ab)^{-1} = b^{-1}a^{-1}$. Therefore, the multiplication in R, restricted to $U(R)$, is a binary operation on $U(R)$ and is clearly associative. Also, the unity 1 will be the identity in $(U(R), \cdot)$. Further, if a is a unit, then so is its inverse and $(a^{-1})^{-1} = a$. Thus, $(U(R), \cdot)$ is a group. ◀

Worked Exercise 9.2.1. Let $1 < n \in \mathbb{Z}$ and $\mathbb{Z}_n$ be the ring of integers modulo n. For any $a \in \mathbb{Z}_n$, prove that a is a unit in $\mathbb{Z}_n$ if and only if a and n are relatively prime.

Answer: Consider the following

$$\begin{aligned}
(a, n) = 1 &\Leftrightarrow xa + yn = 1 \quad \text{for some integers } x \text{ and } y \\
&\Leftrightarrow ba + zn = 1 \quad \text{for some } b, z \in \mathbb{Z} \text{ with } 0 < b < n \\
&\quad\ \text{(use division algorithm to write } x = qn + b) \\
&\Leftrightarrow b \cdot_n a = 1, b \in \mathbb{Z}_n \\
&\Leftrightarrow a \text{ is a unit in } (\mathbb{Z}_n, +_n, \cdot_r).
\end{aligned}$$

Worked Exercise 9.2.2. For any $n > 1$, prove that any nonzero element in the ring $\mathbb{Z}_n$ is either a unit or a zero divisor.

Answer: Let $0 < a < n$. Suppose that a is not a unit in $\mathbb{Z}_n$. Then, $(a, n) > 1$ and let $d = (a, n)$. Then, d is a common divisor for a and n and hence both $\frac{a}{d}$ and $\frac{n}{d}$ are integers and

$$a\left(\frac{n}{d}\right) = n\left(\frac{a}{d}\right) \equiv_n 0.$$

Therefore, $a \cdot_n \left(\frac{n}{d}\right) = 0$ in $\mathbb{Z}_n$ and $\frac{n}{d}$ is nonzero in $\mathbb{Z}_n$. Therefore, a is a zero divisor in $\mathbb{Z}_n$.

Worked Exercise 9.2.3. Let a and b be elements in a ring R such that $ab = ba$. Prove the following:

1. $a + b$ is a nilpotent if a and b are nilpotents.
2. ab is a nilpotent if a or b is a nilpotent.

Answer:

1. Suppose that a and b are nilpotents. Then, there exist positive integers n and m such that $a^n = 0 = b^m$. Now, since $ab = ba$, we have

$$\begin{aligned}(a + b)^{m+n} &= a^{m+n} + (m + n)c_1 a^{m+n-1}b + \cdots + (m + n)c_{m+n} b^{m+n} \\ &= \sum_{r=0}^{m+n} (m+n)C_r a^{m+n-r} b^r.\end{aligned}$$

since $a^s = 0$ for all $s \geq n$ and $b^t = 0$ for all $t \geq m$ and since, for any $0 \leq r \leq m + n$, either $m + n - r \geq n$ or $r \geq m$ (otherwise $(m + n - r) + r < n + m$, an absurd), we get that $a^{m+n-r} = 0$ or $b^r = 0$ for all $0 \leq r \leq m + n$ and therefore $(a + b)^{m+n} = 0$, so that $a + b$ is a nilpotent.

2. If $a^n = 0$, then $(ab)^n = a^n b^n = 0b^n = 0$. Therefore, if a is a nilpotent, then ab $(= ba)$ is also a nilpotent.

Worked Exercise 9.2.4. Prove that $1, -1, i$ and $-i$ are the only units in the ring $\mathbb{Z}[i]$ of Gaussian integers.

Answer: Let $x = a + ib$ be a unit in $\mathbb{Z}[i]$. Then, there exists $y = c + id$ in $\mathbb{Z}[i]$ such that $xy = 1$. Here a, b, c and d are integers. Then,

$$1 = |xy|^2 = |x|^2|y|^2 = (a^2 + b^2)(c^2 + d^2)$$

Therefore, $a^2 + b^2 = 1 = c^2 + d^2$ (since $a^2 + b^2 \geq 1$). Again, since $a^2 \geq 0$ and $b^2 \geq 0$, we get that

$$\begin{aligned} & a = 1 \text{ or } -1 \quad \text{and} \quad b = 0 \\ \text{(or)} \quad & a = 0 \quad \text{and} \quad b = 1 \text{ or } -1 \\ \text{so that} \quad & x = 1 \text{ or } -1 \quad \text{or} \quad x = i \text{ or } -i. \end{aligned}$$

EXERCISE 9(B)

1. Determine all the zero divisors, nilpotents, idempotents and units in each of the following rings
   (i) The ring $\mathbb{R}$ of real numbers.
   (ii) $\mathbb{Z} \times \mathbb{R}$, where $\mathbb{Z}$ is the ring of integers.

(iii) $\mathbb{Z} \times \mathbb{Z}$.

(iv) The ring $\mathbb{Z}_{24}$ of integers modulo 24.

(v) $\mathbb{Z}_{15}$.

(vi) $\mathbb{Z}_{12} \times \mathbb{Z}$.

(vii) $(\mathbb{P}(X), +, \cap)$, for any set X.

(viii) $\mathbb{Z}_4 \times \mathbb{Z}_9$.

2. In any nontrivial ring with unity, prove that no zero divisor is a unit.

3. Let a be a nonzero element in a commutative ring R. Prove that a is not a zero divisor if and only if a satisfies the following cancellation law for any b and c in R:

$$ab = ac \Rightarrow b = c.$$

4. Let R be a Boolean ring with unity. Prove that the unity is the only unit in R_0 and the zero is the only nilpotent in R.

5. Let n be any integer greater than 1. The content of n is defined to be the product of all distinct primes dividing n and is denoted by $c(n)$. Prove that $a \in \mathbb{Z}_n$ is a nilpotent if and only if $c(n)$ divides a.

6. Using 5 above, derive a formula for the number of nilpotents in the ring $\mathbb{Z}_n$ of integers modulo n.

7. For any integers a and n with $0 \leq a < n$, prove that a is an idempotent in $\mathbb{Z}_n$ if and only if $a(a - 1)$ is a multiple of n.

8. Let $R_1, R_2, \ldots, R_n$ be rings and $R = R_1 \times R_2 \times \cdots \times R_n$. For any $a = (a_1, a_2, \ldots, a_n) \in R$, prove that a is a nilpotent (idempotent) in R if and only if each a_i is a nilpotent (respectively idempotent) in R_i for $1 \leq i \leq n$.

9. In 8 above, when each R_i is a ring with unity, prove that $(a_1, a_2, \ldots, a_n)$ is a unit in $R_1 \times R_2 \times \cdots \times R_n$ if and only if a_i is a unit in R_i for each $1 \leq i \leq n$.

10. If a and b are idempotents in a commutative ring R, prove that ab is also an idempotent. Can $a + b$ be an idempotent? If not, give a counter example.

11. Find all the solutions of $x^2 + 2x + 4 = 0$ in the ring $\mathbb{Z}_6$.

12. Prove that 0 is the only solution of $x^2 = 0$ in a ring R if and only if R has no nonzero nilpotents.

13. For any prime number p, prove that the set of all nonzero elements in $\mathbb{Z}_p$ forms a group under multiplication modulo p.

14. Prove that any nonzero nilpotent in any ring R is a zero divisor in R.

15. Let ϕ be the Euler-totient function. Prove that the multiplicative group of units in $\mathbb{Z}_n$ is of order $\phi(n)$, for any integer $n > 1$.

16. Let R be a ring with unity and $a \in R$ be a nilpotent. Then prove that $1 + a$ is a unit.

17. Let R be a ring with unity and without zero divisors. For any a and b in R, prove that $ab = 1$ if and only if $ba = 1$ and that $a^2 = 1$ if and only if $a = 1$ or -1.

18. Let R be a ring without nonzero nilpotents and a be an idempotent in R. Prove that $ax = xa$ for all $x \in R$.

9.3 THE CHARACTERISTIC OF A RING

It is well known that $na \neq 0$ for any positive integer n and any nonzero integer a. That is, there is no positive integer n such that $na = 0$ for all elements a in the ring $\mathbb{Z}$ of integers. However, there are rings for which there exists a positive integer n such that $na = 0$ for all elements a in the ring. For example, consider a positive integer m and the ring $\mathbb{Z}_m$ of integers modulo m. Then, $ma = 0$ for all $a \in \mathbb{Z}_m$. In fact if n is any positive integral multiple of m, then $na = 0$ for all $a \in \mathbb{Z}_n$. Recall that, for any element a in a ring R, the order of a in the group $(R, +)$ is precisely the smallest positive integer (if exists) n such that $na = 0$. In this section, we discuss the existence of a common positive integer n such that $na = 0$ for all elements a in the ring. First recall that, for any positive integer n and for any element a in a ring R, we have defined na inductively by

$$na = \begin{cases} a & \text{if } n = 1 \\ (n-1)a + a & \text{if } n \neq 1 \end{cases}.$$

Definition 9.3.1. Let $(R, +, \cdot)$ be a ring. If there is no positive integer n such that $na = 0$ for all $a \in R$, then the *characteristic* of R is defined to be zero. Otherwise, the smallest positive integer n such that $na = 0$, for all $a \in R$, is called the *characteristic* of R and is denoted by char(R).

Note that $na = 0$ if n is the integer zero or a is the zero element in the ring R. If char(R) $= 0$, then, for each positive integer n, there exists an element a in the ring R such that $na = 0$. If char(R) $= n > 0$, then $na = 0$ for all $a \in R$ and n is the least such positive integer.

Theorem 9.3.1. Let $(R, +, \cdot)$ be ring and char(R) $= n > 0$. Then, for any integer m,

$$ma = 0 \quad \text{for all } a \in R \quad \text{if and only if } n \text{ divides } m.$$

Proof: Let m be any integer. If n divides m, then $m = nr$ for some integer r and hence, for any $a \in R$,

$$ma = (nr)a = r(na) = r0 = 0.$$

Conversely suppose that $ma = 0$ for all $a \in R$. By the division algorithm, we can express m as

$$m = qn + r, \quad \text{where } q \text{ and } r \in \mathbb{Z} \text{ and } 0 \leq r < n.$$

Then, for any $a \in R$, we have

$$0 = ma = (qn + r)a = q(na) + ra = 0 + ra = ra.$$

Since n is the least positive integer such that $na = 0$ for all $a \in R$ and since $r < n$, it follows that r cannot be positive. Since $0 \leq r$, we get that $r = 0$ and hence $m = qn$. Thus, n divides m. ◀

Theorem 9.3.2. Let $(R, +, \cdot)$ be a ring with unity 1. Then, the characteristic of R is precisely the order of the unity in the group $(R, +)$.

Proof: This follows from the fact that, for any $a \in R$ and $n \in \mathbb{Z}$,

$$na = n(1a) = (n1)a$$

and that $na = 0$ for all $a \in R$ if and only if $n1 = 0$. ◀

Example 9.3.1

1. The characteristic of each of the rings $\mathbb{Z}$, $\mathbb{Q}$, $\mathbb{R}$ and $\mathbb{C}$ is zero, since for any integer $n > 0$ and for any nonzero real or complex number a, $na \neq 0$.
2. $\text{char}(\mathbb{Z}_n) = n$ for any positive integer n, where $\mathbb{Z}_n$ is the ring of integers modulo n.
3. $\text{char}(\mathbb{Z}_n \times \mathbb{Z}_m)$ is the least common multiple of m and n for positive integers m and n.

Worked Exercise 9.3.1. Let R and S be rings of characteristic m and n, respectively. Then prove that the characteristic of the product ring $R \times S$ is the least common multiple of m and n.

Answer: We have $\text{char}(R) = m$ and $\text{char}(S) = n$. First let us assume that $m > 0$ and $n > 0$. Let r be the least common multiple of m and n and

$$r = ms \quad \text{and} \quad r = nt$$

for some positive integers s and t. Then, for any element (a, b) in $R \times S$, we have

$$\begin{aligned} r(a, b) &= (ra, rb) \\ &= ((ms)a, (nt)b) \\ &= (s(ma), t(nb)) \\ &= (s0, t0) = (0, 0) \end{aligned}$$

Therefore, $0 < \text{char}(R \times S) \leq r$. Put $\text{char}(R \times S) = k$.
Then, for any $a \in R$,

$$(ka, 0) = k(a, 0) = (0, 0) \quad \text{and hence } ka = 0.$$

By Theorem 9.3.1, char(R) divides k. Similarly, we can prove that char(S) divides k. Therefore, k is a common multiple of m and n and hence $r \leq k$. Thus,

$$\text{char}(R \times S) = r = \text{l.c.m.}\{m, n\}.$$

On the other hand, if $\text{char}(R \times S) = p > 0$, then

$$(pa, pb) = p(a, b) = (0, 0)$$

and hence $pa = 0$ and $pb = 0$ for all $a \in R$ and $b \in S$ so that $\text{char}(R) > 0$ and $\text{char}(S) > 0$.

Worked Exercise 9.3.2. Determine the characteristic of $\mathbb{Z}_{12} \times \mathbb{Z}$.

Answer: If n is any positive integer, then

$$n(0, a) = (n0, na) = (0, na) \neq (0, 0)$$

for any $a \in \mathbb{Z}$ and $(0, a) \in \mathbb{Z}_{12} \times \mathbb{Z}$. Therefore, $\text{char}(\mathbb{Z}_{12} \times \mathbb{Z}) = 0$.

EXERCISE 9(C)

1. Find the characteristic of each of the following rings.
   (i) The ring E of even integers
   (ii) $5\mathbb{Z}$
   (iii) $\mathbb{Z}_{12} \times \mathbb{Z}_{18}$
   (iv) $\mathbb{Z} \times \mathbb{Z}_{24}$
   (v) $(\mathbb{P}(X), +, \cap)$ for any set X
   (vi) $\mathbb{Z}_{30}^{X}$ under the point-wise operations, where X is any set.

(vii) The trivial ring $\mathbb{R} = \{0\}$.

(viii) The ring $\mathbb{Z}[i]$ of Gaussian integers.

(ix) $\mathbb{Z}^X$ under the point-wise operations, where X is any set.

(x) $\mathbb{Z}_6 \times \mathbb{Z}_{20} \times \mathbb{Z}_{14}$

2. Let R be a commutative ring with unity and char(R) $= 3$. For any a and $b \in R$, compute the following and simplify.

   (i) $(a + b)^9$

   (ii) $(a + b)^3$

   (iii) $(a + b)^6$

   (iv) $(a + b)^{12}$

3. Prove that a ring R is trivial if and only if char(R) $= 1$.

4. Prove that the characteristic of any finite ring is positive.

5. Let R be a commutative ring of characteristic 2 and $E(R)$ be the set of idempotents in R. Prove that $E(R)$ is a ring under the operations on R.

6. Give an example of a ring R of characteristic 5 such that every nonzero element in R is a unit.

7. Let R be a commutative ring with unity in which each nonzero element is a unit. If char(R) $= 2$ and R has atleast three elements, then prove that there exist elements a and b in R such that

$$(a + b)^3 \neq a^3 + b^3.$$

8. In Exercise 7 above, suppose that char(R) is a prime number p and

$$A = \{a \in R : a^p = a\}.$$

   Then prove that A is a ring under the operations on R and that every nonzero element in A is a unit in A.

9. Let R be a ring with identity and char(R) $= n > 0$. If n is not prime, prove that R has zero divisors.

10. Let R be a finite ring and $R = \{a_1, a_2, \ldots, a_r\}$. Let $O(a_i)$ be the order of a_i in the group $(R, +)$. Prove that char(R) is the least common multiple of $O(a_1), O(a_2), \ldots, O(a_r)$.

11. Prove the characteristic of a finite ring R that divides $|R|$.

12. Let R be a finite ring with unity and a and $b \in R$. Prove that $ab = 1$ if and only if $ba = 1$.

9.4 SUBRINGS

In this section, we deal with the situation where a subset of a ring constitutes a ring again. Recall the set $\mathbb{Z}$ of integers is a subset of the ring $(\mathbb{R}, +, \cdot)$ of real numbers and $\mathbb{Z}$ itself is a ring under the addition and multiplication of real numbers. This is abstracted in the following definition.

Definition 9.4.1. Let $(R, +, \cdot)$ be a ring. A nonempty subset S of R is called a *subring* of R if S is itself is a ring under the operations $+$ and $\cdot$ on R restricted to S.

In other words, S is a subring of R if S is a subgroup of $(R, +)$ and S is a subsemigroup of $(R, \cdot)$ (that is, $ab \in S$ whenever a and $b \in S$). The reason for this is that the distributive laws and the associativity of the multiplication $\cdot$ hold automatically for the elements of S as a consequence of their validity in the ring R. The following is a simpler characterization of a subring and whose proof is trivial.

Theorem 9.4.1. Let S be a nonempty subset of a ring R. Then, S is a subring of R if and only if

$$a \text{ and } b \in S \Rightarrow a - b \in S \quad \text{and} \quad ab \in S.$$

Clearly $\{0\}$ and the whole of R are subrings of any ring R and are called *trivial subrings* and all other subrings (if they exist) are called *nontrivial subrings*. A subring S of R is called a *proper subring* if $S \neq R$.

Example 9.4.1

1. $\mathbb{Z}$ is a subring of the ring $(\mathbb{Q}, +, \cdot)$ of rational numbers, $\mathbb{Q}$ is a subring of the ring $(\mathbb{R}, +, \cdot)$ of real numbers and $\mathbb{R}$ is a subring of the ring $(\mathbb{C}, +, \cdot)$ of complex numbers.
2. Let Y be a subset of a set X. Then, $\mathbb{P}(Y)$, the power set of Y, is a subring of $(\mathbb{P}(X), +, \cap)$.
3. For any nonnegative integer n, the set $n\mathbb{Z}$ of all integral multiples of n, is a subring of the ring $(\mathbb{Z}, +, \cdot)$ of integers. In particular, the set E of even integers is a subring of $(\mathbb{Z}, +, \cdot)$.

Note that a ring R may possess the unity (multiplicative identity) while a subring may not possess and, even when a subring possesses unity then it may be different from that of R. Consider the following examples.

Example 9.4.2

1. The set E of even integers is a subring of the $(\mathbb{Z}, +, \cdot)$ of integers. $\mathbb{Z}$ has unity, while E has no unity (there is no even integer e such that $ea = a$ for all even integers).
2. Let X be a set and Y be a proper subset of X. Then, $\mathbb{P}(Y)$ is a subring of $(\mathbb{P}(X), +, \cap)$. X and Y are unit elements in $\mathbb{P}(X)$ and $\mathbb{P}(Y)$, respectively and $Y \neq X$.

The following is a routine verification. Observe that any subring S of a ring R contains the zero element 0, since S is a subgroup of $(R, +)$.

Theorem 9.4.2. The intersection of any class of subrings of a ring R is again a subring of R.

Worked Exercise 9.4.1. Let $(R, +, \cdot)$ be a ring and define

$$C(R) = \{a \in R : ax = xa \quad \text{for all } x \in R\}.$$

Then prove that $C(R)$ is a subring of R. $C(R)$ is called the *centre* of R.

Answer: Since $0x = 0 = x0$ for all $x \in R$, $0 \in C(R)$ and therefore $C(R)$ is a nonempty subset of R. For any a and $b \in C(R)$ and $x \in R$, we have

$$(a-b)x = ax - bx = xa-xb = x(a-b)$$
$$\text{and} \quad (ab)x = a(bx) = a(xb) = (ax)b = (xa)b = x(ab)$$

and hence $a - b \in C(R)$ and $ab \in C(R)$. Thus, $C(R)$ is a subring of R.

Worked Exercise 9.4.2. Let S be a subring of a ring R and $\text{char}(R) > 0$. Then prove that $\text{char}(S)$ is a positive divisor of $\text{char}(R)$.

Answer: Let $\text{char}(R) = n$, we are given that $n > 0$ and $na = 0$ for all $a \in R$ and, in particular, $na = 0$ for all $a \in S$. Therefore, $\text{char}(S) > 0$ and, by Theorem 9.3.1, $\text{char}(S)$ divides n.

Worked Exercise 9.4.3. Let R be a ring For any $a \in R$, let

$$C(a) = \{x \in R : ax = xa\}.$$

Prove that $C(a)$ is a subring of R and that $C(R) = \bigcap_{a\in R} C(a)$.

Answer: Clearly a and $0 \in C(a)$ and hence $C(a)$ is a nonempty subset of R. For any x and $y \in C(a)$, we have

$$a(x - y) = ax - ay = xa - ya = (x - y)a$$
$$\text{and} \quad a(xy) = (ax)y = (xa)y = x(ay) = x(ya) = (xy)a$$

and there $x - y \in C(a)$ and $xy \in C(a)$. Thus, $C(a)$ is a subring of R. Also,

$$x \in C(R) \Leftrightarrow xa = ax \quad \text{for all } a \in R$$
$$\Leftrightarrow x \in C(a) \quad \text{for all } a \in R.$$

Therefore, $C(R) = \bigcap_{a\in R} C(a)$.

EXERCISE 9(D)

1. Determine all the subrings of each of the following rings.
   (i) $(\mathbb{Z}_2, +_2, \cdot_2)$
   (ii) $(\mathbb{Z}_7, +_7, \cdot_7)$
   (iii) $(\mathbb{Z}_{10}, +_{10}, \cdot_{10})$
   (iv) $(\mathbb{Z}, +, \cdot)$
2. If S is a subring of a ring R and T is a subring of S, then prove that T is a subring of R.
3. Prove that S is a subring of the ring $\mathbb{Z}$ of integers if and only if $S = n\mathbb{Z}$ for some nonnegative integer n.
4. Let n be a positive integer and $\mathbb{Z}_n$ be the ring of integers modulo n. Prove that S is a subring of $\mathbb{Z}_n$ if and only if $S = \{0, m, 2m, \ldots, (r-1)m\}$ for some divisor m of n and $rm = n$, $r > 0$.
5. Let X be a subset of a ring R and (X) be the intersection of all subrings of R containing X. Prove that (X) is the smallest subring of R containing X. (X) is called the *subring of R generated by X*.
6. In Exercise 5 above, describe all the elements of (X).
7. Let S be a subring of a ring with unity 1 and $1 \in S$. If $a \in C(R)$, prove that
$$(S \cup \{a\}) = \{s_0 + s_1 a + \cdots + s_n a^n : n \geq 0 \text{ and } s_i \in S\}.$$
8. Give an example of a ring R without unity and of a subring S of R such that S is with unity.
9. Let S be a nontrivial subring of a ring R and $1'$ be the unity in S such that $1'$ is not the unity in R. Then prove that $1'$ is a zero divisor in R.
10. Let S and T be subrings of a ring R. Then prove that $S \cup T$ is a subring of R if and only if either $S \subseteq T$ or $T \subseteq S$.
11. Let $\mathscr{C}$ be a class of subrings of a ring R such that, for any S_1 and $S_2 \in \mathscr{C}$, there exist $S_3 \in \mathscr{C}$ containing both S_1 and S_2. Then prove that $\bigcup_{S \in \mathscr{C}} S$ is a subring of R.
12. Let S be a nonempty subset of a finite ring R. Prove that S is a subring of R if and only if
$$a \text{ and } b \in S \Rightarrow a + b \quad \text{and} \quad ab \in S.$$
13. Let R be a ring which has no nonzero nilpotent elements. Prove that every idempotent in R is in the centre $C(R)$.
14. Let R be a ring such that $a^2 + a \in C(R)$ for all $a \in R$. Prove that R is a commutative ring.

15. Let $(R, +, \cdot)$ be a ring of characteristic $n > 0$ and $(\mathbb{Z}_n, +_n, \cdot_n)$ the ring of integers modulo n. Define the operations $+$ and $\cdot$ on $R \times \mathbb{Z}_n$ by

$$(x, a) + (y, b) = (x + y, a +_n b)$$
$$\text{and} \quad (x, a) \cdot (y, b) = (xy + ay + bx, a \cdot_n b).$$

Prove that $(R \times \mathbb{Z}_n, +, \cdot)$ is a ring in which $R \times \{0\}$ is a subring.

9.5 HOMOMORPHISMS OF RINGS

A homomorphism from a ring R into a ring R' is, as one might guess, a function $f : R \to R'$ which preserves both the ring operations. This amounts to applying the familiar homomorphism concept to the underlying additive group $(R, +)$ and the multiplicative semigroup $(R, \cdot)$. In the following, we give a precise definition.

Definition 9.5.1. Let R and R' be rings. *A* function $f : R \to R'$ is called a *homomorphism* of R into R' if

$$f(a + b) = f(a) + f(b)$$
$$\text{and} \quad f(a \cdot b) = f(a) \cdot f(b) \quad \text{for all } a \text{ and } b \in R.$$

Note that the symbols $+$ and $\cdot$ occurring on the left sides of the above equations denote the addition and multiplication in R where as $+$ and $\cdot$ occurring on the right sides denote those in R'. This use of the same symbols for the operations of addition and multiplication in two different rings need cause no confusion provided the reader gives careful attention to the context if the notation is employed. The following is the usual terminology we apply, as in the case of group theory.

Definition 9.5.2

1. An injective homomorphism is called a *monomorphism* or an *embedding*.
2. A surjective homomorphism is called an *epimorphism*.
3. A bijective homomorphism is called an *isomorphism*.
4. A homomorphism of a ring R into itself is called an *endomorphism* of R.
5. An isomorphism of a ring R onto itself is called an *automorphism* of R.
6. A ring R is said to be isomorphic with a ring R' and denote this by $R \cong R'$ if there is an isomorphism of R onto R'. The following examples should help us for a better understanding of the above concepts.

Example 9.5.1

1. Let R and R' be any rings and define $f: R \to R'$ by $f(x) = 0$ for all $x \in R$. Then, for any a and b in R,

$$f(a + b) = 0 = 0 + 0 = f(a) + f(b)$$
$$f(ab) = 0 = 0 \cdot 0 = f(a) \cdot f(b)$$

and therefore f is a homomorphism, which is called the *trivial* or *zero homomorphism*. Note that this is not a monomorphism unless $R = \{0\}$ and is not an epimorphism unless $R' = \{0\}$.

2. The identity mapping I_R of a ring R onto itself is an automorphism of R.
3. Let R be any ring and X be any nonempty set. Consider the ring R^X of all mappings of X into R under the point-wise operations (refer Example 9.1.1 (7)). For any $x \in X$, define $\alpha_x : R^X \to R$ by

$$\alpha_x(f) = f(x) \quad \text{for all } f \in R^X.$$

Then, for any f and $g \in R^X$, we have

$$\alpha_x(f + g) = (f + g)(x) = f(x) + g(x) = \alpha_x(f) + \alpha_x(g)$$
$$\text{and} \quad \alpha_x(f \cdot g) = (f \cdot g)(x) = f(x) \cdot g(x) = \alpha_x(f) \cdot \alpha_x(g)$$

and hence α_x is a homomorphism of R^X into R which is called the *evaluation homomorphism* at x. It can be verified that α_x is an epimorphism.

4. Let n be a positive integer and consider the ring $\mathbb{Z}$ of integers and the ring $\mathbb{Z}_n$ of integers modulo n. Define $f: \mathbb{Z} \to \mathbb{Z}_n$ by

$$f(a) = r, \quad \text{where } a = qn + r, q \text{ and } r \in \mathbb{Z}^+ \text{ and } 0 \leq r < n.$$

That is, $f(a)$ is precisely the remainder obtained by dividing a with n. Then, f is an epimorphism (see Worked Exercise 9.5.1).

In the following, we exhibit a few elementary properties of homomorphisms of rings and prove that some of the structural features are preserved under homomorphisms of ring. First of all, the following is a simple consequence of the fact at a homomorphism of rings is a homomorphism of the underlying additive groups.

Theorem 9.5.1. Let R and R' be rings and $f: R \to R'$ a homomorphism of rings. Then, the following holds.

1. $f(0) = 0$.
2. $f(-a) = -f(a)$ for all $a \in R$.
3. $f(a - b) = f(a) - f(b)$ for all a and $b \in R$.

Theorem 9.5.2. Let $f: R \to R'$ be an epimorphism of rings with unity. Then, $f(1) = 1$ and, for any unit a in R, $f(a)$ is a unit in R' and $f(a^{-1}) = f(a)^{-1}$.

Proof: Let $x' \in R'$. Since f is an epimorphism, we can choose $x \in R$ such that $f(x) = x'$. Now,

$$\begin{aligned} f(1)x' &= f(1)f(x) = f(1x) = f(x) = x' \\ \text{and} \quad x'f(1) &= f(x)f(1) = f(x1) = f(x) = x' \end{aligned}$$

and hence $f(1)$ is the multiplicative identity in R' so that $f(1) = 1$, the unity in R'. Next, let a be a unit in R. Then, there is an element a^{-1} in R such that $aa^{-1} = 1 = a^{-1}a$. By applying f to these, we get that

$$\begin{aligned} f(a)f(a^{-1}) &= f(aa^{-1}) = f(1) = 1 \\ \text{and} \quad f(a^{-1})f(a) &= f(a^{-1}a) = f(1) = 1 \end{aligned}$$

and therefore $f(a)$ is multiplicatively invertible in R' and $f(a)^{-1} = f(a^{-1})$. ◀

Theorem 9.5.3. Let $f: R \to R'$ be a homomorphism of rings. Then, $f(S)$ is a subring of R' for any subring S of R and $f^{-1}(S')$ is a subring of R for any subring S' of R'.

Proof: Let S and S' be subrings of R and R', respectively since $f(0) = 0$ and S and S' contain zero elements, it follows that $f(S) \neq \phi$ and $f^{-1}(S') \neq \phi$. Now,

$$\begin{aligned} x \text{ and } y \in f(S) &\Rightarrow x = f(a) \text{ and } y = f(b) \quad \text{for some } a \text{ and } b \in S \\ &\Rightarrow x - y = f(a) - f(b) = f(a-b) \quad \text{and} \quad a - b \in S \\ \text{and} \qquad & xy = f(a)f(b) = f(ab) \quad \text{and} \quad ab \in S \\ &\Rightarrow x - y \quad \text{and} \quad xy \in f(S) \end{aligned}$$

and therefore $f(S)$ is a subring of R'. Also,

$$\begin{aligned} a \text{ and } b \in f^{-1}(S') &\Rightarrow f(a) \quad \text{and} \quad f(b) \in S' \\ &\Rightarrow f(a-b) = f(a) - f(b) \in S' \end{aligned}$$

and $f(ab) = f(a)f(b) \in S'$

$$\Rightarrow a - b \quad \text{and} \quad ab \in f^{-1}(S')$$

and therefore $f^{-1}(S')$ is a subring of R. ◀

We discuss some more important properties of homomorphisms of rings after introducing the concepts of ideals and quotient rings later.

Worked Exercise 9.5.1. Let n be a positive integer and define $f: \mathbb{Z} \to \mathbb{Z}_n$ by $f(a) = r$, where r is the remainder obtained by dividing a with n. Then prove that f is an epimorphism of rings.

Answer: For any $a \in \mathbb{Z}$, we have

$$f(a) = r, \quad \text{where } a = qn + r, q, r \in \mathbb{Z} \text{ and } 0 \leq r < n.$$

If $r \in \mathbb{Z}_n$, then $0 \leq r < n$ and clearly $f(r) = r$.
Therefore, f is a surjection.
Let a and $b \in \mathbb{Z}$ and

$$a = qn + r \quad \text{and} \quad b = q'n + s,$$

where $q, q', r, s \in \mathbb{Z}$, $0 \leq r < n$ and $0 \leq s < n$. Then, $f(a) = r$ and $f(b) = s$. Now,

$$\begin{aligned} a + b &= qn + r + q'n + s = (q + q')n + (r + s) \\ &= \begin{cases} (q+q')n+(r+_n s) & \text{if } r+s<n \\ (q+q'+1)n+(r+_n s) & \text{if } r+s \geq n \end{cases} \end{aligned}$$

and therefore $f(a + b) = r +_n s = f(a) +_n f(b)$. Also,

$$\begin{aligned} ab &= (qn + r)(q'n + s) \\ &= (qq'n + qs + q'r)n + rs \end{aligned}$$

If $rs = un + t$, $0 \leq t < n$, then

$$ab = (qq'n + qs + q'r + u)n + t$$

and hence $f(ab) = t = r \cdot_n s = f(a) \cdot_n f(b)$.

Thus, f is a homomorphism of rings. Since f is a surjection also, it follows that f is an epimorphism.

Worked Exercise 9.5.2. Prove that the composition of homomorphism of rings is again a homomorphism.

Answer: Let $f: R \to R'$ and $g: R' \to R''$ be homomorphisms of rings. Then, $g \circ f: R \to R''$ is a function and, for any a and $b \in R$, we have

$$(g \circ f)(a + b) = g(f(a + b)) = g(f(a) + f(b)) = g(f(a)) + g(f(b))$$

and $$(g \circ f)(ab) = g(f(a)f(b)) = g(f(a))g(f(b))$$

and therefore $g \circ f$ is a homomorphism.

Worked Exercise 9.5.3. Define $f: \mathbb{Z} \rightarrow \mathbb{Z}_2$ by

$$f(a) = \begin{cases} 0 & \text{if } a \text{ is even} \\ 1 & \text{if } a \text{ is odd} \end{cases}.$$

Then prove that f is an epimorphism of rings.

Answer: This is a special case of Worked Exercise 9.5.1 by taking $n = 2$. However, the following is an independent proof.
Since $f(0) = 0$, $f(1) = 1$ and $\mathbb{Z}_2 = \{0, 1\}$, f is a surjection. Also, note that, for any integers a and b, $a + b$ is even if and only if both a and b are even or both a and b are odd.
Since $0 + 0 = 0 = 1 + 1$, it follows that

$$f(a + b) = f(a) + f(b) \quad \text{for all } a, b \in \mathbb{Z}.$$

Further, ab is even if and only if atleast one of a and b is even. Since $0 \cdot 1 = 0 = 1 \cdot 0 = 0 \cdot 0$ and $1 \cdot 1 = 1$, it follows that

$$f(ab) = f(a)f(b) \quad \text{for all } a \text{ and } b \in \mathbb{Z}.$$

Thus, f is an epimorphism of rings.

Worked Exercise 9.5.4. Determine all the endomorphisms of the ring $\mathbb{Z}$ of integers.

Answers: Let $f: \mathbb{Z} \rightarrow \mathbb{Z}$ be a homomorphism of rings.
Since

$$f(1) = f(1 \cdot 1) = f(1)f(1),$$

we have $f(1)(f(1) - 1) = 0$ and hence $f(1) = 0$ or $f(1) = 1$. If $f(1) = 0$, then, for any $a \in \mathbb{Z}$,

$$f(a) = f(a \cdot 1) = f(a) \cdot f(1) = f(a)0 = 0$$

and hence f is the zero homomorphism. On the other hand, if $f(1) = 1$, then for any $0 < a \in \mathbb{Z}$,

$$\begin{aligned} f(a) &= f(1 + 1 + \cdots + 1) \ (a \text{ times}) \\ &= f(1) + f(1) + \cdots + f(1) \ (a \text{ times}) \\ &= 1 + 1 + \cdots + 1 \ (a \text{ times}) \\ &= a \end{aligned}$$

and for $0 > a \in \mathbb{Z}$, $-a > 0$ and

$$f(a) = f(-(-a)) = -f(-a) = -(-a) = a.$$

Therefore, if $f(1) = 1$, then $f(a) = a$ for all $a \in \mathbb{Z}$.
Thus, the zero homomorphism and the identity homomorphism are the only homomorphisms of $\mathbb{Z}$ into itself. These two are the only endomorphisms of the ring $\mathbb{Z}$. In this context, recall that there are several endomorphisms of the group $(\mathbb{Z}, +)$.

EXERCISE 9(E)

For any rings R and R', let Hom(R, R') denote the set of all ring homomorphisms of R into R' and End(R) denote the set of all endomorphisms of the ring R.

1. Determine all the members of each of the following:
   (i) End($\mathbb{Q}$), where $\mathbb{Q}$ is the ring of rational numbers.
   (ii) End($\mathbb{Z}_n$), for any positive integer n.
   (iii) Hom($\mathbb{R}$, $\mathbb{Z}$), where $\mathbb{R}$ is the ring of real numbers.
   (iv) Hom($\mathbb{Z}$, $\mathbb{R}$)
   (v) Hom($\mathbb{Z}_n$, $\mathbb{Q}$)
   (vi) Hom($\mathbb{Q}$, $\mathbb{Z}_n$)
2. State which of the following are true. Substantiate your answers.
   (i) Every monomorphism of $\mathbb{Z}_n$ into $\mathbb{Z}_n$ is an isomorphism.
   (ii) For any integers $0 < n < m$, there exists a monomorphism of $\mathbb{Z}_n$ into $\mathbb{Z}_m$.
   (iii) If there is a monomorphism of $\mathbb{Z}_n$ into $\mathbb{Z}_m$, then n divides m.
   (iv) If n divides m, then there is an epimorphism of $\mathbb{Z}_m$ onto $\mathbb{Z}_n$.
   (v) For any ring R, End(R) has atleast two members.
   (vi) The zero map is the only homomorphism of $\mathbb{R}$ into $\mathbb{Q}$.
   (vii) Hom($\mathbb{Q}$, $\mathbb{R}$) has exactly two members.
   (viii) For any prime number p, End($\mathbb{Z}_p$) has exactly two members.
3. For any ring R, prove that (End(R), $+$, o) is a ring, where $+$ is the point-wise addition and o is the composition of mappings.
4. Let R be a ring with unity. Prove that $f: R \to R$ is an endomorphism of the ring R if and only if there exists an idempotent e in the centre of R (that is, $ee = e$ and $ex = xe$ for all $x \in R$) such that $f(a) = ea$ for all $a \in R$.
5. Let $f: R \to R'$ be a homomorphism of rings and $a \in R$. Then prove that

$$a + \ker f = \{x \in R : f(x) = f(a)\},$$
$$\text{where } \ker f = \{y \in R : f(y) = 0\}.$$

6. If $f: R \to R'$ is an isomorphism of rings, then prove that $f^{-1} : R' \to R$ is also an isomorphism of rings.

7. For any homomorphism $f: R \to R'$ of rings, prove that $\ker f$ is a subring of R such that ax and $xa \in \ker f$ for all $a \in \ker f$ and $x \in R$.
8. Let R be a ring with unity and define $f: \mathbb{Z} \to R$ by $f(n) = n1$ for all $n \in \mathbb{Z}$. Prove that f is a homomorphism of rings and $\ker f = m\mathbb{Z}$ if $\text{char}(R) = m > 0$ and $\ker f = \{0\}$ if $\text{char}(R) = 0$.
9. Prove the following for any rings R, R' and R''.

   (i) $R \cong R$

   (ii) $R \cong R' \Rightarrow R' \cong R$

   (iii) $R \cong R'$ and $R' \cong R'' \Rightarrow R \cong R''$

10. Let R be a ring with unity. Prove that there is a subring S of R such that S is isomorphic to $\mathbb{Z}$ or to $\mathbb{Z}_m$ depending on whether $\text{char}(R) = 0$ or m.
11. Prove that the rings $\mathbb{R}$ and $\mathbb{C}$ are not isomorphic.
12. Determine all the ring epimorphisms of $\mathbb{Z}$ onto $\mathbb{Z}$.

9.6 CERTAIN SPECIAL TYPES OF RINGS

In almost every occasion where there is a need for an example of a ring, we used to refer till now to the ring of integers or of real numbers or of complex numbers or the ring $\mathbb{Z}_n$ of integers modulo n. Notice that all these are commutative rings. It is not that noncommutative rings are unimportant. In fact the knowledge of noncommutative rings is very important in the study of linear algebra, in particular, of linear transformations from a vector space into itself. In this section, we discuss mainly three types of noncommutative rings.

Theorem 9.6.1 (Ring of Endomorphisms of an Abelian Group). Let $(G, +)$ be an abelian group and End(G) be the set of all endomorphisms of the group G. Then, (End(G), +, o) is a ring with unity, where + is the point-wise addition and o is the composition of mappings.

Proof: Note that, for any f and $g \in \text{End}(G)$, $f + g$ and $f \circ g$ are defined by

$$\begin{aligned} (f+g)(x) &= f(x) + g(x) \\ \text{and}\quad (f \circ g)(x) &= f(g(x)). \end{aligned}$$

For any f and $g \in \text{End}(G)$ and x and $y \in G$, we have

$$\begin{aligned} (f+g)(x+y) &= f(x+y) + g(x+y) \\ &= f(x) + f(y) + g(x) + g(y) \\ &= f(x) + g(x) + f(y) + g(y) \quad \text{(since } G \text{ is abelian)} \\ &= (f+g)(x) + (f+g)(y) \end{aligned}$$

$$\begin{aligned} \text{and} \quad (f \circ g)(x + y) &= f(g(x + y)) \\ &= f(g(x) + g(y)) \\ &= f(g(x)) + f(g(y)) = (f \circ g)(x) + (f \circ g)(y) \end{aligned}$$

and therefore $f + g$ and $f \circ g$ are endomorphisms of the group $(G, +)$. Thus, $+$ and o are binary operations on End(G). The associativity of $+$ in End(G) follows from that of $+$ in G. The zero endomorphism will be the zero element in End(G). Also, for any $f \in$ End(G), the map $-f$, defined by $(-f)(x) = -f(x)$ for all $x \in G$, is the additive inverse of f in End(G). Thus, (End(G), $+$) is an abelian group. Also, clearly the composition o is associative and,

$$\begin{aligned} (f \circ (g + h))(x) &= f(g + h)(x) \\ &= f(g(x) + h(x)) \\ &= f(g(x)) + f(h(x)) \\ &= (f \circ g + f \circ h)(x) \end{aligned}$$

$$\begin{aligned} \text{and} \quad ((f + g) \circ h)(x) &= (f + g)(h(x)) \\ &= f(h(x)) + g(h(x)) \\ &= (f \circ h + g \circ h)(x) \end{aligned}$$

for all $x \in G$ and hence

$$\begin{aligned} f \circ (g + h) &= f \circ g + f \circ h \\ \text{and} \quad (f + g) \circ h &= f \circ h + g \circ h \end{aligned}$$

for any f, g and $h \in$ End(G) Thus, (End(G), $+$, o) is a ring. Further, the identity homomorphism I_G of the group G is the multiplicative identity in the ring End(G). Thus, (End(G), $+$, o) is a ring with unity. ◀

In general the ring End(G) of endomorphisms of an abelian group G is not commutative. For, consider the following example.

Example 9.6.1. Let G be the product group $\mathbb{Z} \times \mathbb{Z}$ under coordinate wise addition and define f and $g : G \to G$ by

$$f(a_1, a_2) = (a_1, a_1 - a_2) \quad \text{and} \quad g(a_1, a_2) = (-a_2, a_1).$$

It can be easily verified that f and g are endomorphisms of $G(= \mathbb{Z} \times \mathbb{Z})$. Now consider

$$\begin{aligned} (f \circ g)(1, 1) &= f(g(1, 1)) = f(-1, 1) = (-1, -1 - 1) = (-1, 2) \\ \text{and} \quad (g \circ f)(1, 1) &= g(f(1, 1)) = g(1, 1 - 1) = g(1, 0) = (0, 1). \end{aligned}$$

Therefore, $(f \circ g)(1, 1) \neq (g \circ f)(1, 1)$ and hence $f \circ g \neq g \circ f$. Thus, End(G) is a noncommutative ring.

Next let us consider the familiar concept of a real matrix; that is, an array in which all the entries are real numbers. In fact, we can replace the real number system here with an abstract ring. In the following, we define addition and multiplication of $n \times n$ matrices in such a way that the set of all $n \times n$ matrices becomes a noncommutative ring.

Definition 9.6.1. Let $(R, +, \cdot)$ be a ring and n a positive integer. By an $n \times n$ matrix over R, we mean an array of n^2 elements of the ring R, not necessarily distinct, arranged in n rows and n columns as given below.

$$\begin{pmatrix} a_{11} & a_{12} & a_{13} & \cdots & a_{1n} \\ a_{21} & a_{22} & a_{23} & \cdots & a_{2n} \\ \vdots & \vdots & \vdots & \cdots & \vdots \\ a_{n1} & a_{n2} & a_{n3} & \cdots & a_{nn} \end{pmatrix}$$

where each a_{ij}, $1 \leq i \leq n$ and $1 \leq j \leq n$, is an element of the given ring. The elements a_{ij}, $1 \leq j \leq n$, constitute the ith row and the elements a_{ij}, $1 \leq i \leq n$ constitute the jth column. As such a_{ij} is the element in both the ith row and j^{th} column. The matrix itself will be denoted by (a_{ij}). The set of all $n \times n$ matrices over a ring R will be denoted by $M_n(R)$. Two matrices (a_{ij}) and (b_{ij}) are said to be equal if $a_{ij} = b_{ij}$ for all $1 \leq i, j \leq n$.

Definition 9.6.2. Let $(R, +, \cdot)$ be a ring and $n \in \mathbb{Z}^+$. For any matrices

$$A = (a_{ij}) \quad \text{and} \quad B = (b_{ij}) \text{ in } M_n(R),$$

we define $A + B$ and $A \cdot B$ as follows.

$$A + B = (c_{ij}), \quad \text{where } c_{ij} = a_{ij} + b_{ij}$$

and $A \cdot B = (d_{ij})$, where $d_{ij} = a_{i1} \cdot b_{1j} + a_{i2} \cdot b_{2j} + \cdots + a_{in} \cdot b_{nj}$ for any $1 \leq i$, $j \leq n$. Note that the $+$ and $\cdot$ on the right sides are those in the ring R.

Theorem 9.6.2. Let n be a positive integer and R be an arbitrary ring. Then, the set $M_n(R)$ of all $n \times n$ matrices over R is a ring under the operations $+$ and $\cdot$ defined above.

Proof: Let $A = (a_{ij})$, $B = (b_{ij})$ and $C = (c_{ij})$ be arbitrary $n \times n$ matrices over R. Since the addition $+$ on R is commutative and associative, we have

$$A + B = (a_{ij} + b_{ij}) = (b_{ij} + a_{ij}) = B + A$$

$$\text{and} \quad (A + B) + C = ((a_{ij} + b_{ij}) + c_{ij})$$
$$= (a_{ij} + (b_{ij} + c_{ij})) = A + (B + C).$$

Therefore, $+$ is commutative and associative on $M_n(R)$. Let us denote the $n \times n$ matrix, all of whose entries are 0, by 0 itself. Then, clearly

$$A + 0 = A = 0 + A$$
$$\text{and} \quad A + (-A) = 0 = (-A) + A,$$

where $-A = (-a_{ij})$. Therefore, $(M_n(R), +)$ is an abelian group. To prove the associativity of $\cdot$, let

$$A \cdot B = (s_{ij}) \quad \text{and} \quad (A \cdot B) \cdot C = (t_{ij}).$$

Then, $s_{ij} = \sum_{k=1}^{n} a_{ik}b_{kj}$ and $t_{ij} = \sum_{k=1}^{n} s_{ik}c_{kj}$. Now,

$$t_{ij} = \sum_{k=1}^{n}\left(\sum_{r=1}^{n} a_{ir}b_{rk}\right)c_{kj}$$

$$= \sum_{k=1}^{n}\sum_{r=1}^{n}(a_{ir}b_{rk})c_{kj}$$

$$= \sum_{k=1}^{n}\sum_{r=1}^{n}a_{ir}(b_{rk}c_{kj})$$

$$= \sum_{r=1}^{n}a_{ir}\left(\sum_{k=1}^{n}b_{rk}c_{kj}\right)$$

$$= \text{the } ij^{\text{th}} \text{ entry in } A \cdot (B \cdot C).$$

Thus, $(A \cdot B) \cdot C = A \cdot (B \cdot C)$. Also,

$$A \cdot (B + C) = \left(\sum_{r=1}^{n} a_{ir}(b_{rj} + c_{rj})\right)$$

$$= \left(\sum_{r=1}^{n} a_{ir}b_{rj}\right) + \left(\sum_{r=1}^{n} a_{ir}c_{rj}\right)$$

$$= A \cdot B + A \cdot C$$

and, similarly $(A + B) \cdot C = A \cdot C + B \cdot C$. Thus, $(M_n(R), +, \cdot)$ is a ring.

Note:

1. If R is a ring with unity 1 and E_r is the $n \times n$ matrix defined by

$$E_n = (e_{ij}), \quad \text{where } e_{ij} = \begin{cases} 1 & \text{if } i = j \\ 0 & \text{if } i \neq j \end{cases}$$

then $E_n \cdot A = A = A \cdot E_n$ for any matrix A in $M_n(R)$. Therefore, if R is with unity, then $M_n(R)$ is a ring with unity.
2. If $n = 1$, then a 1×1 matrix (a) can be identified with a itself and, hence $M_1(R)$ is isomorphic with R.
3. If $n > 1$, then $M_n(R)$ may not be commutative even when R is commutative; for consider the following theorem.

Theorem 9.6.3. For any $n > 1$, the ring $M_n(\mathbb{R})$ of $n \times n$ matrices over the real numbers is a noncommutative ring with unity.

Proof: Let $n > 1$. Since the real number system $\mathbb{R}$ forms a ring with unity, by Theorem 9.6.2 (1), $M_n(\mathbb{R})$ is a ring with unity. Let $A = (a_{ij})$ and $B = (b_{ij})$ be the matrices defined by

$$a_{ij} = \begin{cases} 1 & \text{if } i = 1 = j \\ 0 & \text{otherwise} \end{cases}$$

and $$b_{ij} = \begin{cases} 1 & \text{if } j = 1 \text{ and } i = 1 \text{ or } 2 \\ 0 & \text{otherwise} \end{cases}.$$

Then, $$AB = \begin{pmatrix} 1 & 0 & \dots & 0 \\ 0 & 0 & \dots & 0 \\ 0 & 0 & \dots & 0 \\ \vdots & \vdots & \vdots & \vdots \\ 0 & 0 & \dots & 0 \end{pmatrix} \begin{pmatrix} 1 & 0 & \dots & 0 \\ 1 & 0 & \dots & 0 \\ 0 & 0 & \dots & 0 \\ \vdots & \vdots & \vdots & \vdots \\ 0 & 0 & \dots & 0 \end{pmatrix} = \begin{pmatrix} 1 & 0 & 0 & \dots & 0 \\ 0 & 0 & 0 & \dots & 0 \\ \vdots & \vdots & \vdots & \vdots & \vdots \\ 0 & 0 & 0 & \dots & 0 \end{pmatrix}$$

and $$BA = \begin{pmatrix} 1 & 0 & \dots & 0 \\ 1 & 0 & \dots & 0 \\ 0 & 0 & \dots & 0 \\ \vdots & \vdots & \vdots & \vdots \\ 0 & 0 & \dots & 0 \end{pmatrix} \begin{pmatrix} 1 & 0 & \dots & 0 \\ 0 & 0 & \dots & 0 \\ 0 & 0 & \dots & 0 \\ \vdots & & \dots & \vdots \\ 0 & 0 & \dots & 0 \end{pmatrix} = \begin{pmatrix} 1 & 0 & \dots & 0 \\ 1 & 0 & \dots & 0 \\ 0 & 0 & \dots & 0 \\ \vdots & \vdots & \dots & \vdots \\ 0 & 0 & \dots & 0 \end{pmatrix}$$

and hence $AB \neq BA$. Thus, $M_n(\mathbb{R})$ is a noncommutative ring. ◀

In the following, we give another important example of a noncommutative ring, namely the ring of real quaternions.

Definition 9.6.3. The algebraic system $(\mathbb{R}^4, +, \cdot)$, where $\mathbb{R}^4$ is the set of all quadruples of real numbers and $+$ and $\cdot$ are the binary operations defined as follows, is called the *system of real quaternions* and each element of this system is called a *real quaternion*.
For convenience, let us write a quadruple by (a_0, a_1, a_2, a_3). For any

$$a = (a_0, a_1, a_2, a_3) \quad \text{and} \quad b = (b_0, b_1, b_2, b_3) \text{ in } \mathbb{R}^4,$$

we define

$$a + b = (a_0 + b_0, a_1 + b_1, a_2 + b_2, a_3 + b_3)$$

$$\text{and} \quad a \cdot b = (c_0, c_1, c_2, c_3), \text{ where}$$

$$\begin{aligned} c_0 &= a_0b_0 - a_1b_1 - a_2b_2 - a_3b_3 \\ c_1 &= a_0b_1 + a_1b_0 + a_2b_3 - a_3b_2 \\ c_2 &= a_0b_2 + a_2b_0 + a_3b_1 - a_1b_3 \\ \text{and} \quad c_3 &= a_0b_3 + a_3b_0 + a_1b_2 - a_2b_1. \end{aligned}$$

Even though $+$ is defined coordinate wise, the multiplication $\cdot$ is not coordinate wise and needs special attention. An easy way of remembering the rule for multiplication is given below. Let us represent a quadruple (a_0, a_1, a_2, a_3) by

$$(a_0, a_1, a_2, a_3) = a_0 + a_1 i + a_2 j + a_3 k$$

(as a complex number (a, b) is represented by $a + bi$).

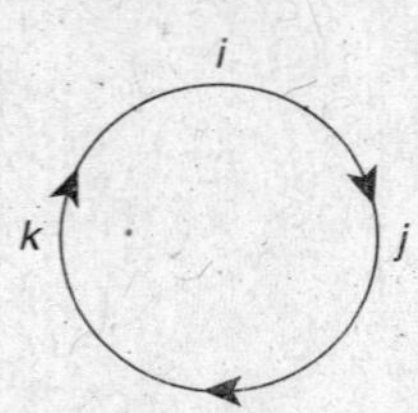

As we go around clockwise, we read off the product; for example,

$$i \cdot j = k, j \cdot k = i, k \cdot i = j.$$

Going around anticlockwise, we read off the negatives; for example,

$$i \cdot k = -j, j \cdot i = -k, k \cdot j = -i.$$

The multiplication in $\mathbb{R}^4$ is now defined as if we multiply two sums of real numbers obeying the following rules.

$$\begin{aligned} &i \cdot i = j \cdot j = k \cdot k = -1 \\ &i \cdot j = k, j \cdot k = i, k \cdot i = j \\ &j \cdot i = -k, k \cdot j = -i, i \cdot k = -j \end{aligned} \qquad \text{(A)}$$

Now, the multiplication of real quaternions can be formally defined as follows.

$$(a_0 + a_1 i + a_2 j + a_3 k) \cdot (b_0 + b_1 i + b_2 j + b_3 k) = c_0 + c_1 i + c_2 j + c_3 k,$$

$$\begin{aligned} \text{where} \quad c_0 &= a_0 b_0 - a_1 b_1 - a_2 b_2 - a_3 b_3 \\ c_1 &= a_0 b_1 + a_1 b_0 + a_2 b_3 - a_3 b_2 \\ c_2 &= a_0 b_2 + a_2 b_0 + a_3 b_1 - a_1 b_3 \\ \text{and} \quad c_3 &= a_0 b_3 + a_3 b_0 + a_1 b_2 - a_2 b_1. \end{aligned} \qquad \text{(B)}$$

To multiply $a_0 + a_1 i + a_2 j + a_3 k$ by $b_0 + b_1 i + b_2 j + b_3 k$ on the right, we first multiply each 'term' in the first quaternion with each term in the second on the right, use the laws given in (A) and collect the terms with each of i, j and k and without any of them. Recall that the elements $1, -1, i, -i, j, -j, k$ and $-k$ form a nonabelian group of order 8 under the above multiplication rules (A) and is called the *group of quaternions*. If we write

$$\begin{gathered} (1, 0, 0, 0) = 1, (0, 1, 0, 0) = i \\ (0, 0, 1, 0) = j \quad \text{and} \quad (0, 0, 0, 1) = k, \end{gathered}$$

then, by (B) above,

$$i^2 = (0, 1, 0, 0) \cdot (0, 1, 0, 0) = -(1, 0, 0, 0) = -1$$

and, similarly $j^2 = -1 = k^2$ and other rules of multiplication in (A) can be derived from (B). Now, the following is a routine verification.

Theorem 9.6.4. The real quaternions form a noncommutative ring with unity under the addition and multiplication given above. This ring is denoted by $Q_{\mathbb{R}}$ and is called the *ring of real quaternions*.

Worked Exercise 9.6.1. Give an example of a noncommutative ring with exactly 16 elements.

Answer: Consider the ring $\mathbb{Z}_2$ of integers modulo 2. Then, $\mathbb{Z}_2 = \{0, 1\}$. Then, the ring $M_2(\mathbb{Z}_2)$ of 2×2 matrices over $\mathbb{Z}_2$ has exactly $2^{2\times2}$ (=16) elements and is not commutative. For, consider

$$\begin{pmatrix} 1 & 1 \\ 0 & 0 \end{pmatrix} \cdot \begin{pmatrix} 1 & 0 \\ 1 & 0 \end{pmatrix} = \begin{pmatrix} 1+1 & 0 \\ 0 & 0 \end{pmatrix} = \begin{pmatrix} 0 & 0 \\ 0 & 0 \end{pmatrix}$$

and $$\begin{pmatrix} 1 & 0 \\ 1 & 0 \end{pmatrix} \cdot \begin{pmatrix} 1 & 1 \\ 0 & 0 \end{pmatrix} = \begin{pmatrix} 1 & 1 \\ 1 & 1 \end{pmatrix}.$$

Worked Exercise 9.6.2. Prove that every nonzero element in the ring of real quaternions is a unit.

Answer: Let $0 \neq a = a_0 + a_1 i + a_2 j + a_3 k \in Q_{\mathbb{R}}$. Put $s = a_0^2 + a_1^2 + a_2^2 + a_3^2$. Since $a \neq 0$, atleast one a_i must be nonzero and hence $s > 0$. Now, consider

$$b = \frac{a_0}{s} - \frac{a_1}{s} i - \frac{a_2}{s} j - \frac{a_3}{s} k.$$

Then, $ab = ba = \frac{s}{s} = 1$. Therefore, a is a unit in $Q_{\mathbb{R}}$.

EXERCISE 9(F)

1. Evaluate the following products $a \cdot$ b and $b \cdot a$ in the rings mentioned against them.
   (i) $a, b \in \text{End}(\mathbb{R}^3)$ defined by
   $a(r_1, r_2, r_3) = (r_1 + r_2, r_1 - r_2, 0)$ and
   $b(r_1, r_2, r_3) = (r_3 - r_1, r_3 - r_2, r_2 - r_3)$
   (ii) $a = \begin{pmatrix} 1 & 0 & 0 \\ 0 & 2 & 3 \\ 2 & 1 & 0 \end{pmatrix}$ and $b = \begin{pmatrix} 0 & 2 & 1 \\ 1 & 2 & 3 \\ 3 & 1 & 2 \end{pmatrix}$ in $M_3(\mathbb{Z})$.
   (iii) $a = 2 + 3i + 4j + 5k$ and $b = 1 + 2i + 3j + k$ in $Q_{\mathbb{R}}$.
   (iv) $a = 1 + 2i + 5j - 3k$ and $b = 3 + 2i - 2j + k$ in $Q_{\mathbb{R}}$.
2. Give an example of a noncommutative ring with unity having exactly 81 elements.
3. Describe the ring $\text{End}(\mathbb{Z}_6)$, where $\mathbb{Z}_6$ is the group of integers modulo 6.
4. Prove or disprove that the set

$$X = \left\{ \begin{pmatrix} a & a+b \\ a+b & b \end{pmatrix} : a \text{ and } b \text{ are integers} \right\}$$

   is a subring of the ring $M_2(\mathbb{R})$ of 2×2 matrices over $\mathbb{R}$.

5. Give examples of two matrices A and B in $M_4(\mathbb{R})$ such that $AB = 0$ and $BA \neq 0$.
6. Let $a = a_0 + a_1 i + a_2 j + a_3 k$ and $b = a_0 - a_1 i - a_2 j - a_3 k$, where a_0, a_1, a_2, a_3 are real numbers. Evaluate the products $a \cdot b$ and $b \cdot a$ in $Q_{\mathbb{R}}$.
7. Prove that $\mathbb{Q}_{\mathbb{R}} - \{0\}$ forms a group under the multiplication of real quaternions.
8. Determine all the nilpotents and all the idempotents in the ring $Q_{\mathbb{R}}$ of real quaternions.
9. If $A = \begin{pmatrix} 0 & 0 & 1 \\ 0 & 0 & 1 \\ 0 & 0 & 1 \end{pmatrix}$, then compute A^n in $M_3(\mathbb{R})$ for any positive integer m.
10. Prove that $\left\{ \begin{pmatrix} a & a \\ b & b \end{pmatrix} : a \text{ and } b \text{ are integers} \right\}$ is a subring of $M_2(\mathbb{R})$.
11. Prove in detail that $\mathbb{Q}_{\mathbb{R}}$ is a ring under the addition and multiplication of real quaternions.
12. Find the centre of the ring of $M_n(\mathbb{R})$ of all $n \times n$ matrices over the real number system $\mathbb{R}$.
13. Extend the above Exercise 12 for an arbitrary ring R in place of the ring $\mathbb{R}$ of real numbers.
14. Determine the centre of the ring of real quaternions.
15. If S is a subring of a ring R, prove that $M_n(S)$ is a subring of $M_n(R)$ for any positive integer n.
16. If R is a ring such that $M_n(R)$ is a ring with unity, then prove that R is a ring with unity.

9.7 INTEGRAL DOMAINS AND FIELDS

Now we turn our attention to certain important special types of commutative rings. One of the motives of inventing the abstract concept of a ring is to put the algebraic properties of the integers into an abstract setting. A ring is not the appropriate abstraction of the integers, because much is lost in the process. Besides the two obvious properties of commutativity and the existence of unity, there is one other essential feature of the integers that rings in general do not satisfy, namely cancellation property for multiplication. In this section, we introduce a special class of rings, known as integral domains, which have all the three properties, namely the commutativity, the existence of unity and the cancellation law for multiplication. Integral domains play a major role in Algebraic Number Theory and various other areas of mathematics. Also, a special kind of integral domains, namely fields, are introduced in this section and several elementary properties of these are discussed. Fields are abstractions of the rational or real or complex number systems.

First let us recall that a nonzero element a in a ring R is called a zero divisor if there exists a nonzero element b in R such that $ab = 0 = ba$.

Definition 9.7.1. A nontrivial commutative ring with unity and without zero divisors is called an *integral domain.*

Recall that a ring R with unity 1 is nontrivial or nonzero (that is, $R \neq \{0\}$) if and only if the additive identity 0 and the multiplicative identity 1 are different. In the following, we obtain some other simple equivalent conditions for a ring to be an integral domain.

Theorem 9.7.1. The following are equivalent to each other for any nontrivial commutative ring $(R, +, \cdot)$ with unity.

1. $(R, +, \cdot)$ is an integral domain.
2. For any elements a, b and c in R,

$$ab = ac \Rightarrow a = 0 \quad \text{or} \quad b = c.$$

3. For any elements a and b in R,

$$ab = 0 \Rightarrow a = 0 \quad \text{or} \quad b = 0.$$

Proof: (1) $\Rightarrow$ (2): If $ab = ac$ and $a \neq 0$, then

$$a(b - c) = ab - ac = 0$$

and, since a is not a zero divisor, $b - c = 0$ or $b = c$.
(2) $\Rightarrow$ (3): If $ab = 0$, then $ab = a0$ and therefore, by (2), $a = 0$ or $b = 0$.
(3) $\Rightarrow$ (1) is clear. ◀

Example 9.7.1

1. The ring $\mathbb{Z}$ of integers, the ring $\mathbb{Q}$ of rational numbers, the ring $\mathbb{R}$ of real numbers and the ring $\mathbb{C}$ of complex numbers are all integral domains with respect to usual addition and multiplication. In each of these, the product of any two nonzero elements is again nonzero and hence there are no zero divisors.
2. The ring $\mathbb{Z}[i]$ of Gaussian integers is an integral domain with respect to the addition and multiplication of complex numbers. Note that $\mathbb{Z}[i]$ is a subring of the ring $\mathbb{C}$ of complex numbers.
3. $(\mathbb{Z}_p, +_p, \cdot_p)$ is an integral domain for any prime number p.

4. $(\mathbb{Z}_6, +_6, \cdot_6)$ is not an integral domain, since $2 \neq 0$, $3 \neq 0$ and $2 \cdot_6 3 = 0$ in $\mathbb{Z}_6$. This is a commutative ring with unity and zero divisors.
5. For any $n > 1$, $(n\mathbb{Z}, +, \cdot)$ is a commutative ring without zero divisors and with no unity and hence not an integral domain.
6. The ring $Q_{\mathbb{R}}$ of real quaternions is a ring with unity and without zero divisors and not commutative and hence not an integral domain.

The examples in (4), (5) and (6) above substantiate that the three defining properties of an integral domain, namely, the nonexistence of zero divisors, the existence of unity and the commutativity are all independent of others, in the sense that no two imply the other. In the following, we introduce a special class of integral domains.

Definition 9.7.2. A nontrivial commutative ring with unity is called a *field* if every nonzero element of it is a unit (that is, multiplicatively invertible).

Example 9.7.2

1. The rings $\mathbb{Q}$, $\mathbb{R}$ and $\mathbb{C}$ are all fields, since for any nonzero number a, there is $1/a$ for which $a \cdot \frac{1}{a} = 1$ and hence a is a unit.
2. The ring $\mathbb{Z}$ of integers is not a field, since 2 is not a unit in $\mathbb{Z}$. In fact, 1 and -1 are the only units in $\mathbb{Z}$.
3. For any prime number p, $(\mathbb{Z}_p, +_p, \cdot_p)$ is a field, since any $0 < a < p$ is relatively prime with p and hence a unit in $\mathbb{Z}_p$.
4. The ring $Q_{\mathbb{R}}$ of real quaternions is not a field, even though every nonzero element is a unit in it; because it is not commutative.

Theorem 9.7.2. Every field is an integral domain and the converse is not true.

Proof: Let $(\mathbb{R}, +, \cdot)$ be a field. Then, R is a nontrivial commutative ring with unity and, if $0 \neq a \in R$, a has multiplicative inverse a^{-1} in R and therefore, for any $b \in R$,

$$ab = 0 \Rightarrow a^{-1}(ab) = 0 \Rightarrow (a^{-1}a)b = 0 \Rightarrow b = 0.$$

Therefore, R has no zero divisors and hence R is an integral domain. The converse is not true, since the ring $\mathbb{Z}$ of integers is an integral domain but not a field. ◀

However, certain special type of integral domains are fields. In this context, we have the following theorem.

Theorem 9.7.3. Every finite integral domain is a field.

Proof: Let $(R, +, \cdot)$ be a finite integral domain. Then, R is a nontrivial commutative ring with unity and without zero divisors. Let $0 \neq a \in R$. Since R is finite, we can write

$$R - \{0\} = \{a_1, a_2, \ldots, a_n\},$$

where $a_1, a_2, \ldots, a_n$ are all the distinct nonzero elements in R. Since R is an integral domain, $a \neq 0$ and each $a_i \neq 0$, we get that $aa_i \neq 0$ for each $1 \leq i \leq n$. Therefore, the set

$$S = \{aa_1, aa_2, \ldots, aa_n\}$$

is a subset of $R - \{0\}$. Further, $aa_i \neq aa_j$ for all $i \neq j$ (since $a \neq 0$ and $a_i \neq a_j$). S is an n-element subset of $R - \{0\}$, which also has n elements. Therefore,

$$R - \{0\} = S = \{aa_1, aa_2, \ldots, aa_n\}.$$

In particular, $1 \in R - \{0\}$ and hence $aa_i = 1$ for some i. Therefore, a is a unit in R. Thus, R is a field. ◀

Corollary 9.7.1. The following are equivalent to each other for any positive integer n.

1. n is a prime number.
2. $(\mathbb{Z}_n, +_n, \cdot_n)$ is an integral domain.
3. $(\mathbb{Z}_n, +_n, \cdot_n)$ is a field.

Proof: First note that, for each of these, n must be necessarily greater than 1 (for, if $n = 1$, $\mathbb{Z}_n$ is trivial).
(1) ⇒ (2) follows from the fact that, for any a and $b \in \mathbb{Z}_n$,

$$ab = 0 \text{ in } \mathbb{Z}_p \Rightarrow n \text{ divides } ab$$

and that, if n is prime,

$$n \text{ divides } ab \Leftrightarrow n \text{ divides } a \text{ or } b.$$

(2) ⇒ (3) follows from Theorem 9.7.3.
(3) ⇒ (1) follows from the fact that, for any $0 < a < n$, a is a unit in $\mathbb{Z}_n$ if and only if a is relatively prime with n.

Corollary 9.7.2. For any prime number p, $(\mathbb{Z}_p, +_p, \cdot_p)$ is a field.

Definition 9.7.3. A nontrivial ring with unity is called a *division ring* if every nonzero element is a unit.

Note that any field is a division ring and the converse is not true, since the ring $Q_{\mathbb{R}}$ of real quaternions is a division ring, but not a field (see Worked Exercise 9.6.2). However, any commutative division ring is a field.

Theorem 9.7.4. The characteristic of any integral domain is either 0 or a prime number.

Proof: Let R be an integral domain and char(R) > 0. Since R is nontrivial, char(R) $\neq 1$. Let char(R) $= n > 1$. Suppose that n is not a prime. Then, there exist positive integers r and s such that $n = rs$, $r > 1$ and $s > 1$.

Now, consider

$$(r1) \cdot (s1) = rs1 = n1 = 0.$$

Since $r < n =$ char(R), $r1 \neq 0$ (otherwise $rx = 0$ for all $x \in R$). Similarly, $s1 \neq 0$. This is a contradiction, since R is an integral domain. Thus, n is prime. ◀

Corollary 9.7.3. The characteristic of any field is either 0 or a prime.

Worked Exercise 9.7.1. Prove that any integral domain has exactly two idempotents.

Answer: Let R be an integral domain. Then, $0 \neq 1$ and clearly these two are idempotents. If a is any idempotent in R, then $a^2 = a$ and hence

$$a(a - 1) = 0 \quad \text{so that } a = 0 \text{ or } a = 1$$

Thus, 0 and 1 are the only idempotents of R.

Worked Exercise 9.7.2. Prove that $\mathbb{Z}_3[i] = \{a + bi : a \text{ and } b \in \mathbb{Z}_3\}$ is a field under addition and multiplication modulo 3 by writing the tables representing the operations $+_3$ and $\cdot_3$ on $\mathbb{Z}_3[i]$. $\mathbb{Z}_3[i]$ is called the *ring of Gaussian integers modulo 3*.

Answer: We have

$$\mathbb{Z}_3[i] = \{0, 1, 2, i, 1 + i, 2 + i, 2i, 1 + 2i, 2 + 2i\}.$$

Here, the elements are added and multiplied as in the complex number system, except that the coefficients are reduced modulo 3. In particular, note that

$$-1 = 2, -i = 2i, \quad 2 \cdot_3 2 = 1, \quad 2i \cdot_3 2i = 2$$

The table representing $+_3$ and $\cdot_3$ on $\mathbb{Z}_3[i]$ are given below.

$+_3$	0	1	2	i	$1+i$	$2+i$	$2i$	$1+2i$	$2+2i$
0	0	1	2	i	$1+i$	$2+i$	$2i$	$1+2i$	$2+2i$
1	1	2	0	$1+i$	$2+i$	i	$1+2i$	$2+2i$	$2i$
2	2	0	1	$2+i$	i	$1+i$	$2+2i$	$2i$	$1+2i$
i	i	$1+i$	$2+i$	$2i$	$1+2i$	$2+2i$	0	1	2
$1+i$	$1+i$	$2+i$	i	$1+2i$	$2+2i$	$2i$	1	2	0
$2+i$	$2+i$	i	$1+i$	$2+2i$	$2i$	$1+2i$	2	0	1
$2i$	$2i$	$1+2i$	$2+2i$	0	1	2	i	$1+i$	$2+i$
$1+2i$	$1+2i$	$2+2i$	$2i$	1	2	0	$1+i$	$2+i$	i
$2+2i$	$2+2i$	$2i$	$1+2i$	2	0	1	$2+i$	i	$1+i$

$\cdot_3$	0	1	2	i	$1+i$	$2+i$	$2i$	$1+2i$	$2+2i$
0	0	0	0	0	0	0	0	0	0
1	0	1	2	i	$1+i$	$2+i$	$2i$	$1+2i$	$2+2i$
2	0	2	1	$2i$	$2+2i$	$1+2i$	i	$2+i$	$1+i$
i	0	i	$2i$	2	$2+i$	$2+2i$	1	$1+i$	$1+2i$
$1+i$	0	$1+i$	$2+2i$	$2+i$	$2i$	1	$1+2i$	2	i
$2+i$	0	$2+i$	$1+2i$	$2+2i$	1	i	$1+i$	$2i$	2
$2i$	0	$2i$	i	1	$1+2i$	$1+i$	2	$2+2i$	$2+i$
$1+2i$	0	$1+2i$	$2+i$	$1+i$	2	$2i$	$2+2i$	i	1
$2+2i$	0	$2+2i$	$1+i$	$1+2i$	i	2	$2+i$	1	$2i$

It is a straight forward verification that $\mathbb{Z}_3[i]$ is a commutative ring with unity. By looking at the multiplication table given above (zero does not appear in any row and column corresponding to nonzero elements), we can infer that $\mathbb{Z}_3[i]$ is an integral domain. Since it is finite (with 9 elements), it is a field also.

Worked Exercise 9.7.3. Prove that $\mathbb{Z}_2[i] = \{a + bi : a \text{ and } b \in \mathbb{Z}_2\}$ is a commutative ring with unity which is not an integral domain under addition and multiplication of complex numbers modulo 2.

Answer: We have

$$\mathbb{Z}_2[i] = \{0, 1, i, 1 + i\}.$$

The tables for $+_2$ and $\cdot_2$ on $\mathbb{Z}_2[i]$ are given below.

$+_2$	0	1	i	$1+i$
0	0	1	i	$1+i$
1	1	0	$1+i$	i
i	i	$1+i$	0	1
$1+i$	$1+i$	i	1	0

$\cdot_2$	0	1	i	$1+i$
0	0	0	0	0
1	0	1	i	$1+i$
i	0	i	1	$1+i$
$1+i$	0	$1+i$	$1-i$	0

It can be easily verified that $(\mathbb{Z}_2[i], +_2, \cdot_2)$ is a commutative ring with unity. Since

$$(1 + i)(1 + i) = 1 + 2i + (-1) = 0,$$

$1 + i$ is a zero divisor and hence $\mathbb{Z}_2[i]$ is not an integral domain.

Worked Exercise 9.7.4. Let R be a nontrivial ring such that, for each $0 \neq a \in R$, there exists unique element x in R such that $axa = a$. Prove that R is a division ring.

Answer: We first prove that R has no zero divisors. Suppose that a and $b \in R$ such that $ab = 0$ and $a \neq 0$. Choose $x \in R$ such that $axa = a$. Then,

$$a(x + b)a = axa + aba = axa + 0 = axa = a.$$

By the uniqueness of x, it follows that $x + b = x$ or $b = 0$. Thus, $ab = 0$ implies that $a = 0$ or $b = 0$. From this we get $R - \{0\}$ is closed under multiplication. Let $0 \neq a \in R$ and x be the unique element in R such that $axa = a$. Then, $ax \cdot ax = ax$ and $xa \cdot xa = xa$ and hence ax and xa are idempotents and are nonzero (since $axa = a \neq 0$). Next, we shall prove that there is only one nonzero idempotent in R. Let e and f be nonzero idempotents in R. Let g be the unique element such that

$$(ef) \cdot g \cdot (ef) = ef.$$

Then, $g \neq 0$, since $ef \neq 0$. Now,

$$(ef)(ge)(ef) = ef \quad \text{and} \quad (ef)(fg)(ef) = ef.$$

By the uniqueness of g, we get that $g = ge = fg$. Also, $(ef)(ge \cdot fg)(ef) = ef$ and hence $ge \cdot fg = g$ which implies that $g^2 = g$. Therefore, $geg = g = gfg$ and hence $e = f$. Thus, $R - \{0\}$ has exactly one idempotent, say e.

In particular, $ax = xa = e$ and hence

$$ae = axa = a \quad \text{and} \quad ea = axa = a.$$

Thus, e is the unity in R and, since $ax = e = xa$, x is the multiplicative inverse of a. Thus, R is a division ring.

EXERCISE 9(G)

1. Consider the following classes of rings.

   FF = The class of all finite fields.

   F = The class of all fields.

   ID = The class of all integral domains.

   R = The class of all rings.

   RU = The class of all rings with unity.

   CR = The class of all commutative rings.

   CRU = The class of all commutative rings with unity.

   NCR = The class of all noncommutative rings.

   Draw a Venn diagram representing the above classes.

2. Which of the following are fields or integral domains? Substantiate your answers.

   (i) $\mathbb{R} \times \mathbb{R}$ under coordinate wise addition and multiplication.

   (ii) $\mathbb{Q}^n$, where $\mathbb{Q}$ is the field of rationals and $n \in \mathbb{Z}^+$ under coordinate wise operations.

(iii) $\mathbb{Z}[\sqrt{2}] = \{a + b\sqrt{2} : a \text{ and } b \in \mathbb{Z}\}$ under the addition and multiplication of real numbers.

(iv) $\mathbb{Z}[i]$, the ring of Gaussian integers.

(v) $(\mathbb{Z}_5[i], +_5, \cdot_5)$.

(vi) $(\mathbb{Z}_4[i], +_4, \cdot_4)$.

(vii) $(\mathbb{Z}_3[i], +_3, \cdot_3)$.

(viii) $(\mathbb{Z}_2[i], +_2, \cdot_2)$.

(ix) $\mathbb{Z}_5 \times \mathbb{Z}_3$.

(x) $\mathbb{Z}_{19}$.

3. Prove that $\mathbb{Q}[i] = \{a + bi : a \text{ and } b \text{ are rationals}\}$ is a field under the addition and multiplication of complex numbers.

4. Let R be a field. Prove that R is a Boolean ring if and only if R has exactly two elements.

5. Give an example of a field with exactly 30 nonzero elements.

6. Let n be a positive integer and

$$\mathbb{Z}_n[i] = \{a + bi : a \text{ and } b \in \mathbb{Z}_n\}.$$

Prove that $\mathbb{Z}_n[i]$ is a ring under addition and multiplication modulo n.

7. Give an example of a positive integer n for which $\mathbb{Z}_n[i]$ is not an integral domain.

8. Prove that $\mathbb{Z}_n[i]$ is an integral domain if and only if it is a field.

9. Let $\mathbb{Z}[\sqrt{2}] = \{a + b\sqrt{2} : a \text{ and } b \textit{ are } \text{integers}\}$. Prove that $\mathbb{Z}[\sqrt{2}]$ is an integral domain under the addition and multiplication of real numbers.

10. Prove that $\mathbb{Z}[\sqrt{2}]$ is not a field.

11. Let p be a prime number and

$$R = \left\{\frac{a}{b} : a \text{ and } b \in \mathbb{Z} \text{ and } p \text{ does not divide } b\right\}.$$

Prove that R is an integral domain under the usual addition and multiplication of rational numbers.

12. Is the above R a field?

13. Let R be a nontrivial finite ring without zero divisors. Then prove that R is with unity and that $(R - \{0\}, \cdot)$ is a group.

14. Prove that any finite commutative ring without zero divisors is a field.

15. For any simple abelian group $(G, +)$, prove that the ring End(G) of all endomorphisms of $(G, +)$ is a division ring

16. Prove that the characteristic of a finite field is a prime number.

17. Is there an integral domain having exactly 6 elements?

18. Let R be a nontrivial finite ring without zero divisors. Then prove that R is a division ring.

19. Let R and S be two rings. Then prove that the product ring $R \times S$ is an integral domain if and only if one of R and S is an integral domain and the other is the trivial ring.

20. Let R be an integral domain, $0 \neq a \in R$ and $n \in \mathbb{Z}^+$ such that $na = 0$. Prove that $\text{char}(R) > 0$.

10

Ideals and Quotient Rings

In the study of finite groups, we have proved several results using the concept of a normal subgroup, quotient construction and induction on the group order. Homomorphic images of groups are identified with quotient groups with the help of the kernel of the homomorphism which is a normal subgroup. The role of normal subgroups in groups is played by ideals in rings. The concepts of ideal and quotient rings are important in the structure theory of rings. A special kind of subrings, which are most suitable (ideal) for the study of the structure of rings, are popularly called ideals.

10.1 IDEALS

In this section, we introduce the notion of an ideal in a ring and discuss several important elementary properties of ideals.

Definition 10.1.1. Let $(R, +, \cdot)$ be a ring and I be a subgroup of $(R, +)$. Then, I is called

1. a *left ideal* of R if $ra \in I$ for all $a \in I$ and $r \in R$.
2. a *right ideal* of R if $ar \in I$ for all $a \in I$ and $r \in R$.
3. an *ideal* of R if it is both a left ideal and a right ideal of R.

Clearly any left ideal or right ideal of a ring R is a subring of R. But a subring of R may be neither a left ideal nor a right ideal. For example, the set $\mathbb{Z}$ of integers is a subring of the ring $\mathbb{R}$ of real numbers and $\mathbb{Z}$ is not an ideal of $\mathbb{R}$, since $\frac{1}{2}\cdot 1 \notin \mathbb{Z}$. If R is a commutative ring, there is no difference between a left ideal, a right ideal and an ideal. Sometimes, we refer to an ideal as a *two-sided ideal*.

Example 10.1.1

1. For any ring R, clearly $\{0\}$ and R are ideals of R and are called *trivial ideals*. $\{0\}$ is called the *zero ideal*. Ideals other than $\{0\}$ and R are called *proper ideals*.
2. If $(R, +, \cdot)$ is a ring with trivial multiplication, that is, $ab = 0$ for any a and b in R, then every subgroup of $(R, +)$ is an ideal of R.
3. Consider the ring $M_2(\mathbb{R})$ of 2×2 matrices over the real $\mathbb{R}$ and let

$$I = \left\{ \begin{pmatrix} 0 & a \\ 0 & b \end{pmatrix} : a \text{ and } b \text{ are real numbers} \right\}.$$

Then, I is a left ideal of $M_2(\mathbb{R})$, since

$$\begin{pmatrix} r & s \\ t & u \end{pmatrix}\begin{pmatrix} 0 & a \\ 0 & b \end{pmatrix} = \begin{pmatrix} 0 & ra+sb \\ 0 & ta+ub \end{pmatrix} \in I \quad \text{for all} \quad \begin{pmatrix} r & s \\ t & u \end{pmatrix} \in M_2(\mathbb{R}).$$

It can be easily verified that I is a subgroup of $M_2(\mathbb{R})$. However, I is not a right ideal of $M_2(\mathbb{R})$, since

$$\begin{pmatrix} 0 & 2 \\ 0 & 3 \end{pmatrix}\begin{pmatrix} 1 & 1 \\ 1 & 1 \end{pmatrix} = \begin{pmatrix} 2 & 2 \\ 3 & 3 \end{pmatrix} \notin I.$$

4. Let $J = \left\{ \begin{pmatrix} a & b \\ 0 & 0 \end{pmatrix} : a \text{ and } b \text{ are real numbers} \right\}$. Then, J is a right ideal of $M_2(\mathbb{R})$ and is not a left ideal.
5. For any nonnegative integer n, let

$$n\mathbb{Z} = \{na : a \text{ is an integer}\}.$$

Then, $n\mathbb{Z}$ is an ideal of the ring $\mathbb{Z}$ of integers. In fact, any ideal of $\mathbb{Z}$ is of this form $n\mathbb{Z}$ for some $n \geq 0$.

6. Let $K = \left\{ \begin{pmatrix} a & b \\ c & d \end{pmatrix} : a, b, c \text{ and } d \text{ are even integers} \right\}$. Then, K is an ideal of $M_2(\mathbb{Z})$. Later, we shall prove that $M_2(\mathbb{R})$ has no nontrivial ideals, since $\mathbb{R}$ is so. However, $\mathbb{Z}$ has several ideals and so is $M_2(\mathbb{Z})$.
7. Let X be any set and consider the ring $(\mathbb{P}(X), +, \cap)$ of all subsets of X, where $+$ and $\cap$ are defined by

   $A + B = (A - B) \cup (B - A)$ and $A \cap B =$ The intersection of A and B.

   Let $I = \{Y : Y \text{ is a finite subset of } X\}$. Then, I is an ideal of $(\mathbb{P}(X), +, \cap)$.
8. Let R be any ring and X be any nonempty set and consider the ring R^X of all mappings of X into R. For any $Y \subseteq X$, let

   $$I_Y = \{f \in R^X : f(y) = 0 \text{ for all } y \in Y\}.$$

   Then, I_Y is an ideal of R^X.

Theorem 10.1.1. Let R be a ring with unity and $U(R)$ be the set of all units in R. Then, the following are equivalent to each other for any left (right or two-sided) ideal I of R.

1. $I = R$
2. $U(R) \subseteq I$
3. $I \cap U(R) \neq \emptyset$
4. $1 \in I$

Proof: Let I be a left ideal of R. (1) $\Rightarrow$ (2) and (2) $\Rightarrow$ (3) are trivial.
(3) $\Rightarrow$ (4): Suppose that $a \in I \cap U(R)$. Then, a has multiplicative inverse a^{-1} in R and $1 = a^{-1} \cdot a \in I$ (since $a \in I$ and I is a left ideal of R).
(4) $\Rightarrow$ (1): If $1 \in I$, then, for any $r \in R$, $r = r \cdot 1 \in I$ and therefore $R \subseteq I$ so that $I = R$.

In the following, our discussion is restricted to ideals of rings. Some of the results proved for ideals can be extended to left or right ideals easily. However, we are more interested in two-sided ideals, since these lead to the construction of quotient rings. First, we discuss certain standard methods of constructing new ideals from given ones.

Theorem 10.1.2. Let $\{I_\alpha\}_{\alpha \in \Delta}$ be a nonempty class of ideals of a ring R. Then, $\bigcap_{\alpha \in \Delta} I_\alpha$ is also an ideal of R.

Proof: First note that every ideal contains the zero element of the ring; for, an ideal I is nonempty and hence there exists $a \in I$ so that $0 = 0a \in I$. Therefore, $0 \in I_\alpha$ for all $\alpha \in \Delta$. Put $I = \bigcap_{\alpha \in \Delta} I_\alpha$ Then, $0 \in I$ and hence I is a nonempty subset of R. Since the intersection of any family of subgroups of $(R, +)$ is again a subgroup, it follows that I is a subgroup of $(R, +)$. Also,

$$\begin{aligned} a \in I \text{ and } r \in R &\Rightarrow a \in I_\alpha \quad \text{for all } \alpha \in \Delta \text{ and } r \in R \\ &\Rightarrow ra \text{ and } ar \in I_\alpha \quad \text{for all } \alpha \in \Delta \\ &\Rightarrow ra \quad \text{and} \quad ar \in I. \end{aligned}$$

Thus, I is an ideal of R.

We have proved above that the intersection of ideals is again an ideal. However, the union of ideals may not be an ideal. If I and J are ideals, then they are subgroups of $(R, +)$. Therefore, $I \cup J$ is a subgroup of $(R, +)$ if and only if $I \subseteq J$ or $J \subseteq I$ (see Theorem 4.1.6). Thus, for any ideals I and J, $I \cup J$ is an ideal if and only if $I \subseteq J$ or $J \subseteq I$. For certain special classes of ideals, union of the class of ideals is again an ideal.

Theorem 10.1.3. Let $\{I_\alpha\}_{\alpha \in \Delta}$ be a class of ideals of a ring. Suppose that, for any α and $\beta \in \Delta$, there exists $\gamma \in \Delta$ such that $I_\alpha \subseteq I_\gamma$ and $I_\beta \subseteq I_\gamma$ and $I_\beta \subseteq I_\gamma$ (such classes are called directed above). Then, $\bigcup_{\alpha \in \Delta} I_\alpha$ is an ideal of R.

Proof: Let $I = \bigcup_{\alpha \in \Delta} I_\alpha$. Then, clearly I is a nonempty subset of R. Now,

$$\begin{aligned} a \text{ and } b \in I &\Rightarrow a \in I_\alpha \text{ and } b \in I_\beta \quad \text{for some } \alpha \text{ and } \beta \in \Delta \\ &\Rightarrow \text{there exists } \gamma \in \Delta \text{ such that } a \in I_\alpha \subseteq I_\gamma \quad \text{and} \quad b \in I_\beta \subseteq I_\gamma \\ &\Rightarrow a \text{ and } b \in I_\gamma, \gamma \in \Delta \\ &\Rightarrow a - b \in I_\gamma \subseteq I \\ &\Rightarrow a - b \in I. \end{aligned}$$

Therefore, I is a subgroup of $(R, +)$. Also,

$$\begin{aligned} a \in I \text{ and } r \in R &\Rightarrow a \in I_\alpha \quad \text{for some } \alpha \in \Delta \text{ and } r \in R \\ &\Rightarrow ra \quad \text{and} \quad ar \in I_\alpha \subseteq I \\ &\Rightarrow ra \quad \text{and} \quad ar \in I. \end{aligned}$$

Thus, I is an ideal of R. ◀

Corollary 10.1.1. Let $\{I_\alpha\}_{\alpha \in \Delta}$ be a chain of ideals of a ring R (that is, given any two members in the class, one of them is contained in the other). Then, $\bigcup_{\alpha \in \Delta} I_\alpha$ is again an ideal of R.

Definition 10.1.2. Let R be a ring and $S \subseteq R$. Let

$$<S> = \cap\{I : I \text{ is an ideal of } R \text{ and } S \subseteq I\}.$$

By Theorem 10.1.2, $<S>$ is an ideal of R and is called the *ideal generated by* S. Note that, for any ideal I of R, $S \subseteq I$ if and only if $<S> \subseteq I$. For this reason, we say that $<S>$ is the smallest ideal of R containing S. If $I = <S>$, then we say that I is the ideal generated by S or S generates I. An ideal I is said to be *finitely generated* if $I = <S>$ for some finite set S. If $S = \{a\}$, then $<S>$ will be denoted by $<a>$ and is called a *principal ideal* generated by a.

A natural question that arises in one's mind is about the precise form of elements in the ideal $<S>$ generated by S. Answer to such a question will be clear if we can determine the precise form of the elements in a principal ideal $<a>$. We do this in the following theorem.

Theorem 10.1.4. Let R be a ring and $a \in R$. Then, any element of $<a>$ is of the form

$$ra + as + na + \sum_{i=1}^{m} x_i a y_i,$$

where m is a nonnegative integer, n is an integer and $r, s, x_1, \ldots, x_m, y_1, \ldots, y_m \in R$.

Proof: Let A be the set of all elements of the form given in the theorem. We shall prove that A is the smallest ideal of R containing a. By taking $r = 0 = s$; $m = 0$ and $n = 1$, we have $a \in A$. If I is any ideal of R containing a, then ra, as, xay and $na \in I$ for any $r, s, x, y \in R$ and $n \in \mathbb{Z}$ and hence $A \subseteq I$. Thus, we are left with only verifying that A is an ideal of R. Using the commutativity of $+$ and the distributivity of the multiplication over the addition, one can easily prove that A is an ideal of R. Thus, $A = <a>$.

In certain special cases, $<a>$ turns out to be much simpler.

Corollary 10.1.2. Let R be a ring with unity and $a \in R$. Then,

$$<a> = \left\{\sum_{i=1}^{m} x_i a y_i : 0 \leq m \in \mathbb{Z},\ x_i \text{ and } y_i \in R\right\}.$$

Proof: Let B be the set given on the right hand side. Then, by taking $m = 1$, $x_1 = 1 = y_1$ (the unity in R), we get that $a \in B$. Since B is closed under $+$, it follows that $na \in B$ for all $n \in \mathbb{Z}$. Also, by taking $m = 1$ and $x_1 = 1$, we get that $ay \in B$ for all $y \in R$. Similarly, $xa \in B$ for all $x \in R$. Thus, by Theorem 10.1.4, $<a> \subseteq B \subseteq <a>$ and hence $<a> = B$.

Corollary 10.1.3. Let R be a commutative ring and $a \in R$. Then,

$$<a> = \{ra + na : r \in R \text{ and } n \in \mathbb{Z}\}.$$

Proof: By Theorem 10.1.4, $ra + na \in <a>$ for any $r \in R$ and $n \in \mathbb{Z}$. Also, since R *is* commutative,

$$ra + as + na + \sum_{i=1}^{m} x_i a y_i = \left(r + s + \sum_{i=1}^{m} x_i a y_i \right) a + na.$$

Thus, any element of $<a>$ is of the form $ra + na$ for some $r \in R$ and $n \in \mathbb{Z}$ and hence

$$<a> = \{ra + na : r \in R \text{ and } n \in \mathbb{Z}\}.$$

Corollary 10.1.4. Let R be a commutative ring with unity and $a \in R$. Then,

$$<a> = \{ra : r \in R\} = Ra = aR.$$

Proof: For any $r \in R$ and $n \in \mathbb{Z}$, we have

$$ra + na = ra + (n \cdot 1)a = (r + n1)a \in Ra,$$

where 1 is the unity in R. Therefore, by Corollary 10.1.3, $<a> = Ra = aR$.

For any ideals I and J of a ring R, clearly $I \cap J$ is the largest ideal contained in both I and J. In the following, we describe the smallest ideal containing both I and J. This may not be $I \cup J$, since $I \cup J$ may not be an ideal at all, unless $I \subseteq J$ or $J \subseteq I$.

Theorem 10.1.5. Let I and J be ideals of a ring R and

$$I + J = \{a + b : a \in I \text{ and } b \in J\}.$$

Then, $I + J$ is the smallest ideal of R containing both I and J; that is, $I + J = <I \cup J>$.

Proof: Since $0 \in I \cap J$, we have

$$I = I + \{0\} \subseteq I + J \text{ and } J = \{0\} + J \subseteq I + J.$$

Since I and J are subgroups of $(R, +)$, so is $I + J$. Also, for any $r \in R$, $a \in I$ and $b \in J$.

Also, for any $r \in R$, $a \in I$ and $b \in J$,

$$r(a + b) = ra + rb \in I + J$$
$$\text{and} \quad (a + b)r = ar + br \in I + J$$

and hence $I + J$ is an ideal of R. Further, if K is any ideal of R containing both I and J, then clearly $I + J \subseteq K$. Thus, $I + J$ is the smallest ideal of R containing both I and J.

Corollary 10.1.5. For any ideals $I_1, I_2, \ldots, I_n$ of a ring R, let

$$\sum_{i=1}^{n} I_i = I_1 + I_2 + \cdots + I_n = \{a_1 + a_2 + \cdots + a_n : a_i \in I_i\}.$$

Then, $\sum_{i=1}^{n} I_i$ is the smallest ideal containing $\bigcup_{i=1}^{n} I_i$.

Corollary 10.1.6. Let $\{I_\alpha\}_{\alpha \in \Delta}$ be a nonempty class of ideals of a ring R and

$$\sum_{\alpha \in \Delta} I_\alpha = \left\langle \bigcup_{\alpha \in \Delta} I_\alpha \right\rangle.$$

Then, $\sum_{\alpha \in \Delta} I_\alpha = \{a_1 + a_2 + \cdots + a_n : a_i \in I_{\alpha_i}$ for some $\alpha_i \in \Delta\}$.

Corollary 10.1.7. Let S be a nonempty subset of a ring R. Then, any element of $\langle S \rangle$ is a finite sum of elements of the form

$$ra + as + na + \sum_{i=1}^{m} x_i a y_i,$$

where $a \in S$, n and $m \in \mathbb{Z}$, $m \geq 0$, and $r, s, x_i, y_i \in R$.

Corollary 10.1.8. Let R be a commutative ring with unity and S be a nonempty subset of R. Then,

$$\langle S \rangle = \left\{ \sum_{i=1}^{m} r_i a_i : m \geq 0, r_i \in R \text{ and } a_i \in S \right\}.$$

For any ideals $I_1, I_2, \ldots, I_n$ of a ring R, we have proved that any element of $\left\langle \bigcup_{i=1}^{n} I_i \right\rangle$ can be expressed as a sum $a_1 + a_2 + \cdots + a_n$ with $a_i \in I_i$, $1 \leq i \leq n$.

However, there is no guarantee that this expression is unique, unless the ideals $I_1, I_2, \ldots, I_n$ satisfy certain additional conditions.

Theorem 10.1.6. Let $I_1, I_2, \ldots, I_n$ be ideals of a ring R and $I = I_1 + I_2 + \cdots + I_n$. Then, any element of I can be uniquely expressed as $a_1 + a_2 + \cdots + a_n$, with $a_i \in I_i$, if and only if

$$I_i \cap \left(\sum_{j \neq i} I_j \right) = \{0\} \quad \text{for all } 1 \leq i \leq n.$$

Proof: Suppose that any element a of I can be uniquely expressed as $a = a_1 + a_2 + \cdots + a_n$, with $a_i \in I_i$. Fix $1 \leq i \leq n$ and let $a \in I_i \cap \left(\sum_{j \neq i} I_j \right)$. Then, $a \in I$ and

$$a = a_1 + \cdots + a_{i-1} + a_{i+1} + \cdots + a_n, a_j \in I_j \quad \text{for } j \neq i$$

$$\text{or} \quad a_1 + \cdots + a_{i-1} - a + a_{i+1} + \cdots + a_n = 0 = 0 + 0 + \cdots + 0.$$

By the uniqueness, $a = 0$. Thus, $I_i \cap \left(\sum_{j \neq i} I_j \right) = \{0\}$. Conversely, suppose that the given condition is satisfied. Since $I = I_1 + I_2 + \cdots + I_n$, any element of I can be expressed as $a_1 + a_2 + \cdots + a_n$, with $a_i \in I_i$. Now, suppose that

$$a_1 + a_2 + \cdots + a_n = b_1 + b_2 + \cdots + b_n,$$

where $a_i, b_i \in I_i$ for $1 \leq i \leq n$. Now, for each i,

$$a_i - b_i = \sum_{j \neq i} (b_j - a_j) \in I_i \cap \left(\sum_{j \neq i} I_j \right) = \{0\}$$

and hence $a_i - b_i = 0$ or $a_i = b_i$. Thus, any element of I can be uniquely expressed as $a_1 + a_2 + \cdots + a_n$, with $a_i \in I_i$.

Corollary 10.1.9. Let I and J be ideals of a ring R. Then, any element of R can be uniquely expressed as $a + b$ with $a \in I$ and $b \in J$ if and only if

$$I + J = R \quad \text{and} \quad I \cap J = \{0\}.$$

Definition 10.1.3. An ideal I of a ring R is said to be a *direct summand* of R if there is an ideal J of R such that $I + J = R$ and $I \cap J = \{0\}$. In this case, R is said to be the *direct sum* of I and J and denote this by $R = I \oplus J$. Also, I and J are called *direct complements* to each other if $R = I \oplus J$.

In a ring with unity, we can have yet another beautiful description of direct summands. Before going to this, let us define the following definition.

Definition 10.1.4. Let R be a ring and $a \in R$. Then, a is called a *central idempotent* if $a^2 = a$ and $ax = xa$ for all $x \in R$; that is, a is an idempotent belonging to the centre of R.

Theorem 10.1.7. Let R be a ring with unity and I be an ideal of R. Then, I is a direct summand of R if and only if I is the principal ideal generated by a central idempotent of R.

Proof: First note that, for any central idempotent e in R, the principal ideal generated by e is of the form

$$<e> = eR = \{ex : x \in R\}.$$

Suppose that $I = <e>$ for some central idempotent e. Then, put $J = (1 - e)R = \{(1 - e)x : x \in R\}$. Since

$$(1 - e)^2 = 1 - e - e + e^2 = 1 - e - e + e = 1 - e$$
$$\text{and} \quad (1 - e)x = x - ex = x - xe = x(1 - e)$$

for all $x \in R$, we get that $1 - e$ is also a central idempotent in R and $J = <1 - e>$, the principal ideal generated by $1 - e$. Now, for any $x \in R$, we can write

$$x = ex + (1 - e)x \in I + J$$

and hence $I + J = R$, Also,

$$a \in I \cap J \Rightarrow a = ex = (1 - e)y \quad \text{for some } x, y \in R$$
$$\Rightarrow a = ex = e^2x = ea = e(1 - e)y = 0$$

and therefore $I \cap J = \{0\}$. Thus, $R = I \oplus J$ and I is a direct summand of R.

Conversely, suppose that I is a direct summand of R. Then, there is an ideal J or R such that $R = I \oplus J$; that is, $I + J = R$ and $I \cap J = \{0\}$. Since R is with unity, $1 \in R = I + J$ and hence $1 = e + s$ for some $e \in I$ and $s \in J$, clearly $s = 1 - e$. Now,

$$es \in I \cap J = \{0\} \quad \text{and hence } es = 0.$$

From $1 = e + s$, we have

$$e = e(e + s) = e^2 + es = e^2 \quad (\text{since } es = 0)$$

and therefore e is an idempotent. Also, for any $x \in R$,

$$ex + sx = (e + s)x = x = x(e + s) = xe + xs$$

and therefore $ex - xe = xs - sx \in I \cap J = \{0\}$ (since $e \in I$ and $s \in J$), so that $ex - xe = 0$ or $ex = xe$. Thus, e is a central idempotent in R. Now, since $e \in I$, we get that $<e> \subseteq I$. Also,

$$\begin{aligned} x \in I &\Rightarrow x = (e + s)x = ex + sx = ex \quad (\text{since } sx \in I \cap J = \{0\}) \\ &\Rightarrow x \in <e>. \end{aligned}$$

Thus, $I = eR = <e>$.

The central idempotents in any ring have certain nice properties. They form a Boolean ring under suitable operations, defined in the following theorem.

Theorem 10.1.8. Let $(R, +, \cdot)$ be a ring and $B(R)$ be the set of all central idempotents in R. For any a and $b \in B(R)$, define

$$a * b = a + b - 2ab\ (= (a - ab) + (b - ab)).$$

Then, $(B(R), *, \cdot)$ is a Boolean ring. Also, if R is with unity, then so is $B(R)$.

Proof: First observe that, for any a and $b \in B(R)$, $a * b$ and $a \cdot b$ are central idempotents of R; i.e.,

$$\begin{aligned} (a * b)^2 &= (a + b - 2ab)^2 = a + b - 2ab \\ (a \cdot b)^2 &= abab = ab \\ (a * b)x &= (a + b - 2ab)x = x(a + b - 2ab) = x(a * b) \\ \text{and} \quad (a \cdot b)x &= abx = axb = xab = x(a \cdot b). \end{aligned}$$

It is a straight forward verification to prove that all the axioms of rings are satisfied in $B(R)$ with $*$ as addition and $\cdot$ as multiplication. Note that $a \cdot a = a$ and $a * a = 0$ for all $a \in B(R)$ and 0 is the zero element in $B(R)$ also. If R has unity 1, then $1 \in B(R)$ and 1 is the unity in $B(R)$ also.

In addition to the two binary operations $\cap$ and $+$ on the set of ideals of a ring, we introduce yet another binary operation, which is denoted by juxtaposition, in the following definition.

Definition 10.1.5. For any ideals I and J of a ring R, define

$$IJ = \left\{ \sum_{i=1}^{n} a_i b_i : n \in \mathbb{Z}^+,\ a_i \in I \text{ and } b_i \in J \right\}.$$

Theorem 10.1.9. Let I and J be ideals of a ring R. Then, IJ is an ideal of R and $IJ \subseteq I \cap J$.

Proof: If x and $y \in IJ$, then $x = \sum_{i=1}^{n} a_i b_i$ and $y = \sum_{j=1}^{m} c_j d_j$, where $n, m \in \mathbb{Z}^+$, a_i and $c_j \in I$ and b_i and $d_j \in J$. Then,

$$x - y = a_1 b_1 + \cdots + a_n b_n + (-c_1)d_1 - \cdots + (-c_m)d_m$$

and hence $x - y \in IJ$, so that IJ is a subgroup of $(R, +)$.
Also, if $r \in R$ and $x = \sum_{i=1}^{n} a_i b_i \in IJ$, then

$$rx = \sum_{i=1}^{n} (ra_i) b_i \quad \text{and} \quad xr = \sum_{i=1}^{n} a_i (b_i r)$$

and ra_i, $a_i \in I$ and b_i, $b_i r \in J$ and hence rx and xr belong to IJ. Thus, IJ is an ideal of R. Clearly $IJ \subseteq I$ and J.

There is an important observation that an ideal of an ideal need not be an ideal; that is, if I is an ideal of a ring R, then I can be treated as a ring on its own (since I is a subring of R) and, if J is an ideal of I, then J need not be an ideal of the ring R. This is illustrated in the following example.

Example 10.1.2. Let $\mathbb{R}$ be the ring of real numbers under the usual addition and multiplication. Consider the ring $\mathbb{R}^{\mathbb{R}}$ of all mapping of $\mathbb{R}$ into $\mathbb{R}$ under the point-wise addition and multiplication. Let

$$S = \{f \in \mathbb{R}^{\mathbb{R}} : f \text{ is continuous}\},$$

where $\mathbb{R}$ is with the usual topology. Then, S is a subring of $\mathbb{R}^{\mathbb{R}}$ and hence S is a commutative ring with unity (the constant map 1 is the unity in S). Let i be the identity map defined by $i(x) = x$ for all $x \in \mathbb{R}$. Then, $i \in S$. Now, let

$$I = \{if : f \in S \text{ and } f(0) = 0\}$$
$$\text{and} \quad J = \{i^2 f + ni^2 : f \in S, f(0) = 0 \text{ and } n \in \mathbb{Z}\}.$$

Then, I is an ideal of the ring S and J is an ideal of I. However, J fails to be an ideal of S, since $i^2 \in J$, $\frac{1}{2} \in S$ and $\frac{1}{2} i^2 \notin J$, where $\frac{1}{2}$ denotes the constant map of $\mathbb{R}$ which maps every element of $\mathbb{R}$ onto the real number $\frac{1}{2}$. The assumption $\frac{1}{2}i^2 \in J$ leads to a contradiction; for, let $\frac{1}{2}i^2 \in J$. Then,

$$\frac{1}{2}i^2 = i^2f + ni^2 \text{ for some } f \in S \text{ with } f(0) = 0 \quad \text{and} \quad n \in \mathbb{Z}.$$

Therefore, $fi^2 = (\frac{1}{2} - n)i^2$; that is, $f(x)x^2 = (\frac{1}{2} - n)x^2$ for all $x \in \mathbb{R}$. If $x \neq 0$, $f(x) = \frac{1}{2} - n$. Therefore, f is a nonzero constant function on $\mathbb{R} - \{0\}$ and $f(0) = 0$. This is a contradiction to the fact that f is continuous.

Next we discuss a characterization theorem for ideals of a matrix ring $M_n(R)$, where R is an arbitrary ring with unity.

Theorem 10.1.10. Let R be a ring with unity, n be a positive integer and $M_n(R)$ be the ring of $n \times n$ matrices over R. For any $I \subseteq M_n(R)$, I is an ideal of $M_n(R)$ if and only if $I = M_n(J)$ for some ideal J of R.

Proof: It can be easily verified that $M_n(J)$ is an ideal of $M_n(R)$ for any ideal J of R. Conversely suppose that I is an ideal of $M_n(R)$.
For any $1 \leq i, j \leq n$, let E_{ij} be the $n \times n$ matrix over R such that the ij^{th} entry is 1 and all other entries are 0. Then, any $n \times n$ matrix (a_{ij}) can be expressed as

$$(a_{ij}) = \sum_{i,j=1}^{n} a_{ij}E_{ij}$$

and $E_{ij}E_{kr} = \delta_{jk}E_{ir}$, where $\delta_{jk} = \begin{cases} 1 & \text{if } j = k \\ 0 & \text{if } j \neq k \end{cases}$.

Now, consider the given ideal I of $M_n(R)$ and define

$$J = \{a \in R : a = a_{11} \text{ for some } (a_{ij}) \in I\}.$$

That is, J is the set of all 11 entries (entries in the first column and first row) of the matrices belonging to I. Since $I \neq \varnothing$, J is a nonempty subset of R. Clearly $a - b \in J$ for any a and $b \in J$. Next, suppose that $a \in J$ and $r \in R$. Then, $a = a_{11}$ for some $(a_{ij}) \in I$. Since I is an ideal of $M_n(R)$, we have

$$ra\,E_{11} = r\,E_{11}\left(\sum_{i,j=1}^{n} a_{ij}E_{ij}\right)E_{11} = rE_{11}(a_{ij})E_{11} \in I$$

and $$ar\,E_{11} = E_{11}\left(\sum_{i,j=1}^{n} a_{ij}E_{ij}\right)rE_{11} = E_{11}(a_{ij})rE_{11} \in I$$

and hence ra and $ar \in J$ (since ra is the 11th entry of raE_{11} and ar is the 11th entry of arE_{11}). Thus, J is an ideal of R. We shall prove that $I = M_n(J)$.
Let $A = (a_{ij}) \in I$. Then, $A = \sum_{i,j=1}^{n} a_{ij}E_{ij}$. For any $1 \leq i, j \leq n$, consider

$$E_{1i}AE_{j1} = E_{1i}\left(\sum_{r,s=1}^{n} a_{rs}E_{rs}\right)E_{j1} = \sum_{r,s=1}^{n} a_{rs}\delta_{ir}E_{1s}E_{j1}$$
$$= \sum_{r,s=1}^{n} a_{rs}\delta_{ir}\delta_{sj}E_{11} = a_{ij}E_{11}.$$

Since $A \in I$ and $a_{ij}E_{11} = E_{1i}AE_{j1} \in I$, we get that $a_{ij} \in J$. Therefore, $I \subseteq M_n(J)$.
On the other hand, let $A = (a_{ij}) \in M_n(J)$. Then, $a_{ij} \in J$ for all i and j. Now, for any $1 \leq i, j \leq n$, $a_{ij} \in J$ and hence there exists $B = (b_{rs}) \in I$ such that $b_{11} = a_{ij}$. Now,

$$E_{i1}BE_{1j} = E_{i1}\left(\sum_{r,s=1}^{n} b_{rs}E_{rs}\right)E_{1j} = \sum_{r,s=1}^{n} b_{rs}E_{i1}E_{rs}E_{1j}$$
$$= \sum_{r,s=1}^{n} b_{rs}\delta_{1r}E_{is}E_{1j}$$
$$= \sum_{r,s=1}^{n} b_{rs}\delta_{1r}\delta_{s1}E_{ij} = b_{11}E_{ij} = a_{ij}E_{ij}.$$

Therefore, $a_{ij}E_{ij} = E_{i1}BE_{1j} \in I$, since I is an ideal and $B \in I$. Now, $A = \sum_{i,j=1}^{n} a_{ij}E_{ij} \in I$. Therefore, $M_n(J) \subseteq I$. Thus, $I = M_n(J)$ and J is an ideal of R.

The above theorem is false for rings without unity. This is illustrated in the following example.

Example 10.1.3. Consider the ring $2\mathbb{Z}$ of even integers. Note that $2\mathbb{Z}$ has no unity. Consider the ring $M_2(2\mathbb{Z})$ of 2×2 matrices over $2\mathbb{Z}$. Let

$$I = \{(a_{ij}) \in M_2(2\mathbb{Z}) : a_{12} \in 4\mathbb{Z}\}.$$

It can be easily checked that I is an ideal of $M_2(2\mathbb{Z})$. Note that $I \neq M_2(J)$ for any ideal J of $2\mathbb{Z}$.

Theorem 10.1.10 is useful only to the extent that we can describe the ideals of a ring $\mathbb{R}$, which is not usually easy to do, although it is easy for the ring $\mathbb{Z}$ of integers (recall that any ideal of $\mathbb{Z}$, being a subgroup of $(\mathbb{Z}, +)$, is generated by a nonnegative integer). However, the ideals of fields are easy to describe.

Theorem 10.1.11. Let R be a nontrivial commutative ring with unity. Then, R is a field if and only if $\{0\}$ and R are the only ideals of R.

Proof: Clearly $\{0\}$ and R are ideals of R and these are distinct, since R is nontrivial. If R is a field and $I \neq \{0\}$ is an ideal of R, then there exists $0 \neq a \in I$ and hence $1 = a^{-1}a \in I$, so that $I = R$. Conversely suppose that $\{0\}$ and R are the only ideals of R. Let $0 \neq a \in R$ and $I = aR = \langle a \rangle$. Then, I is a nonzero ideal of R and hence $I = R$. In particular, $1 \in R = I = aR$ and hence $1 = ab$ for some $b \in R$. Therefore, a is a unit. Thus, R is a field. ◀

Corollary 10.1.10. The ring $\mathbb{R}$ of real numbers (or the ring $\mathbb{C}$ of complex numbers) has only two ideals, namely $\{0\}$ and the whole ring.

Corollary 10.1.11. For any positive integer n, the ring $M_n(\mathbb{R})$ of $n \times n$ matrices over $\mathbb{R}$ has exactly two ideals namely $\{0\}$ and the whole ring $M_n(\mathbb{R})$.

Since $M_n(R)$ is a noncommutative ring for $n > 1$, $M_n(\mathbb{R})$ is not a field (and not a division ring) even though it has only two ideals. This says that the commutativity of the ring R in Theorem 10.1.11 cannot be dropped from the hypothesis. Nontrivial rings having only two ideals are called *simple rings*. $M_n(\mathbb{R})$ is a simple ring for all $n \in \mathbb{Z}^+$.

Worked Exercise 10.1.1. Let I and J be ideals of a ring R with unity. Then prove that $R = I \oplus J$ if and only if there are central idempotents a and b in R such that $a + b = 1$, $ab = 0$, $I = aR$ and $J = bR$.

Answer: Suppose that $R = I \oplus J$. Then, $I + J = R$ and $I \cap J = \{0\}$. Since $1 \in R = I + J$, we have

$$1 = a + b \quad \text{for some } a \in I \text{ and } b \in J.$$

Then, since $ab \in I \cap J = \{0\}$, we have $ab = 0$. Therefore, $a = a \cdot 1 = a(a + b) = a^2 + ab = a^2$. Similarly, $b = b^2$. Also, for any $x \in R$,

$$ax + bx = (a + b)x = x = x(a + b) = xa + xb$$

$$\text{and hence} \quad ax - xa = xb - bx \in I \cap J = \{0\}$$

so that $ax = xa$ and $bx = xb$. Therefore, a and b are central idempotents in R, Also, since $a \in I$, $aR \subseteq I$. Further, if $x \in I$, then $bx \in I \cap J = \{0\}$ and hence

$$x = (a + b)x = ax + bx = ax + 0 = ax \in aR.$$

Thus, $I = aR$. Similarly, $J = bR$.
Conversely suppose that a and b are central idempotents such that $a + b = 1$, $ab = 0$, $aR = I$ and $bR = J$. Then, for any $x \in R$,

$$x = (a+b)x = ax + bx \in I + J$$

and hence $I + J = R$. Also,

$$\begin{aligned} x \in I \cap J &\Rightarrow x = ay = bz \quad \text{for some } y \text{ and } z \in R \\ &\Rightarrow x = a^2y = ax = a(bz) = abz = 0 \end{aligned}$$

and hence $I \cap J = \{0\}$. Thus, $R = I \oplus J$.

Worked Exercise 10.1.2. Let I, J and K be ideals of a ring R such that $I \subseteq K$. Then prove that

$$I + (J \cap K) = (I + J) \cap K.$$

This is known as *modular law*.

Answer: Since $I \subseteq I + J$ and $I \subseteq K$, we have $I \subseteq (I + J) \cap K$. Also, $J \cap K \subseteq (I + J) \cap K$. Therefore, $I + (J \cap K) \subseteq (I + J) \cap K$. On the other hand, let $x \in (I + J) \cap K$. Then, $x \in K$ and $x = a + b$ for some $a \in I$ and $b \in J$. Now,

$$b = x - a \in K \quad (\text{since } x \in K \text{ and } a \in I \subseteq K)$$

and hence $b \in J \cap K$ and $x = a + b \in I + (J \cap K)$. Therefore, $(I + J) \cap K \subseteq I + (J \cap K)$ Thus, $I + (J \cap K) = (I + J) \cap K$.

Worked Exercise 10.1.3. Consider the ring Z of integers. For any positive integers n and m, let $I = n\mathbb{Z}$ and $J = m\mathbb{Z}$. Then, compute $I + J$, $I \cap J$ and IJ.

Answer: Let (m, n) and $[m, n]$ be the greatest common divisor and least common multiple of m and n, respectively. Note that $a \in n\mathbb{Z}$ if and only if a is a multiple of n (or n divides a). Therefore,

$$\begin{aligned} a \in I + J &\Leftrightarrow a \in n\mathbb{Z} + m\mathbb{Z} \\ &\Leftrightarrow a = nx + my \quad \text{for some } x, y \in \mathbb{Z} \\ &\Leftrightarrow (m, n) \text{ divides } a \\ &\Leftrightarrow a \in (m, n)\mathbb{Z} \\ a \in I \cap J &\Leftrightarrow a \in n\mathbb{Z} \cap m\mathbb{Z} \end{aligned}$$

$$\Leftrightarrow a \text{ is a common multiple of } n \text{ and } m$$
$$\Leftrightarrow a \text{ is a multiple of } [m, n]$$
$$\Leftrightarrow a \in [m, n]\mathbb{Z}$$

$$a \in IJ \Leftrightarrow a = \sum_{i=1}^{r} x_i y_i,\ x_i \in n\mathbb{Z} \quad \text{and} \quad y_i \in m\mathbb{Z}$$

$$\Leftrightarrow a = \sum_{i=1}^{r} x_i y_i,\ n \text{ divides } x_i \text{ and } m \text{ divides } y_i$$

$$\Leftrightarrow mn \text{ divides } a$$
$$\Leftrightarrow a \in mn\mathbb{Z}.$$

Thus, $I + J = (m, n)\mathbb{Z}$, $I \cap J = [m, n]\mathbb{Z}$ and $IJ = mn\mathbb{Z}$.

Worked Exercise 10.1.4. Let $f : R \to S$ be a homomorphism of rings. If J is an ideal of S, then prove that $f^{-1}(J)$ is an ideal of R. Also, if f is an epimorphism and I is an ideal of R, prove that $f(I)$ is an ideal of S.

Answer: Let J be an ideal of S. Then,

$$f^{-1}(J) = \{a \in R : f(a) \in J\} \neq \emptyset, \quad \text{since } f(0) = 0 \in J.$$

For any a and $b \in R$, we have

$$a \text{ and } b \in f^{-1}(J) \Rightarrow f(a) \quad \text{and} \quad f(b) \in J$$
$$\Rightarrow f(a - b) = f(a) - f(b) \in J$$
$$\Rightarrow a - b \in f^{-1}(J)$$

and $\quad a \in f^{-1}(J) \text{ and } x \in R \Rightarrow f(a) \in J \quad \text{and} \quad f(x) \in S$

$$\Rightarrow f(ax) = f(a)f(x) \in J$$

and $\quad f(xa) = f(x)f(a) \in J \Rightarrow ax \quad \text{and} \quad xa \in f^{-1}(J).$

Thus, $f^{-1}(J)$ is an ideal of R.
Next, let f be an epimorphism and I be an ideal of R. Then,

$$f(I) = \{f(a) : a \in I\} \neq \emptyset, \quad \text{since } I \neq \emptyset.$$

Now, we have

$$x \text{ and } y \in f(I) \Rightarrow x = f(a) \quad \text{and} \quad y = f(b),\ a \text{ and } b \in I$$
$$\Rightarrow x - y = f(a) - f(b) = f(a - b) \in f(I)$$

$$\begin{aligned} \text{and} \quad x \in f(I) \text{ and } s \in S &\Rightarrow x = f(a) \quad \text{for some } a \in I \text{ and} \\ s &= f(r) \quad \text{for some } r \in R \\ &\Rightarrow xs = f(a)f(r) = f(ar) \in f(I) \\ \text{and} \quad sx &= f(r)f(a) = f(ra) \in f(I). \end{aligned}$$

Thus, $f(I)$ is an ideal of S.

Worked Exercise 10.1.5. Let R be a ring with unity and $I \subseteq R \times R$. Prove that I is an ideal of the ring $R \times R$ if and only if $I = I_1 \times I_2$ for some ideals I_1 and I_2 of R.

Answer: Suppose that I is an ideal of $R \times R$. Put

$$I_1 = \{a \in R : (a, b) \in I \text{ for some } b \in R\}$$
$$\text{and} \quad I_2 = \{b \in R : (a, b) \in I \text{ for some } a \in R\}.$$

Then, I_1 and I_2 are ideals of R, since $I_1 = p_1(I)$ and $I_2 = p_2(I)$, where p_1 and $p_2 : R \times R \to R$ are the first and second projections, respectively and since p_1 and p_2 are epimorphisms of rings. Also,

$$\begin{aligned} (a, b) \in I &\Rightarrow a \in I_1 \quad \text{and} \quad b \in I_2 \\ &\Rightarrow (a, b) \in I_1 \times I_2. \end{aligned}$$

Therefore, $I \subseteq I_1 \times I_2$. On the other hand,

$$\begin{aligned} (a, b) \in I_1 \times I_2 &\Rightarrow a \in I_1 \quad \text{and} \quad b \in I_2 \\ &\Rightarrow (a, c) \in I_1 \quad \text{and} \quad (d, b) \in I_2 \quad \text{for some } c, d \in R. \\ &\Rightarrow (a, 0) = (1, 0)(a, c) \in I \quad \text{and} \quad (0, b) = (0, 1)(d, b) \in I \\ &\Rightarrow (a, b) = (a, 0) + (0, b) \in I \end{aligned}$$

and therefore $I_1 \times I_2 \subseteq I$. Thus, $I = I_1 \times I_2$.
Converse can be easily proved.

EXERCISE 10(A)

1. Determine all the ideals in each of the following rings under the operations.
   (i) The ring $\mathbb{Z}$ of integers.
   (ii) The ring $\mathbb{Q}$ of rational numbers.

(iii) The ring $\mathbb{R}$ of real numbers.

(iv) The ring $\mathbb{C}$ of complex numbers.

(v) The ring $\mathbb{Q}_{\mathbb{R}}$ of real quaternions.

(vi) The ring $\mathbb{Z}_n$ of integers modulo n for any $n > 0$.

(vii) $M_2(\mathbb{R})$, the ring of 2×2 real matrices over $\mathbb{R}$.

(viii) The ring $M_n(\mathbb{Z})$ of $n \times n$ matrices over $\mathbb{Z}$, for any $n > 0$.

(ix) $\mathbb{Z}_{12}$

(x) $\mathbb{Z}_{13} \times \mathbb{Z}_{13}$

2. Which of the following are true? Substantiate your answers.
   (i) $\mathbb{Z}$ is an ideal of $\mathbb{Q}$.
   (ii) $\mathbb{Q}$ is an ideal of $\mathbb{R}$.
   (iii) Every subring of a ring R is an ideal of R.
   (iv) For any ideal I of a ring R, $I + I = I$.
   (v) For any ideal I of a ring R, $I = \{a - b : a, b \in I\}$.
   (vi) There is a finite ideal in any ring.
   (vii) There can be infinitely many ideals in a finite ring.
   (viii) There is a ring with exactly three ideals.

3. For any subsets S and T of a ring R, prove that

$$<S \cup T> = <S> + <T>.$$

Is $<S \cap T> = <S> \cap <T>$ true?

4. Let R be a ring with unity. If R is a division ring, prove that R has only two ideals. Is the converse true?

5. Let R be a commutative ring and $A \subseteq R$. Prove that

$$A^* = \{x \in R : xa = 0 \quad \text{for all } a \in A\}$$

is an ideal of R. A^* is called the *annihilator* of A in R.

6. For any ideals I and J of a commutative ring R, prove the following:
   (i) $(I + J)^* = I^* \cap J^* = (I \cup J)^*$
   (ii) $I \subseteq J \Rightarrow J^* \subseteq I^*$
   (iii) $I^* + J^* \subseteq (I \cap J)^*$

7. Let I be an ideal of the ring $\mathbb{Z}$ of integers and $13 \in I$. Then prove that $I = 13\mathbb{Z}$ or $I = \mathbb{Z}$.

8. Prove the Exercise 7 above with an arbitrary prime number in place of 13.

9. Let R be a commutative ring and I be an ideal of R. Let

$$r(I) = \{a \in R: a^n \in I \text{ for some } n \in \mathbb{Z}^+\}$$

Prove that $r(I)$ is an ideal of R containing I.

10. For any ideals I and J of a commutative ring R, prove the following:
    (i) $r(I \cap J) = r(I) \cap r(J)$
    (ii) $I \subseteq J \Rightarrow r(I) \subseteq r(J)$
    (iii) $r(I) + r(J) \subseteq r(I + J)$
    (iv) $r(I + J) = r(r(I) + r(J))$

11. For any ideas I and J of a ring R, let

$$(I : J) = \{x \in R : xa \in J \text{ for all } a \in I\}.$$

Prove that $(I : J)$ is a left ideal of R.

12. Express each of the following in the form of $n\mathbb{Z}$ for a suitable n.
    (i) $8\mathbb{Z} \cap 12\mathbb{Z}$
    (ii) $6\mathbb{Z} + 9\mathbb{Z}$
    (iii) $(6\mathbb{Z} : 9\mathbb{Z})$
    (iv) $r(12\mathbb{Z})$
    (v) $(12\mathbb{Z})^*$
    (vi) $(9\mathbb{Z} : 6\mathbb{Z})$

13. For any positive integers m and n, express each of the following in the form of $d\mathbb{Z}$ for a suitable integer d:
    (i) $m\mathbb{Z} \cap n\mathbb{Z}$
    (ii) $m\mathbb{Z} + n\mathbb{Z}$
    (iii) $r(m\mathbb{Z})$
    (iv) $(n\mathbb{Z} : m\mathbb{Z})$
    (v) $(m\mathbb{Z})^*$

14. Consider the ring $\mathbb{Z}[\sqrt{3}] = \{a + b\sqrt{3} : a \text{ and } b \in \mathbb{Z}\}$, under the usual addition and multiplication of real numbers. Prove that the set

$$I = \{a + b\sqrt{3} : a,\ b \in \mathbb{Z} \text{ and } a - b \text{ is even}\}$$

is an ideal of $\mathbb{Z}[\sqrt{3}]$.

15. Let R be a ring without zero divisions. If every subring of R is an ideal of R, then prove that R is commutative.

16. Let S be a subset of a ring R and define

$$\text{Ann}_l(S) = \{x \in R : xs = 0 \text{ for all } s \in S\}.$$

Then prove that $\text{Ann}_l(S)$ is a left ideal of R. If S is a left ideal of R, then prove that $\text{Ann}_l(S)$ is an ideal of R. $\text{Ann}_l(S)$ is called the *left annihilator* of S.

17. Define the notion of the right annihilator $\text{Ann}_r(S)$ of S and formulate and prove a statement similar to the Exercise 16 for right annihilators.

18. Prove that every ideal of $\mathbb{Z}_n$ is a principal ideal.

19. Let R be commutative ring in which $\{0\}$ and R are the only ideals. Then prove that R is field or R is a finite ring with $|R|$ as a prime and $ab = 0$ for all a and $b \in R$.

20. Prove that the set

$$I = \{a + b\sqrt{2} : a \text{ and } b \in \mathbb{Z} \text{ and } a \text{ is even}\}$$

is an ideal of the ring $\mathbb{Z}[\sqrt{2}]$.

21. In the ring $\mathbb{Z}[i]$ of Gaussian integers, prove that the set

$$I = \{a + bi : a \text{ and } b \text{ are even}\}$$

is an ideal and find the annihilator I^*.

22. If R is a simple ring with unity, then prove that the ring $M_n(R)$ of all $n \times n$ matrices is a simple ring.

10.2 QUOTIENT RINGS

The concept and construction of quotient rings are same as for quotient groups. For an ideal I of a ring R, I is a subgroup of the abelian group $(R, +)$ and hence I is a normal subgroup of the group $(R, +)$ and therefore, as in Theorem 4.6.1, we can construct the quotient group $(R, +)/I$ as the group of all cosets $a + I$, $a \in R$ under the operation defined by

$$(a + I) + (b + I) = (a + b) + I.$$

Then, $(R, +)/I$ is an abelian group. Since R is a ring, we have the multiplication in R and it is natural to ask whether the quotient group $(R, +)/I$ has a corresponding ring structure using $(a + I) \cdot (b + I) = ab + I$ as the multiplication. This is answered positively in the following theorem.

Theorem 10.2.1. Let I be an ideal of a ring $(R, +, \cdot)$ and

$$R/I = \{a + I : a \in R\}.$$

For any $a + I$ and $b + I$ in R/I, define

$$(a + I) + (b + I) = (a + b) + I$$
$$\text{and } (a + I) \cdot (b + I) = ab + I.$$

Then, $(R/I, +, \cdot)$ is a ring.

Proof: Since I is a subgroup of $(R, +)$, which is an abelian group, I becomes a normal subgroup of $(R, +)$. Therefore, by Theorem 4.6.1, $(R/I, +)$ is an abelian group. Next, with regard to the multiplication in R/I, we should first prove that the operation on R/I is well defined.

To do this, for any a, b, a' and $b' \in R$, we have

$$\left.\begin{matrix} a+I=a'+I \\ b+I=b'+I \end{matrix}\right\} \Rightarrow a-a' \in I \quad \text{and} \quad b-b' \in I$$

$$\begin{aligned} &\Rightarrow (a-a')\,b \in I \quad \text{and} \quad a'(b-b') \in I \\ &\Rightarrow ab-a'b' = (a-a')b + a'(b-b') \in I \\ &\Rightarrow ab+I = a'b'+I. \end{aligned}$$

Therefore, the multiplication on R/I depends on the cosets, but not on their representatives. For any a, b and $c \in R$, we have

$$\begin{aligned} ((a+I)\cdot(b+I))\cdot(c+I) &= (ab)c + I = a(bc)I \\ &= (a+I)\cdot((b+I)\cdot(c+I)). \end{aligned}$$

Therefore, $\cdot$ is associative. Also,

$$\begin{aligned} (a+I)\cdot((b+I)+(c+I)) &= a(b+c)+I \\ &= (ab+ac)+I \\ &= (ab+I)+(ac+I) \\ &= ((a+I)(b+I)) + ((a+I)(c+I)). \end{aligned}$$

Therefore, $\cdot$ distributes over $+$ from left and, similarly from right also. Thus, $(R/I, +, \cdot)$ is a ring.

Definition 10.2.1. For any ideal I of a ring R, the ring $(R/I, +, \cdot)$ constructed above is called *quotient ring of R by I* or *factor ring of R by I*.

Note 10.2.1

1. The zero element in the quotient ring R/I is $0 + I = I$, where 0 is the zero element in R.
2. If the ring R has unity 1, then R/I also has unity, namely $1 + I$. The converse may not hold good. That is, R/I may have unity while R has no unity. For example, R is an ideal of R and the quotient R/R is the trivial ring which obviously has unity (when a ring has only one element, then that element is the additive identity as well as the multiplicative identity).
3. If R is a commutative ring, then the quotient R/I is also commutative ring for any ideal I or R.

In group theory, we have defined the concept of the kernel of a homomorphism and proved that it is a normal subgroup of the domain group and conversely, any normal subgroup is the kernel of a homomorphism. These results are extended to the case of rings in the following definition.

Definition 10.2.2. Let $f : R \to S$ be a homomorphism of rings. Then, the *Kernel* of f is defined to be the set

$$\ker f = \{a \in R : f(a) = 0 \text{ in } S\}.$$

Theorem 10.2.2. The kernel of any homomorphism of rings is an ideal of the domain ring.

Proof: Let $f : R \to S$ be a homomorphism of rings. Then, f is a homomorphism of the group $(R, +)$ into the group $(S, +)$ and hence $\ker f$ is a subgroup of $(R, +)$. Also,

$$\begin{aligned} a \in \ker f \text{ and } r \in R &\Rightarrow f(a) = 0 \quad \text{and} \quad r \in R \\ &\Rightarrow f(ra) = f(r)f(a) = f(r)0 = 0 \end{aligned}$$

and $\quad f(ar) = f(a)f(r) = 0f(r) = 0 \Rightarrow ra \quad$ and $\quad ar \in \ker f.$

Thus, $\ker f$ is an ideal of R.

We prove the converse of the above result; that is, any ideal I of a ring R is the kernel of a homomorphism of R into some ring S.

Theorem 10.2.3. Let I be an ideal of a ring R. Then, there exists a ring S and a homomorphism $f : R \to S$ such that $I = \ker f$.

Proof: Consider the quotient ring R/I and define $f : R \to R/I$ by $f(a) = a + I$ for any $a \in R$. Then, f is a homomorphism of rings; for,

$$\begin{aligned} f(a + b) &= (a + b) + I = (a + I) + (b + I) = f(a) + f(b) \\ \text{and} \quad f(ab) &= ab + I = (a + I)(b + I) = f(a)f(b) \end{aligned}$$

for any a and $b \in R$. Also,

$$\begin{aligned} \ker f &= \{a \in R : f(a) = 0 \text{ in } R/I\} \\ &= \{a \in R : a + I = I\} = I \end{aligned}$$

Definition 10.2.3. For any ideal I of a ring R, the homomorphism $f: R \to R/I$, defined by $f(a) = a + I$ for any $a \in R$, is called the *canonical homomorphism* or *natural homomorphism*. Actually, f is an epimorphism, since any element of R/I is of the form $a + I$ for some $a \in R$.

The fundamental theorem of homomorphisms of groups proved in Theorem 5.2.1 is extended for rings in the following theorem.

Theorem 10.2.4 (Fundamental Theorem of Homomorphism for Rings). Let $f: R \to S$ be a homomorphism of rings. Then, $f(R)$ is a subring of S and

$$R/\ker f \cong f(R).$$

In particular, if f is an epimorphism, then $R/\ker f \cong S$.

Proof: The proof is same as that of Theorem 5.2.1, except that the map $g : R/K \to f(R)$ defined by

$$g(a + k) = f(a)$$

is a ring homomorphism also, where $k = \ker f$.
This is clear from

$$g((a+k)(b+k)) = g(ab + k) = f(ab) = f(a)f(b).$$

The above fundamental theorem can be restated as 'any homomorphic image of a ring R is isomorphic to a quotient ring of R'. This is not only a fundamental result but also an important tool in proving several isomorphism theorems for quotient ring. Some of these are listed below and their proofs are similar to those proved in Section 5.3.

Theorem 10.2.5. For any ideals I and J of a ring R,

$$\frac{I}{I \cap J} \cong I + \frac{J}{J}.$$

Theorem 10.2.6. Let $f: R \to S$ be an epimorphism of rings and I be an ideal of R such that $\ker f \subseteq I$. Then, $f(I)$ is an ideal of S and

$$\frac{R}{I} \cong \frac{S}{f(I)}.$$

Theorem 10.2.7. Let $f: R \to S$ be an epimorphism of rings and J be an ideal of S. Then, $f^{-1}(J)$ is an ideal of R and

$$\frac{R}{f^{-1}(J)} \cong \frac{S}{J}.$$

Theorem 10.2.8. Let I and J be ideals of a ring and $I \subseteq J$. Then, J/I is an ideal of R/I and

$$(R/I)/(J/I) \cong R/J.$$

Further, the correspondence between the subgroups of a quotient group G/N and the subgroups of G containing N can be easily extended to rings as given in the following theorem whose proof is routine.

Theorem 10.2.9. Let I be an ideal of a ring R. Then,

$$J \mapsto J/I$$

is a one-to-one correspondence between the ideals of R containing I and the ideals of R/I.

Example 10.2.1. Let n be a positive integer and $\mathbb{Z}_n$ be the ring of integers modulo n. We shall prove that $\mathbb{Z}_n$ is isomorphic to a quotient of the ring $\mathbb{Z}$ of integers, by using the fundamental theorem of homeomorphisms, as in the case where we have treated these as groups alone (see Example 5.2.1). As usual, define

$$f: \mathbb{Z} \to \mathbb{Z}_n \quad \text{by} \quad f(a) = r,$$

where r is the remainder obtained by dividing a with n; that is, $a = qn + r$, $0 \leq r < n$. Then, f is a epimorphism of rings (see Example 9.5.1 (4)) and ker $f = n\mathbb{Z}$ and hence

$$\mathbb{Z}/n\mathbb{Z} \cong \mathbb{Z}_n.$$

Worked Exercise 10.2.1. Prove that any nontrivial homomorphic image of a field is again a field.

Answer: Let R be a nontrivial homomorphic image of a field F. That is, R is a nontrivial ring and there is an epimorphism $f: F \to R$ of rings. Then, consider

$\ker f$ which is an ideal of F. Since F is a field, $\ker f = \{0\}$ or F. Also, $F/\ker f \cong f(F) = R$. If $\ker f = F$, then $F/\ker f$ is trivial and hence R is trivial, which is a contradiction to the hypothesis that R is nontrivial. Therefore, $\ker f = \{0\}$ and hence

$$F \cong F/\ker f \cong R.$$

Since F is a field, R is also a field.

Worked Exercise 10.2.2. For any ideal I and J of a ring R, prove that $R/I \cap J$ is isomorphic to a subring of $R/I \times R/J$.

Answer: Define $f : R \to R/I \times R/J$ by

$$f(a) = (a + I, a + J) \quad \text{for any } a \in R.$$

It can be easily verified that f is a homomorphism of the ring R into the product ring $R/I \times R/J$. By the fundamental theorem of homomorphisms, $f(R)$ is a subring of $R/I \times R/J$ and $R/\ker f \cong f(R)$. Since $I = (0 + I)$ and $J = (0 + J)$ are the zero elements of R/I and R/J, respectively, (I, J) is the zero element in the ring $R/I \times R/J$. Therefore,

$$\begin{aligned}\ker f &= \{a \in R : f(a) = \text{zero element in } R/I \times R/J\}\\ &= \{a \in R : (a + I, a + J) = (I, J)\}\\ &= \{a \in R : a + I = I \text{ and } a + J = J\}\\ &= \{a \in R : a \in I \text{ and } a \in J\}\\ &= I \cap J.\end{aligned}$$

Thus, $R/I \cap J \cong f(R)$, which is a subring of $R/I \times R/J$.

Worked Exercise 10.2.3. Let $I = 3\mathbb{Z}/12\mathbb{Z}$. Prove that I is isomorphic to an ideal J of $\mathbb{Z}_{12}$ such that $\mathbb{Z}_{12}/J \cong \mathbb{Z}_3$.

Answer: Recall that $12\mathbb{Z} \subseteq 3\mathbb{Z} \subseteq \mathbb{Z}$ and $12\mathbb{Z}$ and $3\mathbb{Z}$ are ideas of $\mathbb{Z}$. By Theorem 10.2.8,

$$(\mathbb{Z}/12\mathbb{Z})/(3\mathbb{Z}/12\mathbb{Z}) \cong \mathbb{Z}/3\mathbb{Z} \cong \mathbb{Z}_3.$$

Since $3\mathbb{Z}/12\mathbb{Z}$ is an ideal of $\mathbb{Z}/12\mathbb{Z} \cong \mathbb{Z}_{12}$ (by Example 10.2.1), there must be an ideal J of $\mathbb{Z}_{12}$ such that

$$I = 3\mathbb{Z}/12\mathbb{Z} \cong J \quad \text{and} \quad \mathbb{Z}_{12}/J \cong \mathbb{Z}_3.$$

Worked Exercise 10.2.4. Let I and J be ideals of array R such that $I \cap J = \{0\}$ and $I + J = R$. Then prove that

$$J \cong R/I, I \cong \mathrm{R}/J.$$

Answer: Define $f: J \to R/I$ by $f(a) = a + I$ for any $a \in J$. Then, clearly f is a homomorphism of rings. Also, for any $x + I \in R/I$, $x \in R$, we can write $x = b + a$, for some $b \in I$ and $a \in J$ (since $R = I + J$) and therefore $x + I = (b + a) + I = a + (b + I) = a + I$ (since $b \in I$) and hence f is an epimorphism. Also, for any $a \in J$,

$$\begin{aligned} f(a) = 0 &\Rightarrow a + I = I, \quad \text{the zero in } R/I \\ &\Rightarrow a \in I \\ &\Rightarrow a \in I \cap J = \{0\} \\ &\Rightarrow a = 0. \end{aligned}$$

Therefore, f is an injection also and hence f is an isomorphism of J onto R/J. Thus, $J \cong R/I$. Similarly, $I \cong R/J$.

Worked Exercise 10.2.5. Let I be an ideal of a ring R and the characteristic of R be $n > 0$. Prove that the characteristic of the quotient ring R/I is a divisor of n.

Answer: We are given that char(R) $= n$ and hence n is the least positive integer such that $na = 0$ for all $a \in R$. Now, for any $a + I \in R/I$, $a \in R$, we have

$$n(a + I) = na + I = 0 + I = I, \quad \text{the zero in } R/I$$

and therefore char(R/I) > 0 and, by Theorem 9.3.1, char(R/I) is a divisor of n.

Worked Exercise 10.2.6. Let $\mathbb{P}(X)$ be the power set of any set X. Let Y be a non-empty proper subset of X. Prove that $\mathbb{P}(Y)$ is an ideal of the ring $(\mathbb{P}(X), +, \cap)$ and describe the quotient ring $\mathbb{P}(X)/\,\mathbb{P}(Y)$.

Answer: We are given that $\mathbb{P}(Y) = \{A : A \subseteq Y\}$. We have

$$\begin{aligned} A \text{ and } B \in \mathbb{P}(Y) &\Rightarrow A + B = (A - B) \cup (B - A) \subseteq A \cup B \subseteq Y \\ \text{and} \quad & A \cap Z \subseteq Y \quad \text{for all } Z \in \mathbb{P}(X). \end{aligned}$$

Therefore, $\mathbb{P}(Y)$ is an ideal of the ring $(\mathbb{P}(X), +, \cap)$ Any element of the quotient ring $\mathbb{P}(X)/\mathbb{P}(Y)$ is of the form $Z + \mathbb{P}(Y)$ for some $Z \in \mathbb{P}(X)$; that is, $Z \subseteq X$. Now, we can write

$$\begin{aligned} Z &= (Z \cap Y) \cup (Z \cap (X - Y)) \\ &= (Z \cap Y) + (Z \cap (X - Y)) \end{aligned}$$

and hence
$$\begin{aligned} Z + \mathbb{P}(Y) &= ((Z \cap Y) - \mathbb{P}(Y)) + ((Z \cap (X - Y)) + \mathbb{P}(Y)) \\ &= (Z \cap (X - Y)) + \mathbb{P}(Y), \quad \text{since } Z \cap Y \in \mathbb{P}(Y). \end{aligned}$$

Therefore, $\mathbb{P}(X)/\mathbb{P}(Y) = \{A + \mathbb{P}(Y) : A \subseteq X - Y\}$.
Note that, for any A and $B \subseteq X - Y$,

$$\begin{aligned} A + \mathbb{P}(Y) = B + \mathbb{P}(Y) &\Rightarrow A - B \in \mathbb{P}(Y) \\ &\Rightarrow A + B \subseteq Y \\ &\Rightarrow A + B \subseteq Y \cap (X - Y) = \varnothing \\ &\Rightarrow A + B = \varnothing \\ &\Rightarrow A = -B = B \end{aligned}$$

Thus, $A \mapsto A + \mathbb{P}(Y)$ is a bijective map of $\mathbb{P}(X - Y)$ onto $\mathbb{P}(X)/\mathbb{P}(Y)$. It can be easily verified that this map preserves the ring operations in $\mathbb{P}(X - Y)$ and $\mathbb{P}(X)/\mathbb{P}(Y)$. Thus, the factor ring $\mathbb{P}(X)/\mathbb{P}(Y)$ is isomorphic to $\mathbb{P}(X - Y)$. One can consider the map $f: \mathbb{P}(X) \to \mathbb{P}(X - Y)$ defined by $f(A) = A \cap (X - Y)$. It can be verified that f is an epimorphism of rings and $\ker f = \mathbb{P}(Y)$ and hence, by the fundamental theorem of homomorphisms,

$$\mathbb{P}(X)/\mathbb{P}(Y) \cong \mathbb{P}(X - Y).$$

EXERCISE 10(B)

1. Determine all the elements of each of the following quotient rings.
   (i) $\mathbb{Z}/5\mathbb{Z}$
   (ii) $3\mathbb{Z}/6\mathbb{Z}$
   (iii) $2\mathbb{Z}/10\mathbb{Z}$
   (iv) $3\mathbb{Z}/9\mathbb{Z}$
   (v) R/I, where $R = \left\{\begin{pmatrix} a & 2b \\ 3c & d \end{pmatrix} : a, b, c \text{ and } d \in \mathbb{Z}\right\}$ and
   $I = \left\{\begin{pmatrix} a & 2b \\ 3c & 3d \end{pmatrix} : a, b, c \text{ and } d \in \mathbb{Z}\right\}$
   (vi) $\mathbb{P}(X)/I$, where $X = \{1, 2, 3, 4, 5\}$ and $I = \mathbb{P}(\{2,4\})$.

2. State whether each of the following is true or false. Substantiate your answers.
   (i) For any ring R, R has unity if and only if R/I has unity for all ideals I of R.
   (ii) For any ideal I of a commutative ring, R/I is commutative.
   (iii) A ring R is commutative if and only if R/I is commutative for all ideals I of R.
   (iv) For any integral domain R, R/I is an integral domain for any ideal I of R.
   (v) A ring R is an integral domain if and only if R/I is an integral domain for any ideal I of R.
   (vi) Any quotient ring of a field is a field.
   (vii) For any ideal I of a ring R, R/I is a field if and only if R is a field.
   (viii) A nontrivial ring R is a field if R/I is a field for all proper ideal, I of R.
3. Prove that a homomorphism $f : R \to S$ of rings is an injection if and only if $\ker f = \{0\}$.
4. For any subring S of a ring R, prove that the multiplication of additive cosets of S in R is well defined by the equation

$$(a + S)(b + s) = ab + S$$

   if and only if S is an ideal of R.
5. For any ideals I and J of a ring R, prove that the set

$$\overline{J} = \{a + I : a \in J\}$$

   is an ideal of R/I.
6. Determine all the idempotents, nilpotents and units in each of the following quotient rings
   (i) $\mathbb{Z}/6\mathbb{Z}$
   (ii) $\mathbb{Z}/8\mathbb{Z}$
   (iii) $\mathbb{Z}/7\mathbb{Z}$
7. Let I be an ideal of the ring $\mathbb{Z}$ such that $\mathbb{Z}/I$ is an integral domain. Then prove that $I = \{0\}$ or $I = p\mathbb{Z}$ for some prime number p.
8. Prove that the following are equivalent to each other for any positive integer n.
   (i) n is a prime number.
   (ii) $\mathbb{Z}/n\mathbb{Z}$ is a field.
   (iii) $\mathbb{Z}/n\mathbb{Z}$ is an integral domain.
9. Let R be a commutative ring and N be the set of all nilpotents in R. Then prove that N is an ideal of R and the quotient ring R/N has no nonzero nilpotents.
10. Prove that $\mathbb{Z}_n$'s, $\{0\}$ and $\mathbb{Z}$ are the only homomorphic images of the ring $\mathbb{Z}$ of integers.

11. Let S be a subring and I be an ideal of a ring R such that $S \cap I = \{0\}$. Prove that S is isomorphic to a subring of the quotient ring R/I.

12. For any two subsets A and B of a ring R, let the product of A and B be defined by the set

$$AB = \{ab : a \in A \text{ and } b \in B\}.$$

Give an example of an ideal I of a ring R such that the product $(x + I)(y + I)$ of two cosets $(x + I)$ and $(y + I)$ is properly contained in the coset $(xy + I)$.

13. For any ideal I of a ring R, prove that

$$M_n(R)/M_n(I) \cong M_n(R/I)$$

for any positive integer n, where $M_n(S)$ denotes the ring of $n \times n$ matrices over S.

14. For any pair of relatively prime positive integers m and n, prove that $\mathbb{Z}_m/nm\mathbb{Z} \cong \mathbb{Z}_n$.

15. Let $R_1, \ldots, R_n$ be rings and $I_1, \ldots, I_n$ be ideals of $R_1, \ldots, R_n$, respectively. Then prove that

$$R_1 \times \cdots \times R_n/I_1 \times \cdots \times I_n \cong R_1/I_1 \times \cdots \times R_n/I_n.$$

16. For any positive integer n, determine all the ideals of the quotient ring $\mathbb{Z}/n\mathbb{Z}$ and all the homomorphic images of $\mathbb{Z}/n\mathbb{Z}$.

10.3 CHINESE REMAINDER THEOREM

In this section, we extend a remarkable result known as 'Chinese Remainder Theorem' in the theory of numbers to the ideals of a ring. Recall that the set $\mathbb{Z}$ of integers forms a ring under the usual addition and multiplication of integers and that the ideals of $\mathbb{Z}$ are of the form $n\mathbb{Z}$ for some nonnegative integer n. Note that, if $I = n\mathbb{Z}$ and $a \in \mathbb{Z}$, then

$$a \in I = n\mathbb{Z} \Leftrightarrow n \text{ divides } a.$$

The classical version of the Chinese Remainder Theorem is that 'given distinct primes $p_1, p_2, \ldots, p_n$ and integers $a_1, a_2, \ldots, a_n$, one can always find an integer a such that

$$a \equiv a_i (\text{mod } p_i) \quad \text{for all } 1 \le i \le n;$$

that is, $a - a_i \in p_i\mathbb{Z}$ or $a + p_i\mathbb{Z} = a_i + p_i\mathbb{Z}$ for all $1 \le i \le n$.' If we take $I_i = p_i\mathbb{Z}$, then the Chinese Remainder Theorem states that, for any elements $a_1, a_2, \ldots, a_n$ in the ring $\mathbb{Z}$, there exists an element a in $\mathbb{Z}$ such that

$$a + I_i = a_i + I_i \quad \text{for all } 1 \le i \le n.$$

In the following theorem, we arrive at a necessary and sufficient condition on ideals $I_1, I_2, \ldots, I_n$ of an arbitrary ring for the validity of the above results.

Theorem 10.3.1. Let R be a ring with identity and $I_1, I_2, \ldots, I_n$ be ideals of R. If $I_i + I_j = R$ for any $i \neq j$, then, for any $x_1, x_2, \ldots, x_n \in R$ there exists $x \in R$ such that

$$x - x_i \in I_i; \text{ that is, } x + I_i = x_i + I_i \quad \text{for all } 1 \leq i \leq n.$$

Proof: Recall that an ideal I of R is the whole of R if and only if the unity 1 belongs to I. Suppose that

$$I_i + I_j = R \quad \text{for all } i \neq j.$$

First, we shall prove that, for each $1 \leq i \leq n$,

$$I_i + \left(\bigcap_{j \neq i} I_j\right) = R.$$

To prove this, fix $1 \leq i \leq n$ and put $K_i = \bigcap_{j \neq i} I_j$. For each $j \neq i$, we have

$$1 \in R = I_i + I_j \quad \text{and hence } 1 = a_j + b_j$$

for some $a_j \in I_i$ and $b_j \in I_j$. Now, consider

$$1 = \prod_{j \neq i} (a_j + b_j) = s_i + t_i,$$

where $t_i = \prod_{j \neq i} b_j$ and $s_i \in I_i$, since the expansion of $\prod_{j \neq i} (a_j + b_j)$ gives a sum in which all the summands, except $\prod_{j \neq i} b_j$, are products involving atleast one a_j which is in I_i and I_i is an ideal. Also, $t_i = \prod_{j \neq i} b_j \in I_j$ for all $j \neq i$ and hence $t_i \in \bigcap_{j \neq i} I_j = K_i$.

Therefore,

$$1 = s_i + t_i, s_i \in I_i \quad \text{and} \quad t_i \in K_i \tag{*}$$

and hence $I_i + (\bigcap_{j \neq i} I_j) = I_i + K_i = R$ for all $1 \leq i \leq n$. Now, let $x_1, x_2, \ldots, x_n$ be any elements in R. Put

$$x = x_1 t_1 + x_2 t_2 + \cdots + x_n t_n.$$

Then, for each $1 \leq i \leq n$,

$$x - x_i = \sum_{j=i}^{n} x_j t_j - x_i(s_i + t_i) \qquad \text{(by (*))}$$

$$= \sum_{j \neq i} x_j t_j - x_i s_i \in I_i$$

since $t_j \in I_i$ for all $j \neq i$ and $s_i \in I_i$.

A converse of the above result is also true and this is proved in the following theorem.

Theorem 10.3.2 Let $I_1, I_2, \ldots, I_n$ be ideals in a ring R such that, for any elements $x_1, x_2, \ldots, x_n$ in R, there exists $x \in R$ with $x - x_i \in I_i$ for all $1 \leq i \leq j$. Then, $I_i + I_j = R$ for all $i \neq j$.

Proof: Fix $i \neq j$. Let $a \in R$. Define $x_1, x_2, \ldots, x_n$ by $x_i = a$ and $x_k = 0$ for all $k \neq i$. Then, by hypothesis, there exists $x \in R$ such that

$$x - a = x - x_i \in I_i \quad \text{and} \quad x = x - x_j \in I_j$$

and therefore $a = (a - x) + x \in I_i + I_j$. Thus, $I_i + I_j = R$ for all $i \neq j$. ◀

Theorem 10.3.3. Let $I_1, I_2, \ldots, I_n$ be ideals of a ring with unity and R/I_1, $R/I_2, \ldots, R/I_n$ be the corresponding quotient rings. Define $f: R \to R/I_1 \times R/I_2 \times \cdots \times R/I_n$ by

$$f(x) = (x + I_1, x + I_2, \ldots, x + I_n)$$

for any $x \in R$. Then, f is an epimorphism if and only if $I_i + I_j = R$ for all $i \neq j$ and, in this case $R / \bigcap_{i=1}^{n} I_i$ is isomorphic to $R/I_1 \times R/I_2 \times \cdots \times R/I_n$.

Proof: Clearly f is a homomorphism of rings and

$$\begin{aligned}
\ker f &= \{a \in R : f(a) = \text{zero in } R/I_1 \times \cdots \times R/I_n\} \\
&= \{a \in R : (a + I_1, \ldots, a + I_n) = (I_1, \ldots, I_n)\} \\
&= \{a \in R : a \in I_i \text{ for } 1 \leq i \leq n\} \\
&= \bigcap_{i=1}^{n} I_i.
\end{aligned}$$

Also, f is a surjection if and only if, for any $x_1, x_2, \ldots, x_n$ in R, there exists $x \in R$ such that

$$x + I_i = x_i + I_i \quad \text{for all } 1 \leq i \leq n$$

which in turn, by Theorems 10.3.1 and 10.3.2, is equivalent to saying that $I_i + I_j = R$ for all $i \neq j$. In this case, we have by the fundamental theorem of homomorphisms, that

$$f : \frac{R}{\bigcap_{i=1}^{n} I_i} \to \frac{R}{I_1} \times \frac{R}{I_2} \times \cdots \times \frac{R}{I_n} \text{ is an isomorphism.}$$

Corollary 10.3.1. Let R be a ring with unity and $I_1, I_2, \ldots, I_n$ be ideals of R such that $I_i + I_j = R$ for all $i \neq j$ and $\bigcap_{i=1}^{n} I_i = \{0\}$. Then, $R \cong R/I_1 \times R/I_2 \times \cdots \times R/I_n$.

Corollary 10.3.2. Let R be a ring with unity and R be the direct sum of ideals I and J. Then,

$$R \cong R/I \times R/J.$$

Corollary 10.3.3 (Chinese Remainder Theorem). Let $m_1, m_2, \ldots, m_n$ be positive integers which are pair-wise relatively prime. For any integers $a_1, a_2, \ldots, a_n$, there exists an integer a such that

$$a \equiv a_i (\text{mod } m_i) \quad \text{for all } 1 \leq i \leq n.$$

Further, a is unique modulo the l.c.m. of $\{m_1, m_2, \ldots, m_n\}$.

Proof: Put $I_i = m_i\mathbb{Z}$ for $1 \leq i \leq n$. Then, I_i is an ideal of $\mathbb{Z}$ and $I_i + I_j =$ g.c.d. $\{m_i, m_j\}\mathbb{Z} = 1\mathbb{Z} = \mathbb{Z}$ for all $i \neq j$. Also, $\bigcap_{i=1}^{n} I_i = m\mathbb{Z}$, where $m =$ l.c.m. of $\{m_1, \ldots, m_n\}$. Now, let $a_1, a_2, \ldots, a_n$ be any integers. Then, by Theorem 10.3.1, there exist $a \in \mathbb{Z}$ such that

$$9.5a - a_i \in I_i = m_i\mathbb{Z} \quad \text{for all } 1 \leq i \leq n$$

and hence m_i divides $a - a_i$, so that $a \equiv a_i(\text{mod } m_i)$. Also, if b is any other integer with this property, then

$$a - b = (a - a_i) - (b - a_i) \in I_i \text{ for all } 1 \leq i \leq n$$

and hence $a - b \in \bigcap_{i=1}^{n} I_i = m\mathbb{Z}$, so that $a \equiv b(\text{mod } m)$. Thus, a is unique modulo m.

Corollary 10.3.4 (Classical Chinese Remainder Theorem). Let $p_1, p_2, \ldots, p_n$ be distinct prime numbers and $a_1, a_2, \ldots, a_n$ be any integers. Then, there exists an integer a such that

$$a \equiv a_i (\text{mod } p_i) \quad \text{for all } 1 \le i \le n$$

and this a is unique modulo the product $p_1 p_2 \cdots p_n$.

Corollary 10.3.5. Let m be a positive integer greater than 1 and

$$m = p_1^{r_1} p_2^{r_2} \ldots p_n^{r_n},$$

where $p_1, p_2, \ldots, p_n$ are distinct primes and $r_1, r_2, \ldots, r_n$ are positive integers. Then,

$$\mathbb{Z}_m \cong \mathbb{Z}_{p_1^{r_1}} \times \mathbb{Z}_{p_2^{r_2}} \times \cdots \times \mathbb{Z}_{p_n^{r_n}}.$$

Proof: $\mathbb{Z}_m \cong \mathbb{Z}/m\mathbb{Z} \cong \mathbb{Z}/p_1^{r_1}\mathbb{Z} \times \cdots \times \mathbb{Z}/p_n^{r_n}\mathbb{Z} \cong \mathbb{Z}_{p_1^{r_1}} \times \cdots \times \mathbb{Z}_{p_n^{r_n}}.$

Worked Exercise 10.3.1. Prove that $\mathbb{Z}_{12} \cong \mathbb{Z}_4 \times \mathbb{Z}_3$.

Answer: Since $12 = 2^2 \times 3^1$ and 2 and 3 are distinct primes, it follows from Corollary 10.3.5 that

$$\mathbb{Z}_{12} \cong \mathbb{Z}_{2^2} \times \mathbb{Z}_{3^1} = \mathbb{Z}_4 \times \mathbb{Z}_3.$$

Worked Exercise 10.3.2. Can $\mathbb{Z}_{12} \cong \mathbb{Z}_2 \times \mathbb{Z}_2 \times \mathbb{Z}_3$?

Answer: No; for, char($\mathbb{Z}_2$) = 2 and char($\mathbb{Z}_3$) = 3 and hence char($\mathbb{Z}_2 \times \mathbb{Z}_2 \times \mathbb{Z}_3$) = l.c.m {2, 2, 3} = 6. But char($\mathbb{Z}_{12}$) = 12 and hence $\mathbb{Z}_{12}$ cannot be isomorphic with $\mathbb{Z}_2 \times \mathbb{Z}_2 \times \mathbb{Z}_3$.

EXERCISE 10(C)

1. Which of the following are true? Substantiate your answer.
   (i) $\mathbb{Z}_{16} \times \mathbb{Z}_5 \cong \mathbb{Z}_{80}$
   (ii) $\mathbb{Z}_6 \times \mathbb{Z}_5 \cong \mathbb{Z}_2 \times \mathbb{Z}_{15}$
   (iii) $\mathbb{Z}_{16} \times \mathbb{Z}_{20} \cong \mathbb{Z}_{320}$

(iv) $\mathbb{P}(X) \cong \mathbb{P}(Y) \times \mathbb{P}(X - Y)$ for any subset Y of a set X, where $\mathbb{P}(X)$ is the ring. $(\mathbb{P}(X), +, \cap)$

(v) $\mathbb{Z} \cong \mathbb{Z}_3 \times \mathbb{Z}_2$

(vi) $\mathbb{Z} \times \mathbb{Z} \cong \mathbb{Z}$

2. Let R be any ring with unity and

$$S = \left\{ \begin{pmatrix} a & b \\ 0 & c \end{pmatrix} : a, b \text{ and } c \in R \right\}.$$

Then prove that S is a ring under the usual addition and multiplication of matrices over R.

3. Let I and J be ideals of a ring R with unity and S be the ring given in Exercise 2. Let

$$K = \left\{ \begin{pmatrix} a & b \\ 0 & c \end{pmatrix} : a \in I, c \in J, b \in R \right\}.$$

Then prove that K is an ideal of S and that

$$S/K \cong R/I \times R/J.$$

4. Let $\{R_j\}_{j\in J}$ be a class of rings and

$$\prod_{j\in J} R_j = \{f : J \to \bigcup_{j\in J} R_j : f(j) \in R_j \text{ for all } j \in J\}.$$

Then prove that $\prod_{j\in J} R_j$ is a ring under the point-wise addition and multiplication.

5. Let $\bigoplus_{j\in J} R_j = \{f \in \prod_{j\in J} R_j : f(j) = 0 \text{ for all but finite } j\text{'s in } J\}$. Then prove that $\bigoplus_{j\in J} R_j$ is a subring of the $\prod_{j\in J} R_j$ given in 4 above.

6. For any nonempty proper subset K of J, prove that there is a subring S of $\bigoplus_{j\in J} R_j$ given in 5 above such that $S \cong \bigoplus_{j\in K} R_j$, S is an ideal of $\bigoplus_{j\in J} R_j$ and $\bigoplus_{j\in J} R_j \big/ S \cong \bigoplus_{j\in J-K} R_j$.

7. For any ideals $I_1, I_2, \ldots, I_n$ of a ring R, prove that $R / \bigcap_{j=1}^{n} I_j$ is isomorphic to a subring of $\bigoplus_{j=1}^{n} R/I_j$.

8. For any class $\{I_j\}_{j\in J}$ of ideals of a ring R, prove that $R / \bigcap_{j\in J} I_j$ is isomorphic to a subring of $\prod_{j\in J} R/I_j$.

10.4 PRIME IDEALS

We have noticed earlier that a quotient ring of a ring may be an integral domain irrespective of whether the given ring is so. For example, the ring $\mathbb{Z}$

of integers is an integral domain and the quotient ring $\mathbb{Z}/6\mathbb{Z}(\cong \mathbb{Z}_6)$ is not an integral domain, while the quotient ring $\mathbb{Z}/5\mathbb{Z}(\cong \mathbb{Z}_5)$. is an integral domain. In this section, we discuss ideals with respect to which the quotient ring becomes an integral domain.

Definition 10.4.1. Let R be a ring. A proper ideal P of R is said to be a *prime ideal* if, for any ideals I and J of R,

$$IJ \subseteq P \Rightarrow I \subseteq P \quad \text{or} \quad J \subseteq P.$$

Example 10.4.1

1. $\{0\}$ is a prime ideal of the ring $\mathbb{Z}$ of integers.
2. A nonzero ideal I of $\mathbb{Z}$ is prime if and only if $I = p\mathbb{Z}$ for some prime number p.
3. Consider the ring $\mathbb{Z} \times \mathbb{Z}$ under co-ordinate wise addition and multiplication. Then, $\mathbb{Z} \times \{0\}$ and $\{0\} \times \mathbb{Z}$ are prime ideals of $\mathbb{Z} \times \mathbb{Z}$.
4. $\{0\}$ is not a prime ideal in the ring $M_2(\mathbb{R})$ of 2×2 matrices over the real number system.

Theorem 10.4.1. Let P be a proper ideal of a ring R such that, for any a and b in R,

$$ab \in P \Rightarrow a \in P \quad \text{or} \quad b \in P.$$

Then, P is a prime ideal. The converse holds if R is a commutative ring.

Proof: Let I and J be ideals of R such that $I \not\subseteq P$ and $J \not\subseteq P$. Then, there exist elements a and b such that $a \in I$, $a \notin P$, $b \in J$ and $b \notin P$. Then, by the hypothesis, $ab \notin P$. Since $ab \in IJ$, it follows that $IJ \not\subseteq P$. Thus, P is a prime ideal of R. Conversely suppose that R is a commutative ring and P is a prime ideal of R. Let a and $b \in R$ such that $a \notin P$ and $b \notin P$. Consider the ideals $<a>$ and $<b>$. We have $<a> \not\subseteq P$ and $<b> \not\subseteq P$ and hence $<a><b> \not\subseteq P$. Therefore, there exists an element $x \in <a><b>$ such that $x \notin P$. x is a finite sum of elements of the form

$$(ra + na)(sb + mb) = rsab + mrab + nsab + nmab$$

(see Corollary 10.1.3) and, since $x \notin P$, it follows that $ab \notin P$.

Corollary 10.4.1. A proper ideal P of a commutative ring R is prime if and only if, for any a and b in R,

$$ab \in P \Leftrightarrow a \in P \quad \text{or} \quad b \in P.$$

Theorem 10.4.2. Let P be an ideal of a commutative ring R with unity. Then, P is prime if and only if the quotient R/P is an integral domain.

Proof: Suppose that P is a prime ideal of R. Since P is proper ideal of R, R/P is a nontrivial ring. Also, since R is a commutative ring with unity, so is R/P. Now, for only $a + P$ and $b + P$ in R/P,

$$\begin{aligned}(a + P)(b + P) = \text{zero in } R/P &\Rightarrow ab + P = P \\ &\Rightarrow ab \in P \\ &\Rightarrow a \in P \quad \text{or} \quad b \in P \\ &\Rightarrow a + P = P \quad \text{or} \quad b + P = P.\end{aligned}$$

Thus, R/P is an integral domain.
Conversely, suppose that R/P is an integral domain. Then, R/P is nontrivial and hence P is a proper ideal of R. Now, for any a and b in R,

$$\begin{aligned}ab \in P &\Rightarrow (a + P)(b + P) = ab + P = P \\ &\Rightarrow (a + P)(b + P) = \text{The zero in } R/P \\ &\Rightarrow a + P = P \quad \text{or} \quad b + P = P \\ &\Rightarrow a \in P \quad \text{or} \quad b \in P.\end{aligned}$$

Thus, P is a prime ideal of R.
Let us recall that, for any ideals I and J of a ring R, the set

$$\frac{J}{I} = \{a + I : a \in J\}$$

is an ideal of the quotient ring R/I and that $J \mapsto J/I$ is a one-to-one correspondence between the ideals of R containing I and the ideals of R/I (see Theorem 10.2.9). This correspondence can be carried to prime ideals also, as proved in the following theorem.

Theorem 10.4.3. Let I be an ideal of a ring R and P be an ideal of R containing I. Then, P is a prime ideal of R if and only if P/I is a prime ideal of R/I.

Proof: Suppose that P is a prime ideal of R. Then, P/I is a proper ideal of R/I, since P is proper in R. Let A and B be ideals of R/I such that $AB \subseteq P/I$. Then, $A = J/I$ and $B = K/I$ for some ideals J and K of R containing I. Also,

$$JK/I = (J/I)(K/I) = AB \subseteq P/I$$

and hence $JK \subseteq P$, so that $J \subseteq P$ or $K \subseteq P$, since P is prime. Therefore, $A = J/I \subseteq P/I$ or $B = K/I \subseteq P/I$. Thus, P/I is a prime ideal of R/I.

Conversely suppose that P/I is a prime ideal of R/I. Then, P/I is proper in R/I and hence P is a proper ideal of R. Let J and K be ideals of R such that $JK \subseteq P$, Then, J/I and K/I are ideals of R/I and

$$(J/I)(K/I) = JK/I \subseteq P/I.$$

Since P/I is a prime ideal of R/I, it follows that $J/I \subseteq P/I$ or $K/I \subseteq P/I$ and hence $J \subseteq P$ or $K \subseteq P$. Thus, P is a prime ideal of R.

Corollary 10.4.2. Let I be an ideal of a ring R. Then, $P \mapsto P/I$ is a one-to-one correspondence between the prime ideals of R containing I and the prime ideals of R/I.

Example 10.4.2. Consider the ring $\mathbb{Z}$ of integers in which $\{0\}$ is a prime ideal and any nonzero prime ideal is precisely of the form $p\mathbb{Z}$ for some prime number p. Recall that any ideal of $\mathbb{Z}$ is a principal ideal. Let $I = \langle n \rangle = n\mathbb{Z}$ and $n = p_1^{a_1} p_2^{a_2} \cdots p_r^{a_r}$, where $r > 0$, $p_1, p_2, \ldots, p_r$ are distinct prime numbers and $a_1, a_2 \ldots, a_r$ are positive integers. Then, P is a prime ideal of $\mathbb{Z}$ containing $n\mathbb{Z}$ if and only if $P = p_i\mathbb{Z}$, for some $1 \leq i \leq r$. Thus, there are exactly r prime ideals in $\mathbb{Z}/n\mathbb{Z}(\cong \mathbb{Z}_n)$ and these are $P_i\mathbb{Z}/n\mathbb{Z}$, $1 \leq i \leq r$.

In the following definition, we introduce a concept which plays an important role in many aspects of the theory of ideals in commutative rings.

Definition 10.4.2. Let R be a commutative ring and I be an ideal of R. The *nil radical* of I is defined as the set

$$\sqrt{I} = \{x \in R : x^n \in I \text{ for some } n \in \mathbb{Z}^+\}.$$

Using the commutativity of the ring R, one can easily prove that $\sqrt{I}$ is an ideal of R containing I. In fact, we prove in the following that $\sqrt{I}$ is the intersection of all prime ideals of R containing I. Before going to the proof of this, let us recall an axiom of the theory of sets, which is popularly known as the *Zorns lemma*. Though it is called a lemma, it is actually an axiom. There are several equivalent formulations of this axiom. We present a convenient form of this in the following lemma.

Zorn's Lemma 10.4.1. Let X be any set and $\mathcal{T}$ be a nonempty class of subsets of X. Suppose that, for any subclass $\mathcal{C}$ of $\mathcal{T}$ in which any two members of $\mathcal{C}$ are comparable (that is, for any A and $B \in \mathcal{C}$, either $A \subseteq B$ or $B \subseteq A$), the union of all the members of $\mathcal{C}$ is a member of $\mathcal{T}$. Then, $\mathcal{T}$ has a maximal member; that is, there exists $M \in \mathcal{T}$ such that M is not properly contained in any member of $\mathcal{T}$.

Theorem 10.4.4. Let I be an ideal of a commutative ring R. Then, the nil radical of I is equal to the intersection of all prime ideals of R containing I.

Proof: Let $\sqrt{I}$ be the nil radical of I and J be the intersection of all prime ideals of R containing I. We shall prove that $\sqrt{I} = J$. If P is any prime ideal of R containing I, then

$$\begin{aligned} x \in \sqrt{I} &\Rightarrow x^n \in I \subseteq P \quad \text{for some } n \in \mathbb{Z}^+ \\ &\Rightarrow x^n \in P, n \in \mathbb{Z}^+ \\ &\Rightarrow x \in P \text{ (since } P \text{ is prime)} \end{aligned}$$

and hence $\sqrt{I} \subseteq P$. Therefore, $\sqrt{I} \subseteq J$.
On the other hand, let $x \in R$ such that $x \notin \sqrt{I}$. Then, $x^n \notin I$ for all $n \in \mathbb{Z}^+$. Put

$$S = \{x^n : n \in \mathbb{Z}^+\}.$$

Then, $x \in S$ and $S \cap I = \varnothing$. Let
$\mathcal{T} = \{K : K \text{ is an ideal of } R, I \subseteq K \text{ and } S \cap K = \varnothing\}$.
Since $I \in \mathcal{T}$, $\mathcal{T}$ is a nonempty class of subsets of R. We shall verify that the hypothesis in the Zorn's Lemma is satisfied for $\mathcal{T}$. Let $\mathcal{C}$ be a subclass of $\mathcal{T}$ such that, for any A and $B \in \mathcal{C}$, either $A \subseteq B$ or $B \subseteq A$. If $K = \bigcup_{A \in \mathcal{C}} A$, then K is an ideal of R (by Corollary 10.1.1), $I \subseteq K$ and $S \cap K = \varnothing$ and hence $K \in \mathcal{T}$. Therefore, $\mathcal{T}$ satisfies the hypothesis of the Zorn's Lemma and hence $\mathcal{T}$ has a maximal member, say M. Then, M is an ideal of R, $I \subseteq M$ and $x \notin M$ (since $x \in S$ and $S \cap M = \varnothing$). We shall prove that M is a prime ideal, which implies that $x \notin J$. Let a and $b \in R$ such that $ab \in M$. Suppose, if possible, $a \notin M$ and $b \notin M$. Then, $M \subsetneq M + \langle a\rangle$ and $M \subsetneq M + \langle b\rangle$. By the maximality of M, $M + \langle a\rangle$ and $M + \langle b\rangle$ cannot be members of $\mathcal{T}$. Therefore,

$$S \cap (M + \langle a\rangle) \neq \varnothing \quad \text{and} \quad S \cap (M + \langle b\rangle) \neq \varnothing.$$

and hence there exist positive integers n and m such that

$$x^n = y + r_1 a + sa \quad \text{and} \quad x^m = z + r_2 b + tb$$

for some $y, z \in M$, $r_1, r_2 \in R$ and $s, t \in \mathbb{Z}$. Now $x^{n+m} \in S$ and

$$\begin{aligned} x^{n+m} &= x^n x^m = (y + r_1 a + sa)(z + r_2 b + tb) \\ &= y(z + r_2 b + tb) + (r_1 a + sa)z + r_1 r_2 ab + tr_1 ab + sr_2 ab + stab. \end{aligned}$$

Since y, z and $ab \in M$, we get that $x^{n+m} \in S \cap M$, which is a contradiction to the fact that $S \cap M = \emptyset$ (since $M \in \mathfrak{T}$). Therefore, $a \in M$ or $b \in M$. Thus, M is a prime ideal of R. Since $x \notin M$ and $I \subseteq M$, we get that $x \notin J$. Therefore, we get that $J \subseteq \sqrt{I}$. Thus, $\sqrt{I} = J$.

Example 10.4.3. In the ring $\mathbb{Z}$ of integers, let $n = p_1^{a_1} p_2^{a_2} \cdots p_r^{a_r}$, where $p_1, p_2, \ldots, p_r$ are distinct primes and $a_1, a_2, \ldots, a_r$ are positive integers. Then, $p_1\mathbb{Z}$, $p_2\mathbb{Z}, \ldots, p_r\mathbb{Z}$ are all the prime ideals containing $n\mathbb{Z}$ and hence

$$\sqrt{n\mathbb{Z}} = (p_1\mathbb{Z}) \cap (p_2\mathbb{Z}) \cap \cdots \cap (p_r\mathbb{Z}) = (p_1 p_2 \cdots p_r)\mathbb{Z}.$$

As concrete illustrations of this, we have

$$\sqrt{24\mathbb{Z}} = 6\mathbb{Z} \text{ (since } 24 = 2^3 \cdot 3^1)$$

$$\text{and} \quad \sqrt{100\mathbb{Z}} = 10\mathbb{Z} \text{ (since } 100 = 2^2 \cdot 5^2).$$

Corollary 10.4.3. The nil radical of $\{0\}$ in any commutative ring R is precisely the set of nilpotents in R; that is,

$$\sqrt{\{0\}} = \{x \in R : x^n = 0 \text{ for some } n \in \mathbb{Z}^+\}.$$

$\sqrt{\{0\}}$ is usually denoted by $N(R)$ and is called the *prime radical* of R.

Corollary 10.4.4. For any ideal I of a commutative ring R,

$$N\left(\frac{R}{I}\right) = \frac{\sqrt{I}}{I}.$$

Worked Exercise 10.4.1. Prove that the following are equivalent to each other for any ideal I of a commutative ring R.

1. $I = \sqrt{I}$
2. $I = \sqrt{J}$ for some ideal J of R.
3. I is the intersection of a class of prime ideals of R.

Answer: (1) $\Rightarrow$ (2) is trivial.
(2) $\Rightarrow$ (3) follows from the fact that $\sqrt{J}$ is equal to the intersection of prime ideals of R containing J (by Theorem 10.4.4).
(3) $\Rightarrow$ (1): Suppose that $\{P_\alpha\}_{\alpha \in \Delta}$ is a class of prime ideals of R such that $I = \bigcap_{\alpha \in \Delta} P_\alpha$, we always have $I \subseteq \sqrt{I}$. On the other hand,

$$x \in \sqrt{I} \Rightarrow x^n \in I = \bigcap_{\alpha \in \Delta} p_\alpha \quad \text{for some } n \in \mathbb{Z}^+$$
$$\Rightarrow x^n \in p_\alpha \quad \text{for all } \alpha \in \Delta$$
$$\Rightarrow x \in p_\alpha \quad \text{for all } \alpha \in \Delta \text{ (since } p_\alpha \text{ is prime)}$$
$$\Rightarrow x \in \bigcap_{\alpha \in \Delta} p_\alpha = I$$

and therefore $\sqrt{I} \subseteq I$. Thus, $I = \sqrt{I}$.

Worked Exercise 10.4.2. Let I be an ideal of a commutative ring R. Prove that $I = \sqrt{I}$ if and only if the quotient ring R/I has no nonzero nilpotent elements.

Answer: Suppose that $I = \sqrt{I}$. Let $a + I$ be a nilpotent element in R/I. Then, $(a + I)^n = I$ for some $n \in \mathbb{Z}^+$ and hence $a^n + I = I$. This implies that $a^n \in I$ and hence $a \in \sqrt{I} = I$, so that $a + I = I$. Therefore, zero element is the only nilpotent in R/I.

Conversely suppose that R/I has no nonzero nilpotents. Then,

$$a \in \sqrt{I} \Rightarrow a^n \in I \text{ for some } n \in \mathbb{Z}^+$$
$$\Rightarrow (a+I)^n = a^n + I = I, \; n \in \mathbb{Z}^+$$
$$\Rightarrow a+I \text{ is a nilpotent in } \frac{R}{I}$$
$$\Rightarrow a+I = I$$
$$\Rightarrow a \in I.$$

Therefore, $\sqrt{I} \subseteq I \subseteq \sqrt{I}$ and hence $I = \sqrt{I}$.

Worked Exercise 10.4.3. Let P and Q be prime ideals of a ring R. Prove that $P \cap Q$ is a prime ideal if and only if $P \subseteq Q$ or $Q \subseteq P$.

Answer: If $P \subseteq Q$ or $Q \subseteq P$, then $P \cap Q = P$ or Q and hence $P \cap Q$ is a prime ideal, conversely suppose that $P \cap Q$ is a prime ideal and $P \nsubseteq Q$. Choose an element $a \in P$ such that $a \notin Q$. Then,

$$b \in Q \Rightarrow ab \in P \cap Q$$
$$\Rightarrow a \in P \cap Q \quad \text{or} \quad b \in P \cap Q$$
$$\Rightarrow b \in P \cap Q \quad (\text{since } a \notin Q)$$
$$\Rightarrow b \in P$$

and hence $Q \subseteq P$.

Worked Exercise 10.4.4. Let R and S be commutative rings with unities and $f: R \to S$ be an epimorphism of rings. Prove that S is an integral domain if and only if $\ker f$ is a prime ideal of R.

Answer: By the fundamental theorem of homomorphisms (Theorem 10.2.4), $R/\ker f \cong S$. Note that S is nontrivial $\Leftrightarrow$ $\ker f$ is a proper ideal and therefore, S is an integral domain if and only if $\ker f$ is a prime ideal of R.

EXERCISE 10(D)

1. Determine all the prime ideals of each of the following rings
   (i) $\mathbb{Z}$
   (ii) $\mathbb{Z}_{180}$
   (iii) $\mathbb{R}$
   (iv) $\mathbb{Q}$
   (v) $\mathbb{R} \times \mathbb{Q}$
   (vi) $\mathbb{Q} \times \mathbb{Z}$
   (vii) $\mathbb{Z} \times \mathbb{Z}$
   (viii) $\mathbb{Z}_{31}$
   (ix) $M_2(\mathbb{R})$
   (x) $\mathbb{R} \times \mathbb{R} \times \mathbb{R}$
   (xi) $M_2(\mathbb{Z})$
   (xii) $(\mathbb{P}(X), +, \cap)$.

2. Let $\{P_\alpha\}_{\alpha\in\Delta}$ be a class of prime ideals of a ring such that, for any α and $\beta \in \Delta$, there is $\gamma \in \Delta$ such that $P_\gamma \subseteq P_\alpha$ and $P_\gamma \subseteq P_\beta$. Then prove that $\bigcap_{\alpha\in\Delta} P_\alpha$ is a prime ideal of R.

3. Let $\{P_\alpha\}_{\alpha\in\Delta}$ be class of prime ideals of a ring R with unity such that, for any α and $\beta \in \Delta$, there is $\gamma \in \Delta$ such that $P_\alpha \subseteq P_\gamma$ and $P_\beta \subseteq P_\gamma$. Then prove that $\bigcup_{\alpha\in\Delta} P_\alpha$ is a prime ideal of R.

4. Let $\{P_\alpha\}_{\alpha\in\Delta}$ be a chain of prime ideals of a ring R with unity (that is, for any α and $\beta \in \Delta$, $P_\alpha \subseteq P_\beta$ or $P_\beta \subseteq P_\alpha$). Then prove that $\bigcap_{\alpha\in\Delta} P_\alpha$ and $\bigcup_{\alpha\in\Delta} P_\alpha$ are prime ideals of R.

5. Let I and J be ideals of a commutative ring R. Then prove the following.
   (i) $\sqrt{IJ} = \sqrt{I \cap J} = \sqrt{I} \cap \sqrt{J}$
   (ii) $\sqrt{I} + \sqrt{J} \subseteq \sqrt{I+J} = \sqrt{\sqrt{I} + \sqrt{J}}$
   (iii) $\sqrt{I} \subseteq \sqrt{J}$ if $I^n \subseteq J$ for some $n \in \mathbb{Z}^+$.
   (iv) $\sqrt{\sqrt{I}} = \sqrt{I}$

(v) $\sqrt{I^n} = \sqrt{I}$ for all $n \in \mathbb{Z}^+$.

(vi) For any prime ideal P of R, $I \subseteq P \Leftrightarrow \sqrt{I} \subseteq P$.

(vii) If I is prime, then $I = \sqrt{I}$.

(viii) $I = \sqrt{I}$ if and only if $a^2 \in I$ implies $a \in I$ for any $a \in R$.

6. A proper ideal I of a commutative ring R is called *primary* if, for any a and $b \in R$,

$$ab \in I \Rightarrow a \in I \quad \text{or} \quad b^n \in I \quad \text{for some } n \in \mathbb{Z}^+.$$

Prove that every prime ideal of R is primary. Give an example to prove that the converse fails.

7. Prove that a nonzero ideal I of the ring $\mathbb{Z}$ of integers is primary if and only if $I = p^n\mathbb{Z}$ for some prime number p and positive integer n.

8. Let I be a primary ideal of a commutative ring R. Prove that $\sqrt{I}$ is the smallest prime ideal of R containing I.

9. Prove that an ideal I of a commutative ring R is primary if and only if every zero divisor of the quotient ring R/I is nilpotent.

10. For any prime number p and any positive integer n, prove that every zero divisor in the ring $\mathbb{Z}_{p^n}$ is nilpotent.

11. For any ideal I of a ring R, prove that $P \mapsto P/I$ is a one-to-one correspondence between the primary ideals of R containing I and the primary ideals of R/I.

12. A commutative ring R is called *regular* if, for any $a \in R$, there exists $b \in R$ such that $aba = a$. Prove that a commutative ring R is regular if and only if $I = \sqrt{I}$ for all ideals I of R.

13. Prove that the prime radical $N(R)$ of a ring R is $\{0\}$ if and only if, for any ideals I and J of R,

$$IJ = \{0\} \Leftrightarrow I \cap J = \{0\}.$$

14. Prove that the following are equivalent to each other for any commutative ring R, with unity.

(i) R has a unique prime ideal.

(ii) Every nonunit in R is nilpotent.

(iii) $\{0\}$ is a primary ideal and every nonunit nonzero element of R is a zero divisor.

15. Let R be the ring of all mappings of the real number system $\mathbb{R}$ into itself under point-wise addition and multiplication. Let

$$I = \{f \in R : f(1) = 0 = f(-1)\}.$$

Prove that I is an ideal of R. Is I a prime ideal of R?

16. Let R be a commutative ring and $\text{char}(R) = n > 0$. If $N(R) = \{0\}$, then prove that n is square free (that is, a^2 does not divide n for any integer $n > 1$).

10.5 MAXIMAL IDEALS

In this section, we discuss another special type of ideals of rings, namely maximal ideals. Clearly, for any ring R, the whole ring R is the largest ideal of R. We search for ideals of R which are maximal among all proper ideals.

Definition 10.5.1. Let R be a nontrivial ring. A proper ideal M of R is called a *maximal ideal* if M is not properly contained in any proper ideal of R; that is, for any ideal I of R,

$$M \subseteq I \subseteq R \Rightarrow M = I \quad \text{or} \quad I = R.$$

Example 10.5.1

1. Let us recall that any ideal of $\mathbb{Z}$ is of the form $n\mathbb{Z}$ for some nonnegative integer n and that $n\mathbb{Z} \subseteq m\mathbb{Z}$ if and only if m divides n. Therefore, $n\mathbb{Z}$ is a maximal ideal of $\mathbb{Z}$ if and only if n is a prime number.
2. In the ring $\mathbb{R}$ of real numbers, $\{0\}$ is a maximal ideal. In fact, $\{0\}$ is a maximal ideal in any field, since a field has two ideals, namely $\{0\}$ and the whole field.
3. Let $R = \mathbb{R} \times \mathbb{R}$ under co-ordinate wise addition and multiplication. Then, $\mathbb{R} \times \{0\}$ and $\{0\} \times \mathbb{R}$ are maximal ideals of R.
4. Consider the ring $M_n(\mathbb{R})$ of $n \times n$ matrices over the real number system $\mathbb{R}$. Then, $\{0\}$ is a maximal ideal of $M_n(\mathbb{R})$, since, by Theorem 10.1.10, $M_n(\mathbb{R})$ is the only nonzero ideal of $M_n(\mathbb{R})$.

Let us recall that a commutative ring R with unity is a field if and only if R has exactly two ideals, namely $\{0\}$ and R. This, together with the fact that the ideals of R/I are in one-to-one correspondence with the ideals of R containing I, imply the following important result. However, we prefer to give an independent proof in view of its technicality.

Theorem 10.5.1. Let M be an ideal of a commutative ring R with unity. Then, M is a maximal ideal of R if and only it the quotient ring R/M is a field.

Proof: First note that M is a proper ideal of R if and only if R/M is nontrivial. Suppose that M is a maximal ideal of R. Then, M is a proper ideal and hence R/M is a nontrivial commutative ring with unity (since so is R). Now, let $a + M$ be a nonzero element of R/M. Then, $a + M \neq M$ and hence $a \notin M$. Let I be the ideal defined by

$$I = M + <a> = \{x + ra : x \in R\}.$$

Then, $M \subsetneq I$ (since $a \in I$ and $a \notin M$). By the maximality of M, it follows that $I = R$. In particular, $1 \in I = x + ra$ for some $x \in M$ and $r \in R$.

Now, we have $1 - ra = x \in M$ and hence

$$(r + M)(a + M) = ra + M = 1 + M, \text{ the unity in } R/M.$$

Therefore, $r + M$ is the multiplicative inverse of $a + M$ in R/M. Thus, every nonzero element in R/M is a unit and hence R/M is a field.

Conversely suppose that R/M is a field. Then, R/M is nontrivial and hence M is a proper ideal of R. Let J be any ideal of R such that $M \subseteq I \subseteq R$. Suppose that $M \neq I$. Then, there exists $a \in I$ such that $a \notin M$. Now, $a + M$ is a nonzero element in the field R/M and hence $a + M$ is a unit in R/M. Therefore, there exists $b \in R$ such that

$$ab + M = (a + M)(b + M) = 1 + M$$

and hence $1 - ab \in M \subseteq I$. Also, since $a \in I$, $ab \in I$ and therefore

$$1 = (1 - ab) + ab \in I$$

which implies that $I = R$. Thus, M is a maximal ideal of R.

Corollary 10.5.1. A commutative ring R with unity is a field if and only if $\{0\}$ is a maximal ideal of R.

Proof: This follows from the fact that $R/\{0\} \cong R$ and from Theorem 10.5.3.

Theorem 10.5.2. Let R be a ring with unity. Then, every maximal ideal of R is a prime ideal and the converse is not true.

Proof: Let M be a maximal ideal of R. Then, M is a proper ideal of R. Let I and J be ideals of R such that $IJ \subseteq M$. Suppose that $I \not\subseteq M$. Choose $a \in I$ such that $a \notin M$. Recall that

$$<a> = \left\{ \sum_{i=1}^{n} x_i a y_i + ar + sa : x_i, y_i, r, s \in R, \; n \geq 0 \right\}.$$

Put $K = M + <a>$. Then, K is an ideal of R containing M properly, since $a \in K$ and $a \notin M$. By the maximality of M, we get that $K = R$. In particular, $1 \in K = M + <a>$ and hence

$$1 = x + \sum_{i=1}^{n} x_i a y_i + ar + sa$$

for some $x \in M$ and $x_i, y_i, r, s \in R$, $n \geq 0$. Now,

$$\begin{aligned} b \in J \Rightarrow b &= 1b = \left(x + \sum_{i=1}^{n} x_i a y_i + ar + sa \right) b \\ \Rightarrow b &= xb + \sum_{i=1}^{n} (x_i a)(y_i b) + (ar)\, b + (sa)\, b \\ \Rightarrow b &\in M + IJ, \quad \text{since } x \in M, a \in I, b \in J \\ \Rightarrow b &\in M, \quad \text{since } IJ \subseteq M \quad \text{and} \quad M + IJ = M. \end{aligned}$$

Therefore, $J \subseteq M$. Thus, $I \subseteq M$ or $J \subseteq M$ and hence M is a prime ideal. The converse is not true; that is, an ideal can be prime without being maximal. For example, $\{0\}$ is a prime ideal of the ring $\mathbb{Z}$ of integers and $\{0\}$ is not a maximal ideal.

Note that, in proving the above theorem, the existence of unity in the ring is essential. For consider the following example.

Example 10.5.2. Consider the ring $2\mathbb{Z}$ of even integers. Let $M = 4\mathbb{Z}$. Then, M is a maximal ideal of $2\mathbb{Z}$; for, let I be an ideal of $2\mathbb{Z}$ such that $M \subseteq I \subseteq 2\mathbb{Z}$. Suppose that $M \neq I$. Then, there exists $2a \in I$ such that $2a \notin M = 4\mathbb{Z}$. Now, a must be odd and hence $a = 2n + 1$ for some $n \in \mathbb{Z}$. Consider

$$2 = 2 \cdot 1 = 2(a - 2n) = 2a - 4n \in I$$

(since $2a \in I$ and $4n \in M \subseteq I$). This implies that $2\mathbb{Z} \subseteq I$ and hence $I = 2\mathbb{Z}$. Thus, M is a maximal ideal of $2\mathbb{Z}$. However, M is not prime, since

$$2 \cdot 2 \in M \quad \text{and} \quad 2 \notin M.$$

Note that $2\mathbb{Z}$ is a commutative ring without unity.

Theorem 10.5.3. Let M be a proper ideal of a ring R. Then, M is a maximal ideal if and only if $M + \langle a \rangle = R$ for all $a \in R - M$.

Proof: Suppose that M is a maximal ideal of R and $a \in R - M$. Then, $M + \langle a \rangle$ is an ideal of R containing M properly. Since M is maximal, we get that $M + \langle a \rangle = R$. Conversely, if M is not maximal, then there exists an ideal

I of R such that $M \subsetneqq I \subsetneqq R$ and therefore there exists $a \in I - M \subseteq R - M$, so that $M + \langle a \rangle \subseteq I$ and hence $M + \langle a \rangle \neq R$. ◀

Corollary 10.5.2. Let R be a commutative ring with unity and M be a proper ideal of R. Then, M is maximal if and only if, for each $a \in R - M$, $1 - ar \in M$ for some $r \in R$.

Proof: This is a consequence of the above and of the fact that $M + \langle a \rangle = R$ if and only if $1 = x + ar$ for some $x \in M$ and $r \in R$. ◀

Next, we obtain a general result which assures the existence of suitably many maximal ideals. The crucial step here is again the Zorn's Lemma; an equivalent form of which is given in Zorn's Lemma 10.4.1.

Theorem 10.5.4. Let R be a ring with unity. Then, any proper ideal of R is contained in a maximal ideal of R.

Proof: Recall that an ideal of R is proper if and only if it does not contain the unity of R. Let I be a proper ideal of R. Consider the class

$$\mathcal{T} = \{J : J \text{ is a proper ideal of } R \text{ and } I \subseteq J\}$$

since $I \in \mathcal{T}$, $\mathcal{T}$ is a nonempty class of subsets of R. If $\{J_\alpha\}$ is a chain in $\mathcal{T}$ (that is, any two members of it are comparable), then by Corollary 10.1.1, $\bigcup_{\alpha \in \Delta} J_\alpha$ is an ideal of R and, since each J_α is a proper ideal of R, $1 \notin J_\alpha$ for each $\alpha \in \Delta$ and hence $1 \notin \bigcup_{\alpha \in \Delta} J_\alpha$, so that $\bigcup_{\alpha \in \Delta} J_\alpha$ is a proper ideal of R. Also, clearly $\bigcup_{\alpha \in \Delta} J_\alpha$ is a member of $\mathcal{T}$ containing each I_α. Therefore, the hypothesis of the Zorn's Lemma is satisfied for $\mathcal{T}$ and hence $\mathcal{T}$ has a maximal member, which is clearly a maximal ideal containing I. ◀

Corollary 10.5.3. Let R be a commutative ring with unity and $a \in R$. Then, a is a unit in R if and only if a does not belong to any maximal ideal of R.

Proof: a is a nonunit in R if and only if $\langle a \rangle$ $(= aR)$ is a proper ideal of R and hence, by the above theorem, $\langle a \rangle$ is contained in a maximal ideal of R. ◀

Theorem 10.5.5. Let R be a commutative ring with unity. Suppose that R has exactly one maximal ideal. Then, 0 and 1 are the only idempotents in R.

Proof: Let M be the unique maximal ideal of R. Let a be an idempotent in R; that is, $a^2 = a \in R$. Suppose that $a \neq 0$ and $a \neq 1$. Then,

$$a(a - 1) = 0$$

and hence a and $a - 1$ are zero divisors in R. Therefore, a and $a - 1$ are both non-units and, by Corollary 10.5.3, $a \in M$ and $a - 1 \in M$. From this, we get that

$$1 = a - (a - 1) \in M$$

which is a contradiction to the fact that $M \neq R$. Thus, either $a = 0$ or $a = 1$. ◀

Recall from Corollary 10.4.3, the prime radical of R is defined as the set of all nilpotents in R and, by Theorem 10.4.4, it is precisely the intersection of all prime ideals of R. In the following, we introduce another type of radical, which plays an important role in the structure thecry of commutative rings.

Definition 10.5.2. Let R be a nontrivial commutative ring with unity. The intersection of all maximal ideals of R is called the *Jacobson radical* of R and is denoted by $J(R)$ or by Rad(R). R is said to be *semisimple* if $J(R) = \{0\}$.

The Jacobson radical always exists, since any nontrivial commutative ring with unity has atleast one maximal by ideal. by Theorem 10.5.4. Also, from the definition of $J(R)$, it is immediate that the Jacobson radical of R is an ideal contained in each maximal ideal of R. Two important examples of semisimple rings are given below.

Example 10.5.3

1. Recall that the maximal ideals of the ring $\mathbb{Z}$ are precisely of the form $p\mathbb{Z}$ for some prime number p. Now, the Jacobson radical of $\mathbb{Z}$ is given by

$$J(\mathbb{Z}) = \bigcap_{p \in P} p\mathbb{Z} = \{0\},$$

where P is the set of prime numbers. Therefore, $\mathbb{Z}$ is a semisimple ring.

2. Let X be a nonempty set and F be a field. Then, the set F^X of all mappings of X into F forms a commutative ring with unity under the point-wise operations, with reference to the ring operations in F. For each $x \in X$, let

$$M_x = \{f \in F^X : f(x) = 0\}$$

and define $\alpha_x : F^X \to F$ by $\alpha_x(f) = f(x)$. It can be easily verified that α_x is an epimorphism of rings and ker $\alpha_x = M_x$. By the fundamental theorem of homomorphisms, $F^X / M_x \cong F$. Since F is a field, it follows from Theorem 10.5.1 that M_x is a maximal ideal of F^X. Also,

$$\bigcap_{x \in X} M_x = \{f \in F^X : f(x) = 0 \quad \text{for all } x \in X\} = \{0\}$$

and therefore the Jacobson radical of F^X is $\{0\}$. Thus, F^X is a semisimple ring.

In the following, we obtain a basic connection between the Jacobson radical and multiplicative invertibility of the elements of the ring.

Theorem 10.5.6. Let I be an ideal of a commutative ring R with unity. Then, $I \subseteq J(R)$ if and only if each element of the coset $1 + I$ is a unit in R.

Proof: Suppose that $I \subseteq J(R)$. Let $1 + a \in 1 + I$, $a \in I$. If $1 + a$ is a nonunit, then by Corollary 10.5.3, there exists a maximal ideal M of R such that $1 + a \in M$. Then, since $a \in I \subseteq J(R) \subseteq M$, we get that

$$1 = (1 + a) - a \in M$$

which is a contradiction to the fact that M is a proper ideal. Thus, $1 + a$ is a unit for any $a \in I$.

Conversely suppose that $1 + a$ is a unit in R for each $a \in I$. Suppose, if possible, that I $\not\subseteq J(R)$. We can choose an element $a \in I$ such that $a \notin J(R)$. Then, there exists a maximal ideal M such that $a \notin M$. Then, M is properly contained in $M + aR$. By the maximality of M, we have $M + aR = R$. In particular, $1 \in R = M + aR$ and hence

$$1 = x + ar \text{ for some } x \in M \quad \text{and} \quad r \in R.$$

Now, since $a \in I$, $-ar \in I$ and $1 - ar \in 1 + I$. Therefore, $x\ (= 1 - ar)$ is a unit in R, which is a contradiction since $x \in M$ and M is a proper ideal. Thus, $I \subseteq J(R)$.

The above theorem gives us a characterization of elements of the Jacobson radical $J(R)$, if we replace I above a principal ideal $\langle a \rangle$.

Theorem 10.5.7. Let R be a commutative ring with unity. Then, the Jacobson radical $J(R)$ is given by

$$J(R) = \{a \in R : 1 + ar \text{ is a unit in } R \text{ for all } r \in R\}.$$

Proof: If $a \in J(R)$, then $\langle a \rangle \subseteq J(R)$ and hence every element of $1 + \langle a \rangle$ is a unit in R; that is, $1 + ar$ is a unit for all $r \in R$. Conversely suppose that $1 + ar$ is a unit for all $r \in R$. Then, every element of $1 + \langle a \rangle$ is a unit and therefore, by the above theorem, $\langle a \rangle \subseteq J(R)$ which is equivalent to saying that $a \in J(R)$.

Corollary 10.5.4. The following holds for any commutative ring R with unity.

1. 0 is the only idempotent in the Jacobson radical $J(R)$.
2. An element $a \in R$ is a unit in R if and only if the coset $a + J(R)$ is a unit in the quotient ring $R/J(R)$.
3. The prime radical $N(R)$ is contained in the Jacobson radical $J(R)$.

Proof:

1. Let $a \in J(R)$ be an idempotent. Then, by Theorem 10.5.7, $1 - a$ is a unit in R and hence there exists $b \in R$ such that $(1 - a)b = 1$. Now,

$$a = a1 = a(1 - a)b = (a - a^2)b = 0 \text{ (since } a^2 = a)$$

   Thus, 0 is the only idempotent in $J(R)$.
2. Let $a \in R$. Suppose that $a + J(R)$ is a unit in $R/J(R)$. Then, there exists $b \in R$ such that

$$(a + J(R))(b + J(R)) = 1 + J(R)$$

   and therefore $1 - ab \in J(R)$. By Theorem 10.5.7, $1 - (1 - ab)$ is a unit in R and therefore ab is a unit. Thus, a is a unit in R. The converse is trivial.
3. $a \in N(R) \Rightarrow a$ belongs to every prime ideal of R
   $\Rightarrow a$ belongs to every maximal ideal of R (since every maximal ideal is prime)
   $\Rightarrow a \in J(R)$.

   Thus, $N(R) \subseteq J(R)$. ◀

Theorem 10.5.8. Let R be a commutative ring with unity and $J(R)$ be the Jacobson radical of R. Then, the quotient ring $R/J(R)$ is semisimple.

Proof: We have to prove that the Jacobson radical of $R/J(R)$ is trivial. Let $a + J(R)$ be an element in the Jacobson radical of $R/J(R)$. Then, by Theorem 15.5.15,

$$1 + J(R) + (a + J(R))(r + J(R))(=(1 + ar) + J(R))$$

is a unit in $R/J(R)$ for all $r + J(R) \in R/J(R)$ and therefore, by Corollary 10.5.4 (2), $1 + ar$ is a unit in R for all $r \in R$. This implies that $a \in J(R)$ and hence $a + J(R)$ is the zero in $R/J(R)$. Thus, $J(R/J(R))$ is trivial and hence $R/J(R)$ is semisimple. ◀

Worked Exercise 10.5.1. Let $n \in \mathbb{Z}^+$ and $R = \mathbb{R} \times \mathbb{R} \times \cdots \times \mathbb{R}$ (n factors). Then, R is a commutative ring with unity under the co-ordinate wise operations. Determine all maximal ideals of the ring R.

Answer: Let any element of R be denoted by

$$a = (a_1, a_2, \ldots, a_n)$$

where $a_1, a_2, \ldots, a_n$ are real numbers. For each $1 \le i \le n$, let

$$M_i = \{a \in R : a_i = 0\}.$$

Let $p_i : R \to \mathbb{R}$ be the ith projection; that is,

$$p_i(a) = a_i \quad \text{for all } a \in R.$$

Then, p_i is an epimorphism of rings and $\ker p_i = M_i$. Therefore, $R/M_i \cong \mathbb{R}$. Since $\mathbb{R}$ is a field, so is R/M_i and hence M_i is a maximal ideal of R. We shall prove that $M_1, M_2, \ldots, M_n$ are the only maximal ideals of R.

Let M be any maximal ideal of R. We shall prove that $M \subseteq M_i$ for some $1 \le i \le n$ and hence $M = M_i$. On the contrary, suppose that $M \not\subseteq M_i$ for all i. Then, we can choose, for each $1 \le i \le n$, $x^i \in M$ such that $x^i \notin M_i$. Then, $x_i^i \neq 0$ for each $1 \le i \le n$. Put

$$a = (x^1)^2 + (x^2)^2 + \cdots + (x^n)^2.$$

Then, $a \in M$, since $x^i \in M$ for all $1 \le i \le n$. Since the ith co-ordinate of a is

$$a_i = (x_i^1)^2 + (x_i^2)^2 + \cdots + (x_i^n)^2.$$

Since $x_i^i \neq 0$, we get that $(x_i^i)^2 > 0$ and hence

$$a_i > 0 \text{ for all } 1 \le i \le n.$$

Now,

$$(1, 1, \ldots, 1) = (a_1, a_2, \ldots, a_n)\left(\frac{1}{a_1}, \frac{1}{a_2}, \ldots, \frac{1}{a_n}\right)$$

and hence a is a unit in R. This is a contradiction to the fact that $a \in M$ and M is a maximal ideal. Thus, $M \subseteq M_i$ and, by the maximality of M, $M = M_i$ for some $1 \le i \le n$. Thus, $M_1, M_2, \ldots, M_n$ are all the maximal ideals of R.

Worked Exercise 10.5.2. Let I be a proper ideal of a Boolean ring B (see Definition 9.2.2). Prove that the following are equivalent to each other.

1. I is a prime ideal of B.
2. I is a maximal ideal of B.
3. For any $a \in B$, either $a \in I$ or $b - ab \in I$ for all $b \in B$, but not both.

Answer: (1) $\Rightarrow$ (2): Suppose that I is a prime ideal of B. Let J be an ideal of B containing I properly. Choose $a \in J$ such that $a \notin I$. Then,

$$a(1 - a) = a - a^2 = 0 \in I.$$

Since I is a prime ideal and $a \notin I$, we get that $1 - a \in I \subseteq J$. Now, since $a \in J$, we have

$$1 = a + (1 - a) \in J$$

and hence $J = B$.
(2) $\Rightarrow$ (3): Suppose that I is a maximal ideal of B and $a \in B$. Suppose that $a \notin I$. Then, $I + \langle a \rangle = B$ and hence, for any $b \in B$,

$$b = x + ar \quad \text{for some } x \in I \text{ and } r \in B$$

$$\text{and} \quad ab = ax + a^2r = ax + ar$$

$$\text{and hence} \quad b - ab = x - ax \in I.$$

If both $a \in I$ and $b - ab \in I$ for all $b \in B$, then

$$b = (b - ab) + ab \in I \quad \text{for all } b \in B$$

and hence $I = B$, which is a contradiction. Thus, either $a \in I$ or $b - ab \in I$ for all $b \in B$ and not both.
(3) $\Rightarrow$ (1): Suppose (3) is satisfied. Let a and $b \in B$ such that $ab \in I$. If $a \notin I$, then $b - ab \in I$ (by (3)) and hence

$$b = (b - ab) + ab \in I.$$

Therefore, $a \in I$ or $b \in I$. Already we are given that I is a proper ideal of B. Thus, I is a prime ideal of B.

Worked Exercise 10.5.3. Consider the ring $\mathbb{Z}[i]$ of Gaussian integers. Let

$$I = \{a + bi \in \mathbb{Z}[i] : a \text{ and } b \text{ are both even}\}$$

Prove that I is an ideal of $\mathbb{Z}[i]$, which is not a maximal ideal.

Answer: If $x = a + bi$ and $y = c + di \in I$, then a, b, c and d are even and hence $a - c$ and $b - d$ are even and therefore $x - y = (a - c) + (b - d)i \in I$. Also, if $x = a + bi \in I$ and $z = s + ti \in \mathbb{Z}[i]$, then

$$xz = (as - bt) + (at + bs)i \in I$$

since a and b are even and hence as, bt, at and bs are all even. Thus, I is an ideal of $\mathbb{Z}[i]$. Note that $\mathbb{Z}[i]$ is a commutative ring with unity. We shall prove that I is not a maximal ideal of $\mathbb{Z}[i]$. Let

$$J = \{a + bi \in \mathbb{Z}[i] : a^2 + b^2 \text{ is even}\}.$$

Observe that $a^2 + b^2$ is even if and only if either both a and b are even or both a and b are odd. We verify that J is an ideal of $\mathbb{Z}[i]$.

Let $x = a + bi$ and $y = c + di \in J$. Then, $a^2 + b^2$ and $c^2 + d^2$ are even and

$$x - y = (a - c) + (b - d)i \in J$$

since $(a - c)^2 + (b - d)^2 = (a^2 + b^2) + (c^2 + d^2) - 2(ac + bd)$, which is even. Also, for any $z = s + ti \in \mathbb{Z}[i]$,

$$xz = (a + bi)(s + ti) = (as - bt) + (at + bs)i$$

and $(as - bt)^2 + (at + bs)^2 = (a^2 + b^2)s^2 + (a^2 + b^2)t^2$
which is even, since $a^2 + b^2$ is even. Therefore, $xz \in J$. Thus, J is an ideal of $\mathbb{Z}[i]$. Further,

$$I \underset{\neq}{\subset} J \underset{\neq}{\subset} \mathbb{Z}[i],$$

since $1 + i \in J - I$ and $1 + 2i \notin J$. Thus, I is not a maximal ideal of $\mathbb{Z}[i]$.

Worked Exercise 10.5.4. Let $I = \{a + bi \in \mathbb{Z}[i]$: 3 divides both a and $b\}$. Prove that I is a maximal ideal of the ring $\mathbb{Z}[i]$ of Gaussian integers.

Answer: Clearly $3 + 3i \in I$ and hence I is a nonempty subset of $\mathbb{Z}[i]$. Let $x = a + bi$ and $y = c + id \in I$. Then, 3 divides a, b, c and d and hence 3 divides $a - c$ and $b - d$, so that

$$x - y = (a - c) + (b - d)i \in I$$

Also, for any $z = s + ti \in \mathbb{Z}[i]$,

$$\begin{aligned} xz &= (a + bi)(s + ti) \\ &= (as - bt) + (at + bs)i \end{aligned}$$

which belongs to I, since 3 divides a and b and hence 3 divides $as - bt$ and $at + bs$. Thus, I is an ideal of $\mathbb{Z}[i]$. To prove that I is maximal, let J be any ideal of $\mathbb{Z}[i]$ such that $I \subsetneqq J \subseteq \mathbb{Z}[i]$. Choose $a + bi \in J$ such that $a + bi \notin I$. Then, 3 does not divide a or b or both. We shall distinguish these cases separately and prove that $1 \in J$ in each case.

1. Suppose that 3 divides a and 3 does not divide b.

   Then, $a = a+0i \in I \subseteq J$ and hence
   $$bi = (a + bi) - a \in J.$$
   Now, $b^2 = (bi)(-bi) \in J$, since J is an ideal. Since 3 does not divide b, g.c.d. $(3, b^2) = 1$ and therefore there exist c and $d \in \mathbb{Z}$ such that $3c + b^2d = 1$. Now, $b^2d \in J$ and $3c \in I \subseteq J$ and hence
   $$1 = 3c + b^2d \in J$$
   which implies that $J = \mathbb{Z}[i]$.
2. Suppose that 3 does not divide a and 3 divides b. In this case, using the technique of (1) above we can prove that $J = \mathbb{Z}[i]$.
3. Suppose that 3 divides neither a nor b. Then, $a = 3k + 1$ or $3k + 2$ for some $k \in \mathbb{Z}$ and $b = 3s + 1$ or $3s + 2$ for some $s \in \mathbb{Z}$. Then,
   $$a^2 = 3(3k^2 + 2k) + 1 \quad \text{or} \quad 3(3k^2 + 4k + 1) + 1$$
   $$\text{and} \quad b^2 = 3(3s^2 + 2s) + 1 \quad \text{or} \quad 3(3s^2 + 4s + 1) + 1$$
   which imply that $a^2 + b^2 = 3t + 2$ for some $t \in \mathbb{Z}$ and hence 3 does not divide $a^2 + b^2$. Put $c = a^2 + b^2$. Then, g.c.d.$(3, c) = 1$ and hence there exist integers n and m such that $3m + cn = 1$. Now,
   $$3 \in I \subseteq J \text{ and } c = a^2 + b^2 = (a + bi)(a - bi) \in J$$
   since $a + bi \in J$. Therefore, 3 and $c \in J$ and hence
   $$1 = 3m + cn \in J$$
   which implies that $J = \mathbb{Z}[i]$.

Thus, in all cases, we have proved that $J = \mathbb{Z}[i]$. Therefore, I is a maximal ideal of $\mathbb{Z}[i]$.

Worked Exercise 10.5.5. Let M and N be two distinct maximal ideals of a commutative ring R with unity. Then prove that $MN = M \cap N$.

Answer: Since M and N are distinct maximal ideals, $M \nsubseteq N$ and $N \nsubseteq M$ and hence $M + N \neq M$ and $M + N \neq N$. In particular, $M+N$ is an ideal of R containing M properly. By the maximality of M, it follows that $M + N = R$. In particular, $1 \in R = M + N$ and hence

$$1 = a + b \quad \text{for some } a \in M \text{ and } b \in N.$$

Now, clearly $MN \subseteq M \cap N$. Also,

$$x \in M \cap N \Rightarrow x = x1 = xa + xb \in MN$$

since $x \in N$, $a \in M$, $x \in M$ and $b \in N$. Therefore, $M \cap N \subseteq MN$. Thus, $MN = M \cap N$.

EXERCISES 10(E)

1. Determine all maximal ideals and the Jacobson radicals in each of the following rings.
   (i) $\mathbb{R}$
   (ii) $\mathbb{Q}$
   (iii) $\mathbb{Z}$
   (iv) $\mathbb{Z}_{120}$
   (v) $\mathbb{Z} \times \mathbb{Z}$
   (vi) $\mathbb{Q} \times \mathbb{Z}$
   (vii) $M_3(\mathbb{R})$
   (viii) $\mathbb{Z}_n$, $n \in \mathbb{Z}^+$.

2. Give an example of a maximal ideal in the ring $3\mathbb{Z}$ which is not a prime.

3. Let I be a proper ideal of a ring R. Prove that $M \mapsto M/I$ is a one-to-one correspondence between the maximal ideals of R containing I and the maximal ideals of R/I.

4. Let $n \in \mathbb{Z}^+$. Prove that an ideal of $\mathbb{Z}_n$ is prime if and only if it is maximal.

5. Let R be a finite commutative ring with unity. Prove that an ideal of R is prime if and only if it is maximal.

6. Let X be a nonempty finite set with n elements. Prove that there are exactly n maximal ideals in the ring $(P(X), +, \cap)$.

7. Let $\mathscr{C}(X, \mathbb{R})$ be the set of all real valued continuous functions defined on a topological space X. Then prove that $\mathscr{C}(X, \mathbb{R})$ is a commutative ring with unity under the point-wise addition and multiplication.

8. In the ring $\mathscr{C}(X, \mathbb{R})$ given in 7 above, prove that the set

$$M_x = \{f \in \mathscr{C}(X, \mathbb{R}) : f(x) = 0\}$$

   is a maximal ideal for each $x \in X$.

9. Let X be a Compact Hausdorff space and $\mathscr{C}(X, \mathbb{R})$ be the ring given in 7 above. Prove that $x \mapsto M_x$ is a bijection of X onto the set of maximal ideals of $\mathscr{C}(X, \mathbb{R})$, where M_x is the set given in 8 above.

10. Let $\mathbb{Z}[i]$ be the ring of Gaussian integers. Prove that the set

$$I = \{a + bi \in \mathbb{Z}[i] : a - b \text{ is even}\}.$$

is a maximal ideal of $\mathbb{Z}[i]$ and find the number of elements of the quotient ring $\mathbb{Z}[i]/I$.

11. Let $\mathbb{Z}[\sqrt{2}] = \{a + b\sqrt{2} : a \text{ and } b \in \mathbb{Z}\}$. Prove that $\mathbb{Z}[\sqrt{2}]$ is a commutative ring with unity under the usual addition and multiplication of real numbers. Prove that the set

$$I = \{a + b\sqrt{2} \in \mathbb{Z}\sqrt{2} : 5 \text{ divides both } a \text{ and } b\}$$

is a maximal ideal of $\mathbb{Z}[\sqrt{2}]$ and find the number of elements in $\mathbb{Z}[\sqrt{2}]/I$.

12. Prove that the set

$$I = \{a + bi \in \mathbb{Z}[i] : 5 \text{ divides both } a \text{ and } b\}$$

is an ideal of $\mathbb{Z}[i]$ and is not maximal. Is this a prime ideal? Estimate the number of elements in the quotient ring $\mathbb{Z}[i]/I$.

13. Prove that a proper ideal M of a ring R is maximal if and only if, for any ideal I of R, either $I \subseteq M$ or $I + M = R$.

14. Let $f: R \rightarrow S$ be an epimorphism of rings.
    (i) If M is a maximal (prime) ideal of R containing $\ker f$, prove that $f(M)$ is a maximal (prime) ideal of S.
    (ii) If M' is a maximal (prime) ideal of S, then prove that $f^{-1}(M')$ is a maximal (prime) ideal of R.
    (iii) Prove that $M \mapsto f(M)$ is a one-to-one correspondence between the maximal (prime) ideals of R containing $\ker f$ and the maximal (prime) ideals of S.

15. A nonzero ideal I of a ring R is called *minimal* if there is no ideal properly in between $\{0\}$ and I. Prove that a nonzero ideal I of R is minimal if and only if $I = \langle a \rangle$ for all $0 \neq a \in I$. Show that the ring $\mathbb{Z}$ of integers has no minimal ideals.

16. Let P be a prime ideal of a commutative ring R such that the quotient ring R/P is finite. Then prove that P is a maximal ideal.

17. Let R be a commutative ring with unity such that, for each $a \in R$, there exists an integer $n > 1$ such that $a^n = a$. Prove that an ideal of R is prime if and only if it is maximal.

18. Let M be a maximal ideal of a commutative ring R with unity and $n \in \mathbb{Z}^+$. Prove that R/M^n has exactly one prime ideal.

19. Prove that the following are equivalent to each other for any commutative ring with unity.

(i) R has a unique prime ideal.

(ii) R has a unique maximal ideal and the Jacobson radical of R is equal to the prime radical of R.

(iii) Every nonunit in R is nilpotent.

20. Prove the following for any commutative ring R with unity.

(i) R is semisimple if and only if, for each $a \in R$, $1 - ra$ is a nonunit for some $r \in R$.

(ii) If R is regular, then it is semisimple.

(iii) If I is an ideal of R such that R/I is semisimple, then the Jacobson radical of R is contained in I.

10.6 EMBEDDINGS OF RINGS

Rings without unity lack certain important properties. However, we shall prove in this section that any ring can be treated as a subring of a ring with unity. It is well known that any subring with unity of a field is an integral domain. We prove a converse of this, in the sense that any integral domain can be treated as a subring of a field. First, let us have the following definition.

Definition 10.6.1. A ring R is said to be *embedded* in a ring S if R is isomorphic to a subring of S.

It can be easily proved that R is embedded in S if and only if there is a monomorphism of R into S. In the following, we prove that any ring can be embedded in a ring with unity.

Theorem 10.6.1. Let R be any ring. Then, there exists a ring S with unity satisfying the following properties:

1. R is embedded in S.
2. R is isomorphic to an ideal of S.
3. R is commutative if and only if S is commutative.
4. $\text{char}(R) = \text{char}(S)$.

Proof: First, we assume that R is of characteristic zero. Let

$$S = \mathbb{Z} \times R.$$

For any (m, a) and $(n, b) \in S$, define

$$(m, a) + (n, b) = (m + n, a + b)$$
$$\text{and} \quad (m, a) \cdot (n, b) = (mn, mb + na + ab).$$

Then, $+$ and $\cdot$ are binary operations on S. Since $+$ is precisely co-ordinate wise addition and $(\mathbb{Z}, +)$ and $(R, +)$ are abelian groups, it follows that $(S, +)$ is also an abelian group in which $(0, 0)$ is the identity, where the first 0 is the integer 0 and the second 0 is the zero element in the ring R. For any (m, a), (n, b) and (r, c) in S, we have

$$\begin{aligned}((m, a) \cdot (n, b)) \cdot (r, c) &= (mn, mb + na + ab) \cdot (r, c) \\ &= ((mn)r, mnc + r(mb + na + ab) + \\ &\quad (mb + na + ab)c) \\ &= (m(nr), m(nc + rb + bc) + nra + \\ &\quad a(nc + rb + bc)) \\ &= (m, a) \cdot ((n, b) \cdot (r, c)).\end{aligned}$$

Therefore, $\cdot$ is associative on S. Also,

$$\begin{aligned}(m, a) \cdot ((n, b) + (r, c)) &= (m, a) \cdot (n + r, b + c) \\ &= (m(n + r), m(b + c) + (n + r)a + a(b + c)) \\ &= (mn + mr, (mb + na + ab) + (mc + ra + ac)) \\ &= (mn, mb + na + ab) + (mr, mc + ra + ac) \\ &= (m, a) \cdot (n, b) + (m, a) \cdot (r, c).\end{aligned}$$

Similarly, we can prove the other distributive law.
Thus, $(S, +, \cdot)$ is a ring. Consider the element $(1, 0)$ in S. For any $(m, a) \in S$, we have

$$\begin{aligned}&(m, a) \cdot (1, 0) = (m1, m0 + 1a + a0) = (m, a) \\ \text{and} \quad &(1, 0) \cdot (m, a) = (1m, 1a + m0 + 0a) = (m, a)\end{aligned}$$

and therefore $(1, 0)$ is the unity (multiplicative identity) in S. Thus, S is a ring with unity. We shall prove that this ring S satisfies all the required three properties.

1. Define $f: R \to S$ by $f(a) = (0, a)$ for any $a \in R$. It can be easily verified that f is a monomorphism of rings. Therefore, R is embedded in S.
2. Put $I = f(R) = \{(0, a) : a \in R\}$. Then, for any $(0, a) \in I$ and $(n, b) \in S$, we have

$$\begin{aligned}&(0, a) \cdot (n, b) = (0n, 0b + na + ab)\ (0, na + ab) \in I \\ \text{and} \quad &(n, b) \cdot (0, a) = (n0, na + 0b + ba) = (0, na + ab) \in I.\end{aligned}$$

Also, $(0, a) + (0, b) = (0, a + b) \in I$.

Thus, I is an ideal of S. Clearly the map f defined in (1) above is an isomorphism of R onto $f(R) = I$.

3. If S is commutative, then $f(R)$, being a subring of S, is commutative and hence R is commutative. Conversely suppose that R is commutative. Then, for any (m, a) and (n, b) in S, we have

$$\begin{aligned}(m, a) \cdot (n, b) &= (mn, mb + na + ab)\\ &= (nm, na + mb + ba)\\ &= (n, b) \cdot (m, a)\end{aligned}$$

and hence S is commutative.

4. Since $\mathbb{Z}$ is of characteristic zero, so is S and hence $\text{char}(R) = 0 = \text{char}(S)$.

Next, we assume that $\text{char}(R) = n > 0$. In this case, the above construction of S is of no use, since $\mathbb{Z}$ is of characteristic zero and hence so is S, irrespective of whether $\text{char}(R)$ is zero or not. Therefore, we slightly change the construction at S. Let

$$S = \mathbb{Z}_n \times R,$$

where $\mathbb{Z}_n$ is the ring of integers modulo n. Define addition and multiplication in S as follows

$$(i, a) + (j, b) = (i +_n j, a + b)$$
$$\text{and} \quad (i, a)(j, b) = (i \cdot_n j, ib + ja + ab).$$

Then, similar to the above case, we can easily prove that $(S, +, \cdot)$ is a ring with unity satisfying the properties (1) through (4). Note that $\text{char}(S) = n = \text{char}(R)$.

We have proved earlier that any field is an integral domain and that any finite integral domain is a field. Also, any subring (with unity) of a field is an integral domain. In the following theorem, we prove that integral domains arise as subrings of fields only.

Theorem 10.6.2. Any integral domain can be embedded in a field.

Proof: Let R be an integral domain; that is, R is a nontrivial commutative ring with unity in which product of any two nonzero elements is again nonzero. Let

$$S = \{(a, b) : a \text{ and } b \in R \text{ and } b \neq 0\}$$

and define *a* binary relation Θ on S by

$$(a, b)\ \Theta\ (c, d) \Leftrightarrow ad = bc.$$

We shall verify that Θ is an equivalence relation on S. Since R is commutative, $ab = ba$ and hence $(a, b)\ \Theta\ (a, b)$ for any $(a, b) \in S$. Therefore, Θ is reflexive. Also,

$$(a, b)\ \Theta\ (c, d) \Rightarrow ad = bc \Rightarrow cb = da \Rightarrow (c, d)\ \Theta\ (a, b)$$

and therefore Θ is symmetric. Further,

$$\begin{aligned}(a, b)\ \Theta\ (c, d) \text{ and } (c, d)\ \Theta\ (s, t) &\Rightarrow ad = bc \text{ and } ct = ds \\ &\Rightarrow adt = bct = bds \\ &\Rightarrow (at)d = (bs)d \\ &\Rightarrow at = bs \quad \text{since } d \neq 0 \\ &\Rightarrow (a, b)\ \Theta\ (s, t).\end{aligned}$$

Therefore, Θ is transitive also. Thus, Θ is an equivalence relation on S. For any $(a, b) \in S$, let

$$\begin{aligned}[a, b] &= \text{The equivalence class containing } (a, b) \\ &= \{(c, d) \in S : (a, b)\ \Theta\ (c, d)\}\end{aligned}$$

Recall that $[a, b] = [c, d] \Leftrightarrow (a, b)\ \Theta\ (c, d) \Leftrightarrow ad = bc$. Let

$$F = \{[a, b] : (a, b) \in S\}$$

we shall define addition and multiplication on F by

$$\begin{aligned}[a, b] + [c, d] &= [ad + bc, bd] \\ \text{and} \quad [a, b] \cdot [c, d] &= [ac, bd].\end{aligned}$$

First note that, since R is an integral domain and $b \neq 0$ and $d \neq 0$, we have $bd \neq 0$ and hence the above definitions of $+$ and $\cdot$ make sense. Next, we have to prove that $+$ and $\cdot$ are well defined, in the sense that they depend on the classes $[a, b]$ and $[c, d]$, but not on the representative elements a, b, c and d. Suppose that

$$[a, b] = [a', b'] \quad \text{and} \quad [c, d] = [c', d'].$$

Then, $ab' = ba'$ and $cd' = dc'$ and hence

$$\begin{aligned}(ad + bc)b'd' &= adb'd' + bcb'd' \\ &= (ab')dd' + (cd')bb' \\ &= (ba')dd' + (dc')bb' \\ &= bd(a'd' + b'c')\end{aligned}$$

and hence $[ad + bc, bd] = [a'd' + b'c', b'd']$. Thus, $+$ is well defined. Similarly, we can prove that $\cdot$ is well defined. In the following, let $[a, b]$, $[c, d]$ and $[s, t]$ be arbitrary elements of F.

$$\begin{aligned}([a, b] + [c, d]) + [s, t] &= [ad + bc, bd] + [s, t]\\ &= [(ad + bc)t + bds, bdt]\\ &= [a(dt) + b(ct + ds), bdt]\\ &= [a, b] + ([c, d] + [s, t]).\end{aligned}$$

Therefore, $+$ is associative.

$$\begin{aligned}[a, b] + [c, d] &= [ad + bc, bd]\\ &= [cb + da, db]\\ &= [c, d] + [a, b].\end{aligned}$$

Therefore, $+$ is commutative.

$$[a, b] + [0, 1] = [a1 + b0, b1] = [a, b]$$

Therefore, $[0, 1]$ is the additive identity in F

$$[a, b] + [-a, b] = [ab + b(-a), b^2] = [0, b^2] = [0, 1].$$

Therefore, $[-a, b]$ is the additive inverse of $[a, b]$ in F. Thus, $(F, +)$ is an abelian group. One can easily verify that $\cdot$ is associative and commutative. Also,

$$[a, b] \cdot [1, 1] = [a1, b1] = [a, b]$$

and therefore $[1, 1]$ is the multiplicative identity in F.

$$\begin{aligned}[a, b] \cdot ([c, d] + [s, t]) &= [a, b] \cdot [ct + ds, dt]\\ &= [a(ct + ds), bdt]\\ &= [b(act + ads), b(bdt)]\\ &= [acbt, + bdas, bdbt]\\ &= [ac, bd] + [as, bt]\\ &= [a, b] \cdot [c, d] + [a, b] \cdot [s, t]\end{aligned}$$

Therefore, $\cdot$ distributes over $+$. Thus, $(F, +, \cdot)$ is a commutative ring with unity. Also, since R is nontrivial, $1 \neq 0$ in R and $[1, 1] \neq [0, 1]$ in F.

Therefore, F is nontrivial. Also for any $[a, b] \neq [0, 1]$, we have $a \neq 0$ and hence $[b, a] \in F$ and

$$[a, b] \cdot [b, a] = [ab, ba] = [1, 1].$$

Therefore, $[b, a]$ is the multiplicative inverse of $[a, b]$. Thus, $(F, +, \cdot)$ is a field.
Now, define $f: R \to F$ by $f(a) = [a, 1]$ for all $a \in R$. One can easily verify that f is a monomorphism of rings. Thus, R is embedded in the field F.

Definition 10.6.2. For any integral domain R, the field F constructed above is called the *field of quotients* of R.

Example 10.6.1. The ring $\mathbb{Z}$ of integers is an integral domain and the field of quotients of $\mathbb{Z}$ is precisely the field $\mathbb{Q}$ of rational numbers. A rational number is usually written as a/b which is precisely $[a, b]$. Recall from the high school mathematics that two rational numbers a/b and c/d are equal if and only if $ad = bc$ and that a/b represents a class of pairs (c, d) for which $ad = bc$.

By means of the monomorphism f of an integral domain R into the field of quotients F defined by $f(a) = [a, 1]$, we can identify an element a in R with the element $[a, 1]$ in F. With this identification, we can treat R as a subring of F. Also, any element $[a, b]$ of F can be expressed as ab^{-1} with $a \in R$ and $0 \neq b \in R$, since

$$[a, b] = [a, 1] \cdot [1, b] = [a, 1][b, 1]^{-1}.$$

Corollary 10.6.1. If R is a field, then the field of quotients of R is isomorphic to R.

Proof: Let R be a field and F be its field of quotients.
Let $f: R \to F$ be the monomorphism defined by

$$f(a) = [a, 1] \quad \text{for any } a \in R.$$

Then, for any $[a, b] \in F$, we can write

$$[a, b] = [ab^{-1}, 1] = f(ab^{-1})$$

and $ab^{-1} \in R$ (Note that $b \neq 0$ and hence b is a unit in R).
Therefore, f is a surjection also and hence f is an isomorphism. Thus, $F \cong R$.

Theorem 10.6.3. Let R be an integral domain and K be a field containing R as a subring. Let

$$Q = \{ab^{-1} \in K: a \text{ and } b \in R \text{ and } b \neq 0\}.$$

Then, Q is a subfield of K and is isomorphic to the field of quotients of R.

Proof: Let F be the field of quotients of R and define $g : F \to Q$ by

$$g([a, b]) = ab^{-1} \quad \text{for any } [a, b] \in F.$$

For any $[a, b]$ and $[c, d] \in F$, we have

$$\begin{aligned}[a, b] = [c, d] &\Leftrightarrow ad = bc \\ &\Leftrightarrow ab^{-1} = cd^{-1}\end{aligned}$$

This shows that g is well defined and is an injection. By the definition of Q, g is a surjection also. Further,

$$\begin{aligned}g([a, b] + [c, d]) &= g([ad + bc, bd]) \\ &= (ad + bc)(bd)^{-1} \\ &= (ad + bc)b^{-1}d^{-1} \\ &= ab^{-1} + cd^{-1} \\ &= g([a, b]) + g([c, d]) \\ \text{and} \quad g([a, b]\,[c, d]) &= g([ac, bd]) = ac(bd)^{-1} \\ &= ab^{-1}cd^{-1} = g([a, b])g([c, d]).\end{aligned}$$

Therefore, g is a homomorphism also. Thus, g is an isomorphism of F onto Q. This implies that Q is a subfield of K and $F \cong Q$.

Corollary 10.6.2. Let R be an integral domain and F be its field of quotients. Then, F is the smallest field containing R, in the sense that any field containing R as a subring should contain an isomorphic copy of F.

Worked Exercise 10.6.1. Let F be the field of quotients of an integral domain R. Then prove that R and F are of same characteristic.

Answer: We can treat R as a subring of F and the unities in R and F are the same. In Theorem 9.3.2, we have proved that the characteristic of any ring with unity is precisely the order of the unity in the additive group of the ring. Therefore,

$$\text{char}(R) = \text{O}(1) = \text{char}(F),$$

where O(1) is the order of 1 in the group $(R, -)$.

Worked Exercise 10.6.2. Let $\mathbb{Z}[i]$ be the ring of Gaussian integers. Determine the field of quotients of $\mathbb{Z}[i]$.

Answer: Recall that

$$\mathbb{Z}[i] = \{a + bi : a \text{ and } b \text{ are integers}\}$$

and that $\mathbb{Z}[i]$ is an integral domain under the addition and multiplication of complex numbers. Let $\mathbb{C}$ be the field of complex numbers. Then, $\mathbb{Z}[i]$ is a subring of $\mathbb{C}$. By Theorem 10.6.3, the field of quotients of $\mathbb{Z}[i]$ is equal (isomorphic) to

$$Q = \{st^{-1} : s \text{ and } t \in \mathbb{Z}[i] \text{ and } t \neq 0\}$$

If $s = a + bi$ and $t = c + di \neq 0$, then $c \neq 0$ or $d \neq 0$ and hence $c^2 + d^2 > 0$. Now,

$$\begin{aligned} st^{-1} &= \frac{a+bi}{c+di} = \frac{(a+bi)(c-di)}{(c+di)(c-di)} \\ &= \frac{ac+bd+(bc-ad)i}{c^2+d^2} \\ &= \frac{ac+bd}{c^2+d^2} + \frac{bc-ad}{c^2+d^2} i \end{aligned}$$

Since a, b, c and d are all integers, it follows that st^{-1} belongs to the set

$$\mathbb{Q}[i] = \{p + qi : p \text{ and } q \text{ are rational}\}$$

and hence $Q \subseteq \mathbb{Q}[i]$.
On the other hand, let $p + qi \in \mathbb{Q}[i]$ and $p = \frac{a}{b}$ and $q = \frac{c}{d}$, where $a, b, c, d \in \mathbb{Z}$, $b \neq 0$ and $d \neq 0$. Then,

$$p = ab^{-1} \quad \text{and} \quad q = cd^{-1}, a, b, c, d \in \mathbb{Z} \subseteq \mathbb{Z}[i]$$

and hence p and $q \in Q$. Since $i \in \mathbb{Z}[i] \subseteq Q$, it follows that $p + qi \in Q$. Therefore, $\mathbb{Q}[i] \subseteq Q$. Thus, $Q = \mathbb{Q}[i]$. That is, the field of quotients of $\mathbb{Z}[i]$ is $\mathbb{Q}[i]$. Recall that $\mathbb{Q}$ is the field of quotients of $\mathbb{Z}$.

Worked Exercise 10.6.3. Prove that any isomorphism between two integral domains can be extended to their fields of quotients.

Answer: Let R and R' be integral domains and $f: R \to R'$ be an isomorphism. Let F and F' be fields of quotients of R and R', respectively. Then,

$$\begin{aligned} F &= \{ab^{-1} : a \text{ and } b \in R \text{ and } b \neq 0\} \\ \text{and} \quad F' &= \{xy^{-1} : x \text{ and } y \in R' \text{ and } y \neq 0\}. \end{aligned}$$

Define $g : F \to F'$ by $g(ab^{-1}) = f(a)f(b)^{-1}$. Note that $b \neq 0$ in R if and only if $f(r) \neq 0$ in R', since f is an isomorphism. Also,

$$\begin{aligned} ab^{-1} = cd^{-1} &\Leftrightarrow ad = bc \\ &\Leftrightarrow f(ad) = f(bc) \\ &\Leftrightarrow f(a)f(d) = f(b)f(c) \\ &\Leftrightarrow f(a)f(b)^{-1} = f(c)f(d)^{-1} \end{aligned}$$

This shows that g is well defined and is an injection. Also, if $xy^{-1} \in F'$ with x, $y \in R'$ and $y \neq 0$, we can choose elements a and b in R such that $f(a) = x$ and $f(b) = y$ (since f is a bijection). Then, $b \neq 0$ and $g(ab^{-1}) = f(a)f(b)^{-1} = xy^{-1}$. Therefore, g is a surjection also. Further, for any ab^{-1}, cd^{-1}, $\in F$,

$$\begin{aligned} g(ab^{-1} + cd^{-1}) &= g((ad + bc)(bd)^{-1}) \\ &= f(ad + bc)f(bd)^{-1} \\ &= (f(a)f(d) + f(b)f(c))(f(b)f(d))^{-1} \\ &= f(a)f(b)^{-1} + f(c)f(d)^{-1} \\ &= g(ab^{-1}) + g(cd^{-1}) \\ \text{and} \quad g((ab^{-1}) \cdot (cd^{-1})) &= g(ac(bd)^{-1}) \\ &= f(ac)f(bd)^{-1} \\ &= f(a)f(c)(f(b)f(d))^{-1} \\ &= f(a)f(b)^{-1}f(c)f(d)^{-1} \\ &= g(ab^{-1})g(cd^{-1}). \end{aligned}$$

Therefore, g is a homomorphism also. Thus, g is an isomorphism of F onto F'. Also, g is an extension of f, since $g(a) = g(a1^{-1}) = f(a)f(1)^{-1} = f(a)1 = f(a)$ for all $a \in R$.

Worked Exercise 10.6.4. Let F be the field of quotients of an integral domain R and S be a subring of F such that $R \subseteq S \subseteq F$. Prove that the field of quotients of S is isomorphic to F.

Answer: Since F is a field, it is an integral domain. Being a subring of F, S is also an integral domain.

Note that the unity in R is same unity in S as well as in F. Now, by Theorem 10.6.3, the field of quotients of S is given by

$$Q = \{ab^{-1}: a \text{ and } b \in S \text{ and } b \neq 0\}.$$

Since F is a field containing S, we get from Corollary 10.6.2 that $Q \subseteq F$. On the other hand,

$$\begin{aligned} x \in F &\Rightarrow x = ab^{-1}, \quad \text{where } a \text{ and } b \in R \text{ and } b \neq 0 \\ &\Rightarrow x = ab^{-1}, a, b \in S, b \neq 0 \quad (\text{since } R \subseteq S). \\ &\Rightarrow x \in Q. \end{aligned}$$

Thus, $F \subseteq Q$ and hence $F = Q$.

EXERCISE 10(F)

1. Determine the field of quotients of each of the following integral domains.
   (i) $\mathbb{Z}$
   (ii) $\mathbb{Z}_7$
   (iii) $\mathbb{Z}_{11}$
   (iv) $\mathbb{Z}_2[i] = \{a + bi : a \text{ and } b \in \mathbb{Z}_2\}$ under addition and multiplication modulo 2.
   (v) $\mathbb{R}$
   (vi) $\mathbb{Q}[i]$
2. Prove in detail that $\mathbb{Z}_n \times R$ in the proof of Theorem 10.6.1 is a ring with unity and is of characteristic n, where R is a ring of characteristic n.
3. Determine the field of quotients of the integral domain
   $$\mathbb{Z}[\sqrt{3}] = \{a + b\sqrt{3} : a \text{ and } b \in \mathbb{Z}\}.$$
4. Let $R = \{\frac{a}{b} : a \text{ and } b \in \mathbb{Z} \text{ and } 5 \text{ does not divide } b\}$.
   Prove that R is a subring of the ring Q of rational numbers and deduce that R is an integral domain. Determine the field of quotients of R.
5. Let F be a field containing no subfield properly (such fields are called *prime fields*). Prove that F is isomorphic either to the field $\mathbb{Q}$ of rational numbers or to the field $\mathbb{Z}_p$ of integers modulo a prime p.

6. Prove that every field contains a subfield, that is, isomorphic to $\mathbb{Q}$ or $\mathbb{Z}_p$ (such subfields are called *prime subfields*).

7. Prove that any automorphism of a field F fixes every element of the prime subfield.

8. Determine the prime subfields of each of the following fields.
   (i) $\mathbb{R}$
   (ii) $\mathbb{Q}$
   (iii) $\mathbb{Z}_{79}$
   (iv) $\mathbb{Q}[\sqrt{5}] = \{a + b\sqrt{5} : a \text{ and } b \in \mathbb{Q}\}$
   (v) $\mathbb{Q}[\sqrt{2}] = \{a + b\sqrt{2} : a \text{ and } b \in \mathbb{Q}\}$
   (vi) $\mathbb{Q}[i] = \{a + bi : a \text{ and } b \in \mathbb{Q}\}$

9. If F is a subfield of a field K, then prove that F and K have the same prime subfields.

10. If a field F has exactly 9 elements, then prove that the prime subfield of F is isomorphic to $\mathbb{Z}_3$.

11. If $\mathbb{Z}_5$ is the prime subfield of a field F, then prove that there exists $a \in F$ such that $a + a \neq 0$.

12. Prove that any automorphism of an integral domain can be extended uniquely to an automorphism of its field of quotients.

13. Let R be a commutative ring with no zero divisions. Then prove that R can be embedded in an integral domain.

14. Prove that a commutative ring can be embedded in a field if and only if it has no zero divisors.

15. Let R be a commutative ring with unity. A subset S of R is said to be *multiplicative* if $1 \in S$, $0 \notin S$ and $ab \in S$ for any a and $b \in S$. Define a binary relation $\boldsymbol{\Theta}$ on $R \times S$ by

$$(a, s)\ \boldsymbol{\Theta}\ (b, t) \Leftrightarrow \text{there exists } u \in S \text{ such that } u(at - bs) = 0.$$

Prove that $\boldsymbol{\Theta}$ is an equivalence relation on $R \times S$. Let the equivalence containing (a, s) be denoted by a/s and let

$$S^{-1}R = \left\{\frac{a}{s} : a \in R \text{ and } s \in S\right\}.$$

Define addition and multiplication on $S^{-1}R$ by

$$\frac{a}{s} + \frac{b}{t} = \frac{at + bs}{st} \quad \text{and} \quad \frac{a}{s} \cdot \frac{b}{t} = \frac{ab}{st}$$

Then prove that $(S^{-1}R, +, \cdot)$ is a commutative ring with unity. This ring is called the *ring of fractions* of R by S.

16. Prove that the field of quotients of an integral domain R is the ring of fractions of R by $R - \{0\}$.

17. Let P be a prime ideal of a commutative ring R with unity. Then prove that $(R - P)^{-1}R$ has a unique maximal ideal.

18. Let S be a multiplicative subset of a commutative ring R with unity. Prove that there is a one-to-one correspondence between the prime ideals of R disjoint with S and the prime ideals of the ring $S^{-1}R$ of fractions of R by S.

11

Polynomial Rings

We are very familiar with polynomials which are introduced to us very early in our mathematical education, in fact, in high school itself, we are thoroughly drilled in adding, multiplying, dividing, factoring and simplifying them. We have learnt the remainder theorem in eighth or ninth standard. Later, at higher level, polynomials appear as functions and we were concerned with their continuity, derivatives and integrals and their maxima and minima. Now, we are interested in polynomials, but from neither of the above view points. Here, polynomials will simply be elements of a certain ring and we shall be concerned with the algebraic properties of this ring.

At the secondary school level, we have studied polynomials with integer coefficients, rational coefficients, real coefficients and, may be even complex coefficients. Notice that, in each case, the set of coefficients is a ring and the set of polynomials also forms a ring under suitable addition and multiplication, with which we are all familiar. In this chapter, we make an abstraction of these cases and study polynomials with coefficients from a given abstract ring.

11.1 RINGS OF POLYNOMIALS

The word 'polynomial' reminds us an expression or symbol of the form

$$a_0 + a_1x + a_2x^2 + \cdots + a_nx^n.$$

We used to quickly add, multiply and divide such expressions. Now, it is appropriate to clarify what different terms of such an expression mean. Of course, $a_0, a_1, a_2, \ldots, a_n$ are from some number system. Instead, we can take these a_i's as elements of an abstract ring. What are $a_1x, a_2x^2, \ldots, a_nx^n$? What is the + in between them? We used to write

$$3x + x^3 + 2x^5 \quad \text{for} \quad 0 + 3x + 0x^2 + 1x^3 + 0x^4 + 2x^5.$$

Also, we used to treat two polynomials

$$a_0 + a_1x + a_2x^2 + \cdots + a_nx^n \quad \text{and} \quad b_0 + b_1x + b_2x^2 + \cdots + b_mx^m$$

equal if and only if $a_i = b_i$ for all $i \geq 0$.

In essence, the coefficients $a_0, a_1, a_2, \ldots, a_n$ are only important things and the arithmetics of polynomials only depend on these a_i's. In this section, we shall formulise these intuitive ideas and arrive at an exact idea of how the classical arithmetic of polynomials fit into our ring theory. To begin with, we have following definition.

Definition 11.1.1. Let $(R, +, \cdot)$ be any ring. *A polynomial over R* is defined to be an infinite sequence

$$(a_0, a_1, a_2, \ldots, a_n, \ldots)$$

of elements of R such that a_n's are zero for all but finite number of n's; equivalently, there exists a nonnegative integer k such that $a_n = 0$ for all $n \geq k$. The set of all polynomials over R will be denoted by Poly(R).

Recall that an infinite sequence of elements of R can be viewed as a mapping of the set of nonnegative integers into R. Let us agree that two polynomials

$$f = (a_0, a_1, a_2, \ldots) \quad \text{and} \quad g = (b_0, b_1, b_2, \ldots)$$

are considered to be equal if and only if $a_n = b_n$ for all $n \geq 0$.

Often it is convenient to use the notation $(a_0, a_1, a_2, \ldots, a_n, 0, 0, \ldots)$ for a polynomial with a_n as the last nonzero term; when $n = 0$, we allow the possibility that $a_0 = 0$ in order to include the zero polynomial $(0, 0, 0, \ldots)$ each of whose term is zero. With this notation, we have

$$\text{Poly}(R) = \{(a_0, a_1, a_2, \ldots, a_n, 0, 0, \ldots) : a_i \in R \text{ and } n \geq 0\}.$$

Therefore, (0, 1, 1, 0, 1, 0, 0, 0, ...) is a polynomial over $\mathbb{Z}$, where as (0, 1, 0, 1, 0, 1, 0, 1, ...) is not. In the following, we introduce suitable operations on Poly(R) to make it a ring.

Definition 11.1.2. For any polynomials $a = (a_0, a_1, a_2, \ldots)$ and $b = (b_0, b_1, b_2, \ldots)$ over a ring $(R, +, \cdot)$, define

$$a + b = (a_0 + b_0,\ a_1 + b_1,\ a_2 + b_2,\ \ldots)$$

$$\text{and} \quad a \cdot b = (c_0,\ c_1,\ c_2,\ \ldots),$$

where $c_n = a_0 b_n + a_1 b_{n-1} + \cdots + a_n b_0 = \sum_{r=0}^{n} a_r b_{n-r} = \sum_{r+s=n} a_r b_s$.

Note that the additive operation $+$ and the multiplicative operation $\cdot$ on the right sides of the above defining equations are those in the $(R, +, \cdot)$. Since a and b are in Poly(R), there exist nonnegative integers m and n such that

$$a_i = 0 \text{ for all } i \geq m \quad \text{and} \quad b_i = 0 \text{ for all } i \geq n$$

and hence, for any $i \geq \max\{m, n\}$, $a_i = 0 = b_i$ so that

$$a_i + b_i = 0$$

$$\text{and} \quad c_i = \sum_{r=0}^{i} a_r b_{i-r} = 0 \quad \text{for all } i \geq \max\{m, n\}$$

(since $i \geq m + n$ implies $r \geq m$ or $i - r \geq n$ for any $0 \leq r \leq i$ and therefore, $a_r = 0$ or $b_{i-r} = 0$). This leads to the fact that $a + b$ and $a \cdot b$ are well defined.

Theorem 11.1.1. For any ring $(R, +, \cdot)$, (Poly(R), $+$, $\cdot$) is a ring, where $+$ and $\cdot$ are the operations defined above.

Proof: Let $(R, +, \cdot)$ be a ring and a, b and c be arbitrary elements of Poly(R). Then, a, b and c are polynomials given by

$$a = (a_0,\ a_1,\ a_2,\ \ldots)$$

$$b = (b_0,\ b_1,\ b_2,\ \ldots)$$

$$\text{and} \quad c = (c_0,\ c_1,\ c_2,\ \ldots)$$

with a_i's, b_i's and c_i's are elements in the given ring R. Then, using the associatively and commutativity of $+$ in R, we can prove that

$$(a + b) + c = a + (b + c) \quad \text{and} \quad a + b = b + a.$$

Therefore, + is associative and commutative in Poly(R). Let us write 0 for the polynomial (0, 0, 0, …). Then,

$$a+0=a=0+a \quad \text{for all } a\in \text{Poly}(R).$$

Therefore, 0 is the identity element for +. Also, for any $a = (a_0, a_1, a_2, \ldots)$ in Poly(R), the polynomial $-a$ defined by

$$-a=(-a_0,\ -a_1,\ -a_2,\ \ldots)$$

satisfies the property

$$a+(-a)=(0,\ 0,\ 0,\ \ldots)=0=-a+a$$

and hence $-a$ is the inverse of a with respect to +. Thus, (Poly(R), +) is an abelian group.
To prove the associativity of multiplication, let

$$\begin{aligned} a\cdot b &= (d_0,\ d_1,\ d_2,\ \ldots) \\ \text{and} \quad (a\cdot b)\cdot c &= (x_0,\ x_1,\ x_2,\ \ldots). \end{aligned}$$

Then,

$$d_n=\sum_{r=0}^{n} a_r b_{n-r}=\sum_{r+s}^{n} a_r b_s$$

$$\begin{aligned} \text{and} \quad x_n &= \sum_{s=0}^{n} d_s c_{n-s} = \sum_{s+t=n} d_s c_t \\ &= \sum_{s+t=n}\left(\sum_{r+u=s} a_r b_u\right)c_t \\ &= \sum_{r+u+t=n} a_r b_u c_t \\ &= \sum_{r+s=n} a_r\left(\sum_{u+t=s} b_u c_t\right) \end{aligned}$$

which is precisely the nth term in $a \cdot (b \cdot c)$. Thus,

$$(a\cdot b)\cdot c=a\cdot(b\cdot c).$$

Also, if $a \cdot (b + c) = (y_0, y_1, y_2, \ldots)$, then

$$\begin{aligned} y_n &= \sum_{r+s=n} a_r(b_s + c_s) \\ &= \sum_{r+s=n} a_r b_s + \sum_{r+s=n} a_r c_s \end{aligned}$$

which is precisely the nth term in $(a \cdot b) + (a \cdot c)$. Thus, $a \cdot (b + c) = a \cdot b + a \cdot c$. Similarly, one can prove that $(a + b) \cdot c = a \cdot c + b \cdot c$. Thus, (Poly($R$), $+$, $\cdot$) is a ring.

Definition 11.1.3. For any ring R, (Poly(R), $+$, $\cdot$) is called the *ring of polynomials over R* and is simply denoted by Poly(R).

In the following, we prove that any given ring R is isomorphic to a subring of the ring of polynomials over R.

Theorem 11.1.2. Any ring R can be embedded in Poly(R).

Proof: Let $(R, +, \cdot)$ be a ring and Poly(R) be the ring of polynomials over R. Define $f : R \to \text{Poly}(R)$ by $f(a) = (a, 0, 0, 0, \ldots)$. Then, for any a and $b \in R$,

$$\begin{aligned} f(a+b) &= (a+b,\ 0,\ 0,\ \ldots) \\ &= (a+b,\ 0+0,\ 0+0,\ \ldots) \\ &= (a,\ 0,\ 0,\ \ldots)+(b,\ 0,\ 0,\ \ldots) \\ &= f(a)+f(b) \end{aligned}$$

and

$$\begin{aligned} f(ab) &= (ab,\ 0,\ 0,\ \ldots) \\ &= (a,\ 0,\ 0,\ \ldots)\cdot(b,\ 0,\ 0,\ \ldots) \\ &= f(a)\cdot f(b). \end{aligned}$$

Therefore, f is a homomorphism. Also, for any a and $b \in R$,

$$\begin{aligned} f(a)=f(b) &\Rightarrow (a,\ 0,\ 0,\ \ldots)=(b,\ 0,\ 0,\ \ldots) \\ &\Rightarrow a=b. \end{aligned}$$

Thus, f is an injection also and hence f is a monomorphism of R into Poly(R). This says that R can be embedded in Poly(R) and R is isomorphic to the subring $f(R)$ of Poly(R).

Theorem 11.1.3. For any ring R, the ring Poly(R) is commutative if and only if R is commutative.

Proof: Let R be a ring. If Poly(R) is commutative, then clearly R is commutative, since R is isomorphic to a subring of Poly(R). Conversely, suppose that R is commutative. Then, for any

$$a = (a_0, a_1, a_2, \ldots) \quad \text{and} \quad b = (b_0, b_1, b_2, \ldots),$$

in Poly(R), we have, for any $n \geq 0$,

$$\begin{aligned} n^{\text{th}} \text{ term in } a \cdot b &= \sum_{r+s=n} a_r b_s \\ &= \sum_{s+r=n} b_s a_r \\ &= n^{\text{th}} \text{ term in } b \cdot a \end{aligned}$$

and hence $a \cdot b = b \cdot a$. Thus, Poly(R) is commutative.

Theorem 11.1.4. A ring R is with unity if and only if the ring Poly(R) is with unity.

Proof: Let R be a ring. Suppose that R has unity 1.
Let $e = (1, 0, 0, 0, \ldots)$. Then, for any $a = (a_0, a_1, a_2, \ldots)$ in Poly(R), we have, for any $n \geq 0$,

$$\begin{aligned} n^{\text{th}} \text{ term in } a \cdot e &= \sum_{r+s=n} a_r e_s \\ &= a_n \text{ (since } e_s = 0 \quad \text{for all } s > 0 \text{ and } e_1 = 1) \\ &= \sum_{r+s=n} e_r a_s \\ &= n^{\text{th}} \text{ term in } e \cdot a \end{aligned}$$

and hence $a \cdot e = a = e \cdot a$. Thus, e is the unity element in Poly(R).

Note that a ring R is trivial if and only if the ring Poly(R) is trivial. Recall that a nontrivial commutative ring with unity and without zero divisors is called an integral domain.

Theorem 11.1.5. Let $(R, +, \cdot)$ be a ring R. Then, R is an integral domain if and only if Poly(R) is an integral domain.

Proof: First note that, from Theorems 11.1.1 and 11.1.4, R is a nontrivial commutative ring with unity if and only if so is Poly(R). Therefore, we can

assume that R is a nontrivial commutative ring with unity, which is a necessary qualification for R (and hence for Poly(R)) to be an integral domain.
Suppose that R is an integral domain. Let $a = (a_0, a_1, a_2, \ldots)$ and $b = (b_0, b_1, b_2, \ldots)$ be any nonzero elements in Poly(R). Since $a \neq 0$ and $b \neq 0$, there exist nonnegative integers n and m such that $a_n \neq 0$, $b_m \neq 0$ and $a_i = 0$ for all $i > n$ and $b_j = 0$ for all $j > m$. Now, consider the $(n + m)^{\text{th}}$ term in the product $a \cdot b$. It is given by

$$\sum_{r+s=n+m} a_r b_s = a_n b_m \neq 0,$$

since R is an integral domain, $a_n \neq 0$ and $b_m \neq 0$. (Note that $r + s = n + m \Rightarrow r \geq n$ or $s \geq m \Rightarrow (r = n$ or $a_r = 0)$ or $(s = m$ or $b_s = 0)$. Therefore, the $(n + m)^{\text{th}}$ term of $a \cdot b$ is nonzero and hence $a \cdot b \neq 0$. Therefore, Poly(R) has no zero divisors. Thus, Poly(R) is an integral domain.
Conversely, suppose that Poly(R) is an integral domain. Since R is isomorphic to a subring of Poly(R), it follows that R is an integral domain.

In the following result, we demonstrate that Poly(R) can never be a field, even when R is a field.

Theorem 11.1.6. For any ring R, Poly(R) can never be a field.

Proof: Let R be a ring and suppose that Poly(R) is a field. Then, Poly(R) is an integral domain and hence, by Theorem 11.1.5, R is also an integral domain. Consider the element

$$x = (0, 1, 0, 0, \ldots) \text{ in Poly}(R).$$

Then, x is a nonzero element in Poly(R) and hence x is a unit. Therefore, there exists $a = (a_0, a_1, a_2, \ldots)$ in Poly(R) such that

$$x \cdot a = a \cdot x = 1 = (1, 0, 0, \ldots)$$
$$\text{hence } (a, a_0, a_1, a_2, \ldots) = (1, 0, 0, \ldots).$$

Therefore, $0 = 1$, which is a contradiction to the fact that R is nontrivial. Thus, Poly(R) is not a field.
The element x given in the above proof is of special importance. Even from the high school days, we are well aware that a polynomial over $\mathbb{R}$ is an expression of the form

$$a_0 + a_1 x_1 + a_2 x^2 + \cdots + a_n x^n,$$

where $a_0, a_1, a_2, \ldots, a_n$ are real numbers and x is an indeterminate. Though we are familiar with this, we did not know what x is what ax is for any $a \in \mathbb{R}$. Further, we should give a mathematically valid explanation for the operation symbol $+$ in the above expression of a polynomial. We shall give satisfactory answers to these questions in the following result. Recall that R can be identified with a subring of Poly(R) and any element a in R can be identified with the polynomial $(a, 0, 0, \ldots)$.

Theorem 11.1.7. Let $(R, +, \cdot)$ be a ring with unity and x be the polynomial over R given by

$$x = (0, 1, 0, 0, \ldots).$$

Then, any polynomial over R can be expressed uniquely as

$$a_0 + a_1 \cdot x + a_2 \cdot x^2 + \cdots + a_n \cdot x^n,$$

where $a_0, a_1, \ldots, a_n$ are elements of R, identified with the elements of Poly(R).

Proof: Note that, in the expression $a_0 + a_1 \cdot x + a_2 \cdot x^2 + \cdots + a_n \cdot x^n$, the operation symbols $+$ and $\cdot$ denote the addition and multiplication in Poly(R). Using the definitions of x and the multiplication in Poly(R), the following can be proved easily.

$$\begin{aligned}
x &= (0, 1, 0, 0, \ldots) \\
x^2 &= x \cdot x = (0, 0, 1, 0, 0, \ldots) \\
x^3 &= x^2 \cdot x = (0, 0, 0, 1, 0, 0, \ldots) \\
&\vdots \\
x^n &= x^{n-1} \cdot x = (0, 0, \ldots, 0, \underset{(n+1)^{\text{th}} \text{ term}}{1}, 0, 0, \ldots)
\end{aligned}$$

for any positive integer n. Also, for any $a \in R$, by identifying a with $(a, 0, 0, \ldots)$ in Poly(R), we have

$$\begin{aligned}
a \cdot x &= (0, a, 0, 0, \ldots) \\
a \cdot x^2 &= (0, 0, a, 0, 0, \ldots) \\
a \cdot x^3 &= (0, 0, 0, a, 0, 0, \ldots) \\
&\vdots \\
a \cdot x^n &= (0, 0, \ldots, 0, \underset{(n+1)^{\text{th}} \text{ term}}{\mathbf{a}}, 0, 0, \ldots)
\end{aligned}$$

for any positive integer n. If p is a polynomial over R, then p is a sequence $(a_0, a_1, a_2, \ldots, a_n, 0, 0, \ldots)$ with a_i's are in R. These a_i's are unique in R such that

$$p = (a_0, a_1, a_2, \ldots, a_n, 0, 0, \ldots).$$

Now, we have

$$\begin{aligned} p &= (a_0, 0, 0, \ldots) + (0, a_1, 0, 0, \ldots) + \cdots + (0, 0, \ldots, 0, a_n, 0, 0, \ldots) \\ &= a_0 + a_1 \cdot x + a_2 \cdot x^2 + \cdots + a_n \cdot x^n . \end{aligned}$$

Remarks 11.1.1

1. If we identify $a \in R$ with $(a, 0, 0, \ldots)$ in Poly(R) and identify R as a subring of Poly(R), then Poly(R) is the subring generated by R and x. For this reason, we prefer to write $R[x]$ for Poly(R).
2. The expression $a_0 + a_1x + a_2x^2 + \cdots + a_nx^n$ for a polynomial looks simple and elegant for two reasons: The first one is that we are familiar with this right from our school days. The second is that we can straight away multiply two polynomials by treating x also as a real number or as an element in the ring containing a_i's.
3. As mentioned in the above theorem and its proof, x is not an element in the ring R and it is an element in Poly(R); that is, x is a polynomial over R.
4. x is usually called an *indeterminate.*
5. When we are completely aware of what Poly(R) and $R[x]$ are, we prefer to use familiar notation for polynomials over a given ring R and for the ring of polynomials over R. $R[x]$ is the most standard notation used to denote the ring polynomials over R.
6. The expression $a_0 + a_1x + \cdots + a_nx^n$ for a polynomial is known as *polynomial in an indeterminate form.* Though x is called an indeterminate, it is actually a polynomial by itself and the operations + and · in the above expression are the addition and multiplication of polynomials only.
7. Often, for convenience, a polynomial $a_0 + a_1x + \cdots + a_nx^n$ is also written as $\sum_{i=1}^{n} a_ix^i$, with the assumption that $x^0 = 1$.

To sum up, the polynomials in an indeterminate form are expressions of the form $a_0 + a_1x + a_2x^2 + \cdots + a_nx^n$, where $a_0, a_1, \ldots, a_n$ are elements of a given ring R and the set of all such expressions will be denoted by $R[x]$. The elements of $R[x]$ can be added and multiplied as we do in the number systems.

Example 11.1.1

1. $\mathbb{Z}[x] = \{a_0 + a_1x + \cdots + a_nx^n : n \geq 0, a_i \in \mathbb{Z}\}$. $3 + 2x + x^2 + 4x^3$ is in $\mathbb{Z}[x]$, while $2 + \sqrt{3}x + 3x^2$ is not in $\mathbb{Z}[x]$. However, $2 + \sqrt{3}x + 3x^2$ is an element of $\mathbb{R}[x]$.
2. Let us compute $(3 + 2x + x^3) + (5x + 3x^2 + 2x^4)$ and $(3 + 2x + x^3) \cdot (5x + 3x^2 + 2x^4)$.
   First write $p = 3 + 2x + x^3 = a_0 + a_1x + a_2x^2 + a_3x^3$ and $q = 5x + 3x^2 + 2x^4 = b_0 + b_1x + b_2x^2 + b_3x^3 + b_4x^4$, where $a_0 = 3, a_1 = 2, a_2 = 0$ and $a_3 = 1$; and $b_0 = 0, b_1 = 5, b_2 = 3, b_3 = 0$ and $b_4 = 2$.
   $$p+q=\sum_{i=0}^{3}a_ix^i+\sum_{i=0}^{4}b_ix^i=\sum_{i=0}^{4}(a_i+b_i)x^i$$
   where $a_i = 0$ for all $i > 3$ and $b_i = 0$ for all $i > 4$.
   $$\begin{aligned}\therefore p + q &= (3 + 0) + (2 + 5)x + (0 + 3)x^2 + (1 + 0)x^3 + (0 + 2)x^4\\ &= 3 + 7x + 3x^2 + x^3 + 2x^4.\end{aligned}$$
   Also, $pq = c_0 + c_1x + c_2x^2 + c_3x^3 + c_4x^4 + c_5x^5 + c_6x^6 + c_7x^7$,
   $$\begin{aligned}\text{where } c_0 &= a_0 \cdot b_0 = 3 \cdot 0 = 0\\ c_1 &= a_0 \cdot b_1 + a_1 \cdot b_0 = 3 \cdot 5 + 2 \cdot 0 = 15\\ c_2 &= a_0 \cdot b_2 + a_1 \cdot b_1 + a_2 \cdot b_0 = 3 \cdot 3 + 2 \cdot 5 + 0 \cdot 0 = 19\\ c_3 &= \Sigma_{i=0}^{3} a_i\, b_{3-i} = 3 \cdot 0 + 2 \cdot 3 + 0 \cdot 5 + 1 \cdot 0 = 6\\ c_4 &= 3 \cdot 2 + 2 \cdot 0 + 0 \cdot 3 + 1 \cdot 5 + 0 \cdot 0 = 11\\ c_5 &= 3 \cdot 0 + 2 \cdot 2 + 0 \cdot 0 + 1 \cdot 3 + 0 \cdot 5 + 0 \cdot 0 = 7\\ c_6 &= 3 \cdot 0 + 2 \cdot 0 + 0 \cdot 2 + 1 \cdot 0 + 0 \cdot 2 + 0 \cdot 5 + 0 \cdot 0 = 0\\ c_7 &= 3 \cdot 0 + 2 \cdot 0 + 0 \cdot 0 + 3 \cdot 2 + 0 \cdot 0 + 0 \cdot 3 + 0 \cdot 5 +\\ &\quad 0 \cdot 0 = 6.\end{aligned}$$
   $$\begin{aligned}\text{Thus, } p \cdot q &= 0 + 15x + 19x^2 + 6x^3 + 11x^4 + 7x^5 + 0x^6 + 6x^7\\ &= 15x + 19x^2 + 6x^3 + 11x^4 + 7x^5 + 6x^7.\end{aligned}$$
3. Let us compute $(2 + x + x^3) + (1 + 2x + x^2 + x^3)$ in $\mathbb{Z}_3[x]$. The required sum is
   $$(2 +_3 1) + (1 +_3 2)x + (0 +_3 1)x^2 + (1 +_3 1)x^3 = x^2 + 2x^3$$
   since $2 +_3 1 = 0$ in $\mathbb{Z}_3$.
4. $1 + 2x$ is a unit in $\mathbb{Z}_4[x]$; for
   $$(1 + 2x)(1 + 2x) = 1 + (2 +_4 2)x + (2 \cdot_4 2)x^2 = 1$$
   since $2 +_4 2 = 0 = 2 \cdot_4 2$. Here, $1 + 2x$ is the multiplicative inverse of itself.

Definition 11.1.4. Let $p = (a_0, a_1, a_2, \ldots)$ be a nonzero polynomial over a ring R. Then, $a_i \neq 0$ for some i. The largest n for which $a_n \neq 0$ is called the *degree* of p and is denoted by deg(p).

Note that we have not defined the degree of the zero polynomial (0, 0, 0, ...). Also, if p is a polynomial of degree n, then p can expressed as

$$p = a_0 + a_1x + \cdots + a_nx^n$$

The a_i's involved in this expression are called the *coefficients* in the polynomial p.

Example 11.1.2

1. The degree of $2 + 3x^4$ in $\mathbb{R}[x]$ is 4, since

$$2 + 3x^4 = 2 + 0x + 0x^2 + 0x^3 + 4x^4$$

2. In $\mathbb{Z}_4[x]$, the degree of $(1 + 2x)^2$ is 0, since

$$\begin{aligned}(1 + 2x)^2 &= 1 + (2 +_4 2)x + (2 \cdot_4 2)x^2 \\ &= 1 + 0x + 0x^2 = 1.\end{aligned}$$

Definition 11.1.5. The zero polynomial and the polynomials of degree 0 are called *constant polynomials*.

For any ring R, the set of constant polynomials over a ring R form a subring of $R[x]$ and is isomorphic to R. In the following, we discuss how the degrees of polynomials vary when we take sums and products.

Definition 11.1.6. Let R be a ring and $R[x]$ be the ring of polynomials over R. The following holds for any nonzero polynomials f and $g \in R[x]$:

1. Either $f + g = 0$ or $\deg(f + g) \leq \max\{\deg(f), \deg(g)\}$.
2. Either $f \cdot g = 0$ or $\deg(f \cdot g) \leq \deg(f) + \deg(g)$.
3. If R is an integral domain, then $f \cdot g \neq 0$ and $\deg(f \cdot g) = \deg(f) + \deg(g)$.

Proof: Let $\deg(f) = m$ and $\deg(g) = n$. Then,

$$\begin{aligned} f &= a_0 + a_1x + a_2x^2 + \cdots + a_mx^m, a_m \neq 0 \\ \text{and} \quad g &= b_0 + b_1x + b_2x^2 + \cdots + b_nx^n, b_n \neq 0. \end{aligned}$$

Let $k = \max\{m, n\}$. Then, we can write

$$\begin{aligned} f &= a_0 + a_1x + a_2x^2 + \cdots + a_kx^k \\ \text{and} \quad g &= b_0 + b_1x + b_2x^2 + \cdots + b_kx^k, \end{aligned}$$

where $a_i = 0$ for $i > m$ and $b_j = 0$ for $j > n$.

1. We have $f + g = (a_0 + b_0) + (a_1 + b_1)x + \cdots + (a_k + b_k)x^k$ and therefore either $f + g = 0$ or

$$\deg(f + g) \leq k = \max\{\deg(f), \deg(g)\}.$$

2. Suppose that $f \cdot g \neq 0$. Let

$$f \cdot g = c_0 + c_1 x + c_2 x^2 + \cdots + c_s x^s,$$

where $c_r = \sum_{i=0}^{r} a_i b_{r-i} = \sum_{i+j=r} a_i b_j$.
If $i + j = r > m + n$, then either $i > m$ or $j > n$ and hence $a_i = 0$ or $b_j = 0$ and, in either case $a_i b_j = 0$. Therefore,

$$c_r = 0 \quad \text{for all } r > m + n$$

which implies that $\deg(f \cdot g) \leq m + n = \deg(f) + \deg(n)$.

3. Suppose that R is an integral domain. Then,

$$C_{m+n} = \sum_{i+j=m+n} a_i b_j = a_m b_n \neq 0 \quad \text{(since } R \text{ is an integral domain)}$$

and, as in (2), $c_r = 0$ for all $r > m + n$. Therefore,

$$\deg(f \cdot g) = m + n = \deg(f) + \deg(g).$$

In fact, the converse of (3) above is also true, in the sense of the following corollary.

Corollary 11.1.1. Let R be a nontrivial commutative ring with unity. Then, the following are equivalent to each other.

1. $\deg(f \cdot g) = \deg(f) + \deg(g)$ for all nonzero f and $g \in R[x]$.
2. R is an integral domain.
3. $R[x]$ is an integral domain.

Proof: We have proved (2) $\Leftrightarrow$ (3) in Theorem 11.1.5 and (2) $\Rightarrow$ (1) by Definition 11.1.6 (3).

(1) $\Rightarrow$ (3) follows from the fact that the degree is defined for nonzero polynomials only and hence, by (1), $f \cdot g \neq 0$ for all nonzero f and g in $R[x]$. ◀

Worked Exercise 11.1.1. In $\mathbb{Z}_6[x]$, let $f = 2 + x + 4x^2 + 3x^3$

$$\text{and} \quad g = 4 + 3x + 5x^2 + 3x^3.$$

Compute $f + g$ and $f \cdot g$ and their degrees.

Answer:
$$\begin{aligned} f + g &= (2 +_6 4) + (1 +_6 3)x + (4 +_6 5)x^2 + (3 +_6 3)x^3 \\ &= 0 + 4x + 3x^2 + 0 \cdot x^3 \\ &= 4x + 3x^2 \end{aligned}$$

and therefore, $\deg(f + g) = 2 < \max\{\deg(f), \deg(g)\}$.

Also,
$$\begin{aligned} f \cdot g &= (2 + x + 4x^2 + 3x^3)(4 + 3x + 5x^2 + 3x^3) \\ &= (2 \cdot_6 4) + (2 \cdot_6 3 +_6 1 \cdot_6 4)x + (2 \cdot_6 5 +_6 1 \cdot_6 3 +_6 4 \cdot_6 4)x^2 \\ &\quad + (2 \cdot_6 3 +_6 1 \cdot_6 5 +_6 4 \cdot_6 3 +_6 3 \cdot_6 4)x^3 \\ &\quad + (1 \cdot_6 3 +_6 4 \cdot_6 5 +_6 3 \cdot_6 3)x^4 + (4 \cdot_6 3 +_6 3 \cdot_6 5)x^5 \\ &\quad + (3 \cdot_6 3)x^6 \\ &= 2 + 4x + (4 +_6 3 +_6 4)x^2 + (5 +_6 0 +_6 0)x^3 \\ &\quad + (3 +_6 2 +_6 3)x^4 + (0 +_6 3)x^5 + 3x^6 \\ &= 2 + 4x + 5x^2 + 5x^3 + 2x^4 + 3x^5 + 3x^6 \end{aligned}$$

and therefore, $\deg(f \cdot g) = 6 = \deg(f) + \deg(g)$.

Worked Exercise 11.1.2. Give examples of two polynomials f and g over a ring R for which $f \cdot g \neq 0$ and

$$\deg(f \cdot g) < \deg(f) + \deg(g).$$

Answer: Let $f = 1 + 2x$ and $g = 1 + 3x^2$ in $\mathbb{Z}_6[x]$.

Then,
$$\begin{aligned} f \cdot g &= (1 + 2x)(1 + 3x^2) \\ &= 1 + 2x + 3x^2 + (2 \cdot_6 3)x^3 \\ &= 1 + 2x + 3x^2 + (\text{since } 2 \cdot_6 3 = 0). \end{aligned}$$

Therefore, $\deg(f \cdot g) = 2 < \deg(f) + \deg(g)$.

Worked Exercise 11.1.3. Let R be a commutative ring with unity. Then, prove that R and $R[x]$ have the same characteristic.

Answer: First note that both R and $R[x]$ have the same unity, namely 1. Now, it follows from Theorem 9.3.2 that

$$\begin{aligned} \text{char } R &= O(1) \text{ in } (R, +) \\ &= O(1) \text{ in } (R[x], +) \\ &= \text{char } R[x]. \end{aligned}$$

EXERCISE 11(A)

1. Evaluate the following:
   (i) $(1, 2, 0, 4, 0, 0, \ldots) + (2, 0, 1, 4, 0, 0, \ldots)$ in Poly($\mathbb{Z}_5$).
   (ii) $(1, 2, 0, 3, 0, 0, \ldots) \cdot (3, 1, 0, 4, 2, 0, 0, \ldots)$ in Poly($\mathbb{Z}$).
   (iii) $(4, 3, 2, 1, 0, 0, \ldots) \cdot (1, 2, 3, 4, 5, 0, 0, \ldots)$ in Poly($\mathbb{Z}_6$).
   (iv) $(1, 2, 3, 4, 0, 0, \ldots) + (4, 3, 2, 1, 0, 0, \ldots)$ in Poly($\mathbb{Z}_5$).
   (v) $(0, 1, 0, 0, \ldots)^n$ in Poly($\mathbb{R}$) for any $n \in \mathbb{Z}^+$.
   (vi) $(1, 1, 0, 0, \ldots)^n$ in Poly($\mathbb{Z}_n$), for each $n \in \mathbb{Z}^+$.
2. State whether each of the following is true and substantiate your answer.
   (i) $\mathbb{Z}_2[x]$ is an integral domain.
   (ii) $\mathbb{Z}_5[x]$ is a field.
   (iii) $\mathbb{Z}_6[x]$ is an integral domain.
   (iv) A ring $\mathbb{R}$ is finite if and only if $R[x]$ is finite.
   (v) $1 + x$ is a unit in $\mathbb{Z}[x]$.
   (vi) $2 + 2x$ is a nilpotent in $\mathbb{Z}_4[x]$.
   (vii) $1 + x$ is a zero divisor in $\mathbb{Z}_4[x]$.
   (viii) $\mathbb{Z}_2[x]$ is a finite integral domain.
3. If S is a subring of a ring R, then prove that $S[x]$ is a subring of $R[x]$.
4. If I is an ideal of a ring R, then prove that $I[x]$ is an ideal of $R[x]$.
5. Determine all the units in the ring $\mathbb{Z}[x]$.
6. Prove that a polynomial f over $\mathbb{R}$ is a unit in $\mathbb{R}[x]$ if and only if f is a nonzero constant polynomial.
7. For any positive integer n, prove that $\mathbb{Z}_n[x]$ is an integral domain if and only if n is a prime number.
8. Determine the number of nonzero polynomials of degree ≤ 5 in $\mathbb{Z}_2[x]$.
9. For any positive integers m and n, derive a formula for the number of polynomials of degree less then m in $\mathbb{Z}_n[x]$.
10. For any ring R, prove that the set
$$I = \{a_0 + a_1x + \cdots + a_nx^n \in R[x] : a_0 = 0\}$$
is an ideal of $R[x]$.
11. For any $i > 0$, let
$$J_i = \{a_0 + a_1x + \cdots + a_nx^n \in R[x] : a_i = 0\}.$$
Then, for any ring R, is J_i an ideal of $R[x]$.

12. Let R be a commutative ring with unity. Prove that $a_0 + a_1x + \cdots + a_nx^n \in R[x]$ is a unit in $R[x]$ if and only if a_0 is a unit in R and $a_1, a_2, \ldots, a_n$ are nilpotents in $\mathbb{R}$.

13. Deduce Exercise 6 above from Exercise 12 above.

14. Let R be a commutative ring with unity and I be a proper ideal of R. Then prove that I is a prime ideal of R if and only if $I[x]$ is a prime ideal of $R[x]$.

15. Can we replace 'prime ideal' in Exercise 14 above by 'maximal ideal'?

16. Let R be a commutative ring with unity. Then prove that $R[x]/\langle x\rangle \cong R$, where $\langle x\rangle$ is the ideal generated by x in $R[x]$.

11.2 THE DIVISION ALGORITHM

It is well known that there is an algorithm through which, by dividing any integer by any nonzero integer, we get the quotient and the remainder, precisely, if a is any integer and b is a nonzero integer, then there exist unique integers q and r such that $a = qb + r$ and $|r| < |b|$. The algorithm through which we get q and r is called the division algorithm in $\mathbb{Z}$. In this section, we extend this algorithm to polynomials over commutative rings with unity. The degree of a polynomial is used in the derivation of the division algorithm in as much the same way as the absolute value is employed among integers.

First, let us have a small change in the notation for polynomials. The elements of $R[x]$ are the polynomials over R in the indeterminable form and these will be denoted by $f(x)$, $g(x)$, etc., in order to mention the indeterminate x also. This will also be helpful later in treating polynomials as functions.

Definition 11.2.1. Let R be a ring and $f(x) = a_0 + a_1x + \cdots + a_nx^n$ be a nonzero polynomial over R of degree n. Then, $a_n \neq 0$ and a_n is called the *leading coefficient* of $f(x)$. If a_n is the unity in R, then $f(x)$ is called a *monic polynomial*.

Example 11.2.1

1. If $f(x) = 2 + 3x + x^2 + 6x^3 \in \mathbb{Z}[x]$, then degree of $f(x)$ is 3 and 6 is the leading coefficient in $f(x)$.
2. If $f(x) = 2 - x^2 \in \mathbb{Z}[x]$, then -1 is the leading coefficient of $f(x)$.
3. $3 + x - 2x^3 + x^4$ is a monic polynomial over $\mathbb{Z}$.

Theorem 11.2.1 (Division Algorithm for Polynomials). Let $f(x)$ and $g(x)$ be polynomials over a commutative ring R with unity such that $g(x) \neq 0$ and the leading coefficient of $g(x)$ is a unit in R. Then, there exists unique polynomials $q(x)$ and $r(x)$ in $R[x]$ such that

$$f(x) = q(x)g(x) + r(x)$$
$$\text{and either} \quad r(x) = 0 \text{ or } \deg(r(x)) < \deg(g(x)).$$

Proof: If $f(x) = 0$ or $f(x) \neq 0$ such that $\deg(f(x)) < \deg(g(x))$, then we can take $q(x) = 0$ and $r(x) = f(x)$. Therefore, we can assume that $f(x) \neq 0$ and $\deg(f(x)) \geq \deg(g(x))$.

We apply induction on the degree of $f(x)$. First, let $\deg(f(x)) = 0$. Then, since $\deg(f(x)) \geq \deg(g(x)) \geq 0$, it follows that $\deg(g(x)) = 0$ and hence both $f(x)$ and $g(x)$ are constant polynomials, so that $f(x)$ and $g(x)$ are elements of R. In particular, the leading coefficient of $g(x)$ is $g(x)$ itself and is invertible in R (by the hypothesis). Now, put

$$q(x) = f(x)g(x)^{-1} \quad \text{and} \quad r(x) = 0.$$

Then, $q(x)$ and $r(x)$ satisfy the required properties.

Next, let $\deg(f(x)) = n > 0$ and assume that the theorem is true for all polynomials $f_1(x)$ of degree less than n. Let $\deg(g(x)) = m$. Then, we have $n \geq m$.

Let $f(x) = a_0 + a_1x + \cdots + a_nx^n, a_n \neq 0$

and $g(x) = b_0 + b_1x + \cdots + b_mx^m, b_m \neq 0$.

By hypothesis, b_m is a unit in R. Put

$$f_1(x) = f(x) - a_n b_m^{-1} x^{n-m} g(x).$$

Then, $f_1(x) \in R[x]$ and $\deg(f_1(x)) \leq n$. Since the coefficient of x^n in $f_1(x)$ is $a_n - (a_n b_m^{-1}) b_m = 0$, it follows that $\deg(f_1(x)) < n$. By the induction hypothesis, there exists polynomials $q_1(x)$ and $r(x) \in R[x]$ such that

$$f_1(x) = q_1(x)g(x) + r(x),$$

where $r(x) = 0$ or $\deg(r(x)) < \deg(g(x))$.

From this, by substituting for $f_1(x)$, we get

$$\begin{aligned} f(x) &= (q_1(x) + a_n b_m^{-1} x^{n-m}) g(x) + r(x) \\ &= q(x)g(x) + r(x), \end{aligned}$$

$$\text{where} \quad q(x) = q_1(x) + a_n b_m^{-1} x^{n-m}.$$

To prove the uniqueness of $q(x)$ and $r(x)$, suppose that

$$f(x) = q(x)g(x) + r(x) = q'(x)g(x) + r'(x),$$

where $r(x)$ and $r'(x)$ satisfy the requirements of the theorem. Then, we get that

$$r(x) - r'(x) = (q'(x) - q(x))g(x).$$

Since the leading coefficient b_m of $g(x)$ is a unit in R, it is not a zero divisor. If $q'(x) - q(x) \neq 0$, then

$$\begin{aligned}\deg[(q'(x) - q(x))g(x)] &= \deg(q'(x) - q(x)) + \deg(g(x)) \\ &\geq \deg(g(x)) > \deg(r(x) - r'(x))\end{aligned}$$

which is a contradiction. The last in equality is based on the fact that the degrees of both $r(x)$ and $r'(x)$ are less than $\deg(g(x))$. Thus, it is necessary that $q'(x) - q(x) = 0$ and hence $r(x) - r'(x) = 0$. Therefore, $q(x) = q'(x)$ and $r(x) = r'(x)$.

Definition 11.2.2. The polynomials $q(x)$ and $r(x)$ in the above theorem and called, respectively, the *quotient* and *remainder* on dividing $f(x)$ by $g(x)$. The proof of the above theorem actually provides an algorithm to find the quotient and remainder and hence the theorem is called the division algorithm. Let us take up an example.

Example 11.2.2. In $\mathbb{Z}[x]$, let $f(x) = 2 + 3x - 4x^2 + x^3 - 3x^4$ and $g(x) = 3 + x - x^2$. The leading coefficient of $g(x)$ is -1 which is a unit in the ring $\mathbb{Z}$. Put

1. $f_1(x) = f(x) - a_n b_m^{-1} x^{n-m} g(x)$ (as in the above proof).
   Then, $f_1(x) = (2 + 3x - 4x^2 + x^3 - 3x^4) - (-3)(-1)^{-1}x^{4-2}(3 + x - x^2)$
   $= 2 + 3x - 4x^2 + x^3 - 3x^4 - 3x^2(3 + x - x^2)$
   $= 2 + 3x - 13x^2 - 2x^3$

2. Put $f_2(x) = f_1(x) - ab_m^{-1}x^{3-2}g(x)$, where a is the leading coefficient of $f_1(x)$. Then, $f_2(x) = (2 + 3x - 13x^2 - 2x^3) - (-2)(-1)^{-1}x(3 + x - x^2)$
   $= (2 + 3x - 13x^2 - 2x^3) - 2x(3 + x - x^2)$
   $= 2 - 3x - 15x^2$

3. Put $f_3(x) = f_2(x) - bb_m^{-1}x^{2-2}g(x)$, where b is the leading coefficient of $f_2(x)$. Then, $f_3(x) = (2 - 3x - 15x^2) - (-15)(-1)^{-1}(3 + x - x^2)$
   $= (2 - 3x - 15x^2) - 15(3 + x - x^2)$
   $= -43 - 18x$

Now, $\deg(f_3(x)) = 1 < \deg(g(x))$ and the process stops, we have, from (1), (2) and (3),

$$\begin{aligned}
f(x) &= f_1(x) + a_n b_m^{-1} x^{n-m} g(x) && \text{(by (1))}\\
&= f_1(x) + (-3)(-1)^{-1} x^{4-2} g(x)\\
&= f_2(x) + a\, b_m^{-1} x^{3-2}\, g(x) + 3x^2 g(x) && \text{(by (2))}\\
&= f_2(x) + (-2)(-1)^{-1} x g(x) + 3x^2 g(x)\\
&= f_3(x) + b\, b_m^{-1} x^{2-2}\, g(x) + 2x\, g(x) + 3x^2 g(x) && \text{(by (3))}\\
&= -43 - 18x + (-15)(-1)^{-1} g(x) + 2x g(x) + 3x^2 g(x)\\
&= (15 + 2x + 3x^2) g(x) + (-43 - 18x)\\
&= q(x) g(x) + r(x),
\end{aligned}$$

where $q(x) = 15 + 2x + 3x^2$ and $r(x) = -43 - 18x$.

Algorithm 11.2.1 (The Process of Division Algorithm). Consider $f(x)$ and $g(x)$ as in Theorem 11.2.1. Let $\deg(f(x)) = n$, $\deg(g(x)) = m \leq n$. Let

$$f_1(x) = f(x) - a_n b_m^{-1} x^{n-m} g(x),$$

where a_n and b_m are leading coefficients of $f(x)$ and $g(x)$, respectively. Let c_1 be the leading coefficient of $f_1(x)$ and $n_1 = \deg(f_1(x))$. Let

$$f_2(x) = f_1(x) - c_1 b_m^{-1} x^{n_1 - m} g(x)$$

and continue the process of obtaining polynomials $f_0(x) = f(x), f_1(x), f_2(x), f_3(x), \ldots$ of degrees $n_0 = n, n_1, n_2, n_3, \ldots$ with leading coefficients $c_0 = a_n, c_1, c_2, c_3, \ldots$, respectively. Then,

$$\begin{aligned}
&f_{r+1}(x) = f_r(x) - c_r b_m^{-1} x^{n_r - m} g(x) \quad \text{for all } r \geq 0\\
&\text{and} \quad \deg(f(x)) > \deg(f_2(x)) > \deg(f_3(x)) > \ldots
\end{aligned}$$

At some stage, we should get $f_s(x)$ such that $\deg(f_s(x)) < m = \deg(g(x))$ and let $f_s(x)$ be first such stage. Then,

$$\begin{aligned}
f(x) &= a_n b_m^{-1} x^{n-m} g(x) + f_1(x)\\
&= c_0 b_m^{-1} x^{n_0 - m} g(x) + c_1 b_m^{-1} x^{n_1 - m} g(x) + f_2(x)\\
&= \ldots
\end{aligned}$$

$$= \left(\sum_{t=0}^{s-1} b_m^{-1} c_t x^{n_t - m} \right) g(x) + f_s(x)$$

$$\therefore f(x) = q(x)g(x) + r(x),$$

where $q(x) = \sum_{t=0}^{s-1} b_m^{-1} c_t x^{n_t - m}$ and $r(x) = f_s(x)$ and these are the quotient and remainder, respectively.

The following are an immediate consequences of the division algorithm (Theorem 11.2.1).

Corollary 11.2.1. Let $f(x)$ and $g(x)$ be polynomials over a commutative ring R with unity. If $g(x)$ is a nonzero monic polynomial, then there exist unique $q(x)$ and $r(x)$ in $R[x]$ such that

$$f(x) = q(x)g(x) + r(x)$$
$$\text{and either} \quad r(x) = 0 \quad \text{or} \quad \deg(r(x)) < \deg(g(x)).$$

Corollary 11.2.2. Let F be a field and $f(x)$ and $g(x) \in F[x]$ with $g(x) \neq 0$. Then, there exists unique $q(x)$ and $r(x)$ in $R[x]$ such that

$$f(x) = q(x)g(x) + r(x)$$
$$\text{and either} \quad r(x) = 0 \quad \text{or} \quad \deg(r(x)) < \deg(g(x)).$$

In the following, we introduce the concept of an evaluation homomorphism which is an important tool in the study of solutions of polynomial equations.

Theorem 11.2.2. Let S be a commutative ring with unity and R be a subring of S containing the unity of S. For any $\alpha \in S$, define

$$\phi_\alpha : R[x] \to S$$

by $\phi_\alpha(a_0 + a_1 x + a_2 x^2 + \cdots + a_n x^n) = a_0 + a_1\alpha + a_2\alpha^2 + \cdots + a_n\alpha^n$ for all $a_0 + a_1 x + \cdots + a_n x^n \in R\,[x]$. Then, ϕ_α is a homomorphism of $R[x]$ into S such that $\phi_\alpha(x) = \alpha$ and $\phi_\alpha(a) = a$ for all $a \in R$.

Proof: First of all, observe that $\alpha \in S$ and $a_0, a_1, \ldots, a_n \in R \subseteq S$ and hence $a_0 + a_1\alpha + \cdots + a_n\alpha^n \in S$. Also, $a_0 + a_1 x + \cdots + a_n x^n = b_0 + b_1 x + \cdots + b_n x^n$ implies that $a_i = b_i$ for all $1 \leq i \leq n$ and hence $a_0 + a_1\alpha + \cdots + a_n\alpha^n = b_0 + b_1\alpha + \cdots + b_n\alpha^n$. Therefore, ϕ_α is a well-defined mapping of $R[x]$ into S.

Now, let $f = a_0 + a_1x + \cdots + a_mx^m$
and $g = b_0 + b_1x + \cdots + b_nx^n$
be arbitrary elements in $R[x]$, where $a_i, b_i \in R$.
Put $r = \max\{m, n\}$, $a_i = 0 = b_j$ for $i > m$ and $j > n$. Then, we have

$$\begin{aligned}\phi_\alpha(f + g) &= \phi_\alpha((a_0 + b_0) + (a_1 + b_1)x + \cdots + (a_r + b_r)x^r) \\ &= (a_0 + b_0) + (a_1 + b_1)\alpha + \cdots + (a_r + b_r)\alpha^n \\ &= (a_0 + a_1\alpha + \cdots + a_r\alpha^r) + (b_0 + b_1\alpha + \cdots + b_r\alpha^r) \\ &= (a_0 + a_1\alpha + \cdots + a_m\alpha^m) + (b_0 + b_1\alpha + \cdots + b_n\alpha^n) \\ &= \phi_\alpha(f) + \phi_\alpha(g).\end{aligned}$$

Also, $f \cdot g = c_0 + c_1x + \cdots + c_{m+n}x^{m+n}$, where $c_r = \sum_{i=0}^{r} a_ib_{r-i}$
Now, $\phi_\alpha(f \cdot g) = c_0 + c_1\alpha + \cdots + c_{m+n}\alpha^{m+n}$
$= (a_0 + a_1\alpha + \cdots + a_m\alpha^m)\,(b_0 + b_1\alpha + \cdots + b_n\alpha^n)$
(by the commutativity in S)
$= \phi_\alpha(f) \cdot \phi_\alpha(g)$.

Thus, ϕ_α is a homomorphism of $R[x]$ into S. By the very definition of ϕ_α, it follows that $\phi_\alpha[x] = \alpha$ and $\phi_\alpha[a] = a$ for all $a \in R$. ◀

Definition 11.2.3. The homomorphism ϕ_α defined above is called the *evaluation homomorphism* at α.

The evaluation homomorphism at α is actually unique with respect to its defining properties, namely $\phi_\alpha(x) = \alpha$ and $\phi_\alpha(a) = a$ for all $a \in R$, for the simple reason that $R[x]$ is generated by R and x; that is, $R[x]$ is the only subring of $R[x]$ containing R and x. This justifies the notation $R[x]$ for the ring of polynomials over R.

Theorem 11.2.3. Let S be a commutative ring with unity 1 and R be a subring of S containing 1. For each $\alpha \in S$, there exists a unique homomorphism ϕ : $R[x] \to S$ such that $\phi(x) = \alpha$ and $\phi(a) = a$ for all $a \in R$.

Proof: Let $\alpha \in S$. We have the existence of the required homomorphism, namely ϕ_α, in Definition 11.2.1. To prove the uniqueness, let $\phi : R[x] \to S$ be a homomorphism such that $\phi(x) = \alpha$ and $\phi(a) = a$ for all $a \in R$. Then, for any $f = a_0 + a_1x + \cdots + a_nx^n$ in $R[x]$, we have

$$\begin{aligned}\phi(f) &= \phi(a_0 + a_1x + \cdots + a_nx^n) \\ &= \phi(a_0) + \phi(a_1)\phi(x) + \cdots + \phi(a_n)\phi(x)^n \\ &= a_0 + a_1\alpha + \cdots + a_n\alpha^n \\ &= \phi_\alpha(f).\end{aligned}$$

Thus, $\phi = \phi_\alpha$. ◀

For any $\alpha \in R$ and the evaluation homomorphism ϕ_α, we often write $f(\alpha)$ for $\phi_\alpha(f)$, where $f = f(x)$ is a polynomial over R. This is to say that $\phi_\alpha(f)$ is just the element of R obtained by substituting α for x in the polynomial $f = f(x)$. The following is an important result to which we were exposed at school level itself and now, we supplement a proof of the most general version of the remainder theorem.

Theorem 11.2.4 (Remainder Theorem). Let $f(x)$ be a polynomial over a commutative ring R with unity and $a \in R$. Then, there exists unique $q(x) \in R[x]$ such that

$$f(x) = q(x)(a - x) + f(a),$$

where $f(a)$ is the element in R obtained by substituting a for x in $f(x)$; that is, $f(a) = \phi_a(f(x))$.

Proof: Let $f(x) = a_0 + a_1x + \cdots + a_nx^n$. Then,

$$f(a) = a_0 + a_1a + \cdots + a_na^n \in R.$$

Consider the polynomial $a - x$ whose leading coefficient is -1, which is a unit in R. Therefore, by Theorem 11.2.1 (the division algorithm), there exist unique $q(x)$ and $r(x)$ in $R[x]$ such that

$$f(x) = q(x)(a - x) + r(x) \tag{*}$$

where $r(x) = 0$ or $\deg(r(x)) < \deg(a - x) = 1$. Since $\deg(r(x))$ is always a nonnegative integer, it follows that $\deg(r(x)) = 0$ and hence $r(x)$ is a constant polynomial over R. Let $r(x) = b \in R$.

$$f(a) = q(a)(a - a) + b = 0 + b = b = r(x).$$

Therefore, again by (*),

$$f(x) = q(x)(a - x) + f(a).$$

Definition 11.2.4. Let $f(x)$ and $g(x)$ be polynomials over a commutative ring R. Then, $g(x)$ is said to be a *divisor* (*or a factor*) of $f(x)$ if $g(x)h(x) = f(x)$ for some $h(x) \in R[x]$; in this case, we also say that $f(x)$ is a *multiple* of $g(x)$.

Corollary 11.2.3. Let $f(x)$ be a polynomial over a commutative ring R and $a \in R$. Then, $a - x$ divides $f(x)$ if and only if $f(a) = 0$.

Definition 11.2.5. For any polynomial $f(x)$ over R and $a \in R$, we say that a is a *root* (or zero) of $f(x)$ if $f(a) = 0$.

In other words, a is a root of $f(x)$ if and only if $f(x) = (a - x)g(x)$ for some $g(x) \in R[x]$. The next theorem is about the number of roots of a polynomial in a integral domain.

Theorem 11.2.5. Let $f(x)$ be a nonzero polynomial of degree n over an integral domain R. Then, $f(x)$ can have at most n distinct roots in R.

Proof: We apply induction on the degree of $f(x)$, If $\deg(f(x)) = 0$, then the theorem is trivial, since $f(x)$, being nonzero, cannot have any root in R. Next, suppose that $\deg(f(x)) = n > 0$ and assume that any nonzero polynomial of degree $m < n$ can have at most m distinct roots in R. If $f(x)$ has a root in R; that is, if $a \in R$ is such that $f(a) = 0$, then, by Theorem 11.2.5,

$$f(x) = (a - x)g(x) \quad \text{for some } g(x) \in R[x].$$

Now, b is a root of $f(x)$ in R implies that

$$0 = f(b) = (a - b)g(b)$$

and, since R is an integral domain, it follows that $b = a$ or $g(b) = 0$. Therefore, the number of roots of $f(x)$ in R other than a cannot exceed the number of roots of $g(x)$ in R. Since R is an integral domain, we have

$$n = \deg(f(x)) = \deg(a - x) + \deg(g(x)) = 1 + \deg(g(x))$$

and hence $\deg(g(x)) = n - 1$. By the induction hypothesis, it follows that $g(x)$ can have at most $n - 1$ district roots in R. Thus, $f(x)$ can have at most n distinct roots in R.

Corollary 11.2.4. Let $f(x)$ and $g(x)$ be polynomials of degree n over an integral domain R and $a_1, a_2, \ldots, a_{n+1}$ be distinct elements of R such that $f(a_i) = g(a_i)$ for $1 \le i \le n$. Then, $f(x) = g(x)$.

Proof: Consider $h(x) = f(x) - g(x) \in R[x]$. If $h(x) \neq 0$, then $\deg(h(x)) \le n$ and $a_1, \ldots, a_{n+1}$ are distinct roots of $h(x)$ in R, which is a contradiction to Theorem 11.2.5. Therefore, $h(x) = 0$ and $f(x) = g(x)$.

Corollary 11.2.5. Let $f(x)$ be a polynomial over an integral domain R. Suppose that A is an infinite subset of R such that $f(a) = 0$ for all $a \in A$. Then, $f(x)$ is the zero polynomial.

Proof: If $f(x)$ is a nonzero polynomial, then $\deg(f(x)) = n$ and $f(x)$ can have at most n district roots in R which is a contradiction to the assumption that each element of the infinite set A is a root of $f(x)$.

Worked Exercise 11.2.1. Find the quotient and remainder when $f(x) = 3 + 4x + 3x^2 + 2x^3 + 2x^4 + x^5$ is divided by $g(x) = 5 + 3x + 4x^2 + x^3$ in $\mathbb{Z}_6[x]$.

Answer: We follow the process of division algorithm as given in Algorithm 11.2.1 (the process of division algorithm). Note that the addition and multiplication in $\mathbb{Z}_6$ are modulo 6.

$$
\begin{array}{r|l|l}
5+3x+4x^2+x^3 & 3+4x+3x^2+2x^3+2x^4+x^5 & x^2 \\
 & -(5x^2+3x^3+4x^4+x^5) & \\
\hline
 & 3+4x+4x^2-5x^3+4x^4 & 4x \\
 & -(2x+0+4x^3+4x^4) & \\
\hline
 & 3+2x+4x^2+x^3 & 1 \\
 & -(5+3x+4x^2+x^3) & \\
\hline
 & 4+5x &
\end{array}
$$

Therefore, the quotient is $1 + 4x + x^2$ and the remainder is $4 + 5x$.

Worked Exercise 11.2.2. Let R be a commutative ring with unity and $\alpha \in R$. Then prove that $R[x]/<\alpha-x> \cong R$, where $<\alpha - x>$ is the principal ideal generated by $\alpha - x$ in $R[x]$.

Answer: Consider the evaluation homomorphism ϕ_α from $R[x]$ into R (by taking $S = R$ in Definition 11.2.1) defined by $\phi_\alpha(a_0 + a_1x + \cdots + a_nx^n) = a_0 + a_1\alpha + \cdots + a_n\alpha^n$. If $a \in R$, then, for the constant polynomial a in $R[x]$,we have $\phi_\alpha(a) = a$. Therefore, ϕ_α is an epimorphism and hence, by the fundamental theorem of homomorphism,

$$R[x]\big/\ker \phi_\alpha \cong R.$$

If $f \in <\alpha - x> = (\alpha - x)R[x]$, then $f = (\alpha - x)g$ for some $g \in R[x]$ and hence

$$\phi_\alpha(f) = \phi_\alpha(\alpha - x)\phi_\alpha(g) = (\alpha - \alpha)\phi_\alpha(g) = 0$$

so that $f \in \ker \phi_\alpha$. Using the division algorithm (whose proof is given in the next section), one can prove that $\phi_\alpha(f) = 0$ implies $f = (\alpha - x)g$ for some $g \in R[x]$. Therefore,

$$\ker \phi_\alpha = <\alpha - x>$$

Thus, $R[x]/<\alpha - x> \cong R$.

EXERCISE 11(B)

1. Find the remainder and quotient when $f(x)$ is divided by $g(x)$ in the rings mentioned in each of the following:
   (i) $f(x) = 2 + 3x + x^2 + x^4 + 2x^5$ and $g(x) = 2 + x^2 - x^3$ in $\mathbb{Z}[x]$.
   (ii) $f(x) = 1 + x + x^2 + x^4 + x^6$ and $g(x) = 1 + x + x^2$ in $\mathbb{Z}_2[x]$.
   (iii) $f(x) = \frac{1}{2} + x + \frac{2}{3}x^2 + \frac{3}{4}x^3$ and $g(x) = \frac{1}{3} + \frac{1}{2}x$ in $\mathbb{Q}[x]$.
   (iv) $f(x) = 2 + 3x + 4x^2 + 5x^3$ and $g(x) = 1 - x$ in $\mathbb{Z}[x]$.
   (v) $f(x) = 1 + 2x + 3x^2 + 4x^3$ and $g(x) = 1 + x^2$ in $\mathbb{Z}_5[x]$.
   (vi) $f(x) = (1 + i) + (2 + 3i)x + (1 - 2i)x^2 + (1 + 3i)x^3$ and $g(x) = i + (2 + i)x - 2x^2$ in $\mathbb{C}[x]$.
2. Evaluate each of the following for the indicated evaluation homomorphism $\phi_a : \mathbb{Z}_5[x] \to \mathbb{Z}_5$.
   (i) $\phi_0(2 + 3x + 4x^2 + x^3)$
   (ii) $\phi_1(1 + x + x^2 + x^3 + x^4 + x^5)$
   (iii) $\phi_2(1 + 2x + 3x^2 + 4x^3 + x^4)$
   (iv) $\phi_3(2 + 3x + 4x^2 + x^5 + x^6)$
   (v) $\phi_4(2 + x + 3x^2 + x^4 + x^7)$
3. Find eight elements in the Kernel of the evaluation homomorphism $\phi_5: \mathbb{Q}[x] \to \mathbb{R}$.
4. For any subfield F of any field E, prove that the Kernel of the evaluation homomorphism $\phi_a: F[x] \to E$ is an infinite set for each $a \in E$.
5. Find all the roots in $\mathbb{Z}_5$ of the polynomial $2 + 4x + 3x^2 + 4x^3 + x^4$ in $\mathbb{Z}_5[x]$.
6. Prove that $1 + 4x$ is a unit in $\mathbb{Z}_8[x]$.
7. Let F be an infinite field and $f(x) \in F[x]$. Prove that $f(a) = 0$ for infinitely many elements a in F if and only if $f(x) = 0$.
8. Let R be an integral domain and $f(x)$ and $g(x) \in R[x]$. Prove that $\{a \in R : f(a) = g(a)\}$ is infinite if and only if $f(x) = g(x)$.
9. Determine all the roots of $x - x^5$ in $\mathbb{Z}_5$.
10. For any prime number p, prove that every element of $\mathbb{Z}_p$ is a root of the polynomial $x - x^p$ in $\mathbb{Z}_p[x]$.

11.3 POLYNOMIALS OVER A FIELD

The ring $F[x]$ of polynomials over a field F has a rich structure theory. In particular, $F[x]$ satisfies most of the ring theoretic properties that are satisfied by the ring $\mathbb{Z}$ of integers. For example, we will be proving that any ideal of $F[x]$ is generated by a single element, as in the case of $\mathbb{Z}$. Also, we have unique factorizations in $F[x]$. More so, we have the division algorithm, as mentioned in Corollary 11.2.2 and Theorem 11.2.4 (Remainder Theorem). Throughout our discussions in this section, F always denotes an arbitrary field, unless otherwise stated.

Let us recall from Theorem 11.2.4 (Remainder Theorem) that, for any $a \in F$ and $0 \neq f(x) \in F[x]$, $f(a)$ is the remainder obtained by dividing $f(x)$ with $a - x$ and that a is a root of $f(x)$ if and only if $a - x$ divides $f(x)$ (or $a - x$ is a factor of $f(x)$).

Definition 11.3.1. Let F be a field, $f(x) \in F[x]$ and $a \in F$. If $(a - x)^n$ is a factor of $f(x)$ for some $n > 1$, then a is called a *multiple root* of $f(x)$, and the least such n is called the *multiplicity* of the root a.

The following is a slight variation of Theorem 11.2.5, where we have proved that any nonzero polynomial of degree n over an integral domain R can have at most n distinct roots.

Theorem 11.3.1. Any nonzero polynomial of degree n over a field F can have at most n roots in F, including the multiplicity of the roots.

Proof: Let F be a field, $0 \neq f(x) \in F[x]$ and $\deg(f(x)) = n$. We shall use induction on n. If $n = 0$, there is nothing to prove. If $n = 1$, then $f(x) = a_0 + a_1 x$ and $a_1 \neq 0$ and hence $-a_0 a_1^{-1}$ is the only root of $f(x)$.
Suppose that $n > 1$ and that the theorem is true for all polynomials of degree less that n. If $f(x)$ has no roots in F, then we are done. Let a be a root of $f(x)$ in F. Then,

$$f(x) = (a - x)^m g(x) \quad \text{for some } g(x) \in F[x] \text{ and } m \in \mathbb{Z}^+.$$

Comparing the degrees both the sides, we get that

$$n = \deg(f(x)) = m + \deg(g(x))$$

and hence $\deg(g(x)) = \deg(f(x)) - m < n - m < n$.
If $f(x)$ has no roots other than a in F, then we are done, since $m \leq n$. On the other hand, if $b \neq a$ is a root of $f(x)$ in F, then

$$0 = f(b) = (a - b)^m g(b)$$

and hence $g(b) = 0$, so that b is a root of $g(x)$. By the induction hypothesis, $g(x)$ has at most $\deg(g(x))$ roots in F, including the multiplicity of the roots. Thus, $f(x)$ has at most $n - m$ ($= \deg(g(x))$) roots in F other than a. Therefore, $f(x)$ has at most $m + (n - m)$ roots in F, including the multiplicity of the roots.

Let us recall that an ideal I of a ring R is called a principal ideal if $I = <a>$ for some $a \in R$; that is, I is generated by a single element of R. It is well known that every ideal of the ring $\mathbb{Z}$ of integers is a principal ideal. This property is satisfied by the rings of polynomials over fields also.

Theorem 11.3.2. Let F be a field and $F[x]$ be the ring of polynomials over F. Then, $F[x]$ is an integral domain in which every ideal is principal.

Proof: By Theorem 11.1.5, $F[x]$ is an integral domain since F is so. Let I be an ideal of $F[x]$. If $I = \{0\}$, then there is nothing to prove, since $I = <0> = \{0\}$. Suppose that I is a nonzero ideal; that is, I contains atleast one nonzero polynomial. Consider the set

$$S = \{\deg(f(x)) : 0 \neq f(x) \in I\}.$$

Since S is a nonempty set of nonnegative integers, S has a least member, say n. Then, $n = \deg(f(x))$ for some $f(x) \in I$ and $n \leq \deg(g(x))$ for all $0 \neq g(x) \in I$. We shall prove that I is generated by $f(x)$. Since $f(x) \in I$, we have $<f(x)> \subseteq I$. On the other hand, suppose that $g(x) \in I$. Then, by the division algorithm, there exist $q(x)$ and $r(x) \in F[x]$ such that

$$g(x) = q(x)f(x) + r(x),$$

where $r(x) = 0$ or $\deg(r(x)) < \deg(f(x)) = n$. Now,

$$r(x) = g(x) - q(x)f(x) \in I$$

since $g(x)$ and $f(x) \in I$ and I is an ideal. By the least property of n, it follows that $r(x) = 0$ and hence

$$g(x) = q(x)f(x) \in <f(x)>.$$

Therefore, $I \subseteq <f(x)>$. Thus, $I = <f(x)>$.

The converse of the above result is also true in the sense of the following theorem.

Theorem 11.3.3. Let R be a ring such that $R[x]$ is an integral domain in which every ideal is principal, Then, R is a field.

Proof: Since $R[x]$ is given to be an integral domain, it follows from Theorem 11.1.5 that R is an integral domain. To prove that R is a field, let $0 \neq a \in R$. Consider the ideal $<a, x>$ generated by a and x in $R[x]$. By hypothesis, this ideal must be principal and hence there exists $f(x) \in R[x]$ such that

$$<a, x> = <f(x)>.$$

First of all, note that $f(x) \neq 0$, since $0 \neq a \in <f(x)>$. Since both a and $x \in <f(x)>$, we get that

$$a = f(x)g(x) \tag{1}$$

$$\text{and} \quad x = f(x)h(x) \tag{2}$$

for some $g(x)$ and $h(x) \in R[x]$. From (1), we have

$$0 = \deg(a) = \deg(f(x)g(x)) = \deg(f(x)) + \deg(g(x))$$

and hence $\deg f(x) = 0$. Let $f(x) = a_0 \in R$.
Then, from (2), we have $x = a_0\, h(x)$ and hence

$$1 = \deg(x) = \deg(a_0) + \deg(h(x)) = \deg(h(x)).$$

Therefore, $h(x) = b_0 + b_1x$ for some b_0 and $0 \neq b_1$ in R and hence, from (2), we have

$$x = a_0(b_0 + b_1 x).$$

By comparing the coefficient of x on both sides, we get that

$$1 = a_0b_1 \in <f(x)> = <a, x>.$$

Therefore, there exist $f_1(x)$ and $f_2(x)$ in R$[x]$ such that

$$1 = af_1(x) + xf_2(x)$$

which implies that $1 = ac_0$ where c_0 is the constant term in $f_1(x)$. Thus, a is a unit in R. Therefore, R is a field.

Worked Exercise 11.3.1. Give an example of an ideal of $\mathbb{Z}[x]$ which is not principal.

Answer: Since $\mathbb{Z}$ is not a field, it follows from Definition 13.3.2 that there must be an ideal of $\mathbb{Z}[x]$ which is not principal. Let

$$I = \{a_0 + a_1x + \cdots + a_nx^n \in \mathbb{Z}[x] : a_0 \text{ is even}\}.$$

Then, one can easily verify that I is an ideal of $\mathbb{Z}[x]$. We prove that I is not a principal ideal. On the contrary, suppose that

$$I = <f(x)>, f(x) \in \mathbb{Z}[x].$$

Then, $2 \in I = <f(x)>$ and hence

$$2 = f(x)g(x), \quad \text{for some } g(x) \in \mathbb{Z}[x].$$

Then, $0 = \deg(2) = \deg(f(x)g(x)) = \deg(f(x)) + \deg(g(x))$, so that $\deg(f(x)) = 0 = \deg(g(x))$. Let $f(x) = b$ and $g(x) = c \in \mathbb{Z}$. Then, $2 = bc$, we can assume that $b > 0$ and $c > 0$ (since $<-f(x)> = <f(x)>$). Then, $b = 1$ or 2. But $b \neq 1$, since $1 \notin I$. Therefore, $b = 2$. This implies that $I = <2>$, which is a contradiction, since $2 + x \in I$ and $2 + x \notin <2>$. Thus, $I \neq <f(x)>$ for any $f(x) \in \mathbb{Z}[x]$. That is, I is not a principal ideal in $\mathbb{Z}[x]$.

Worked Exercise 11.3.2. Let F be a field. For any $a \in F$, let

$$M_a = \{f(x) \in F[x] : f(a) = 0\}.$$

Then, prove that M_a is a maximal ideal of $F[x]$ and that $F[x]/M_a \cong F$. If F is infinite, prove that $\bigcap_{a\in F} M_a = \{0\}$.

Answer: Let $a \in F$. Consider the evaluation homomorphism $\phi_a : F[x] \to F$ defined by $\phi_a(f(x)) = f(a)$. Then, ϕ_a is an epimorphism (for, if $r \in F$, then $r \in F[x]$ and $\phi_a(r) = r$). Also,

$$\begin{aligned}\ker \phi_a &= \{f(x) \in F[x] : \phi_a(f(x)) = 0\} \\ &= \{f(x) \in F[x] : f(a) = 0\} \\ &= M_a.\end{aligned}$$

By the fundamental theorem of homomorphisms, M_a is an ideal of $F[x]$ and $F[x]/M_a \cong F$. Since F is a field, so is $F[x]/M_a$ and hence M_a is a maximal ideal of $F[x]$. Further, suppose that F is infinite. If $f(x) \in \bigcap_{a\in F} M_a$, then $f(a) = 0$ for infinitely many a and therefore, by Corollary 11.2.5, $f(x) = 0$. Thus,

$$\bigcap_{a\in F} M_a = \{0\}.$$

Worked Exercise 11.3.3. Let F be a field and M_a be as defined in Worked Exercise 11.3.2 above for any $a \in F$. Then prove that

$$\bigcap_{a \in F} M_a = \{0\} \Leftrightarrow F \text{ is infinite.}$$

Answer: Suppose that F is finite and $|F| = n$. Consider the group $F^* = F - \{0\}$ under the multiplication in F. Then, F^* is a finite group of order $n - 1$ and hence

$$a^{n-1} = 1 \quad \text{for all } a \in F^*$$

Therefore, $a^n = a$ for all $a \in F$. Put $f(x) = x - x^n$. Then, $0 \neq f(x) \in F[x]$ and $f(a) = 0$ for all $a \in F$. Therefore, $0 \neq f(x) \in \bigcap_{a \in F} M_a$. Thus, $\bigcap_{a \in F} M_a \neq \{0\}$. Converse of this is proved in Worked Exercise 11.3.2.

Worked Exercise 11.3.4. Let R be an integral domain. Prove that R is a field if and only if $<x>$ is a maximal ideal of $R[x]$.

Answer: Suppose that R is a field. For any $a \in R$. Let

$$M_a = \{f(x) \in R[x] : f(a) = 0\}.$$

Then, by Worked Exercise 11.3.2, M_c is a maximal ideal of $R[x]$. In particular, M_0 is a maximal ideal of $R[x]$. For any $f(x) \in R[x]$, we have

$$\begin{aligned} f(x) \in M_0 &\Leftrightarrow f(0) = 0 \\ &\Leftrightarrow x \text{ divides } f(x) \quad \text{(by Corollary 11.2.3)} \\ &\Leftrightarrow f(x) \in <x>. \end{aligned}$$

Therefore, $M_0 = <x>$ and hence $<x>$ is a maximal ideal of $R[x]$. Conversely, suppose that $<x>$ is a maximal ideal of $R[x]$. Since $<x> = M_0 = \ker \phi_0$, where ϕ_0 is the evaluation homomorphism at 0, it follows that

$$R[x]/<x> \cong R.$$

since $<x>$ is a maximal ideal of $R[x]$, $R[x]/<x>$ and hence R is a field.

We close this section with a remark that several other properties of rings of polynomials over fields will be discussed in the next section and in the next chapter.

EXERCISE 11(C)

1. Let F be field and $f(x)$ be a polynomial of degree n $(n > 0)$ over F. Then prove that the quotient ring $f[x]/<f(x)>$ is bijective with F^n.

2. Let R be a commutative ring with unity in which every ideal is principal. Then prove that an ideal in R is maximal if and only if it is prime.

3. Let F be a field, $0 \neq a \in F$ and $f(x) \in F[x]$, prove that $<f(x)> = <af(x)>$.

4. Let F be a subfield of a field E and $a \in E$. Then prove that $\{f \in F[x] : a \text{ is a root of } f(x)\}$ is an ideal of $F[x]$ and describe a generator of this ideal.

5. Let F be a field and I be the set of all polynomials over F for each of which the sum of the coefficients is zero. Then prove that I is an ideal of $F[x]$ and determine a generator of I.

6. Prove that there are infinitely many polynomials $f(x)$ in $\mathbb{Z}_3[x]$ for each of which every element of $\mathbb{Z}_3$ is a root.

7. Prove that the rings $\mathbb{Q}[\sqrt{2}]$ and $\mathbb{Q}[x]/\sqrt{2-x^2}$ are isomorphic.

8. Let p be a prime number. Then prove that $(x - 1)(x - 2)(x - 3) \ldots (x - (p - 1)) = x^{p-1} - 1$ in $\mathbb{Z}_p[x]$.

9. Let $f(x) \in \mathbb{R}[x]$, $a \in \mathbb{R}$ and $f'(x)$ be the derivative of f with respect to x. Then prove that $f(a) = 0 = f'(a)$ if and only if $(a - x)^2$ divides $f(x)$.

10. In $\mathbb{Z}[x]$, prove that $<x>$ is a prime ideal but not a maximal ideal.

11. Let P be a prime ideal of a commutative ring R with unity. Then prove that $P[x]$ is a prime ideal of $R[x]$. If M is a maximal ideal of R, is $M[x]$ a maximal ideal of $R[x]$?

12. Prove that $\mathbb{R}[x]/<1 + x^2>$ is isomorphic to the field of complex numbers.

13. Prove that $\mathbb{Z}[x]/<1 + x^2>$ is isomorphic with the ring of Gaussian integers $\mathbb{Z}[i]$.

14. Let R be an integral domain. Then prove that the set

$$I = \{f(x) \in R[x] : f(x) = 0 \text{ or } \deg(f(x)) > 0\}$$

is an ideal of the ring $R[x]$. Also, for any $n > 0$, prove that

$$I^n = \{f(x) \in R[x] : f(x) = 0 \text{ or } \deg(f(x)) \geq n\}.$$

15. In Exercise 14 above, determine $\bigcap_{n=1}^{\infty} I^n$.

16. Let $f(x) \in \mathbb{Z}[x]$ be a monic polynomial and a be a rational number such that $f(a) = 0$. Then prove that a must be integer.

11.4 IRREDUCIBLE POLYNOMIALS

Several interesting questions on factorization of polynomials are based on the idea of irreducibility. In this section, we discuss irreducible polynomials over mainly integral domains.

Definition 11.4.1. Let R be an integral domain. A polynomial $f(x)$ of positive degree over R is said to be *irreducible* over R if $f(x)$ cannot be expressed as a product of two polynomials of positive degree over R. That is, $f(x) \in R[x]$ is said to be *irreducible* over R if $\deg(f(x)) > 0$ and, *for* any $g(x)$ and $h(x) \in R[x]$.

$$f(x) = g(x)h(x) \Rightarrow \deg(g(x)) = 0 \quad \text{or} \quad \deg(h(x)) = 0.$$

If $f(x)$ is not irreducible and $\deg(f(x)) > 0$, we say that $f(x)$ is *reducible*.

Note that the above definition applies only to polynomials of positive degree and as such the constant polynomials are neither reducible nor irreducible. Also, the irreducibility of a polynomial $f(x) \in R[x]$ depends much on the integral domain R; that is, a given polynomial may be irreducible when viewed as a polynomial over one domain, yet reducible when viewed as a polynomial over another domain. For, consider the following example.

Example 11.4.1

1. The polynomial $1 + x^2$ is irreducible over $\mathbb{R}$ the field $\mathbb{R}$ of real numbers; but it is reducible over the field $\mathbb{C}$ of complex numbers, since $1 + x^2 = (1 + ix)(1 - ix)$ and $1 + ix$ and $1 - ix \in \mathbb{C}[x]$.
2. $1 + x^2$ is reducible in $\mathbb{Z}_2[x]$, since $1 + x^2 = (1 + x)(1 + x)$ in $\mathbb{Z}_2[x]$; but $1 + x^2$ is irreducible in $\mathbb{Z}_3[x]$.

Thus, to ask merely whether a polynomial is irreducible, without specifying the coefficient ring involved, is incomplete and meaningless. More often, it is a formidable task to decide when a given polynomial is irreducible over a specific ring. The following provide certain simple tips in finding a given polynomial to be irreducible over a given field or an Integral Domain.

Theorem 11.4.1

1. Let R be an integral domain and $f(x) \in R[x]$ with $\deg(f(x)) = 1$. Then, $f(x)$ is irreducible over R.
2. Let F be a field and $f(x) \in F[x]$ with $\deg(f(x)) = 2$ or 3. Then, $f(x)$ is irreducible if and only if $f(x)$ has no root in F.

Proof:

1. Recall that the degree of any nonzero polynomial is a nonnegative integer. If $f(x) = g(x)h(x)$ and $g(x);h(x) \in R[x]$, then

$$1 = \deg(f(x)) = \deg(g(x)) + \deg(h(x))$$

   and hence $\deg(g(x)) = 0$ or $\deg(h(x)) = 0$. Therefore, $f(x)$ cannot be expressed as a product of two polynomials of positive degree. Thus, $f(x)$ is irreducible over R.

2. We shall prove that $f(x)$ is reducible over F if and only if $f(x)$ has a root in F. First note that any polynomial $a_0 + a_1x$, $a_1 \neq 0$ of degree over F one has a root, namely, $-a_1^{-1}a_0$ in F. Suppose that $f(x)$ is reducible over F. Then,

$$f(x) = g(x)h(x) \text{ for some } g(x) \quad \text{and} \quad h(x) \in F[x]$$

   with $\deg(g(x)) > 0$ and $\deg(h(x)) > 0$. Then, either $\deg(g(x)) = 1$ or $\deg(h(x)) = 1$ (otherwise, if $\deg(g(x)) \geq 2$ and $\deg(h(x)) \geq 2$, then $\deg(f(x)) = \deg(g(x)) + \deg(h(x)) \geq 4$, which is a contradiction to the hypothesis that $\deg(f(x)) = 2$ or 3). If $\deg(g(x)) = 1$, then, by (1) above, $g(x)$ has a root in F and hence $f(x)$ has a root in F. Similarly, if $\deg(h(x)) = 1$, then $h(x)$ and hence $f(x)$ has a root in F.

Conversely, suppose that $f(x)$ has a root in F. Let a be a root of $f(x)$ in F. Then, by Corollary 11.2.3, $a - x$ divides $f(x)$. Therefore, there exists $g(x) \in F[x]$ such that $f(x) = (a - x)g(x)$. Since $\deg f(x) = 2$ or 3, it follows that $\deg(g(x)) > 0$ and hence $f(x)$ is reducible over F.

Theorem 11.4.2. Let $f(x)$ be a nonzero polynomial over a field F. Then, the following are equivalent

1. $f(x)$ is irreducible over F.
2. $<f(x)>$ is a maximal ideal of $F[x]$.
3. $<f(x)>$ is a prime ideal of $F[x]$.

Proof: (1) $\Rightarrow$ (2): Suppose that $f(x)$ is irreducible over F. Then, $f(x)$ is not a constant polynomial and hence $f(x)$ is not a unit in $F[x]$, so that $<f(x)>$ is a proper ideal of $F[x]$. Let I be any ideal of $F[x]$ containing $<f(x)>$. By Theorem 11.3.2, $I = <g(x)>$ for some $g(x) \in F[x]$. Since $<f(x)> \subseteq I = <g(x)>$, we get that $f(x) = g(x)h(x)$ for some $h(x) \in F[x]$. Since $f(x)$ is irreducible, it follows that either $g(x)$ or $h(x)$ is a constant, If $g(x)$ is a constant, then $g(x)$ is a unit (note that $f(x) \neq 0$ and $h(x) \neq 0$, since $\deg(f(x)) > 0$ and hence $f(x) \neq 0$) so that $I = <g(x)> = F[x]$. If $h(x)$ is a constant, then $h(x)$ is

a unit in $F[x]$ and $g(x) = f(x)h(x)^{-1} \in \langle f(x) \rangle$ so that $I \subseteq \langle f(x) \rangle$ and hence $I = \langle f(x) \rangle$. Thus, $\langle f(x) \rangle$ is a maximal ideal of $F[x]$.
(2) ⇒ (3): This is trivial, since any maximal ideal of a commutative ring with unity is a prime ideal.
(3) ⇒ (1): Suppose that $\langle f(x) \rangle$ is a prime ideal of $F[x]$. Then, $\langle f(x) \rangle$ is a proper ideal and hence $f(x)$ is not a unit. Therefore, $\deg(f(x)) > 0$. For any $g(x)$ and $h(x) \in F[x]$,

$$\begin{aligned} f(x) = g(x)h(x) &\Rightarrow g(x)h(x) \in \langle f(x) \rangle \\ &\Rightarrow g(x) \in \langle f(x) \rangle \quad \text{or} \quad h(x) \in \langle f(x) \rangle \\ &\Rightarrow g(x) = f(x)g_1(x) \quad \text{or} \quad h(x) = f(x)h_1(x) \quad \text{for some} \\ & g_1(x) \text{ and } h_1(x) \in F[x] \\ &\Rightarrow f(x) = f(x)g_1(x)h(x) \quad \text{or} \quad f(x) = g(x)f(x)h_1(x) \\ &\Rightarrow g_1(x)h(x) = 1 \text{ or } = g(x)h_1(x) = 1 \\ &\Rightarrow h(x) \quad \text{or} \quad g(x) \text{ is a unit} \\ &\Rightarrow \deg(h(x)) = 0 \quad \text{or} \quad \deg(g(x)) = 0. \end{aligned}$$

Thus, $f(x)$ is irreducible over $F[x]$.

When we deal with polynomials over the field of complex numbers, the crucial tool is the fundamental theorem of algebra. There are several proofs of this theorem but none of these come under the topics covered in this book and hence proof is omitted here and the reader can simply assume the validity of the following theorem.

Theorem 11.4.3 (Fundamental Theorem of Algebra). Any nonconstant polynomial over the field $\mathbb{C}$ of complex numbers has atleast one root in $\mathbb{C}$.

Corollary 11.4.1. Let $f(x) \in \mathbb{C}[x]$ be a polynomial of degree $n > 0$. Then, $f(x)$ can be expressed as a product of n (not necessarily distinct) polynomials, each of degree one.

Proof: This follows from Theorem 11.4.3 (fundamental theorem of algebra) and from the fact that a is a root of $f(x)$ if and only if $a - x$ is a factor of $f(x)$ and also by using induction on the $\deg(f(x))$.

Corollary 11.4.2. The only irreducible polynomials over $\mathbb{C}$ are those of degree one. Since the field $\mathbb{R}$ of real numbers can be considered as a subfield of the field $\mathbb{C}$ of complex numbers, $\mathbb{R}[x]$ can be considered as a subring of $\mathbb{C}[x]$. Therefore, any nonzero polynomial over $\mathbb{R}$ can be considered as a polynomial over $\mathbb{C}$ and hence has a root in $\mathbb{C}$. This observation leads to the following corollary.

Corollary 11.4.3. Let $f(x)$ be a nonconstant monic polynomial over the real number system $\mathbb{R}$. Then, $f(x)$ can be expressed as a product of polynomials over $\mathbb{R}$, each of degree 2 or 1.

Proof: Let $\deg(f(x)) = n$. Since $f(x) \in \mathbb{R}[x] \subseteq \mathbb{C}[x]$, it follows from Corollary 11.4.1 that

$$f(x) = g_1(x)g_2(x) \dots g_n(x),$$

where $g_j(x)$ is a polynomial of degree one over $\mathbb{C}$, for each $1 \le j \le n$. Since $f(x)$ is monic, we can assume that each $g_j(x)$ is monic (the leading coefficient is a unit) and hence

$$g_j(x) = s - x, s \in \mathbb{C}.$$

If $s \in \mathbb{R}$, then $g_j(x) \in \mathbb{R}[x]$. If $s \notin \mathbb{R}$, then $s = a + bi$, where a and $b \in \mathbb{R}$ and $b \neq 0$. Note that s is a root of $f(x)$ (since $g_j(x) = s - x$ is a divisor of $f(x)$).

Also, since $f(x) \in \mathbb{R}[x]$, all the coefficients of the $f(x)$ are real numbers. Now, consider the complex conjugate $\bar{s} = a - bi$ of s.

If $f(x) = a_0 + a_1x + \cdots + a_nx^n$, $a_i \in \mathbb{R}$, then

$$\begin{aligned} f(\bar{s}) &= a_0 + a_1\bar{s} + \cdots + a_n\bar{s}^n \\ &= \overline{a_0} + \overline{a_1 s} + \cdots + \overline{a_n s^n} \\ &= \overline{f(s)} = \bar{0} = 0. \end{aligned}$$

and hence $\bar{s}$ is a root of $f(x)$, so that $\bar{s} - x$ is also a factor of $f(x)$, so that $\bar{s} - x = g_k(x)$ for some $1 \le k \le n$. Now,

$$\begin{aligned} g_j(x)g_k(x) &= (s - x)(\bar{s} - x) \\ &= (a + bi - x)(a - bi - x) \\ &= (a^2 + b^2) - (a + bi + a - bi)x + x^2 \\ &= a^2 + b^2 - 2ax + x^2 \in \mathbb{R}[x]. \end{aligned}$$

Therefore, $g_j(x)g_k(x)$ is a factor of $f(x)$ and is a polynomial of degree 2 over $\mathbb{R}$. Thus, $f(x)$ is a product of polynomials over $\mathbb{R}$, each of degree 1 or 2.

In the next chapter, we prove some more important properties of irreducible polynomials and, in particular, we prove the Eisenstein criterion to find the irreducibility of certain polynomials. ◀

Worked Exercise 11.4.1. Prove that the polynomial $f(x) = 1 + x + x^3$ is irreducible over $\mathbb{Z}_2$.

Answer: If there are any factors $f(x)$, then atleast one factor must be of degree one, say $a + bx$ (with a and $b \in \mathbb{Z}_2$ and $b \neq 0$). In this case, $-b^{-1}a$ is a root of $a + bx$ and hence of $f(x)$. Therefore, if $f(x)$ is reducible, then $f(x)$ must have a root in $\mathbb{Z}_2$; but it can be easily verified that $f(0) = 1 = f(1)$ (sense $1 + 1 = 0$ in $\mathbb{Z}_2$) . Therefore, $f(x)$ is irreducible over $\mathbb{Z}_2$.

Worked Exercise 11.4.2. Let $f(x) \in \mathbb{R}[x]$ and s be a complex number. Then, prove that s is a root of $f(x)$, if and only if $\bar{s}$ is a root of $f(x)$, where $\bar{s}$ is the complex conjugate of s.

Answer: Let $s = a + bi$, where a and $b \in \mathbb{R}$. Then, $\bar{s} = a - bi$. Let

$$f(x) = a_0 + a_1x + \cdots + a_nx^n,$$

where $a_0, a_1, \ldots, a_n \in \mathbb{R}$. Note that r is a real number if and only if $\bar{r} = r$. Now, suppose that s is a root of $f(x)$. Then, $f(s) = 0$ and

$$f(\bar{s}) = a_0 + a_1\bar{s} + \cdots + a_n(\bar{s})^n$$

$$= \frac{a_0 + \overline{as} + \cdots + \bar{a}_n\bar{s}^n}{a_0 + a_1s + \cdots + a_ns^n}$$

$$= \overline{f(s)} = \bar{0} = 0.$$

Thus, $\bar{s}$ is a root of $f(x)$. The converse follows from the fact that $\bar{\bar{s}} = s$.

EXERCISE 11(D)

1. Which of the following are true?
   (i) The degree of any irreducible polynomial is positive.
   (ii) The degree of any reducible polynomial is greater that one.
   (iii) $1 + x + x^2 + x^3$ is irreducible over $\mathbb{Z}_2$.
   (iv) $1 + x + x^2 + x^3$ is reducible over $\mathbb{Z}_3$.
   (v) $3 - x^2$ is irreducible over $\mathbb{Q}$.
   (vi) The degree of any irreducible polynomial is less that 4.
   (vii) The number of irreducible polynomial in $\mathbb{Z}_2[x]$ is finite.
   (viii) $<x>$ is a maximal ideal in $\mathbb{R}[x]$.
   (ix) $<x>$ is a maximal ideal in $\mathbb{Z}[x]$.
   (x) $<x>$ is a prime ideal in $\mathbb{Z}[x]$.

2. Let $f(x)$ be a nonconstant polynomial over a field F. Then, prove that $f(x)$ is irreducible if and only if $F[x]/<f(x)>$ is a field.

3. Prove that $\mathbb{Z}_2[x]/1 + x + x^2$ is a field and determine the number of elements in it.
4. Give an example of a field with exactly nine elements.
5. Let p be a prime number and $f(x)$ be an irreducible polynomial of degree n over $\mathbb{Z}_p$, then prove that $\mathbb{Z}_p[x]/<f(x)>$ has exactly p^n elements.
6. Let F be a field, $f(x) \in F[x]$ and $0 \neq a \in F$. Then prove that $f(x)$ is irreducible over F if and only if so is $f(x)$.
7. Is $1 + x + x^2 + x^3$ reducible over $\mathbb{Z}_5[x]$?
8. Prove that Theorem 11.4.1 (2) fails if $\deg(f(x)) \geq 4$.
9. Prove that $1 + x + x^3 + x^4$ is reducible over any field F.
10. Let $R[x]$ be the ring of polynomials over a commutative ring R with unity. Then, the ring of polynomials over $R[x]$ will be denoted by $R[x, y]$; that is, $R[x, y] = R[x][y]$. By induction, we define

$$R[x_1, x_2, \ldots, x_n] = R[x_1, \ldots, x_{n-1}][x_n]$$

for any $n > 1$. This is called the ring of polynomials in n indeterminates over R. Prove that $R[x_1, x_2, \ldots, x_n]$ is an integral domain if and only if so is R.
11. Prove that there is an ideal in $R[x, y]$ which is not principal.
12. For any field F, prove that

$$F[x,y]/<x+y> \cong F[x] \cong F[y].$$

13. For any field F, prove that the ideal $<x, y>$ generated by $\{x, y\}$ in $F[x, y]$ is a maximal ideal of $F[x, y]$.
14. Prove that $\mathbb{Z}_4[x]$ has infinitely many units and infinitely many nilpotent elements.

12

Factorization in Integral Domains

This chapter is concerned with the problem of factoring elements of an integral domain. The motivation for this lies in the ring $\mathbb{Z}$ of integers, where the Fundamental Theorem of Arithmetic states that every integer $n > 1$ can be written, in an essentially unique way, as a product of prime numbers; for example,

$$6{,}300 = 2 \times 2 \times 3 \times 3 \times 5 \times 5 \times 7$$

and 2, 3, 5 and 7 are prime numbers. In this chapter, we extend the factorization theory of the ring $\mathbb{Z}$ and, in particular, the above-mentioned Fundamental Theorem of Arithmetic, to a more general setting. Naturally, any reasonable abstraction of these number theoretic ideals depends on a suitable interpretation of prime elements (the building blocks for the study of divisibility problems in $\mathbb{Z}$). All the topics discussed in this chapter are more concerned with integral domains. We proceed from the most general results about divisibility, prime elements and factorization to stronger results concerning certain specific classes of integral domains.

First, let us recall that an integral domain is a nontrivial commutative ring with unity and without zero divisors (or equivalently, product of two nonzero elements is again nonzero).

12.1 DIVISIBILITY IN INTEGRAL DOMAINS

In this section, we extend the concepts of divisibility, greatest common divisor (g.c.d.), least common multiple (l.c.m.) and primes in the ring $\mathbb{Z}$ of integers to arbitrary integral domains. Let us begin with the following definition.

Definition 12.1.1. Let R be an integral domain and a and $b \in R$. If $a = bu$ for some unit u in R, then we say that a is an associate of b and denote this by $a \sim b$.

Since $a = a1$, it follows that $a \sim a$ for each $a \in R$. Also, if $a \sim b$, then $a = bu$ for some unit u in R and hence $b = au^{-1}$ so that $b \sim a$. Further, if $a = bu$ and $b = cv$ for some units u and v in R, then uv is a unit in R and $a = c(uv)$ and hence $a \sim c$. All these arguments say that $\sim$ is an equivalence relation. If $\tilde{a}$ stands for the set of all associates of a, then the following can be easily proved.

1. $\tilde{0} = \{0\}$
2. $\tilde{a} = \{au : u \text{ is a unit in } R\}$, for any $a \in R$.
3. $\tilde{1}$ = The set of all units in R.
4. $\tilde{a} = \tilde{b}$ if and only if $a \sim b$.
5. $\tilde{a}$'s form a partition of R.

Example 12.1.1

1. In the ring $\mathbb{Z}$ of integers, 1 and -1 are the only units and hence $a \sim b$ if and only if $|a| = |b|$, for any a and $b \in \mathbb{Z}$, where $|a|$ is the absolute value of a.
2. In a field F, each nonzero element is a unit and hence $a \sim b$ (since $a = b(b^{-1}a)$) for any nonzero elements a and b in F. Therefore, $\tilde{a} = R - \{0\}$ for any $0 \neq a \in R$.
3. Consider the ring $\mathbb{Z}[i]$ of Gaussian integers in which $1, -1, i$ and $-i$ are the only units. For any $x = a + bi \in \mathbb{Z}[i]$,
$$\begin{aligned}\tilde{x} &= \{x, -x, ix, -ix\}\\ &= \{a + bi, -a - bi, -b + ai, b - ai\}.\end{aligned}$$
4. Let $R[x]$ be the ring of polynomials over an integral domain R. Then, the units of $R[x]$ are precisely the units in R. For any $f(x) \in R[x]$,
$$\widetilde{f(x)} = \{uf(x) : u \text{ is a unit in } R\}.$$

Definition 12.1.2. Let a and b be any elements of an integral domain R. If there exists $x \in R$ such that $ax = b$, then we say that a *divides* b (or a is a *divisor* of b or b is a *multiple* of a) and denote this by $a|b$.

Virtually all statements about divisibility can be phrased in terms of principal ideals. Let us recall that, for any element a in an integral domain R, the principal ideal generated by a in R is given by

$$<a> = aR = \{ar : r \in R\}.$$

Theorem 12.1.1. The following holds for any elements a, b and c in an integral domain R.

1. $a|b \Leftrightarrow b \in <a> \Leftrightarrow <b> \subseteq <a>$
2. $a|b$ and $b|a \Leftrightarrow a \sim b \Leftrightarrow <a> = <b>$
3. $a|0$
4. $a|1 \Leftrightarrow a$ is a unit in $R \Leftrightarrow <a> = R$

Proof:

1. $a|b \Leftrightarrow ax = b$ for some $x \in R$
   $\Leftrightarrow b \in aR = <a>$
   $\Leftrightarrow <b> \subseteq <a>$.

2. By (1), $a|b$ and $b|a \Leftrightarrow <a> = <b>$.
   Suppose that $a \sim b$. Then, $a = bu$ and $au^{-1} = b$ for some unit u in R and hence $a|b$ and $b|a$. Conversely, suppose that $a|b$ and $b|a$. Then, $ax = b$ and $by = a$ for some x and $y \in R$ and hence

   $$a = by = (ax)y = a(xy).$$

   First note that $a = 0$ if and only if $b = 0$. Now, if $a \neq 0$, then $xy = 1$ (since $a = a(xy)$ and R is an integral domain) and hence x and y are units. Since $ux = b$ and x is a unit, we get that $a \sim b$.
3. $a0 = 0$ and hence $a|0$.
4. $a|1 \Leftrightarrow ax = 1$ for some $x \in R$.
   $\Leftrightarrow a$ is a unit in R
   $\Leftrightarrow <a> = R$.

When $a|b$, we often use the phrases 'a is a factor of b' or 'b is divisible by a' or 'b is a multiple of a'. When a does not divide b, we write $a \nmid b$. The next result is a routine verification and hence its proof is left as an exercise to the reader.

Theorem 12.1.2. The following holds for any elements a, b and c of an integral domain R.

1. $a|a$ and $1|a$
2. $a|b$ and $b|c \Rightarrow a|c$
3. $a|b \Rightarrow ac|bc$; the converse holds if $c \neq 0$.
4. $a|c$ and $a|b \Rightarrow a|cx + by$ for all x and $y \in R$.

In the following, we introduce the notion of the g.c.d. for a given finite set of nonzero elements in an integral domain.

Definition 12.1.3. Let $a_1, a_2, \ldots, a_n$ be nonzero elements in an integral domain R. An element $d \in R$ is called a g.c.d. of $a_1, a_2, \ldots, a_n$ if the following are satisfied.

(i) $d|a_i$ for all $1 \leq i \leq n$.

(ii) If $c \in R$ and $c|a_i$ for all $1 \leq i \leq n$, then $c|d$.

The use of the superlative adjective 'greatest' in the above definition does not imply that d has greatest magnitude than any other common divisor c of $a_1, a_2, \ldots, a_n$; but only that d is a multiple of any such c. A natural question that arises is whether the elements $a_1, a_2, \ldots, a_n$ can possess two different g.c.d.s. The answer is affirmative; for, in the ring $\mathbb{Z}$ of integers, both 2 and -2 are g.c.d.s of 6 and 10, as per the above definition. However, 2 and -2 are associates to each other. The same is true in a general integral domain. If d and d' are both g.c.d.s of $a_1, a_2, \ldots, a_n$, then, by (ii) above, $d|d'$ and $d'|d$ and hence d and d' are associates to each other. Thus, the g.c.d. of $a_1, a_2, \ldots, a_n$ is unique up to associates, whenever it exists and is usually denoted by $(a_1, a_2, \ldots, a_n)$. The following theorem deals with the existence of g.c.d.

Theorem 12.1.3. Let $a_1, a_2, \ldots, a_n$ be any nonzero elements in an integral domain R. Then, $a_1, a_2, \ldots, a_n$ have g.c.d. d expressible in the form

$$d = a_1r_1 + a_2r_2 + \cdots + a_nr_n \ (r_i \in R)$$

if and only if the ideal $<a_1, a_2, \ldots, a_n>$ generated by the set $\{a_1, a_2, \ldots, a_n\}$ in R is the principal ideal.

Proof: Suppose that $d = a_1r_1 + a_2r_2 + \cdots + a_nr_n$ $(r_i \in R)$ is a g.c.d. of $a_1, a_2, \ldots, a_n$ in R. Then, $d|a_i$ and hence $a_i \in <d>$ for all $1 \leq i \leq n$. Therefore,

$$<a_1, a_2, \ldots, a_n> \subseteq <d>,$$

where $<a_1, a_2, \ldots, a_n>$ is the ideal generated by $a_1, a_2, \ldots, a_n$ in R. Also, since

$$d = a_1r_1 + a_2r_2 + \cdots + a_nr_n \in <a_1, a_2, \ldots, a_n>,$$

we get that $<d> \subseteq <a_1, a_2, \ldots, a_n>$. Thus,

$$<a_1, a_2, \ldots, a_n> = <d>.$$

Conversely, suppose that $<a_1, a_2, \ldots, a_n> = <d>$ for some $d \in R$. Then, $a_i \in <d>$ and hence $d|a_i$ for all $1 \leq i \leq n$. If c is a common divisor of $a_1, a_2, \ldots, a_n$, then $c|a_i$ and hence $a_i \in <c>$ for all $1 \leq i \leq n$, so that

$$<d> = <a_1, a_2, \ldots, a_n> \subseteq <c>.$$

Therefore, $c|d$. Thus, d is a g.c.d. of $a_1, a_2, \ldots, a_n$. Also, since

$$d \in <d> = <a_1, a_2, \ldots, a_r> = a_1R + a_2R + \cdots + a_nR,$$

it follows that $d = a_1r_1 + a_2r_2 + \cdots + a_nr_n$ for some $r_1, r_2, \ldots, r_n \in R$.

It is well known that the ring $\mathbb{Z}$ of integers is an integral domain in which every ideal is a principal ideal. This together with the above theorem implies the following corollary.

Corollary 12.1.1. Any nonzero integers $a_1, a_2, \ldots, a_n$ have g.c.d. and

$$\text{g.c.d. } \{a_1, a_2, \ldots, a_n\} = a_1r_1 + a_2r_2 + \cdots + a_nr_n$$

for some integers $r_1, r_2, \ldots, r_n$.

Dual to the notation of g.c.d., we have the concept of l.c.m. which is defined in the following definition.

Definition 12.1.4. Let a_1 $a_2, \ldots, a_n$ be any nonzero elements in an integral domain R. An element $d \in R$ is called l.c.m of a_1 $a_2, \ldots, a_n$ if the following are satisfied.

(i) $a_i|d$ for all $1 \leq i \leq n$.
(ii) If $c \in R$ and $a_i|c$ for all $1 \leq i \leq n$, then $d|c$.

In other words, a common multiple of a_1 $a_2, \ldots, a_n$ is called l.c.m. if it divides any other common multiple Note that a l.c.m., if it exists, is unique apart from the distinction between associates, for, if d and d' are l.c.m.'s of a_1 $a_2, \ldots, a_n$ in R, then by (ii) above, $d| d'$ and $d'|d$ and hence d and d' are associates to each other.

Theorem 12.1.4. For any nonzero elements a_1 $a_2, \ldots, a_n$ in an integral domain R, $a_1, a_2, \ldots, a_n$ have l.c.m. if and only if the ideal $\bigcap_{i=1}^{n} < a_i >$ is principal.

Proof: This follows from the definition of l.c.m. and from the fact that, for any a and $b \in R$, $a|b$ if and only if $\langle b \rangle \subseteq \langle a \rangle$. Note that $\bigcap_{i=1}^{n} \langle a_i \rangle$ is the largest ideal contained in each of $\langle a_i \rangle$.

Next, we introduce two new classes of elements, namely prime and irreducible elements in an arbitrary integral domain. When we consider the ring $\mathbb{Z}$ of integers, these two concepts become equivalent and yield the usual notion of a prime number.

Definition 12.1.5. Let p be a nonzero and nonunit element in an integral domain R. Then,

1. p is called a *prime element* if, for any a and $b \in R$,

$$p|ab \Rightarrow p|a \quad \text{or} \quad p|b.$$

2. p is called an *irreducible element* if, for any a and $b \in R$,

$$p = ab \Rightarrow a \text{ is a unit} \quad \text{or} \quad b \text{ is a unit.}$$

In other words, a nonzero and nonunit element p is called irreducible if it cannot be factored in R in a nontrivial way; that is, the only factors of p are its associates and units in R. Note that any unit u is a factor of every element, since $u(u^{-1}a) = a$. In fields, where each nonzero element is a unit, the concepts of prime elements and irreducible elements are of no significance.

Theorem 12.1.5. Let R be an integral domain. Then, every prime element in R is irreducible. The converse is false.

Proof: Let p be a prime element in R. Then, p is nonzero and nonunit. To prove the irreducibility of p, let a and $b \in R$ such that $p = ab$. Then, $p|ab$ and, since p is prime, $p|a$ or $p|b$. Now,

$$\begin{aligned} p|a &\Rightarrow ps = a \quad \text{for some } s \in R \\ &\Rightarrow abs = a \\ &\Rightarrow bs = 1 \quad (\text{since } p \neq 0 \text{ and hence } a \neq 0) \\ &\Rightarrow b \text{ is a unit in } R. \end{aligned}$$

Similarly, $p|b \Rightarrow a$ is a unit in R. Thus, either a or b is a unit in R. Therefore, p is irreducible. The converse fails; for, consider the following example in which we exhibit an irreducible element which is not prime.

Example 12.1.2. Let $\mathbb{Z}[\sqrt{-5}] = \{a + b\sqrt{-5}\,;\, a \text{ and } b \in \mathbb{Z}\}$. Then, $\mathbb{Z}[\sqrt{-5}]$ is an integral domain under the usual addition and multiplication of complex numbers. For any $x=a+b\sqrt{-5}$, let $|x|$ be the usual modulus of the complex number $x=a+ib\sqrt{-5}$; that is,

$$|x| = \sqrt{a^2+5b^2}.$$

The following can be easily verified for any x and y in $\mathbb{Z}[\sqrt{-5}]$.

1. $|xy| = |x||y|$
2. $|x| = 0 \Leftrightarrow x = 0$
3. x is a unit in $\mathbb{Z}[\sqrt{-5}] \Leftrightarrow |x| = 1 \Leftrightarrow x = \pm 1$.

Now, we shall prove that $2+\sqrt{-5}$ is irreducible, but not prime in $\mathbb{Z}[\sqrt{-5}]$.
Let $p=2+\sqrt{-5}$. Then, $|p| = \sqrt{9} = 3$.
Suppose that x and $y \in \mathbb{Z}[\sqrt{-5}]$ such that $p = xy$.
Then, $3 = |p| = |xy| = |x||y|$ and hence $|x| = 1$ or $|y| = 1$ so that x or y is a unit in $\mathbb{Z}[\sqrt{-5}]$.
Thus, p is irreducible. Now, consider

$$3\times 3 = 9 = (2+\sqrt{-5})(2-\sqrt{-5}) = p(2-\sqrt{-5}).$$

Therefore, p divides 3×3. But p does not divide 3, since we cannot find integers a and b such that $(2+[\sqrt{-5}])(a+b\sqrt{-5}) = 3$. Thus, p is not prime in $\mathbb{Z}[\sqrt{-5}]$.

Recall that we have introduced the notion of an irreducible polynomial over an integral domain R as a nonconstant polynomial over R which cannot be expressed as a product of two nonconstant polynomials. In the integral domain $R[x]$, an irreducible element may be a constant and hence not an irreducible polynomial. However, if F is a field, then the ring $F[x]$ of polynomials over F is an integral domain in which units are precisely nonzero constant polynomials and hence irreducible polynomials over F and irreducible elements in $F[x]$ are same.

In the following, we establish a relation between primeness (irreducibility) of an element p and the primeness (maximality) of the ideal $<p>$ generated by p. First, let us define a principal ideal of R to be a maximal principal ideal if it is maximal (with respect to the inclusion relation) in the set of proper principal ideals of R.

Theorem 12.1.6. The following holds for any nonzero and nonunit p in an integral domain R.

1. p is a prime element in R if and only if $<p>$ is a prime ideal of R.
2. p is an irreducible element in R if and only if $<p>$ is a maximal principal ideal of R.

Proof:

1. is trial because of the fact that $x \in \langle p\rangle$ if and only if p divides x.
2. Suppose that p is an irreducible element in R and $\langle p\rangle \subseteq \langle x\rangle$, $x \in R$. Then, $\langle p\rangle \neq R$ (since p is a nonunit) and $p \in \langle x\rangle$ and hence $p = xy$ for some $y \in R$. Since p is irreducible, either x or y is a unit in R. If x is a unit, then $\langle x\rangle = R$. If y is a unit, then $x = py^{-1} \in \langle p\rangle$ and hence $\langle x\rangle \subseteq \langle p\rangle$, so that $\langle x\rangle = \langle p\rangle$. Thus, $\langle p\rangle$ is a maximal principal ideal of R.

Conversely, suppose that $\langle p\rangle$ is a maximal principal ideal of R. Then, $\langle p\rangle \neq R$ and hence p is not a unit. Suppose a and $b \in R$ such that $p = ab$. Then, $p \in \langle a\rangle$ and hence $\langle p\rangle \subseteq \langle a\rangle$. By the maximality of $\langle p\rangle$, either $\langle p\rangle = \langle a\rangle$ or $\langle a\rangle = R$. If $\langle a\rangle = R$, then a is a unit. If $\langle p\rangle = \langle a\rangle$, then $a = pc$ for some $c \in R$ and hence

$$p = ab = (pc)b = p(cb)$$

since R is an integral domain and $p \neq 0$, it follows that $1 = cb$ and hence b is a unit. Therefore, in any case, either a or b is a unit. Thus, p is an irreducible element in R. ◀

Worked Exercise 12.1.1. Let p and q be associates to each other in an integral domain R. Then, prove the following:

1. p is prime if and only if q is prime.
2. p is irreducible if and only if q is irreducible.

Answer: Since $p \sim q$, we have $p = qu$ for some unit u in R and hence $q = pu^{-1}$. Therefore, $p = 0$ if and only if $q = 0$ and p is a unit if and only if q is a unit. Also, note that $p|q$ and $q|p$.

1. Suppose that p is prime. Then, p is nonzero and nonunit and hence so is q. For any a and $b \in R$,

$$\begin{aligned} q|ab &\Rightarrow p|ab \quad (\text{since } p|q) \\ &\Rightarrow p|a \text{ or } p|b \quad (\text{since } p \text{ is prime}) \\ &\Rightarrow q|a \text{ or } q|b \quad (\text{since } q|p). \end{aligned}$$

   Thus, q is prime. Converse follows from the fact that $p \sim q$ if and only if $q \sim p$.
2. Suppose that p is irreducible. Then, p is nonzero and nonunit and hence so is q. Let a and $b \in R$ such that $q = ab$. Then, $p = qu = abu$. Since p is irreducible, either a or bu is a unit. Therefore, a or b is a unit. Thus, q is irreducible.

Worked Exercise 12.1.2. Prove that an integral domain is a field if and only if there are exactly two associate classes.

Answer: Let R be an integral domain. If R is a field, then every nonzero element of R is a unit and hence $a \sim 1$ for all $a \neq 0$ in R, so that $\tilde{0}$ and $\tilde{1}$ $(= R - \{0\})$ are the only associate classes in R. Conversely, suppose that there are exactly two associate classes in R. Since $\tilde{0} = \{0\}$, we get that $\tilde{a} = R - \{0\}$ for any $0 \neq a \in R$. In particular, $1 \sim a$ for all $0 \neq a \in R$ and hence, every nonzero element in R is a unit. Therefore, R is a field.

EXERCISE 12(A)

1. State whether the following are true and justify your answers.
   - (i) 5 is an irreducible element in $\mathbb{Z}$.
   - (ii) 10 is an irreducible element in $\mathbb{Z}[i]$.
   - (iii) 13 is a prime element in $\mathbb{Z}[i]$.
   - (iv) Any prime element in $\mathbb{Z}$ is a prime element in $\mathbb{Z}[i]$.
   - (v) $2+\sqrt{2}$ is an associate of $\sqrt{2}$ in $\mathbb{Z}[\sqrt{2}]$.
   - (vi) $\sqrt{2}$ is irreducible in $\mathbb{R}$.
   - (vii) 5 is a prime element in $\mathbb{R}$.
   - (viii) -5 is a prime element in $\mathbb{Z}$.

2. Determine all the units in each of the following:
   - (i) $\mathbb{Z}$
   - (ii) $\mathbb{Z}[i]$
   - (iii) $\mathbb{Z}[\sqrt{2}]$
   - (iv) $\mathbb{Z}[x]$
   - (v) $\mathbb{R}[x]$
   - (vi) $\mathbb{Z}_5[x]$.

3. Determine all the associates of each of the following in the rings mentioned against them
   - (i) 4 in $\mathbb{Z}$
   - (ii) $1 + x$ in $\mathbb{R}[x]$
   - (iii) $1 + i$ in $\mathbb{Z}[i]$
   - (iv) $2 + x$ in $\mathbb{Z}_5[x]$
   - (v) $1 + 2x$ in $\mathbb{Z}_3[x]$
   - (vi) $1 + x + x^2$ in $\mathbb{Z}_2[x]$.

4. Consider the ring $\mathbb{Z}[\sqrt{3}] = \{a + b\sqrt{3} : a \text{ and } b \in \mathbb{Z}\}$. In this, prove that $2 + \sqrt{3}$ is a unit and $3 + 2\sqrt{3}$ is an associate $\sqrt{3}$.

5. Let a and $b \in \mathbb{Z}$ such that $a^2 + b^2$ is prime in $\mathbb{Z}$. Then prove that $a + bi$ is prime in $\mathbb{Z}[i]$.

6. Prove that 2 and $1 + \sqrt{5}$ are irreducible in $\mathbb{Z}[\sqrt{5}]$ but not prime.

7. Let a and $b \in \mathbb{Z}$ such that $|a^2 - 10b^2|$ is prime in $\mathbb{Z}$. Then prove that $a + b\sqrt{10}$ is irreducible in $\mathbb{Z}[\sqrt{10}]$.

8. Let R be an integral domain and $a_1, a_2, \ldots, a_n \in \mathbb{R}$. If p is a prime element in R and p divides the product $a_1, a_2, \ldots, a_n$, then prove that p divides atleast one a_i.

9. Let a and $b \in \mathbb{Z}$ such that $a^2 + 3b^2$ is prime in $\mathbb{Z}$. Then prove that $a + b\sqrt{-3}$ is irreducible in $\mathbb{Z}[\sqrt{-3}]$.

10. Let a and b be nonzero elements in an integral domain R and $a \sim b$. If $c \in R$ and $a = bc$, prove that c is a unit in R.

12.2 PRINCIPAL IDEAL DOMAINS

It is well known that every ideal of the ring $\mathbb{Z}$ of integers is a principal ideal. In this, we discuss integral domains in which every ideal is principal.

Definition 12.2.1. An integral domain in which every ideal is a principal is called a principal ideal domain (PID).

Example 12.2.1

1. $\mathbb{Z}$ is a PID.
2. Any field is a PID, since $<0>$ and $<1>$ are the only ideals of a field.
3. The ring $F[x]$ of polynomials over a field F is a PID (refer Theorem 11.3.2).
4. The ring $\mathbb{Z}[x]$ of polynomials over $\mathbb{Z}$ is an integral domain, but not a PID; for, we have exhibited, in Worked Exercise 11.3.1 an ideal of $\mathbb{Z}[x]$ which is not principal.

The following is an immediate consequence of Theorems 12.1.3 and 12.1.4.

Theorem 12.2.1. In a PID, any finite number of nonzero elements have both g.c.d. and l.c.m.

Corollary 12.2.1. Let R be a PID and $a_1, a_2, \ldots, a_n \in R$, Then,

$$\text{g.c.d. } \{a_1, a_2, \ldots, a_n\} = a_1r_1 + a_2r_2 + \cdots + a_nr_n$$

for suitable elements $r_1, r_2, \ldots, r_n$ in R.

Definition 12.2.2. Let $a_1, a_2, \ldots, a_n$ be nonzero elements of an integral domain such that g.c.d. $\{a_1, a_2, \ldots, a_n\}$ is a unit in R. Then, $a_1, a_2, \ldots, a_n$ are said to be *relatively prime* and denote this by g.c.d. $\{a_1, a_2, \ldots, a_n\} \sim 1$.

Note that any nonzero elements $a_1, a_2, \ldots, a_n$ of a PID R are relatively prime if and only if there exist elements $r_1, r_2, \ldots, r_n$ in R such that

$$a_1r_1 + a_2r_2 + \cdots + a_nr_n = 1.$$

This identity is known as *Bezout's identity*. The following is one of the most useful applications of Bezout's identity.

Theorem 12.2.2. Let R be a PID and a, b and $c \in R$ such that a and b are relatively prime and a divides bc. Then, a divides c.

Proof: Since a and b are relatively prime, there exist r and $s \in R$ such that

$$ar + bs = 1$$

(by Bezout's identity). Now,

$$c = c \cdot 1 = car + cbs = a(cr) + (bc)s.$$

Since a divides bc, $ax = bc$ for some $x \in R$. Therefore,

$$c = acr + bcs = a(cr + xs)$$

and hence a divides c.

Although prime elements are irreducible in a general integral domain (by Theorem 12.1.5), we have observed that the converse is not true. However, in a PID, any irreducible element is prime, as proved in the following theorem.

Theorem 12.2.3. The following are equivalent to each other for any nonzero element p in a PID R.

1. p is a prime element.
2. p is an irreducible element.
3. $<p>$ is a maximal ideal of R.
4. $<p>$ is a prime ideal of R.

Proof: First observe that, to satisfy any of the conditions (1) through (4) above, it is necessary that p is a nonunit in R. (1) $\Rightarrow$ (2) follows from Theorem 12.1.5.

(2) ⇔ (3): By Theorem 12.1.6 (2), p is irreducible if and only if $<p>$ is maximal among proper principal ideals of R. But R being a PID, every ideal of R is principal. Therefore, p is irreducible if and only if $<p>$ is a maximal ideal of R.

(3) ⇒ (4) is trivial and (4) ⇒ (1) follows from Theorem 12.1.6 (1). ◀

Corollary 12.2.2. A nonzero ideal of a PID is maximal if and only if it is a prime ideal.

The next results are concerned with the ideal structure of a PID. A sequence $\{I_n\}$ of ideals of a ring R is said to be an *ascending* (or *increasing*) sequence if $I_n \subseteq I_{n+1}$ for all n. A sequence $\{I_n\}$ is said to *terminate* at a finite stage if there exists $n \in \mathbb{Z}^+$ such that $I_n = I_{n+1} = I_{n+2} = I_{n+k}$ for all $k \in \mathbb{Z}^+$.

Theorem 12.2.4. The following holds in any PID R.

1. Every ascending sequence of ideals of R terminates at a finite stage.
2. Any nonempty class of ideals of R has a maximal member.

Proof:

1. Let $I_1 \subseteq I_2 \subseteq I_3 \subseteq \ldots$ be an ascending sequence of ideals of R. Put $I = \bigcup_{n=1}^{\infty} I_n$. Then, I is an ideal of R. Since R is a PID, there exists $a \in R$ such that

$$\bigcup_{n=1}^{\infty} I_n = I = <a>.$$

Then, for some $n \in \mathbb{Z}^+$, $a \in I_n$ and hence

$$I_{n+k} \subseteq I = <a> \subseteq I_n \subseteq I_{n+k}$$

for all $k \in \mathbb{Z}^+$, so that $I_n = I_{n+k}$ for all $k \in \mathbb{Z}^+$. Thus, the sequence terminates at a finite stage.

2. Let $\mathscr{C}$ be a nonempty class of ideals of R and suppose, if possible, that $\mathscr{C}$ has no maximal member. Since $\mathscr{C}$ is nonempty, choose $I_1 \in \mathscr{C}$ Then, I_1 is not maximal in $\mathscr{C}$ and hence there exists $I_2 \in \mathscr{C}$ such that $I_1 \subsetneq I_2$. Again, since I_2 is not maximal, there exists $I_3 \in \mathscr{C}$ such that $I_2 \subsetneq I_3$. Continuing this procedure, we get an ascending sequence

$$I_1 \subsetneq I_2 \subsetneq I_3 \subsetneq \ldots$$

of ideals of R which does not terminate at any finite stage. This is a contradiction to (1.) above. Thus, $\mathscr{C}$ must contain a maximal member. ◀

Corollary 12.2.3. Let R be a PID and $\{a_n\}$ be a sequence of elements in R such that a_n divides a_{n-1} for all $n > 1$. Then, there exists $n \in \mathbb{Z}^+$ such that

$$a_n \sim a_{n+1} \sim a_{n+k} \quad \text{for all } k \in \mathbb{Z}^+.$$

That is, a_n and a_{n+k} are associates for all $k \in \mathbb{Z}^+$.

Proof: This is an immediate consequence of Theorem 12.2.4 above and of the facts that, for any a and $b \in R$,

$$<a> \subseteq <b> \text{ if and only if } b \text{ divides } a$$
$$\text{and} \quad <a> = <b> \text{ if and only if } a \text{ and } b \text{ are associates.}$$ ◀

Theorem 12.2.5. Let R be a PID and a be a nonzero nonunit element in R. Then, there exists a prime element p in R such that p divides a.

Proof: Since a is a nonunit, the principal ideal $<a>$ is a proper ideal and hence $<a>$ is contained in a maximal ideal M of R. Since R is a PID, there exists $p \in R$ such that $M = <p>$. Also, since $a \neq 0$. We have

$$<0> \neq <a> \subseteq M = <p> \quad \text{and hence } p \neq 0.$$

Since M is a maximal ideal, we get from Theorem 12.2.3, that p is a prime element of R. Since $<a> \subseteq <p>$, we get that p divides a. ◀

Theorem 12.2.6. Any nonzero nonunit in a PID can be expressed as a finite product of prime elements.

Proof: Let R be a PID and a be a nonzero nonunit element in R. By the above theorem, there exists a prime element p_1 in R such that p_1 divides a and hence

$$a = p_1 a_1 \quad \text{for some } a_1 \in R.$$

Then, $a_1 \neq 0$ (since $a \neq 0$). If a_1 is a unit, then a is an associate of p_1 and hence a itself is prime. If a_1 is not a unit, then again by the above theorem, there exists a prime element p_2 dividing a_1 and hence $a_1 = p_2 a_2$ for some $a_2 \in R$. Repeating this process with a_2 and so on, we get prime elements $p_1, p_2, \ldots$ in R and elements $a_1, a_2, \ldots$ in R such that $a_n = p_{n+1} a_{n+1}$. Now,

$$a = p_1 a_1 = p_1 p_2 a_2 = \cdots = p_1 p_2 p_3 \cdots p_n a_n.$$

Then, $\{a_n\}$ is a sequence of nonzero elements such that a_{n+1} divides a_n for all n. By Corollary 12.2.3, there exists n such that $a_n \sim a_{n+k}$ for all $k \in \mathbb{Z}^+$.

Let n be the least positive integer such that $a_n \sim a_{n+k}$ for all $k \in \mathbb{Z}^+$. We claim that a_n (and hence all a_{n+k}) is a unit. For, otherwise, we have the prime element p_{n+1} such that $a_n = p_{n+1}a_{n+1}$ and, since $a_n \sim a_{n+1}$, it follows that p_{n+1} is a unit which is a contradiction to the primeness of p_{n+1}. Thus, a_n is a unit and hence $p_n a_n$ is also prime. Now, we have

$$a = p_1 p_2 \cdots p_{n-1} p_n a_n$$

and $p_1, p_2, \ldots, p_{n-1}$ and $(p_n a_n)$ are primes in R. ◀

Worked Exercise 12.2.1. Prove that the ring $\mathbb{Z}[i]$ of Gaussian integers is a PID.

Answer: We have $\mathbb{Z}[i] = \{a + bi : a \text{ and } b \in \mathbb{Z}\}$. $\mathbb{Z}[i]$ is an integral domain under the usual addition and multiplication of complex members. Let I be an ideal of $\mathbb{Z}[i]$. If $I = \{0\} = \langle 0 \rangle$, we are done. Suppose $I \neq \{0\}$. Let

$$A = \{a^2 + b^2 : 0 \neq a + bi \in I\}.$$

Then, A is a nonempty subset of $\mathbb{Z}^+$ and hence A has a least member. Let $0 \neq a + bi \in I$ be such that $a^2 + b^2$ is least in A. Now, we shall prove that I is the principal ideal generated by $a + bi$. Put $x = a + bi$. Since $x \in I$, we have $\langle x \rangle \subseteq I$. On the other hand, let $y = c + di \in I$. Consider the complex number

$$\frac{y}{x} = \frac{(c+di)(a-bi)}{(a+bi)(a-bi)} = \frac{(ac+bd)+(ad-bc)i}{a^2+b^2} = \alpha + \beta i,$$

where α and β are rational numbers given by

$$\alpha = \frac{ac+bd}{a^2+b^2} \quad \text{and} \quad \beta = \frac{ad-bc}{a^2+b^2}.$$

Choose integers m and n such that

$$|m-\alpha| \leq \frac{1}{2} \quad \text{and} \quad |n-\beta| \leq \frac{1}{2}.$$

Now, $y = (\alpha + \beta i)x = (m + ni)x + ((\alpha - m) + (\beta - n)i)x$

Put $r = ((\alpha - m) + (\beta - n)i)x$. Then,

$$r = y - (m + ni)x \in I \quad (\text{since } x \text{ and } y \in I).$$

Also,

$$\begin{aligned} |r|^2 &= ((\alpha - m)^2 + (\beta - n)^2)\,|x|^2 \\ &\leq \left(\frac{1}{4}+\frac{1}{4}\right)|x|^2 = \frac{1}{2}|x|^2 < a^2 + b^2. \end{aligned}$$

Since $a^2 + b^2$ is least in A, it follows that $r = 0$ and hence

$$y = (m + ni)x \in \langle x \rangle.$$

Therefore, $I \subseteq \langle x \rangle$. Thus, $I = \langle x \rangle$. Therefore, $\mathbb{Z}[i]$ is a PID.

Worked Exercise 12.2.2. Let R be a PID. Then prove that any nonzero proper ideal of R can be expressed as a finite product of maximal ideals of R.

Answer: Let I be a nonzero proper ideal of R. Since R is a PID, $I = \langle a \rangle$ for some nonzero nonunit a in R. By Theorem 12.2.6, a can be expressed as a product of prime elements. Let

$$a = p_1 p_2 \cdots p_n,$$

where $p_1, p_2, \ldots, p_n$ are prime elements. Put $M_i = \langle p_i \rangle$ for each $1 \leq i \leq n$. By Theorem 12.2.3, each M_i is a maximal ideal of R. Now,

$$\begin{aligned} I = \langle a \rangle &= \langle p_1 p_2 \cdots p_n \rangle \\ &= \langle p_1 \rangle \langle p_2 \rangle \cdots \langle p_n \rangle \\ &= M_1 M_2 \cdots M_n. \end{aligned}$$

Worked Exercise 12.2.3. Determine all the units of the ring $\mathbb{Z}[i]$ of Gaussian integers.

Answer: Let $a + bi$ be a unit in $\mathbb{Z}[i]$. Then, there exist integers c and d such that

$$(a + bi)(c + di) = 1.$$

By taking the absolute values, we have

$$(a^2 + b^2)(c^2 + d^2) = 1.$$

Since a, b, c and d are all integers, it follows that

$$a^2 + b^2 = 1 = c^2 + d^2$$

and ($a = 0$ and $b = \pm 1$) or ($b = 0$ and $a = \pm 1$) and hence $a + bi = 1$ or -1 or i or $-i$. Thus, $1, -1, i$ and $-i$ are all the units in $\mathbb{Z}[i]$.

Worked Exercise 12.2.4. Let I be a nonzero ideal of the ring $\mathbb{Z}[i]$. Then prove that the quotient ring $\mathbb{Z}[i]/I$ is finite.

Answer: Since $\mathbb{Z}[i]$ is a PID, I is a nonzero principal ideal and hence $I = \langle x \rangle$ for some $0 \neq x \in \mathbb{Z}[i]$. Let $x = a + bi$. Then, $a \neq 0$ or $b \neq 0$ and hence $a^2 + b^2$ is a positive integer. Consider an element $y + I \in \mathbb{Z}[i]/I$ with $y \in \mathbb{Z}[i]$. As in Worked Exercise 12.2.1, we can write

$$y = (m + ni)x + r$$

for some integers m and n and $r \in \mathbb{Z}[i]$ such that

$$|r|^2 < |x|^2 = a^2 + b^2.$$

Now, $y - r = (m + ni)x \in \langle x \rangle = I$ and hence $y + I = r + I$, where $r \in \mathbb{Z}[i]$ such that $|r|^2 < a^2 + b^2$. Since r must be of the form $c + di$ with c and d integers and $c^2 + d^2 < a^2 + b^2$ and since there can be only finitely many pairs (c, d) of integers such that $c^2 + d^2 < a^2 + b^2$, it follows that

$$\mathbb{Z}[i]/I = \{(c + di) + I : c \text{ and } d \in \mathbb{Z} \text{ and } c^2 + d^2 < a^2 + b^2\}$$

is finite.

EXERCISE 12(B)

1. Which of the following are PIDs? Justify your answers.
   (i) $\mathbb{Z}$
   (ii) $\mathbb{Q}$
   (iii) $\mathbb{R}[x]$
   (iv) $\mathbb{C}[x]$
   (v) $\mathbb{Z}_{10}$
   (vi) $\mathbb{Z}_{31}$
   (vii) $\mathbb{Z}_2$
   (viii) $\mathbb{Z}_6$

(ix) $F[x]$, where F is a field.

(x) $F[x, y]$ $(= F[x][y])$, where F is a field.

2. Prove that $\mathbb{Z}[x]$ is not a PID.

3. For any integral domain R, prove that the polynomial ring $R[x]$ is a PID if and only if R is a field.

4. Prove that any homomorphic image of a PID is a PID.

5. Prove that $\mathbb{Z}[\sqrt{-3}] = \{a + b\sqrt{-3} : a \text{ and } b \in \mathbb{Z}\}$ is an integral domain under the usual addition and multiplication of complex numbers and that $\mathbb{Z}[\sqrt{-3}]$ is not a PID.

6. Prove that $\mathbb{Z}[\sqrt{-19}]$ is not a PID.

7. For any odd prime number p, prove that $\mathbb{Z}[\sqrt{-p}]$ is not a PID.

8. Prove that $\mathbb{Z}[\sqrt{-2}]$ is a PID

9. Let $D = \{a_0 + a_1x + \cdots + a_nx^n \in R[x] : a_0 \text{ is rational}\}$. Prove that D is an integral domain under the usual addition and multiplication of polynomials.

10. Let R be a PID and S be a multiplicative subset of R. Prove that the ring of fractions of R by S (refer Exercise 15 of 10(f)) is a PID.

11. Let I be a nonzero ideal of a PID R. Then prove that any descending chain of ideals of R/I terminates at a finite stage.

12. Prove that an integral domain R is a PID if and only if R/I is a PID for each ideal I of R.

13. Let R be an integral domain in which any descending chain of ideals terminates at a finite stage. Then prove that R is a field.

14. Prove that a PID R has a unique maximal ideal if and only if any two irreducible elements of R are associates.

15. Prove that a nontrivial commutative ring with identity is a field if and only if every proper ideal is prime.

16. Let $R_1, R_2, \ldots, R_n$ be PIDs and $R = R_1 \times R_2 \times \cdots \times R_n$. Then prove that every ideal of R is principal. Is R a PID?

17. Prove the following for any ideals I, J and K of a PID R.

(i) $I(J \cap K) = (IJ) \cap (IK)$

(ii) $I \cap (J + K) = (I \cap J) + (I \cap K)$

(iii) $I + (J \cap K) = (I + J) \cap (I + K)$

(iv) If $I + J = R$, then $IJ = I \cap J$.

18. Let R be a PID. Prove that a nonzero ideal P of R is primary (refer Exercise 6 of 10(d)) if and only if $P = \langle p^n \rangle$ for some prime element p in R and $n \in \mathbb{Z}^+$.

12.3 UNIQUE FACTORIZATION DOMAINS

We have developed the analogues, in an arbitrary integral domain, of the concepts of the divisibility and prime numbers in the ring $\mathbb{Z}$ of integers. Recall that any nonzero nonunit in $\mathbb{Z}$ is a product of finite number of prime numbers (or their associates) and that this factorization is unique, except for the order of occurrences of the primes and their associates. We shall formalise this in the following definition.

Definition 12.3.1. Let R be an integral domain and $a \in R$ such that

$$a = p_1 p_2 \cdots p_n,$$

where each p_i is an irreducible element in R. Then, the equation $a = p_1 p_2 \cdots p_n$ is called a *factorization* of a in R.

Examples 12.3.1

1. $6 = 2 \cdot 3$ and $6 = (-2)(-3)$ are factorizations of 6 in $\mathbb{Z}$, since 2, 3, -2 and -3 are irreducible in $\mathbb{Z}$.
2. $20 = 2 \cdot 2 \cdot 5$, $20 = (-2)2(-5)$ and $20 = 2(-2)(-5)$ are factorizations of 20 in $\mathbb{Z}$.
3. $1 + 2x + x^2 = (1 + x)(1 + x)$ is a factorization of $1 + 2x + x^2$ in $\mathbb{Z}[x]$, since $1 + x$ is irreducible in $\mathbb{Z}[x]$.
4. $1 + x^2 = (1 + x)(1 + x)$ is a factorization of $1 + x^2$ in $\mathbb{Z}_2[x]$, since $1 + x$ is an irreducible element in $\mathbb{Z}_2[x]$.

Definition 12.3.2. An integral domain R is called a *factorization domain* (FD) if every nonzero nonunit in R has a factorization in R.

Examples 12.3.2

1. $\mathbb{Z}$ is a FD.
2. Any PID is a FD (recall Theorem 12.2.6). Note that an element in a PID is prime if and only if it is irreducible.
3. The ring $\mathbb{Z}[i]$ of Gaussian integers is a FD, since it is a PID (see Worked Exercise 12.2.1).
4. The ring $F[x]$ of polynomials over a field F is a FD, since $F[x]$ is a PID.

In the following, we give sufficient condition on an integral domain for it to be a FD. This helps us as a tool to quickly ascertain that a given integral domain is a FD.

Theorem 12.3.1. Let R be an integral domain such that there is a map $\delta : R - \{0\} \to \mathbb{Z}$ satisfying the following for any elements a and b in $R - \{0\}$.

1. $\delta(a) \geq 0$
2. $\delta(ab) \geq \delta(a)$
3. $\delta(ab) = \delta(a)$ if and only if b is a unit in R.

Then, R is a FD.

Proof: We have to prove that every nonzero nonunit in R has a factorization in R. Let S be the set of all nonzero nonunits in R which have no factorizations in R. It is enough if we prove that S is empty. On the contrary, suppose that S is not empty. Consider the set

$$A = \{\delta(a) : a \in S\}.$$

Then, A is a nonempty set of nonnegative (by (1)) integers. By the well-ordering principle, A has a least member. Let n be the least in A. Then, $n = \delta(a)$ for some $a \in S$. Then, a is not irreducible (since every irreducible element gives a factorization of itself) Therefore, $a = bc$ for some nonzero nonunits b and c in R. Then, by (2) and (3),

$$\delta(a) = \delta(bc) > \delta(b) \quad \text{and} \quad \delta(a) > \delta(c).$$

By the least property of $\delta(a)$, it follows that $b \notin S$ and $c \notin S$ and hence both b and c have factorizations in R. But then a $(= bc)$ also has a factorization in R, which is a contradiction to the fact that $a \in S$. Thus, S is empty and hence any nonzero nonunit in R has a factorization in R. Thus, R is a FD. ◀

Corollary 12.3.1. $\mathbb{Z}$ is a FD.

Proof: The map $a \mapsto |a|$ from $\mathbb{Z} - \{0\}$ into $\mathbb{Z}$ satisfies the properties mentioned in Theorem 12.3.1 and hence $\mathbb{Z}$ is a FD. ◀

Corollary 12.3.2. The ring $\mathbb{Z}[i]$ of Gaussian integers is a FD.

Proof: Define $\delta : \mathbb{Z}[i] - \{0\} \to \mathbb{Z}$ by

$$\delta(a + bi) = a^2 + b^2.$$

Then, δ satisfies (1), (2) and (3) of Theorem 12.3.1 and hence $\mathbb{Z}[i]$ is a FD. ◀

Corollary 12.3.3. Let n be a positive integer greater than 1 and

$$\mathbb{Z}\left[\sqrt{-n}\right]=\left\{a+b\sqrt{-n} : a \text{ and } b\in\mathbb{Z}\right\}.$$

Then, $\mathbb{Z}\left[\sqrt{-n}\right]$ is a FD.

Proof: Define $\delta : \mathbb{Z}\left[\sqrt{-n}\right] \rightarrow \mathbb{Z}$ by

$$\delta(a+b\sqrt{-n})=a^2+nb^2.$$

Then, δ satisfies the conditions in Theorem 12.3.1 and hence $\mathbb{Z}[\sqrt{-n}]$ is a FD. ◀

Definition 12.3.3. A FD R is called an *unique factorization domain* (*UFD*) if the following is satisfied:

If $p_1p_2\cdots p_n = q_1q_2\cdots q_m$, where p_i's and q_j's are irreducible elements in R, then $n = m$ and $a_i \sim b_{\sigma(i)}$ for some permutation σ on $\{1,2, \ldots, n\}$.

In other words, an integral domain R is called a UFD if every nonzero and nonunit in R has a factorization in R which is unique, except for the associates and order of occurrences of the irreducible factors.

Before going to certain examples of an UFD, we first prove two important properties of UFD's which are tools in determining whether a given FD is a UFD.

Theorem 12.3.2. Let R be an UFD. Then, an element p in R is irreducible if and only if it is prime.

Proof: We know that every prime element in any integral domain (and hence in R) is irreducible. Conversely, suppose that p is an irreducible element in R. Let b and $c \in R$ such that $p|bc$. We can assume that b and c are both nonzero (since $p|0$). If b is a unit, then $p|c$. Similarly, if c is a unit, then $p|b$. Suppose that neither b nor c is a unit. Since $p|bc$, there exists $a \in R$ such that $pa = bc$. Then, $a \neq 0$ (since $b \neq 0$ and $c \neq 0$) and a is nonunit (otherwise, if a is a unit, then $p = b(ca^{-1})$ and, since p is irreducible, b or ca^{-1} is a unit which is not true). Thus, a, b and c are nonzero nonunits in R. Since R is a UFD, we get that

$$a = p_1p_2\cdots p_n,\ b = q_1q_2\cdots q_m \quad \text{and} \quad c = r_1r_2\cdots r_t,$$

where p_i's, q_i's and r_i's are irreducible elements in R. Therefore, we have

$$pp_1p_2\cdots p_n = q_1q_2\cdots q_m\, r_1r_2\cdots r_t$$

By the uniqueness of the factorizations in R, p is an associate of q_i or r_j and hence p divides b or c. Thus, p is a prime element in R.

Example 12.3.3. Consider the integral domain $\mathbb{Z}[\sqrt{-5}]$. We have seen in Corollary 12.3.3 that $\mathbb{Z}[\sqrt{-5}]$ is a FD. Also, in Example 12.1.2, we have proved that $2+\sqrt{-5}$ is irreducible, but not prime in $\mathbb{Z}[\sqrt{-5}]$. Therefore, by the above theorem, $\mathbb{Z}[\sqrt{-5}]$ is not an UFD.

Next result is a converse of Theorem 12.3.2 in the sense that a FD is an UFD if every irreducible element is prime.

Theorem 12.3.3. Let R be a FD. Then, R is an UFD if and only if every irreducible element in R is prime.

Proof: Suppose that every irreducible element in R is prime. Since R is a FD, any nonzero nonunit in R can be expressed as a product of finite number of irreducible elements. We have to prove only the uniqueness of the factorizations. Let $p_1, p_2, \ldots, p_n$ and $q_1, q_2, \ldots, q_m$ be irreducible elements in R such that

$$p_1p_2\cdots p_n = q_1q_2\cdots q_m.$$

Suppose, if possible, that $n \neq m$. Without loss of generality, we can assume that $n > m$. Since all the p_i's and q_j's are irreducible, they are primes. Since p_1 is a prime and p_1 divides $q_1q_2\cdots q_m$, p_1 should divide some q_j. Let $\sigma(1)$ be such a j. That is, $1 \leq \sigma(1) \leq m$ and p divides $q_{\sigma(1)}$. Since $q_{\sigma(1)}$ is irreducible, it follows that p_1 is an associate of $q_{\sigma(1)}$. Therefore, there exists a unit u_1 in R such that $p_1u_1 = q_{\sigma(1)}$. Now, we have

$$p_1p_2\cdots p_n = q_{\sigma(1)}\left(\prod_{j\neq\sigma(1)} q_j\right) = p_1u_1\left(\prod_{j\neq\sigma(1)} q_j\right).$$

Since R is an integral domain and $p_1 \neq 0$, we can cancel p_1 on both the sides. Then, we have

$$p_2\cdots p_n = u_1\left(\prod_{j\neq\sigma(1)} q_j\right).$$

We can repeat the above process, with p_2 in place of p_1, to get $\sigma(2) \in \{1, 2, \ldots, m\} - \{q(1)\}$ such that $p_2u_2 = q_{\sigma(2)}$ for some unit u_2 in R. Then,

$$p_2\cdots p_n = u_1p_2u_2 \prod_{j\neq\sigma(1),\sigma(2)} q_j$$

and hence $$p_3\cdots p_n = u_1u_2 \prod_{j\neq\sigma(1),\sigma(2)} q_j$$

This process can be continued for m steps (since $n > m$) to exhaust all q_j's and then we get

$$p_{m+1}\cdots p_n = u_1 u_2 \cdots u_m,$$

where $u_1, u_2, \ldots, u_m$ are units and hence their product $u_1 u_2 \cdots u_m$ is also a unit. Now, p_{m+1} divides the unit $u_1 u_2 \cdots u_m$ and hence p_{m+1} itself is a unit which is a contradiction to the fact that an irreducible element is necessarily a nonunit. Thus, $n = m$ and we have permutation σ on $\{1, 2, \ldots, n\}$ such that $p_i \sim q_{\sigma(i)}$ for all $1 \leq i \leq n$. Thus, R is an UFD, converse is proved in Theorem 12.3.2. ◀

Corollary 12.3.4. Every PID is an UFD.

Proof: Let R be a PID. In Theorem 12.2.3, we have proved that an element in R is irreducible if and only if it is prime. Also, in Theorem 12.2.6, we have proved that R is a FD. Therefore, by the above Theorem 12.3.3, R is an UFD. ◀

The converse of the above result fails. That is, there are UFDs which are not PIDs. For example, $\mathbb{Z}[x]$ is not a PID (refer Worked Exercise 11.3.1). However, we prove in the next section that $\mathbb{Z}[x]$ is an UFD.

Corollary 12.3.5. The ring $F[x]$ of polynomials over a field F is an UFD in which the irreducible elements are precisely the irreducible polynomials over F.

Proof: We have proved in Theorem 11.3.2 that, for any field F, $F[x]$ is a PID and hence, by the above corollary, $F[x]$ is an UFD. ◀

Unique factorizations in an UFD help us in determining the g.c.d. and l.c.m. of two elements. In this direction, we have the following theorem whose proof is a routine verification.

Theorem 12.3.4. Let R be an UFD and a and b be nonzero nonunits in R. Then, we can express a and b by

$$a = p_1^{\alpha_1} p_2^{\alpha_2} \cdots p_n^{\alpha_n}$$

$$\text{and} \quad b = p_1^{\beta_1} p_2^{\beta_2} \cdots p_n^{\beta_n},$$

where $p_1, p_2, \ldots, p_n$ are pair-wise nonassociate irreducible elements in R (i.e., p_i is not an associate to any p_j, $j \neq i$) and α_i's and β_i's are nonnegative integers. Also, we have the following:

1. a divides b if and only if $\alpha_i \leq \beta_i$ for all $1 \leq i \leq n$.
2. g.c.d.$\{a, b\} = \prod_{i=1}^{n} p_i^{\min\{\alpha_i, \beta_i\}}$
3. l.c.m.$\{a, b\} = \prod_{i=1}^{n} p_i^{\max\{\alpha_i, \beta_i\}}$.

We have proved earlier (see Theorem 12.2.3) that, in a PID, the principal ideal generated by an irreducible element is a maximal ideal. We prove the converse in the following theorem.

Theorem 12.3.5. Let R be an UFD. Then, R is a PID if and only if $<p>$ is a maximal ideal of R for any irreducible element p of R.

Proof: Suppose that $<p>$ is a maximal ideal for any irreducible p in R. Let I be an ideal of R. If $I = \{0\}$, then I is principal. Therefore, we can assume that $I \neq \{0\}$. Also, we can assume that $I \neq R$. Put

$$A = \left\{ \begin{array}{l} m \in \mathbb{Z}^+ : m = \sum_{i=1}^{r} \alpha_i \text{ and} \prod_{i=1}^{r} p_i^{\alpha_i} \in I, \\ \qquad\qquad \text{where } p_1, \ldots, p_r \text{ are irreducible in } R \end{array} \right\}.$$

Since I has atleast one nonzero nonunit, if follows that A is a nonempty set of positive integers. Let m be the least member in A. Then, there exist irreducible elements $p_1, p_2, \ldots, p_r$ in R and positive integers $\alpha_1, \alpha_2, \ldots, \alpha_r$ such that

$$d = \prod_{i=1}^{r} p_i^{\alpha_i} \in I \quad \text{and} \quad m = \sum_{i=1}^{r} \alpha_i.$$

Now, we claim that $I = <d>$. Since $d \in I$, we have $<d> \subseteq I$. On the other hand, suppose that $0 \neq x \in I$. We can assume that each p_i is not an associate of any other p_j, $j \neq i$. It is enough if we can prove that each $p_i^{\alpha_i}$ divides x. Suppose that $p_1^{\alpha_1}$ does not divide x. Then, we have

$$x = p_1^n y, \quad \text{where } 0 \leq n < \alpha_1 \quad \text{and} \quad p_1 \nmid y.$$

Then, $y \notin <p_1>$ and $<p_1>$ is a maximal ideal of R. Therefore, $<y> + <p_1> = R$ and hence

$$1 = ay + bp_1 \quad \text{for some } a \text{ and } b \in R.$$

Put $Z = p_1^{\alpha_1 - 1} p_2^{\alpha_2} \ldots p_r^{\alpha_r}$. Then,

$$\begin{aligned} z = z1 &= zay + zbp_1 \\ &= p_1^n y(ap_1^{\alpha_1 - 1 - n} p_2^{\alpha_2} \ldots p_r^{\alpha_r}) + db \\ &= x(ap_1^{\alpha_1 - 1 - n} p_2^{\alpha_2} \ldots p_r^{\alpha_r}) + db \in I \end{aligned}$$

since $x \in I$ and $d \in I$. Therefore, $z \in I$ which is a contradiction to the least property of m. Therefore, $p_1^{\alpha_1} \nmid x$. Similarly, $p_1^{\alpha_1} \nmid x$ for all $1 \leq i \leq n$ and hence $d|x$. Therefore, $x \in \langle d \rangle$. Thus, $I = \langle d \rangle$. Therefore, R is a PID. Converse is proved in Theorem 12.2.3. ◀

Worked Exercise 12.3.1. Let $\mathbb{Z}[\sqrt{3}] = \{a + b\sqrt{3} : a \text{ and } b \in \mathbb{Z}\}$. Then prove that $\mathbb{Z}[\sqrt{3}]$ is a FD.

Answer: Define $\delta : \mathbb{Z}[\sqrt{3}] - \{0\} \to \mathbb{Z}$ by

$$\delta(a + b\sqrt{3}) = |a^2 - 3b^2|.$$

Then, $\delta(x) \geq 0$ for all $0 \neq x \in \mathbb{Z}[\sqrt{3}]$. Also, if $x = a + b\sqrt{3}$ and $y = c + d\sqrt{3}$, then

$$\begin{aligned}
\delta(xy) &= \delta((ac + 3bd) + (ad + bc)\sqrt{3}) \\
&= |(ac + 3bd)^2 - 3(ad + bc)^2| \\
&= |a^2c^2 + 9b^2d^2 - 3a^2d^2 - 3b^2c^2| \\
&= |a^2 - 3b^2|\,|c^2 - 3d^2| = \delta(x)\delta(y) \\
&\geq |a^2 - 3b^2| = \delta(x)
\end{aligned}$$

$$\begin{aligned}
\text{and} \quad \delta(x) = 1 &\Leftrightarrow |a^2 - 3b^2| = 1 \\
&\Leftrightarrow |(a + b\sqrt{3})(a - b\sqrt{3})| = 1 \\
&\Leftrightarrow (a + b\sqrt{3})(a - b\sqrt{3}) = \pm 1 \\
&\Leftrightarrow a + b\sqrt{3} \text{ is a unit.}
\end{aligned}$$

$\therefore \delta(xy) = \delta(x)$ if and only if y is a unit. Therefore, by Theorem 12.3.1, $\mathbb{Z}[\sqrt{3}]$ is a FD.

Worked Exercise 12.3.2. Let R be a UFD and P be a nonzero prime ideal of R. Then prove that there exists an irreducible element in P.

Answer: Since $P \neq \{0\}$, we can choose $0 \neq x \in P$. Also since $P \neq R$, x is a nonunit. Now, we have

$$x = p_1^{\alpha_1} p_2^{\alpha_2} \ldots p_r^{\alpha_r},$$

where $p_1, p_2, \ldots, p_r$ are irreducible elements in R and $\alpha_1, \alpha_2, \ldots, \alpha_r$ are positive integers. Since $x \in P$ and P is a prime ideal, $p_i \in P$ for some i.

EXERCISE 12(C)

1. Which of the following ring are UFDs? Justify your answers.
   (i) $\mathbb{Z}$
   (ii) $\mathbb{Q}$
   (iii) $\mathbb{R}$
   (iv) $\mathbb{Z}_6$
   (v) $\mathbb{Z}[i]$
   (vi) $\mathbb{Z}\left[\sqrt{-5}\right]$
   (vii) $\mathbb{R}[x]$
   (viii) $\mathbb{Q}[x]$.

2. Prove that the l.c.m. of any finite subset of an UFD exists and is unique up to associates.

3. Let R be an UFD and F be its field of quotients. For any prime p in R, let

$$R_{(p)} = \left\{\frac{a}{b} \in F : \text{g.c.d.}\{a, b\} = 1 \text{ and } p \nmid b\right\}.$$

   Prove that $R_{(p)}$ is a subring of F and that $R_{(p)}$ is a FID and hence an UFD.

4. Let S be a multiplicative set in an UFD R and $S^{-1}R$ the ring of fractions of R by S. Prove that $S^{-1}R$ is an UFD.

5. Let R be an UFD and $\{a_n\}$ be a sequence of elements in R such that a_{n+1} divides a_n for all $n \in \mathbb{Z}^+$. Then prove that there exists $n \in \mathbb{Z}^+$ such that a_n is an associate of a_{n+k} for all $k \in \mathbb{Z}^+$.

6. Let p be a prime element in an UFD R such that any prime element in R is an associate of p. Prove that every nonzero proper ideal of R is of the form $\langle p^n \rangle$ for some $n \in \mathbb{Z}^+$.

7. Let R be an UFD and P be the only nonzero prime ideal of R. Then prove that any nonzero proper ideal I of R is of the form P^n for some $n \in \mathbb{Z}$.

8. Prove that any increasing sequence of principal ideals in an UFD terminates at a finite stage.

9. If P is a nonzero prime ideal of an UFD R, then is R/P a UFD?

10. Let R be an UFD and (a, b) denote the g.c.d. $\{a, b\}$ for any a and $b \in R$. Prove the following for any nonzero elements a, b and $c \in R$.
    (i) $(a, (b, c)) \sim ((a, b), c)$
    (ii) $(a, 1) \sim 1$
    (iii) $(ca, cb) \sim c(a, b)$

(iv) $(a, ab) \sim a$

(v) $(a, b) \sim 1 \sim (a, c) \Rightarrow (a, bc) \sim 1$

(vi) $(a, b) \sim 1$, $a|c$ and $b|c \Rightarrow ab|c$

(vii) $(a, b) \sim 1$ and $a|bc \Rightarrow a|c$

(viii) $ab \sim (a, b)[a, b]$, where $[a, b] =$ l.c.m. $\{a, b\}$.

12.4 POLYNOMIALS OVER UFDS

Polynomials over an UFD are of special importance in view of the famous Gauss theorem which states that such polynomials again form an UFD. In this section, we introduce few more concepts in polynomials which ultimately lead to the proof of the Gauss theorem. Let us recall that, in any UFD, any finite number of nonzero elements have g.c.d.

Definition 12.4.1. Let R be an UFD and $f(x) = a_0 + a_1x + a_2x^2 + \cdots + a_nx^n$ be a nonzero polynomial over R. Then, the *content* of f is defined to be the g.c.d. of the coefficients $a_0, a_1, a_2, \ldots, a_n$ and is denoted by $c(f(x))$; that is,

$$c(f(x)) = \text{g.c.d. } \{a_0, a_1, a_2, \ldots, a_n\}.$$

A polynomial $f(x)$ is said to be *primitive* over R if $c(f(x))$ is a unit in R.

The content and the primitivity of a polynomial depends on the UFD over which the polynomial is defined. Consider the following example.

Example 12.4.1. Let $f(x) = 4 + 8x + 6x^2$. If we consider $f(x)$ as a polynomial over $\mathbb{Z}$, then

$$c(f(x)) = \text{g.c.d. } \{4, 8, 6\} = 2$$

and therefore, $f(x)$ is not primitive over $\mathbb{Z}$. However, if we consider $f(x)$ as a polynomial over the field $\mathbb{Q}$ of rational numbers, then

$$c(f(x)) = 1,$$

since any nonzero in $\mathbb{Q}$ is a unit. Therefore, $f(x)$ is primitive over $\mathbb{Q}$.

Note 12.4.1

1. Any monic polynomial over any UFD is primitive.
2. If R is an UFD and $0 \neq f(x) \in R[x]$, then

$$f(x) = c(f(x))g(x)$$

for some primitive polynomial $g(x)$ in $R[x]$.

In the following, we prove that primitive polynomials over any UFD are closed under multiplication.

Theorem 12.4.1. Let R be an UFD and $f(x)$ and $g(x)$ be primitive polynomials over R. Then, $f(x)g(x)$ is primitive.

Proof: Let $f(x) = a_0 + a_1x + \cdots + a_nx^n, a_n \neq 0$
and $g(x) = b_0 + b_1x + \cdots + b_mx^m, b_m \neq 0.$

If $\deg(f(x)) = n = 0$, then $c(f(x)) = a_0$ and hence $a_0(= f(x))$ is a unit so that $c(f(x)g(x)) \sim c(g(x)) \sim 1$. Therefore, we can assume that $n > 0$ and $m > 0$.

We prove that there is no prime element in R that divides all the coefficients in $f(x)g(x)$. To prove this, let p be a prime in R. Since $f(x)$ is primitive, p does not divide some coefficient a_i in $f(x)$. Let a_i be the first coefficient in $f(x)$ which is not divisible by p. Similarly, let b_j be the first coefficient in $g(x)$ which is not divisible by p. Let c_{i+j} be the coefficient of x^{i+j} in $f(x)g(x)$. Then,

$$\begin{aligned} c_{i+j} &= a_0b_{i+j} + a_1b_{i+j-1} + \cdots + a_{i-j}b_0 \\ &= \sum_{r+s=i+j} a_rb_s. \end{aligned}$$

Since $p|a_r$ for all $0 \leq r < i$ and $p|b_s$ for all $0 \leq s < j$, it follows that $c_{i+j} - a_rb_s$ is divisible by p. Since p does not divide a_ib_j, we get that p does not divide c_{i+j}. Therefore, there is no prime dividing all the coefficients in $f(x)g(x)$ and hence

$$c(f(x)g(x)) \text{ is a unit.}$$

Thus, $f(x)g(x)$ is primitive. ◀

Corollary 12.4.1. For any nonzero polynomials $f(x)$ and $g(x)$ over an UFD, $c(f(x)g(x)) = c(f(x))c(g(x))$.

Proof: Let R be an UFD and $f(x)$ and $g(x) \in R[x] - \{0\}$. Then, there exist primitive polynomials $f_1(x)$ and $g_1(x)$ in $R[x]$ such that

$$f(x) = c(f(x))f_1(x) \quad \text{and} \quad g(x) = c(g(x))g_1(x).$$

Thus, by the above theorem, $f_1(x)g_1(x)$ is primitive and

$$f(x)g(x) = c(f(x))c(g(x))f_1(x)g_1(x)$$

and therefore, $c(f(x)g(x)) = c(f(x))c(g(x))$.

Let us recall that an element p in an integral domain R is called irreducible if it is nonzero, nonunit and not a product of two nonunits. (See Definition 12.1.5). Also, recall that a polynomial over R is called irreducible over R if it is nonconstant and not a product of two nonconstant polynomials over R. In the case of polynomial rings over an integral domain R, there is a subtle difference between irreducible elements in $R[x]$ and irreducible polynomials over R. However, if R is a field, both these concepts coincide. In the case of general integral domains, we distinguish these two. Irreducible elements in $R[x]$ and irreducible polynomials over R are two different concepts. Irreducible elements in $R[x]$ are often called irreducible in $R[x]$.

Before going further, let us observe that, if F is the field of quotients of an UFD R and $0 \neq f(x) \in F[x]$, then we can write $f(x) = ab^{-1}g(x)$ for some a and $b \in R$ and $g(x)$ is primitive in $R[x]$; for, if

$$f(x)=\frac{a_0}{b_0}+\frac{a_1}{b_1}x+\cdots+\frac{a_n}{b_n}x^n$$

and $b = b_0b_1 \ldots b_n \in R$ and $f(x) = b^{-1}h(x) = b^{-1}c(h(x))g(x)$ for some primitive $g(x)$ in $R[x]$. ◀

Theorem 12.4.2 (Gauss Lemma). Let R be an UFD and F be the field of quotients of R. Let $f(x) \in R[x]$ be a primitive polynomial of positive degree. Then, $f(x)$ is irreducible in $R[x]$ if and only if $f(x)$ is irreducible in $F[x]$.

Proof: Suppose that $f(x)$ is irreducible in $F[x]$. Suppose that $f(x) = g(x)h(x)$ where $g(x)$ and $h(x) \in R[x]$. If $g(x)$ and $h(x)$ are both of positive degree then they are nonunits in $R[x]$ and in $F[x]$, which is a contradiction to the irreducibility of $f(x)$ in $F[x]$. Therefore, $g(x)$ or $h(x)$ is of degree 0. Let $\deg(g(x)) = 0$. Then, $g(x) \in R$. Since $f(x)$ is primitive in $R[x]$, we get that $g(x)$ is a unit in R. Similarly, if $\deg(h(x)) = 0$, then $h(x)$ is a unit in R. Thus, $f(x)$ is irreducible in $R[x]$.

Conversely, suppose that $f(x)$ is reducible in $F[x]$. Then, $f(x) = g(x)h(x)$ where $g(x)$ and $h(x)$ are nonunits in $F[x]$ and hence $g(x)$ and $h(x)$ are polynomials of positive degree (since F is a field). Let

$$g(x)=\frac{a_0}{b_0}+\frac{a_1}{b_1}x+\cdots+\frac{a_n}{b_n}x^n,\ a_n \neq 0,\ n>0$$

and $$h(x)=\frac{c_0}{d_0}+\frac{c_1}{d_1}x+\cdots+\frac{c_m}{d_m}x^m,\ c_m \neq 0,\ m>0$$

where $a_i, b_i, c_j, d_j \in R$, $b_i \neq 0$ and $d_j \neq 0$.

Put $b = b_0 b_1 b_2 \ldots b_n$ and $d = d_0 d_1 d_2 \ldots d_m$.
Then, $bg(x)$ and $dh(x) \in R[x]$. We can write

$$bg(x) = ag_1(x) \quad \text{and} \quad dh(x) = ch_1(x)$$

for some primitive polynomials $g_1(x)$ and $h_1(x)$ in $R[x]$. Now, we have

$$bdf(x) = bg(x)dh(x) = acg_1(x)h_1(x).$$

By Theorem 12.4.1, $g_1(x)h_1(x)$ is primitive in $R[x]$. Since $f(x)$ is also primitive, it follows that bd is an associate of ac and hence $bdu = ac$ for some unit u in R. Now,

$$bdf(x) = bdug_1(x)h_1(x).$$

and hence $f(x) = ug_1(x)h_1(x)$, which implies that $f(x)$ is reducible in $R[x]$. ◄

Theorem 12.4.3 (Gauss Theorem). The ring $R[x]$ of polynomials over an UFD R is also an UFD.

Proof: Let R be an UFD. Then, clearly $R[x]$ is an integral domain. Let $0 \neq f(x) \in R[x]$. If $\deg(f(x)) = 0$, then $f(x) \in R$ and, since R is an UFD, $f(x)$ has a factorization in R and hence in $R[x]$. Therefore, we can assume that $\deg(f(x)) > 0$.
Let F be the quotient field of R. Then, R is a subring of F. We can write $f(x) = c(f(x))g(x)$, where $g(x)$ is a primitive polynomial over R and hence over F. Now, recall that $F[x]$ is an UFD (by Corollary 12.3.5). Since $g(x) \in F[x]$ and $\deg(g(x)) > 0$, we get that

$$g(x) = g_1(x)g_2(x)\cdots g_n(x)$$

for some irreducible polynomials $g_1(x), g_2(x), \ldots, g_n(x)$ in $F[x]$. Then, $\deg(g_i(x)) > 0$ for each $1 \leq i \leq n$. We can write $g_i(x) = a_i b_i^{-1} h_i(x)$ where $a_i, b_i \in R$ $(b_i \neq 0)$ and $h_i(x)$ is primitive in $R[x]$. Then, we have

$$b_1 b_2 \cdots b_n g(x) = a_1 a_2 \cdots a_n h_1(x)h_2(x)\cdots h_n(x).$$

Since each $h_i(x)$ is primitive, so is their product. Also, since $g(x)$ is primitive, it follows, by taking contents both sides that

$$b_1 b_2 \cdots b_n \sim a_1 a_2 \cdots a_n \text{ in } R$$

and hence $g(x) = uh_1(x)h_2(x)\cdots h_n(x)$,

where u is a unit in R. Further, since each $h_i(x)$ is irreducible in $F[x]$ and primitive in $R[x]$, it follows that each $h_i(x)$ is irreducible in $R[x]$ (by the Gauss Lemma (Theorem 12.4.2)). Thus,

$$f(x) = c(f(x))uh_1(x)h_2(x)\cdots h_n(x).$$

If $c(f(x))$ is a unit in R, $c(f(x))uh_1(x), h_2(x), \ldots, h_n(x)$ are irreducible elements in $R[x]$ and hence we have a factorization of $f(x)$ in $R[x]$. If $c(f(x))$ is not a unit in R, then, by the factorization property in R,

$$c(f(x)) = p_1 p_2 \cdots p_m,$$

where $p_1, p_2, \ldots, p_m$ are irreducible elements in R and hence in $R[x]$. Then,

$$f(x) = p_1 p_2 \cdots p_{m-1}(p_m u)h_1(x)h_2(x)\cdots h_n(x)$$

is a factorization of $f(x)$ in $R[x]$. Thus, $R[x]$ is a FD.
To prove the uniqueness of the factorizations in $R[x]$, first observe that any irreducible polynomial in $R[x]$ of positive degree must be primitive. Any factorization of $f(x)$ in $R[x]$ must be of the form

$$f(x) = c_1 c_2 \cdots c_n g_1(x)g_2(x)\cdots g_m(x) \tag{1}$$

where $c_1, c_2, \ldots, c_n$ $(n \geq 0)$ are irreducible elements in R and $g_1(x), g_2(x), \ldots, g_m(x)$ $(m \geq 0)$ are irreducible in $R[x]$, each of positive degree, and hence $g_1(x), g_2(x), \ldots, g_m(x)$ are primitive. Now, suppose that

$$f(x) = d_1 d_2 \cdots d_r h_1(x)h_2(x)\cdots h_s(x) \tag{2}$$

is another factorization of $f(x)$ in $R[x]$. Since each of $g_i(x)$ and $h_i(x)$ are irreducible polynomials of positive degree in $R[x]$, they are primitive and hence these are irreducible in $F[x]$. Also, by taking contents in (1) and (2), we get that

$$c_1 c_2 \cdots c_n = ud_1 \cdots d_r \tag{3}$$

for some unit u in R. Since R is an UFD, $n = r$ and each c_i is an associate of some d_j. Further, from (1), (2) and (3), we get that

$$h_1(x)h_2(x)\cdots h_s(x) = ug_1(x)g_2(x)\cdots g_m(x).$$

Since $F[x]$ is an UFD, it follows that $s = m$ and each $h_i(x)$ is an associate of $g_j(x)$. Thus, the factorization (1) is unique. Therefore, $R[x]$ is an UFD.

Corollary 12.4.2. If R is an UFD, then so is $R[x_1, x_2, \ldots, x_n]$, where $R[x_1, x_2, \ldots, x_n] = R[x_1, \ldots, x_{n-1}][x_n]$.

Corollary 12.4.3. $\mathbb{Z}[x]$ is an UFD but not a PID.

Corollary 12.4.4. For any field F, $F[x_1, x_2, \ldots, x_n]$ is an UFD.

Though we have proved in the above theorem that any polynomial over an UFD R can expressed as a product of irreducible elements in $R[x]$, it is a difficult task to find such a factorization for a given polynomial. Gauss theorem only ensures the existence of a factorization. There is no general explicit method for obtaining such a factorization, not even for deciding whether a given polynomial is irreducible or not. However, we have certain sufficient conditions for the irreducibility of a polynomial over an UFD as given below.

Theorem 12.4.4 (Eisenstein's Criterion). Let R be an UFD and F be the field of quotients of R. Let

$$f(x) = a_0 + a_1x + \cdots + a_nx^n, n > 0, a_n \neq 0$$

be a nonconstant polynomial in $R[x]$. Suppose that there exists a prime element p in R such that

1. p divides a_i for $0 \leq i < n$.
2. p does not divide a_n.
3. p^2 does not divide a_0.

Then, $f(x)$ is irreducible in $F[x]$. Also, if $f(x)$ is primitive, then $f(x)$ is irreducible in $R[x]$.

Proof: First, we assume that $f(x)$ is primitive and prove that $f(x)$ is irreducible in $R[x]$. Suppose, if possible, $f(x)$ is not irreducible. Since $f(x)$ is a nonconstant primitive polynomial, there exist two nonzero nonunit polynomials $f_1(x)$ and $f_2(x)$ in $R[x]$ such that

$$f(x) = f_1(x)f_2(x).$$

Let $f_1(x) = c_0 + c_1x + \cdots + c_rx^r$ and $f_2(x) = d_0 + d_1x + \cdots + d_sx^s$. Then,

$$n = \deg(f(x)) = \deg(f_1(x)) + \deg(f_2(x)) = r + s.$$

If $r = 0$, then $f_1(x) = c_0$ and $f(x) = c_0f_2(x)$ and hence c_0 is a nonzero nonunit (since so is $f_1(x)$) divisor of $c(f(x))$ which is a unit. Therefore, $r \neq 0$.

Similarly, $s \neq 0$. Since $n = r + s$, it follows that $0 < r < n$ and $0 < s < n$. Now, $f(x) = f_1(x)f_2(x)$ implies that $a_0 = c_0d_0$. Since $p|a_0$ and $p^2\nmid a_0$, it follows that p divides exactly one of c_0 and d_0. Without loss of generality, we can assume that $p|c_0$ and $p\nmid d_0$. Also, we have $a_n = c_rd_s$ and $p\nmid a_n$ and hence $p\nmid c_r$ and $p\nmid d_s$. Therefore, we have $p|c_0$ and $p\nmid c_r$. Let i be the least positive integer such that $p\nmid c_i$. Then, $p|c_j$ for all $0 \leq j < i$. Equating the coefficients of x^i in $f(x)$ and $f_1(x)$ $f_2(x)$, we get that

$$a_i = c_0d_i + c_1d_{i-1} + \cdots + c_id_0,$$

where $c_j = 0$ for all $j > r$ and $d_j = 0$ for all $j > s$. Now, since $p|c_0, p|c_1, \ldots, p|c_{i-1}$, we get that

$$p \mid a_i - c_id_0.$$

Also, since $0 < i \leq r < n, p|a_i$, by (1). Therefore, $p|c_id_0$ which is a contradiction since p is prime, $p\nmid c_i$ and $p\nmid d_0$.

Thus, $f(x)$ is irreducible in $R[x]$. Also, since $f(x)$ is primitive, the Gauss Lemma (Theorem 12.4.2) implies that $f(x)$ is irreducible in $F[x]$.

Now, let us take up the general case. There exists a primitive polynomial $g(x)$ in $R[x]$ such that

$$f(x) = c(f(x))g(x),$$

where $c(f(x))$ is the content of $f(x)$. If $c(f(x))$ is a unit, then $f(x)$ is primitive and hence, by the first case, $f(x)$ is irreducible in $R[x]$ and in $F[x]$. Suppose that $c(f(x))$ is a nonunit. Let

$$c(f(x)) = d \quad \text{and} \quad g(x) = b_0 + b_1x + \cdots + b_nx^n.$$

Then, $\deg(f(x)) = \deg(g(x))$ and hence $b_n \neq 0$. Since $f(x) = dg(x)$, we get that

$$a_i = db_i \quad \text{for all } 0 \leq i \leq n.$$

Since $p\nmid an$, we get that $p\nmid d$ and $p\nmid b_n$. Also, for any $1 \leq i < n$, $p|a_i$ implies that $p|b_i$. Further, $p^2\nmid b_0$, since $p^2\nmid a_0$. Thus, by the first case, $g(x)$ is irreducible in $R[x]$ and hence in $F[x]$. Since d is a unit in F, we get that $f(x)$ is an associate of $g(x)$ in $F[x]$. Thus, $f(x)$ is irreducible in $F[x]$. ◀

Since $\mathbb{Z}$ is a UFD and the rational number field $\mathbb{Q}$ is the field of quotients of $\mathbb{Z}$, the following is an immediate consequence of the Eisenstein's Criterion.

Corollary 12.4.5. Let $f(x) = a_0 + a_1x + \cdots + a_nx^n$ be a constant polynomial in $\mathbb{Z}[x]$ and p be a prime number satisfying the following:

1. p divides a_i for all $0 \leq i < n$.
2. p does not divide a_n.
3. p^2 does not divide a_0.

Then, $f(x)$ is irreducible in $\mathbb{Q}[x]$. Further, if $f(x)$ is primitive, then $f(x)$ is irreducible in $\mathbb{Z}[x]$.

Corollary 12.4.6. Let p be a prime number and $f(x) = 1 + x + x^2 + \cdots + x^{p-1}$. Then, $f(x)$ is irreducible in $\mathbb{Q}[x]$ and in $\mathbb{Z}[x]$. This polynomial $f(x)$ is called the *cyclotomic polynomial.*

Proof: Since $f(x)$ is a monic polynomial, it is primitive in $\mathbb{Z}[x]$.Consider the polynomial $f(x + 1)$. We are given that

$$f(x)=1+x+x^2+\cdots+x^{p-1}=\frac{x^p-1}{x-1}.$$

Therefore, $$f(x+1)=\frac{(x+1)^p-1}{(x+1)-1}$$

$$=\frac{1}{x}(x^p+px^{p-1}+(p_2)x^{p-2}+\cdots+px)$$

$$=\sum_{r=0}^{p-1}(p_r)x^{p-r-1}$$

$$=x^{p-1}+px^{p-2}+\cdots+(p_i)x^{p-i-1}+\cdots+p.$$

By the Eisenstein's Criterion (Theorem 12.4.4), $f(x + 1)$ is irreducible in $\mathbb{Q}[x]$ and therefore, $f(x)$ is irreducible in $\mathbb{Q}[x]$. Since $f(x)$ is primitive in $\mathbb{Z}[x]$, it is irreducible in $\mathbb{Q}[x]$. ◀

The following theorem is a very useful tool to determine the irreducibility of certain polynomials in $\mathbb{Z}[x]$ and their rational roots.

Theorem 12.4.5. Let $f(x) = a_0 + a_1x + \cdots + a_ix^n \in \mathbb{Z}[x]$ with $a_0 \neq 0$ and $a_n \neq 0$. Let p and q be relatively prime integers and $q > 0$. Suppose that p/q is a root of $f(x)$ in $\mathbb{Q}$. Then, p divides a_0 and q divides a_n.

Proof: We are given that $f(p/q)=0$ and therefore

$$a_0+a_1\left(\frac{p}{q}\right)+a_2\left(\frac{p}{q}\right)^2+\cdots+a_n\left(\frac{p}{q}\right)^n=0.$$

By multiplying with q^n, we have

$$a_0q^n + a_1pq^{n-1} + a_2p^2q^{n-2} + \cdots + a_{n-1}p^{n-1}q + a_np^n = 0$$

and hence

$$a_0q^n = p\left(-\sum_{r=1}^{n} a_r p^r q^{n-r}\right)$$

and

$$a_np^n = q\left(-\sum_{r=0}^{n-1} a_r p^r q^{n-r}\right).$$

Therefore, p divides a_0q^n and q divides a_np^n. Since p and q are relatively prime, it follows that p divides a_0 and q divides a_n. ◀

Worked Exercise 12.4.1. Let $f(x) = 5 + 11x - 7x^2 + 9x^3 \in \mathbb{Z}[x]$. Prove that $f(x)$ is irreducible over $\mathbb{Q}$ as well as over $\mathbb{Z}$.

Answer: Since $\deg(f(x)) = 3$, $f(x)$ is irreducible over $\mathbb{Q}$ if and only if $f(x)$ has a root in $\mathbb{Q}$. Suppose that p/q is a root of $f(x)$, we can assume that p and q are relatively prime integers and $q > 0$. Then, by the above Theorem 12.4.5, p should divide 5 and q should divide 9. Therefore, $p = \pm 1$ or ± 5 and $q = \pm 1$ or ± 3 or ± 9 and hence

$$\frac{p}{q} \in \left\{\pm 1,\ \pm 5,\ \pm\frac{1}{3},\ \pm\frac{5}{3},\ \pm\frac{1}{9},\ \pm\frac{5}{9}\right\}.$$

But, none of the elements in this set is a root of $f(x)$. Therefore, $f(x)$ has no root in $\mathbb{Q}$. Thus, $f(x)$ is irreducible over $\mathbb{Q}$. Since $f(x)$ is primitive in $\mathbb{Z}[x]$, it follows from Gauss Lemma (Theorem 12.4.2), that $f(x)$ is irreducible over $\mathbb{Z}$ also.

Worked Exercise 12.4.2. Let $f(x) = -2 + 15x - 9x^2 + x^3 \in \mathbb{Z}[x]$. Prove that $f(x)$ is irreducible over neither $\mathbb{Q}$ nor $\mathbb{Z}$.

Answer: Suppose, if possible, $(p/q) \in \mathbb{Q}$ is a root of $f(x)$. We can assume that p and q are relatively prime integers and $q > 0$. Then, by the above Theorem 12.4.5, p should divide -2 and q should divide 1 and hence $p = \pm 1$ or ± 2 and $q = \pm 1$. Therefore,

$$\frac{p}{q} = \pm 1 \quad \text{or} \quad \pm 2.$$

By a physical verification, we can see that 2 is a root of $f(x)$; that is, $f(2) = 0$. Thus, $x - 2$ is a factor of $f(x)$; in fact,

$$f(x) = (x - 2)(x^2 - 7x + 1).$$

Thus, $f(x)$ is reducible over $\mathbb{Q}$ as well as over $\mathbb{Z}$.

EXERCISE 12(D)

1. Which of the following are UFDs? Justify your answers.
   (i) $\mathbb{Z}[x]$
   (ii) $\mathbb{Q}[x]$
   (iii) $\mathbb{R}[x]$
   (iv) $\mathbb{C}[x]$
   (v) $\mathbb{Z}_3[x]$
   (vi) $\mathbb{Z}_6[x]$
   (vii) $\mathbb{Z}_9[x]$
   (viii) $\mathbb{Z}_{13}[x]$.
2. Prove that every field is an UFD.
3. Let R be an UFD and $f(x) \in R[x]$. Prove that $f(x)$ is irreducible over $\mathbb{R}$ if and only if $f(x + a)$ is irreducible over $\mathbb{R}$ for some $a \in R$.
4. Which of the following polynomials are irreducible over the UFDs mentioned against them?
   (i) $15 - 9x^2 + 6x^3 + 2x^4$ over $\mathbb{Z}$
   (ii) $3 + 2x^2 + x^3$ over $\mathbb{Q}$
   (iii) $4 + 2x + x^3$ over $\mathbb{Z}_5$
   (iv) $1 + x^2 + x^5$ over $\mathbb{Z}_2$
   (v) $9 - x^3$ over $\mathbb{Z}_{31}$
   (vi) $1 + x^3 + x^6$ over $\mathbb{Q}$
   (vii) $5 + 10x + 15x^3 + 2x^5$ over $\mathbb{Z}$
   (viii) $2 + 2x + x^4$ over $\mathbb{Q}$
   (ix) $9 - x^3$ over $\mathbb{Z}_{11}$
   (x) $14 - 7x + 10x^4$ over $\mathbb{Q}$.
5. Prove that any polynomial over $\mathbb{R}$ of degree ≥ 3 is reducible over $\mathbb{R}$.
6. For any prime number p, prove that $p - x^n$ is irreducible over $\mathbb{Q}$ for any positive integer n.

7. Prove that $1 + x^4$ is irreducible over $\mathbb{Q}$ and reducible over $\mathbb{Z}_p$ for any prime number p.
8. Determine all irreducible polynomials of degree 2 in $\mathbb{Z}_2[x]$.
9. Give an example of a polynomial which is irreducible over $\mathbb{Z}$ but not irreducible over $\mathbb{Z}_2$.
10. For any prime p, prove that there are exactly $(p(p - 1))/2$ irreducible monic polynomials of degree 2 in $\mathbb{Z}_p[x]$.
11. Let p be a prime number and

$$f(x) = 1 - x + x^2 - x^3 + \cdots + (-1)^{p-1}x^{p-1}.$$

Then prove that $f(x)$ is irreducible over $\mathbb{Z}$.

12. Let R be an UFD and F be its field of quotients. Prove Theorem 12.4.5 with R and F in place of $\mathbb{Z}$ and $\mathbb{Q}$, respectively.

12.5 EUCLIDEAN DOMAINS

Another important class of integral domains, about which we discuss in this section, is the class of Euclidean domains. These arose out of attempts to generalize the familiar Euclidean division algorithm for integers to elements of arbitrary rings. Let us begin our discussion with the following.

Definition 12.5.1. An integral domain R is said to be an *Euclidean domain* if there exists a function $g : R - \{0\} \to \mathbb{Z}^+$ satisfying the following conditions:

1. $g(ab) = g(a)g(b)$ for all a and $b \in R - \{0\}$.
2. For any a and $b \in R$ with $b \neq 0$, there exist elements q and $r \in R$ such that

$$a = qb + r \quad \text{and either } r = 0 \quad \text{or} \quad g(r) < g(b).$$

The function g is called the *guage function* (or *Euclidean valuation*).

In other words, an integral domain R is called an Euclidean domain if, to each nonzero element a of R, there is an associated positive integer $g(a)$, called the *guage of a*, satisfying the conditions (1) and (2) above. Condition (2) is called the *Euclidean division algorithm* and q and r in (2) are called *quotient* and *remainder*, respectively. We first mention the following simple examples of Euclidean domains.

Example 12.5.1

1. The ring $\mathbb{Z}$ of integers is an Euclidean domain, in which the gauge function is defined by

$$g(a) = |a|, \text{ the absolute value of } a,$$

for any $a \in \mathbb{Z} - \{0\}$. Clearly, $|ab| = |a||b|$. To prove the division algorithm, let a and $b \in \mathbb{Z}$ and $b \neq 0$. Without loss of generality, we can assume that $b > 0$. Let q be the integral part of the rational number a/b; that is, q is an integer such that

$$q \leq \frac{a}{b} \quad \text{and} \quad q+1 > \frac{a}{b}.$$

Then, $bq \leq a$ and $bq + b > a$. Now, put $r = a - bq$. Since a, b and q are all integers, we get that $r \in \mathbb{Z}$ and

$$a = qb + r, \quad \text{where } r = 0 \text{ or } |r| = r < b = |b|.$$

Thus, $\mathbb{Z}$ is an Euclidean domain.

2. Every field F is an Euclidean domain; for, define $g : F - \{0\} \to \mathbb{Z}^+$ by $g(a) = 1$ for all $a \in F - \{0\}$. Then, $g(ab) = 1 = g(a)g(b)$ for all a and $b \in F - \{0\}$ and

$$a = (ab^{-1})b + 0 = qb + r, \quad \text{where } q = ab^{-1} \quad \text{and} \quad r = 0.$$

In the following, we exhibit certain elementary properties of the gauge function of an Euclidean domain.

Theorem 12.5.1. Let R be an Euclidean domain with gauge function g. Then, the following holds

1. $g(1) = 1$.
2. For any $0 \neq a \in R$, a is a unit in R if and only if $g(a) = 1$.
3. If a and b are associates in $R - \{0\}$, then $g(a) = g(b)$.

Proof:

1. follows from the facts that $g(1)$ is a positive integer and $g(1) = g(1 \cdot 1) = g(1)g(1)$.
2. Let $0 \neq a \in R$. If a is a unit in R, then there exists $b \in R$ such that $ab = 1$. Then, $b \neq 0$ and

$$1 = g(1) = g(ab) = g(a)g(b).$$

Since $g(a)$ and $g(b)$ are positive integers, we get that $g(a) = 1 = g(b)$. Conversely, suppose that $g(a) = 1$. By the Euclidean division algorithm, there exist q and $r \in R$ such that

$$1 = qa + r, \quad \text{where } r = 0 \quad \text{or} \quad g(r) < g(a) = 1.$$

If $r \neq 0$, then $g(r) \in \mathbb{Z}^+$ which is not true, since $g(r) < 1$. Therefore, necessarily $r = 0$ and $1 = qa$. Thus, a is a unit in R.

3. Suppose a and b are associates. Then, $au = b$ for some unit u in R. Now,

$$g(b) = g(au) = g(a)g(u) = g(a) \quad (\text{since } g(u) = 1). \qquad ◀$$

Theorem 12.5.2. Every Euclidean domain is a PID.

Proof: Let R be an Euclidean domain with gauge function g. Let I be an ideal of R. If $I = \{0\}$, then $I = \langle 0 \rangle$ and hence I is principal. Suppose that $I \neq \{0\}$.

Consider the set

$$A = \{g(a) : 0 \neq a \in I\}.$$

Then, I is a nonempty set of positive integers and, by the well-ordering principal, A has a least member, say $g(a)$ with $0 \neq a \in I$. Now we prove that $A = \langle a \rangle$. Since $a \in I$, we have $\langle a \rangle \subseteq I$. On the other hand, let $x \in I$. Then, by Euclidean division algorithm, there exist q and $r \in R$ such that

$$x = qa + r, \quad \text{where } r = 0 \quad \text{or} \quad g(r) < g(a).$$

Now, $r = x - qa \in I$ (since x and $a \in I$). Since $g(a)$ is least in A, it follows that $r = 0$ and hence $x = qa \in \langle a \rangle$. Thus, $I \subseteq \langle a \rangle$ and hence $I = \langle a \rangle$. Thus, R is a PID. ◀

Corollary 12.5.1. Every Euclidean domain is an UFD.

Proof: This follows from the fact that every PID is an UFD and from the above theorem. ◀

Theorem 12.5.3. The ring $F[x]$ of polynomials over a field F is an Euclidean domain and hence a PID and an UFD.

Proof: Let F be a field. For any $0 \neq f(x) \in F[x]$, define

$$g(f(x)) = 2^{\deg(f(x))}.$$

Since $\deg(f(x)) \geq 0$, $g : F[x] - \{0\} \to \mathbb{Z}^+$ is a function.
Also, for any nonzero $f(x)$ and $h(x)$ in $F[x]$,

$$\begin{aligned} g(f(x)h(x)) &= 2^{\deg(f(x)h(x))} \\ &= 2^{\deg(f(x)+\deg(h(x))} \\ &= 2^{\deg(f(x))} \cdot 2^{\deg(h(x))} \\ &= g(f(x))g(h(x)). \end{aligned}$$

Also, for any $f(x)$ and $h(x) \in F[x]$ with $h(x) \neq 0$, by the division algorithm for polynomials, there exist $q(x)$ and $r(x) \in F[x]$ such that

$$f(x) = q(x)h(x) + r(x),$$

where $r(x) = 0$ or $\deg(r(x)) < \deg(h(x))$ and hence $r(x) = 0$ or $g(r(x)) < g(h(x))$. Thus, $F[x]$ is an Euclidean domain. ◀

The following is a generalization of the well-known algorithm to find the g.c.d. of any two positive integers.

Theorem 12.5.4. Let R be an Euclidean domain with gauge function g and a and $b \in R - \{0\}$. Let $\{q_n\}$ and $\{r_n\}$ be sequences of elements in R satisfying the following:

$$\begin{aligned} b &= q_1 a + r_1, \text{ where } r_1 = 0 \text{ or } g(r_1) < g(a) \\ a &= q_2 r_1 + r_2, \text{ where } r_2 = 0 \text{ or } g(r_2) < g(r_1) \\ r_1 &= q_3 r_2 + r_3, \text{ where } r_3 = 0 \text{ or } g(r_3) < g(r_2) \\ r_2 &= q_4 r_3 + r_4, \text{ where } r_4 = 0 \text{ or } g(r_4) < g(r_3) \\ &\vdots \quad \vdots \quad \vdots \quad \vdots \end{aligned}$$

Then, there exists n such that $r_{n+1} = 0$ and $r_{n-1} = q_{n+1} r_n$. If n is the least such integer, then

$$r_n = \text{g.c.d. } \{a, b\}.$$

Proof: First observe that, since $g(a) > g(r_1) > g(r_2) > \ldots$, the above process of getting q_n's and r_n's should terminate (at most after $g(a)$ number of steps) and hence $r_{n+1} = 0$ for some n. Writing from bottom to top of the above equations, we have

$$\begin{aligned} r_{n-1} &= q_{n+1} r_n \\ r_{n-2} &= q_n r_{n-1} + r_n, \ 0 \neq r_n = r_{n-2} - q_n r_{n-1} \\ r_{n-3} &= q_{n-1} r_{n-2} + r_{n-1}, \ 0 \neq r_{n-1} = r_{n-3} - q_{n-1} r_{n-2} \\ &\vdots \quad \vdots \quad \vdots \quad \vdots \\ r_1 &= q_3 r_2 + r_3, \ 0 \neq r_3 = r_1 - q_3 r_2 \\ a &= q_2 r_1 + r_2, \ 0 \neq r_2 = a - q_2 r_1 \\ b &= q_1 a + r_1, \ 0 \neq r_1 = b - q_1 a. \end{aligned}$$

Tracing from top to bottom of the left hand side equations, we get that $r_n | r_{n-1}, r_n | r_{n-2}, \ldots, r_n | a$ and $r_n | b$. Therefore, r_n is a common divisor of a and b.

Also, if d is any common divisor of a and b, then tracing from bottom to top of the equations on the right hand side above, we get that

$$d|a,\ d|b,\ d|r_1,\ d|r_2,\ d|r_3,\ \ldots,\ d|r_{n-1} \quad \text{and} \quad d|r_n.$$

Thus, r_n is the g.c.d. of a and b. ◀

Worked Exercise 12.5.1. Let R be an Euclidean domain or a PID and a and $b \in R - \{0\}$. Then prove that the g.c.d. $\{a, b\}$ exists and is of the form $ax + by$ for some x and $y \in R$.

Answer: Since every Euclidean domain is a PID, we prove this result in a PID. Consider the ideal $aR + bR$, Then, there exists $d \in R$ such that

$$aR + bR = \langle d \rangle = dR.$$

Then, it can be easily verified that d is the g.c.d.$\{a, b\}$ and $d = ax + by$ for some x and $y \in R$.

Worked Exercise 12.5.2. Let R be an Euclidean domain with gauge function g and $a \in R - \{0\}$. Then prove that a is a unit in R if and only if $g(ab) = g(b)$ for some $0 \neq b \in R$.

Answer: If a is a unit in R and $0 \neq b \in R$, then

$$g(ab) = g(a)g(b) = 1g(b) = g(b).$$

Also, if $g(ab) = g(b)$ for some $0 \neq b \in R$, then

$$g(a)g(b) = g(ab) = g(b).$$

and hence $g(a) = 1$ (since $g(b) > 0$), so that a is a unit in R.

Worked Exercise 12.5.3. Let $\mathbb{Z}[\sqrt{2}] = \{a + b\sqrt{2} : a \text{ and } b \in \mathbb{Z}\}$. Then prove that $\mathbb{Z}[\sqrt{2}]$ is an Euclidean domain under the usual addition and multiplication of real numbers.

Answer: It can be easily proved that $\mathbb{Z}[\sqrt{2}]$ is a nontrivial subring of the integral domain $\mathbb{R}$ and hence $\mathbb{Z}[\sqrt{2}]$ is an integral domain. Define

$$g : \mathbb{Z}[\sqrt{2}] - \{0\} \to \mathbb{Z}^+ \text{ by } g(a + b\sqrt{2}) = |\, a^2 - 2b^2 \,|.$$

Since $a+b\sqrt{2} \neq 0$ implies $a \neq 0$ or $b \neq 0$ and hence $|a^2 - 2b^2| > 0$ (for, $a^2 = 2b^2$ has no integral solutions). Thus,

$$g(a+b\sqrt{2}) > 0 \quad \text{for all } a+b\sqrt{2} \in \mathbb{Z}[\sqrt{2}].$$

Also, for any $x = a + b\sqrt{2}$ and $y = c + d\sqrt{2}$ in $\mathbb{Z}[\sqrt{2}] - \{0\}$,

$$\begin{aligned} g(xy) &= g((a+b\sqrt{2})\,(c+d\sqrt{2})) \\ &= g(ac+2bd+(ad+bc)\sqrt{2}) \\ &= |(ac+2bd)^2 - 2(ad+bc)^2| \\ &= |a^2c^2 + 4b^2d^2 + 4acbd - 2a^2d^2 - 2b^2c^2 - 4adbc| \\ &= |(a^2-2b^2)||(c^2-2d^2)| \\ &= g(x)g(y). \end{aligned}$$

Next, let $x = a + b\sqrt{2}$ and $0 \neq y = c + b\sqrt{d} \in \mathbb{Z}[\sqrt{2}]$.
Then, $c \neq 0$ or $d \neq 0$. Now,

$$\frac{x}{y} = \frac{(a+b\sqrt{2})\,(c-d\sqrt{2})}{c^2-2d^2} = \alpha + \beta\sqrt{2}, \text{ say}$$

where α and β are rational numbers. Choose integers m and n such that

$$|m-\alpha| \leq \frac{1}{2} \quad \text{and} \quad |n-\beta| \leq \frac{1}{2}.$$

Then, $x = y(\alpha+\beta\sqrt{2} = (m+n\sqrt{2})y + [(\alpha-m)+(\beta-n)\sqrt{2}]y$.
Now, $[(\alpha-m)+(\beta-n)\sqrt{2}]\,y = x-(m+n\sqrt{2})y \in \mathbb{Z}[\sqrt{2}]$
Put $r = [(\alpha-m)+(\beta-n)\sqrt{2}\,]\,y$. Then, $x = yq + r$,

where $q = m+n\sqrt{2}$ and $r = 0$ or

$$\begin{aligned} g(r) &= |(\alpha-m)^2 - 2(\beta-n)^2 \,||\, c^2 - 2d^2| \\ &\leq \left|\frac{1}{2}+\frac{2}{4}\right| |c^2-2d^2| < |c^2-2d^2| = g(y). \end{aligned}$$

Thus, $\mathbb{Z}[\sqrt{2}]$ is an Euclidean domain.

Worked Exercise 12.5.4. Determine the g.c.d. of $1 + x + x^2$ and $1 + 2x + 3x^3 + x^5$ in $\mathbb{R}[x]$.

Answer: We know that $\mathbb{R}[x]$ is an Euclidean domain. Let us follow the algorithm given in Theorem 12.5.4.

Let $f(x) = 1 + x + x^2$ and $g(x) = 1 + 2x + 3x^3 + x^5$. Then,

$$
\begin{aligned}
g(x) &= (-2 + 3x - x^2 + x^3)(1 + x + x^2) + x + 3 \\
1 + x + x^2 &= (x - 2)(x + 3) + 7 \\
x + 3 &= \left(\frac{1}{7}x\right)7 + 3 \\
7 &= 2 \cdot 3 + 1 \\
3 &= 3 \cdot 1 + 0.
\end{aligned}
$$

Therefore, 1 is the g.c.d. of $1 + x + x^2$ and $1 + 2x + 3x^3 + x^5$. This is better understood by the following method.

$$
\begin{array}{r|l|l}
x^2+x+1 & x^5+3x^3+2x+1 & x^3-x^2+3x-2 \\
 & \underline{x^5+3x^3+x+3} &
\end{array}
$$

$$
\begin{array}{r|l|l}
x+3 & x^2+x+1 & x-2 \\
 & \underline{x^2+x-6} &
\end{array}
$$

$$
\begin{array}{r|l|l}
7 & x+3 & \frac{1}{7}x \\
 & \underline{x} &
\end{array}
$$

$$
\begin{array}{r|l|l}
3 & 7 & 2 \\
 & \underline{6} &
\end{array}
$$

$$
\begin{array}{r|l|l}
1 & 3 & 3 \\
 & \underline{3} & \\
 & & 0.
\end{array}
$$

Before we close this section, let us summarize various classes of integral domains introduced in this section and discuss their inter relationships. Let us fix notation for these classes as given below.

ID = The class of integral domains
FD = The class of factorization domains
UFD = The class of unique factorization domains
PID = The class of principle ideal domains
ED = The class of Euclidean domains
F = The class of fields.

Theorem 12.5.5. We have the following inclusions among the above classes

$$\mathrm{F} \subset \mathrm{ED} \subset \mathrm{PID} \subset \mathrm{UFD} \subset \mathrm{FD} \subset \mathrm{ID}.$$

and these are strict inclusions.

Proof:

1. In Example 12.5.1 (2), we have proved that every field is an Euclidean domain and therefore $\mathrm{F} \subseteq \mathrm{ED}$.
   $\mathbb{Z}$ is an Euclidean domain, but not a field and hence $\mathrm{F} \subsetneq \mathrm{ED}$.
2. In Theorem 12.5.2, we have proved that every Euclidean domain is a PID and therefore $\mathrm{ED} \subseteq \mathrm{PID}$. Consider the ring R given by

$$R = \left\{ a + \frac{b}{2}(1 + i\sqrt{19}) : a \text{ and } b \in \mathbb{Z} \right\}.$$

   Then, R is a PID, but not an Euclidean domain (the proof of this is little bit involved and hence we skip the proof). Therefore, $\mathrm{ED} \subsetneq \mathrm{PID}$.
3. In Corollary 12.3.4, we have proved that every PID is an UFD and hence $\mathrm{PID} \subseteq \mathrm{UFD}$. The ring $\mathbb{Z}[x]$ is an UFD but not a PID (see Theorem 12.4.3). Therefore, $\mathrm{PID} \subsetneq \mathrm{UFD}$.

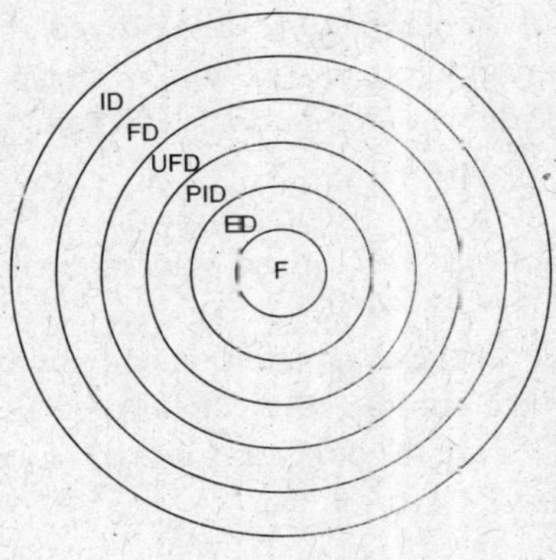

4. Clearly, every unique factorization is a FD. We have seen in 12.3…. that $\mathbb{Z}\sqrt{-5}$ is a FD which is not an UFD. Therefore, $\mathrm{UFD} \subsetneq \mathrm{FD}$.
5. Clearly, every FD is an integral domain. There are integral domains which are not FDs. That is, $\mathrm{FD} \subsetneq \mathrm{ID}$. ◀

EXERCISE 12(E)

1. Let R be an Euclidean domain with gauge function g. For any a and $b \in R - \{0\}$, prove that a and b are associates if and only if a divides b and $\phi(a) = \phi(b)$.

2. Prove that $\mathbb{Z}[x]$ is not an Euclidean domain.

3. Prove that $\mathbb{Q}[x]$, $\mathbb{R}[x]$ and $\mathbb{C}[x]$ are all Euclidean domains.

4. If R is an integral domain which is not a field, prove that $R[x]$ is not an Euclidean domain.

5. Prove that any nonzero prime ideal in an Euclidean domain is maximal.

6. In any integral domain with gauge function g, prove that $g(a) = g(-a)$ for any nonzero element a.

7. Prove that, for any $n \in \mathbb{Z}^+$, $g : \mathbb{Z} - \{0\} \to \mathbb{Z}^+$ defined by $g(a) = |a|^n$ is a gauge function on $\mathbb{Z}$.

8. Prove that $\mathbb{Z}[\sqrt{3}] = \{a + b\sqrt{3} : a \text{ and } b \in \mathbb{Z}\}$ is a Euclidean domain.

9. Let R be an Euclidean domain and $a, b \in R$ with $b \neq 0$. Let $q, r \in R$ such that $a = bq + r$ with $r \neq 0$. Then prove that g.c.d. $\{a, b\}$ = g.c.d. $\{b, r\}$.

10. Let $x = -102 + 10\sqrt{3}$ and $y = 1 + 7\sqrt{3}$. Find q and r in $\mathbb{Z}[\sqrt{3}]$ such that $x = yq + r$ where either $r = 0$ or $r = a + b\sqrt{3}$ with $|a^2 - 3b^2| < 146$.

12.6 SOME APPLICATIONS TO NUMBER THEORY

In this section, we apply the general results proved about Euclidean rings to the ring of Gaussian integers and obtain a relatively difficult theorem about prime numbers due to the famous mathematician, Fermat. First recall that

$$\mathbb{Z}[i] = \{a + bi : a \text{ and } b \text{ are integers}\}.$$

is an integral domain under the usual addition and multiplication of complex numbers. In fact $\mathbb{Z}[i]$ is a subring of the field of complex numbers. Also, recall that $1, -1, i$ and $-i$ are the only units in $\mathbb{Z}[i]$ and that $\mathbb{Z}[i]$ is called the ring of Gaussian integers.

Theorem 12.6.1. The ring of Gaussian integers $\mathbb{Z}[i]$ is an Euclidean domain.

Proof: Define $g : \mathbb{Z}[i] - \{0\} \to \mathbb{Z}^+$ by

$$g(a + bi) = a^2 + b^2.$$

Note that $a + bi \neq 0$ in $\mathbb{Z}[i]$ implies that $a \neq 0$ or $b \neq 0$ and hence $a^2 + b^2$ is a positive integer.

For any $x = a + bi$ and $y = c + di \in \mathbb{Z}[i]$

$$\begin{aligned} g(xy) &= g((a + bi)(c + di)) \\ &= g((ac - bd) + (ad + bc)i) \\ &= (ac - bd)^2 + (ad + bc)^2 \\ &= a^2c^2 + b^2d^2 + a^2d^2 + b^2c^2 \\ &= (a^2 + b^2)(c^2 + d^2) \\ &= g(x)g(y). \end{aligned}$$

To prove the Euclidean division algorithm, let $x = c + bi$ and $0 \neq y = c + di \in \mathbb{Z}[i]$. Then, $c^2 + d^2 > 0$. Consider the complex number

$$\frac{x}{y} = \frac{(a+bi)(c-di)}{(c+di)(c-di)} = \frac{(ac+bd)+(bc-ad)i}{c^2+d^2} = \alpha + \beta i$$

for some rational numbers α and β. Now, choose integers m and n such that

$$|\alpha - m| \leq \frac{1}{2} \quad \text{and} \quad |\beta - n| \leq \frac{1}{2}$$

$$\text{we have} \quad \begin{aligned} x &= (\alpha + \beta i)y \\ &= (m + ni)y + ((\alpha - m) + (\beta - n)i)y. \end{aligned}$$

Put $q = m + ni$ and $r = [(\alpha - m) + (\beta - n)\, i]y$.
Then, clearly $q \in \mathbb{Z}[i]$ and

$$r = x - (m + ni)y \in \mathbb{Z}[i].$$

Now, $x = qy + r$ and either $r = 0$

$$\begin{aligned} \text{or} \quad g(r) &= ((\alpha - m)^2 + (\beta - n)^2)g(y) \\ &\leq \left(\frac{1}{4} + \frac{1}{4}\right) g(y) < g(y). \end{aligned}$$

Thus, $\mathbb{Z}[i]$ is an Euclidean domain. ◀

Corollary 12.6.1. $\mathbb{Z}[i]$ is a PID and an UFD.

Now, we are free to apply the properties of PIDs and UFDs to $\mathbb{Z}[i]$ to prove the following purely number theoretic results.

Theorem 12.6.2. Let $p \in \mathbb{Z}^+$ be prime and $n \in \mathbb{Z}$ such that p does not divide n. Suppose that we can find integers x and y such that $np = x^2 + y^2$. Then, p can be expressed as a sum of two squares of integers; that is, $p = a^2 + b^2$ for some a and $b \in \mathbb{Z}$.

Proof: First observe that any integer m can be treated as a Gaussian integer $m + 0i$ and that $\mathbb{Z}$ is a subring of $\mathbb{Z}[i]$. Next, we prove that p cannot be a prime element in $\mathbb{Z}[i]$. Suppose, if possible, that p is prime in $\mathbb{Z}[i]$. Since

$$pn = x^2 + y^2 = (x + iy)(x - iy),$$

we get that p divides either $x + iy$ or $x - iy$. If p divides $x + iy$ then

$$p(s + it) = x + iy$$

for some s and $t \in \mathbb{Z}$ and hence $ps = x$ and $pt = y$, so that $p(s - it) = x - iy$. Therefore,

$$pn = (x + iy)(x - iy) = p(s + it)p(s - it) = p^2(s^2 + t^2)$$

from which it follows that p^2 divides pn and hence p divides n, which is a contradiction to our hypothesis. Therefore, p is not a prime in $\mathbb{Z}[i]$. Since $\mathbb{Z}[i]$ is an UFD, an element in $\mathbb{Z}[i]$ is prime if and only if it is irreducible. Therefore, p is not an irreducible element in $\mathbb{Z}[i]$. Therefore, there exists two nonunits $a + bi$ and $c + di$ in $\mathbb{Z}[i]$ such that

$$p = (a + bi)(c + di)$$
$$\text{and hence } p^2 = g(p) = g(a + bi)g(c + di),$$

where g is the gauge function on $\mathbb{Z}[i]$ defined in Theorem 12.6.1. Therefore,

$$p^2 = (a^2 + b^2)(c^2 + d^2)$$

and hence $a^2 + b^2$ is a divisor of p^2. Therefore, $a^2 + b^2 = 1$ or p or p^2. But, since $a + bi$ and $c + di$ are nonunits in $\mathbb{Z}[i]$, $a^2 + b^2 > 1$ and $c^2 + d^2 > 1$. Again, since $c^2 + d^2 > 1$, if follows from $p^2 = (a^2 + b^2)(c^2 + d^2)$ that $a^2 + b^2 = p$. ◀

The following is a famous theorem in elementary number theory and we state this without proof. First recall that, for integers a, b and n, we write $a \equiv b \pmod{n}$ when n divides $a - b$.

Theorem 12.6.3 (Wilson's Theorem). For any prime number p in $\mathbb{Z}^+$,

$$(p-1)! \equiv -1 \pmod{p}.$$

Note that any positive integer m must be of the form $4n$ or $4n+1$ or $4n+2$ or $4n+3$, for some integer $n \geq 0$. In particular, when m is a prime > 2, then $m = 4n+1$ or $4n+3$ for some $n \geq 0$. In the following two results, we prove that primes of the form $4n+1$ have certain special properties; in particular, we prove that each such prime is a sum of two perfect squares.

Theorem 12.6.4. Let p be a prime number of the form $4n+1$. Then, there exists an integer x such that

$$x^2 \equiv -1 \pmod{p}.$$

Proof: First note that, since $p = 4n+1$, $(p-1)/2$ is an even integer. Now, put $x = ((p-1)/2)!$. Then,

$$x = 1\cdot 2\cdot 3 \cdots \left(\frac{p-1}{2}\right) = (-1)(-2)(-3) \cdots \left(-\frac{p-1}{2}\right)$$

since, $(p-1)/2$ is even. Also, since $p - a \equiv -a \pmod{p}$ for any $a \in \mathbb{Z}$, we have

$$\begin{aligned}
x^2 &= \left[1\cdot 2\cdot 3 \cdots \left(\frac{p-1}{2}\right)\right]\left[(-1)(-2)(-3) \cdots \left(-\frac{p-1}{2}\right)\right]\\
&\equiv \left[1\cdot 2\cdot 3 \cdots \left(\frac{p-1}{2}\right)\right]\left[(p-1)(p-2)(p-3) \cdots \left(p-\frac{p-1}{2}\right)\right] \pmod{p}\\
&= 1\cdot 2\cdot 3 \cdots \frac{p-1}{2}\cdot\frac{p+1}{2}\cdot\frac{p+3}{2} \cdots (p-2)(p-1)\\
&= (p-1)! \equiv -1 \pmod{p}.
\end{aligned}$$

Thus, $x^2 \equiv -1 \pmod{p}$. ◀

Theorem 12.6.5 (Fermat's Theorem). Any prime number of the form $4n+1$ can be expressed as a sum of two perfect squares.

Proof: Let p be a prime number and $p = 4n+1$. By the above theorem, there exists an integer x such that $x^2 \equiv -1 \pmod{p}$. x can be chosen in such a way that $0 \leq x \leq p - 1$; for, consider the remainder of x on division by p. That is, let

$$x = qp + r, \quad \text{where } 0 \leq r < p.$$

Then, $r^2 = (x - qp)^2 = x^2 + p(q^2p - 2xq) \equiv -1 \pmod{p}$. Further, we can assume that $r \leq (p/2)$; for, if $r > (p/2)$, then $s = r - p$ satisfies the property that $s^2 \equiv -1 \pmod{p}$ and $0 < s \leq (p/2)$. Thus, we can find an integer s such that

$$s^2 \equiv -1 \pmod{p} \quad \text{and} \quad 0 \leq s \leq (p/2).$$

Now, p divides $s^2 + 1$ and hence $pn = s^2 + 1^2$.
Also, since $pn = s^2 + 1 \leq (p^2/4) + 1 < p^2$, we get that p does not divide n. Thus, by Theorem 12.6.2, it follows that $p = a^2 + b^2$ for some integers a and b. ◀

Worked Exercise 12.6.1. Prove that any integer of the form $4n + 3$ cannot be expressed as the sum of two perfect squares.

Answer: Suppose, if possible, that

$$4n + 3 = a^2 + b^2$$

for some $n, a, b \in \mathbb{Z}$. Since $4n + 3$ is odd, one of a^2 and b^2 must be odd and the other must be even. Therefore, one of a and b must be odd and the other must be even. Without loss of generality, we can suppose that $a = 2r$ and $b = 2s + 1$ for some r and $s \in \mathbb{Z}$. Now, we have

$$4n + 3 = (2r)^2 + (2s + 1)^2 = 4r^2 + 4s^2 + 4s + 1$$

and therefore, $2 = 4(r^2 + s^2 - s - n)$ so that $1 = 2(r^2 + s^2 - s - n)$, which is absurd. Thus, $4n + 3$ cannot be expressed as a sum of two squares of integers.

Worked Exercise 12.6.2. Let I be a nonzero ideal of the ring $\mathbb{Z}[i]$ of Gaussian integers. Then prove that the quotient ring $\mathbb{Z}[i]/I$ is finite.

Proof: By Corollary 12.6.1, $\mathbb{Z}[i]$ is a PID and hence $I = \langle x \rangle$ for some $x \in \mathbb{Z}[i]$. Let $x = a + bi$ with a and $b \in \mathbb{Z}$. Since I is nonzero, we have $x \neq 0$ and hence $a \neq 0$ or $b \neq 0$, so that $g(x) = a^2 + b^2 > 0$, where g is the gauge function on $\mathbb{Z}[i]$. Any element of $\mathbb{Z}[i]/I$ is of the form $y + I, y \in \mathbb{Z}[i]$. Now, if $y \in \mathbb{Z}[i]$, then by Euclidean division algorithm, there exist elements q and $r \in \mathbb{Z}[i]$ such that

$$y = qx + r, \quad \text{where } r = 0 \quad \text{or} \quad g(r) < g(x)$$

and hence $y - r = qx \in \langle x \rangle = I$, so that

$$y + I = r + I, r = 0 \quad \text{or} \quad g(r) < g(x).$$

Therefore, $\frac{\mathbb{Z}[i]}{I} = \{r + I : r = 0 \text{ or } g(r) < g(x)\}$.
Since there can be only finitely many pairs (c, d) of integers such that $c^2 + d^2 < g(x) = a^2 + b^2$, it follows that $\mathbb{Z}[i]/I$ is a finite set. ◀

EXERCISE 12(F)

1. In each of the following two elements x and y of $\mathbb{Z}[i]$ are given. Find q and r in $\mathbb{Z}[i]$ such that $x = qy + r$; with $r = 0$ or $g(r) < g(y)$.
   (i) $x = 3 + 2i$ and $y = 2 - 3i$
   (ii) $x = 5$ and $y = 2i$
   (iii) $x = 1 + i$ and $y = 2 + i$
   (iv) $x = 2 + 3i$ and $y = 1 - i$
   (v) $x = 4 - 5i$ and $y = 5 + 4i$.
2. Prove the neither 2 nor 17 is a prime element in $\mathbb{Z}[i]$.
3. Prove that $\mathbb{Z}\sqrt{2}$ and $\mathbb{Z}\sqrt{-2}$ are Euclidean domains.
4. Prove that any nonzero prime ideal in $\mathbb{Z}[i]$ is maximal.
5. Prove that $\mathbb{Z}\sqrt{-6}$ is not a Euclidean domain.
6. Prove that $\mathbb{Z}\sqrt{-5}$ is not a Euclidean domain.
7. Determine all prime elements in $\mathbb{Z}[i]$.
8. Prove that $2 - 7i$ and $2 + 11i$ are relatively prime in $\mathbb{Z}[i]$.
9. Find the g.c.d. of $-5 + 10i$ and $3 + i$ in $\mathbb{Z}[i]$.
10. Find x and $y \in \mathbb{Z}[i]$ such that g.c.d. $\{-5 + 10i, 3 + i\} = (-5 + 10i)x + (3 + i)y$.

13

Modules and Vector Spaces

Another important algebraic structure is that of a module over a ring, in particular, a vector space over a division ring or a field. Till now, we have come across groups and rings, where one or two binary operations are involved. Modules are concerned with one binary operation and several binary operations, one corresponding to each element in the ring. Consider an abelian group $(M, +)$ and let $R = \text{End}(M, +)$, the set of all endomorphisms of $(M, +)$. It is well known that R is a ring under point-wise addition and composition of mappings (as multiplication). For each $f \in R$ and $x \in M$, let fx be the image of x under f in M. Then, one can easily see that the following conditions are satisfied for any x and $y \in M$ and f and $g \in R$:

1. $f(x + y) = fx + fy$
2. $(f + g)x = fx + gx$
3. $(fg)(x) = f(gx)$
4. $1x = x$, where 1 is the unity in R.

This situation is abstracted in this chapter to introduce the concept of a module over a ring and prove certain elementary properties of modules over a given ring.

13.1 MODULES AND SUBMODULES

In this section, we define the notions of modules and submodules and discuss certain examples and properties of these. Let us begin with the following definition.

Definition 13.1.1. Let $(R, +, \cdot)$ be a ring and $(M, +)$ be an abelian group. Then, M is called a *left R-module* (or a *left module over R*) if there exists a mapping $(a, x) \mapsto ax$ of $R \times M$ into M satisfying the following for any x and $y \in M$ and a and $b \in R$:

1. $a(x + y) = ax + ay$
2. $(a + b)x = ax + bx$
3. $a(bx) = (ab)x$
4. $1x = x$, if 1 is the unity in R.

Example 13.1.1

1. Let $(M, +)$ be any abelian group and $R = \text{End}(M)$, the ring of all endomorphisms of $(M, +)$, then M is a left R-module, where the map $(f, x) \mapsto fx$ of $R \times M$ into M is defined simply by $fx = f(x)$, the image of x under f.
2. Let $(M, +)$ be any abelian group and consider the ring $\mathbb{Z}$ of integers. For any $n \in \mathbb{Z}$ and $x \in M$, define

$$nx = \begin{cases} 0 & \text{if } n = 0 \\ (n-1)\,x + x & \text{if } n > 0. \\ (-n)(-x) & \text{if } n < 0 \end{cases}$$

   That is, $nx = 0$ if $n = 0$ and,

$$nx = x + x + \cdots + x \ (n \text{ times}) \text{ if } n > 0$$
$$\text{and} \quad nx = (-x) + (-x) + \cdots + (-x) \ (-n \text{ times}) \text{ if } n < 0.$$

   Then, under this map M is a left $\mathbb{Z}$-module.
3. Any ring R can itself be treated as a left R-module by defining ax to be the product of a and x as elements of R, for any $a \in R$ and $x \in R$.
4. Let m and n be any positive integers and R be any ring. Let $M_{m\times n}(R)$ be the set of all $m \times n$ matrices over R. Then, clearly $M_{m\times n}(R)$ is an abelian group under the usual addition of matrices. Now, for any $a \in R$ and $A = (a_{ij}) \in M_{m\times n}(R)$, define

$$aA = (aa_{ij}).$$

   Then, under this map $(a, A) \mapsto aA$, $M_{m\times n}(R)$ is a left R-module.

5. Let n be any positive integer and R be a ring. Then, $M = R \times R \times \cdots \times R$ (n factors) is an abelian group under the coordinate-wise addition. Also, for any $a \in R$ and $x = (x_1, x_2, \ldots, x_n) \in M$, define
$$ax = (ax_1, ax_2, \ldots, ax_n).$$
Under this map, M is a left R-module. Actually, this is a special case of (4) above, since $R \times \cdots \times R$ is precisely the set of all $1 \times n$ matrices over R. However, this needs a special mention, in view of its importance.

When M is a left R-module, then we often call the map $(a, x) \mapsto ax$ as the *scalar multiplication* and ax is called the *scalar multiplication* or simply the *multiplication of x by a on the left*. One can define a right R-module similarly by considering the mappings $(x, a) \mapsto xa$ from $M \times R$ into M satisfying the properties similar to (1) through (4) of Definition 13.1.1. When R is a commutative ring and M is a left R-module, then M can be made into a right R-module by defining $xa = ax$ (since $(xa)b = b(xa) = b(ax) = (ba)x = (ab)x = x(ab)$, the multiplication in R being commutative). In this case, left R-modules are same as right R-modules. When R is commutative, we do not distinguish between left R-modules and right R-modules and simply call them R-modules. The following are certain elementary properties of R-modules.

Theorem 13.1.1. Let M be a left R-module. Then, the following holds for any $x \in M$ and $a \in R$:

1. $0x = 0$, where 0 on the left is the zero element in the ring R and 0 on the right is the zero element (identity element) in the group $(M, +)$.
2. $a0 = 0$, where 0 is the zero element in M
3. $(-a)x = -(ax) = a(-x)$.

Proof:

1. Since $0x + 0 = 0x = (0 + 0)x = 0x + 0x$, it follows that $0x = 0$.
2. Also, since $a0 + 0 = a0 = a(0 - 0) = a0 + a0$, we get that $a0 = 0$.
3. We have $0 = 0x = (a - a)x = ax - (-a)x$ and hence $(-a)x = -(ax)$. Also,
$$0 = a0 = a(x + (-x)) = ax + a(-x)$$
and therefore $a(-x) = -(ax)$.

Note that we are using the same symbol 0 to denote the additive identity (zero element) in the ring as well as the identity element in the group M. This need not create any ambiguity and we should take it as per the context. Further, throughout our discussions in this chapter, all modules are assumed to be left R-modules, unless otherwise stated. Further, a module over R will be denoted by $(M, +, R)$ or simply by M when there is no ambiguity about the group operation $+$ on M and about the ring R over which M is a module.

Definition 13.1.2. Let $R = (R, +, \cdot)$ be a ring and $M = (M, +, R)$ be an R-module. A nonempty subset N of M is called an *R-submodule* (or simply, a *submodule*) of M if N is a subgroup of $(M, +)$ and $rx \in N$ for all $r \in R$ and $x \in N$.

Before going for certain examples of submodules of modules, let us mention the following whose proof is an easy exercise.

Theorem 13.1.2. Let R be a ring with unity and M be an R-module. Then, the following are equivalent to each other for any subset N of M:

1. N is an R-submodule of M.
2. $N \neq \emptyset$ and $ax - by \in N$ for all $a, b \in R$ and $x, y \in N$.
3. $0 \in N$ and $ax - by \in N$ for all $a, b \in R$ and $x, y \in N$.

Example 13.1.2

1. As in Example 13.1.1 (3), any ring R can be treated as an R-module and the R-submodules of R are precisely the left ideals of R.
2. Let R be any ring and $R[x]$ be the set of all polynomials over R. Then, $R[x]$ is an R-module under the usual addition and multiplication of polynomials (recall that elements of R can be treated as polynomials of degree zero). For any $n \geq 0$, let
$$R_n[x] = \{f(x) \in R[x] : f(x) = 0 \text{ or } \deg f(x) \leq n\}.$$
Then, $R_n[x]$ is an R-submodule of $R[x]$.
3. Let M be an R-module, $x \in M$ and
$$Rx = \{ax : a \in R\}.$$
Then, Rx is an R-submodule of M, since $0 = 0x \in Rx$ and, for any r, s, a and $b \in R$, we have
$$r(ax) - s(bx) = (ra - sb)x \in Rx.$$
4. Let M be an R-module, $x \in M$ and
$$\langle x\rangle = \{ax + nx : a \in R \text{ and } n \in \mathbb{Z}\}.$$
Then, $\langle x\rangle$ is an R-submodule of M, since
$$0 = 0x + 0x \in \langle x\rangle$$
and, for any $r, s, a, b \in R$ and $n, m \in \mathbb{Z}$, we have
$$\begin{aligned} r(ax + nx) - s(bx + mx) &= rax + rnx - sbx - smx \\ &= (ra + nr - sb - ms)x \\ &= (ra + nr - sb - ms)x + 0x \in \langle x\rangle. \end{aligned}$$

5. If R is a ring with unity and M is an R-module, then, for any $x \in M$,

$$\langle x \rangle = Rx.$$

Theorem 13.1.3. The intersection of any class of R-submodules of an R-module M is again an R-submodule of M.

Proof: Let $\{M_i\}_{i\in I}$ be a class of R-submodules of an R-module M and let $N = \bigcap_{i\in I} M_i$. Since any submodule should contain 0, it follows that $0 \in M_i$ for all $i \in I$ and hence $0 \in N$. Also,

$$\begin{aligned} a, b \in R \text{ and } x, y \in N &\Rightarrow a, b \in R \text{ and } x, y \in M_i \text{ for all } i \in I \\ &\Rightarrow ax - by \in M_i \text{ for all } i \in I \\ &\Rightarrow ax - by \in N. \end{aligned}$$

Thus, N is a submodule of M.

However, as usual, the union of submodules may not be a submodule. In fact, as in the case of subgroups of a group, for any submodules N and K of an R-module M, $N \cup K$ is a submodule of M if and only if either $N \subseteq K$ or $K \subseteq N$.

Definition 13.1.3. Let M be an R-module and X be a subset of M. Then, the intersection of all submodules of M containing X is called the *submodules generated by* X and is denoted by $\langle X \rangle$. Clearly, $\langle X \rangle$ is the smallest submodule of M containing X. If $X = \{x\}$, then

$$\langle X \rangle = \langle x \rangle = \{ax + nx : a \in R \text{ and } n \in \mathbb{Z}\}.$$

Theorem 13.1.4. Let $\{N_i\}_{i\in I}$ be a class of R-submodules of an R-module M. Then

$$\left\langle \bigcup_{i\in I} N_i \right\rangle = \left\{x_1 + x_2 + \cdots + x_n : x_j \in N_{i_j},\ \ i_1, \ldots, i_n \in I\right\}.$$

and this will be denoted by $\sum_{i\in I} N_i$. In particular, for any R-submodules $N_1, \ldots, N_m$ of M,

$$\langle N_1 \cup \ldots \cup N_m \rangle = \{x_1 + \cdots + x_m : x_j \in N_j \text{ for } 1 \le j \le m\}.$$

Proof: Let $N = \left\{x_1 + x_2 + \cdots + x_n : x_j \in N_{i_j},\ i_1, i_2, \ldots, i_n \in I\right\}$.

We shall prove that N is the smallest R-submodule of M containing $\bigcup_{i\in I} N_i$. Clearly, $N_i \subseteq N$ for all $i \in I$ and hence $\bigcup_{i\in I} N_i \subseteq N$. Also,

$$\begin{aligned} a, b \in R \text{ and } x, y \in N &\Rightarrow a, b \in R, x = x_1 + \cdots + x_n \text{ and } y = y_1 + \cdots + y_m, \\ &\quad \text{where } x_j \in N_{i_j}, y_k \in N_{i_k}, i_j \text{ and } i_k \in I \\ &\Rightarrow ax - by = ax_1 + \cdots + ax_n + b(-y_1) + \cdots + b(-y_m) \\ &\Rightarrow ax - by \in N. \end{aligned}$$

Therefore, N is an R-submodule of M containing $\underset{i \in I}{\cup} N_i$. On the other hand, let P be any submodule of M containing $\underset{i \in I}{\cup} N_i$. Then, $N_i \subseteq P$ for all $i \in I$ and hence $N \subseteq P$. Thus,

$$N = \left\langle \bigcup_{i \in I} N_i \right\rangle.$$

In particular, if $N_1, N_2, \ldots, N_m$ are submodules of M, then $0 \in N_j$ for all $1 \leq j \leq n$ and hence

$$<N_1 \cup \ldots \cup N_m> = \{x_1 + \cdots + x_m : x_j \in N_j \text{ for } 1 \leq j \leq m\}.$$

Corollary 13.1.1. Let N and K be R-submodules of an R-module M and

$$N + K = \{x + y : x \in N \text{ and } y \in K\}.$$

Then, $<N \cup K> = N + K$, which is the smallest submodule of M containing both N and K.

Definition 13.1.4. Let M be an R-module. If $X = \{x_1, x_2, \ldots, x_n\}$ is a finite subset of M, then the submodule of M generated by X is denoted by $<x_1, x_2, \ldots, x_n>$; that is,

$$<X> = <x_1, x_2, \ldots, x_n>.$$

It can be easily verified that

$$<x_1, x_2, \ldots, x_n> = <x_1> + <x_2> + \cdots + <x_n> = \sum_{i=1}^{n} <x_i>.$$

M is said to be *finitely generated* if $M = <x_1, x_2, \ldots, x_n>$ for some $x_1, x_2, \ldots, x_n \in M$. The elements $x_1, x_2, \ldots, x_n$ are said to be *generators* of M and the set $X = \{x_1, x_2, \ldots, x_n\}$ is said to generate M.

Definition 13.1.5. An R-module M is called *cyclic* if $M = \langle x \rangle$ for some $x \in M$. Recall from Example 13.1.2 (4) that M is a cyclic R-module if and only if there exists $x \in M$ such that

$$M = \{ax + nx : a \in R \text{ and } n \in \mathbb{Z}\}.$$

If R is a ring with unity, then an R-module M is cyclic if and only if

$$M = Rx = \{ax : a \in R\}$$

for some $x \in M$.

Note that the set of generators of a module need not be unique. For example, the $\mathbb{Z}$-module $\mathbb{Z}$ is cyclic, since $\mathbb{Z} = \langle 1 \rangle$ and also $\mathbb{Z} = \langle 2, 3 \rangle$. In fact, we have following exercise.

Worked Exercise 13.1.1. Consider the ring $\mathbb{Z}$ of integers. Then, $\mathbb{Z}$ is a module over itself. For each positive integer n, prove that the $\mathbb{Z}$-module $\mathbb{Z}$ has an n-element generating set X such that no proper subset of X generates $\mathbb{Z}$.

Answer: Let n be a given positive integer. If $n = 1$, then $\mathbb{Z} = \langle 1 \rangle = \langle -1 \rangle$ and hence we are through. Let $n > 1$. Consider any distinct primes $p_1, p_2, \ldots, p_n$. For each $1 \leq i \leq n$, let

$$x_i = p_1 p_2 \cdots p_{i-1} p_{i+1} \cdots p_n = \prod_{j \neq i} p_j$$

and $X = \{x_1, x_2, \ldots, x_n\}$. Then, $x_1, x_2, \ldots, x_n$ are relatively prime (since there is no prime dividing all x_i's).
Hence, there exist integers $y_1, y_2, \ldots, y_n$ such that

$$1 = y_1 x_1 + y_2 x_2 + \cdots + y_n x_n.$$

Therefore, any $a \in \mathbb{Z}$ can be expressed as

$$a = (ay_1)x_1 + (ay_2)x_2 + \cdots + (ay_n)x_n.$$

Thus, $\mathbb{Z} = \langle x_1, x_2, \ldots, x_n \rangle$. Further, for each $1 \leq i \leq n$, p_i divides all x_j, $j \neq i$ and, in fact

$$\text{g.c.d. } \{x_j : j \neq i\} = p_i$$

and hence $\sum_{j \neq i} \langle x_j \rangle = \langle p_i \rangle \neq \mathbb{Z}$. Therefore, no proper subset of X generates $\mathbb{Z}$.

Worked Exercise 13.1.2. Let F be any field and n be a positive integer. Let M be the set of all polynomials over F of degree less than n. Then, prove that M is a finitely generated F-module and exhibit two distinct sets of generators of M, each with n elements.

Answer: Under the usual addition of polynomials and scalar multiplication, M is clearly an F-module. Also, any element of M can be expressed as

$$f(x) = a_0 + a_1x + \cdots + a_{n-1}x^{n-1}$$

and therefore $\{1, x, x^2, \ldots, x^{n-1}\}$ is a generating set for M and has n elements. Also, for any $a \neq 0$ in F,

$$\{a, a + x, x^2, \ldots, x^{n-1}\}$$

is a generating set for M; since $x = (a + x) - a$ and $1 = a^{-1}a$, we get that

$$<a, a + x, x^2, \ldots, x^{n-1}> = <1, x, x^2, \ldots, x^{n-1}> = M.$$

EXERCISE 13(A)

1. State whether each of the following is true or false and justify your answer:
   (i) Any ring R is a finitely generated R-module.
   (ii) Any ring R with unity is a finitely generated R-module.
   (iii) If R is a finite ring, then any R-module is finitely generated.
   (iv) Any finite R-module is finitely generated.
   (v) Any module over a finite ring is finite.
   (vi) Any finite R-module has only a finite number of R-submodules.
   (vii) Any finitely generated R-module has only a finite number of R-submodules.
   (viii) If an R-module M is cyclic and $M = <x>$, then $M = <ax>$ for each $0 \neq a \in R$.
   (ix) If a is a unit in a ring R and M is an R-module, then $<x> = <ax>$ for each $x \in M$.
   (x) If M is a finite R-module, then R is finite.
2. Consider the real member field $\mathbb{R}$ and the $\mathbb{R}$-module $\mathbb{R}^4$. Which of the following are $\mathbb{R}$-submodules of $\mathbb{R}^4$?
   (i) $\{(a_1, a_2, a_3, a_4) \in \mathbb{R}^4 : a_2 = 0\}$
   (ii) $\{(a_1, a_2, a_3, a_4) \in \mathbb{R}^4 : a_1 = a_2 = a_3 = a_4\}$
   (iii) $\{(a_1, a_2, a_3, a_4) : a_1 + a_2 = 0\}$
   (iv) $\{(a_1, a_2, a_3, a_4) : a_1 = a_2\}$

(v) $\{(a_1, a_2, a_3, a_4) : a_1 \neq 0\}$

(vi) $\{(a_1, a_2, a_3, a_4) : a_1a_2 = 0\}$

(vii) $\{(a_1, a_2, a_3, a_4) : a_1 + a_2 \leq 0\}$

(viii) $\{(a_1, a_2, a_3, a_4) : a_1 + a_2 + a_3 + a_4 \geq 0\}$.

3. For any R-module M, prove that the set

$$\text{Ann}(M) = \{a \in R : ax = 0 \text{ for all } x \in M\}$$

is an ideal of R.

4. An R-module M is called *faithful* if $\text{Ann}(M) = \{0\}$. Give an example of a faithful R-module and of an R-module which is not faithful.

5. Let R be a ring and M be the set of all mappings of R into itself. Prove that M is an R-module under the operations defined by

$$(f + g)(a) = f(a) + g(a)$$
$$\text{and } (af)(b) = af(b)$$

for all f and $g \in M$ and a and $b \in R$.

6. For any R-submodules P and Q of an R-module M, prove that

$$P + Q = Q \Leftrightarrow P \subseteq Q \Leftrightarrow P = P \cap Q.$$

7. Let N, P and Q be R-submodules of an R-module M, such that $N \subseteq P$. Then, prove that

$$N + (Q \cap P) = (N + Q) \cap P$$

(This is called the *modular law.*)

8. Give an example of three R-submodules N, P and Q of an R-module M such that

$$N \cap (P + Q) \neq (N \cap P) + (N \cap Q).$$

9. Let M be an R-module. Prove that the set

$$RM = \{a_1x_1 + a_2x_2 + \cdots + a_ix_i : a_i \in R \text{ and } x_i \in M\}$$

is an R-submodule of M.

10. Let M be an R-module and N_0 be the intersection of all nonzero submodules of M. If $N_0 \neq \{0\}$, prove that N_0 is cyclic and $N_0 = \langle x \rangle$ for all $0 \neq x \in N_0$.

11. Let M be an R-module. An equivalence relation θ on M is called an *R-congruence on M* if,

$$(x, y) \in \theta \text{ and } (s, t) \in \theta \Rightarrow (x + s, y + t) \in \theta$$

and $(x, y) \in \theta \Rightarrow (ax, ay) \in \theta$ for all $a \in R$.

If θ is an R-congruence on M and $x \in M$, then the set

$$\theta(x) = \{y \in M : (x, y) \in \theta\}$$

is called the *congruence class of* x relative to θ on M. Prove the following for any R-congruence θ on M:

(i) $\theta(0)$ is an R-submodule of M.

(ii) $\theta(x) = x + \theta(0)$ for all $x \in M$.

(iii) $\theta(x) = \theta(y) \Leftrightarrow x - y \in \theta(0)$.

12. Let M be an R-module. For any R-submodule N of M, let

$$\theta_N = \{(x, y) \in M \times M : x - y \in N\}.$$

Then, prove that θ_N is an R-congruence on M and every R-congruence on M is of the form θ_N for some R-submodule N of M. Also prove that, for any R-submodules A and B of M,

$$A \subseteq B \Leftrightarrow \theta_A \subseteq \theta_B.$$

13.2 HOMOMORPHISMS AND QUOTIENTS OF MODULES

For a fixed ring R, a function from one R-module into another R-module which preserves the operations is called a homomorphism. Note that with each element a of the ring R there corresponds a binary operation on an R-module M given by $x \mapsto ax$. A homomorphism should preserve these binary operations also in addition to the group operation $+$. To be more precise, we have the following definition.

Definition 13.2.1. Let R be a ring and M and N be R-modules. A function $f : M \to N$ is called an *R-homomorphism* (or simply, a *homomorphism* when there is no ambiguity about the ring R) if the following are satisfied:

1. $f(x + y) = f(x) + f(y)$ for all x and $y \in M$.
2. $f(ax) = af(x)$ for all $a \in R$ and $x \in M$.

An R-homomorphism of R-modules is also called an *R-linear mapping* or, simply, *linear mapping*. An R-homomorphism $f : M \to N$ is necessarily a homomorphism of the group $(M, +)$ into the group $(N, +)$ and hence we have following theorem.

Theorem 13.2.1. The following holds for any R-homomorphism $f : M \to N$ of R-modules:

1. $f(0) = 0$
2. $f(-x) = -f(x)$ for all $x \in M$.
3. $f(x - y) = f(x) - f(y)$ for all x and $y \in M$.

Definition 13.2.2. The set of all R-homomorphisms of an R-module M into an R-module N is denoted by $\text{Hom}_R(M, N)$. An R-homomorphism of M into itself is called an *R-endomorphism*, or simply, an *endomorphism* of M and the set of all R-endomorphisms of M is denoted by $\text{End}_R(M)$; that is, $\text{Hom}_R(M, N) = \text{End}_R(M)$.

As usual an R-homomorphism is called an *R-monomorphism* if it is injective, an *R-epimorphism* if it is surjective and an *R-isomorphism* if it is bijective. A bijective R-endomorphism is called an *R-automorphism*. If there is an R-isomorphism of M onto N, then we say that M is *R-isomorphic* to N and denote this by $M \cong N$. Being R-isomorphic is clearly an equivalence relation on the set of R-modules, for any given ring R.

Definition 13.2.3. Let $f: M \to N$ be an R-homomorphism.

1. The set $\{x \in M : f(x) = 0\}$ is called the *kernel* of f and is denoted by $\ker f$.
2. The set $\{f(x) : x \in M\}$ is called the *image of M under f* and is denoted by $f(M)$.

Theorem 13.2.2. The following holds for any R-homomorphism $f: M \to N$.

1. $\ker f$ is an R-submodule of M.
2. $f(M)$ is an R-submodule of N.

Proof:

1. Since $f(0) = 0$, $0 \in \ker f$. If a and $b \in R$ and x and $y \in \ker f$, then $f(x) = 0 = f(y)$,

$$\begin{aligned} & f(x - y) = f(x) - f(y) = 0 - 0 = 0 \\ \text{and} \quad & f(ax) = af(x) = a0 = 0. \end{aligned}$$

   Therefore, $\ker f$ is an R-submodule of M.
2. Clearly, $f(M)$ is a nonempty subset of N. If s and $t \in f(M)$, then $s = f(x)$ and $t = f(y)$ for some x and $y \in M$ and hence

$$s - t = f(x) - f(y) = f(x - y) \in f(M)$$

   and $as = af(x) = f(ax) \in f(M)$ for all $a \in R$ and therefore $f(M)$ is an R-submodule of N.

Example 13.2.1

1. Let R be any ring and M and N be R-modules. Define $f: M \to N$ by $f(x) = 0$ for all $x \in M$. Then, f is an R-homomorphism and is called the *zero homomorphism*. Note that $\ker f = M$ and $f(M) = \{0\}$.

2. Let M be any R-module and define $f: M \to M$ by $f(x) = x$ for all $x \in M$. Then, f is an R-endomorphism of M and is called the *identity endomorphism* of M. Here, $\ker f = \{0\}$ and $f(M) = M$.
3. Let R be any ring. For any positive integers m and n, the set $M_{m\times n}(R)$ of all $m \times n$ matrices over R is an R-module under the usual addition and scalar multiplication of matrices. For any $m \times n$ matrix A, define $f_A : M_{1\times m}(R) \to M_{1\times n}(R)$ by

$$f_A(B) = BA \text{ for any } 1 \times m \text{ matrix } B.$$

   Then, f_A is an R-homomorphism. Note that $M_{1\times m}(R) = R^m$ and $M_{1\times n}(R) = R^n$. Here, $\ker f_A = \{B \in R^m : BA = (0)\}$ and $f_A(R^m) = \{BA : B \in R^m\}$.
4. Let R be a commutative ring and M be an R-module. Let a be a fixed elements of R and define $f_a : M \to M$ by

$$f_a(x) = ax \text{ for all } x \in M.$$

   Then, f_a is an R-endomorphism of M.

Fundamental theorem of homomorphisms and other isomorphism theorems are analogous to those of groups and rings. Before going to these, we formally define the notion of quotient module.

Theorem 13.2.3. Let N be an R-submodule of an R-module M and, for any $x \in M$, let

$$x + N = \{x + s : s \in N\}.$$

Then, the set

$$\frac{M}{N} = \{x + N : x \in M\}$$

forms an R-module under the operations defined by

$$(x + N) + (y + N) = (x + y) + N$$
$$\text{and} \quad a(x + N) = ax + N$$

for any x and $y \in M$ and $a \in R$.

Proof: As in the case of groups (since N is a subgroup of the abelian group $(M, +)$), two cosets $x + N$ and $y + N$ are equal if and only if $x - y \in N$. Also, any two cosets are either equal or disjoint. The operation $+$ defined on M/N is well defined and $(M/N, -)$ is an abelian group. Also, for any x and $y \in M$ and $a \in R$,

$$\begin{aligned} x + N = y + N &\Rightarrow x - y \in N \\ &\Rightarrow ax - ay = a(x - y) \in N \\ &\Rightarrow ax + N = ay + N \end{aligned}$$

and therefore the scalar multiplication defined in M/N is also well defined. It is routine to verify all the axioms of an R-module for M/N. Thus, M/N is an R-module.

Definition 13.2.4. The R-module M/N defined above is called the *quotient R-module* (or simply *quotient module*) of M by N.

The proofs of the following two theorems are similar to those in groups and rings.

Theorem 13.2.4 (Fundamental Theorem of *R*-homomorphisms) Let $f: M \to N$ be an R-homomorphism of R-modules.

Then

$$\frac{M}{\ker f} \cong f(M).$$

If f is an R-epimorphism, then $(M/\ker f) \cong N$.

Theorem 13.2.5. Let N be an R-submodule of an R-module M and M/N be the quotient of M by N. Then, any R-submodule of M/N of the form A/N for some R-submodule A of M containing N.

Worked Exercise 13.2.1. For any R-module M, prove that $\mathrm{End}_R(M)$ is a ring with unity under the point-wise addition and the composition of mappings as multiplication.

Answer: $\mathrm{End}_R(M)$ is the set of all R-endomorphisms of M (R-homomorphisms of M into itself). For any f and g in $\mathrm{End}_R(M)$, we define $f + g$ and $f \cdot g$ by

$$\begin{aligned} (f + g)(x) &= f(x) + g(x) \\ \text{and} \quad (f \cdot g)(x) &= f(g(x)) \end{aligned}$$

for all $x \in M$. Since $+$ is commutative in M, we have

$$\begin{aligned}(f+g)(x+y) &= f(x+y)+g(x+y)\\ &= f(x)+f(y)+g(x)+g(y)\\ &= f(x)+g(x)+f(y)+g(y)\\ &= (f+g)(x)+(f+g)(y)\end{aligned}$$

$$\text{and}\quad \begin{aligned}(f+g)(ax) &= f(ax)+g(ax)\\ &= af(x)+ag(x)\\ &= a(f(x)+g(x))\\ &= a(f+g)(x)\end{aligned}$$

for all x and $y \in M$ and $a \in R$. Therefore, $f + g$ is an R-endomorphism. Similarly $f \cdot g \in \text{End}_R(M)$. The zero homomorphism acts as identity for $+$ in $\text{End}_R(M)$. It can be easily verified that $(\text{End}_R(M), +)$ is an abelian group. Also clearly the composition o of mappings is associative. Further,

$$f \circ (g+h) = f \circ g + f \circ h$$
$$\text{and}\quad (f+g) \circ h = f \circ h + g \circ h$$

for all f, g and $h \in \text{End}_R(M)$. Also, the identity homomorphism of M is the identity for o in $\text{End}_R(M)$. Thus, $\text{End}_R(M)$ is a ring with unity.

Worked Exercise 13.2.2. Let A and B be R-submodules of an R-module M such that $A \subseteq B$, Then, prove that B/A is an R-submodule of M/A and

$$\frac{\left(M/A\right)}{\left(B/A\right)} \cong \frac{M}{B}.$$

Proof: By Theorem 13.2.5, B/A is an R-submodule of M/A and hence we can form the quotient module $(M/A)/(B/A)$. Define $f: M/A \to M/B$ by $f(x + A) = x + B$ for any $x + A \in M/A$, $x \in M$. For any x and $y \in M$,

$$\begin{aligned}x + A = y + A &\Rightarrow x - y \in A \subseteq B\\ &\Rightarrow x - y \in B\\ &\Rightarrow x + B = y + B.\end{aligned}$$

Therefore, f is well defined. It can be easily verified that f is an R-homomorphism. Also,

$$\begin{aligned}\ker f &= \left\{x+A \in \frac{M}{A} : f(x+A) = \text{The zero in } \frac{M}{B}\right\}\\ &= \left\{x+A \in \frac{M}{A} \ : x+B = B\right\}\\ &= \left\{x+A \in \frac{M}{A} \ : x \in B\right\}\\ &= \frac{B}{A}.\end{aligned}$$

Also, clearly f is a surjection. Therefore, by the Fundamental Theorem of R-homomorphisms, it follows that

$$\frac{\left(M/A\right)}{\left(B/A\right)} \cong \frac{M}{B}.$$

Worked Exercise 13.2.3. For any R-submodules A and B of an R-module M, prove that $(A+B)/A \cong B/(A \cap B)$.

Proof: Clearly, $A + B$ is an R-module (being an R-submodule of M) and A is an R-submodule of $A + B$. Also, $A \cap B$ is an R-submodule of B. Define $f : B \to (A+B)/A$ by $f(b) = b + A$ for any $b \in B$.

Since $B \subseteq A + B$, f is well defined. Clearly, f is an R-homomorphism. Also, for any $x = a + b \in A + B$ with $a \in A$ and $b \in B$, we have

$$x - b = a \in A$$

and hence $x + A = b + A = f(b)$, $b \in B$. Therefore, f is an R-epimorphism. Further

$$\begin{aligned}\ker f &= \left\{b \in B : f(b) = \text{zero in } \frac{A+B}{A}\right\}\\ &= \{b \in B : b + A = A\}\\ &= \{b \in B : b \in A\}\\ &= A \cap B.\end{aligned}$$

By the fundamental theorem of R-homomorphisms,

$$\frac{B}{A \cap B} = \frac{B}{\ker f} \cong \frac{A+B}{A}.$$

Thus, $(A+B)/A \cong B/(A \cap B)$.

Worked Exercise 13.2.4. Let R be a ring with unity and M an R-module. Prove that M is cyclic if and only if $M \cong (R/I)$ for some left ideal I of R.

Proof: Let I be a left ideal of R. Then, clearly R/I is a left R-module. Put $x = 1 + I$. Then, for any $a \in R$,

$$a + I = a(1 + I) = ax \in <x>$$

Therefore, $(R/I) = <x>$ and hence R/I is a cyclic R-module. If $M \cong (R/I)$, then M is also cyclic.

Conversely, suppose that M is cyclic and $M = <x>$. Define $f: R \to M$ by $f(a) = ax$ for all $a \in R$. Then, f is an R-epimorphism and let $I = \ker f$.

$$\begin{aligned}\ker f &= \{a \in R : f(a) = 0\}\\ &= \{a \in R : ax = 0\}.\end{aligned}$$

It can be easily verified that $\ker f$ is a left ideal of R. By the fundamental theorem of R-homomorphisms,

$$\frac{R}{I} = \frac{R}{\ker f} \cong M.$$

EXERCISE 13(B)

1. Which of the following are homomorphisms of modules?
   (i) $f: \mathbb{R}^3 \to \mathbb{R}; f(a_1, a_2, a_3) = a_1 + a_2$
   (ii) $f: \mathbb{R} \to \mathbb{R}; f(a) = 2a$
   (iii) $f: \mathbb{R}^3 \to \mathbb{R}; f(a_1, a_2, a_3) = a_3$
   (iv) A and B are R-submodules of an R-module M and $f : A \times B \to M$; $f(a, b) = a + b$
   (v) I is a left ideal of a ring R and $f: R \to R/I; f(a) = a + I$.
   (vi) For any ring R and $r \in R, f: R \to R; f(a) = ra$.
2. Let $M_1, M_2, \ldots, M_n$ be R-modules. Prove that $M_1 \times M_2 \times \cdots \times M_n$ is an R-module under the coordinate-wise addition and the scalar multiplication defined by

$$a\,(x_1, x_2, \ldots, x_n) = (ax_1, ax_2, \ldots, ax_n).$$

3. Let M and N be R-modules and P and Q be R-submodules of M and N, respectively. Prove that $M \times N$ is an R-module and $P \times Q$ is an R-submodule of $M \times N$ and $(M \times N)/(P \times Q) \cong (M/P) \times (N/Q)$.
4. For any R-module M, prove that the set $\text{End}_R(M)$ of all R-endomorphisms of M is a subring of the ring $\text{End}(M)$ of all endomorphisms of the group $(M, +)$.

5. Let R be a ring with unity and consider R as a right R-module. Then, prove that the ring R is isomorphic to the ring $\text{End}_R(R)$ of all R-endomorphisms of R.

6. Let $f: M \to N$ be an R-homomorphism of R-modules and A and B be R-submodules of M and N, respectively. Then, prove that $f(A)$ and $f^{-1}(B)$ are R-submodules of N and M, respectively.

7. Prove that an R-homomorphism of R-modules is an R-monomorphism if and only if its kernel is trivial (zero).

8. Let M be the set of all differentiable real valued functions defined on the set $\mathbb{R}$ of real numbers. Then, prove that M is an $\mathbb{R}$-module under the point-wise addition and scalar multiplication. Prove that the derivative operator is an $\mathbb{R}$-homomorphism of M into $\mathbb{R}^{\mathbb{R}}$ and determine its kernel.

9. If M is a cyclic R-module, prove that the quotient M/N is also a cyclic R-module for any R-submodule N of M.

10. Let M and N be R-module and $M \cong N$. Prove that M is cyclic (finitely generated) if and only if N is so.

11. A sequence (finite or infinite) of R-modules and R-homomorphisms $\cdots \xrightarrow{f_{n-1}} M_{n-1} \xrightarrow{f_n} M_n \xrightarrow{f_{n+1}} M_{n+1} \xrightarrow{f_{n+2}} \cdots$ is called *exact* if $f_i(M_{i-1}) = \ker f_{i+1}$ for all i.

    Suppose that the following diagram of R-modules and R-homomorphisms is commutative.

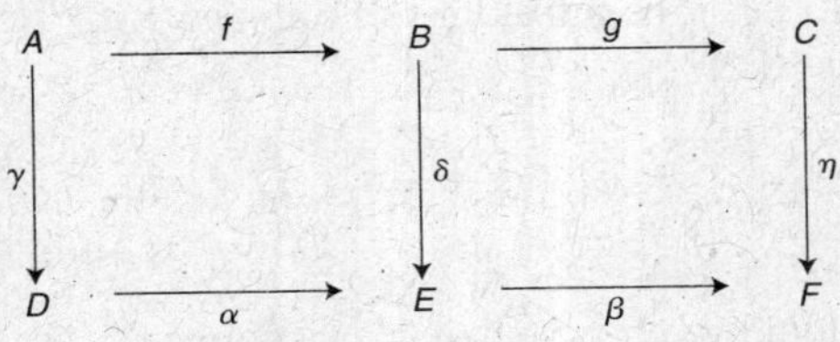

    That is, $\alpha \circ \gamma = \delta \circ f$ and $\beta \circ \delta = \eta \circ g$. Prove the following:

    (i) If γ, η and α are injections, then so is δ.

    (ii) If γ, η and g are surjections, then so is δ.

12. Consider the group $(\mathbb{Q}, +)$ of rational numbers as a $\mathbb{Z}$-module. Prove that the ring $\text{End}_{\mathbb{Z}}(\mathbb{Q})$ of $\mathbb{Z}$-endomorphisms of $\mathbb{Q}$ is isomorphic to the ring $\mathbb{Q}$ of rational numbers.

13. If R is a ring with unity and I is a left ideal of R such that $R/I \cong R$ as R-modules, then prove that there is an idempotent e in R such that $\text{Re} = I$. Is the converse true?

14. For any R-submodules A and B of an R-module M such that $A + B = M$, prove that

$$\frac{M}{A \cap B} \cong \frac{M}{A} \times \frac{M}{B}.$$

15. Let M be an R-module and $A_1, A_2, \ldots, A_n$ be R-submodules of M such that

$$A_i + \left(\bigcap_{j \neq i} A_j\right) = M \text{ for all } 1 \leq i \leq n.$$

Prove that $\dfrac{M}{\bigcap_{i=1}^{n} A_i} \cong \dfrac{M}{A_1} \times \cdots \times \dfrac{M}{A_n}$.

16. Let $f: M \to N$ be an R-homomorphism of R-module and let

$$\theta_f = \{(x, y) \in M \times M : f(x) = f(y)\}.$$

Then, prove the following:

(i) θ_f is an R-congruence on M.

(ii) $\theta_f(0) = \ker f$

(iii) For any $s \in f(M)$, $\{x \in M : f(x) = s\}$ is a congruence class relative to θ_f in M.

(iv) $\theta_f = \{(x, y) \in M \times N : x - y \in \ker f\}$.

13.3 DIRECT PRODUCTS AND SUMS

For any given in R, the class of R-modules exhibits a special property that has direct products and sums of arbitrary subclasses. First, let us introduce the following definition.

Definition 13.3.1. Let R be a fixed given ring and $\{M_i\}_{i \in I}$ be any nonempty class of R-modules. Let

$$M = \left\{\alpha : I \to \bigcup_{i \in I} M_i \ : \ \alpha(i) \in M_i \text{ for all } i \in I\right\}.$$

By the choice axiom, we observe that M is a nonempty set. For any α and $\beta \in M$ and $a \in R$, define

$$(\alpha + \beta)(i) = \alpha(i) + \beta(i)$$
$$\text{and} \quad (a\alpha)(i) = a\alpha(i)$$

for all $i \in I$. Then, M is an R-module under these addition and scalar multiplication. This M is called the *direct product* of $\{M_i\}_{i \in I}$ and is denoted by $\prod_{i \in I} M_i$. Each M_i is called a *direct factor* of the direct product M.

The direct product of any class of R-modules satisfies the following categorical property.

Theorem 13.3.1. Let M be the direct product of a given class of R-modules $\{M_i\}_{i \in I}$ over a given ring R. Then, M satisfies the following properties:

1. There exist R-homomorphisms $p_i : M \to M_i$, for each $i \in I$.

2. If N is any R-module and $q_i : N \to M_i$, $i \in I$, are R-homomorphisms, then there exists a unique R-homomorphism $q : N \to M$ such that

$$p_i \circ q = q_i \text{ for all } i \in I.$$

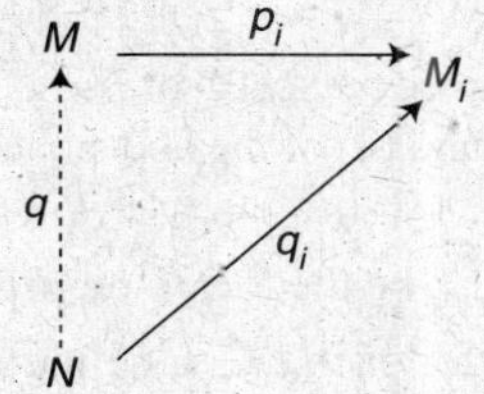

Proof:

1. For each $i \in I$, define $p_i : M \to M_i$ by

$$p_i(\alpha) = \alpha(i) \text{ for all } \alpha \in M.$$

Then, $p_i(\alpha + \beta) = (\alpha + \beta)(i) = \alpha(i) + \beta(i) = p_i(\alpha) + p_i(\beta)$

$$\text{and } p_i(a\alpha) = (a\alpha)(i) = a\alpha(i) = ap_i(\alpha)$$

for all α and $\beta \in M$ and $a \in R$. Therefore, each p_i, $i \in I$, is an R-homomorphism.

2. Let N be any R-module and, for each $i \in I$, let $q_i : N \to M_i$ be an R-homomorphism. Then, define $q : N \to M$ by

$$q(x)(i) = q_i(x)$$

for all $x \in N$ and $i \in I$. Then, $q(x) \in M$ for all $x \in N$ and q is a well-defined mapping of N into M. For any x and $y \in N$ and $a \in R$, we have

$q(x + y)(i) = q_i(x + y) = q_i(x) + q_i(y) = q(x)(i) + q(y)(i) = (q(x) + q(y))(i)$

and $\quad q(ax)(i) = q_i(ax) = aq_i(x) + aq(x)(i) = (aq(x))(i)$

for all $i \in I$ and hence

$$q(x + y) = q(x) + q(y) \text{ and } q(ax) = aq(x).$$

Therefore, q is an R-homomorphism. Also, for any $x \in N$,

$$(p_i \circ q)(x) = p_i(q(x)) = q(x)(i) = q_i(x)$$

and therefore $p_i \circ q = q_i$ for all $i \in I$.

To prove the uniqueness of q, let $q' : N \to M$ be any R-homomorphism such that $p_i \circ q' = q_i$ for all $i \in I$. Then, for any $x \in N$,

$$(q'(x))(i) = p_i(q'(x)) = q_i(x) = q(x)(i)$$

for all $i \in I$, and hence $q'(x) = q(x)$ for all $x \in N$. Therefore, $q' = q$. Thus, q is a unique R-homomorphism of N into M such that $p_i \circ q = q_i$ for all $i \in I$.

The converse of the above theorem is also true in the sense of the following.

Theorem 13.3.2. Let $\{M_i\}_{i\in I}$ be any class of R-modules and M be an R-module satisfying the properties (1) and (2) of the above theorem. Then, $M \cong \prod_{i\in I} M_i$.

Proof: Let $p_i : M \to M_i$, $i \in I$, be the R-homomorphisms satisfying (1) and (2) above. Let $N \cong \prod_{i\in I} M_i$ and, for each $i \in I$, define $q_i : N \to M_i$ by $q_i(\alpha) = \alpha(i)$ for all $\alpha \in N$. Then, by (2) above, there exists a unique R-homomorphism $q : N \to M$ such that $p_i \circ q = q_i$ for all $i \in I$.

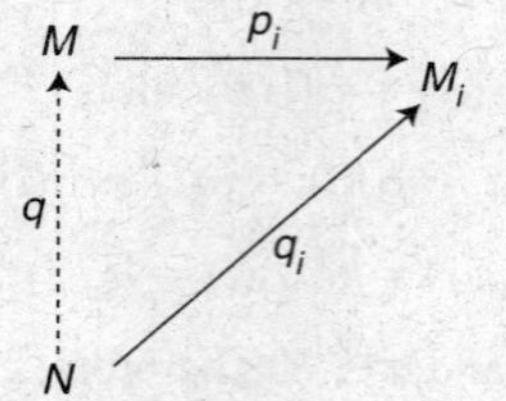

Also, since the direct product N satisfies the properties (1) and (2) (from Theorem 13.3.1), there exists a unique R-homomorphism $p : M \to N$ such that

$$q_i \circ p = p_i \text{ for all } i \in I.$$

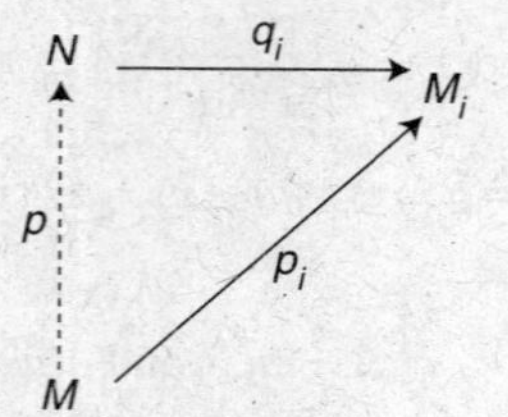

Now, consider $q \circ p : M \to M$. We have

$$\begin{aligned} p_i \circ (q \circ p) &= (p_i \circ q) \circ p \\ &= q_i \circ p \\ &= p_i = p_i \circ (\mathrm{Id}_M) \end{aligned}$$

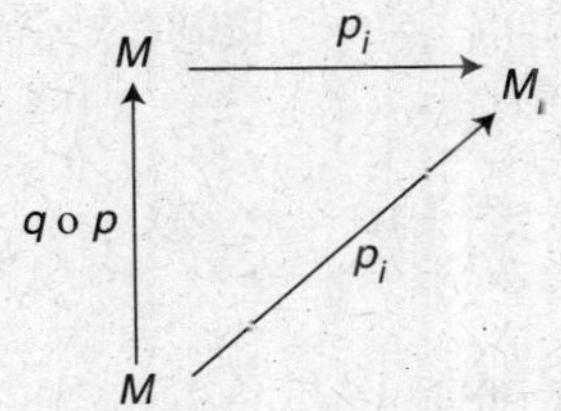

By the uniqueness of the R-homomorphism in (2), we get that

$$q \circ p = \mathrm{Id}_M.$$

Similarly, we can prove that $p \circ q = \mathrm{Id}_N$. Therefore, p and q are bijections and inverses to each other. In particular, $p : M \to N$ is an R-isomorphism. Thus,

$$M \cong N = \prod_{i \in I} M_i.$$

Definition 13.3.2. Let $\{M_i\}_{i \in I}$ be a class of R-modules and $M = \prod_{i \in I} M_i$. For each $i \in I$, the map $p_i : M \to M_i$, defined by

$$p_i(\alpha) = \alpha(i) \quad \text{for any } \alpha \in M$$

is called the ith projection.

Theorem 13.3.3. Let $M = \prod_{i \in I} M_i$ be the direct product of R-modules $\{M_i\}_{i \in I}$. Then, each projection $p_i : M \to M_i$ is an R-epimorphism and $(M / \ker p_i) \cong M_i$.

Proof: Choose some element $\alpha \in M$. Fix $i \in I$ and let $x_i \in M_i$.

Define $\beta : I \to \bigcup_{j \in I} M_j$ by

$$\beta(j) = \begin{cases} x_i & \text{if } j = i \\ \alpha(j) & \text{if } j \neq i \end{cases}.$$

Then, $\beta \in M$ and $p_i(\beta) = \beta(i) = x_i$. Therefore, p_i is an R-epimorphism. By the fundamental theorem of R-homomorphisms,

$$\frac{M}{\ker p_i} \cong M_i.$$

If $I = \{1, 2, \ldots, n\}$ is a finite set and $M_1, M_2, \ldots, M_n$ are R-modules, then it can be easily observed that

$$\prod_{i\in I} M_i = \prod_{i=1}^{n} M_i = M_1 \times M_2 \times \cdots \times M_n$$

which is the usual set of n-tuples $(x_1, x_2, \ldots, x_n)$ with $x_i \in M_i$.

Next let us turn our attention to the concept of direct sum of a given family of R-modules over a given fixed ring R.

Definition 13.3.3. Let $\{M_i\}_{i\in I}$ be a nonempty family of R-modules, where R is a given fixed ring. An R-module M is called an *external direct sum* if the following are satisfied:

1. For each $i \in I$, these is an R-homomorphism $f_i : M_i \to M$.
2. If N is an R-module and, for each $i \in I$, $g_i : M_i \to N$ is an R-homomorphism, then there exists unique R-homomorphism $g : M \to N$ such that $g \circ f_i = g_i$ for all $i \in I$.

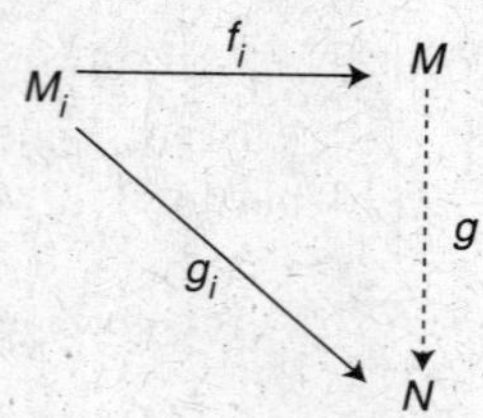

In the following, we prove uniqueness (up to isomorphism) of the external direct sum of a given family of R-modules.

Theorem 13.3.4. Let $\{M_i\}_{i\in I}$ be a nonempty family of R-modules, M and M' be R-modules and $\{f_i\}_{i\in I}$ and $\{f_i'\}_{i\in I}$ be R-homomorphisms satisfying the properties (1) and (2) above. Then, there exists an R-isomorphism $f' : M \to M'$ such that $f' \circ f_i = f_i'$ for all $i \in I$.

Proof: Since M and $\{f_i\}_{i\in I}$ satisfy (2) with M' in place of N and f_i' in place of g_i, there exists an R-homomorphism $f' : M \to M'$ such that $f' \circ f_i = f_i$ for all $i \in I$.

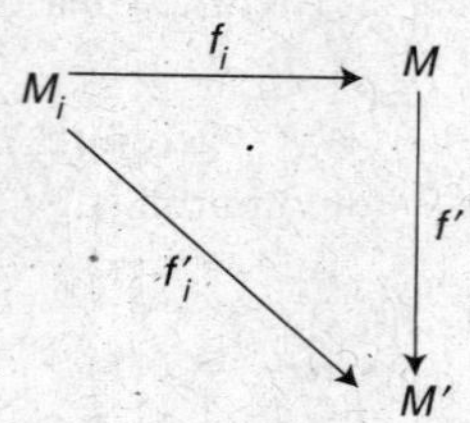

Also, since M' and $\{f_i'\}_{i\in I}$ satisfies (2) with M in place of N and f_i in place of g_i, there exists an R-homomorphism $f : M' \to M$ such that

$$f \circ f_i' = f_i \text{ for all } i \in I.$$

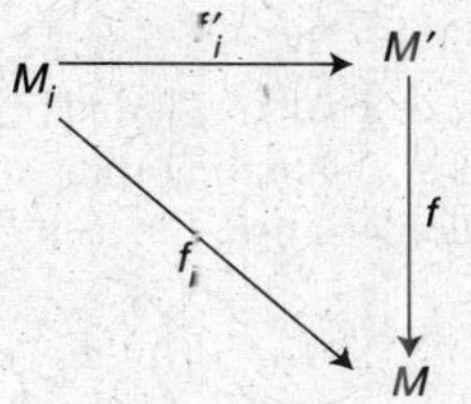

Now, consider $f \circ f' : M \to M$. We have

$$(f \circ f') \circ f_i = f \circ f_i' = f_i = \mathrm{Id}_M \circ f_i \text{ for all } i \in I.$$

From the uniqueness of the R-homomorphism in (2), with M in place of N and f_i in place of g_i, we get that $f \circ f' = \mathrm{Id}_M$. Similarly, by interchanging the roles of M and M', we can prove that $f' \circ f = \mathrm{Id}_{M'}$. Therefore, f and f' are bijections and are inverses to each other. Thus, $f' : M \to M'$ is an R-isomorphism and $f' \circ f_i = f_i'$ for all $i \in I$.

Before taking up the proof of the existence of the external direct sums, let us have the following notation.

Definition 13.3.4. Let $\{M_i\}_{i\in I}$ be a nonempty family of R-modules and $\alpha \in \prod_{i\in I} M_i$. Then, the set

$|\alpha| = \{i \in I : \alpha(i) \neq 0 \text{ in } M_i\}$ is called the *support* of α.

Theorem 13.3.5. Let $\{M_i\}_{i\in I}$ be a nonempty family of R-modules and $\prod_{i\in I} M_i$ be the direct product of M_i's. Let

$$M = \left\{\alpha \in \prod_{i\in I} M_i : |\alpha| \text{ is finite}\right\}.$$

Then, M is the external direct sum of $\{M_i\}_{i\in I}$.

Proof: First observe that, for any α and $\beta \in \prod_{i\in I} M_i$,

$$|-\alpha| = |\alpha|,\ |\alpha + \beta| \subseteq |\alpha| \cup |\beta| \text{ and } |a\alpha| \subseteq |\alpha|$$

for all $a \in R$ and hence M is an R-submodule of the direct product $\prod_{i\in I} M_i$. Therefore, M is an R-module.

For each $i \in I$, define $f_i : M_i \to M$ by

$$f_i(x)(j) = \begin{cases} x & \text{if } j = i \\ 0 & \text{if } j \neq i \end{cases}$$

for any $x \in M_i$ and $j \in I$. Note that $|f_i(x)| \subseteq \{i\}$ and hence $f_i(x) \in M$ for all $x \in M_i$. It can be easily verified that f_i is an R-homomorphism. Now, let N be any R-module and $g_i : M_i \to N$ be an R-homomorphism for each $i \in I$. Define $g : M \to N$ by

$$g(\alpha) = \sum_{i \in I} g_i\left(\alpha(i)\right) \text{ for any } \alpha \in M.$$

Since $|\alpha|$ is finite, $\alpha(i) = 0$ and hence $g_i(\alpha(i)) = 0$ for all but finite number of i's. Therefore,

$$g(\alpha) = \sum_{i \in |\alpha|} g_i\left(\alpha(i)\right)$$

and the summation in the definition of $g(\alpha)$ is meaningful. For any α and $\beta \in M$ and $a \in R$, we have

$$\begin{aligned} g(\alpha + \beta) &= \sum_{i \in I} g_i\left((\alpha + \beta)(i)\right) \\ &= \sum_{i \in I} g_i\left(\alpha(i) + \beta(i)\right) \\ &= \sum_{i \in I} g_i\left(\alpha(i)\right) + \sum_{i \in I} g_i\left(\beta(i)\right) = g(\alpha) + g(\beta) \end{aligned}$$

and
$$\begin{aligned} g(a\alpha) &= \sum_{i \in I} g_i\left((a\alpha)(i)\right) \\ &= \sum_{i \in I} g_i\left(a\alpha(i)\right) \\ &= \sum_{i \in I} a g_i\left(\alpha(i)\right) \\ &= a \sum_{i \in I} g_i\left(\alpha(i)\right) = a g(\alpha). \end{aligned}$$

Therefore, g is an R-homomorphism. Also, for any $x \in M_i$, we have

$$(g \circ f_i)(x) = g\left(f_i(x)\right) = \sum_{j \in I} g_j\left(f_i(x)(j)\right) = g_i(x),$$

since $f_i(x)(j) = 0$ for $j \neq i$ and $f_i(x)(i) = x$. Therefore, $g \circ f_i = g_i$ for all $i \in I$. To prove the uniqueness of g, let us take an R-homomorphism $g' : M \to N$ such that $g' \circ f_i = g_i$ for all $i \in I$. Then, for any $\alpha \in M$,

$$\begin{aligned} g(\alpha) &= \sum_{i\in I} g_i\,(\alpha(i)) \\ &= \sum_{i\in I} (g' \circ f_i)(\alpha(i)) \\ &= \sum_{i\in I} g'\left(f_i\,(\alpha(i))\right) \\ &= g'\left(\sum_{i\in I} f_i\,(\alpha(i))\right) = g'(\alpha), \end{aligned}$$

since $\sum_{i\in I} f_i(\alpha(i)) = \sum_{i\in|\alpha|} f_i(\alpha(i)) = \alpha$. Thus, $g = g'$ and hence g is the unique R-homomorphism such that $g \circ f_i = g_i$ for all $i \in I$. Thus, M is the external direct sum of $\{M_i\}_{i\in I}$.

Corollary 13.3.1. Let M be the external direct sum of a family $\{M_i\}_{i\in I}$ of R-modules. Then, there exists R-submodules $\{N_i\}_{i\in I}$ of M satisfying the following:

1. $M_i \cong N_i$ for all $i \in I$.
2. Each nonzero element of M can be uniquely expressed as a sum $x_1 + x_2 + \cdots + x_n$ where $0 \neq x_j \in N_{i_j}$ for $1 \leq j \leq n$ and $i_1, i_2, \ldots, i_n$ are distinct members of the index set I.

Proof: From Theorems 13.3.4 and 13.3.5, we can take M as the R-submodule of the direct product $\prod_{i\in I} M_i$ given by

$$M = \left\{\alpha \in \prod_{i\in I} M_i \ : \ |\alpha| \text{ is finite}\right\}$$

and $f_i : M_i \to M$ as the R-homomorphism defined by

$$f_i(x)(j) = \begin{cases} x & \text{if } j = i \\ 0 & \text{if } j \neq i \end{cases}$$

for any $x \in M_i$ and $j \in I$.

Now, put $N_i = f_i(M_i)$ for each $i \in I$. Then, each N_i is an R-submodule of M. It can be easily observed that each f_i is an R-monomorphism and hence f_i can be treated as an R-isomorphism of M_i onto N_i. Therefore, $M_i \cong N_i$ for each $i \in I$. Let $0 \neq \alpha \in M$. Then, the support $|\alpha|$ is a nonempty finite subset of I. Let

$$|\alpha| = \{i_1, i_2, \ldots, i_n\}.$$

Put $x_j = f_{i_j}\left(\alpha(i_j)\right) \in N_{i_j}$ for $1 \leq j \leq n$. Then,

$$\alpha = x_1 + x_2 + \cdots + x_n$$

(by evaluating both sides at each $k \in I$) and clearly this expression of α is unique.

Definition 13.3.5. Let M be an R-module and $\{N_i\}_{i \in I}$ be a nonempty family of R-submodules of M. If each element x of M can be uniquely expressed as a sum

$$x = \sum_{i \in I} x_i,$$

with $x_i \in N_i$ for all $i \in I$ and $x_i = 0$ for all but a finite number i's, then M is called the *internal direct sum* of $\{N_i\}_{i \in I}$ and denote this by $M = \bigoplus_{i \in I} N_i$. In this case, each N_i is called a *direct summand* of M. If I is a finite set, say $I = \{1, 2, \ldots, n\}$, then $\bigoplus_{i \in I} N_i$ will be written as $N_1 \oplus N_2 \oplus \cdots \oplus N_n$.

If M is the external direct sum of R-modules $\{M_i\}_{i \in I}$, then we have proved in Corollary 13.3.1 that there are R-submodules $\{N_i\}$ of M such that $N_i \cong M_i$ and M is the internal direct sum of $\{N_i\}_{i \in I}$. On the other hand, if M is the internal direct sum of R-submodules $\{M_i\}_{i \in I}$, then we prove below that M is (isomorphic to) the external direct sum of $\{M_i\}_{i \in I}$.

Theorem 13.3.6. Let M be an R-module and M be the internal direct sum of R-submodules $\{M_i\}$. Then, M is isomorphic to the external direct sum of $\{M_i\}_{i \in I}$.

Proof: Let N be the external direct sum of $\{M_i\}_{i \in I}$. That is, by Theorem 13.3.5,

$$N = \left\{\alpha \in \prod_{i \in I} M_i \ : \ |\alpha| \text{ is finite}\right\}.$$

Define $f: M \to N$ as follows.

$$f(0) = 0 \text{ and, if } 0 \neq x \in M, \text{ then } x = x_1 + x_2 + \cdots + x_n$$

where $x_j \in M_{i_j}$, $i_1, i_2, \ldots, i_n \in I$ and define

$$f(x)(i) = \begin{cases} x_j & \text{if } i = i_j,\ 1 \leq j \leq n \\ 0 & \text{otherwise} \end{cases}.$$

Then, it can be verified that f is an R-isomorphism.

Theorem 13.3.7. Let $\{M_i\}_{i \in I}$ be a family of R-submodules of an R-module M such that $M = \sum_{i \in I} M_i$. Then, the following are equivalent to each other:

1. $M = \bigoplus_{i \in I} M_i$.
2. $x_1 + x_2 + \cdots + x_n = 0$, $x_j \in M_{i_j}$ and $i_1, i_2, \ldots, i_n \in I$ imply that $x_1 = x_2 = \ldots = x_n = 0$.
3. $M_i \cap \left(\sum_{i \neq j \in I} M_j \right) = \{0\}$ for each $i \in I$.

Proof: (1) ⇒ (2): If $M = \bigoplus_{i \in I} M_i$, then any element x of M can be uniquely expressed as $x = x_1 + x_2 + \cdots + x_n$ with $x_j \in M_{i_j}$ for $1 \leq j \leq n$ and $i_1, i_2, \ldots, i_n \in I$. Therefore, if $x_1 + x_2 + \cdots + x_n = 0 = 0 + 0 + \cdots + 0$, it follows from the uniqueness that $x_1 = 0 = x_2 = \cdots = x_n$.
(2) ⇒ (3): Let $i \in I$ be fixed

$$x_i \in M_i \cap \left(\sum_{i \neq j \in I} M_j \right) \Rightarrow x_i = x_{j_1} + \cdots + x_{j_n}, j_k \neq i \text{ for } 1 \leq k \leq n \text{ and } x_{j_k} \in M_{j_k}$$

$$\Rightarrow (-x_i) + x_{j_1} + \cdots + x_{j_n} = 0.$$

$$\Rightarrow x_i = 0 \text{ (by (2))}.$$

Therefore, $M_i \cap \left(\sum_{i \neq j \in I} M_j \right) = \{0\}$ for all $i \in I$.

(3) ⇒ (1): We are given that $M = \sum_{i \in I} M_i$.
If $x_1 + x_2 + \cdots + x_n = y_1 + y_2 + \cdots + y_n$ and $x_j, y_j \in M_{i_j}$, then

$$x_1 - y_1 = (y_2 - x_2) + \cdots + (y_n - x_n) \in M_{i_1} \cap \left(\sum_{i \neq i_1} M_i \right) = \{0\}$$

and hence $x_1 = y_1$. Similarly, $x_2 = y_2, \ldots, x_n = y_n$. Thus, $M = \bigoplus_{i \in I} M_i$.

Observe that, if $N_1, N_2, \ldots, N_n$ are finite number of R-submodules of an R-module M such that $M = \bigoplus_{i=1}^{n} N_i$, then $M \cong \prod_{i=1}^{n} N_i$. That is, for any finite number of R-modules, their direct product and direct sum are equal (isomorphic).

Worked Exercise 13.3.1. Let $M = \mathbb{R}^3$ be consider M as an $\mathbb{R}$-module. Let $x_1 = (2, 0, 0)$, $x_2 = (0, 1, 0)$ and $x_3 = (0, 0, 3)$. Then, prove that $M = \langle x_1 \rangle \oplus \langle x_2 \rangle \oplus \langle x_3 \rangle$.

Answer: If $x = (a_1, a_2, a_3) \in M$, then

$$x = \frac{a_1}{2} x_1 + a_2 x_2 + \frac{a_3}{3} x_3 \in \langle x_1 \rangle + \langle x_2 \rangle + \langle x_3 \rangle$$

Therefore, $M = \langle x_1 \rangle + \langle x_2 \rangle + \langle x_3 \rangle$.
Suppose $y_i \in \langle x_i \rangle$ for $i = 1, 2, 3$ such that $y_1 + y_2 + y_3 = 0$.
Then $y_i = a_i x_i$ for some $a_i \in \mathbb{R}$ for $i = 1, 2, 3$ and

$$\begin{aligned}(0, 0, 0) &= y_1 + y_2 + y_3 = a_1 x_1 + a_2 x_2 + a_3 x_3 \\ &= a_1(2, 0, 0) + a_2(0, 1, 0) + a_3(0, 0, 3) \\ &= (2a_1, a_2, 3a_3)\end{aligned}$$

and hence $2a_1 = 0 = a_2 = 3a_3$ which imply that $a_1 = 0 = a_2 = a_3$ so that $y_1 = 0 = y_2 = y_3$. Therefore, by Theorem 13.3.7, $M = \langle x_1 \rangle + \langle x_2 \rangle + \langle x_3 \rangle$.

Worked Exercise 13.3.2. Let e_1 and e_2 be idempotents in a ring R and regard R as a left R-module. Prove that

$$Re_1 \oplus R(e_2 - e_2 e_1) = Re_1 + Re_2$$

Answer: Let $M = Re_1 + Re_2$, $M_1 = Re_1$ and $M_2 = R(e_2 - e_2 e_1)$. Then, M is an R-module and M_1 and M_2 are R-submodules of M. Since

$$\begin{aligned}& e_1 = e_1 e_1 \in Re_1 \subseteq M \\ \text{and} \quad & e_2 = e_2 e_1 + e_2(e_2 - e_2 e_1) \in M_1 + M_2,\end{aligned}$$

it follows that $Re_1 + Re_2 \subseteq M_1 + M_2$ and therefore
$M = M_1 + M_2$. Also,

$$\begin{aligned}x \in M_1 \cap M_2 &\Rightarrow x = re_1 \text{ and } x = s(e_2 - e_2 e_1) \text{ for some } r, s \in R \\ &\Rightarrow x = re_1 e_1 = s(e_2 - e_2 e_1)e_1 = s(e_2 e_1 - e_2 e_1) = 0\end{aligned}$$

and therefore $M_1 \cap M_2 = \{0\}$. By Theorem 13.3.7, it follows that $M = M_1 \oplus M_2$.

Worked Exercise 13.3.3. Let R be a ring with unity and I be a left ideal of R. Consider R as a left R-module. Then, prove that I is a direct summand of R if and only if $I = Re$ for some idempotent e in R.

Answer: Suppose that $I = Re$ and $e^2 = e \in R$. Then, put $J = R(1 - e)$. Then, I and J are R-submodules of R and, any $x \in R$ can be written as

$$x = xe + x(1 - e) \in I + J$$

and hence $I + J = R$. Also,

$$\begin{aligned} x \in I \cap J &\Rightarrow x = re = s(1 - e) \text{ for some } r \text{ and } s \in R \\ &\Rightarrow x = xe = s(1 - e)e = s0 = 0 \end{aligned}$$

and therefore $I \cap J = \{0\}$. Thus, $R = I \oplus J$ and I is a direct summand of R. Conversely suppose that I is a direct summand of R. Then, there exists an R-submodule J of R such that $R = I \ominus J$. Then, I and J are left ideals of R, $R = I + J$ and $I \cap J = \{0\}$. Since $I \in R = I + J$, we get that

$$1 = e + f \text{ for some } e \in I \text{ and } f \in J.$$

Now, $e = e(e + f) = e^2 + ef$ and hence

$$e - e^2 = ef \in I \cap J = \{0\}.$$

Therefore, $e - e^2 = 0$ or e is an idempotent of R. Also, since I is a left ideal of R and $e \in I$, we have $Re \subseteq I$. Further,

$$\begin{aligned} x \in I &\Rightarrow x = x(e + f) = xe + xf \\ &\Rightarrow x - xe = xf \in I \cap J = \{0\} \\ &\Rightarrow x - xe = 0 \\ &\Rightarrow x = xe \in Re. \end{aligned}$$

Thus, $I \subseteq Re$ and hence $I = Re$.

EXERCISE 13(C)

1. Consider the $\mathbb{R}$-module $\mathbb{R}^4$ and determine whether $\mathbb{R}^4$ is the direct sum of $<x_1>$, $<x_2>$, $<x_3>$ and $<x_4>$ in each of the following cases:
   (i) $x_1 = (0, 2, 0, 3)$, $x_2 = (0, 3, 0, 4)$, $x_3 = (2, 0, 0, 5)$ and $x_4 = (0, 6, 0, 9)$

(ii) $x_1 = (0, 0, 1, 0)$, $x_2 = (0, 2, 0, 0)$, $x_3 = (3, 0, 0, 0)$ and $x_4 = (0, 0, 0, 4)$

(iii) $x_1 = (3, 0, 0, 0)$, $x_2 = (3, 3, 0, 0)$, $x_3 = (3, 3, 3, 0)$ and $x_4 = (3, 3, 3, 3)$

(iv) $x_1 = (0, 2, 2, 2)$, $x_2 = (0, 3, 3, 3)$, $x_3 = (0, 1, 1, 1)$ and $x_4 = (0, 4, 4, 4)$

2. Consider $\mathbb{R}^4$ as $\mathbb{Z}$-module and determine whether $\mathbb{R}^4$ is the direct sum of $<x_1>$, $<x_2>$, $<x_3>$ and $<x_4>$ in each of the above cases in Exercise 1.

3. Let I be any nonempty set and R be any ring. Then, R^I, the set of all mappings of I into R, is an R-module under the usual point-wise operations. Prove that there exists a family $\{M_i\}_{i\in I}$ of R-modules such that M_i is R-isomorphic to R for each $i \in I$ and $R^I \cong \prod_{i\in I} M_i$.

4. In the above exercise, let

$$R^{(I)} = \{\alpha \in R^I : |\alpha| \text{ is finite}\}.$$

Then, prove that there exists R-submodules $\{N_i\}_{i\in I}$ of $R^{(I)}$ such that N_i is R-isomorphic to R for each $i \in I$ and $R^{(I)} = \bigoplus_{i\in I} N_i$.

5. Let $\{M_i\}_{i\in I}$ be a family of R-modules and $\phi \neq J \subsetneq I$. If $M = \prod_{i\in I} M_i$, $A = \prod_{j\in J} M_j$ and $B = \prod_{i\in I-J} M_i$, then prove that A and B are (isomorphic to) R-submodules of M and $M \cong A \oplus B$.

6. Let $\{M_i\}_{i\in I}$ be class of R-modules and M be an R-module. For each $i \in I$, let $p_i : M \to M_i$ be an R-homomorphism satisfying the properties (1) and (2) of Theorem 13.3.1. Then, prove that each p_i, $i \in I$, is an R-epimorphism.

7. Let $M = \prod_{i\in I} M_i$. Prove that there exists an R-submodule N_i of M such that $(M/N_i) \cong M_i$ for each $i \in I$ and $N_i \cong \prod_{i\neq j\in I} M_j$.

8. Let M be the external direct sum of $\{M_i\}_{i\in I}$ and $f_i : M_i \to M$ be the R-homomorphism as in Definition 13.3.3 (1). Then, prove that f_i is a R-monomorphism for each $i \in I$.

9. Let $\{M_i\}_{i\in I}$ and $\{N_i\}_{i\in I}$ be two families of R-modules and, for each $i \in I$, let $f_i : M_i \to N_i$ be an R-homomorphism. Then, prove that there is a unique R-homomorphism $f : \prod_{i\in I} M_i \to \prod_{i\in I} N_i$ such that $f(\alpha)(i) = f_i(\alpha(i))$ for all $\alpha \in \prod_{i\in I} M_i$ and $i \in I$.

10. In the above exercise, if each f_i is an R-isomorphism, then prove that f is an R-isomorphism.

11. State results similar to the above two exercises for external or internal direct sums of R-modules and prove them.

12. Let R be a ring with unity and I be a left ideal of R such that $(R/I) \cong R$ regarded as R-modules. Then, prove that $I = Re$ for idempotent e in R and deduce that I is a direct summand of R.

13.4 SIMPLE AND COMPLETELY REDUCIBLE MODULES

A module with nontrivial scalar multiplication and without nontrivial submodules is called a simple module. A module which can be expressed as a sum of simple submodules is called completely reducible. In this section, we briefly discuss about the simple modules and completely reducible modules. First recall that, for an R-module M, RM is an R-submodule of M defined by

$$RM = \left\{ \sum_{i=1}^{n} r_i x_i : r_i \in R \text{ and } x_i \in M \right\}.$$

Definition 13.4.1. An R-module M is called *simple* or *irreducible* if $RM \neq \{0\}$ and M has no R-submodules except $\{0\}$ and M.

Note that a simple module M is necessarily nonzero and the scalar multiplication is nontrivial, in the sense that, $rx \neq 0$ for some $r \in R$ and $x \in M$.

Example 13.4.1

1. Let R be a field or a division ring and consider R as left R-module. Then, R has no left ideals except $\{0\}$ and R. Therefore, R is a simple R-module.
2. Let F be a field and $R = M_n(F)$, the ring of all $n \times n$ matrices over F. For each $1 \leq i, j \leq n$, let e_{ij} be the matrix whose ij^{th} entry is 1 and all the other entries are zero. Fix $1 \leq k \leq n$ and consider $M = Re_{kk}$. Then, M is a left R-module. We prove that M is a simple R-module. Clearly, $RM \neq \{0\}$. Let $N \neq \{0\}$ be an R-submodule of M and let $0 \neq (a_{ij}) \in N$.

   Then, since

   $$(a_{ij}) = \sum_{i=1}^{n} a_{ik} e_{ik},$$

   $a_{jk} \neq 0$ for some $1 \leq j \leq n$ and a_{jk} has multiplicative inverse in F. Now,

   $$e_{kk} = a_{jk}^{-1} e_{kj} \left(\sum_{i=1}^{n} a_{ik} e_{ik} \right) \in N$$

   and hence $Re_{kk} \subseteq N$ so that $N = M$. Thus, M is a simple R-module. Note that M is precisely the set of all $n \times n$ matrices over F in which every entry is zero except possibly in the kth column.
3. Let R be a ring with unity and M be a minimal left ideal of R. Then, clearly M is a simple R-module.

   In general, a minimal left ideal of a ring R need not be a simple R-module; for consider an abelian group $(R, +)$ of order p, where p is a prime number, and define $ab = 0$ for all a and b in R. Then, R is a ring (without unity). R is a minimal left ideal (in fact, it is the only nonzero left ideal) and not a simple R-module, since $RR = \{0\}$. The following is an useful characterization of simple modules over rings with unity.

Theorem 13.4.1. The following are equivalent to each other for any module M over a ring R with unity:

1. M is a simple R-module.
2. There exists a maximal left ideal I of R such that $M \cong (R/I)$.
3. $M \neq \{0\}$ and $M = Rx$ for any $0 \neq x \in M$.

Proof: (1) $\Rightarrow$ (2): Suppose that M is a simple R-module. Since R is a ring with unity, $M = RM \neq \{0\}$. Choose $0 \neq x \in M$. Then, Rx is a nonzero R-submodule of M and, since M is simple, $Rx = M$. Define

$$f : R \to M \text{ by } f(a) = ax \text{ for all } a \in R.$$

Then, f is an R-epimorphism. Put $I = \ker f = \{a \in R : ax = 0\}$. Then, I is a left ideal of R and, by the fundamental theorem of R-homomorphisms, $(R/I) \cong M$. Now, since M is simple R-module, so is R/I. The R-submodules of R/I are in one-to-one correspondence with the left ideals of R containing I. Therefore, I and R are the only left ideals of R containing I. Thus, I is a maximal left ideal of R and $(R/I) \cong M$.

(2) $\Rightarrow$ (3): Let I be a maximal left ideal of R such that $M \cong (R/I)$; without loss of generality, we can assume that $M = R/I$. Since $I \neq R$, it follows that $M \neq \{0\}$ and $RM = M \neq \{0\}$. If $0 \neq x \in M$, then

$$x = a + I \quad \text{for some } a \in R - I$$

Then, $Ra + I$ is a left ideal of R containing I properly. By the maximality of I, it follows that $Ra + I = R$. Therefore, $Rx = R(a+I) = Ra+I = (R/I) = M$.

(3) $\Rightarrow$ (1): If N is any nonzero R-Submodule of M and $0 \neq x \in N$, then $M = Rx \subseteq N$ and hence $N = M$. Also, $RM = M \neq \{0\}$. Thus, M is a simple R-module.

The following result, which is popularly known as Schur's lemma, is an important property of simple R-modules. Let us first recall that, for any R-module M, the set $\text{End}_R(M)$ of all R-endomorphisms of M forms a ring with unity under the point-wise addition and the composition of mappings as multiplication. Also note that an R-module M can also be viewed as a module over the ring $\text{End}_R(M)$, where the scalar multiplication is defined by

$$fx = f(x)$$

for any $f \in \mathrm{End}_R(M)$ and $x \in M$.

Theorem 13.4.2. (Schur's lemma) $\mathrm{End}_R(M)$ is a division ring for any simple R-module M.

Proof: Let M be a simple R-module. We have already observed that the set $\mathrm{End}_R(M)$ of all R-endomorphisms of M is a ring with unity under the pointwise addition and composition of mappings as multiplication. Note that the identity map is the unity element in the ring $\mathrm{End}_R(M)$. We have to only prove that every nonzero element in $\mathrm{End}_R(M)$ has multiplicative inverse in $\mathrm{End}_R(M)$. Let $0 \neq f \in \mathrm{End}_R(M)$. Then, $\ker f \neq M$ and $f(M) \neq \{0\}$. Both $\ker f$ and $f(M)$ are R-submodules of M. Since M is simple, it follows that $\ker f = \{0\}$ and $f(M) = M$. These imply that f is a bijection and hence f is an R-isomorphism, so that the inverse map f^{-1} exists and is an R-endomorphism. Therefore, f has multiplicative inverse in $\mathrm{End}_R(M)$. Thus, $\mathrm{End}_R(M)$ is a division ring.

Definition 13.4.2. An R-module M is called *completely reducible* if there exists a family $\{M_\alpha\}_{\alpha\in\Delta}$ of simple R-submodules of M such that

$$M = \sum_{\alpha\in\Delta} M_\alpha = \left\{x_1 + \cdots + x_n \ : \ x_i \in M_{\alpha_i},\ \alpha_i \in \Delta \text{ for } 1 \le i \le n\right\}.$$

Example 13.4.2

1. Clearly any simple R-module is completely reducible.
2. Let $\{M_\alpha\}_{\alpha\in\Delta}$ be any family of simple R-modules and M be the (external) direct sum of $\{M_\alpha\}_{\alpha\in\Delta}$. Then, M is completely reducible.

   In fact, we prove below that any completely reducible module is necessarily a direct sum of a family of simple R-modules.

Theorem 13.4.3. Let N be an R-submodule of a *completely reducible* R-module M and $\{M_\alpha\}_{\alpha\in\Delta}$ be a family of simple R-submodules of M such that $M = \sum\limits_{\alpha\in\Delta} M_\alpha$. Then, there exists a subset I of Δ such that $\sum\limits_{\alpha\in I} M_\alpha$ is a direct sum and $M = N \oplus (\bigoplus\limits_{\alpha\in I} M_\alpha)$.

Proof: Here, we use the Zorn's lemma. Let

$$\mathbb{P} = \left\{J \subseteq \Delta : \sum_{\alpha\in J} M_\alpha \text{ is a direct sum and } N \cap \left(\sum_{\alpha\in J} M_\alpha\right) = \{0\}\right\}.$$

If $J = \emptyset$, then by $\sum_{\alpha \in J} M_\alpha$ we mean $\{0\}$. Since the empty set $\emptyset$ is a member of $\mathbb{P}$, we get that $\mathbb{P}$ is a nonempty class of subsets of Δ. It can be proved that $\mathbb{P}$ is closed under unions of chains (totally ordered subsets of $\mathbb{P}$, with respect to the inclusion ordering). Therefore, by Zorn's lemma, $\mathbb{P}$ has a maximal member. Let I be a maximal member of $\mathbb{P}$ and let

$$A = N \oplus \left(\sum_{\alpha \in I} M_\alpha \right) = N \oplus \left(\bigoplus_{\alpha \in I} M_\alpha \right).$$

We prove that $A = M$. For this it is enough if we can prove that $M_\beta \subseteq A$ for all $\beta \in \Delta$. Let $\beta \in \Delta$ be arbitrarily fixed. If $\beta \in I$, then $M_\beta \subseteq \sum_{\alpha \in I} M_\alpha \subseteq A$. Suppose that $\beta \notin I$. Now $M_\beta \cap A$ is an R-submodule of M_β. Since M_β is simple, we get that $M_\beta \cap A = \{0\}$ or M_β. Suppose, if possible, $M_\beta \cap A = \{0\}$. Then

$$M_\beta \bigcap \left(\bigoplus_{\alpha \in I} M_\alpha \right) = \{0\}$$

and hence $\sum_{\alpha \in I \cup \{\beta\}} M_\alpha$ is a direct sum and has zero intersection with N. This implies that $I \cup \{\beta\} \in \mathbb{P}$, which is a contradiction to the fact that I is a maximal member of $\mathbb{P}$, since $\beta \notin I$ and $I \subsetneq I \cup \{\beta\}$. Therefore, $M_\beta \cap A \neq \{0\}$ and hence $M_\beta \cap A = M_\beta$ so that $M_\beta \subseteq A$. Thus, $\sum_{\beta \in \Delta} M_\beta \subseteq A$ and hence $A = M$. Thus, $\sum_{\alpha \in I} M_\alpha$ is a direct sum and $M = N \oplus (\bigoplus_{\alpha \in I} M_\alpha)$.

Corollary 13.4.1. Let $\{M_\alpha\}_{\alpha \in \Delta}$ be a family of simple R-submodules of an R-module M. Then, there exists a subset I of Δ such that M is the direct sum of $\{M_\alpha\}_{\alpha \in I}$.

Proof: This follows from the Theorem 13.4.7 by taking $N = \{0\}$.

Corollary 13.4.2. An R-module M is completely reducible if and only if it is a direct sum of a family of simple R-submodules of M.

Worked Exercise 13.4.1. Prove that any nonzero R-submodule of a completely reducible R-module M is completely reducible and is a direct summand of M.

Answer: Let M be a completely reducible R-module and N be a nonzero R-submodule of M. Let $\{M_\alpha\}_{\alpha \in \Delta}$ be a family of simple R-submodules of M such that $M = \bigoplus_{\alpha \in \Delta} M_\alpha$. (by Corollary 13.4.2). By Theorem 13.4.3, there exists a subset I of Δ such that

$$M = N \oplus \left(\underset{\alpha \in I}{\oplus} M_\alpha \right).$$

Then,

$$N \cong \frac{M}{\left(\underset{\alpha \in I}{\oplus} M_\alpha \right)} = \frac{\left(\left(\underset{\alpha \in I}{\oplus} M_\alpha \right) \ominus \left(\underset{\alpha \in \Delta - I}{\oplus} M_\alpha \right) \right)}{\left(\underset{\alpha \in I}{\oplus} M_\alpha \right)} \cong \underset{\alpha \in \Delta - I}{\oplus} M_\alpha .$$

Thus, N is completely reducible and N is a direct summand of M.

EXERCISE 13(D)

1. Which of the following are simple modules and which of them are completely reducible?
   - (i) $\mathbb{Z}$, as a $\mathbb{Z}$-module
   - (ii) $\mathbb{R}$, as a $\mathbb{R}$-module
   - (iii) $\mathbb{R}$, as a $\mathbb{Q}$-module
   - (iv) $\mathbb{Q}$, as a $\mathbb{Z}$-module
   - (v) $\mathbb{Q}$, as a $\mathbb{Q}$-module
   - (vi) For any field F, $F[x]$ as an F-module.
2. Let X be any nonempty set and $\mathbb{R}$ be the ring of all real numbers. Let
   $$M = \{\alpha : X \rightarrow \mathbb{R} : |\alpha| \text{ is finite}\}.$$
   Prove that M is a completely reducible $\mathbb{R}$-module and the M is simple if and only if X is a singleton set.
3. Let R be ring of 2×2 matrices over a field F. Prove that R is a completely reducible R-module.
4. Let M be a completely reducible R-module and N be a proper R-submodule of M. Then, prove that the quotient M/N is a completely reducible R-module.
5. Let R be a ring with unity. Prove that R as an R-module is completely reducible if and only if every R-module M is completely reducible.
6. Prove that the direct sum of any family of completely reducible R-modules is completely reducible.

13.5 FREE MODULES

Modules over fields are best examples of free modules and are called vector spaces. In the next section, we are going to have a detailed discussion on vector spaces. In this section, we consider a generalization of vector spaces; that is, a module over a general ring with unity, which satisfies one of the most important properties of a vector space. Throughout this section, R stands for a nonzero ring with unity, unless otherwise stated.

Definition 13.5.1. Let M be an R-module. A finite sequence $\{x_1, x_2, \ldots, x_n\}$ of distinct elements in M is said to be *linearly independent* if, for any $a_1, a_2, \ldots, a_n \in R$,

$$a_1x_1 + a_2x_2 + \cdots + a_nx_n = 0 \Rightarrow a_1 = a_2 = \cdots = a_n = 0.$$

A finite sequence of distinct elements of M is said to be *linearly dependent* if it is not linearly independent. A subset X of M is said to be *linearly independent* if every finite sequence of distinct elements in X is linearly independent; otherwise X is called linearly dependent. Clearly, any linearly independent set does not contain 0.

Example 13.5.1

1. Consider R as an R-module. Then, clearly $\{1\}$ is linearly independent.
2. For any positive integer n, consider R^n as an R-module. For each $1 \leq i \leq n$, let e_i denote the n-tuple in which the ith coordinate is 1 and all other coordinates are 0. Then, $\{e_1, e_2, \ldots, e_n\}$ is linearly independent.
3. Let X be any nonempty set and consider R^X as an R-module. For each $x \in X$, define $e_x : X \to R$ by

$$e_x(y) = \begin{cases} 1 & \text{if } y = x \\ 0 & \text{if } y \neq x \end{cases}.$$

   Then, the set $X' = \{e_x : x \in X\}$ is linearly independent in the R-module R^X.
4. Consider the set $R[x]$ of polynomials over R. Then, $R[x]$ is an R-module in which $\{1, x, x^2, \ldots\}$ is a linearly independent set, since for any polynomial $f(x) = a_0 + a_1x + \cdots + a_nx^n$,

$$f(x) = 0 \Rightarrow a_i = 0 \text{ for all } i.$$

Definition 13.5.2. An R-module M is said to be a *free R-module* if there exists a subset B of M such that

1. B is linearly independent and
2. M is generated by B as an R-module.

In this case, B is called a basis for M.

In other words, M is called a free R-module if M has a basis. Of course, there can be more than one basis. For example, in any field F, as an F-module, every nonzero element constitutes a basis.

Example 13.5.2

1. Any ring R is a free R-module, since $\{1\}$ is a basis.
2. Also, as in Example 13.5.1 (2), R^n is a free R-module. The set $\{e_1, e_2, \ldots, e_n\}$ is a basis for R^n. Any element $x = (a_1, a_2, \ldots, a_n)$ can be expressed as

$$x = a_1e_1 + a_2e_2 + \cdots + a_ne_n.$$

   This basis $\{e_1, e_2, \ldots, e_n\}$ is called the *standard basis* of R^n.
3. The set $R[x]$ of polynomials over a ring R is a free R-module, since $\{1, x, x^2, \ldots\}$ is a basis for $R[x]$.
4. Consider the R-module R^X as in Example 13.5.1 (3). Then, R^X is not a free R-module. However, the R-submodule M of R^X given by

$$M = \{f \in R^X : |f| \text{ is finite}\}$$

   is a free R-module, since $\{e_x : x \in X\}$ is a basis for M.

The following is an interesting result on its own and it serves as a good example for free $\mathbb{Z}$-modules. Recall that any abelian group $(G, +)$ can be regarded as a $\mathbb{Z}$-module.

Theorem 13.5.1. Consider a cyclic group $(G, +)$ and regard it as a $\mathbb{Z}$-module. Then, G is a free $\mathbb{Z}$-module if and only if G is infinite (i.e., G is isomorphic to the group $\mathbb{Z}$ of integers).

Proof: Suppose that G is finite. Then, there exists a positive integer n such that $na = 0$ for all $a \in G$ (for example, we can take $n = |G|$) and hence there is no linearly independent set in G; in particular, G has no basis. Therefore, G is not a free $\mathbb{Z}$-module. Conversely, suppose that G is an infinite group. Since G is given to be cyclic, we get G is isomorphic to $\mathbb{Z}$ as $\mathbb{Z}$-module. Since $\mathbb{Z}$ is a free $\mathbb{Z}$-module, it follows that G is also a free $\mathbb{Z}$-module.

The following result suggests an alternate proof of the above theorem.

Theorem 13.5.2. Let M be a free R-module with a basis $\{x_1, x_2, \ldots, x_n\}$. Then, $M \cong R^n$, as R-modules.

Proof: Define $f: R^n \to M$ by

$$f(a_1, a_2, \ldots, a_n) = a_1x_1 + a_2x_2 + \cdots + a_nx_n.$$

Since M is generated $\{x_1, x_2, \ldots, x_n\}$, any element of M can be expressed as $a_1x_1 + a_2x_2 + \cdots + a_nx_n = f(a_1, a_2, \ldots, a_n)$ and hence f is a surjection. It can be easily verified that f is an R-homomorphism. Also,

$$\begin{aligned}(a_1, a_2, \ldots, a_n) \in \ker f &\Rightarrow f(a_1, a_2, \ldots, a_n) = 0\\ &\Rightarrow a_1x_1 + a_2x_2 + \cdots + a_nx_n = 0\\ &\Rightarrow a_1 = a_2 = \cdots = a_n = 0.\end{aligned}$$

Therefore, f is an R-monomorphism. Thus, $f : R^n \to M$ is an R-isomorphism and $M \cong R^n$, as R-modules.

In the next few results, we exhibit certain special properties of finitely generated free modules over commutative rings.

Theorem 13.5.3. Let M be a finitely generated free module over a commutative ring R. Then, all basis of M are finite and have the same number of elements.

Proof: Let $B = \{e_i\}_{i\in I}$ be a basis of M. Since M is finitely generated, there is finite set $\{x_1, x_2, \ldots, x_n\}$ generating M. For each $1 \le j \le n$, we have

$$x_j = \sum_{i\in I} a_{ji}e_i,\ a_{ji} \in R$$

in which all but finite number of a_{ji} are zero. For each $1 \le j \le n$, let

$$S_j = \{i \in I : a_{ji} \ne 0\}$$

and $S = \bigcap_{j=1}^{n} S_j$. Then, each S_j and hence S are finite subsets of I.
Let $D = \{e_i : i \in s\}$.
Then, D is a linearly independent finite set and generates M and hence D is a basis. Since $D \subseteq B$ and B is also a basis of M, if follows that $D = B$. Thus, B is a finite basis of M.

Let C be any other basis of M and $|C| = m$ and $|B| = n$. Then, by Theorem 13.5.2, $M \cong R^m$ and $M \cong R^n$. Therefore, there exists an R-isomorphism $g : R^m \to R^n$ and let $h = g^{-1}$. Suppose, if possible, that $m \ne n$. Without loss of generality, we can suppose that $m < n$. Let $\{e_1, e_2, \ldots, e_m\}$ and $\{f_1, f_2, \ldots, f_n\}$ be the standard basis of R^m and R^n, respectively. Let

$$g(e_i) = \sum_{j=1}^{n} a_{ji} f_j \quad \text{for each } 1 \le i \le m$$

and

$$g^{-1}(f_j)=\sum_{i=1}^{m} b_{ij}e_i \text{ for each } 1\le j\le n.$$

and let $A = (a_{ji})$ and $B = (b_{ij})$ be the corresponding $n \times m$ and $m \times n$ matrices, respectively. Then, for each i,

$$\begin{aligned}(e_i)=g^{-1}g(e_i)&=\sum_{j=1}^{n} a_{ji}g^{-1}(f_j)\\&=\sum_{j=1}^{n} a_{ji}\sum_{k=1}^{m} b_{kj}e_k.\end{aligned}$$

Therefore, $e_i = \sum_{k=1}^{m}\left(\sum_{j=1}^{n} b_{kj}a_{ji}\right)e_k$ for each $1 \le i \le m$. Since $\{e_i\}$ are linearly independent, we get that

$$\sum_{j=1}^{n} b_{kj}a_{ji} = \delta_{ki} = \begin{cases}1 & \text{if } k = i\\ 0 & \text{if } k \ne i\end{cases}.$$

Thus, the matrix product $BA = I_m$.

Consider the augmented matrices

$$A' = \begin{bmatrix}A & 0\end{bmatrix} \text{ and } B' = \begin{bmatrix}B\\0\end{bmatrix}.$$

These are $n \times n$ matrices, where each of the 0 blocks is a matrix of appropriate size. Then,

$$A'B' = I_n \text{ and } B'A' = \begin{bmatrix}I_m & 0\\ 0 & 0\end{bmatrix}.$$

This implies that $\det(A'B') = 1$ and $\det(B'A') = 0$. This is a contradiction, since A' and B' are $n \times n$ matrices over the commutative ring R and $\det(A'B') = \det(B'A')$. Thus, $m < n$ is impossible. Similarly, $n < m$ is impossible. Thus, $m = n$.

Definition 13.5.3. Let M be a finitely generated free module over a commutative ring R with unity. The number of elements in any basis of M is called the *rank of* M and denoted by rank(M).

Worked Exercise 13.5.1. Let e be an idempotent in a commutative ring with unity, $e \neq 0$ and $e \neq 1$. Then, prove that Re is not a free R-module.

Answer: Clearly, Re is an R-module. If $0 \neq x \in Re$, then $x = re$ for some $r \in R$ and

$$(1 - e)x = (1 - e)re = r(1 - e)e = r(e - e^2) = 0$$

and $1 - e \neq 0$. Therefore, there are no linearly independent sets in Re. Thus, the R-module Re is not a free R-module.

Worked Exercise 13.5.2. Let R be a ring with unity, M an R-module and $X \subseteq M$. Then, prove that X is a basis of M if and only if, for any R-module N and for any mapping $f : X \to N$, there exists unique R-homomorphism $\overline{f} : M \to N$ such that $\overline{f}(x) = f(x)$ for all $x \in X$.

Answer: Suppose that X is basis of M. Let N be any R-module and $f : X \to N$ any mapping. Define $\overline{f} : M \to N$ by

$$\overline{f}\,(y) = \sum_{i=1}^{n} a_i f(x_i) \text{ if } y = \sum_{i=1}^{n} a_i x_i \quad a_i \in R,\ x_i \in X.$$

Since any element of M can be uniquely expressed as $a_1x_1 + \cdots + a_nx_n$ with $a_i \in R$ and $x_i \in X$, $\overline{f}$ is well defined. It can be easily verified that $\overline{f}$ is an R-homomorphism and $\overline{f}(x) = f(x)$ for all $x \in X$.

Conversely, suppose that the given condition is satisfied. First, we prove that X generates M. Put

$$M' = \langle X \rangle = \left\{ \sum_{i=1}^{n} a_i x_i : a_i \in R \text{ and } x_i \in X \right\}.$$

Then, M' is an R-submodule of M and consider the quotient module M/M'. Define

$$f : X \to \frac{M}{M'} \text{ by } f(x) = M', \text{ the zero element in } \frac{M}{M'}.$$

Then, the natural map $g : M \to \frac{M}{M'}$ defined by $g(y) = y + M'$ and the zero homomorphism $h : M \to \frac{M}{M'}$ defined by $h(y) = M'$ for all $y \in M$ are both R-homomorphisms such that

$$g(x) = f(x) = h(x) \quad \text{for all } x \in X.$$

By the uniqueness of the R-homomorphism, it follows that $g = h$; that is,

$$y + M' = g(y) = h(y) = M' \text{ for all } y \in M$$

so that $y \in M'$ for all $y \in M$ and hence $M = M' = \langle x \rangle$.
Thus, X generates M.
Next, we prove that X is linearly independent. Let $x_1, x_2, \ldots, x_n$ be distinct elements in X and $a_1, a_2, \ldots, a_n \in R$ such that

$$a_1x_1 + a_2x_2 + \cdots + a_nx_n = 0.$$

We have to prove that $a_i = 0$ for each $1 \le i \le n$. Let $1 \le i \le n$ be fixed. Consider R as an R-module and define

$$f: X \to R \text{ by } f(x) = \begin{cases} 1 & \text{if } x = x_i \\ 0 & \text{if } x \neq x_i. \end{cases}$$

Then, there exists an R-homomorphism $\overline{f} : M \to N$ such that

$$\overline{f}(x) = f(x) \text{ for all } x \in X.$$

Now, we have
$0 = \overline{f}(a_1x_1 + \cdots + a_nx_n) = a_1\overline{f}(x_1) + \cdots + a_n\overline{f}(x_n) = a_i$ since $\overline{f}(x_i) = 1$ and $\overline{f}(x_j) = 0$ for all $j \neq i$. Therefore, $a_i = 0$ for each $1 \le i \le n$. Thus, X is linearly independent. Thus, X is a basis of M.

EXERCISE 13(E)

1. Which of the following R-modules M are free? Justify your answers.
   (i) $R = \mathbb{R} = M$
   (ii) $R = \mathbb{R}$ and $M = \mathbb{R}^n$, $n \in \mathbb{Z}^+$
   (iii) $R = \mathbb{R}^2$ and $M = \mathbb{R} \times \{0\}$
   (iv) $R = \mathbb{Z}$ and $M = \mathbb{Z}_n$, $n \in \mathbb{Z}^+$
   (v) $R = \mathbb{Z} = M$
   (vi) $R = \mathbb{Z}^2$ and $M = \{0\} \times \mathbb{Z}$
   (vii) $R = \mathbb{Z}^3$ and $M = \mathbb{Z} \times \{0\} \times \mathbb{Z}$
   (viii) $R = \mathbb{Q}$ and $M = \mathbb{Q}^n$, $n \in \mathbb{Z}^+$.
2. Let B be a basis of a free R-module M. Then, prove that $M = \bigoplus_{x \in X} Rx$.
3. Prove that the direct sum of a family of free R-modules is again a free R-module. Is this true for direct products?

4. If $M_1, M_2, \ldots, M_n$ are free R-modules, prove that the direct product $M_1 \times M_2 \times \cdots \times M_n$ is a free R-module.

5. Prove that $\mathbb{Q}$ is not a free $\mathbb{Z}$-module.

6. Prove that every ideal of $\mathbb{Z}$ is a free $\mathbb{Z}$-module.

7. Let R be an integral domain and $x \in R$. Then, prove that Rx is a free R-module.

8. Prove that every R-module is a homomorphic image of a free R-module.

9. Let M and N be R-modules and $M \cong N$. Prove that M is a free R-module if and only if N is a free R-module.

10. Prove that any finitely generated R-module is isomorphic to a quotient of a free R-module.

11. If B is a basis for an R-module M, then prove that B is a minimal generating set; that is, no proper subset of B generates M.

12. Prove that any basis B of a free R-module M is a maximal linearly independent set in M; that is, any subset of M containing B properly is linearly dependent.

13. Let R be a ring with unity, M be an R-module and $X \subseteq M$. Prove that X is a basis of M if and only if any element y of M can be uniquely expressed as

$$y = a_1x_1 + \cdots + a_nx_n$$

where $a_i \in R$ and $x_1, x_2, \ldots, x_n$ are distinct elements of X.

14. Let F be a free R-module and M be another R-module. If N is an R-submodule of M and $f : F \to (M/N)$ is an R-homomorphism, then prove that there exists an R-homomorphism $g : F \to M$ such that

$$f(x) = g(x) + N \text{ for all } x \in F.$$

15. If $f : M \to N$ is an R-epimorphism of R-modules and N is a free R-module, then prove that $\ker f$ is a direct summand of M.

13.6 VECTOR SPACES

Modules over a field are called vector spaces and these play a vital role in any branch of mathematics. In particular, the homomorphisms between vector spaces exhibit rich structural properties. In this, we discuss briefly about the vector spaces and homomorphisms between these.

Definition 13.6.1. Let F be a field. Then, any F-module V is called a *vector space over F* or a *F-vector space* or, simply, a *vector space*. The elements of V are called *vectors* and elements of the field F are called *scalars*.

Definition 13.6.2. Let V and W be vector spaces over a field F. Then, an F-homomorphism $f : V \to W$ is called a *linear transformation* of V into W.

Definition 13.6.3. Let V be a vector space over a field F. Then, a F-submodule W of V is called a *subspace* of V. For any vectors $x_1, x_2, \ldots, x_n$ and for any scalars $a_1, a_2, \ldots, a_n$, the sum

$$a_1x_1 + a_2x_2 + \cdots + a_nx_n$$

is called a *linear combination* of $x_1, x_2, \ldots, x_n$. For any subset X, the subspace generated by X is called *linear span* of X in V and is denoted, as usual, by $<X>$. Recall that

$$<X> = \{a_1x_1 + a_2x_2 + \cdots + a_nx_n : x_i \in X, a_i \in F\}.$$

The following is a special property of vector spaces.

Theorem 13.6.1. Any nonzero vector space over any field F is free (i.e., a free F-module).

Proof: Let F be a field and V be a nonzero vector space over F. Let

$$\mathbb{P} = \{X \subseteq V : X \text{ is linearly independent}\}.$$

First note that for any $0 \neq a \in F$ and $x \in V$,

$$ax = 0 \Rightarrow a^{-1}(ax) = 0 \Rightarrow (a^{-1}a)x = 0 \Rightarrow x = 0$$

and hence, for any $0 \neq x \in V$ and $a \in F$,

$$ax = 0 \Rightarrow a = 0.$$

Therefore, $\{x\}$ is linearly independent for any $0 \neq x \in V$ and hence $\mathbb{P}$ is a nonempty class of subsets of V. Let $\{X_i\}_{i\in I}$ be a chain in $\mathbb{P}$ and $X = \bigcup_{i\in I} X_i$. Then, X is linearly independent; for, if $a_1, a_2, \ldots, a_n \in F$ and $x_1, x_2, \ldots, x_n \in X$ such that $a_1x_1 + a_2x_2 + \cdots + a_nx_n = 0$, then $x_j \in X_{i_j}$ for some $i_j \in I$ and, since $\{X_i\}_{i\in I}$ is a chain, there exists $i \in I$ such that, $X_{i_j} \subseteq X_i$ for all $1 \leq j \leq n$ and hence $x_1, x_2, \ldots, x_n \in X_i$ and by the linear independence of X_i, it follows that $a_1 = a_2 = \cdots = a_n = 0$. Therefore, $X \in \mathbb{P}$. That is, $\mathbb{P}$ is closed under unions of chains. By the Zorn's lemma, $\mathbb{P}$ has a maximal member, say B. Now, we prove that B is a basis of V. Since $B \in \mathbb{P}$, B is linearly independent. Let $0 \neq x \in V$ and $x \notin B$. Then, by the maximality of B, $B \cup \{x\} \notin \mathbb{P}$; that is, $B \cup \{x\}$ is linearly dependent and hence there exist $x_1, x_2, \ldots, x_n \in B$ and $a_1a_2, \ldots, a_n, a \in F$, with $a \neq 0$, such that

$$a_1x_1 + a_2x_2 + \cdots + a_nx_n + ax = 0.$$

Since $0 \neq a \in F$ and F is a field, a^{-1} exists in F and

$$x = (-a^{-1}a_1)x_1 + (-a^{-1}a_2)x_2 + \cdots + (-a^{-1}a_n)x_n$$

Therefore, $x \in \langle B \rangle$ for all $x \in V$, so that $V = \langle B \rangle$. Thus, B generates V. Thus, B is a basis of V and hence V is a free F-module.

Corollary 13.6.1. Let V be a vector space over a field F. Then, any linearly independent subset of V can be extended to a basis of V.

Proof: Let S be a linearly independent subset of V and consider the class

$$\mathbb{P} = \{X \subseteq V : X \text{ is linearly independent and } S \subseteq X\}.$$

On the lines of the proof given above, we can prove that $\mathbb{P}$ has a maximal member, say B. Then, B is a basis of V and $S \subseteq B$.

Theorem 13.6.2. Let V be a vector space over a field F. Then, any subspace of V is a direct summand of V.

Proof: Let W be a subspace of V. Then, W is a vector space over F and, by Theorem 13.6.1, W has a basis B. Now, since B is a linearly independent subset of V and hence, by Corollary 13.6.1, there exists a basis C of V containing B. Now, put $W' = \langle C - B \rangle$. We prove that $V = W \oplus W'$

$$x \in W \cap W' \Rightarrow x = \sum_{i=1}^{n} a_i x_i = \sum_{j=1}^{m} b_j y_j,$$

where $a_i, b_j \in F$ and $x_1, \ldots, x_n \in B$ and $y_1, \ldots, y_m \in C - B$

$$\begin{aligned}
&\Rightarrow a_1x_1 + \cdots + a_nx_n - b_1y_1 - b_2y_2 - \cdots - b_my_m = 0 \\
&\Rightarrow a_1 = \ldots = a_n = b_1 = \cdots = b_m = 0 \\
&\quad \text{(since } C \text{ is linearly independent)} \\
&\Rightarrow x = 0.
\end{aligned}$$

Therefore, $W \cap W' = \{0\}$. Also, clearly

$$V = \langle C \rangle = \langle B \rangle + \langle C - B \rangle = W + W'.$$

Thus, $V = W \oplus W'$ and hence W is a direct summand of V.

Theorem 13.6.3. Any vector space is completely reducible.

Proof: Let V be a vector space over a field and let B be a basis of V. Then, it can be proved that

$$V = \bigoplus_{x \in B} Fx.$$

Also, for any $0 \neq x \in V$, $Fx \cong F$ as vector spaces over F and, since F is a field, F is a single F-module. Therefore, each Fx is a simple F-module. Thus, V is completely reducible.

Definition 13.6.4. Let V be a finitely generated vector space over a field F. Then, the rank of V (that is, the number of elements in any basis of V) is called the *dimension of V over F* and is denoted by $\dim_F V$. If V is not finitely generated, then V is said to be infinite dimensional. Recall that, if $\dim_F V = n < \infty$, then $V \cong F^n$, as vector spaces over F.

Recall that, in F^n, the elements $e_1, e_2, \ldots, e_n$ form a basis, where e_i is the n-tuple in which the ith coordinate is 1 and the other coordinates are 0. This is called an ordered basis for F^n. Now, let us turn our attention to linear transformations (F-homomorphisms) of finite dimensional vector spaces.

Theorem 13.6.4. Let V and W be vector spaces of dimensions m and n, respectively, over a field F. Then,

$$\text{Hom}_F(V, W) \cong F^{m \times n}$$

as vector spaces over F, where $\text{Hom}_F(V, W)$ is the vector space of all linear transformations of V into W.

Proof: Fix bases $B = \{e_1, e_2, \ldots, e_m\}$ and $C = \{f_1, f_2, \ldots, f_n\}$ for V and W, respectively. If $\alpha : V \to W$ is a linear transformation, then, for each $1 \leq i \leq m$, $\alpha(e_i) \in w = <C>$ and hence

$$\alpha(e_i) = \sum_{j=1}^{n} a_{ij} f_j.$$

Then, the $m \times n$ matrix $A_\alpha = (a_{ij})$ is called the matrix of α with respect to the bases B and C. Conversely, if $A = (a_{ij})$ is a $m \times n$ matrix over F, then we can define $\alpha : V \to W$ by

$$\alpha(e_i) = \sum_{j=1}^{n} a_{ij} f_j \quad \text{for each } 1 \leq i \leq n$$

$$\text{and} \quad \alpha(x) = \sum_{i=1}^{m} a_i \alpha(e_i) \ \text{ if } x = \sum_{i=1}^{m} a_i e_i.$$

Then, α is a linear transformation of V into W such that $A_\alpha = A$. It is a straight forward verification to prove that $\alpha \mapsto A_\alpha$ is an isomorphism of $\text{Hom}_F(V, W)$ onto $F^{m\times n}$ as vector spaces over F.

Corollary 13.6.2. If V and W are vector spaces of dimensions m and n, respectively, over a field F, then $\text{Hom}_F(V, W)$ is a vector space over F and

$$\dim_F(\text{Hom}_F(V, W)) = mn.$$

Corollary 13.6.3. For any n-dimensional vector space V over F,

$$\text{Hom}_F(V, V) \cong M_n(F)$$

where $M_n(F)$ is the vector space of all $n \times n$ matrices over F.

In fact, both $\text{Hom}_F(V, V)$ and $M_n(F)$ are algebras over F in the sense of the following definition.

Definition 13.6.5. Let A be a vector space over a field. If there is another binary operation on A such that $(A, +, \cdot)$ is a ring in which the ring multiplication and the scalar multiplication are compatible with each other, then A is called an *algebra* over F.

Corollary 13.6.4. For any n-dimensional vector space over a field F, $\text{Hom}_F(V, V)$ and $M_n(F)$ are isomorphic as algebras over F.

Now, we consider change of basis and their effect on the matrices of linear transformation. First, we have the following theorem.

Theorem 13.6.5. Let V be a n-dimensional vector space over a field F and $\{e_1, e_2, \ldots, e_n\}$ be a basis of V. Let $A = (a_{ij})$ be an $n \times n$ matrix over F and

$$e_i' = a_{i_1} e_1 + a_{i_2} e_2 + \cdots + a_{i_n} e_n \ \text{ for each } 1 \le i \le n.$$

Then, $\{e_1', e_2', \ldots, e'_n\}$ is also a basis of V if and only if A is an invertible matrix (unit) in the ring $M_n(F)$ of all $n \times n$ matrices over F.

Proof: Let $B = \{e_1, e_2, \ldots, e_n\}$ and $B' = \{e_1', e_2', \ldots, e'_n\}$.
First, we suppose that B' is a basis of V. For, each $1 \le i \le n$, we have

$$e_i = \sum_{j=1}^{n} b_{ij} e'_j \text{ for some } b_{ij} \in F.$$

Now, consider

$$e_i = \sum_{j=1}^{n} b_{ij} e'_j = \sum_{j=1}^{n} b_{ij} \left(\sum_{k=1}^{n} a_{jk} e_k \right)$$

$$\therefore \quad e_i = \sum_{k=1}^{n} \left(\sum_{j=1}^{n} b_{ij} a_{jk} \right) e_k.$$

Since $\{e_1, e_2, \ldots, e_n\}$ are linearly independent, we get that

$$\sum_{j=1}^{k} b_{ij} a_{jk} = \delta_{ik} = \begin{cases} 1 & \text{if } i = k \\ 0 & \text{if } i \neq k \end{cases}.$$

This shows that the matrix product

$$A'A = I \text{ where } A' = (b_{ij}) \text{ and } A = (a_{ij}).$$

Similarly, by considering

$$e'_i = \sum_{j=1}^{n} a_{ij} e_j = \sum_{j=1}^{n} a_{ij} \sum_{k=1}^{n} b_{jk} e'_k = \sum_{k=1}^{n} \left(\sum_{j=1}^{n} a_{ij} b_{jk} \right) e'_k,$$

we get that $AA' = I$. Therefore, A is an invertible matrix and A' is the inverse of A.

Conversely suppose that A is an invertible matrix and $A' = (b_{ij})$ be the inverse of A. We prove that B' is a basis of V. Let

$$r_1 e'_1 + r_2 e'_2 + \cdots + r_n e'_n = 0, \text{ where } r_i \in F.$$

Then,

$$\sum_{j=1}^{n} r_j \left(\sum_{i=1}^{n} a_{ji} e_i \right) = 0$$

$$\sum_{i=1}^{n} \left(\sum_{j=1}^{n} r_j a_{ji} \right) e_i = 0.$$

Since $e_1, e_2, \ldots, e_n$ are linearly independent, we get that

$$\sum_{j=1}^{n} r_j a_{ji} = 0 \text{ for all } 1 \le i \le n.$$

Let

$$P = \begin{pmatrix} r_1 \\ r_2 \\ \vdots \\ r_n \end{pmatrix}$$

be the $n \times 1$ matrix. Then,

$$AP = \begin{pmatrix} 0 \\ 0 \\ \vdots \\ 0 \end{pmatrix}.$$

Therefore,

$$A'AP = \begin{pmatrix} 0 \\ 0 \\ \vdots \\ 0 \end{pmatrix}$$

and hence

$$P = \begin{pmatrix} 0 \\ 0 \\ \vdots \\ 0 \end{pmatrix}$$

since $A'A = I_n$.

Thus, $r_1 = r_2 = \cdots = r_n = 0$. Therefore, $e_1', e_2', \ldots, e_n'$ are linearly independent and hence $\{e_1', e_2', \ldots, e_n'\}$ forms a basis of V.

Note: The matrix $A = (a_{ij})$ given in the above theorem is called the *matrix of transformation from the basis B' to the basis B.*

The following describes the effect of a change of a basis on the matrix of a linear transformation. First, recall that if $B = \{e_1, e_2, \ldots, e_m\}$ and $C = \{f_1, f_2, \ldots, f_n\}$ are basis of vector spaces V and W, respectively, and $\alpha : V \to W$ is a linear transformation such that

$$\alpha(e_i) = \sum_{j=1}^{n} a_{ij} f_j \text{ for each } 1 \le i \le m,$$

then the $m \times n$ matrix $A = (a_{ij})$ is called the *matrix of* α *with respect to the bases B and C*.

Theorem 13.6.6. Let V and W be vector spaces of dimensions m and n, respectively, over a field F. Let $B = \{e_1, e_2, \ldots, e_m\}$ and $C = \{f_1, f_2, \ldots, f_n\}$ bases of V and W, respectively, and $A = (a_{ij})$ be the matrix of a linear transformation $\alpha : V \to W$ with respect to the bases B and C.

1. Let $B' = \{e_1', e_2', \ldots, e_m'\}$ and $C' = \{f_1', f_2', \ldots, f_n'\}$ be new bases of V and W, respectively. Then, the matrix of α with respect to the bases B' and C' is of the form PAQ^{-1} where P and Q are matrices of transformations from B' to B and C' to C, respectively.
2. Conversely, if P and Q are $m \times m$ and $n \times n$ invertible matrices, respectively, then there exist bases B' and C' of V and W, respectively, such that PAQ^{-1} is the matrix of α with respect to the bases B' and C'.

Proof: (1) We have $\alpha(e_i) = \sum_{j=1}^{n} a_{ij} f_j$ for each $1 \le i \le m$.
Let $A' = (a_{ij})'$ be the matrix of α with respect to the bases B' and C'. Then, we have

$$\alpha(e_i') = \sum_{j=1}^{n} a_{ij}' f_j' \text{ for each } 1 \le i \le m.$$

Let $P = (p_{ij})$ and $Q = (q_{ij})$ be matrices of transformations from B' to B and C' to C, respectively.
Also let $Q^{-1} = (q'_{ij})$ be the matrix of transformation from C to C' (by Theorem 13.6.5). Now, we have

$$f_j = \sum_{k=1}^{n} q_{jk}' f_k' \text{ for } 1 \le j \le n$$

$$\text{and} \quad e_i' = \sum_{l=1}^{m} p_{il} e_l \text{ for } 1 \le i \le m.$$

Therefore,

$$\begin{aligned}\alpha(e_i') &= \alpha\left(\sum_{l=1}^{m} p_{il} e_l\right) \\ &= \sum_{l=1}^{m} p_{il}\alpha(e_l) \\ &= \sum_{l=1}^{m} p_{il}\left(\sum_{j=1}^{n} a_{lj} f_j\right) \\ &= \sum_{l=1}^{m} p_{il}\left(\sum_{j=1}^{n} a_{lj}\left(\sum_{k=1}^{n} q'_{jk} f'_k\right)\right) \\ &= \sum_{k=1}^{n}\left(\sum_{l=1}^{m} p_{il}\left(\sum_{j=1}^{n} a_{lj} q'_{jk}\right)\right) f'_k.\end{aligned}$$

Thus, the matrix α with respect to the bases B' and C' is equal to PAQ^{-1}. This proves (1). The proof of (2) is similar to the above and to the proof of Theorem 13.6.5.

Worked Exercise 13.6.1. Let F be any field and consider the vector spaces F^3 and F^2 over F. Define $\alpha : F^3 \rightarrow F^2$ by

$$\alpha(a, b, c) = (a + b + c, b + c)$$

Then, prove that α is a linear transformation and determine the matrix of α with respect to the standard basis of F^3 and F^2.

Answer: For any $x = (a, b, c)$ and $y = (a', b', c') \in F^3$ and r and $s \in F$, we have

$$\begin{aligned}\alpha(rx + sy) &= \alpha(ra + sa', rb + sb', rc + rc') \\ &= (ra + sa' + rb + sb' + rc + sc', rb + sb' + rc + sc') \\ &= r(a + b + c, b + c) + s(a' + b' + c', b' + c') \\ &= r\alpha(x) + s\alpha(y).\end{aligned}$$

Therefore, α is a linear transformation of F^3 into F^2. Let $\{e_1, e_2, e_3\}$ and $\{f_1, f_2\}$ be standard bases of F^3 and F^2, respectively. Then

$$e_1 = (1, 0, 0), e_2 = (0, 1, 0) \text{ and } e_3 = (0, 0, 1)$$
$$\text{and } \; f_1 = (1, 0), f_2 = (0, 1).$$

Now,
$$\alpha(e_1) = \alpha(1, 0, 0) = (1, 0) = 1 \cdot f_1 + 0 \cdot f_2$$
$$\alpha(e_2) = \alpha(0, 1, 0) = (1, 1) = 1 \cdot f_1 + 1 \cdot f_2$$
$$\alpha(e_3) = \alpha(0, 0, 1) = (1, 1) = 1 \cdot f_1 + 1 \cdot f_2.$$

Therefore,

$$\begin{pmatrix} 1 & 0 \\ 1 & 1 \\ 1 & 1 \end{pmatrix}$$

is the matrix of α with respect to the bases $\{e_1, e_2, e_3\}$ and $\{f_1, f_2\}$.

EXERCISE 13(F)

1. Let V and W be vector spaces over a field F and $\alpha : V \rightarrow W$ be a mapping. Prove that α is a linear transformation if and only if $\alpha(ax + by) = a\alpha(x) + b\alpha(y)$ for all $x, y \in V$ and $a, b \in F$.

2. Let $V = F^5$ and F be a field. Let

   $$e_1 = (2, 0, 0, 0, 0),\ e_2 = (2, 1, 0, 0, 0) \text{ and } e_3 = (1, 2, 3, 0, 0).$$

   Prove that $\{e_1, e_2, e_3\}$ is linearly independent in V and extend this to a basis of V.

3. Define $\alpha : F^3 \rightarrow F^2$ by $\alpha(a, b, c) = (a + b + c, b + c)$. Determine the matrix of α with respect to the bases $\{(-1, 0, 2), (0, 1, 1), (3, -1, 0)\}$ and $\{(-1, 1), (1, 0)\}$ of F^3 and F^2, respectively.

4. Let F be a field and $V = \{f(x) \in F[x] : \deg(f(x)) \leq 4\}$. Prove that $B = \{1, x, x^2, x^3, x^4\}$ and $C = \{1, 1 + x, 1 + x + x^2, 1 + x + x^2 + x^3, 1 + x + x^2 + x^3 + x^4\}$ are bases of V. If D is the differentiation operator on V, then determine the matrix of D with respect to each of the bases B and C. Also determine the matrix of transformation from B to C and from C to B.

5. For any subset X of a vector space V over a field F, let $\langle X \rangle$ be the linear span of X in V. Prove the following for any subsets X and Y in V:

   (1) $X \subseteq \langle X \rangle$

   (2) $\langle\langle X \rangle\rangle = \langle X \rangle$

   (3) $\langle X \cup Y \rangle = \langle X \rangle + \langle Y \rangle$

   (4) $\langle X \rangle = \cup \{\langle Y \rangle : Y \text{ is finite subset of } X\}$.

PART IV
Field Theory

14

Extension Fields

It is well known that the field $\mathbb{Q}$ of rational numbers is a subfield of the field $\mathbb{R}$ of real numbers and, in this case, we say that $\mathbb{R}$ is an extension field of $\mathbb{Q}$. Likewise, the field $\mathbb{C}$ of complex numbers is an extension of $\mathbb{R}$. Let us recall that the polynomial $1 + x^2$ has no root in $\mathbb{R}$. However, there is an extension field, namely $\mathbb{C}$, containing a root of $1 + x^2$. In this chapter, we discuss in detail about the existence of an extension field containing roots of a given polynomial over a given field.

The field $\mathbb{R}$ of real numbers has a deficit that not all polynomials over $\mathbb{R}$ have roots in $\mathbb{R}$. The field $\mathbb{C}$ of complex number is an extension of $\mathbb{R}$ containing all the roots of any polynomial over $\mathbb{C}$. Such fields like $\mathbb{C}$ are called algebraically closed fields. We discuss these and similar concepts in the present chapter.

Here afterwards F denotes an arbitrary field, unless otherwise stated. Also, a homomorphism of one field F into another field K is always assumed to be a ring homomorphism of F into K carrying the unity in F onto the unity in K.

14.1 EXTENSIONS OF A FIELD

Any field K can be considered as a vector space over any of its subfield and we can discuss their dimensions and related concepts. If a field K has a subfield which is an isomorphic copy of another field F, then we can treat F itself as a subfield of K. Formally, we have the following definition.

Definition 14.1.1. Let F and K be fields. Then, K is said to be a *field extension of F* or, F is said to be a *subfield of K*, if there exists a monomorphism (or *embedding*) $\sigma : F \to K$. Note that this σ is an injective homomorphism of rings such that $\sigma(1) = 1$.

If $\sigma : F \to K$ is a monomorphism, then F is isomorphic to $\sigma(F)$ which is a subfield of K and hence we can consider a field extension K of F to be a field containing F as a subfield. We write $F \subseteq K$ when K is an extension of F. Also, in this case, for any $a \in F$ and $x \in K$, we have $ax \in K$ (note that a is identified with $\sigma(a)$ in K). Therefore, we have mapping $F \times K \to K$ given by $(a, x) \mapsto ax$. With this as the scalar multiplication, we get that the additive group $(K, +)$ becomes a vector space over F.

Definition 14.1.2. If K is a field extension of F, then the dimension of the vector space K over F is called the *degree* of K over F and is denoted by $[K : F]$. K is said to be *finite* or *infinite extension* of F according as the degree of K over F is finite or infinite.

Example 14.1.1

1. For any field F, $[F : F] = 1$.
2. The degree of the field $\mathbb{C}$ of complex numbers over the field $\mathbb{R}$ of real numbers is 2, since $\{1, i\}$ is a basis of $\mathbb{C}$ over $\mathbb{R}$.
3. Let F be any field and $F[x]$ be the ring of polynomials over F. Let K be the field of quotient of $F[x]$. Then, K is a field extension of F. Also, consider the set $\{1, x, x^2, \ldots\}$. If $a_0 1 + a_1 x + a_2 x^2 + \cdots + a_n x^n = 0$, then $a_0 = a_1 = \cdots = a_n = 0$ and hence $\{1, x, x^2, \ldots\}$ is an infinite linearly independent subset of K. Therefore, K is an infinite extension of F.
4. Let $\mathbb{Q}[\sqrt{2}] = \{a + b\sqrt{2} : a \text{ and } b \in \mathbb{Q}\}$. Then, $\mathbb{Q}[\sqrt{2}]$ is a field extension of $\mathbb{Q}$ and is of degree 2 over $\mathbb{Q}$, since $\{1, \sqrt{2}\}$ is a basis of $\mathbb{Q}[\sqrt{2}]$ over $\mathbb{Q}$.

The following theorem about the degrees of finite extensions of fields is very important and useful.

Theorem 14.1.1. Let F be any field, K be a field extension of F and L be a field extension of K (i.e., $F \subseteq K \subseteq L$). Then, $[L : F]$ is finite if and only if both $[L : K]$ and $[K : F]$ are finite and, in this case,

$$[L : F] = [L : K]\,[K : F].$$

Proof: Suppose that $[L : F]$ is finite. Since K is a subspace of L over F, we get that $[K : F] \leq [L : F]$ and hence $[K : F]$ is finite. Also, if B is a basis of L

over K, then B is a linearly independent subset of L over K and hence linearly independent over F also (since $F \subseteq K$) and therefore $|B| \leq [L : F]$. Therefore, $[L : K] \leq [L : F]$ and hence $[L : K]$ is finite.

Conversely suppose that $[L : K] = m$ and $[K : F] = n$, where m and n are positive integers. Let $\{x_1, \ldots, x_m\}$ be a basis of L over K and $\{y_1, \ldots, y_n\}$ be a basis of K over F. Now, let

$$B = \{x_i y_j;\ 1 \leq i \leq m \text{ and } 1 \leq j \leq n\}.$$

Then, B is a subset of L and $|B| = mn$ (since x_i's are independent over K and $y_j \in K$, $x_i y_j = x_r y_s \Rightarrow i = r$ and $j = s$). We prove that B is a basis of L over F. To prove the linear independency of B over F, let $a_{ij} \in F$ for $1 \leq i \leq m$ and $1 \leq j \leq n$ such that

$$\sum_{\substack{1 \leq i \leq m \\ 1 \leq j \leq n}} a_{ij} x_i y_j = 0.$$

For each $1 \leq i \leq m$, let $b_i = \sum_{j=1}^{m} a_{ij} y_j \in K$.

Then, $b_i \in K$, since $a_{ij} \in F \subseteq K$ and $y_j \in K$. Now,

$$\sum_{i=1}^{m} b_i x_i = \sum_{i=1}^{m} \left(\sum_{j=1}^{n} a_{ij} y_j \right) x_i = 0.$$

Since $x_1, x_2, \ldots, x_m$ are linearly independent over K and $b_i \in K$, it follows that $b_i = 0$ for each $1 \leq i \leq m$. Again, for each $1 \leq i \leq m$,

$$\sum_{j=1}^{n} a_{ij} y_j = b_i = 0$$

and, since $\{y_1, y_2, \ldots, y_n\}$ is linearly independent over F and $a_{ij} \in F$, it follows that

$$a_{ij} = 0 \text{ for each } 1 \leq i \leq m \quad \text{and} \quad 1 \leq j \leq n.$$

Thus, B is linearly independent over F.

Next, we prove that B generates L over F. Let $x \in L$. Since x_i's generate L over K, we get that

$$x = k_1 x_1 + k_2 x_2 + \cdots + k_n x_m \quad \text{for some } k_1, \ldots, k_m \in K.$$

Again, since y_j's generate K over F, we get that

$$K_i = \sum_{j=1}^{n} a_{ij} y_j, \text{ for each } 1 \leq i \leq m$$

where $a_{ij} \in F$. Therefore, we have

$$x = \sum_{i=1}^{m} k_i x_i = \sum_{i=1}^{m} \left(\sum_{j=1}^{n} a_{ij} y_j \right) x_i = \sum_{\substack{1 \leq i \leq m \\ 1 \leq j \leq n}} a_{ij} x_i y_j.$$

Therefore, B generates L over F and hence $[L : F]$is finite. Also,

$$[L : F] = |B| = mn = [L : K]\,[K : F].$$

Corollary 14.1.1. If $F \subseteq K \subseteq L$ are fields and $[L : F]$ is finite, then $[L : K]$ and $[K : F]$ are divisors of $[L : F]$.

Corollary 14.1.2. Let $F \subseteq K \subseteq L$ be fields such that $[L : F]$ is a prime number. Then, $K = F$ or $K = L$.

Proof: Let $[L : F] = p$. Then, $[K : F]$ is a divisor of the prime p and hence $[K : F] = 1$ or p. If $[K : F] = 1$, then $K = F$. If $[K : F] = p$, then $[L : K] = 1$ (since $[L : K]\,[K : F] = [L : F] = p$) and hence $K = L$.

Corollary 14.1.3. Let $F_1 \subseteq F_2 \subseteq \ldots \subseteq F_n$ be fields such that $[F_i : F_{i-1}]$ is finite for all $1 \leq i \leq n$. Then, $[F_n : F_1]$ is finite.

Proof: Apply induction on n.

Worked Exercise 14.1.1. Let $F_1 \subseteq F_2 \subseteq \ldots \subseteq F_n$ be fields. Then, prove that $[F_n : F_1]$ is finite if and only if $[F_i : F_j]$ is finite for all $i > j$ and, in this case,

$$[F_n : F_1] = \prod_{i=2}^{n} [F_i : F_{i-1}].$$

Answer: Suppose that $[F_n : F_1]$ is finite. Then, for each $1 < i \leq n$, $[F_i : F_1]$ is finite and $[F_i : F_j]$ is finite for all $j < i$. Conversely, suppose that $[F_i : F_j]$ is finite for all $i > j$. Then, $[F_i : F_{i-1}]$ is finite for all $1 < i \leq n$ and, in particular, $[F_n : F_1]$ is finite. Also, in this case,

$$
\begin{aligned}
[F_n : F_1] &= [F_n : F_{n-1}]\,[F_{n-1} : F_1] \\
&= [F_n : F_{n-1}]\,[F_{n-1} : F_{n-2}]\,[F_{n-2} : F_1] \\
&= \prod_{i=2}^{n}[F_i : F_{i-1}] \text{ (by induction on } n)
\end{aligned}
$$

Worked Exercise 14.1.2. Let p be a prime number and

$$\mathbb{Q}(\sqrt{p}) = \{a + b\sqrt{p} \,:\, a \text{ and } b \in \mathbb{Q}\}.$$

Then, prove that $\mathbb{Q}(\sqrt{p})$ is a field extension of $\mathbb{Q}$ and is of degree 2 over $\mathbb{Q}$.

Proof: First, let us recall that $\sqrt{p}$ is a real number which is not rational. For any a, b, c and $d \in \mathbb{Q}$, we have

$$(a+b\sqrt{p})-(c+d\sqrt{p}) = (a-c)+(b-d)\sqrt{p}$$

$$(a+b\sqrt{p})(c+d\sqrt{p}) = (ac+bdp)+(ad+bc)\sqrt{p}$$

and $$\frac{1}{a+b\sqrt{p}} = \frac{a-b\sqrt{p}}{a^2-pb^2} = \frac{a}{a^2-pb^2}+\left(\frac{-b}{a^2-pb^2}\right)\sqrt{p} \in \mathbb{Q}(\sqrt{p}).$$

Therefore, $\mathbb{Q}(\sqrt{p})$ is a subfield of $\mathbb{R}$, clearly $\mathbb{Q} \subseteq \mathbb{Q}(\sqrt{p})$. ◀

Therefore, $\mathbb{Q}(\sqrt{p})$ is a field extension of $\mathbb{Q}$. Also, since $\{1, \sqrt{p}\}$ is a basis of $\mathbb{Q}(\sqrt{p})$ over $\mathbb{Q}$, it follows that $[\mathbb{Q}(\sqrt{p}) : \mathbb{Q}] = 2$. That is, $\mathbb{Q}(\sqrt{p})$ is of degree 2 over $\mathbb{Q}$.

Worked Exercise 14.1.3. Let K be the smallest subfield of $\mathbb{R}$ containing $\mathbb{Q} \cup \{\sqrt{2}, \sqrt{3}\}$. Then, prove that K is a finite extension of $\mathbb{Q}$ and determine the degree of K over $\mathbb{Q}$.

Answer: Let $F = \mathbb{Q}(\sqrt{2}) = \{a + b\sqrt{2} : a \text{ and } b \in \mathbb{Q}\}$ by Worked Exercise 14.1.2, F is an extension of $\mathbb{Q}$ and $[F : \mathbb{Q}] = 2$.

Now, observe that $\sqrt{3} \notin F$; for, if $\sqrt{3} = a + b\sqrt{2}$ with a and $b \in \mathbb{Q}$, then $a \neq 0$ (otherwise $3 = 2b^2$) and $b \neq 0$ and hence $ab \neq 0$ and therefore $3 = (a + b\sqrt{2})^2 = a^2 + 2b^2 + 2ab\sqrt{2}$ which is not true. Now, consider

$$F(\sqrt{3}) = \{a + b\sqrt{3} \,:\, a \text{ and } b \in F\}.$$

Then, it can be easily proved that $F(\sqrt{3})$ is an extension of F and $\{1, \sqrt{3}\}$ is a basis of $F(\sqrt{3})$ over F and hence $[F(\sqrt{3}) : F] = 2$. Now, we have $\mathbb{Q} \subseteq F \subseteq F(\sqrt{3})$ and hence

$$[F(\sqrt{3}) : \mathbb{Q}] = [F(\sqrt{3}) : F]\,[F : \mathbb{Q}] = 2 \cdot 2 = 4.$$

It can be verified that $F(\sqrt{3})$ is the smallest subfield of $\mathbb{R}$ containing $\mathbb{Q} \cup \{\sqrt{2}, \sqrt{3}\}$, that is, $K = F(\sqrt{3})$. Thus, K is a finite extension of $\mathbb{Q}$ and is of degree 4 over $\mathbb{Q}$.

EXERCISE 14(A)

1. Let K be the smallest subfield of $\mathbb{R}$ containing $\mathbb{Q} \cup \{\sqrt{2}, \sqrt{3}\}$. Then, find a basis of K over $\mathbb{Q}$.

2. Consider two distinct prime numbers p and q and let K be the field given in Exercise 1 above with p and q in place of 2 and 3, respectively. Then, prove that K is a finite extension of $\mathbb{Q}$ and find the degree and basis of K over $\mathbb{Q}$.

3. Let $p_1, p_2, p_3, \ldots$ be the sequence of all prime numbers. Define F_n recursively as follows:

$$F_0 = \mathbb{Q} \text{ and } F_n = F_{n-1}(\sqrt{p_n}) = \{a + b\sqrt{p_n} : a \text{ and } b \in F_{n-1}\}.$$

   Then, prove that, for each $n \in \mathbb{Z}^+$, F_n is a finite extension of F_{n-1} and is of degree 2 over F_{n-1}.

4. Deduce from Exercise 3 above that $\mathbb{R}$ is an infinite extension of $\mathbb{Q}$.

5. Determine each of the following.
   (i) $[\mathbb{C} : \mathbb{R}]$
   (ii) $[\mathbb{C} : \mathbb{Q}]$
   (iii) $[\mathbb{R} : \mathbb{Q}(\sqrt{2})]$
   (iv) $[\mathbb{Q}(\sqrt{2}, \sqrt{3}, \sqrt{5}) : \mathbb{Q}]$, where $\mathbb{Q}(\sqrt{2}, \sqrt{3}, \sqrt{5})$ is the smallest subfield of $\mathbb{R}$ containing $\mathbb{Q} \cup \{\sqrt{2}, \sqrt{3}, \sqrt{5}\}$.
   (v) $[\mathbb{Q}(\sqrt{p_1}, \sqrt{p_2}, \ldots, \sqrt{p_n}) : \mathbb{Q}]$ for any distinct primes $p_1, p_2, \ldots, p_n$.

6. Is $\mathbb{Q}$ a field extension of $\mathbb{Z}_p$, for any prime p?

7. Construct a field extension of $\mathbb{Z}_3$ with exactly 9 elements.

8. If $F \subseteq E$ are fields, prove that $\text{char}(F) = \text{char}(E)$.

14.2 ALGEBRAIC EXTENSIONS

Let us recall that a polynomial $f(x)$ over a field F is called *irreducible* over F if $f(x)$ cannot be expressed as a product of two nonconstant polynomials over F. If E is a field extension of a field F and $f(x) \in F[x]$ such that $f(a) = 0$ for some $a \in E$, then a is a called a *root* of $f(x)$ in E. Note that any polynomial of degree n over a field F can have at most n roots in any field extension E of F.

Theorem 14.2.1. Let F be any field and $p(x) \in F[x]$ be irreducible over F. Then, there exists a field extension E of F such that E contains a root of $p(x)$.

Proof: Since $p(x)$ is given to be irreducible, the principal ideal $<p(x)>$ generated by $p(x)$ in the ring $F[x]$ is a maximal ideal and hence the quotient ring $F[x]/<p(x)>$ is a field. Define $\sigma : F \rightarrow F[x]/<p(x)>$ by

$$\sigma(a) = a + <p(x)> \text{ for any } a \in F.$$

Then, clearly σ is an embedding of F in $F[x]/<p(x)>$. Put $E = F[x]/<p(x)>$. Then, E is a field extension of F. If $p(x) = a_0 + a_1x + a_2x^2 + \cdots + a_nx^n$, $n > 0$, and $a_i \in F$, then $x + <p(x)>$ is a root of $p(x)$ in E, since

$$\begin{aligned} p(x+<p(x)>) &= \sum_{i=0}^{n} a_i(x+<p(x)>)^i \\ &= \sum_{i=0}^{n} a_i(x^i+<p(x)>) \\ &= \left(\sum_{i=0}^{n} a_i(x^i)\right)+<p(x)> \\ &= p(x)+<p(x)>=0 \text{ in } E. \end{aligned}$$

Thus, E is field extension of F containing a root of $p(x)$, namely $x + <p(x)>$.

Corollary 14.2.1. Let F be a field and $f(x)$ be a nonconstant polynomial over F. Then, F has a field extension E containing a root of $f(x)$.

Proof: Since $f(x)$ is a nonconstant polynomial over F, $f(x)$ is a nonunit in $F[x]$. Since $F[x]$ is a unique factorization domain, it follows that there exists an irreducible polynomial $p(x)$ in $F[x]$ such that $p(x)$ divides $f(x)$. By Theorem 14.2.1, there exists a field extension E of F such that E contains a root of $p(x)$.

Let $f(x) = p(x)\, g(x)$. If $a \in E$ is a root of $p(x)$, then
$$f(a) = p(a)\, g(a) = 0.g(a) = 0$$

and therefore a is a root of $f(x)$ in E.

Theorem 14.2.2. Let $f(x)$ be a nonconstant polynomial over a filed F. Then, F has a field extension E containing all the roots of $f(x)$.

Proof: We prove this by using induction on the degree of $f(x)$. Let $\deg(f(x)) = n$. If $n = 1$, then $f(x) = a + bx$, a and $0 \neq b \in F$ and $-ab^{-1}$ is the only root of $f(x)$ and hence F itself is the extension F containing all the roots of $f(x)$. Let $n > 1$ and assume the theorem for all nonconstant polynomials of degree less then n. By the above Corollary 14.2.1, there exists a field extension K of F such that K contains a root a of $f(x)$ in K. Now, $x - a$ divides $f(x)$ in $K[x]$ and hence

$$f(x) = (x - a)\,g(x) \quad \text{for some } g(x) \in K[x].$$

Now, $\deg(g(x)) = n - 1 > 0$ and, by the induction hypothesis, there exists a field extension E of K such that E contains all the roots of $g(x)$. Any root of $f(x)$ other than a must be a root of $g(x)$. Any root of $f(x)$ other than a must be a root of $g(x)$. Therefore, E contains all the roots of $f(x)$ and E is a field extension of F, since $F \subseteq K \subseteq E$.

Corollary 14.2.2. Let $f_1(x), f_2(x), \ldots, f_m(x)$ be nonconstant polynomials over a field F, then F has a field extension E containing all the roots of $f_i(x)$ for all $1 \leq i \leq m$.

Proof: Consider $f(x) = f_1(x), f_2(x), \ldots, f_m(x)$ and use Theorem 14.2.2 above.

Definition 14.2.1. Let $F \subseteq K$ be fields and $a \in K$. Then, a is said to be *algebraic over F* if a is a root of a nonzero polynomial in $F[x]$; that is, if $f(a) = 0$ for some $0 \neq f(x) \in F[x]$. If a is not algebraic over F, then a is called *transcendental over F*.

Example 14.2.1

1. For any field F, every element a of F is algebraic over F, since $x - a \in F[x]$ and a is a root of $x - a$.
2. $\sqrt{2}$ is algebraic over $\mathbb{Q}$, since $x^2 - 2 \in \mathbb{Q}[x]$ and $\sqrt{2}$ is a root of $x^2 - 2$.
3. The complex number i is algebraic over $\mathbb{Q}$ since i is a root of $1 + x^2 \in \mathbb{Q}[x]$.
4. It is known that the real numbers e and π are transcendental over $\mathbb{Q}$. The proofs of these are beyond the scope of this book.

Let us recall that a nonzero polynomial over a field is called monic if its leading coefficient is the unity element of the field.

Theorem 14.2.3. Let $F \subseteq K$ be field and $a \in K$ be algebraic over F. Then, there exists a unique monic irreducible polynomial $p(x)$ over F such that $p(a) = 0$. Also, for any $f(x) \in F[x]$,

$$f(a) = 0 \text{ if and only if } p(x) \text{ divides } f(x) \text{ in } F[x].$$

Proof: Consider the set

$$I = \{f(x) \in F[x] : f(a) = 0\}.$$

It can be easily verified that I is an ideal of the ring $F[x]$. Since $F[x]$ is a principal ideal domain, there exists $p(x) \in I$ such that

$$I = \langle p(x) \rangle = \{p(x)\,g(x) : g(x) \in F[x]\}.$$

Without loss of generality, we can assume that $p(x)$ is monic; for, if $p(x) = a_0 + a_1x + \cdots + a_nx^n$, $a_n \neq 0$ and $g(x) = a_n^{-1}a_0 + a_n^{-1}a_1x + \ldots + a_n^{-1}a_{n-1}x^{n-1} + x^n$, then $p(x) = a_ng(x)$ and hence $p(x)$ and $g(x)$ are associates in $F[x]$ so that $\langle p(x) \rangle = \langle g(x) \rangle$. Thus, we can assume that $p(x)$ is a monic polynomial in $F[x]$ such that

$$\langle p(x) \rangle = \{f(x) \in F[x] : f(a) = 0\} = I.$$

Since a is algebraic over F, there exists a nonconstant polynomial $f(x)$ in $F[x]$ such that $f(x) = 0$ and, in this case, $f(x) = p(x)g(x)$ for some $g(x) \in F[x]$. Therefore, $p(x)$ is also a nonconstant polynomial. Further,

$$\deg(p(x)) \leq \deg(f(x)) \text{ for all } 0 \neq f(x) \in I \qquad (*)$$

From this, it follows that $p(x)$ is irreducible; for, let $p(x) = g(x)h(x)$ for some nonconstant $g(x)$ and $h(x) \in F[x]$. Then, $0 = p(a) = g(a)h(a)$ and hence $g(a) = 0$ or $h(a) = 0$, so that $g(x)$ or $h(x) \in I$. But, since $\deg(p(x)) = \deg(g(x)) + \deg(h(x))$ and both $\deg(g(x))$ and $\deg(h(x))$ are positive, it follows that $\deg(g(x)) < \deg(p(x))$ and $\deg(h(x)) < \deg(p(x))$ which is a contradiction to (*). Thus, $p(x)$ is an irreducible monic polynomial in $F[x]$ such that $p(a) = 0$.
To prove the uniqueness of $p(x)$, let

$$p(x) = a_0 + a_1x + \cdots + a_{n-1}x^{n-1} + x^n$$
$$\text{and} \quad q(x) = b_0 + b_1x + \cdots + b_{m-1}x^{m-1} + x^m$$

be two irreducible monic polynomials in $F[x]$ such that $p(a) = 0 = q(a)$. Then, $q(x) \in I$ and hence $q(x) = p(x)g(x)$ for some $g(x) \in F[x]$. The irreducibility of $q(x)$ implies that $g(x)$ is a unit (that is, $g(x) \in F$) and hence $m = \deg(q(x)) = \deg(p(x)) = n$. Now, $p(a) - q(a) = 0$ and hence $p(x) - q(x) \in I$. This implies, by (*), that $p(x) - q(x) = 0$. Thus, $p(x) = q(x)$.
The last assertion in the theorem follows from the fact that

$$\langle p(x) \rangle = \{f(x) \in F[x] : f(a) = 0\}. \quad ◀$$

Definition 14.2.2. If $F \subseteq K$ are fields, $a \in K$ and a is algebraic over F, then the unique irreducible monic polynomial in $F[x]$, for which a is a root, is called the *minimal polynomial of a over F*.

Example 14.2.2

1. We have $\mathbb{Q} \subseteq \mathbb{R}$ and $\sqrt{2} \in \mathbb{R}$. $x^2 - 2$ is the minimal polynomial of $\sqrt{2}$ over $\mathbb{Q}$.
2. Let ω be a root of $1 + x + x^2 + \cdots + x^{p-1}$ in $\mathbb{C}$, where p is a given prime number. Then, using the Eisenstein criterion, we have proved that $1 + x + x^2 + \cdots + x^{p-1}$ is irreducible over $\mathbb{Q}$ and hence it is the minimal polynomial of ω over $\mathbb{Q}$.

Let $F \subseteq K$ be fields and $a \in K$. Then, the smallest subfield of K containing F and a will be demoted by $F(a)$. Note that $F(a)$ is the intersection of all subfields of K containing $F \cup \{a\}$ and it can be easily verified that

$$F(a) = \{f(a)\, g(a)^{-1} : f(x) \quad \text{and} \quad g(x) \in F[x] \text{ and } g(a) \neq 0\}.$$

The following provides an elegant description of the elements of $F(a)$, when a is algebraic over F.

Theorem 14.2.4. Let $F \subseteq K$ be fields and $a \in K$. Then, a is algebraic over F if and only if $F(a)$ is a finite extension of F and, in this case, $[F(a) : F] = \deg(p(x))$, where $p(x)$ is the minimal polynomial of a over F. Also, if

$$p(x) = a_0 + a_1x + \cdots + a_{n-1}x^{n-1} + x^n,$$

$$\text{then} \quad F(a) = \{b_0 + b_1a + \cdots + b_{n-1}a^{n-1} : b_i \in F\}.$$

Proof: The theorem is trivial if $a \in F$, since, $-a + x$ is the minimal polynomial of a over F. Therefore, we can assume that $a \notin F$.
First suppose that $F(a)$ is a finite extension of F and $[F(a) : F] = m$. Then, $1, a, a^2, a^3, \ldots, a^m$ are all elements of $F(a)$ and these are $m + 1$ in number. Since $\dim_F(F(a)) = m$, $1, a, a^2, \ldots, a^m$ must be linearly dependent. Therefore, there exists $a_0, a_1, a_2, \ldots, a_m$ in F, not all zero, such that

$$a_0 1 + a_1 a + a_2 a^2 + \cdots + a_m a^n = 0.$$

Therefore, a is a root of the nonzero polynomial $a_0 + a_1x + \cdots + a_mx^m$ in $F[x]$. Thus, a is algebraic over F. Conversely, suppose that a is algebraic over F and let

$$p(x) = a_0 + a_1x + \cdots + a_{n-1}x^{n-1} + x^n$$

be the minimal polynomial of a over F. Then,

$$a_0 + a_1 a + \cdots + a_{n-1} a^{n-1} + a^n = p(a) = 0$$

and hence each of $a^n, a^{n+1}, a^{n+2}, \ldots$ can be expressed in the form $b_0 + b_1 a + \cdots + b_{n-1} a^{n-1}$ with $b_i \in F$ and hence so is $f(a)$ for any $f(x) \in F[x]$. Now, put

$$E = \{b_0 + b_1 a + \cdots + b_{n-1} a^{n-1} : b_i \in F\}.$$

Then, clearly E is a subring of K containing F and a. Define $\sigma : F[x] \to E$ by $\sigma(f(x)) = f(a)$. Then, σ is an epimorphism and

$$\begin{aligned} f(x) \in \ker \sigma &\Leftrightarrow f(a) = 0 \\ &\Leftrightarrow p(x) \text{ divides } f(x) \text{ (by Theorem 14.2.3)} \\ &\Leftrightarrow f(x) \in <p(x)> \end{aligned}$$

and therefore, $\ker \sigma = <p(x)>$. By the Fundamental Theorem of Homomorphisms,

$$F[x] \big/ <p(x)> \cong E.$$

Now, since $p(x)$ is irreducible in $F[x]$, $<p(x)>$ is a maximal ideal of $F[x]$ and hence $F[x]/<p(x)>$ is a field. Therefore, E is also a field. Thus, E is a subfield of K containing F and a. It is clear that any subfield of K containing F and a must contain E. Thus, $E = F(a)$. Also, $\{1, a, a^2, \ldots, a^{n-1}\}$ is linearly independent over F (since a is not a root of any polynomial of degree less than n) and it generates $F(a)$. Therefore, $\{1, a, \ldots, a^{n-1}\}$ is a basis of $F(a)$ over F and hence

$$[F(a) : F] = n = \deg(p(x)).$$ ◀

Definition 14.2.3. Let K be a field extension of F. K is said to be an *algebraic extension* of F if every element of K is algebraic over F.

Theorem 14.2.5. Let F be any field. Then, any finite extension of F is an algebraic extension of F.

Proof: Let E be a finite extension of F and $[E : F] = n$. Then, for any $a \in E$, we have $F \subseteq F(a) \subseteq E$ and hence $[F(a) : F]\,[E : F(a)] = [E : F] = n$, so that $[F(a) : F]$ is finite. By Theorem 14.2.4, a is algebraic over F for each $a \in E$. Thus, E is an algebraic extension of F. ◀

The converse of the above theorem is not true. That is, an algebraic extension of a field F need not be a finite extension of F. For, consider the following example.

Example 14.2.3. It is well known that the set of prime numbers is a countably infinite set and hence we can express this set in a sequential form. Let $p_1, p_2, \ldots, p_n, \ldots$ be all the distinct primes. For each $n \geq 0$, E_n be the subfield of $\mathbb{R}$ defined recursively by

$$E_0 = \mathbb{Q} \quad \text{and} \quad E_0 = E_{n-1}(\sqrt{p_n}) \text{ for all } n > 0.$$

Clearly $E_n = \mathbb{Q}(\sqrt{p_1}, \sqrt{p_2}, \ldots, \sqrt{p_n})$ which is the smallest subfield of $\mathbb{R}$ containing $\mathbb{Q} \cup \{\sqrt{p_1}, \sqrt{p_2}, \ldots, \sqrt{p_n}\}$. We first prove that $\sqrt{p_{n+1}} \notin E_n$ for all $n \geq 0$, by using induction on n. Since $\sqrt{p_1}$ is irrational, $\sqrt{p_1} \notin \mathbb{Q} = E_0$. Let $n > 0$ assume that $\sqrt{p_n} \notin E_{n-1}$. If $\sqrt{p_{n+1}} \in E_n$, then

$$\sqrt{p_{n+1}} = a + b\sqrt{p_n} \text{ for some } a \text{ and } b \in E_{n-1}$$

and hence $p_{n+1} = a^2 + p_n b^2 + 2ab\sqrt{p_n}$ which implies that $\sqrt{p_n} = \frac{1}{2ab}(p_{n+1} - a^2 - p_n b^2) \in E_{n-1}$ (note that a and b are nonzero elements of E_{n-1}), which is a contradiction. Thus, $\sqrt{p_{n+1}} \notin E_n$ for all n and hence

$$[E_{n+1} : E_n] = 2 \text{ for all } n.$$

Now, $\mathbb{Q} = E_0 \subset E_1 \subset E_2 \subset \ldots \subset E_n \subset \ldots$ is a strictly ascending chain of subfields of $\mathbb{R}$.

Let $E = \bigcup_{n=0}^{\infty} E_n$. Then, $E = \mathbb{Q}(\sqrt{p_1}, \sqrt{p_2}, \ldots)$ which is the smallest subfield of $\mathbb{R}$ containing $\mathbb{Q}$ and the square roots of all the prime numbers. Now, by Theorem 14.1.1, we have

$$[E_n : \mathbb{Q}] = 2^n \quad \text{for all } n \geq 0$$

and, since each E_n is a subfield of E, it follows that $[E : \mathbb{Q}]$ is infinite; that is, E is an infinite extension of $\mathbb{Q}$. However, we observe that E is an algebraic extension of $\mathbb{Q}$. For, if $a \in E$, then $a \in E_n$ for some n and hence $\mathbb{Q}(a) \subseteq E_n$ so that $[\mathbb{Q}(a) : \mathbb{Q}]$ is a divisor of $[E_n : \mathbb{Q}] = 2^n$ and therefore $\mathbb{Q}(a)$ is a finite extension of $\mathbb{Q}$. By Theorem 14.2.4, a is algebraic over $\mathbb{Q}$. Thus, E is an algebraic extension of $\mathbb{Q}$.

We have proved in Theorem 14.2.2 that, for any nonconstant polynomial $f(x)$ over a field F, there exists a field extension E of F containing all the roots of $f(x)$. In the following, we prove that this extension E can be chosen to be a finite extension containing all the roots of a given $f(x)$.

Theorem 14.2.6 (Kronecker's Theorem). Let $f(x)$ be a nonconstant polynomial over a field F and $\deg(f(x)) = n$. Then there is an extension E of F such that $[E : F] \le n!$ and E contains all the roots of $f(x)$; that is, $f(x)$ can be factored over E as

$$f(x) = a_0(x - b_1)(x - b_2) \cdots (x - b_n),\ b_i \in E \text{ and } a_0 \in F.$$

Proof: We have $f(x) \in F[x]$ and $\deg(f(x)) = n \ge 1$. Let $p(x)$ be an irreducible polynomial $\mathbb{R}$ over F such that $p(x)$ divides $f(x)$. By Theorem 14.2.1, there exists a field extension K of F such that K contains a root of $p(x)$. Let $a \in K$ be a root of $p(x)$ and hence of $f(x)$. By Theorem 14.2.4,

$$[F(a) : F] = \deg(p(x)) \le \deg(f(x)) = n.$$

Let $F(a) = E_1$. We can write

$$f(x) = (x - a) f_1(x) \text{ for some } f_1(x) \in E_1[x].$$

Now $\deg(f_1(x)) = n - 1$. Continuing this process, we get an extension E of E_1, such that $[E : E_1] \le (n - 1)!$ and E_1 contains all the roots of $f_1(x)$. Therefore, E is an extension F such that

$$[E : F] = [E : E_1]\,[E_1 : F] \le (n - 1)!\, n = n!$$

and E contains all the roots of $f(x)$. ◀

Definition 14.2.4. Let $F \subseteq E$ be fields. Then, E is said to be *finitely generated over F* if there exists $a_1, a_2, \ldots, a_n \in E$ such that $E = F(a_1, a_2, \ldots, a_n)$, where

$$F(a_1, a_2, \ldots, a_i) = F(a_1, a_2, \ldots, a_{i-1})(a_i)$$

for each $1 < i \le n$.

Note that any finite extension of F is finitely generated over F; but a finitely generated extension may not be a finite extension. For, if a is a transcendental number, then $\mathbb{Q}(a)$ is finitely generated over $\mathbb{Q}$ and it is not a finite extension of $\mathbb{Q}$. However, we have the following theorem.

Theorem 14.2.7. Let $E = F(a_1, a_2, \ldots, a_n)$ be a finitely generated extension of F and $a_1, a_2, \ldots, a_n$ be algebraic over F. Then, E is a finite extension of F and hence an algebraic extension of F.

Proof: We prove this by using induction of n. If $n = 1$, then the theorem follows from Theorem 14.2.4. Assume the theorem for $n - 1$. That is,

$$[F(a_1, a_2, \dots, a_{n-1}) : F] \text{ is finite.}$$

Since a_n is algebraic over F, it is algebraic over $F(a_1, a_2, \dots, a_{n-1})$ also and hence

$$[F(a_1, a_2, \dots, a_{n-1})(a_n) : F(a_1, a_2, \dots, a_{n-1})] \text{ is finite.}$$

Therefore, $[F(a_1, \dots, a_n) : F]$ is finite, by Theorem 14.1.1. Thus, E is a finite extension of F and hence, by Theorem 14.2.5, E is an algebraic extension of F. ◀

Theorem 14.2.8. Let E be an extension of F and

$$K = \{a \in E : a \text{ is algebraic over } F\}.$$

Then, K is a subfield of E and K is an algebraic extension of F.

Proof: Let a and $b \in K$. Then, by Theorem 14.2.7, $F(a, b)$ is an algebraic extension of F and, since $a \pm b$, ab and a^{-1} (if $a \neq 0$) $\in F(a, b)$, it follows that these $a \pm b$, ab and a^{-1} (if $a \neq 0$) are all algebraic over F and hence these are elements of K. Thus, K is a subfield of F and is clearly an algebraic extension of F. ◀

Theorem 14.2.9. Let $F \subseteq K \subseteq L$ be fields. Let $a \in L$ be algebraic over K and K an algebraic extension of F. Then, a is algebraic over F.

Proof: Since $a \in L$ is algebraic over K, there exists a nonzero polynomial $f(x) = a_0 + a_1x + \cdots + a_nx^n \in K[x]$ for which a is a root. Now,

$$f(x) \in F(a_0, a_1, \dots, a_n)[x]$$

where each a_i is algebraic over F (since $a_i \in K$). Put

$$E = F(a_0, a_1, \dots, a_n).$$

By Theorem 14.2.7, E is a finite extension of F. Also, since $f(x) \in E[x]$ and $f(a) = 0$, a is algebraic over E and hence $[E(a) : E]$ is finite. Now,

$$[E(a) : F] = [E(a) : E]\,[E : F] < \infty.$$

Since $F(a) \subseteq E(a)$, $[F(a) : F] \leq [E(a) : F] < \infty$. Thus, $F(a)$ is a finite extension of F and hence a is algebraic over F. ◀

Corollary 14.2.3. Let $F \subseteq K \subseteq L$ be fields, L be algebraic over K and K be algebraic over F. Then, L is algebraic over F.

Definition 14.2.5. Let $F \subseteq E$ be fields. A homomorphism $\sigma : E \to E$ is said to be an F-homomorphism if $\sigma(a) = a$ for all $a \in F$. We denote this by $\sigma/F = \text{Id}$.

Theorem 14.2.10. Let $K = F(a_1, a_2, \ldots, a_n)$. Then, any F-homomorphism $\sigma : K \to K$ is completely determined by $\sigma(a_1), \sigma(a_2), \ldots, \sigma(a_n)$.

Proof: Let σ and τ be F-homomorphism of K into K such that $\sigma(a_i) = \tau(a_i)$ for all $1 \leq i \leq n$. Put

$$L = \{a \in K : \sigma(a) = \tau(a)\}.$$

Then L is a subfield of K containing F and $a_1, a_2, \ldots, a_n$. Therefore, $F(a_1, a_2, \ldots, a_n) \subseteq L \subseteq K = F(a_1, a_2, \ldots, a_n)$ and hence $L = K$. Thus, $\sigma(a) = \tau(a)$ for all $a \in K$; that is, $\sigma = \tau$. ◀

Theorem 14.2.11. Let $F \subseteq K$ be fields, $p(x)$ a monic irreducible polynomial over F and a and $b \in K$ be roots of $p(x)$. Then, there exists an isomorphism $\sigma : F(a) \to F(b)$ such that $\sigma(a) = b$ and $\sigma(s) = s$ for all $s \in F$.

Proof: As in the proof of Theorem 14.2.4, there exists an isomorphism

$$\sigma_1 : F[x]/<p(x)> \to F(a)$$

defined by $\sigma_1(f(x) + <p(x)>) = f(a)$. Similarly, there exists an isomorphism

$$\sigma_2 : F[x]/<p(x)> \to F(b)$$

defined by $\sigma_2(f(x) + <p(x)>) = f(b)$. Now, put

$$\sigma = \sigma_2 \circ \sigma_1^{-1} : F(a) \to F(b).$$

Then, σ is an isomorphism, $\sigma(a) = \sigma_2(x + <p(x)>) = b$ and, for any $s \in F$, $\sigma(s) = \sigma_2(s + <p(x)>) = s$. ◀

Theorem 14.2.12. Let E be an algebraic extension of F and $\sigma : E \to E$ be an embedding such that $\sigma \mid F = \text{Id}$. Then, σ is an automorphism of E.

Proof: Let $0 \neq a \in E$. Then, a is algebraic over F. Let

$$p(x) = a_0 + a_1x + \cdots + a_{n-1}x^{n-1} + x^n \in F[x]$$

be the minimal polynomial of $p(x)$ over F. Suppose that $a = b_1, b_2, \ldots, b_m \in E$ be all the roots of $p(x)$ in E. Then, by Theorem 14.2.7, $F(b_1, b_2, \ldots, b_m)$ is a finite extension of F.

Since each b_i is a root of $p(x)$, we have

$$0 = p(b_i) = a_0 + a_1b_i + \cdots + a_{n-1}b_i^{n-1} + b_i^n.$$

By applying σ both sides, we get that

$$0 = a_0 + a_1\sigma(b_i) + \cdots + a_{n-1}\sigma(b_i)^{n-1} + \sigma(b_i)^n$$

and hence $\sigma(b_1), \ldots, \sigma(b_m)$ are also roots of $p(x)$ in E.

Since σ is an injection, $\sigma(b_1), \sigma(b_2), \ldots, \sigma(b_m)$ must be the same as $b_1, b_2, \ldots, b_m$ in some order. Now, Let

$$E' = F(b_1, b_2, \ldots, b_m).$$

$$\text{Then, } \sigma(E') = \sigma(F(b_1, b_2, \ldots, b_m)) \cong F(\sigma(b_1), \sigma(b_2), \ldots, \sigma(b_m))$$
$$= F(b_1, b_2, \ldots, b_m) = E'.$$

Therefore, $[\sigma(E') : F] = [E' : F]$, since $\sigma(E') \cong E'$. ◀

But, since $\sigma(E') \subseteq E'$, it follows that $\sigma(E') = E'$. In particular, since $a = b_1 \in E'$, there exists $b \in E'$ such that $\sigma(b) = a$. Thus, σ is a surjection also. Therefore, σ is an isomorphism of E onto E. Thus, σ is an automorphism of E.

Worked Exercise 14.2.1. Prove that the field $\mathbb{C}$ of complex numbers is an algebraic extension of the field $\mathbb{R}$ of real numbers.

Answer: Clearly $\mathbb{C}$ is a field extension of $\mathbb{R}$. Also, any $a \in \mathbb{R}$ is algebraic over $\mathbb{R}$. Since i is a root of $1 + x^2 \in \mathbb{R}[x]$, i is algebraic over $\mathbb{R}$. Thus, any $a + bi \in \mathbb{C}$ is algebraic over $\mathbb{R}$, by Theorem 14.2.8. Thus, $\mathbb{C}$ is an algebraic extension of $\mathbb{R}$.

Worked Exercise 14.2.2. Prove that an algebraic extension E of F is finitely generated over F if and only if E is a finite extension of F.

Answer: Let E be an algebraic extension of F. Suppose that E is finitely generated over F. Then, $E = F(a_1, a_2, \ldots, a_n)$ for some $a_1, a_2, \ldots, a_n \in E$. Since E is an algebraic extension of F, each a_i is algebraic over F. Now, by Theorem 14.2.7, E is a finite extension of F. Conversely suppose that E is a finite extension of F. Let $[E : F] = n$ and $\{ a_1, a_2, \ldots, a_n\}$ be a basis of E over F. Then, $F(a_1, a_2, \ldots, a_n) = E$ and hence E is finitely generated over F.

EXERCISE 14(B)

1. Prove that the polynomial $p(x) = x^2 - x - 1 \in \mathbb{Z}_3[x]$ is irreducible over $\mathbb{Z}_3$ and that there is a field extension E of $\mathbb{Z}_3$ with exactly nine elements and containing all the roots of $p(x)$.

2. Find the smallest extension of $\mathbb{Q}$ having a root of $x^4 - 2 \in \mathbb{Q}[x]$.

3. Determine the degrees of the following:
   (i) $\mathbb{Q}(\sqrt{3})$ over $\mathbb{Q}$
   (ii) $\mathbb{Q}(\sqrt{5})$ over $\mathbb{Q}$
   (iii) $\mathbb{Q}(\sqrt{3}, \sqrt{5})$ over $\mathbb{Q}$
   (iv) $\mathbb{Q}(\sqrt{3}, \sqrt{5})$ over $\mathbb{Q}$

4. Find the degree of $\mathbb{R}$ over $\mathbb{Q}$.

5. Let $F \subseteq K$ be fields and a and $b \in K$ be algebraic over F such that $[F(a) : F] = m$ and $[F(b) : F] = n$, where m and n are relatively prime. Then, prove that $[F(a, b) : F] = mn$.

6. Determine the minimal polynomials over $\mathbb{Q}$ of the following:
   (i) $\sqrt{5}$
   (ii) $\sqrt{2}+5$
   (iii) $5+3\sqrt{2}$
   (iv) $\sqrt{-1+\sqrt{2}}$
   (v) $\sqrt{2}-3\sqrt{3}$
   (vi) $\sqrt{6}$

7. Prove that there is $a \in \mathbb{R}$ such that $\mathbb{Q}(\sqrt{2}, \sqrt{5}) = \mathbb{Q}(a)$

8. Let $F \subseteq E$ be fields such that $[E : F]$ is a prime number. Prove that there are no fields property between F and E.

9. Let D be an integral domain and F be a field contained in D as a subring such that $[D : F]$ is finite. Then prove that D is a field.

10. Let $F \subseteq E$ be fields and $a \in E$ be algebraic over F. If the minimal polynomial of a over F is of odd degree, prove that $F(a) = F(a^2)$.

14.3 ALGEBRAICALLY CLOSED FIELDS

A field F is called algebraically closed if an element a, in any extension of F, is algebraic over F, then $a \in F$. In other words, no element outside F is algebraic over F. In this section, we discuss various properties of algebraically closed fields. Let us begin with a formal definition in the following.

Definition 14.3.1. A field K is said to be *algebraically closed* if it has no proper algebraic extensions; that is, if $K \subseteq L$ and L is algebraic over K, then $K = L$.

The following provides two characterizations of algebraically closed fields.

Theorem 14.3.1. The following are equivalent to each other for any field K.

1. K is algebraically closed.
2. Any nonconstant polynomial over K can be factored completely into linear factors in $K[x]$.
3. If $f(x)$ is a nonconstant polynomial over K, then all the roots of $f(x)$ belong to K.
4. Every nonconstant polynomial over K has atleast one root in K.

Proof: Recall that an element a (in any extension of K) is a root of $F(x) \in K[x]$ if and only if $x - a$ is a factor of $f(x)$ and therefore (2) $\Leftrightarrow$ (3) is clear.

(1) $\Rightarrow$ (2): Let $f(x)$ be a nonconstant polynomial over K. Then, by Theorem 14.2.6, there exists a finite extension E of K such that E contains all the roots of $f(x)$. Since any finite extension is algebraic, E is an algebraic extension of K and hence $E = K$. Thus, K contains all the roots of $f(x)$ and $f(x)$ can be factored completely into linear factors in $K[x]$.

(3) $\Rightarrow$ (4) is trivial.

(4) $\Rightarrow$ (1): Let L be an algebraic extension of K and let $a \in L$. Then, a is algebraic over K. Let $p(x)$ be the minimal polynomial of a over K. Then, $p(x) \in K[x]$ and $\deg(p(x)) \geq 1$. Then, by (4), $p(x)$ has a root, say b, in K. Then, $x - b$ divides $p(x)$ in $K[x]$ and hence $p(x) = x - b$ (since $p(x)$ is irreducible monic polynomial in $K[x]$). Now $a - b = p(a) = 0$ and hence $a = b \in K$.

Thus, $L \subseteq K$ and $L = K$. Thus, K is algebraically closed. ◀

Corollary 14.3.1. A field K is algebraically closed if and only if every irreducible polynomial in $K[x]$ is of degree 1.

Definition 14.3.2. An algebraic extension E of a field F is called an *algebraic closure* of F if E is algebraically closed.

Example 14.3.1. The field $\mathbb{C}$ of complex numbers is an algebraic closure of the field $\mathbb{R}$ of real numbers.

We prove later any two algebraic closures of a field F are isomorphic under an isomorphism that keeps each element of F fixed. The following theorem is in the direction of proving the existence of an algebraic closure of any field.

For any field F, we know that for any given indeterminates $x_1, x_2, \ldots, x_n$, the set $F[x_1, x_2, \ldots, x_n]$ of all polynomials in $x_1, x_2, \ldots, x_n$ with coefficients in F is an integral domain and we identify $x_i x_j$ with $x_j x_i$; that is, the indeterminates are commuting with each other. For any set $S = \{x_\alpha\}_{\alpha \in \Delta}$ of commuting indeterminates, we define

$$F[S] = \bigcup_{\substack{T \subseteq S \\ T \text{ is finite}}} F[T].$$

Then, it can be easily verified that $F[S]$ is an integral domain.

Theorem 14.3.2. Any field has an algebraically closed extension.

Proof: Let F be a field. Let us first construct an extension K_1 of F in which every nonconstant polynomial has a root. For each nonconstant polynomial $f = f(x) \in F[x]$, we correspond an indeterminate x_f and let

$$S = \{x_f : f = f(x) \in F[x] \text{ and } \deg(f(x)) \geq 1\}.$$

Consider the polynomial ring $F[S]$, which is an integral domain. Let A be the ideal in $F[S]$ generated by all polynomials $f(x_f)$ of positive degree in $F[S]$. We claim that A is a proper ideal of $F[S]$. Suppose, if possible that $A = F[S]$. Then, $1 \in A$ and therefore

$$1 = g_1\ f_1(x_{f_1}) + g_2\ f_2(x_{f_2}) + \cdots + g_x\ f_x(x_{f_n}),$$

where $g_1, g_2, \ldots, g_n \in F[S]$. Note that $g_1, g_2, \ldots, g_n$ will involve only a finite number of variable (indeterminates). Write $x_{f_i} = x_i$ for each $f_i \in F[x]$. After reindexing, we can assume that $x_{f_1} = x_1, \cdots, x_{f_n} = x_n$ and the variables occurring in all the g_i, $1 \leq i \leq n$, are in the set $\{x_1, x_2, \ldots, x_n, \ldots, x_m\}$. Therefore, we have

$$1 = \sum_{i=1}^{n} g_i(x_1, \ldots, x_m) f_i(x_i) \qquad (*)$$

Now, let E be an extension of F in which each of the polynomials $f_1, f_2, \ldots, f_n$ has a root and let a_i be a root of f_i in E, for each $1 \leq i \leq n$. If we substitute $x_i = a_i$ for $1 \leq i \leq n$ and $x_{n+1} = \cdots = x_m = 0$ in (*), we get that $1 = 0$, which is absurd.

Thus, A is a proper ideal of $F[S]$ and hence, using the Zorn's lemma, we get a maximal ideal M of $F[S]$ containing A. Then, there is an embedding $a \mapsto a + M$ of F into $F[S]/M$ and hence $F[S]/M$ can be regarded as a field extension of F. Also, each nonconstant polynomial $f = f(x) \in F[x]$ has a root in $F[S]/M$. Thus, we have constructed a field $K_1(= F[S]/M)$ which is an extension of F and in which every nonconstant polynomial in $F[x]$ has a root.

Now, inductively, we can form a chain of fields

$$F = K_0 \subset K_1 \subset K_2 \subset \cdots$$

such that any nonconstant polynomial over K_n has a root in K_{n+1} for all $n \geq 0$.

Put $K = \bigcup_{n=0}^{\infty} K_n$.

Then, K is a field extension of F. If

$$g(x) = b_0 + b_1 x + \cdots + b_m x^m, b_m \neq 0, m > 0$$

is a polynomial over K, then there exists n such that $b_0, b_1, \ldots, b_m \in K_n$ and therefore $g(x) \in K_n[x]$ so that $g(x)$ has a root in $K_{n+1} \subseteq K$. Thus, F has an algebraically closed extension. ◀

The following will be useful in proving the uniqueness (up to isomorphism) of algebraic closure of a given field.

Theorem 14.3.3. Let F be a field and $\sigma : F \to L$ be an embedding of F into an algebraically closed field L. Let $E = F(a)$ be an algebraic extension of F. Then, σ can be extended to an embedding $\eta : E \to L$ and the number of such extensions is equal to the number of distinct roots of the minimal polynomial of a over F.

Proof: Let $p(x) = a_0 + a_1 x + \cdots + a_{n-1} x^{n-1} + x^n$ be the minimal polynomial of a over F and write

$$p^\sigma(x) = \sigma(a_0) + \sigma(a_1)x + \cdots + \sigma(a_{n-1})x^{n-1} + x^n$$

Then, $p^\sigma(x) \in L[x]$. Since L is algebraically closed, $p^\sigma(x)$ has all the roots in L. Let $p \in L$ be a root of $p^\sigma(x)$. By Theorem 14.2.4,

$$E = F(a) = \{b_0 + b_1 a + \cdots + b_{n-1} a^{n-1} : b_i \in F\}.$$

Define $\eta_p : E \to L$ by

$$\eta_p(b_0 + b_1 a + \cdots + b_{n-1} a^{n-1}) = \sigma(b_0) + \sigma(b_1)p + \cdots + \sigma(b_{n-1})p^{n-1}.$$

Since $p(x)$ is the polynomial of least degree for which a is a root, it follows that any element of E ($= F(a)$) can be uniquely expressed as $b_0 + b_1 a + \cdots +$

$b_{n-1}a^{n-1}$ and hence η_p is well defined. It is a routine verification to prove that η_p is an embedding of $F(a)$ into L and that it is an extension of σ; that is, $\eta_p(s) = \sigma(s)$ for all $s \in F$. Further, $p \mapsto \eta_p$ is a bijective correspondence between the distinct roots of $p^\sigma(x)$ and the extensions of σ to E. Therefore, the number of extensions of σ to E is equal to the number of distinct roots of $p^\sigma(x)$ in L. ◀

The above theorem is generalised in the following theorem which is proved by using the Zorn's lemma.

Theorem 14.3.4. Let E be an algebraic extension of F and $\sigma : F \to L$ be an embedding of F into an algebraically closed field F. Then, σ can be extended to an embedding $\eta : E \to L$.

Proof: Consider the set

$$\mathbb{P} = \{(K, \theta) : F \subseteq K \subseteq E, \theta : K \to L \text{ is an embedding and } \theta/F = \sigma\}.$$

Then, $\mathbb{P}$ is not empty, since $(F, \sigma) \in \mathbb{P}$. Define a binary relation $\leq$ on $\mathbb{P}$ by

$$(K_1, \theta_1) \leq (K_2, \theta_2) \Leftrightarrow K_1 \subseteq K_2 \quad \text{and} \quad \theta_2/K_1 = \theta_1.$$

Then, it can be easily verified that $\leq$ is a partial order on $\mathbb{P}$. We prove that the partially ordered set $(\mathbb{P}, \leq)$ satisfies the hypothesis of the Zorn's lemma. Let $\{(K_\alpha, \theta_\alpha)\}_{\alpha \in \Delta}$ be a chain of elements in $\mathbb{P}$. Put

$$K = \bigcup_{a \in \Delta} K_a \text{ and define } \theta : K \to L \text{ by } \theta(a) = \theta_\alpha(a) \text{ if } a \in K_\alpha.$$

If $a \in K_\alpha \cap K_\beta$, then $\theta_\alpha(a) = \theta_\beta(a)$, since either

$$(K_\alpha, \theta_\alpha) \leq (K_\beta, \theta_\beta) \quad \text{or} \quad (K_\beta, \theta_\beta) \leq (K_\alpha, \theta_\alpha).$$

Therefore, θ is a well-defined mapping on K and, clearly K is a subfield of E containing F. It is easy to verify that $\theta : K \to L$ is an embedding and, clearly $(K_\alpha, \theta_\alpha) \leq (K, \theta)$ for all $\alpha \in \Delta$. Therefore, every chain in $(\mathbb{P}, \leq)$ has an upper bound in $\mathbb{P}$ and hence, by the Zorn's lemma, there is a maximal member, say (M, η), in $(\mathbb{P}, \leq)$. We prove that $M = E$. Suppose, if possible, that $M \neq E$. Then, choose $a \in E$ such that $a \notin M$. Since E is an algebraic extension of F, a is algebraic over F and hence over M. By Theorem 14.3.3, there exists an embedding $\eta' : M(a) \to L$ such that $\frac{\eta'}{M} = \eta$. But then $(M, \eta) < (M(a), \eta')$ $\in \mathbb{P}$, which is a contradiction to the maximality of (M, η). Thus, $M = E$ and $\eta : E \to L$ is an embedding and is an extension of σ. ◀

Theorem 14.3.5. Every field has an algebraic closure.

Proof: Let F be a field. By Theorem 14.3.2, there exists an extension E of F which is algebraically closed. Let

$$K = \{a \in E : a \text{ is algebraic over } F\}.$$

Then, K is a subfield of E and is an algebraic extension of F. Now, we prove that K is algebraically closed also. Let $f(x) \in K[x]$ be a nonconstant polynomial and a be a root of $f(x)$. Since $f(x) \in E[x]$ and E is algebraically closed, we get that $a \in E$. Also, since a is a root of $f(x) \in K[x]$, a is algebraic over K. Now, K is algebraic over F and hence, by Theorem 14.2.9, a is algebraic over F. Therefore, $a \in K$. That is, every root of $f(x) \in K[x]$ is in K. Thus, K is an algebraic closure of F. ◀

The following theorem proves the uniqueness (up to isomorphism) of the algebraic closure of a given field.

Theorem 14.3.6. Let F be a field and K and K' be algebraic closures of F. Then, K is isomorphic to K' under an isomorphism which is identity on F.

Proof: We can treat F as a subfield of K and K'. Let $\sigma : F \to K'$ be the inclusion map; that is, $\sigma(a) = a$ for all $a \in F$. Since K is an algebraic extension of F and K' is algebraically closed, we get (from Theorem 14.3.4) an embedding $\sigma^* : K \to K'$ which is an extension of σ. Now

$$K \cong \sigma^*(K) \subseteq K'$$

and K' is an algebraic extension of F and hence an algebraic extension of $\sigma^*(K)$ and therefore $\sigma^*(K) = K'$. Thus, $\sigma^* : K \to K'$ is an isomorphism such that $\sigma^*(a) = \sigma(a) = a$ for all $a \in F$. ◀

Definition 14.3.3. By Theorems 14.3.5 and 14.3.6, any field F has a unique (up to isomorphism) algebraic closure and, henceforth, we denote the algebraic closure of F by $\overline{F}$.

Definition 14.3.4. Let F be a field of $f(x) \in F[x]$ be of degree $n \geq 1$. Then, a field extension E of F is called a *splitting field* of $f(x)$ over F if the following are satisfied:

1. $f(x)$ factors into linear factors in $E[x]$; that is, there exists $a_1, a_2, \ldots, a_n \in E$ such that
$$f(x) = c(x - a_1)(x - a_2) \cdots (x - a_n).$$
2. $E = F(a_1, a_2, \ldots, a_n)$; that is, E is generated by $a_1, a_2, \ldots, a_n$ over F.

For any nonconstant polynomial $f(x)$ over a given field F, by Theorem 14.2.6, there exists a field extension K of F which contains all the roots of $f(x)$

and $[K : F] \leq n!$, where $n = \deg(f(x))$. If E is the intersection of all subfields of K containing F and the roots $a_1, a_2, \ldots, a_n$ of $f(x)$, then $E = F(a_1, a_2, \ldots, a_n)$ which becomes the splitting field of $f(x)$ and $[E : F] \leq n!$. Also, E is a finite extension and hence an algebraic extension of F. The following theorem says that any two splitting fields of a given polynomial over a given field F are isomorphic under an isomorphism which is the identity on F.

Theorem 14.3.7. Let $f(x)$ be a nonconstant polynomial over a field F and let E and K be two splitting fields of $f(x)$ over F. Then, there exists an isomorphism $\sigma : E \rightarrow K$ which is identity on F.

Proof: Let $\overline{K}$ and $\overline{F}$ be the algebraic closures of K and F, respectively. Then, $F \subseteq K \subseteq \overline{K}$, $\overline{K}$ is algebraic over K and K is algebraic over F. Therefore, by Corollary 14.2.3, $\overline{K}$ is algebraic over F and hence $\overline{K} = \overline{F}$. Let $\lambda : F \rightarrow \overline{F} = \overline{K}$ be the inclusion map. Suppose that

$$f(x) = a_0 + a_1x + \cdots + a_nx^n \in F[x], n > 0,$$

and $b_1, b_2, \ldots, b_n \in E$ be all the roots of $f(x)$ (not necessarily distinct). Then, $E = F(b_1, b_2, \ldots, b_n)$ is an algebraic extension of F. By Theorem 14.3.4, there exists an embedding $\lambda^* : E \rightarrow \overline{K}$ such that $\lambda^*/F = \text{Id}$. Put

$$f^{\lambda^*}(x) = \lambda^*(a_0) + \lambda^*(a_1)x + \cdots + \lambda^*(a_n)x^n.$$

Now, $\lambda^*(b_1), \lambda^*(b_2), \ldots, \lambda^*(b_n)$ are the roots of $f^{\lambda^*}(x)$ in $\overline{K}$, and if $\beta_1, \beta_2, \ldots, \beta_n \in K$ are the roots of $f(x)$, then

$$\{\beta_1, \beta_2, \ldots, \beta_n\} = \{\lambda^*(b_1), \lambda^*(b_2), \ldots, \lambda^*(b_n)\},$$

since $K \subseteq \overline{K}$. Also,

$$\begin{aligned} K = F(\beta_1, \beta_2, \ldots, \beta_n) &= F(\lambda^*(b_1), \lambda^*(b_2), \ldots, \lambda^*(b_n)) \\ &= \lambda^*(F(b_1, b_2, \ldots, b_n)) = \lambda^*(E) \end{aligned}$$

Thus, λ^* is an isomorphism of E onto K such that $\lambda^*|F = \text{Id}$. ◀

Worked Exercise 14.3.1. Find the splitting field of $x^3 - 2$ over the field $\mathbb{Q}$ of rational numbers and its degree over $\mathbb{Q}$.

Answer: Let $f(x) = x^3 - 2 \in \mathbb{Q}[x]$. By the Eisenstein's criterion, $f(x)$ is irreducible over $\mathbb{Q}$. In fact, it is the minimal polynomial of $2^{1/3}$. Therefore ,

$$\mathbb{Q}[x] \big/ {<f(x)>} \cong \mathbb{Q}(2^{1/3}) \quad \text{and} \quad [\mathbb{Q}(2^{1/3}) : \mathbb{Q}] = 3.$$

However, $\mathbb{Q}(2^{1/3})$ is not the splitting field of $f(x)$, since

$$f(x) = x^3 - 2 = (x - 2^{1/3})(x^2 + 2^{1/3}x + 2^{2/3})$$

and hence $f(x)$ has two complex roots, say w and $\overline{w}$. Let

$$p(x) = x^2 + 2^{1/3}x + 2^{2/3} \in \mathbb{Q}(2^{1/3})[x].$$

Then, $p(x)$ is irreducible over $\mathbb{Q}(2^{1/3})$ and hence

$$\mathbb{Q}(2^{1/3})[x] \big/ {<p(x)>} \cong \mathbb{Q}(2^{1/3})(w) = \mathbb{Q}(2^{1/3}, w)$$

and $[\mathbb{Q}(2^{1/3}, w) : \mathbb{Q}(2^{1/3})] = 2$, which is the degree of $p(x)$.
Further $\overline{w} \in \mathbb{Q}(2^{1/3}, w)$. Thus, $\mathbb{Q}(2^{1/3}, w)$ is the splitting field of $x^3 - 2$ over $\mathbb{Q}$ and

$$\begin{aligned}[\mathbb{Q}(2^{1/3}, w) : \mathbb{Q}] &= [\mathbb{Q}(2^{1/3}, w) : \mathbb{Q}(2^{1/3})][\mathbb{Q}(2^{1/3}) : \mathbb{Q}] \\ &= 3.2 = 6.\end{aligned}$$

Therefore, the degree of the splitting field of $x^3 - 2$ over $\mathbb{Q}$ is 6.

Worked Exercise 14.3.2. Let $f(x) = x^2 + 3$ and $g(x) = x^2 + x + 1$ be polynomials over $\mathbb{Q}$. Prove that their splitting fields are equal and find its degree over $\mathbb{Q}$.

Answer: $\sqrt{3}i$ and $-\sqrt{3}i$ are the roots of $f(x)$ and hence $\mathbb{Q}(\sqrt{3}i)$ is the splitting field of $f(x)$. Further, $\dfrac{-1 \pm \sqrt{3}\, i}{2}$ are the roots of $g(x)$. Put $w = \dfrac{-1 + \sqrt{-3}}{2}$. Then, $\sqrt{-3} = 2w + 1$ and hence $\sqrt{-3} \in \mathbb{Q}(w)$. Therefore $\mathbb{Q}(\sqrt{-3}) \subseteq \mathbb{Q}(w)$. Also,

$$w = -\frac{1}{2} + \frac{1}{2}\sqrt{-3} \in \mathbb{Q}(\sqrt{-3})$$

and hence $\mathbb{Q}(w) \subseteq \mathbb{Q}(\sqrt{-3})$. Thus, $\mathbb{Q}(w) = \mathbb{Q}(\sqrt{-3})$ is the splitting field of $f(x)$ and $g(x)$. Also, $f(x)$ and $g(x)$ are both irreducible over $\mathbb{Q}$ and $[\mathbb{Q}(w) : \mathbb{Q}] = 2$.

EXERCISE 14(C)

1. Let K be an algebraically closed field and F be a subfield of K. Then prove that the algebraic closure $\overline{F}$ of F in K is also algebraically closed.

2. If $F_1 \subset F_2 \subset \ldots$ is a chain of fields such that F_{n+1} is a field extension of F_n for all $n > 0$, then prove that $\bigcup_{n=1}^{\infty} F_n$ is a field extension of each F_n.

3. Construct splitting fields of the following polynomials over the field $\mathbb{Q}$ of rational numbers.
   (i) $x^4 + 1$
   (ii) $x^3 - 1$
   (iii) $x^6 - 1$

4. Prove that no finite field is algebraically closed.

5. Construct a splitting field for $x^3 + x + 1$ over the field $\mathbb{Z}_2$ and list all its elements.

6. In the proof of Theorem 14.3.4, prove that $K = \bigcup_{\alpha \in \Delta} K_\alpha$ is a subfield of E and that the map $\theta : K \to L$ is an embedding.

7. Prove that the degree of a splitting field of a polynomial of degree n (> 0) over a field F is almost $n!$

8. Let $p(x) \in F[x]$ be an irreducible polynomial over F. If $p(x)$ has one root in a splitting field E of a polynomial $f(x) \in F[x]$, then prove that $p(x)$ has all its roots in E.

9. Let E be any field extension of $\mathbb{Q}$. Prove that the polynomial $x^3 - 3x + 1$ is either irreducible or splits into linear factors over E.

10. For any prime p, find the splitting field of $x^p - 1$ over $\mathbb{Q}$ and its degree over $\mathbb{Q}$.

14.4 DERIVATIVES AND MULTIPLE ROOTS

In this section, we introduce the notion of the derivative of a polynomial as the usual derivative of a function in calculus and relate this to the multiplicity of a polynomial. Let us recall that a is a root of a polynomial if and only if $x - a$ is a divisor of $f(x)$. If $(x - a)^n$ divides $f(x)$ for $x > 1$, then a is called a multiple root of $f(x)$.

The properties of derivatives which are familiar from the calculus are not necessarily valid here. For example. $f'(x) = 0$ does not imply that $f(x)$ is a constant. Consider the polynomial $f(x) = x^3$ over the field $\mathbb{Z}_3$ of integers modulo 3. Then, $f'(x) = 3x^2 = 0$. However, the ordinary rules for operating with derivatives remain the same. Let us formally define the derivative of a polynomial in the following definition.

Definition 14.4.1. Let $f(x) = a_0 + a_1x + a_2x^2 + \cdots + a_nx^n$ be a polynomial over a field F. Then, the derivative of $f(x)$ is defined as

$$f'(x) = a_1 + 2a_2x + 3a_3x^2 + \cdots + na_nx^{n-1}.$$

In particular, the derivative of any constant polynomial is defined to be 0.

Example 14.4.1

1. Let $f(x) = 2 + 3x + x^2 + 4x^3 + 2x^4 \in \mathbb{Q}[x]$. Then,
$$f'(x) = 3 + 2x + 12x^2 + 8x^3$$
2. Let $f(x) = 1 + x + x^2 + x^3 + x^4 \in \mathbb{Z}_2[x]$, where $\mathbb{Z}_2$ is the field of integers modulo 2, Then,
$$f'(x) = 1 + x^2$$
since $2 \equiv 0$, $3 \equiv 1$ and $4 \equiv 0$ in $\mathbb{Z}_2$.
3. Let $f(x) = 2 + 3x + 4x^2 + 2x^3 + 3x^4 + 2x^5 \in \mathbb{Z}_5[x]$.
Then, $f'(x) = 3 + 3x + x^2 + 2x^3$
since $8 \equiv 3$, $6 \equiv 1$, $12 \equiv 2$ and $10 \equiv 0$ in $\mathbb{Z}_5$.

The following are routine verifications.

Theorem 14.4.1. Let F be any field, $f(x)$ and $g(x) \in F[x]$ and $a \in F$. Then the following holds.

1. $(f(x) + g(x))' = f'(x) + g'(x)$
2. $(a\,f(x))' = a\,f'(x)$
3. $(f(x)g(x))' = f'(x)g(x) + f(x)\,g'(x)$
4. If $\text{char}(F) = 0$ and $\deg(f(x)) = n > 0$, then $\deg(f'(x)) = n - 1$
5. If $\text{char}(F) = p$ and $f(x) = x^p$, then $f'(x) = 0$.

Definition 14.4.2. Let F be any field and $f(x) \in F[x]$. Let E be any extension of F and $a \in E$. For any positive integer m, if $(x - a)^m$ divides $f(x)$ in $E[x]$ and $(x - a)^{m+1}$ does not divide $f(x)$, then a is called a *root of $f(x)$ of multiplicity m* and m is called the *multiplicity* of the root a. A root of multiplicity 1 is called a *simple root* and a root of multiplicity $m > 1$ is called a *multiple root.*

Earlier in Ring Theory, we have proved that a polynomial degree n over F can have at most n roots in any extension of F. For this counting purpose, we shall count a root a as m roots if a is a root of multiplicity m and not as one root. In the following, we obtain a necessary and sufficient condition in terms of the derivative of $f(x)$ for a root of $f(x)$ to be a multiple root.

Theorem 14.4.2. Let $f(x) \in F[x]$ be a nonconstant polynomial and a be an element in any field extension E of F. Then, a is a multiple root of $f(x)$ if and only if $f(a) = 0 = f'(a)$; that is, a is a root of both $f'(x)$ and $f'(x)$.

Proof: Suppose that a is a multiple root of $f(x)$ of multiplicity $m > 1$. Then, $(x - a)^m$ divides $f(x)$ in $E[x]$ and hence

$$f(x) = (x - a)^m g(x) \text{ for some } g(x) \in E[x].$$

Now, $f'(x) = m(x - a)^{m-1}g(x) + (x - a)^m g'(x)$, $m - 1 > 0$ and therefore $f'(a) = 0 = f(a)$; that is, a is a common root of $f(x)$ and $f'(x)$.

Conversely suppose that $f'(a) = 0 = f(a)$. Then,

$$f(x) = (x - a)\, p(x)$$

for some $p(x) \in E[x]$, and therefore

$$f'(x) = p(x) + (x - a)\, p'(x)$$

and hence $0 = f'(a) = p(a) + (a - a)\, p'(x) = p(a)$. Therefore, a is a root of $p(x)$ so that $p(x) = (x - a)\, h(x)$ for some $h(x) \in E[x]$. Now,

$$f(x) = (x - a)\, p(x) = (x - a)^2\, h(x).$$

Therefore, $(x - a)^m$ divides $f(x)$ for some $m > 1$. Thus, a is a multiple root of $f(x)$. ◀

Corollary 14.4.1. Let $f(x) \in F[x]$. Then, $f(x)$ has only simple roots in any extension E of F if and only if the g.c.d.$\{f(x), f'(x)\}$ is a unit in $E[x]$.

Corollary 14.4.2. Let $f(x) = x^n - 1 \in F[x]$ where $n > 0$. Suppose that $\text{char}(F) = 0$ or $\text{char}(F) = p$ where p does not divide n. Then, the roots of $f(x)$ are all distinct.

Proof: We have $f'(x) = nx^{n-1}$ and hence, if a is a multiple root of $f(x)$, we have $0 = f'(a) = na^{n-1}$, so that $a = 0$ (since $n > 0$) which is a contradiction (since 0 is not a root of $x^n - 1$). Thus, $f(x)$ has no multiple roots; that is, the roots of $f(x)$ are all distinct. ◀

Theorem 14.4.3. Let $f(x)$ be an irreducible polynomial over F. Then, $f(x)$ has a multiple root in some field extension of F if and only if $f'(x) = 0$.

Proof: Suppose that $f'(x) = 0$. If E is a field extension of F and $a \in E$ is a root of $f(x)$, then $f(a) = 0 = f'(a)$ and hence, by Theorem 14.4.2, a is a multiple root of $f(x)$ in E.

Conversely suppose that $f(x)$ has a multiple root a in a field extension E of F. Then, by Theorem 14.4.2, a is a root of $f'(x)$. Suppose, if possible, that $f'(x) \neq 0$. Let $p(x)$ be the minimal polynomial of a over F. Then, since $f(a) = 0 = f'(a)$, we get that $p(x)$ divides $f(x)$ and $f'(x)$. But, since $f(x)$ is irreducible, if follows that $f(x)$ is an associate of $p(x)$ and hence $f(x)$ divides $f'(x)$, which is a contradiction, since $\deg(f'(x)) < \deg(f(x))$.

Thus, $f'(x) = 0$. ◀

Theorem 14.4.4. The following holds for any irreducible polynomial $f(x)$ over a field F.

1. If $\text{char}(F) = 0$, then $f(x)$ has no multiple roots.
2. When $\text{char}(F) = p \neq 0$, $f(x)$ has a multiple root if and only if $f(x) = g(x^p)$ for some $g(x) \in F[x]$.

Proof:

1. Let $\text{char}(F) = 0$ and $f(x) = a_0 + a_1x + \cdots + a_nx^n$.
   Since $f(x)$ is irreducible, $\deg(f(x)) = n > 0$. Then,
   $$\begin{aligned} f'(x) = 0 &\Rightarrow a_1 + 2a_2x + \cdots + na_nx^{n-1} = 0 \\ &\Rightarrow a_1 = a_2 = \cdots = a_n = 0 \text{ (since char}(F) = 0) \\ &\Rightarrow f(x) = a_0, \text{ which is not true.} \end{aligned}$$
   Therefore $f'(x) \neq 0$ and hence, by the above theorem, $f(x)$ has no multiple roots.
2. Let $\text{char}(F) = p \neq 0$ and $f(x) = a_0 + a_1x + \cdots + a_nx^n$ with $a_n \neq 0$ and $n > 0$. Then, $f'(x) = a_1 + 2a_2x + \cdots + na_nx^{n-1}$. Suppose that $f(x)$ has a multiple root. Then, by Theorem 14.4.3, $f'(x) = 0$ and hence $a_1 = 2a_2 = 3a_3 = \cdots = na_n = 0$. Since $\text{char}(F) = p$, it follows that, for each $1 \leq K \leq n$, either $a_K = 0$ or p divides K (that is, $K = K_1p$ for some integer $K_1 > 0$).
   Thus, we have
   $$f(x) = b_0 + b_1x^p + b_2x^{2p} + \cdots + b_mx^{mp} = g(x^p)$$
   for some positive integer m, where $g(x) = b_0 + b_1x + \cdots + b_mx^m$. Conversely suppose that $f(x) = g(x^p) = b_0 + b_1x^p + \cdots + b_mx^{mp}$. Then, $f'(x) = pb_1x^{p-1} + \cdots + pmb_mx^{mp-1} = 0$ (since $\text{char}(F) = p$) and hence, again by Theorem 14.4.3, $f(x)$ has a multiple root. ◀

Theorem 14.4.5. All the roots of an irreducible polynomial over a field F have the same multiplicity.

Proof: Let $p(x)$ be an irreducible polynomial over F. Let $\overline{F}$ be the algebraic closure of F and let a and b be roots $p(x)$ in $\overline{F}$ with multiplicities m and m', respectively.

Then, we know that,

$$F[x]/<p(x)> \cong F(a) \quad \text{and} \quad F[x]/<p(x)> \cong F(b)$$

under the isomorphisms $\sigma: F[x]/<p(x)> \to F(a)$ and $\tau: F[x]/<p(x)> \cong F(b)$ and given by

$$\begin{aligned} &\sigma(f(x) + <p(x)>) = f(a) \\ \text{and} \quad &\tau(f(x) + <p(x)>) = f(b) \end{aligned}$$

Now, put $\eta = \tau \circ \sigma^{-1}$. Then, $\eta : F(a) \to F(b)$ is an isomorphism and is given by

$$\eta(a_0 + a_1 a + \cdots + a_n a^n) = a_0 + a_1 b + \cdots + a_n b^n.$$

Then, η can be extended to an isomorphism $\eta^* : \overline{F} \to \overline{F(b)} = \overline{F}$.
Let $\alpha : \overline{F}[x] \to \overline{F}[x]$ be the ring homomorphism induced by η^*. Then, α is given by

$$\alpha(a_0 + a_1 x + \cdots + a_s x^s) = \eta^*(a_0) + \eta^*(a_1)x + \cdots + \eta^*(a_s)x^s.$$

Note that $\alpha(p(x)) = p(x)$. Since $\alpha(x - a)^m = (x - b)^m$, we get that $(x - b)^m$ is a factor of $p(x)$ and hence $m \leq m'$. By interchanging the roles of a and b, we get that $m' \leq m$. Thus, $m = m'$ and hence a and b are of same multiplicities. ◀

Worked Exercise 14.4.1. Prove that a polynomial $f(x) \in F[x]$ has a multiple root if and only if $f(x)$ and $f'(x)$ have a nonconstant common factor.

Answer: Suppose that $f(x)$ has a multiple root a in an extension E of F. Then,

$$f(x) = (x - a)^m g(x),\ m > 1 \text{ and } g(x) \in E[x]$$

$$\begin{aligned} \text{Therefore, } f'(x) &= m(x - a)^{m-1} g(x) + (x - a)^m g'(x) \\ &= (x - a)^{m-1} (mg(x) + (x - a)g'(x)),\ m - 1 > 0 \end{aligned}$$

and hence $(x - a)^{m-1}$ is a nonconstant common factor of $f(x)$ and $f'(x)$.

Conversely suppose that $f(x)$ and $f'(x)$ have a nonconstant common factor, say $p(x)$ with $\deg(p(x)) > 0$.
Suppose, if possible, that all the roots of $f(x)$ are distinct.
Let $a_1, a_2, \ldots, a_n$ be all the distinct roots of $f(x)$ in any extension of F. Then,

$$f(x) = a\prod_{i=1}^{n}(x - a_i), \text{ for some } a \in F.$$

$$\text{and} \quad f'(x) = a\sum_{i=1}^{n}\left(\prod_{j\neq i}(x - a_j)\right).$$

Therefore, for each $1 \leq i \leq n$, $f'(a_i) = \prod_{j\neq i}(a_i - a_j) \neq 0$ and hence no root of $f(x)$ is a root of $f'(x)$. Thus, $f(x)$ and $f'(x)$ have no nonconstant common factor, which is a contradiction to the hypothesis. Thus, $f(x)$ has a multiple root.

Worked Exercise 14.4.2. Let F be a finite field and $f(x)$ be an irreducible polynomial over F. Then prove that $f(x)$ has no multiple roots.

Answer: We will be proving later in the next section that $|F| = p^n$, where p is prime, $\text{char}(F) = p$ and $n \in \mathbb{Z}^+$ and $a \mapsto a^p$ is an automorphism of the field F. Suppose, if possible, that $f(x)$ has a multiple root. Then, by Theorem 14.4.4 (2),

$$f(x) = g(x^p) = a_0 + a_1x^p + \cdots + a_nx^{np}, a_i \in F.$$

Since $a \mapsto a^p$ is an automorphism of F, we can choose $b_0, b_1, \ldots, b_n \in F$ such that $a_i = b_i^p$ for each $1 \leq i \leq n$.
Now,

$$\begin{aligned} f(x) &= b_0^p + b_1^p x^p + b_2^p x^{2p} + \cdots + b_n^p x^{np} \\ &= (b_0 + b_1x + b_2x^2 + \cdots + b_nx^n)^p \end{aligned}$$

which implies that $f(x)$ is reducible over F, which is a contradiction to the hypothesis. Thus, $f(x)$ has no multiple roots.

EXERCISE 14(D)

1. Prove Corollary 14.4.1.
2. If $f(x) = \prod_{i=1}^{n} g_i(x)$, then prove that
$$f'(x) = \sum_{i=1}^{n}\left(\prod_{j\neq i} g_j(x)\right)g_i'(x)$$
3. Let $f(x) \in F[x]$ and $\deg(f(x)) = n > 0$. If $\text{char}(F) = p$ and $f'(x) = 0$, then prove that p divides n and $f(x)$ has at most $\frac{n}{p}$ distinct roots.

4. Let a be a root of $f(x) \in F[x]$. Prove that a is a root of multiplicity $m > 1$ if and only if $f^{(r)}(a) = 0$ for each $1 \leq r < m$ and $f^{m}(a) \neq 0$, where $f^{(r)}(a)$ is the rth derivative of $f(x)$ at $x = a$. The rth derivative of $f(x)$ is inductively defined by

$$f^{(r)}(x) = (f^{(r-1)}(x))' \text{ and } f^{(0)}(x) = f(x).$$

5. Let F be a field, a and $b \in F, f(x) \in F[x]$ and $\deg(f(x)) = n$. Prove that

$$f(a+b) = \sum_{r=0}^{n} \frac{1}{r!} f^{(r)}(a) b^r.$$

6. Let F be a field of characteristic 3 and $E = F[x]$, the field of quotients of the integral domain $F[x]$. Prove that the polynomial $y^3 - x \in E[y]$ is irreducible over E and has multiple roots.

14.5 FINITE FIELDS

Recall that a field having no proper subfields is called a *prime field* and that any field E is an extension of a prime field, which is precisely the intersection of all subfields of E. Also recall that $\mathbb{Z}_p$, where p is a prime, and $\mathbb{Q}$ are the only (up to isomorphism) prime fields. If E is a field of characteristic zero, then $\mathbb{Q}$ is the prime subfield of E and, if char$(E) = p$, then $\mathbb{Z}_p$ is the prime subfield of E.

In this section, we mainly discuss about finite fields which have necessarily $\mathbb{Z}_p$ as their prime subfields, where p is the characteristic of the given field. Let us begin with the following definition.

Definition 14.5.1. Let F be a field of characteristic p, where p is a prime. Then the intersection of all subfields of F is called the *prime subfield* of F and is denoted by F_p. Note that F_p has no proper subfields and hence F_p is a prime field. Also $F_p \cong \mathbb{Z}_p$, the field of integers modulo p.

The following theorem determines the number of elements in any given finite field.

Theorem 14.5.1. Let F be a finite field of characteristic p. Then, $|F| = p^n$ for some positive integer n.

Proof: It is well known that the characteristic of a finite field must be a prime number. Therefore, p is a prime number. Consider the prime subfield F_p of F. Then, $F_p \cong \mathbb{Z}_p$ and hence $|F_p| = p$. Also, F becomes a vector space over F_p. Since F is finite, F is a finite dimensional vector space over F_p and hence $F \cong F_p^n$, where $n = [F : F_p]$, the dimension of F over F_p. Thus, $|F| = |F_p^n| = |F_p|^n = p^n, n > 0$. ◀

We have proved above that the number of elements in any finite field is of the form p^n, for some prime p and $n \in \mathbb{Z}^+$. On the other hand, we prove in the next two results below that, for any prime p and a positive integer n, there exists unique (up to isomorphism) finite field with exactly p^n elements. First, we take up the uniqueness.

Theorem 14.5.2. Let F be a finite field with p^n elements, where p is a prime and n is a positive integer. Then F is the splitting field of the polynomial $x^{p^n} - x \in \mathbb{Z}_p[x]$.

Proof: First recall that the prime subfield F_p is isomorphic to $\mathbb{Z}_p$ and hence $\mathbb{Z}_p$ can be considered as the prime subfield of F. Let $F^* = F - \{0\}$. Then, F^* is a group under multiplication and $|F^*| = p^n - 1$. Therefore, $a^{p^n-1} = 1$ for all $a \in F^*$. This implies that $a^{p^n} = a$ for all $a \in F$. Therefore, every element of F is a root of the polynomial $x^{p^n} - x$, which can have at most p^n roots in any extension of F. Thus, F is precisely equal to the set of all roots of $x^{p^n} - x$ and hence F is the splitting field of $x^{p^n} - x$, over $\mathbb{Z}_p$. ◀

Since any two splitting of a polynomial are isomorphic, the following is an immediate consequence of the above theorem.

Corollary 14.5.1. If E and F are finite fields and $|E| = p^n = |F|$, where p is a prime and n is a positive integer, then $E \cong F$.

Theorem 14.5.3. Let p be a prime and n be a positive integer. Then there exists a field F with exactly p^n elements.

Proof: Consider $f(x) = x^{p^n} - x \in \mathbb{Z}_p[x]$. Let K be an extension of $\mathbb{Z}_p$ containing all the roots of $f(x)$. Then,

$$f'(x) = p^n x^{p^n-1} - 1 = -1 \neq 0$$

and hence all the roots of $f(x)$ are distinct. We have to prove that the roots of $f(x)$ form a subfield of K. If a and b are roots of $f(x)$, then

$$(a+b)^{p^n} = a^{p^n} + b^{p^n} = a + b \text{ (since } p \text{ divides } pnc_r\text{)}$$

$$(-a)^{p^n} = -a \text{ (if } p = 2\text{, note that } a = -a\text{)}$$

$$(ab)^{p^n} = a^{p^n} b^{p^n} = ab$$

$$\text{and} \quad (a^{-1})^{p^n} = (a^{p^n})^{-1} = a^{-1}, \text{ if } a \neq 0.$$

Thus, the set E of all roots of $f(x)$ forms a field and, since the roots are distinct, $|E| = \deg(f(x)) = p^n$. Note that E is the splitting field of $f(x)$ over $\mathbb{Z}_p$. ◀

Corollary 14.5.2. For any prime p and for any positive integer n, there exists unique (up to isomorphism) field E such that $|E| = p^n$ and this is the splitting field of $x^{p^n} - x$ over $\mathbb{Z}_p$.

Definition 14.5.2. Let p be a prime and n be a positive integer. Then the field with p^n elements is called the *Galois field of order* p^n and is denoted by $GF(p^n)$.

We have proved in Theorem 14.5.1 that any finite field is a $GF(p^n)$ for some prime p and $n \in \mathbb{Z}^+$. In the following, we prove that any finite field F has an extension of any given finite degree over F.

Theorem 14.5.4. Let $F = GF(p^n)$ and m be a positive integer. Then there exists a field extensions E of F such that $[E : F] = m$ and any two such extensions are isomorphic.

Proof: Consider the algebraic closure $\overline{F}$ of F and the polynomial

$$f(x) = x^{p^{mn}} - x \in F[x].$$

Since the multiplicative group $F^* = F - \{0\}$ is of order $p^n - 1$, we have $a^{p^n - 1} = 1$ for all $a \in F^*$ Also, since

$$(p^n - 1)(p^{n(m-1)} + p^{n(m-2)} + \cdots + p^n + 1) = p^{nm} - 1,$$

we get that $a^{p^{nm} - 1} = 1$ for all $a \in F^*$ and hence

$$a^{p^{nm}} = a \text{ for all } a \in F.$$

Therefore, each element of F is a root of $f(x)$ and, as in the proof of Theorem 14.5.3, the p^{mn} roots of $f(x)$ are distinct and form a field E. Now, we have

$$\mathbb{Z}_p \cong F_p \subset F \subset E \subset \overline{F}$$

where $[F : F_p] = n$ and $[E : F_p] = mn$ and thus $[E : F] = m$. ◀

The following is an important result regarding the multiplicative group of nonzero elements of a finite field.

Theorem 14.5.5. Let F be a field and $F^* = F - \{0\}$, the group of nonzero elements of F under the multiplication. Then, F^* is a cyclic group if and only if F is a finite field.

Proof: Suppose that F^* is a cyclic group and let

$$F^* = \langle a \rangle = \{a^n : n \in \mathbb{Z}\}.$$

That is, F^* is the cyclic group generated by a. Now, if char(F) = 0, then $-1 \neq 1$ and $-1 \in F^* = \langle a \rangle$ and hence $-1 = a^n$ so that $1 = (-1)^2 = a^{2n}$ for some $n \neq 0$, which implies that a is of finite order in the group F^* and hence F^* and F are finite.
Therefore, we can suppose that char(F) = p, a prime. Since $F^* = \langle a \rangle$, we get that $F_p(a) = F$. If $1 + a = 0$, then $a^2 = 1$ and hence O(a) is finite in the group $F^* = \langle a \rangle$, so that F^* and F are finite. Suppose that $1 + a \neq 0$. Then $1 + a = a^r$ for some $r \in \mathbb{Z}$ and hence a is a root of the polynomial $x^r - x - 1 \in F_p[x]$. This implies that a is algebraic over F_p and hence $[F_p(a) : F_p]$ and $[F_p(a) : F]$ are finite. Therefore, $[F : F_p]$ is finite. Thus, F is finite. Conversely suppose that F is a finite field. We know that F^* is abelian group. Let

$$F^* = \{a_1, a_2, \ldots, a_m\}.$$

Then, since F^* is an abelian group, there exists $a \in F^*$ such that

$$\text{O}(a) = \text{l.c.m.}\{\text{O}(a_1), \text{O}(a_2), \ldots, \text{O}(a_m)\}.$$

If O(a) = r, then clearly $r \leq m = |F^*|$. Also, $a_i^r = 1$ for all $1 \leq i \leq m$ and hence $a_1, a_2, \ldots, a_m$ are the roots of the polynomial $x^r - 1$ which has at most r roots in any extension of F. Therefore, $m \leq r$. Thus, $m = r = \text{O}(a)$ so that $\langle a \rangle = F^*$, which implies that F^* is a cyclic group. ◀

Corollary 14.5.3. Any finite extension E of a finite field F is a simple extension; that is, $E = F(a)$ for some $a \in E$.

Proof: Let F be a finite field and E be a finite extension of F. Let $[E : F] = n$. Then E is a n dimensional vector space over F and hence $|E| = |F|^n$, so that E is a finite field. By the above theorem, E^* is a cyclic group generated by $a \in E$. Now, E is the smallest subfield of E containing F and a. Thus, $E = F(a)$. ◀

Corollary 14.5.4. Let F be a finite field and n be a positive integer. Then, there exists an irreducible polynomial of degree n over F.

Proof: By Theorem 14.5.4, there exists an extension E of F such that $[E : F] = n$. Then, by Corollary 14.5.3, $E = F(a)$ for some $a \in E$. Since E is a finite extension of F, a is algebraic over F. Let $p(x)$ be irreducible over F and

$$\deg(p(x)) = [F(a) : F] = [E : F] = n.$$ ◀

For any field F, let us recall that the set Aut(F) of all automorphisms of F is a group under the composition of mappings. In the following, we prove that Aut(F) is a cyclic group for any finite field F.

Theorem 14.5.6. Let F be a finite field of characteristic p and $|F| = p^n$. Then, Aut(F) is a cyclic group of order n.

Proof: Define $\phi : F \to F$ by $\phi(a) = a^p$ for any $a \in F$. By using the fact that p is the characteristic of F, it can be easily verified that ϕ is a homomorphism. Also, for any a and $b \in F$,

$$a^p = b^p \Rightarrow (a-b)^p = 0 \Rightarrow a-b = 0 \Rightarrow a = b.$$

Therefore, ϕ is an injection of F into F. Since F is finite, it follows that ϕ is a surjection also. Thus, ϕ is an automorphism of F. We prove that O(ϕ) = n and |Aut(F)| = n. Since $|F^*| = p^n - 1$, we get that $a^{p^n-1} = 1$ for all $a \in F^*$ and hence

$$\phi^n(a) = a^{p^n} = a \text{ for all } a \in F.$$

Therefore, ϕ^n = Id and hence O(ϕ) $\leq n$. On the other hand, suppose ϕ^d = Id for some $d > 0$. Then, $a^{p^d} = a$, for all $a \in F$ and hence every element of F is a root of the polynomial $x^{p^d} - x$. Therefore, $p^n \leq p^d$ and hence $n \leq d$. Thus, O(ϕ) = n. Now, we have $F_p \leq F \leq \overline{F}$ and by Corollary 14.5.3, $F = F_p(a)$ where $F^* = \langle a \rangle$. Let $p(x)$ be the minimal polynomial of a over F. Then,

$$[F_p(a) : F_p] = \deg(p(x)) = n.$$

Further, by Theorem 14.5.2, F is equal to the set of all roots of the polynomial $x^{p^n} - x \in F_p[x]$. If $\sigma : F_p(a) \to \overline{F}$ is an embedding, then $\sigma(a) \in F = F_\sigma(a)$ (since $\sigma(a)$ is a root of $x^{p^n} - x$), which implies that $\sigma(F) \subseteq F$. Since F is finite and σ is an injection, we get that $\sigma(F) = F$.
By Theorem 14.3.3, we know that the number of extensions $\sigma : F_p(a) \to \overline{F}$ is equal to the degree of the irreducible polynomial satisfied by a. Therefore, Aut(F) contains precisely n elements. Since O(ϕ) = n, ϕ is a generator of Aut(F). Thus, Aut(F) is a cyclic group of order n. ◀

Worked Exercise 14.5.1. Let F be a finite field and $|F| = p^n$, p a prime and $n \in Z^+$. For each divisor m of n, prove that F has exactly one subfield E of F such that $|E| = p^m$.

Answer: Let us recall from group theory that any cyclic group of order n has a unique subgroup of order d for each divisor d of n. Let m be a divisor of n. Consider the cyclic group $F^* = F - \{0\}$ of order $p^n - 1$. Then, $p^m - 1$ divides $p^n - 1$; for, if $md = n$, then

$$p^n - 1 = (p^m - 1)\,(p^{m(d-1)} + p^{m(d-2)} + \cdots + p^m + 1).$$

Therefore, there exists a unique subgroup G of F^* of order $p^m - 1$. Then, for all $a \in G$, $a^{p^m - 1} = 1$ and hence $a^{p^m} = a$ for all $a \in G \cup \{0\}$. Since the roots of $x^{p^m} - x$ form a field, it follows that $H \cup \{0\}$ is the unique subfield of F of order p^m.

EXERCISE 14(E)

1. Construct fields with 4, 8, 9 and 16 elements.
2. Find a generator for the multiplicative groups of nonzero elements of a field with 8 elements.
3. Prove that a finite extension E of a field F is a simple extension of F if and only if there are only a finite number of subfields of E containing F.
4. Let F be a field such that $|F| = 4$. Then find irreducible polynomials over F of degree 2, 3 and 4.
5. Prove that for $n \geq 3$, the polynomial $x^{2^n} + x + 1$ is irreducible over the field $\mathbb{Z}_2$ of integers modulo 2.
6. Let F be a finite field. Prove that any element of F can be expressed as the sum of two squares.
7. Let a and b be two elements of a finite field F. Then prove that there exist elements α and β in F such that $\alpha + a\alpha^2 + b\beta^2 = 0$.
8. Let F be a finite field and char$(F) = p$. Prove that each element a of F has a unique p^{th} root $\sqrt[p]{a}$ in F.

15

Galois Theory

Galois theory of fields is one of the most elegant theories in Abstract Algebra and it is an excellent combination of group theory and the theory of algebraic field extensions. In general, Galois theory and, in particular, the fundamental theorem of Galois theory has several applications to the theory of equations and geometry. Although we are not making a detailed study of Galois theory, we shall discuss its fundamental concepts and give certain simple applications like proving the fundamental theorem of algebra and the nonconstructability of certain geometric figures using straight-edge and compass alone. In this chapter, we introduce the concepts of separable extensions and normal extensions and prove certain important properties of these. Also, we prove the celebrated theorem 'the fundamental theorem of Galois theory' which establishes a one-to-one correspondence between the subfields of a splitting field E of a separable polynomial in $F[x]$ and the set of subgroups of the group of F-automorphisms of E. This one-to-one correspondence transforms certain problems of subfields of fields into more simpler and amenable problems about the subgroups of groups. Certain applications of Galois theory are discussed in the next chapter.

15.1 SEPARABLE AND NORMAL EXTENSIONS

In this section, we discuss certain special extensions of a field, namely separable extensions and simple extensions and prove certain important properties of these. First, we introduce the notion of a separable polynomial in the following definition.

Definition 15.1.1. Let F be any field. An irreducible polynomial in $F[x]$ is called a *separable polynomial* if all its roots are distinct (that is, if it has no multiple roots). Any polynomial $f(x) \in F[x]$ is called *separable* if all the irreducible factors of $f(x)$ are separable. A polynomial that is not separable is called *inseparable*.

Example 15.1.1

1. Clearly, any polynomial of degree one is separable.
2. $x^2 + 1 \in \mathbb{R}[x]$ is an irreducible polynomial and its roots i and $-i$ are distinct and therefore $x^2 + 1$ is a separable polynomial.
3. The polynomial $1 + x + x^2 \in \mathbb{Q}[x]$ is separable, since its roots are $(\frac{-1}{2}) \pm (\frac{\sqrt{3}}{2})i$ and it is an irreducible polynomial.
4. The polynomial $1 + 2x + x^2$ is separable, since its irreducible factors are $1 + x$ alone. In fact, for any $a \in F$, $(x - a)^n$ is a separable polynomial.

Definition 15.1.2. Let E be a field extension of a field F and $a \in E$. Then, a is said to be *separable over* F if a is algebraic over F and the minimal polynomial of a over F is separable. E is said to be a *separable extension* of F if every element of E is separable over F.

Clearly, every separable extension of F is an algebraic extension of F. However, not every algebraic extension is separable. If $\text{char}(F) = 0$, then by Theorem 14.4.4, every algebraic extension of F is a separable extension. Also, by Worked Exercise 14.4.2, every algebraic extension of a finite field F is a separable extension. In the following example, we give an example of an algebraic extension which is not separable.

Example 15.1.2. Let K be the field of quotients of $\mathbb{Z}_3[x]$ and consider

$$f(y) = y^3 - x \in K[y].$$

We prove that $f(y)$ is irreducible over K and has multiple roots. If $f(y)$ is reducible over K, there must exist a root $\big(g(x)/h(x)\big)(h(x) \neq 0)$ in K which implies that

$$\frac{(g(x))^3}{(h(x))^3} = x$$

and hence $3\deg(h(x)) + 1 = 3\deg(g(x))$ which is absurd. Therefore, $f(y)$ is irreducible over K. Also, if b_1 and b_2 are roots of $f(y)$ in its splitting field, then

$$b_1^3 = x = b_2^3$$

which implies that $(b_1 - b_2)^3 = b_1^3 - b_2^3 = 0$ (note that char(K) = 3) and hence $b_1 - b_2 = 0$. This proves that $f(y)$ has only one root whose multiplicity is three. If a is a root of $f(y)$ in any extension of K, then $K(a)$ is an algebraic extension of K and $K(a)$ is not a separable extension.

Definition 15.1.3. A field F is called a *perfect field* if each of its algebraic extension is separable.

Any field of characteristic zero is a perfect field and likewise, any finite field is also perfect. The field K given in Example 15.1.2 is not perfect.

Let us recall that an extension E of F is called simple if $E = F(a)$ for some $a \in E$, where a is algebraic over F. By Corollary 14.5.4, every finite extension of a finite field is simple. This result is extended to separable extensions of any field in the following theorem.

Theorem 15.1.1. Any finite separable extensions of any field F is a simple extension of F.

Proof: Let F be any field and E be a finite separable extension of F. By Corollary 14.5.4, we may assume that F is an infinite field. We first prove that, for any a and $b \in E$, there exists $c \in E$ such that $F(a, b) = c$ and then, by using induction on n, we can prove that $F(a_1, a_2, \ldots, a_n) = F(c)$ for some $c \in E$. Now, let a and $b \in E$. Since E is a separable extension, it is an algebraic extension of F. In particular, a and b are algebraic over F. Let $f(x)$ and $g(x)$ be the minimal polynomials of a and b, respectively, over F. Let $a = a_1, a_2, \ldots, a_n$ be the roots of $f(x)$ in its splitting field and $b = b_1, b_2, \ldots, b_m$ be the roots of $g(x)$ in its splitting field. By hypothesis, a_i's are distinct and b_i's are distinct. Put

$$\alpha_{ij} = \frac{a_i - a}{b - b_j} \quad \text{for } 1 \le i \le n \quad \text{and} \quad 1 < j \le m.$$

Since F is infinite, we can choose $0 \ne \alpha \in F$ such that $\alpha \ne \alpha_{ij}$ for all $1 \le i \le n$ and $1 < j \le m$. Then,

$$\alpha \ne \frac{a_i - a}{b - b_j} \quad \text{and hence } \alpha b - \alpha b_j \ne a_i - a$$

for all $1 \le i \le n$ and $1 < j \le m$. Take $c = a + \alpha b$.
Then, clearly $F(c) \subseteq F(a, b)$ (since $\alpha \in F$).
Now, put $h(x) = f(c - \alpha x) \in F(c)[x]$.
Then, $h(b) = f(c - \alpha b) = f(a) = 0$.
Also, $h(b_j) = f(c - \alpha b_j) \ne 0$ for all $j > 1$, since $c - \alpha b_j \ne a_i$ for all i and $a_1, a_2, \ldots, a_n$ are the roots of $f(x)$. Therefore, $x - b$ is the only common

factor of $g(x)$ and $h(x)$ in $F(c)[x]$. Now, b is algebraic over F and hence over $F(c)$. If $p(x) \in F(c)[x]$ is the minimal polynomial of b over $F(c)$, then $p(x)$ divides $g(x)$. Also, $p(x)$ divides $h(x)$ and hence $p(x)$ divides $x - b$ which implies that $p(x) = x - b$. Therefore, $b \in F(c)$ and hence $\alpha b \in F(c)$, since $\alpha \in F$. Now, $a = c - \alpha b \in F(c)$. Thus, a and $b \in F(c)$ and hence $F(a, b) \subseteq F(c)$. Thus, $F(a, b) = F(c)$. If $E = F(a_1, a_2, \ldots, a_n)$, then by using induction on n, we can prove that $E = F(c)$ for some $c \in E$. Thus, E is a simple extension of F. ◀

Since any finite extension of a field is an algebraic extension and since any algebraic extension of a field of characteristic zero is a separable extension, the following is an immediate consequence of the above theorem.

Corollary 15.1.1. Every finite extension of a field of characteristic zero is a simple extension.

Theorem 15.1.2. A finite extension E of a field F is a simple extension of F if and only if there are only finitely many intermediate fields between F and E.

Proof: Let E be a finite extension of a field F. Suppose that E is a simple extension of F and $a \in E$ such that $E = F(a)$. Let $p(x)$ be the minimal polynomial of a over F. Let K be a subfield of E containing F. Now, a is algebraic over K and let

$$q(x) = a_0 + a_1x + a_2x^2 + \cdots + a_{n-1}x^{n-1} + x^n, \ a_i \in K$$

be the minimal polynomial of a over K. Then, clearly $q(x)$ divides $p(x)$ in $K[x]$ and $[K(a) : K] = n$. Consider $K' = F(a_0, a_1, \ldots, a_{n-1})$. Then,

$$q(x) \in K'[x], K' \subseteq K \quad \text{and} \quad [K'(a) : K'] = n$$

and therefore we have

$$E = F(a) \subseteq K'(a) \subseteq K(a) \subseteq E(a) = E.$$

Therefore, $K'(a) = K(a)$. Also,

$$[K(a) : K] = [K'(a) : K'] = n \quad \text{and} \quad K' \subseteq K$$

and hence we get that $K = K'$. Thus, we have proved that, for any intermediate field K between F and E,

$$K = F(a_0, a_1, \ldots, a_{n-1}),$$

where $q(x) = a_0 + a_1x + \cdots + a_{n-1}x^{n-1} + x^n$ is a divisor of $p(x)$ in $K[x]$. Since the number of monic polynomials dividing $p(x)$ is finite, it follows that the number of intermediate fields between F and E is finite.
Conversely suppose that the number of intermediate fields between F and E is finite. If F is finite field, then E is finite and hence $E - \{0\}$ is a cyclic group and, if $E - \{0\} = \langle a \rangle$, then $F(a) = E$ and hence E is a simple extension of F. Now, suppose that F is infinite. For each $a \in E$, we have $F \subseteq F(a) \subseteq E$. Let

$$A = \{[F(a)\ F] : a \in E\}.$$

Since $[E : F]$ is finite and $[F(a) : F]$ is a positive divisor of $[E : F]$, we get that A is a finite set of positive integers and hence A has the largest member. Choose $b \in E$ such that $[F(b) : F]$ is the largest in A; that is,

$$[F(a) : F] \leq [F(b) : F] \quad \text{for all } a \in E.$$

We prove that $F(b) = E$. Suppose, if possible, that $F(b) \subsetneqq E$. Choose $c \in E$ such that $c \notin F(b)$. Now, since the number of intermediate fields between F and E is finite, we get that

$$\{F(cd + b) : d \in F\} \text{ is finite.}$$

Since F is infinite, we can choose $d_1 \neq d_2 \in F$ such that $F(cd_1 + b) = F(cd_2 + b)$. Put $z = cd_1 + b$. Then, $z \in F(z)$ and

$$cd_2 + b \in F(cd_2 + b) = F(cd_1 + b) = F(z)$$

and therefore $c(d_1 - d_2) \in F(z)$. Since $d_1 - d_2 \neq 0 \in F$, it follows that $c \in F(z)$ and therefore

$$b = z - cd_1 \in F(z) \quad \text{and} \quad F(b) \subseteq F(z).$$

Also, $c \in F(z)$ and $c \notin F(b)$. Therefore, $F(b) \subsetneqq F(z)$ which implies that $[F(z) : F(b)] > 1$. Hence

$$[F(z) : F] = [F(z) : F(b)][F(b) : F] > [F(b) : F]$$

which is a contradiction to the largest property of $[F(b) : F]$. Thus, we have $E = F(b)$ and E is a simple extension of F. ◀

Recall that an element a in an algebraic extension E of a field F is called separable if its minimal polynomial over F is separable. Now, we have the following theorem.

Theorem 15.1.3. Let $F \subseteq E$ be fields and $a \in E$ be algebraic over F. Then, a is separable over F if and only if $F(a)$ is a separable extension of F.

Proof: If $F(a)$ is a separable extension of F, then every element of $F(a)$, in particular, a is separable over F. Conversely suppose that a is separable over F and $b \in F(a)$. We prove that b is separable over F. We have $F \subseteq F(b) \subseteq F(a)$. Let K be an algebraically closed field and $\sigma : F \to K$ be an embedding. Let $p(x)$ be the minimal polynomial of b over F that has m distinct roots. By Theorem 14.3.3, there are m distinct extensions of σ to $F(b)$. Let $\sigma_1, \sigma_2, \ldots, \sigma_m$ be such extensions of σ to $F(b)$. Let $q(x)$ be the minimal polynomial of a over $F(b)$ and suppose that $q(x)$ has n distinct roots. Then, again by Theorem 14.3.3, for each σ_i, $1 \leq i \leq m$, there are exactly n extensions σ_{ij}, $1 \leq j \leq n$ to $F(a)$. Then, clearly the embeddings σ_{ij}, $1 \leq i \leq m$ and $1 \leq j \leq n$, are the only embeddings of $F(a)$ into K which extend $\sigma : F \to K$. Now, let $r(x)$ be the minimal polynomial of a over F. Then,

1. $[F(a) : F]$ = degree of $r(x)$
   = The number of distinct roots of $r(x)$ (since a is separable over F)
   = The number of extensions of σ to $F(a)$.

   Also, since a is separable over F, we get that a is separable over $F(b)$ and hence we have
2. $[F(a) : F(b)]$ = degree of $q(x)$
   = The number of distinct roots of $q(x)$
   = The number of extensions of each σ_i to $F(a)$
   $= n$.

   Also, we have
3. $[F(b) : F]$ = degree of $p(x)$ and
4. The number of distinct roots of $p(x)$ = The number of extensions of σ to $F(b) = m$.

   Now, from (1) to (4), we get that

$$\begin{aligned} n \cdot \deg(p(x)) &= [F(\alpha) : F(b)][F(\beta) : F] \\ &= [F(\alpha) : F] = mn \end{aligned}$$

and hence $m = deg(p(x))$ = The number of distinct roots of $p(x)$.

Therefore, $p(x)$ is a separable polynomial. Thus, b is separable over F. Therefore, every element of $F(a)$ is separable over F. Thus, $F(a)$ is a separable extension of F. ◀

Next, we introduce the notion of a normal extension which plays an important role in the applications of the Galois theory and discuss certain properties of normal extensions.

Definition 15.1.4. An algebraic extension E of a field F is said to be a *normal extension* of F if each irreducible polynomial $f(x)$ over F, having a root in E, splits into linear factors over E.

Note that if E is a normal extension of F and $a \in E$, then all the roots of the minimal polynomial of a over F belong to E.

Example 15.1.3

1. The field $\mathbb{C}$ of complex numbers is a normal extension of the field $\mathbb{R}$ of real numbers and $[\mathbb{C} : \mathbb{R}] = 2$.
2. $\mathbb{R}$ is not a normal extension of the field $\mathbb{Q}$ of rational numbers; for, $x^3 - 2 \in \mathbb{Q}[x]$ is irreducible over $\mathbb{Q}$ and has a root $\sqrt[3]{2}$ in $\mathbb{R}$, but it does not split into linear factors in $\mathbb{R}$, since it has complex roots.

Definition 15.1.5. Let $\{f_\alpha(x)\}_{\alpha \in \Delta}$ be a family of nonconstant polynomials over a field F. An extension E of F is called a *splitting field* of $\{f_\alpha(x)\}_{\alpha \in \Delta}$ if every $f_\alpha(x)$ splits into linear factors in $E(x)$ and E is generated over F by all the roots of the polynomials $f_\alpha(x)$, $\alpha \in \Delta$.

If $\{f_1(x), f_2(x), \ldots, f_n(x)\}$ is a finite family of polynomials over F, then the splitting field of the family $\{f_1(x), \ldots, f_n(x)\}$ over F is same as the splitting field of the simple polynomial $f(x) = \prod_{i=1}^{n} f_i(x)$ over F. On the same lines of the proof of Theorem 14.3.7, we can prove that any two splitting fields of a family $\{f_\alpha(x)\}_{\alpha \in \Delta}$ of polynomials over F are isomorphic under an isomorphism that keeps each element of F fixed. The following gives equivalent statements for an extension to be normal.

Theorem 15.1.4. Let E be an algebraic extension of a field F such that E is a subfield of the algebraic closure $\overline{F}$ of F. Then, the following are equivalent to each other:

1. E is the splitting field of a family of polynomials in $F[x]$.
2. Every embedding $\sigma : E \to \overline{F}$ for which $\sigma(a) = a$ for all $a \in F$ maps E onto E.
3. E is a normal extension of F.

Proof: (1) $\Rightarrow$ (2): Let $\{f_\alpha(x)\}_{\alpha \in \Delta}$ be a family of polynomials in $F[x]$ and E be the splitting filed of $\{f_\alpha(x)\}_{\alpha \in \Delta}$. If a is a root of some $f_\alpha(x)$, $\alpha \in \Delta$, then, for any embedding $\sigma : E \to \overline{F}$ such that $\sigma(b) = b$ for all $b \in F$, $\sigma(a)$ is a root of $f_\alpha(x)$. Since E is generated by all the roots of all the polynomials $f_\alpha(x)$, it follows that $\sigma(d) = d$ for all $d \in E$. Now, by Theorem 14.2.12, σ is an automorphism of E and hence σ maps E onto E.

(2) $\Rightarrow$ (3): Let $p(x)$ be an irreducible polynomial over F that has a root $a \in E$. Let $b \in \overline{F}$ be another root of $p(x)$. We prove that $b \in E$. Since a and b are roots of the same irreducible polynomial $p(x)$, we get F-isomorphisms.

$$\sigma_1 : \frac{F[x]}{<p(x)>} \to F(a)$$

$$\text{and} \quad \sigma_2 : \frac{F[x]}{<p(x)>} \to F(b)$$

given by $\sigma_1(f(x) + <p(x)>) = f(a)$ and $\sigma_2(f(x) + <p(x)>) = f(b)$.
Put $\sigma = \sigma_2 \circ \sigma_1^{-1}$; $F(a) \to F(b)$. Then, σ is an isomorphism such that $\sigma(a) = b$ and $\sigma(c) = c$ for all $c \in F$. Then, by Theorem 14.3.4, σ can be extended to an embedding $\sigma^* : E \to \overline{F}$. Now, by (2), σ^* maps E onto E; that is, σ^* is an automorphism of E. Therefore,

$$b = \sigma(a) = \sigma^*(a) \in E, \quad \text{since } a \in E.$$

Therefore, E contains all the roots of $p(x)$. Thus, E is a normal extension of F.

(3) $\Rightarrow$ (1): Suppose that E is a normal extension of F. For each $a \in E$, let $p_a(x)$ be the minimal polynomial of a over F. Since E is a normal extension of F, $p_a(x)$ splits into linear factors in E, for each $a \in E$. Therefore, it is immediate that E is a splitting field of the family $\{p_a(x)\}_{a\in E}$ of polynomials over F. ◀

Theorem 15.1.5. A finite extension E of F is a normal extension of F if and only if E is the splitting field of a polynomial over F.

Proof: Let E be a finite extension of F. Then, $E = F(a_1, a_2, \ldots, a_n)$, where each $a_i \in E$ and a_i is algebraic over F. For each $1 \le i \le n$, let $p_i(x)$ be the minimal polynomial of a_i over F. Now, suppose that E is a normal extension of F. Since $a_i \in E$, we get that every root of $p_i(x)$ belongs to E. Put

$$p(x) = p_1(x)\, p_2(x) \cdots p_n(x).$$

Then, $E = F(a_1, a_2, \ldots, a_n)$ becomes the splitting field of $f(x)$.
Conversely suppose that E is the splitting field of a polynomial $f(x) \in F[x]$. If $a_1, a_2, \ldots, a_n$ are all the roots of $f(x)$, then $E = F(a_1, a_2, \ldots, a_n)$. If $\sigma : E \to \overline{F}$ is an embedding such that $\sigma/F = \text{Id}$, then $\sigma(a_i)$ is also a root of

$f(x)$ and hence $\sigma(a_i) = a_j$ for some j which implies that $\sigma(a_i) \in E$ for all $1 \leq i \leq n$. Therefore, $\sigma(E) \subseteq E$ and σ is an automorphism of E. By Theorem 15.1.4, E is a normal extension of F. ◀

Worked Exercise 15.1.1. Let $F \subseteq E$ be fields such that $[E : F]=2$. Then prove that E is a normal extension of F.

Answer: Since $[E : F] > 1$, $F \subsetneqq E$ and hence we can choose $a \in E$ such that $a \notin F$. Then, a is algebraic over F and let $p(x)$ be the minimal polynomial of a over F. Then, $F \subsetneqq F(a) \subseteq E$ and hence

$$[E : F(a)][F(a) : F] = [E : F] = 2.$$

Also, since $F \subsetneqq F(a)$, $[F(a) : F] > 1$ and hence it follows that $[F(a) : F] = 2$ and $[E : F(a)] = 1$. Therefore, $F(a) = E$ and

$$2 = [F(a) : F] = \text{degree of } p(x).$$

Since $\deg(p(x)) = 2$ and since E has a root of $p(x)$, it follows that the other root of $p(x)$ must also be in E. Thus, E is the splitting field of $p(x) \in F[x]$ and hence E is a normal extension of F.

Worked Exercise 15.1.2. Let $F \subseteq E \subseteq K$ be fields such that K is a finite separable extension of E and E is a finite separable extension of F. Then prove that K is a finite separable extension of F.

Answer: By Theorem 15.1.1, $E = F(a)$ and $K = E(b)$ for some $a \in E$ and $b \in K$. Let $c \in F(a, b)$ such that $c \notin F(a)$. Then, we have

$$\begin{aligned} &F \subset F(c) \subset F(a, c) \\ \text{and} \quad &F \subset F(a) \subset F(a, c). \end{aligned}$$

Also, $F(a)$ is a finite separable extension of F and c is a separable element over $F(a)$. We prove that c is separable over F. Let $f(x)$, $g(x)$, $p(x)$ and $q(x)$ be the minimal polynomials of a over F, c over $F(a)$, c over F and a over $F(c)$, respectively, and let $\deg(f(x)) = m$, $\deg(g(x)) = n$, $\deg(p(x)) = s$ and $\deg(q(x)) = t$. Let $\sigma : F \to L$ be an embedding of F into an algebraically closed field L. Since a is separable over F, there are exactly m extensions $\sigma_1, \sigma_2, \ldots, \sigma_m$ of σ to $F(a)$. Also, since c is separable over $F(a)$, there are exactly n extensions $\sigma_{i1}, \sigma_{i2}, \ldots, \sigma_{in}$ of each σ_i to $F(a, c)$. Now, there are exactly mn extensions of $\sigma : F \to L$ to $\sigma_{ij} : F(a, c) \to L$, $1 \leq i \leq m$, $1 \leq j \leq n$. Similarly, by considering extensions of $\sigma : F \to L$ to $F(a, c)$ via $F(c)$, we get that there are exactly st

extensions of σ to $F(a, c)$. Therefore, $mn = st$. Now, suppose, if possible, that c is not separable over F. Then, the number of extensions of σ to $F(c)$ is less than s (=degree of the minimal polynomial of c over F), which implies that the number of extensions of σ to $F(a, c)$ is less than $st = mn$, a contradiction. Thus, c is separable over F. Thus, K is a finite separable extension of F.

EXERCISE 15(A)

1. Find $a \neq \sqrt{3}+\sqrt{5}$ such that $\mathbb{Q}(\sqrt{3},\sqrt{5})=\mathbb{Q}(a)$.
2. If z is the complex number such that $z \neq 1$ and $z^3 = 1$, then find a such that $\mathbb{Q}(\sqrt{2}, z)=\mathbb{Q}(a)$.
3. Prove that any extension of $\mathbb{Q}$ is separable.
4. Prove that any finite extension of a finite field is separable.
5. Let K be a field of characteristic $p \neq 0$. Then prove that K is perfect if and only if $K^p = K$.
6. Prove that $\mathbb{Q}(2^{1/4})$ is not a normal extension of $\mathbb{Q}$.
7. If E is an extension of a field F, then prove that the set of all elements in E which are separable over F forms a subfield of E containing F.
8. Let $F \subseteq K \subseteq L$ be fields such that L is a finite normal extension of F. Then prove that L is a finite normal extension of K.
9. Prove that a field F of characteristic $p \neq 0$ is perfect if and only if the mapping $a \mapsto a^p$ is an automorphism of F.

15.2 AUTOMORPHISM GROUPS AND FIXED FIELDS

Recall that any finite separable extension E of a field F is simple and hence $E = F(a)$ for some $a \in E$. Throughout this section, we confine ourselves to finite separable extensions and their groups of automorphisms. First recall that, for any field E, the set Aut(E) of all automorphisms of E forms a group under the composition of mappings.

Definition 15.2.1. Let F be a field and E be an extension of F. An automorphism σ of E is called an *F-automorphism* if σ fixes all elements of F; that is, $\sigma(a) = a$ for all $a \in F$. Let

$$G\left(\frac{E}{F}\right)=\{\sigma \in \mathrm{Aut}(E):\sigma(a)=a \text{ for all } a\in F\}.$$

Then, $G(E/F)$ becomes a group under the composition of mappings and is called the group of F-automorphisms of E. Note that $G(E/F)$ is a subgroup of Aut(E).

Theorem 15.2.1. Let E be a finite separable extension of a field F. Then,

$$\left|G\left(\frac{E}{F}\right)\right| \leq [E:F]$$

Proof: Recall that any finite separable extension E of F is a simple extension of F. Therefore, $E = F(a)$ for some $a \in E$. Let $p(x)$ be the minimal polynomial of a over F and $\deg(p(x)) = n$. Then,

$$[E : F] = [F(a) : F] = \deg(p(x)) = n.$$

By Theorem 14.3.3, we get that

$$\left|G\left(\frac{E}{F}\right)\right| \leq n = [E:F].$$

◄

Definition 15.2.2. Let E be any field and H be a subgroup of Aut(E). Then, the set

$$E_H = \{a \in E : \sigma(a) = a \text{ for all } \sigma \in H\}$$

is called the *fixed field* of H.

It can be easily verified that E_H is a subfield of E for any subgroup H of Aut(E). If E is a field extension of F and H is a subgroup of $G(E/F)$, then

$$F \subseteq E_H \subseteq E.$$

As a simple example, consider $H = G(\mathbb{C}/\mathbb{R})$. Then, $\mathbb{C}_H = \mathbb{R}$.

The following result is an important tool in proving the main result of this section.

Theorem 15.2.2 (Dedikind Theorem). Let F and E be fields and $\sigma_1, \sigma_2, \ldots, \sigma_n$ be distinct embeddings of F into E. Let $a_1, a_2, \ldots, a_n \in E$ such that

$$a_1\sigma_1(a) + a_2\sigma_2(a) + \cdots + a_n\sigma_n(a) = 0 \quad \text{for all } a \in F.$$

Then, $a_1 = a_2 = \cdots = a_n = 0$ (This can also be expressed by saying that distinct embeddings of F into E are linearly independent over E.)

Proof: Suppose, if possible, that there exist $a_1, a_2, \ldots, a_n \in E$, not all zero, such that

$$a_1\sigma_1(a) + a_2\sigma_2(a) + \cdots + a_n\sigma_n(a) = 0 \quad \text{for all } a \in E.$$

Then, we can find such a relation having as few nonzero coefficients as possible. On renumbering, we can assume that this relation is

$$b_1\sigma_1(a) + b_2\sigma_2(a) + \cdots + b_m\sigma_m(a) = 0 \quad \text{for all } a \in E. \tag{1}$$

If $m = 1$, then $b_1\sigma_1(a) = 0$ for all $a \in E$ and, in particular, $b_1 = b_1 1 = b_1\sigma_1(1) = 0$. Therefore, we can assume that $m > 1$. Now, $\sigma_1 \neq \sigma_m$ and hence there exists an element $c \in E$ such that $\sigma_1(c) \neq \sigma_m(c)$. The Equation (1) holds for all $a \in E$ and, in particular, for ca for all $a \in E$. Therefore,

$$b_1\sigma_1(ca) + b_2\sigma_2(ca) + \cdots + b_m\sigma_m(ca) = 0$$

$$\text{and hence} \quad b_1\sigma_1(c)\sigma_1(a) + \cdots + b_m\sigma_m(c)\sigma_m(a) = 0 \tag{2}$$

for any $a \in E$. Multiplying (1) by $\sigma_1(c)$ and subtracting the result from (2), we get that

$$b_2(\sigma_2(c) - \sigma_1(c))\sigma_2(a) + \cdots + b_m(\sigma_m(c) - \sigma_1(c))\sigma_m(a) = 0$$

for all $a \in E$. This is a contradiction to the choice of Equation (1), since $b_m(\sigma_m(c) - \sigma_1(c)) \neq 0$. ◀

Now, we prove the following which is the main theorem in this section.

Theorem 15.2.3. Let E be any field and H be a finite subgroup of Aut(E). Then, E is a finite extension of E_H and

$$[E : E_H] = |H|.$$

Proof: We have $E_H = \{a \in E : \sigma(a) = a \text{ for all } \sigma \in H\}$. Then, clearly E_H is a subfield of E and hence E is an extension of E_H. Let $|H| = n < \infty$ and $H = \{e = g_1, g_2, \ldots, g_n\}$. Suppose, if possible, that $n < [E : E_H]$. Then, there exist elements $a_1, a_2, \ldots, a_{n+1}$ in E such that $\{a_1, a_2, \ldots, a_{n+1}\}$ is linearly independent over E_H. Consider the system of n homogeneous linear equations

$$g_j(a_1)x_1 + g_j(a_2)x_2 + \cdots + g_j(a_{n+1})x_{n+1} = 0,\ 1 \leq j \leq n$$

in $n + 1$ unknowns $x_1, x_2, \ldots, x_{n+1}$. This system should have a nontrivial solution. Therefore, there exists elements $y_1, y_2, \ldots, y_{n+1}$ in E, not all zero, such that

$$g_j(a_1)y_1 + \cdots + g_j(a_{n+1})y_{n-1} = 0 \tag{1}$$

for all $1 \leq j \leq n$. We can choose these $y_1, \ldots, y_{n+1}$ such that as few of them as possible are nonzero and renumber them such that

$$y_i \neq 0 \text{ for } i = 1, \ldots, r \quad \text{and} \quad y_i = 0 \quad \text{for } r < i \leq n + 1.$$

Then, Equation (2) becomes

$$g_j(a_1)y_1 + \cdots + g_j(a_r)y_r = 0 \tag{2}$$

for all $1 \leq j \leq n$. Now, let $g \in H$ and operate on (2) with g. Then, we get the system of equations

$$g(g_j(a_1))g(y_1) + \cdots + g(g_j(a_r))g(y_r) = 0 \tag{3}$$

Since $H = \{gg_1, gg_2, \ldots, gg_n\} = \{g_1, g_2, \ldots, g_n\}$, (3) is equivalent to the system of equations

$$g_j(a_1)g(y_1) + \cdots + g_j(a_r)g(y_r) = 0 \tag{4}$$

By multiplying (2) by $g(y_1)$ and (4) by y_1 and by subtracting, we get

$$g_j(a_2)(y_2g(y_1) - g(y_2)y_1) + \cdots + g_j(a_r)(y_rg(y_1) - g(y_r)y_1) = 0 \tag{5}$$

for all $1 \leq j \leq n$. This is a system of equations like (2), but with fewer terms, which becomes a contradiction to our assumption, unless

$$y_ig(y_1) - g(y_i)y_1 = 0 \text{ for all } 2 \leq i \leq r.$$

If this happens, then $g(y_iy_1^{-1}) = y_iy_1^{-1}$ for all $g \in H$ and hence $y_iy_1^{-1} \in E_H$ for all $2 \leq i \leq r$. Therefore, there exist $z_1, z_2, \ldots, z_r \in E_H$ such that $y_1z_i = y_i$ for $1 \leq i \leq r$ (take $z_1 = 1$ and $z_i = y_iy_1^{-1}$ for $2 \leq i \leq r$). Then, Equation (2) with $j = 1$ becomes

$$g_1(a_1)y_1z_1 + \cdots + g_1(a_r)y_1z_r = 0$$

which implies that $g_1(a_1)z_1 + \cdots + g_1(a_r)z_r = 0$ (since $y_1 \neq 0$). Now, since $z_i \in E_H$, we get that

$$g_1(a_1z_1 + \cdots + a_rz_r) = 0$$

and hence $a_1z_1 + \cdots + a_rz_r = 0$, since g_1 is an embedding. Now, since $\{a_1, \ldots, a_r\}$ are linearly independent over E_H, it follows that

$$z_1 = z_2 = \cdots = z_r = 0$$

and hence $y_1 = y_2 = \cdots = y_r = 0$, which is a contradiction.
Thus, we must have $[E : E_H] \leq n < \infty$ and therefore E is a finite extension of E_H and $[E : E_H] \leq n$.

On the other hand, suppose, if possible, that $[E : E_H] < n$. Let $[E : E_H] = m$ and $\{a_1, a_2, \ldots, a_m\}$ be a basis of E over E_H. Consider the system of m homogeneous linear equations

$$g_1(a_j)x_1 + \cdots + g_n(a_j)x_n = 0, 1 \leq j \leq m,$$

in n unknowns $x_1, \ldots, x_n$. Since $m < n$, this system has a nontrivial solution. Therefore, there exist $y_1, y_2, \ldots, y_n \in E$, not all zero, such that

$$g_1(a_j)y_1 + \cdots + g_n(a_j)y_n = 0 \tag{6}$$

for all $1 \leq j \leq m$. Since $\{a_1, a_2, \ldots, a_m\}$ is a basis of E over E_H, any element $a \in E$ can be uniquely written as $a = s_1a_1 + \cdots + s_ma_m$ with $s_i \in E_H$, and hence

$$\begin{aligned}\sum_{i=1}^{n} g_i(a)y_i &= \sum_{i=1}^{n} g_i\left(\sum_{j=1}^{m} s_ja_j\right)y_i \\ &= \sum_{j=1}^{m}\sum_{i=1}^{n} s_jg_i(a_j)y_i \\ &= \sum_{j=1}^{m} s_j\left(\sum_{i=1}^{n} g_i(a_j)y_i\right) \\ &= 0 \text{ (by (6))}\end{aligned}$$

Therefore, we have $y_1, y_2, \ldots, y_n \in E$, not all zero, and

$$y_1g_1(a) + y_2g_2(a) + \cdots + y_ng_n(a) = 0 \quad \text{for all } a \in E$$

which is a contradiction to the fact that $g_1, g_2, \ldots, g_n$ are distinct embeddings of E into E (refer Dedikind's Theorem 15.2.2). Therefore, $m < n$ is impossible so that $[E : E_H] \geq n$. Thus,

$$[E : E_H] = n = |H|. \qquad ◀$$

Theorem 15.2.4. Let E be a finite separable extension of a field F and let H be a subgroup of $G(E/F)$. Then,

$$G\left(\frac{E}{E_H}\right) = H \quad \text{and} \quad [E : E_H] = \left|G\left(\frac{E}{E_H}\right)\right|$$

Proof: If $\sigma \in H$, then $\sigma(a) = a$ for all $a \in E_H$ and hence $\sigma \in G(E/E_H)$. Therefore, H is a subgroup of $G(E/E_H)$. By Theorem 15.2.3, $|H| = [E : E_H]$. Also, by Theorem 15.2.1,

$$|H| \leq \left|G\left(\frac{E}{E_H}\right)\right| \leq [E : E_H] = |H|$$

and hence $G(E/E_H) = H$, since H is a subgroup of $G(E/E_H)$ and $|H| = |G(E/E_H)|$. Also, we have

$$[E : E_H] = \left|G\left(\frac{E}{E_H}\right)\right|. \qquad ◀$$

Theorem 15.2.5. A finite separable extension E of a field F is a normal extension of F if and only if the fixed field of $G(E/F)$ is F.

Proof: Let E be a finite separable extension of a field F, Then, by Theorem 15.1.1, E is a simple extension of F and hence $E = F(a)$ for some $a \in E$. Let $p(x)$ be the minimal polynomial of a over F and $\deg(p(x)) = n$. Then, we have

$$[E : F] = [F(a) : F] = n.$$

Let E_0 be the fixed field of $G(E/F)$; that is,

$$E_0 = \{s \in E : \sigma(s) = s \text{ for all } \sigma(s) = s \text{ for all } \sigma \in G(E/F)\},$$

Then, $F \subseteq E_0 \subseteq E$ and, by Theorem 15.2.3, we have

$$[E : E_0] = \left| G\left(\frac{E}{F}\right) \right|.$$

Now, we prove that E is a normal extension of F if and only if $E_0 = F$.
First, suppose that $E_0 = F$. Then, $|G(E/F)| = [E : E_0] = [E : F] = n$. Let $G(E/F) = \{e = \sigma_1, \sigma_2, \ldots, \sigma_n\}$, where e is the identity automorphism, consider the polynomial

$$f(x) = \prod_{i=1}^{n} (x - \sigma_i(a)).$$

Each $\sigma_i \in G(E/F)$ induces a natural homomorphism $\sigma_i^* : E[x] \to E[x]$ given by

$$\sigma_i^*(a_0 + a_1 x + \cdots + a_r x^r) = \sigma_i(a_0) + \sigma_i(a_1)x + \cdots + \sigma_i(a_r)x^r.$$

Now, $\sigma_i^* f(x) = \prod_{j=1}^{n} (x - \sigma_i(\sigma_j(a)))$.
Since $\sigma_i\sigma_1, \sigma_i\sigma_2, \ldots, \sigma_i\sigma_n$ are distinct members of $G(E/F)$ and $|G(E/F)| = n$, we get that $G(E/F) = \{\sigma_i\sigma_1, \sigma_i\sigma_2, \ldots, \sigma_i\sigma_n\}$ and hence $\sigma_i^*(f(x)) = f(x)$ for all $1 \leq i \leq n$. By expanding $f(x)$, we have

$$f(x) = x^n - c_1x^{n-1} + c_2x^{n-2} + \cdots + (-1)^n c_n,\ c_i \in E$$

and, from $\sigma_i^* f(x) = f(x)$, we have

$$\sigma_i(c_j) = c_j \quad \text{for all } 1 \leq i, j \leq n.$$

Therefore, $c_j \in E_0$, the fixed field of $G(E/F)$ and, by hypothesis $c_j \in F$ for all $1 \leq j \leq n$. This implies that $f(x) \in F[x]$. Further, all the roots of $f(x)$ lie in $E = F(a)$ and a is one of the roots of $f(x)$. Thus, E is the splitting field of $f(x) \in F[x]$ and hence E is a normal extension of F.
Conversely, suppose that E is a normal extension of F. By Theorem 14.3.3, the number of extensions of the inclusion map $F \to \overline{F}$ to the embedding $F(a) \to \overline{F}$ is equal to the number of distinct roots of $p(x)$. Since E is a separable extension of F and $a \in E$, a is a separable element of E and hence its minimal polynomial $p(x)$ over F has distinct roots. Therefore, the number of distinct roots of $p(x)$ is n. Also, since $E = F(a)$ is a normal extension of F, any embedding $\sigma : F(a) \to \overline{F}$ should map $F(a)$ onto $F(a)$. Further, any

member of $G(E/F)$ is an extension of the inclusion map $F \to \overline{F}$ and hence we have

$$\left|G\left(\frac{E}{F}\right)\right| = \text{The number of distinct roots of } p(x) = n.$$

Therefore, $[E : F] = n = |G(E/F)| = [E : E_0]$. Since $F \subseteq E_0 \subseteq E$, it follows that $[E_0 : F] = 1$ and hence $E_0 = F$. Thus, F is the fixed field of $G(E/F)$. ◀

Corollary 15.2.1. A finite separable extension E of a field F is a normal extension of F if and only if $[E : F] = |G(E/F)|$.

Proof: Let E be a finite separable extension of a field F. Suppose that E is a normal extension of F. Then, by Theorem 15.2.5, F is the fixed field of $G(E/F)$. Then, by Theorem 15.2.4,

$$[E : F] = \left|G\left(\frac{E}{F}\right)\right|.$$

Conversely suppose that $[E : F] = |G(E/F)|$. Let E_0 be the fixed field of $G(E/F)$. Then, $[E : E_0] = |G(E/F)|$ and hence $[E : F] = [E : E_0]$. Since $F \subseteq E_0 \subseteq E$, it follows that $[E_0 : F] = 1$ so that $F = E_0$. Again by Theorem 15.2.5, E is a normal extension of F. ◀

Worked Exercise 15.2.1. Let $f(x) \in F[x]$ has r distinct roots in its splitting field E over F. Then prove that $G(E/F)$ is isomorphic to a subgroup of the symmetric group S_r of degree r.

Answer: Let $a_1, a_2, \ldots, a_r$ be all the distinct roots of $f(x)$ in its splitting field E over F. For any $\sigma \in G(E/F)$, $\sigma(a_i)$ is again a root of $f(x)$ in E. Also, $\sigma(a_i) \neq \sigma(a_j)$ for $a_i \neq a_j$. Thus, $\sigma(a_1), \sigma(a_2), \ldots, \sigma(a_r)$ is a permutation of $a_1, a_2, \ldots, a_r$ and let us denote this permutation by ϕ_σ. Therefore, $\phi_\sigma \in S_r$ for each $\sigma \in G(E/F)$. Define $\phi : G(E/F) \to S_r$ by $\phi(\sigma) = \phi_\sigma$. For any σ and $\eta \in G(E/F)$,

$$\phi(\sigma \circ \eta)(a_i) = (\sigma \circ \eta)(a_i) = \sigma(\eta(a_i)) = (\phi(\sigma) \circ \phi(\eta))(a_i)$$

for all $1 \leq i \leq r$ and hence $\phi(\sigma \circ \eta) = \phi(\sigma) \circ \phi(\eta)$. Therefore, ϕ is a homomorphism of groups. Also, for any $\sigma \in G(E/F)$,

$$\phi(\sigma) = \text{Id} \Rightarrow \sigma(a_i) = a_i \quad \text{for all } 1 \le i \le r$$
$$\Rightarrow \sigma = \text{Id}, \quad \text{since } E = F(a_1, \ldots, a_r)$$

Thus, $\sigma : G(E/F) \to S_r$ is a monomorphism of groups and σ is an isomorphism of $G(E/F)$ onto $\sigma(G(E/F))$, which is a subgroup of the symmetric group S_r.

Worked Exercise 15.2.2. Let $E = \mathbb{Q}(\omega, \sqrt[3]{2})$, where $\omega^3 = 1$ and $\omega \neq 1$ (that is, ω is a cube root of unity in $\mathbb{C}$) and let H be the subgroup of $G(E/\mathbb{Q})$ given by $H = \{\text{Id}, \sigma\}$ where $\sigma : E \to E$ is defined by $\sigma(a) = a$ for all $a \in \mathbb{Q}$, $\sigma(\omega) = \omega^2$ and $\sigma(\sqrt[3]{2}) = \sqrt[3]{2}\ \omega^2$. Then, find the fixed field E_H.

Answer: For simplicity, let $\sqrt[3]{2} = c$. Then, c is a real number such that $c^3 = 2$. We are given that $E = \mathbb{Q}(\omega, c)$ and $H = \{\text{Id}, \sigma\}$, where σ is defined by $(\sigma/\mathbb{Q}) = \text{Id}$, $\sigma(\omega) = \omega^2$ and $\sigma(c) = c\omega^2$. First note that $\mathbb{Q} \subset \mathbb{Q}(c) \subset \mathbb{Q}(\omega, c)$, $\{1, c, c^2\}$ is a basis of $\mathbb{Q}(c)$ over $\mathbb{Q}$ and $\{1, \omega\}$ is a basis of (ω, c) over $\mathbb{Q}(c)$ therefore the basis of E over $\mathbb{Q}$ is $\{1, c, c^2, \omega, \omega c, \omega c^2\}$. Consider any $a \in E$.

Then, $a = r_0 + r_1 c + r_2 c^2 + r_3\omega + r_4 c\omega + r_5 c^2\omega$, with $r_i \in \mathbb{Q}$, and

$$\begin{aligned}
\sigma(a) &= r_0 + r_1 c\omega^2 + r_2 c^2\omega + r_3\omega^2 + r_4 c\omega + r_5 c^2 \\
&= r_0 + r_1 c(-1 - \omega) + r_2 c^2\omega + r_3(-1 - \omega) + r_4 c\omega + r_5 c^2 \\
&\quad (\text{since } 1 + \omega + \omega^2 = 0) \\
&= (r_0 - r_3) + (-r_1)c + r_5 c^2 + (-r_3)\omega + (-r_1 + r_4)c\omega + \\
&\quad r_2 c^2\omega.
\end{aligned}$$

Therefore, $\sigma(a) = a \Rightarrow r_0 - r_3 = r_0, -r_1 = r_1, r_5 = r_2, -r_3 = r_3, -r_1 + r_4 = r_4$ and $r_2 = r_5$

$$\begin{aligned}
&\Rightarrow r_3 = 0, r_1 = 0 \quad \text{and} \quad r_2 = r_5 \\
&\Rightarrow a = r_0 + r_2 c^2 + r_4 c\omega + r_2 c^2\omega \\
&\Rightarrow a = r_0 + r_4 c\omega + r_2 c^2(1 + \omega) \\
&\Rightarrow a = r_0 + r_4 c\omega - r_2(c\omega)^2 \in \mathbb{Q}(c\omega).
\end{aligned}$$

Therefore, $E_H \in \mathbb{Q}(c\omega)$. On the other hand, if $a \in \mathbb{Q}(c\omega)$, then clearly $\sigma(a) = a$ and hence $a \in E_H$. Thus, the fixed field E_H of H is equal to $\mathbb{Q}(c\omega) = \mathbb{Q}(\sqrt[3]{2}\ \omega)$. Note that $(c\omega)^2 = c^2\omega^2 = \sqrt[3]{4}\ \omega^2$ and $(c^2\omega^2)^2 = 2c\omega$ and therefore $E_H = \mathbb{Q}(c\omega) = \mathbb{Q}(c^2\omega^2) = \mathbb{Q}(\sqrt[3]{4}\ \omega^2)$.

EXERCISE 15(B)

1. Let $f(x) = x^4 - 2 \in \mathbb{Q}[x]$ and E be the splitting field of $f(x)$ over $\mathbb{Q}$. Prove that $G(E/\mathbb{Q})$ is isomorphic to the group of symmetries of a square.

2. Prove that $G(\mathbb{C}/\mathbb{R})$ is a cyclic group of order 2.

3. Let F be a field of characteristic 2 and $x^2 - a$ be an irreducible polynomial over F. Prove that $G(E/F)$ is a group of order 2, where E is the splitting field of $x^2 - a$ over F.

4. Let $\omega \neq 1$ be a cube root of unity and $E = \mathbb{Q}(\omega, \sqrt[3]{2})$. Let σ be the automorphism of E defined by $\sigma(r) = r$ for all $r \in \mathbb{Q}$, $\sigma(\omega) = \omega^2$ and $\sigma(\sqrt[3]{2}) = \sqrt[3]{2}\,\omega$. Then prove that $\{\mathrm{Id}, \sigma\}$ is a subgroup of $G(E/\mathbb{Q})$ and find its fixed field.

5. Let E be the splitting field of $x^4 - x^2 + 1$ over the field of rationals $\mathbb{Q}$. Then, determine the group $G(E/\mathbb{Q})$.

6. Let $a \neq 1$ and $a^5 = 1$. Then prove that $\mathbb{Q}(a)$ is a normal extension of $\mathbb{Q}$ and that $G(\mathbb{Q}(a)/\mathbb{Q})$ is isomorphic to $\mathbb{Z}_4$, the group of integers modulo 4.

15.3 FUNDAMENTAL THEOREM OF GALOIS THEORY

In this section, we prove the much awaited main theorem of the present chapter, which is popularly known as the fundamental theorem of Galois theory. Before going to prove this theorem, let us introduce the following terminology.

Definition 15.3.1. Any finite normal and separable extension of a field F is called a *Galois extension* of F.

Definition 15.3.2. Let E be the splitting field of a polynomial $f(x) \in F[x]$ over F. Then, group $G(E/F)$ of F-automorphisms of E is called the *Galois group* of $f(x)$ over F.

Theorem 15.3.1 (Fundamental Theorem of Galois Theory). Let E be a Galois extension of F, ξ be the class of all subfields of E containing F and ζ be the class of all subgroups of $G(E/F)$. Then, the following holds for any $K \in \xi$.

1. $G(E/K)$ is a subgroup of $G(E/F)$ and $K \mapsto G(E/K)$ is a bijection of ξ onto ζ such that, for any K_1 and $K_2 \in \xi$.

$$K_1 \subseteq K_2 \Leftrightarrow G(E/K_2) \text{ is a subgroup of } G(E/K_1).$$

2. K is a normal extension of F if and only if $G(E/K)$ is a normal subgroup of $G(E/F)$.
3. If K is a normal extension of F, then

$$\frac{G\left(\dfrac{E}{F}\right)}{G\left(\dfrac{E}{K}\right)} \cong G\left(\frac{K}{F}\right).$$

Proof:

1. Define $\phi : \xi \to \zeta$ and $\theta : \zeta \to \xi$ by

$$\phi(K) = G(E/K) \text{ and } \theta(H) = E_H, \text{ the fixed field of } H$$

for any $K \in \xi$ and $H \in \xi$. Clearly, $G(E/K)$ is a subgroup of $G(E/F)$ for any field K such that $F \subseteq K \subseteq E$, and E_H is an intermediate field between F and E for any subgroup H of $G(E/F)$. Therefore, ϕ and θ are well-defined mappings. We prove that these mappings ϕ and θ are inverses to each other. If $K \in \xi$, then $F \subseteq K \subseteq E$ and E is a normal extension of K and hence, by Theorem 15.2.5, K is the fixed field of $G(E/K)$; that is,

$$(\theta \circ \phi)(K) = \theta\left(G\left(\frac{E}{K}\right)\right) = K.$$

This is true for any $K \in \xi$ and hence $\theta \circ \phi$ is the identity map of ξ. On the other hand, let $H \in \zeta$. Then, H is a subgroup of $G(E/F)$ and, by Theorem 15.2.4,

$$H = G\left(\frac{E}{E_H}\right) = (\phi \circ \theta)(H).$$

This implies that $\phi \circ \theta$ is the identity map of ζ. Therefore, ϕ is a bijection and $\phi^{-1} = \theta$. Also, for any K_1 and $K_2 \in \xi$,

$$K_1 \subseteq K_2 \Rightarrow G\left(\frac{E}{K_2}\right) \subseteq G\left(\frac{E}{K_1}\right) \Rightarrow \phi(K_2) \subseteq \phi(K_1)$$

and $$G\left(\frac{E}{K_2}\right) \subseteq G\left(\frac{E}{K_1}\right) \Rightarrow E_{G\left(\frac{E}{K_1}\right)} \subseteq E_{G\left(\frac{E}{K_2}\right)} \Rightarrow K_1 \subseteq K_2.$$

2. First we prove that, for any $F \subseteq K \subseteq E$, K is a normal extension of F if and only if $\sigma(K) = K$ for any $\sigma \in G(E/F)$. Let $K \in \xi$. Then, $F \subseteq K \subseteq E$. Suppose that K is a normal extension of F and $\sigma \in G(E/F)$. Then, $\sigma/_K : K \to E$ is an embedding. Since $E \subseteq \overline{F}$, σ/K is an embedding of K into $\overline{F}$. Then, $(\sigma/K)(K) = K$ and hence $\sigma(K) = K$. Conversely, suppose that $\sigma(K) = K$ for any $\sigma \in G(E/F)$. Let $\sigma^* : K \to \overline{F}$ be an embedding such that $(\sigma^*/F) = \text{Id}$. Then, σ^* can be extended to $\lambda : E \to \overline{F}$. Since E is a normal extension of F, we have $\lambda(E) = E$ and hence $\lambda \in G(E/F)$ and $\lambda(K) = K$. Since λ is an extension of σ^* to E,

it follows that $\sigma^*(K) = K$. Thus, K is a normal extension of F. Now, we prove the assertion in (2). Suppose that K is a normal extension of F. Clearly, $G(E/K)$ is a subgroup of $G(E/F)$. Let $\sigma \in G(E/F)$ and $\tau \in G(E/K)$. By the above observation, $\sigma(K) = K$. For any $a \in K$, $\sigma(a) \in K$ and hence $\tau(\sigma(a)) = \sigma(a)$, so that $(\sigma^{-1} \circ \tau \circ \sigma)(a) = a$. This implies that $\sigma^{-1} \circ \tau \circ \sigma \in G(E/K)$ and hence $G(E/K)$ is a normal subgroup of $G(E/F)$.

Conversely suppose that $G(E/K)$ is a normal subgroup of $G(E/F)$. By the above observation, it is enough to prove that $\sigma(K) = K$ for all $\sigma \in G(E/F)$. Let $\sigma \in G(E/F)$ and $a \in K$. For any $\tau \in G(E/K)$, we have $\sigma^{-1}\tau\sigma \in G(E/K)$ and hence

$$(\sigma^{-1}\tau\sigma)(a) = a; \quad \text{that is, } \tau(\sigma(a)) = \sigma(a).$$

This implies that $\sigma(a) \in E_{G(E/K)} = K$. Therefore, we have proved that $\sigma(a) \in K$ for all $a \in K$ and hence $\sigma(K) = K$. Thus, K is a normal extension of F.

3. Let K be a normal extension of F. For any $\sigma \in G(E/F)$, let $\sigma^* = (\sigma/K)$. Then, by the above observation $\sigma^* \in G(K/F)$ (since $\sigma(K) = K$). Now, define a mapping $f : G(E/F) \to G(K/F)$ by

$$f(\sigma) = \sigma^* = \frac{\sigma}{K}$$

It can be easily verified that F is a homomorphism of groups. Also,

$$\ker f = \left\{\sigma \in G\left(\frac{E}{F}\right) : \frac{\sigma}{K} = \text{Id}\right\} = G\left(\frac{E}{K}\right).$$

By the fundamental theorem of homomorphism of groups, we have

$$\frac{G\left(\frac{E}{F}\right)}{G\left(\frac{E}{K}\right)} \cong \text{Im}\, f \subseteq G\left(\frac{K}{F}\right).$$

Also, we have $[E : F] = [E : K][K : F]$, E is a normal extension of F, E is a normal extension of K and K is a normal extension of F and therefore

$$\left|G\left(\frac{E}{F}\right)\right| = [E : F] = [E : K][K : F] = \left|G\left(\frac{E}{K}\right)\right|\left|G\left(\frac{K}{F}\right)\right|$$

Now,

$$|\text{Im } f| = \frac{\left|G\left(\frac{E}{F}\right)\right|}{\left|G\left(\frac{E}{K}\right)\right|} \cong \left|G\left(\frac{K}{F}\right)\right|$$

and hence $\text{Im } f = G(K/F)$. Thus, we have

$$\frac{G\left(\frac{E}{F}\right)}{G\left(\frac{E}{K}\right)} \cong G\left(\frac{K}{F}\right). \qquad \blacktriangleleft$$

Worked Exercise 15.3.1. Prove that the Galois group of $x^3 - 2 \in \mathbb{Q}[x]$ is the group of symmetries of the triangle.

Answer: Let E be the splitting field of $x^3 - 2$ over $\mathbb{Q}$. We have seen earlier that $E = \mathbb{Q}\,(2^{1/3}, \omega)$, where ω is the root of the irreducible polynomial $x^2 + x + 1$ in $\mathbb{Q}(2^{1/3})$ and hence $[E : \mathbb{Q}] = 6$. Notice that

$$x^3 - 2 = (x - 2^{1/3})(x - \omega 2^{1/3})(x - \omega^2 2^{1/3}).$$

Since E is a normal extension of $\mathbb{Q}$ and $[E : \mathbb{Q}] = 6$, there are six automorphisms of E and these are determined by the manner in which they transform the roots of $x^3 - 2$. The root $2^{1/3}$ can have only three images, namely $2^{1/3}$, $2^{1/3}\omega$ and $2^{1/3}\omega^2$ and the root ω can have only two images, namely ω and ω^2. There are six possible combinations, since there are exactly six automorphisms. The Galois group $G(E/\mathbb{Q})$ of $x^3 - 2$ over $\mathbb{Q}$ is given by following table, where

$$G\left(\frac{E}{\mathbb{Q}}\right) = \{\text{Id}, \sigma, \sigma^2, \tau, \sigma\tau, \sigma^2\tau\}$$

	Id	σ	σ^2	τ	$\sigma\tau$	$\sigma^2\tau$
$2^{1/3}$	$2^{1/3}$	$\omega 2^{1/3}$	$\omega^2 2^{1/3}$	$2^{1/3}$	$\omega 2^{1/3}$	$\omega^2 2^{1/3}$
ω	ω	ω	ω	ω^2	ω^2	ω^2

This group $G(E/\mathbb{Q})$ is isomorphic to the group of symmetries of the triangle.

Worked Exercise 15.3.2. Prove that the Galois group of $x^4 - 2 \in \mathbb{Q}[x]$ is the group of symmetries of a square (that is, the octic group).

Answer: First observe that $x^4 - 2$ is irreducible over $\mathbb{Q}$ and we can factorise $x^4 - 2$ by $x^4 - 2 = (x - 2^{1/4})(x + 2^{1/4})(x + i2^{1/4})(x - i2^{1/4})$, where is the square root of -1 in the field $\mathbb{C}$ of complex numbers. Therefore, $E = \mathbb{Q}(2^{1/4}, i)$ is the splitting field of $x^4 - 2$ over $\mathbb{Q}$. Also, $[E : \mathbb{Q}] = 8$ since $[E : \mathbb{Q}(2^{1/4})] = 2$ and $[\mathbb{Q}(2^{1/4}) : \mathbb{Q}] = 4$.
Since E is a normal separable extension of $\mathbb{Q}$, it follows that

$$\left|G\left(\frac{E}{\mathbb{Q}}\right)\right| = [E : \mathbb{Q}] = 8.$$

Therefore, the Galois group $G(E/\mathbb{Q})$ is a group of order 8. Note that $\{1, 2^{1/4}, 2^{1/2}, 2^{3/4}, i, i2^{1/4}, i2^{1/2}, i2^{3/4}\}$ is a basis of E over $\mathbb{Q}$. Now, if $b \in E$, then

$$b = a_0 + a_1 2^{1/4} + a_2 2^{1/2} + a_3 2^{3/4} + a_4 i + a_5 i2^{1/4} + a_6 i2^{1/2} + a_7 2^{3/4} i$$

for some $a_i \in \mathbb{Q}$, $0 \le i \le 7$ and hence, for any $\sigma \in G(E/\mathbb{Q})$,

$$\begin{aligned}\sigma(b) &= a_0 + a_1\sigma(2^{1/4}) + a_2\sigma(2^{1/2}) + a_3\sigma(2^{3/4}) + a_4\sigma(i) \\ &\quad + a_5\sigma(i\,2^{1/4}) + a_6\sigma(i\,2^{1/2}) + a_7\sigma(i\,2^{3/4}) \\ &= a_0 + a_1\sigma(2^{1/4}) + a_2\sigma(2^{1/4})^2 + a_3\sigma(2^{1/4})^3 + a_4\sigma(i) \\ &\quad + a_5\sigma(i)\sigma(2^{1/4}) + a_6\sigma(i)\sigma(2^{1/4})^2 + a_7\sigma(i)\sigma(2^{1/4})^3.\end{aligned}$$

Therefore, σ is completely determined by $\sigma(2^{1/4})$ and $\sigma(i)$. Since $\sigma(2^{1/4}) = 2^{1/4}$ or $-2^{1/4}$ or $i2^{1/4}$ or $-i2^{1/4}$; since $\sigma(i) = i$ or $-i$ and since $|G(E/\mathbb{Q})| = 8$, it follows that $G(E/\mathbb{Q}) = \{\sigma_1, \sigma_2, \sigma_3, \sigma_4, \sigma_5, \sigma_6, \sigma_7, \sigma_8\}$ where σ_i's are given by the following table

	σ_1	σ_2	σ_3	σ_4	σ_5	σ_6	σ_7	σ_8
i	i	i	i	i	$-i$	$-i$	$-i$	$-i$
$2^{1/4}$	$2^{1/4}$	$i2^{1/4}$	$-2^{1/4}$	$-i2^{1/4}$	$2^{1/4}$	$i2^{1/4}$	$-2^{1/4}$	$-i2^{1/4}$

For convenience, let us write

$$a_1 = 2^{1/4},\ a_2 = i2^{1/4},\ a_3 = -2^{1/4} \quad \text{and} \quad a_4 = -i2^{1/4}.$$

As demonstrated in the diagram below, the elements of $G(E/\mathbb{Q})$ permute the roots a_1, a_2, a_3 and a_4 of $x^4 - 2$.

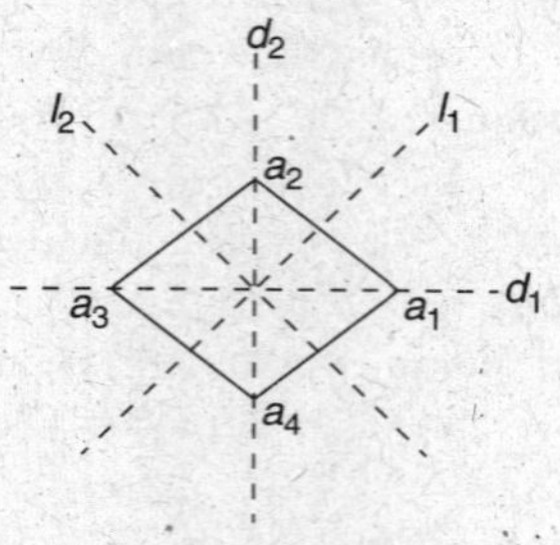

σ_1 : rotation by 0°
σ_2 : rotation by 90°
σ_3 : rotation by 180°
σ_4 : rotation by 270°
σ_5 : rotation by d_1
σ_6 : rotation by l_1
σ_7 : rotation by d_2
σ_8 : rotation by l_2

These are precisely all the symmetries of the square whose vertices are a_1, a_2, a_3 and a_4. Thus, the Galois group $G(E/\mathbb{Q})$ of the polynomial $x^4 - 2$ over $\mathbb{Q}$ is isomorphic with the group of symmetries of a square and hence $G(E/\mathbb{Q})$ is the octic group.

Worked Exercise 15.3.3. Prove that the Galois group of $x^5 - 1 \in \mathbb{Q}[x]$ is a cyclic group of order 4.

Answer: Let $f(x) = x^5 - 1 \in \mathbb{Q}[x]$. Then,

$$f(x) = (x - 1)(x^4 + x^3 + x^2 + x + 1)$$

If ω is a root of $1 + x + x^2 + x^3 + x^4$, then $1, \omega, \omega^2, \omega^3, \omega^4$ are all the roots of $f(x)$ and hence $\mathbb{Q}(\omega)$ is the splitting field of $f(x)$ over $\mathbb{Q}$ and $[\mathbb{Q}(\omega) : \mathbb{Q}] = 4$ and hence $\left|G(\mathbb{Q}(\omega)/\mathbb{Q})\right| = 4$. If σ is an automorphism in $G(\mathbb{Q}(\omega)/\mathbb{Q})$ such that

$$\sigma(1) = 1 \quad \text{and} \quad \sigma(\omega) = \omega^2$$

then $\sigma^2(\omega) = \omega^4$, $\sigma^3(\omega) = \omega^3$ and $\sigma^4(\omega) = \omega$ and hence $\sigma^4 = \text{Id}$ and σ generates $G(\mathbb{Q}(\omega)/\mathbb{Q})$. This implies that $G(\mathbb{Q}(\omega)/\mathbb{Q})$ is a cyclic group of order 4.

EXERCISE 15(C)

1. In each of the following, find all the subgroups of the Galois group of $f(x)$ and the corresponding fixed fields.
    (i) $f(x) = x^3 - 2 \in \mathbb{Q}[x]$
    (ii) $f(x) = x^4 + 1 \in \mathbb{Q}[x]$
    (iii) $f(x) = x^4 - 2 \in \mathbb{Q}[x]$
    (iv) $f(x) = x^2 - a \in F[x]$, where F is a field of characteristic $\neq 2$.

2. Prove that the Galois group of $x^4 + 1 \in \mathbb{Q}[x]$ is the Klein four-group and is isomorphic to $\mathbb{Z}_2 \times \mathbb{Z}_2$.

3. If $E = \mathbb{Q}(\sqrt{3}, \sqrt{5})$, then find the Galois group $G(E/\mathbb{Q})$.

4. Let a be a real number such that $\mathbb{Q}(a)$ is a normal extension of $\mathbb{Q}$ for which $[\mathbb{Q}(a) : \mathbb{Q}] = 2^m$, where $m \geq 0$. Prove that there are fields $E_0 = \mathbb{Q} \subset E_1 \subset E_2 \subset \ldots \subseteq E_m = \mathbb{Q}(a)$ such that $[E_i : E_{i-1}] = 2$ for each $1 \leq i \leq m$.

5. If K is the splitting field of $x^4 - 3x^2 + 4$ over $\mathbb{Q}$, then find the Galois group $G(K/\mathbb{Q})$.

6. Let $a = \cos(2\pi/3) + i\sin(2\pi/3)$. Then, find the Galois group $G(\mathbb{Q}(a)/\mathbb{Q})$ and all its subgroups and the corresponding fixed fields.

16

Selected Applications of Galois Theory

In this chapter, we discuss certain important applications of Galois theory to classical problems. The first is the fundamental theorem of algebra which states that any polynomial over the field of complex numbers $\mathbb{C}$ has all the roots in $\mathbb{C}$ which is equivalent to saying that any polynomial over $\mathbb{C}$ can be factored completely into linear factors over $\mathbb{C}$. We prove this by using various techniques of Galois theory. Also we discuss problems of finding solutions of polynomial equations (that is, finding roots of polynomials) by radicals; that is, expressing the roots of a polynomial in terms of the coefficients using the field operations and the operations of taking square roots, cube roots and so on. We are familiar with the fact that a quadratic polynomial $f(x) = ax^2 + bx + c$ $(a \neq 0)$ with real coefficients has $\frac{-b \pm \sqrt{b^2 - 4ac}}{2a}$ as its roots in $\mathbb{C}$. During the 17th century, similar formulae were found for cubic and biquadratic equations. Later, Abel proved that such formulae were not possible for polynomial of degrees ≥ 5. But the general problems of finding a way of deciding whether a given polynomial could be solvable by radicals was not completed by Abel. Galois and Liouville gave necessary and sufficient conditions for the solvability by radicals for polynomials of degrees ≥ 5 which linked with the group properties of the symmetric group S_n. Finally, we discuss certain impossibilities of geometric constructions by using ruler and compass only. For example, we prove that an angle cannot be trisected by using ruler and compass only.

16.1 FUNDAMENTAL THEOREM OF ALGEBRA

It is well known from elementary analysis that the intermediate value theorem gives us a root of a polynomial $f(x)$ over $\mathbb{R}$ if there are real numbers a and b such that $f(a) < 0 < f(b)$. In this section, we prove that any nonconstant polynomial over the field $\mathbb{C}$ of complex numbers has a root in $\mathbb{C}$. This result is known as the fundamental theorem of algebra. Even though there are several proofs of this fundamental result, most of them use the techniques of topology or real analysis or complex analysis. The proof we are offering here is more elegant and use the techniques of Galois theory. Before taking up the proof of the main theorem, let us have a brief preparation.

Theorem 16.1.1. The field $\mathbb{C}$ has no extension of degree 2.

Proof: Let K be a field extension of $\mathbb{C}$ such that $[K : C] = 2$. Then, there exists $a \in K$ such that $K = \mathbb{C}(a)$. Then, $p(x)$ be the minimal polynomial of a over $\mathbb{C}$. Then, degree of $p(x)$ must be two. Let

$$p(x) = a + 2bx + x^2 \quad \text{where } a \text{ and } b \in \mathbb{C}.$$

Now,

$$p(x) = \left(x + b - \sqrt{b^2 - a}\right)\left(x + b + \sqrt{b^2 - a}\right),$$

where $b - \sqrt{b^2 - a}$ and $b + \sqrt{b^2 - a} \in \mathbb{C}$. This is a contradiction, since $p(x)$ is irreducible over $\mathbb{C}$. ◀

Theorem 16.1.2. Let $f(x) \in \mathbb{R}[x]$ be of odd degree. Then, $f(x)$ has a real root.

Proof: Without loss of generality, we can assume that $f(x)$ is a monic polynomial. Suppose

$$f(x) = a_0 + a_1x + \cdots + a_{n-1}x^{n-1} + x^n, \; n \text{ is odd},$$

where $a_i \in \mathbb{R}$. Put $s = 1 + |a_0| + |a_1| + \cdots + |a_{n-1}|$. Then, $|a_i| \leq s - 1$ for all $0 \leq i \leq n - 1$. Therefore,

$$\begin{aligned} |a_0 + a_1s + \cdots + a_{n-1}s^{n-1}| &\leq (s - 1)(1 + s + \cdots + s^{n-1}) \\ &= s^{n-1} < s^n \end{aligned}$$

and hence $f(s) > 0$. Also

$$\begin{aligned} f(-s) &= a_0 - a_1 s + a_2 s^2 + \cdots + (-1)^{n-1} a_{n-1} s^{n-1} + (-s)^n \\ &= a_0 - a_1 s + a_2 s^2 - \cdots - s^n, \quad \text{since } n \text{ is odd} \\ &\leq s^n - 1 - s^n = -1 < 0. \end{aligned}$$

Therefore, $f(-s) < 0 < f(s)$ and $s \in \mathbb{R}$. By the intermediate value theorem in analysis, there exists a real number a such $f(a) = 0$. Thus, a is a root of $f(x)$ in $\mathbb{R}$. ◀

For any $f(x) = a_0 + a_1 x + \cdots + a_n x^n \in \mathbb{C}[x]$, let us write $\overline{f}(x) = \overline{a}_0 + \overline{a}_1 x + \cdots + \overline{a}_n x^n$, where $\overline{a}_i$ is the complex conjugate of a_i. It is a straight forward verification to prove that $f(x)\overline{f}(x) \in \mathbb{R}[x]$ for any $f(x) \in \mathbb{C}[x]$. Recall that $a\overline{a} = |a|^2 \in \mathbb{R}$ for any $a \in \mathbb{C}$.

Theorem 16.1.3 (Fundamental Theorem of Algebra). Every nonconstant polynomial $f(x)$ over $\mathbb{C}$ completely factors into linear factors in $\mathbb{C}[x]$; that is, any nonconstant polynomial in $\mathbb{C}[x]$ is a product of polynomials of degree one in $\mathbb{C}[x]$.

Proof: Let $f(x) = a_0 + a_1 x + \cdots + a_n x^n$, $n > 0$ and $a_n \neq 0$ be a polynomial in $\mathbb{C}[x]$. Put

$$\begin{aligned} g(x) &= (x^2+1) f(x)\overline{f}(x) \\ &= (x^2+1)(a_0 + a_1 x + \cdots + a_n x^n)(\overline{a}_0 + \overline{a}_1 x + \cdots + \overline{a}_n x^n). \end{aligned}$$

Then, $g(x) \in \mathbb{R}[x]$. Let E be the splitting field of $g(x)$ over $\mathbb{R}$. Then, $\mathbb{R} \subseteq \mathbb{C} \subseteq E$, since i is a root of $g(x)$. We prove that $E = \mathbb{C}$. First observe that $[\mathbb{C} : \mathbb{R}] = 2$ and is a divisor of $[E : \mathbb{R}]$. Therefore, $[E : \mathbb{R}]$ is an even positive integer. Suppose that

$$[E : \mathbb{R}] = 2^m q, \quad \text{where } m \text{ and } q \in \mathbb{Z}^+ \text{ and } q \text{ is odd.}$$

Let G be the Galois group $G(E/\mathbb{R})$. Since E is a normal extension of $\mathbb{R}$, it follows that

$$|G| = |G(E/\mathbb{R})| = [E : \mathbb{R}] = 2^m q.$$

By the Sylow Theorem I in group theory, there exists a subgroup H of G such that $|H| = 2^m$ (H is a 2-Sylow subgroup of G). Let E_H be the fixed field of H. Then, $R \subseteq E_H \subseteq E$ and

$$2^m q = [E : \mathbb{R}] = [E : E_H][E_H : \mathbb{R}] = |H|[E_H : \mathbb{R}] = 2^m [E_H : \mathbb{R}]$$

and hence $[E_H : \mathbb{R}] = q$. Also, since E_H is a finite separable extension of $\mathbb{R}$, E_H is a simple extension of $\mathbb{R}$ and hence $E_H = \mathbb{R}(b)$ for some $b \in \mathbb{R}$. Let

$$q(x) = b_0 + b_1x + \cdots + b_qx^q \in \mathbb{R}[x]$$

be the minimal polynomial of b over $\mathbb{R}$ (note that $\deg(q(x)) = [\mathbb{R}(b) : \mathbb{R}] = [E_H : \mathbb{R}] = q$). By Theorem 16.1.2, $q(x)$ has a real root and hence $q(r) = 0$ for some $r \in \mathbb{R}$. This implies that $x - r$ is a factor of $q(x)$ in $\mathbb{R}[x]$. Being the minimal polynomial of b over $\mathbb{R}$, $q(x)$ must be irreducible. All these imply that $q = 1$ and $[E_H : \mathbb{R}] = 1$ and hence $E_H = R$. Therefore, $[E : \mathbb{R}] = 2^m$. Now, we prove that $m = 1$. Suppose, if possible, that $m > 1$. Then, $[E : \mathbb{C}] = 2^{m-1}$ and hence $|G(E/\mathbb{C})| = 2^{m-1}$. Again, by the Sylow Theorem I, $G(E/\mathbb{C})$ has a subgroup S of order 2^{m-2}. If E_S is the fixed field of S, then

$$[E : E_S] = |S| = 2^{m-2}.$$

Therefore, $[E_S : \mathbb{C}] = 2$, since $[E : \mathbb{C}] = 2^{m-1}$. This is a contradiction to Theorem 16.1.1. Thus, $m = 1$ and $[E : \mathbb{R}] = 2$. Since $\mathbb{R} \subseteq \mathbb{C} \subseteq E$ and $[\mathbb{C} : \mathbb{R}] = 2$, it follows that $[E : \mathbb{C}] = 1$ and hence $E = \mathbb{C}$. Thus, $\mathbb{C}$ is the splitting of $g(x)$ In particular, $\mathbb{C}$ contains all the roots of $g(x)$ and hence of $f(x)$. Thus, $f(x)$ completely factors into linear factors in $\mathbb{C}[x]$. ◀

Corollary 16.1.1. $\mathbb{C}$ is an algebraically closed field.

Worked Exercise 16.1.1. For any $f(x) \in \mathbb{C}[x]$, prove that $\overline{f}(x) \in \mathbb{R}[x]$.

Answer: Let $f(x) \in \mathbb{C}[x]$ be of degree n. We use induction on n. If $n = 0$, then $f(x) = a \in \mathbb{C}$ and $f(x)\overline{f}(x) = a\overline{a} = |a|^2 \in \mathbb{R}$. Let $n > 0$ and assume that the result is true for all polynomials $g(x)$ of degree less than n in $\mathbb{C}[x]$. Let

$$f(x) = a_0 + a_1x + \cdots + a_nx^n = g(x) + a_nx^n$$

$$\text{and} \quad \overline{f}(x) = \overline{g}(x) + \overline{a}_nx^n,$$

where $g(x) = a_0 + a_1x + \cdots + a_{n-1}x^{n-1} \in \mathbb{C}[x]$. Now,

$$\begin{aligned} f(x)\overline{f}(x) &= (g(x) + a_nx^n)(\overline{g}(x) + \overline{a}_nx^n) \\ &= g(x)\overline{g}(x) + (a_n\overline{g}(x) + \overline{a}_ng(x))x^n + a_n\overline{a}_nx^{2n} \\ &= g(x)\overline{g}(x) + \sum_{i=0}^{n-1}(a_n\overline{a}_i + \overline{a}_na_i)x^{n+i} + a_n\overline{a}_nx^{2n} \end{aligned}$$

since $a\overline{b}+\overline{a}b=a\overline{b}+\overline{a\overline{b}}\in\mathbb{R}$, $g(x)\overline{g}(x)\in\mathbb{R}[x]$ and $a_n\overline{a}_n\in\mathbb{R}$ and , it follows that $f(x)\overline{f}(x)\in\mathbb{R}[x]$.

16.2 CYCLIC EXTENSIONS

In this section, we discuss a special type of Galois extensions, namely cyclic extensions which are Galois extensions whose corresponding Galois group is a cyclic group (that is, generated by a single element). Before these, let us first have the following definition.

Definition 16.2.1. Let F be a field and n be a positive integer. An element a in an extension E of F is called a *primitive n^{th} root of unity* in E if $a^n = 1$ and $a^m \neq 1$ for any $0 < m < n$.

Example 16.2.1

1. For any $n \in \mathbb{Z}^+$, $e^{\left(\frac{2\pi i}{n}\right)}=\cos\left(\frac{2\pi}{n}\right)+i\sin\left(\frac{2\pi}{n}\right)$ is primitive n^{th} root of unity in $\mathbb{C}$.
2. Let F be a field of characteristic zero. Then, for any positive integer n, $x^n - 1$ has n distinct roots in its splitting field and form a cyclic group under multiplication. If a is a primitive n^{th} root of unity, then a^m, where $0 < m < n$ and $(m, n) = 1$, is also a primitive n^{th} root of unity and we know that the number of such m is $\phi(n)$.
3. $\dfrac{-1+i\sqrt{3}}{2}$ is a primitive cube root of unity.
4. $\cos\left(\frac{6\pi}{5}\right)+i\sin\left(\frac{6\pi}{5}\right)$ is a primitive 5th root of unity.

Theorem 16.2.1. Let F be a field of characteristic zero and $E = F(a)$, where a is a primitive n^{th} root of unity. Then, $G(E/F)$ is isomorphic to a subgroup of $\mathbb{Z}_n^*$, the group of all multiplicatively invertible elements in $\mathbb{Z}_n$ and hence $G(E/F)$ is abelian.

Proof: Since a is a primitive n^{th} root of unity, $E = F(a)$ becomes the splitting field of $x^n - 1 \in F[x]$. Therefore, E is a finite normal extension of F and $G(E/F)$ is a Galois group.

Any $\sigma \in G(E/F)$ is completely determined by its value $\sigma(a)$ and $\sigma(a)$ is also an nth root of unity and hence $\sigma(a) = a^i$ for some $i < n$ and $(i, n) = 1$. Therefore, the correspondence $\sigma \mapsto i$, where $\sigma(a) = a^i$, is a homomorphism from $G(E/F)$ into $\mathbb{Z}_n^*$. Also, $\sigma \circ \tau = \tau \circ \sigma$, for σ and $\tau \in G(E/F)$. Also the map $\sigma \mapsto i$, is one-to-one. Thus, $G(E/F)$ is an abelian group and is isomorphic to a subgroup of $\mathbb{Z}_n^*$. ◀

In general, $G(E/F)$ need not be a cyclic group. For example, the Galois group of $x^8 - 1$ is isomorphic to the Klein 4-group and hence $G(E/F)$ is not cyclic.

Definition 16.2.2. Let E be Galois extension of a field F. Then, E is called a *cyclic extension* of F if $G(E/F)$ is a cyclic group (that is, generated by a single element).

Examples 16.2.2

1. Recall that any finite extension of a finite field is separable. If E is the splitting field of a polynomial $f(x)$ over a finite field, then E is a Galois extension of F and, by Worked Exercise 14.5.1, $G(E/F)$ is a cyclic group and hence E is a cyclic extension of F.
2. Let p be a prime and a be a primitive p^{th} root of unity. Then, $E = \mathbb{Q}(a)$ is the splitting field of $x^p - 1 \in \mathbb{Q}[x]$ and E is a cyclic extension of $\mathbb{Q}$.

Theorem 16.2.2. Let F be a field and suppose that F contains a primitive n^{th} root of unity. Then, E is a finite cyclic extension of F of degree n if and only it E is the splitting field of an irreducible polynomial $x^n - b \in F[x]$.

Proof: Suppose that E is a finite cyclic extension of F of degree n; that is, $[E : F] = n$ and $G(E/F)$ is a cyclic group of order n. Let σ be a generator of $G(E/F)$. Then, Id, $\sigma, \sigma^2, \ldots, \sigma^{n-1}$ are linearly independent over F. Let ω be a nth root of unity in F. Then, $\omega \neq 0$ and

$$\text{Id} + \omega^{-1}\sigma + \omega^{-2}\sigma^2 + \cdots + \omega^{-n+1}\sigma^{n-1}$$

is a nonzero endomorphism of E as a vector space over F. Therefore, there exists $a \in E$ such that

$$c = \text{Id}(a) + \omega^{-1}\sigma(a) + \omega^{-2}\sigma^2(a) + \cdots + \omega^{-n+1}\sigma^{n-1}(a) \neq 0$$

Now,
$$\begin{aligned}\sigma(c) &= \sigma(a) + \omega^{-1}\sigma^2(a) + \omega^{-2}\sigma^3(a) + \cdots + \omega^{-n+1}\sigma^n(a)\\ &= \omega(\omega^{-1}\sigma(a) + \omega^{-2}\sigma^2(a) + \cdots + \omega^{-n+1}\sigma^{n-1}(a) + \omega^{-n}\sigma^n(a))\\ &= \omega(\omega^{-1}\sigma(a) + \omega^{-2}\sigma^2(a) + \cdots + \omega^{-n+1}\sigma^{n-1}(a) + a)\\ &= \omega c,\end{aligned}$$

since $\omega^{-n} = 1$ and $\sigma^n = \text{Id}$. Recursively, we can prove that

$$\sigma^r(c) = \omega^r c \quad \text{for all } 1 \leq r \leq n - 1.$$

Therefore, $\sigma^r(c^n) = (\omega^r c)^n = c^n$ for all $1 \leq r \leq n - 1$.
This implies that c^n is in the fixed field of $G(E/F)$. But F is the fixed field of $G(E/F)$ and hence $c^n \in F$. Put $b = c^n \in F$. Then, $x^n - b \in F[x]$ and

$c, c\omega, c\omega^2, \ldots, c\omega^{n-1}$ are the roots of $x^n - b$. However $\sigma^r(c) = c\omega^r$ implies that $c, c\omega, c\omega^2, \ldots, c\omega^{n-1}$ are also roots of the minimal polynomial $f(x)$ of c over F. Therefore, $x^n - b$ divides $f(x)$. Since c is a root of $x^n - b$, it follows that $f(x)$ divides $x^n - b$ and hence $f(x) = x^n - b$. Since $f(x)$ is the minimal polynomial of c over $F, f(x)$ is irreducible over F; that is, $x^n - b$ is irreducible over F. Also,

$$[F(c) : F] = n = [E : F] \quad \text{and} \quad F(c) \subseteq E$$

and hence $F(c) = E$. Therefore, E is the splitting field of the irreducible polynomial $x^n - b \in F[x]$ over F.

Conversely suppose that $x^n - b \in F[x]$ is an irreducible polynomial over F and E is its splitting field over F. Let $c \in E$ be a root of $x^n - b$; that is, $b = c^n$. Then, clearly $c, c\omega, c\omega^2, \ldots, c\omega^{n-}$ are the n distinct roots of $x^n - b$, where $\omega \in F$ is a primitive n^{th} root of unity. Therefore, $x^n - b$ is a separable irreducible polynomial and hence $E = F(c)$ is a Galois extension of F. For each $\sigma \in G(E/\sigma)$, let the set A_σ be defined by

$$A_\sigma = \{r \in \mathbb{Z} : \sigma(c) = \omega^r c\}.$$

Then, A_σ is nonempty, since $\sigma(c)$ is also a root of $x^n - b$. Also, for any $r \in A_\sigma$,

$$A_\sigma = r + n\mathbb{Z},$$

since $\omega^r c = \omega^s c$ if and only if $r \equiv s(\text{mod } n)$. Further, for any σ and $\tau \in G(E/F)$,

$$\sigma(c) = \omega^r c \text{ and } \tau(c) = \omega^s c \Rightarrow (\sigma\tau)(c) = \sigma(\omega^s c) = \omega^s \sigma(c) = \omega^{r+s} c.$$

Therefore, $A_{\sigma\tau} = A_\sigma + A_\tau$, where the sum on the right side is interpreted as the binary operation on the additive group $\mathbb{Z}/n\mathbb{Z}$ $(\cong\mathbb{Z}_n)$ of integers modulo n. Finally, if $A_\sigma = n\mathbb{Z}$, the zero in $\mathbb{Z}/n\mathbb{Z}$, then $\sigma(c) = c$. Therefore, σ is the identity on E (since $E = F(c)$ and $\sigma/F = \text{Id}$). Consequently, $\sigma \mapsto A_\sigma$ is an isomomorphism of $G(E/F)$ onto a subgroup of $\mathbb{Z}/n\mathbb{Z}$. Also,

$$\begin{aligned}[F(c) : F] &= \text{The degree of the minimal polynomial of } c \text{ over } F \\ &= \text{The degree of } x^n - b = n.\end{aligned}$$

since $E = F(c)$ is a finite separable and normal extension of F, it follows that $|G(E/F)| = [E : F] = n$ and hence $G(E/F) \cong \mathbb{Z}/n\mathbb{Z}$, so that $G(E/F)$ is a cyclic group of order n. Thus, E is a finite cyclic extension of F degree n. ◀

16.3 SOLVABLE GROUPS

Before we take up certain applications of Galois theory to the theory of equations and geometric constructions in the next two sections, we first introduce the concept of a solvable group and discuss some of their properties.

Definition 16.3.1. Let G be a group. A sequence

$$\{e\} = G_0 \subseteq G_1 \subseteq \cdots \subseteq G_n = G$$

of subgroups of G is called a *normal series* of the group G if each G_i is a normal subgroup of G_{i+1} for $0 \leq i < n$.

Note that, in the above, G_i may not be a normal subgroup of G; it is a normal subgroup of G_{i+1}. Also, $\{e\}$ is always a normal subgroup of any subgroup of G and therefore, for being a normal series, we can simply prescribe that G_i is normal in G_{i+1} for $0 < i < n$ and we can talk about the quotient groups G_{i+1}/G_i.

Definition 16.3.2. A group G is said to be a *solvable group* if there exists a normal series

$$\{e\} = G_0 \subseteq G_1 \subseteq G_2 \subseteq \cdots \subseteq G_n = G$$

such that the quotient G_{i+1}/G_i is an abelian group for $0 \leq r < n$; and, in this case the series $\{e\} = G_0 \subseteq G_1 \subseteq \cdots \subseteq G_n = G$ is called a *solvable series*.

Examples 16.3.1

1. Any abelian group G is a solvable group, since $\{e\} = G_0 \subset G_1 = G$ is a solvable series.
2. Let S_3 be the symmetric group of degree 3 and let $H = \{e, (1\ 2\ 3), (3\ 2\ 1)\}$, where $(1\ 2\ 3)$ is the 3-cycle mapping $1 \to 2$, $2 \to 3$, and $3 \to 1$. Then, H is a subgroup of S_3. Now,

$$\{e\} \subset H \subset S_3$$

is a solvable series in S_3, since H is an abelian normal subgroup and S_3/H is a group of order 2 and hence abelian. Therefore, S_3 is a solvable group.

Recall from group theory that for any elements a and b of a group G, the element $aba^{-1}b^{-1}$ is called a commutator in G and the subgroup G' generated by the set of all commutators in G is called the *derived subgroup* of G. For any positive integer n, we define the nth *derived subgroup* of G, denoted by $G^{(n)}$, is defined recursively as follows:

$$G^{(1)} = G' \quad \text{and} \quad G^{(n)} = (G^{(n-1)})' \quad \text{for } n > 1.$$

Clearly, G is abelian if and only if $G' = \{e\}$.

Theorem 16.3.1. A group G is solvable if and only if $G^{(n)} = \{e\}$ for some positive integer n.

Proof: Suppose that G is solvable and

$$\{e\} = G_0 \subseteq G_1 \subseteq \cdots \subseteq G_n = G$$

is a solvable series in G. For each i, G_{i+1}/G_i is an abelian group and hence its derived subgroup is trivial. This implies that

$$G'_{i+1} \subseteq G_i \quad \text{for all } 0 \leq i < n.$$

In particular, $G' = G'_n \subseteq G_{n-1}$ and hence

$$G^{(2)} \subseteq G'_{n-1} \subseteq G_{n-2}.$$

By induction, we can prove that $G^{(i)} \subseteq G_{n-1}$ and hence $G^{(n)} \subseteq G_0 = \{e\}$. Thus, $G^{(n)} = \{e\}$.

Conversely suppose that $G^{(n)} = \{e\}$ for some $n > 0$. Then,

$$\{e\} = G^{(n)} \subseteq G^{(n-1)} \subseteq \cdots \subseteq G^{(1)} \subseteq G$$

is a solvable series, since H/H_1 is abelian for any group H. Thus, G is a solvable group. ◀

Theorem 16.3.2. Any subgroup and any quotient group of a solvable group is solvable.

Proof: Let G be a solvable group. Then, there exists a positive integer n such that $G^{(n)} = \{e\}$. If H is any subgroup of G, then $H^{(n)} \subseteq G^{(n)} = \{e\}$ and hence $H^{(n)} = \{e\}$ so that H is solvable. Let G/N be a quotient group of G, where N is a normal subgroup of G. Now,

$$(G/N)^{(n)} = G^{(n)} + N = \{N\}, \text{ since } G^{(n)} = \{e\},$$

and hence G/N is solvable. ◀

By the fundamental theorem of homomorphisms, any homomorphic image of a group G is isomorphic to a quotient group of G and hence we have the following corollary.

Corollary 16.3.1. Any homomorphic image of a solvable group is solvable. The following is a converse of the above theorem.

Theorem 16.3.3. Let N be a normal subgroup of a group G such that both N and G/N are solvable groups. Then, G is a solvable group.

Proof: Recall that any subgroup of G/N is of the form H/N, where H is a subgroup of G containing N. Now, since G/N is a solvable group, there exist a sequence of subgroups

$$N \subseteq H_1 \subseteq H_2 \subseteq \cdots \subseteq H_n = G$$

such that $\{N\} \subseteq H_1/N \subseteq H_2/N \subseteq \cdots \subseteq H_n/N = G/N$ is a solvable series in G/N. In particular, each H_i is a normal subgroup of H_{i+1} (since H_i/N is a normal subgroup of H_{i+1}/N). Also,

$$(H_{i+1}/N)/(H_i/N) \cong H_{i+1}/H_i$$

and hence H_{i+1}/H_i is an abelian group. Since N is also solvable, there exists solvable series

$$\{e\} = N_0 \subseteq N_1 \subseteq \cdots \subseteq N_m = N$$

in N. Now, the series

$$\{e\} = N_0 \subseteq N_1 \subseteq \cdots \subseteq N_m \subseteq H_1 \subseteq H_2 \subseteq \cdots \subseteq H_n = G$$

is a solvable series in G. Thus, G is a solvable group. ◀

Theorem 16.3.4. A finite group G is solvable if and only if there exists a sequence $\{e\} = G_0 \subseteq G_1 \subseteq \cdots \subseteq G_n = G$ of subgroups in G such that, for each $0 \leq i < n$, G_i is a normal subgroup of G_{i+1} and G_{i+1}/G_i is a cyclic group of prime order.

Proof: First observe that, if B is a proper normal subgroup of a finite group A such that A/B is abelian, then there exists a maximal normal subgroup N of A containing B and then A/N is a simple abelian group and hence of prime order. Now, suppose that G is a solvable group and $\{e\} = G_0 \subseteq G_1 \subseteq \cdots \subseteq G_n = G$ be a solvable series in G. By the above observation, we can construct a sequence of subgroups.

$$G_i = H_{i,0} \subseteq H_{i,1} \subseteq \cdots \subseteq H_{i,m_i} = G_{i+1}$$

such that $H_{i,j}$ is a normal subgroup of $H_{i,j+1}$ and $H_{i,j+1}/H_{i,j}$ is a group of prime order. This is true for each $0 \leq i < n$ and hence, by clubbing all these sequences we get a required sequence of subgroups in G. The converse is trivial. ◀

Worked Exercise 16.3.1. Prove that the symmetric group S_4 of degree 4 is a solvable group.

Answer: Let A_4 be the alternating group of degree 4 (that is, the subgroup of even permutations in S_4). Then, A_4 is a normal subgroup of S_4. Put

$$a = (1\ 2) \circ (3\ 4),\ b = (1\ 3) \circ (2\ 4) \quad \text{and} \quad c = (1\ 4) \circ (2\ 3)$$

and $H = \{\text{Id}, a, b, c\}$. Then, H is a normal subgroup of A_4 and H is an abelian group (since $ab = c = ba$, etc.). Now, $\{\text{Id}\} \subset H \subset A_4 \subset S_4$ is a solvable series in S_4. Thus, S_4 is a solvable group.

Worked Exercise 16.3.2. Prove that the dihedral group D_n is solvable for any positive integer $n > 1$.

Answer: Let σ be the cycle $(1\ 2\ 3\ \cdots\ n)$, Then,

$$H = \{\text{Id}, \sigma, \sigma^2, \ldots, \sigma^{n-1}\}$$

is a cyclic group of order n and of index 2 (since D_n is of order $2n$). Therefore, H is a normal subgroup of D_n and D_n/H is of order 2 and hence an abelian group. Therefore, $\{\text{Id}\} \subset H \subset D_n$ is a solvable series in D_n. Thus, D_n is a solvable group.

Worked Exercise 16.3.3. Prove that the symmetric group S_n is not solvable for any $n > 4$.

Answer: Let $n > 4$. It is known that the derived subgroup of S_n is A_n, the alternating group of degree n; that is, $S_n' = A_n$. Also, since A_n is simple and A_n' is a normal subgroup of A_n, it follows that $A_n' = \{e\}$ or $A_n' = A_n$. But $A_n' \neq \{e\}$ and hence $A_n' = A_n$. Now, $S_n^{(2)} = A_n' = A_n$ and

$$S_n^{(r)} = A_n \quad \text{for all } r \geq 2.$$

Thus, by Theorem 16.3.1, S_n is not solvable.

16.4 POLYNOMIALS SOLVABLE BY RADICALS

In this section, we use the techniques of Galois theory to find necessary and sufficient conditions for a polynomial over field to be solvable by radicals. First we need some preparation. Throughout this section, we assume that all fields are of characteristic zero.

Definition 16.4.1. Let F be a field and K be an extension of F. Then, K is said to be a *radical extension* of F if there exists a sequence of fields

$$F = F_0 \subseteq F_1 \subseteq \cdots \subseteq F_n = K$$

such that, for each $0 < i \leq n$, $F_i = F_{i-1}(a_i)$ for some $a_i \in F_i$ with the property that $a_i^{r_i} \in F_{i-1}$ for some $r_i \geq 1$.
Observe that, if $F_i = F_{i-1}(a_i)$ and $a_i^{r_i} \in F_{i-1}$ then a_i is a root of the polynomial $x^{r_i} - a_i^{r_i} \in F_{i-1}[x]$ and hence F_i is a simple algebraic extension of F_{i-1} and therefore, $[F_i : F_{i-1}]$ is finite. Also, since

$$[K : F] = [K : F_{n-1}][F_{n-1} : F_{n-2}] \ldots [F_1 : F],$$

K itself is a finite extension of F and

$$K = F(a_1, a_2, \ldots, a_n).$$

Definition 16.4.2. A nonconstant polynomial $f(x)$ over a field F is said to be *solvable by radicals* if the splitting field L of $f(x)$ over F is contained in a radical extension of F.

Example 16.4.1. Let $f(x) = a + bx + x^2$ be a monic quadratic polynomial over the field $\mathbb{Q}$ of rational numbers. Its roots are $a_1 = \frac{-b+\sqrt{b2-4a}}{2}$ and $a_2 = \frac{-b-\sqrt{b2-4a}}{2}$.

Let $L = \mathbb{Q}(\omega)$, where $\omega = \sqrt{b^2 - 4a}$. Then, $\omega^2 = b^2 - 4a \in \mathbb{Q}$ and therefore L is a radical extension of $\mathbb{Q}$. Also, L itself is the splitting field of $f(x)$ over $\mathbb{Q}$. Therefore, $f(x)$ is solvable by radicals.

It is known that polynomial of degree 3 and 4 are always solvable by radicals. In this section, we prove that not all polynomials of degree greater than four are solvable by radicals.

Theorem 16.4.1. Let E be a radical extension of a field F and

$$F = E_0 \subseteq E_1 \subseteq E_2 \subseteq \cdots \subseteq E_r = E$$

be an ascending sequence of intermediate fields between F and E. Then, there exists a radical extension E' of F and an ascending sequence of intermediate fields

$$F = E_0' \subseteq E_1' \subseteq E_2' \subseteq \cdots \subseteq E_s' = E'$$

between F and E' satisfying the following:

1. $E = E_r \subseteq E'_s = E'$
2. E' is a normal extension of F.
3. For each $1 \leq i \leq s$, E'_i is a splitting field of a polynomial of the form $x^{m_i} - b_i \in E'_{i-1}[x]$.

Proof: We can assume that $E_i = E_{i-1}(a_i)$ where a_i is a root of $x^{n_i} - c_i \in E_{i-1}[x]$ for each $1 \leq i \leq r$. Put $n = n_1 n_2 \ldots n_r$. Let ω be a primitive n^{th} root of unity. Consider the sequence

$$F = E_0 \subset E_0(\omega) = F(\omega) \subset E_1(\omega).$$

Clearly, $E_1(\omega)$ is a radical extension of F. Since $F(\omega)$ is a splitting field of $x^n - 1 \in F[x]$, $F(\omega)$ is a normal extension of F. Therefore, F is the fixed field of $G(F(\omega)/F)$ and hence the polynomial

$$f_1(x) = \prod_{\sigma \in G\left(\frac{F(\omega)}{F}\right)} (x^{n_1} - \sigma(c_1))$$

is in $F[x]$. Here, we have $f_1(x) = (x^{n_1} - c_1)^t$, where $t = |G(F(\omega)/F)|$, since $c_1 \in F$ and hence $\sigma(c_1) = c_1$ for all $\sigma \in G(F(\omega)/F)$. Next, let $g_1(x) = (x^n - 1)$ $f_1(x)$. Then, $g_1(x) \in F[x]$. Let K be the splitting field of $g_1(x)$ over F, so that K is a normal extension of F. Clearly, $a_1 \in K$, $\omega \in K$ and $E_1 \subseteq K$. Further, it is clear that there is a finite ascending sequence of intermediate fields between F and K such that each field is a splitting field of a polynomial of the form $x^m - b$ over the preceding field.

Now, we construct a field L such that K and E_2 are subfields of L and L is a normal extension of F. For this, we consider the polynomial

$$g_2(x) = g_1(x) f_2(x), \quad \text{where } f_2(x) = \prod_{\sigma \in G\left(\frac{K}{F}\right)} (x^{n_2} - \sigma(c_2)).$$

Since K is a normal extension of F, $f_2(x) \in F(x)$ and hence $g_2(x) \in F[x]$. Take L to be the splitting field of $g_2(x)$ over F. Then, L contains a_2 and K and hence $E_2 = E_1(a_2) \subseteq L$. Therefore, L is a normal extension of F containing E_2. Also, because of the nature of the polynomial $g_2(x)$, it is clear that there exists a finite ascending sequence of intermediate fields between K and L such that any member of the sequence is a splitting field of a polynomial of the form

$x^m - b$ over the preceding member. Continuing like this, we can construct a radical extension E' of F satisfying the required properties. ◀

Before going to the main theorem, we prove the following two important results which are useful in proving the main theorem.

Theorem 16.4.2. Let n be a positive integer and F be a field containing all the nth roots of unity. Let E be the splitting field of the polynomial $x^n - a \in F[x]$. Then, the Galois group $G(E/F)$ is abelian.

Proof: If b is any root of $x^n - a$, then $b, bc, bc^2, \ldots, bc^{n-1}$ are the roots of $x^n - a$, where c is a primitive n^{th} root of unity and, also $E = F(b)$. For any σ_1 and $\sigma_2 \in G(E/F)$, $\sigma_1(b) = bc^i$ and $\sigma_2(b) = bc^j$ for some i and j and hence

$$(\sigma_1 \circ \sigma_2)(b) = \sigma_1(bc^j) = bc^jc^i = bc^{i+j} = (\sigma_2 \circ \sigma_1)(b).$$

This implies that $\sigma_1 \circ \sigma_2 = \sigma_2 \circ \sigma_1$ for any σ_1 and $\sigma_2 \in G(E/F)$. Thus, $G(E/F)$ is an abelian group. ◀

Theorem 16.4.3. Let n be a positive integer and F be a field. If E is the splitting field of the polynomial $x^n - a \in F[x]$, then $G(E/F)$ is a solvable group.

Proof: Let E be the splitting field of the polynomial $x^n - a \in F[x]$. If F contains a primitive n^{th} root of unity, then by Theorem 16.4.2, $G(E/F)$ is abelian and hence solvable. Suppose that F does not contain any primitive n^{th} root of unity. Let c be a primitive n^{th} root of unity and $b \in E$ be a root of $x^n - a$. Then, cb is a root of $x^n - a$ and hence $cb \in E$. Also, $c = b^{-1}(bc) \in E$ and therefore

$$F \subseteq F(c) \subseteq E \quad \text{and} \quad \{e\} \subseteq G(E/F(c)) \subseteq G(E/F).$$

We prove that the later is a solvable series in $G(E/F)$, so that $G(E/F)$ is a solvable group. Since $F(c)$ is the splitting field of the polynomial $x^n - 1 \in F[x]$, we get that $F(c)$ is a normal extension of F. By the fundamental theorem of Galois theory, $G(E/F(c))$ is a normal subgroup of $G(E/F)$. Also, since $F(c)$ contains the primitive n^{th} root of unity c and since E is the splitting field of the polynomial $x^n - a \in F(c)[x]$, we get from Theorem 16.4.2 that $F(E/F(c))$ is an abelian group, Further, by the fundamental theorem of Galois theory, we have

$$\frac{G\left(\dfrac{E}{F}\right)}{G\left(\dfrac{E}{F(c)}\right)} \cong G\left(\frac{F(c)}{F}\right) \cong \mathbb{Z}_n^*,$$

where $\mathbb{Z}_n^*$ is the multiplicative group $\{r : 0 < r < n \text{ and } (r, n) = 1\}$, which is abelian. Therefore,

$$\{e\} \subseteq G(E/F(c)) \subseteq G(E/F)$$

is a solvable series in $G(E/F)$. Thus, $G(E/F)$ is a solvable group. ◀

Note that a polynomial $f(x)$ over a field F is solvable by radicals if we can extract every root of $f(x)$ by using a finite sequence of operations of addition, subtraction, multiplication, division and taking n^{th} roots, starting with elements of the field F. Now, we prove the following main theorem.

Theorem 16.4.4. Let $f(x)$ be a polynomial over a field F of characteristic zero and E be the splitting field of $f(x)$ over F. Then, $f(x)$ is solvable by radicals over F if and only if the Galois group $G(E/F)$ is a solvable group.

Proof: Suppose that $f(x)$ is solvable by radicals over F. That is, the splitting field E of $f(x)$ over F is contained in a radical extension of F. Hence we can find a sequence of fields

$$F = F_0 \subseteq F_1 = F_0(\omega_1) \subseteq F_2 = F_1(\omega_2) \subseteq \cdots \subseteq F_m = F_{m-1}(\omega_m)$$

such that $\omega_i^{r_i} \in F_{i-1}$ for some integers $r_i \geq 1$ and $E \subseteq F_m$. By Theorem 16.4.1, we can suppose that F_m is a normal extension of F and F_i is the splitting field of $x^{r_i} - \omega_i^{r_i} \in F_{i-1}[x]$. Now,

$$\{e\} = G(F_m/F_m) \subseteq G(F_m/F_{m-1}) \subseteq \cdots \subseteq G(F_m/F)$$

and F_i is a normal extension of F_{i-1} (since it is the splitting field of a polynomial). By the fundamental theorem of Galois theory, we have

$$G(F_m/F_{i-1})/G(F_m/F_i) \cong G(F_i/F_{i-1}).$$

Also, by Theorem 16.4.3, $G(F_i/F_{i-1})$ is a solvable group. Now, put $H_i = G(F_m/F_i)$, we have

$$\{e\} = H_m \subseteq H_{m-1} \subseteq \cdots \subseteq H_0 = G(F_m/F),$$

where H_{i-1}/H_i is a solvable group. Now, H_m $(=\{e\})$ and H_{m-1}/H_m are solvable and hence, by Theorem 16.3.3, H_{m-1} is a solvable group. Also, since H_{m-2}/H_{m-1} is solvable, so is H_{m-2}. Continuing this process, we get that H_0 is a solvable group. That is, $G(F_m/F)$ is a solvable group. Since $E \subseteq F_m$ and

$$G(F_m/F)/G(F_m/E) \cong G(E/F),$$

we get that $G(E/F)$ is a solvable group.

Conversely suppose that $G(E/F)$ is a solvable group, since the characteristic of F is zero, E is a normal separable extension and hence

$$[E : F] = |G(E/F)| = n, \text{ say.}$$

We divide the proof into two parts (1) and (2) as given below.

1. First we assume that F contains a primitive n^{th} root of unity. Then, F contains primitive m^{th} roots of unity for all positive integers m which divide n. since $G(E/F)$ is a finite solvable group, there exists a sequence of subgroups

   $$\{e\} = G_0 \subseteq G_1 \subseteq G_2 \subseteq \cdots \subseteq G_r = G(E/F)$$

   of $G(E/F)$ such that G_i is a normal subgroup of G_{i+1} and G_{i+1}/G_i is a cyclic group. Let F_i be the fixed field of G_i. Then, we have the sequence

   $$E = F_0 \supseteq F_1 \supseteq F_2 \supseteq \cdots \supseteq F_r = F$$

   of intermediate fields between F and E. Then, by the fundamental theorem of Galois theory, we have that $G(E/F_i) = G_i$ and $G_{r-1} = G(E/F_{r-1})$ is a normal subgroup of $G(E/F)$ which implies that F_{r-1} is a normal extension of F. Now, E can be regarded as the splitting field of $f(x)$ over F_{r-1} and hence E is a finite normal extension of F_{r-1}. Since G_{r-2} is a normal subgroup of G_{r-1}, F_{r-2} is a normal extension of F_{r-1}. Continuing this process, we get that F_{i-1} is a normal extension of F_i. By the fundamental theorem of Galois theory, we also have

   $$G(F_{i-1}/F_i) \cong G(E/F_i)/G(E/F_{i-1}) = G_i/G_{i-1}$$

   which is cyclic. Therefore, F_{i-1} is a cyclic extension of F_i. Then, by Theorem 16.2.2, F_{i-1} is the splitting field of an irreducible polynomial $x^{n_i} - b_i$ belonging to $F_i[x]$ and $F_{i-1} = F_i(a_i)$, where $a_i^{n_i} = b_i \in F_i$. Then, we have

   $$E = F(a_r, a_{r-1}, \ldots, a_2, a_1) = F(a_1, a_2, \ldots, a_r),$$

   $a_r^{n_r} \in F_r = F$ and $a_i^{n_i} \in F_i = F(a_r, a_{r-1}, \ldots, a_i)$ for all $1 \leq i < r$. Thus, $f(x)$ is solvable by radicals.

2. Now, we take up the general case and drop the assumption that F contains a primitive n^{th} root of unity. Note that the polynomial $x^n - 1 \in E[x]$ has roots in $\bar{E}$. Let ω be a primitive n^{th} root of unity in $\bar{E}$. Then, $E(\omega)$ is the splitting field of $f(x)$ regarded as a polynomial over $F(\omega)$. If σ is an $F(\omega)$-automorphism of $E(\omega)$, σ leaves the coefficients of $f(x)$ unaltered. Since E is a normal extension of F.

   $$\sigma/E \in G(E/F) \quad \text{for all } \sigma \in G(E(\omega)/F(\omega)).$$

Also, the map $\sigma \mapsto \sigma/E$ is an monomorphism of the group $G(E(\omega)/F(\omega))$ into the group $G(E/F)$, since any subgroup of a solvable group is solvable and since $G(E/F)$ is a solvable group, it follows that $G(E(\omega)/F(\omega))$ is solvable. By part (1) above, $E(\omega)$ is a radical extension of $F(\omega)$ and hence $E(\omega)$ is a radical extension of F. Since the splitting field E of $f(x)$ is contained in the radical extension $E(\omega)$ of F, it follows that $f(x)$ is solvable by radicals over F.

Since the symmetric group S_n is not solvable (by Worked Exercise 16.3.3) for any $n > 4$, the following is an immediate consequence of the above theorem.

Corollary 16.4.1. Let $f(x) \in F[x]$ such that the Galois group of $f(x)$ is isomorphic to S_n for some $n > 4$. Then, $f(x)$ is not solvable by radicals over F.

Recall from Worked Exercise 15.2.1 that the Galois group of a polynomial $f(x) \in F[x]$ having r distinct roots is isomorphic to a subgroup of S_r which is the group of permutations of the r distinct roots of $f(x)$. Before we give some more applications of Theorem 16.4.4, let us have the following definition.

Definition 16.4.3. Let S_n be the symmetric group of degree n. Then, a subgroup H of S_n is called *transitive permutation group* if, for any $i, j \in \{1, 2, \ldots, n\}$, there exists $\sigma \in H$ such that $\sigma(i) = j$.

Theorem 16.4.5. Suppose that a polynomial $f(x) \in F[x]$ has no multiple roots. Then, $f(x)$ is irreducible over F if and only if the Galois group of $f(x)$ is isomorphic to a transitive permutation group.

Proof: Let E be a splitting field of $f(x)$ over F and $a_1, a_2, \ldots, a_n$ be the roots of $f(x)$ in E. Let G be the Galois group of $f(x)$ over F. Then, $G = G(E/F)$. For each $\sigma \in G$, $\sigma(a_1), \sigma(a_2), \ldots, \sigma(a_n)$ are roots of $f(x)$ in E and hence a permutation of $a_1, a_2, a_3, \ldots, a_n$. As in Worked Exercise 15.2.1, we can consider G as a subgroup of S_n.
Now, suppose that G is a transitive permutation group. Let $p(x)$ be the minimal polynomial of a_1 over F. For each root a_i, there exist $\sigma \in G$ such that $\sigma(a_1) = a_i$ (since G is transitive). Then,

$$p(a_i) = p(\sigma(a_1)) = \sigma(p(a_1)) = \sigma(0) = 0.$$

Therefore, each a_i is a root of $p(x)$. Since $p(x)$ divides $f(x)$, it follows that $f(x) = cp(x)$ for some $c \in F$. Since $p(x)$ is irreducible over F, $f(x)$ is also irreducible over F.

Conversely suppose that $f(x)$ is irreducible over F. Then, $<f(x)>$, the ideal generated by $f(x)$ in $F(x)$, is a maximal ideal of $F(x)$ and hence $F[x]/<f(x)>$ is a field and

$$F(a_i) \cong F[x]/<f(x)> \quad \text{for each } 1 \le i \le n.$$

If $\alpha_i : F[x]/<f(x)> \to F(a_i)$ is the isomorphism given by $\alpha_i(g(x) + <f(x)>) = g(a_i)$, then $\alpha_j \circ \alpha_i^{-1}$ is an isomorphism of $F(a_i)$ onto $F(a_j)$ sending a_i to a_j and fixing the elements of F. Since E is a normal extension of F, $\alpha_j \circ \alpha_i^{-1}$ can be extended to an F-automorphism σ of E. Then, $\sigma \in G(E/F) = G$ and $\sigma(a_i) = a_j$. Thus, G is a transitive permutation group.

Theorem 16.4.6. Let p be a prime number and $f(x)$ be a monic irreducible polynomial over $\mathbb{Q}$ of degree p. Suppose that $f(x)$ has exactly two nonreal roots in $\mathbb{C}$. Then, the Galois group of $f(x)$ is isomorphic to the symmetric group S_p.

Proof: Let E be the splitting field of $f(x)$ over $\mathbb{Q}$. Then, $\mathbb{Q} \subseteq E \subseteq \mathbb{C}$. By Theorem 16.4.5, $G(E/\mathbb{Q})$ is isomorphic to a transitive permutation group H, which is a subgroup of S_p. Let $a_1, a_2, \ldots, a_p$ be the roots of $f(x)$ among which a_1 and a_2 are nonreal. Since the coefficients in $f(x)$ are rational numbers, it follows that a complex number α is a root of $f(x)$ if and only if its conjugate $\bar{\alpha}$ is also a root of $f(x)$. Therefore, a_1 and a_2 must be conjugates to each other; that is $\bar{a}_1 = a_2$ and $\bar{a}_2 = a_1$. Consider the embedding $\sigma; E \to \bar{\mathbb{Q}}$ defined by $\sigma(z) = \bar{z}$, where $\bar{\mathbb{Q}}$ is the algebraic closure of $\mathbb{Q}$. Since E is a normal extension of $\mathbb{Q}$, σ maps E onto E and hence $\sigma \in G(E/\mathbb{Q})$. σ takes a_1 to a_2 and a_2 to a_1 and $\sigma(a_r) = a_r$ for all $2 < r \le p$. This implies that H contains the transposition $(a_1\ a_2)$. Also, since $[\mathbb{Q}(a_1) : \mathbb{Q}] = p$ and $[E : \mathbb{Q}] = |G(E/\mathbb{Q})| = |H|$, it follows that p divides the order of the group and hence, by the Cauchy's theorem in group theory, H has an element of order p. That is, there exist $e \neq \tau \in H$ such that $\tau^p = e$. Since p is prime, τ must be a p-cycle. Thus, H contains a p-cycle and a transposition. By the exercise given below in Worked Exercise 16.4.1, $H = S_p$. Thus, $G(E/\mathbb{Q}) \cong S_p$.

Worked Exercise 16.4.1. Let H be a transitive permutation group in S_n containing an n-cycle and a transposition. Then prove that $H = S_n$.

Answer: Since any permutation is a product of transpositions, it is enough if we prove that H contains all transpositions in S_n. Without loss of generality, we can assume that $\sigma = (1\ 2\ 3 \cdots n) \in H$ and $\alpha = (1\ 2) \in H$. Now, it can be verified that

$$(i+1\ i+2) = \sigma^i \circ \alpha \circ \sigma^i \in H \quad \text{for all } 0 \le i \le n-2.$$

Therefore, $(i\ i+1) \in H$ for all $1 \le i < n$. Also,

$$(i\ i+1) \circ (i+1\ i+2) \circ (i\ i-1) = (i\ i+2)$$
$$\text{and} \quad (i\ i+2) \circ (i+2\ i+3) \circ (i\ i+2) = (i\ i+3) \text{ and so on.}$$

Therefore, $(i\,j) \in H$ for all $1 \le i < j \le n$. Thus, H contains all transpositions in S_n and hence $H = S_n$.

Worked Exercise 16.4.2. Prove that the polynomial $x^5 - 9x + 3 \in \mathbb{Q}[x]$ is not solvable by radicals over $\mathbb{Q}$.

Answer: By the Eisenstein's criterion for the irreducibility of polynomial over $\mathbb{Q}$, it follows that $f(x) = x^5 - 9x + 3$ is irreducible over $\mathbb{Q}$. Since $f(0) = 3 > -5 = f(1)$, we get a real root of $f(x)$ in the interval $(0, 1)$ (by the intermediate value theorem in analysis). In the same way, we get a real root in the interval $(1, 2)$ also. Further, by the Descarte's rule of signs, we have

$$\text{the number of positive real roots} \le \text{the number of changes of signs in } f(x)\ (=2)$$
$$\text{and} \quad \text{the number of negative real roots} \le \text{the number of changes of signs in } f(-x)\ (=1).$$

Also 0 is not a root of $f(x)$. Thus, there are at most three real roots of $f(x)$. Also, the roots occur in conjugate pairs. Thus, all these imply that $f(x)$ has exactly two nonreal roots in $\mathbb{C}$. By Theorem 16.4.6, the Galois group of $f(x)$ over $\mathbb{Q}$ is isomorphic with S^5, which is not a solvable group. Thus, by Theorem 16.4.4, $f(x)$ is not solvable by radicals over $\mathbb{Q}$.

Worked Exercise 16.4.3. Let $f(x) \in F[x]$ be an irreducible polynomial over a field F. If $f(x)$ has a root in a radical extension of F, then prove that $f(x)$ is solvable by radicals over F.

Answer: Let E be a radical extension of F containing a root of $f(x)$. By Theorem 16.4.1, there exists a radical extension E' of F such that $E \subseteq E'$ and E' is a normal extension of F. Since $f(x)$ is irreducible over F and has a root in E, it follows that $f(x)$ has a root in E'. Further, since E' is a normal extension of F, E' contains a splitting field of $f(x)$. Thus, $f(x)$ is solvable by radicals over F.

16.5 CONSTRUCTIONS BY RULER AND COMPASS

There are several problems, open for several years, in Euclidean geometry and some of these are the following:

1. Can we construct by ruler and compass a square having the same area as that of a given square?
2. Can we construct by ruler and compass a cube having twice the volume of a given cube?
3. Can an angle be trisected using ruler and compass?
4. Can we construct a regular polygon having n sides using ruler and compass?

Mathematicians proved that all of the above impossible using the techniques of Galois theory. In this section, we present proofs of the impossibilities of the above.

Before we take up the proofs, let us be clear that a *ruler* (or *straight edge*) is an instrument through which we can draw a line segment joining two given points in the Euclidean space and that a *compass* is an instrument by which we can draw a circle with a given point as centre and passing through another given point. Also, let us understand that 'construction by using ruler and compass' means 'construction by using ruler and compass only in finite number of steps'.

Let us imagine that we are given a line segment which we shall define to be one unit in length.

Definition 16.5.1. A real number a is said to be *constructible* if one can construct a line segment of length $|a|$ in a finite number of steps from the given line segment of unit length by using ruler and compass alone.

Recall that by using ruler and compass, it is possible to draw a perpendicular to a given straight line at the given point on the line and to draw a line passing through a given point and parallel to a given line.

Theorem 16.5.1. If a and b are constructible real numbers, then so are $a+b$, $a-b$, ab and a/b if $b \neq 0$.

Proof: Let a and b be constructible real numbers. We can suppose that a and b are not zero. Then, there are line segments of lengths $|a|$ and $|b|$ available to us. Now, extend the line segment of length $|a|$, and lay off on the extension the length b with compass. This constructs a line segment of length $a + b$. Similarly, we construct $a - b$. Notice that, clearly $-a$ and $-b$ are constructible.

Draw a line OA of length $|a|$ and extend it. Draw a line through O not containing A. Suppose OB is of length $|b|$. Take the point P such that OP is of unit length, by using the compass. Draw a line parallel to PA and passing through the point B which cuts the line OA at Q. By the property of similar triangles, we get that

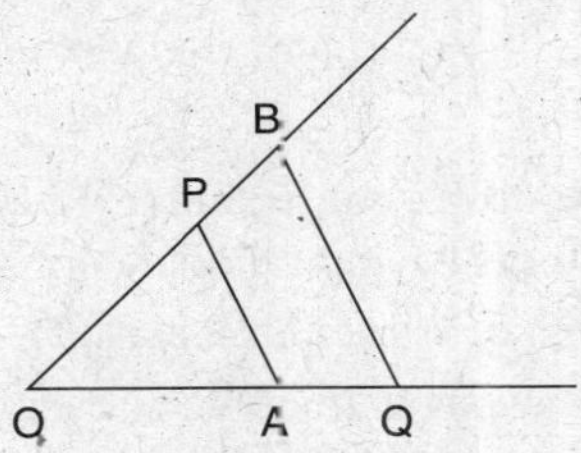

$$\frac{1}{|a|} = \frac{|b|}{|OQ|}$$

and hence $|OQ| = |a||b| = |ab|$. Thus, ab is constructible. Next, let $|OA| = |a|$ and draw a line OB of length $|b|$ through O not containing A. Suppose P is a point on OB such that OP is of unit length. Draw BA and a line parallel to BA passing through P which cuts the line OA at Q, say. Then, again by a property of similar triangles, we have $|OQ| = \frac{|a|}{|b|} = \left|\frac{a}{b}\right|$. Thus, a/b is constructible. ◄

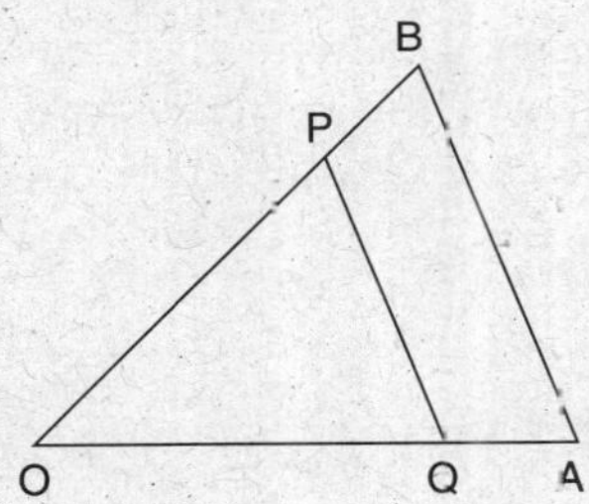

Corollary 16.5.1. The set of all constructible real members forms a subfield F of the field $\mathbb{R}$ of real numbers.

Since $\mathbb{Q}$ is the prime subfield of $\mathbb{R}$, it follows that the field F of constructible real numbers is a field extension of $\mathbb{Q}$. Now, in the two-dimensional Euclidean plane, we can locate any point (q_1, q_2) whose both coordinates are rational numbers.

Definition 16.5.2. A point (a, b) is said to be a *constructible point* if both a and b are constructible real numbers. A line is said to be *constructible* if it passes through two distinct constructible points. A circle is said to be *constructible* if its centre is a constructible point and it passes through another constructible point (or equivalently, its radius is a constructible real number).

Theorem 16.5.2. Let $a \in \mathbb{R}$ be constructible. Then, there exists a subfield K of $\mathbb{R}$ containing a such that $[K : \mathbb{Q}] = 2^n$ for some nonnegative integer n.

Proof: First notice that starting from $\mathbb{Q} \times \mathbb{Q}$, any further point in the plane which can be located by using ruler and compass can be found in one of the following three ways:

1. As an intersection of two constructible lines (that is, lines passing through two given points having rational coordinates).
2. As an intersection of a constructible line and a constructible circle.
3. As an intersection of two constructible circles.

Equations of lines and circles of the type mentioned in (1), (2) and (3) above are of the form.

$$ax + by + c = 0$$
$$\text{and} \quad x^2 + y^2 + dx + ey + f = 0,$$

where a, b, c, d, e and f are all rational numbers. It is clear that, for case (1) above, a simultaneous solution of two linear equations with rational coefficients can only lead to rational values of x and y, which give us a new constructible point. For case (2) above, upon substitution in a quadratic equation using linear equation and when solved by the quadratic formula, gives solutions involving square roots of numbers which are possibly not square in $\mathbb{Q}$. In case (3), the intersection of two circles, whose equations are

$$x^2 + y^2 + d_1x + e_1y + f_1 = 0$$
$$\text{and} \quad x^2 + y^2 + d_2x + e_2y + f_2 = 0,$$

is same as the intersection of the first circle and the line whose equation is

$$(d_1 - d_2)x + (e_1 - e_2)y + (f_1 - f_2) = 0$$

which reduces to case (2). In any case, the new constructed numbers lie in a field $\mathbb{Q}(\sqrt{a})$ for some $0 < a \in \mathbb{Q}$. If E is the smallest field containing those real numbers constructed so far, the above argument shows that the next new number constructed, lie in a field $E(\sqrt{a})$ for some $a \in H$. Therefore, starting from $\mathbb{Q}$, we get

$\mathbb{Q}_1$ = the set of constructible numbers from $\mathbb{Q}$ in the above way.

In a similar way, we get $\mathbb{Q}_2$ from $\mathbb{Q}_1$ and so on. In this way, we get an ascending chain of subfields of $\mathbb{R}$

$$\mathbb{Q} = \mathbb{Q}_0 \subseteq \mathbb{Q}_1 \subsetneq \mathbb{Q}_2 \subseteq \dots$$

It follows from all these that, if a real number a is constructible from $\mathbb{Q}$, then there exists an ascending chain of subfields of $\mathbb{R}$

$$\mathbb{Q} = \mathbb{Q}_0 \subseteq \mathbb{Q}_1 \subseteq \mathbb{Q}_2 \subseteq \cdots \subseteq \mathbb{Q}_m$$

satisfying the following:

1. $a \in \mathbb{Q}_m$
2. $\mathbb{Q}_i = \mathbb{Q}_{i-1}(a_i)$, where $a_i^2 \in \mathbb{Q}_{i-1}$ for $1 \leq i \leq m$
3. $[\mathbb{Q}_i, \mathbb{Q}_{i-1}] \leq 2$.

Thus, it follows that $[\mathbb{Q}_m : \mathbb{Q}] = 2^n$ for some $0 \leq n \leq m$ and $\mathbb{Q}_m$ is the required subfield K of $\mathbb{R}$ containing a. ◀

Corollary 16.5.2. If $a \in \mathbb{R}$ is a constructible number, then

$$[\mathbb{Q}(a) : \mathbb{Q}] = 2^r \quad \text{for some } 0 \leq r \in \mathbb{Z}.$$

In fact, the converse of the above result is also true. But before this, we first prove the following theorem.

Theorem 16.5.3. If $a > 0$ is constructible, then so is $\sqrt{a}$.

Proof: Let OA be a line segment of length a and find a point P on extended OA so that OP is of unit length. Find the midpoint of PA and draw a semicircle with PA as diameter. Draw a perpendicular at O intersecting the semicircle at Q. Now, the triangles OPQ and OQA are similar ($\angle OPQ = \angle OQA$ and $\angle OAQ = \angle PQO$) and hence we have

$$\frac{OQ}{a} = \frac{OQ}{OA} = \frac{OP}{OQ} = \frac{1}{OQ}.$$

Therefore, OQ is of length $\sqrt{a}$. Thus, $\sqrt{a}$ is constructible. ◀

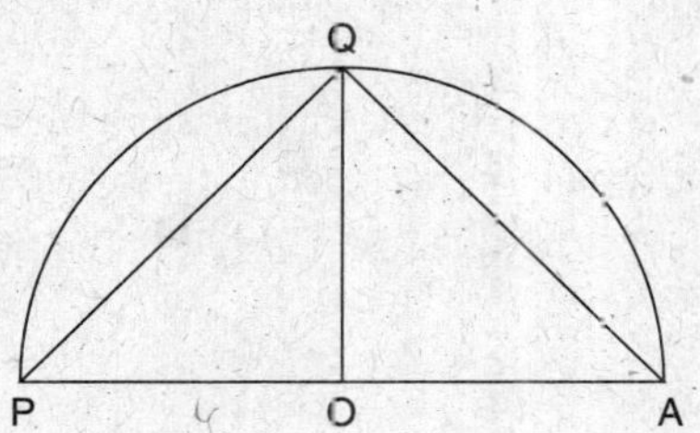

Corollary 16.5.3. Let K be the subset of $\mathbb{R}$ consisting of numbers constructible from $\mathbb{Q}$. Then, K is a subfield of $\mathbb{R}$ containing $\mathbb{Q}$ and square roots of all nonnegative numbers in K.

Theorem 16.5.4. Let $a \in \mathbb{R}$ such that $[\mathbb{Q}(a) : \mathbb{Q}] = 2^r$ for some $0 \leq r \in \mathbb{Z}$. Then, a is constructible.

Proof: Using the fundamental theorem of Galois theory, we can get a_1, a_2, ..., a_r such that

$$\mathbb{Q} \subseteq \mathbb{Q}(a_1) \subseteq \mathbb{Q}(a_1, a_2) \subseteq \cdots \subseteq \mathbb{Q}(a_1, a_2, \ldots, a_r) = \mathbb{Q}(a),$$

where $[\mathbb{Q}(a_1, \ldots, a_i) : \mathbb{Q}(a_1, \ldots, a_{i-1})] = 2$ and $a_1, a_2, \ldots, a_r$ are constructible. Therefore, a is constructible. ◀

Definition 16.5.3. An angle θ is said to be constructible by ruler and compass if the point $(\cos\theta, \sin\theta)$ is constructible from $\mathbb{Q} \times \mathbb{Q}$.

Now, we prove the impossibilities of certain geometric constructions mentioned in the beginning of this section.

Theorem 16.5.5. Doubling the cube is impossible; that is, when a cube is given, it is not always possible to construct with ruler and compass alone the side of a cube which has double the volume of the given cube.

Proof: Without loss of generality, we can assume that the side of the given cube is of unit length and hence its volume is 1. Let the side of the cube to be constructed, if possible, be x. Then, $x^3 = 2$ (double the volume of the given cube) and hence $x^3 - 2 = 0$. Therefore, we have to construct the number $2^{1/3}$. But $x^3 - 2$ is an irreducible polynomial over $\mathbb{Q}$ and hence $[\mathbb{Q}(2^{1/3}) : \mathbb{Q}] = 3$ which is not of the form 2^n. Thus, by Corollary 16.5.2, $2^{1/3}$ is not constructible by ruler and compass alone. ◀

Theorem 16.5.6. Squaring a circle is impossible; that is, when a circle is given, it is not always possible to construct with ruler and compass alone a square having area equal to the area of the given circle.

Proof: Without loss of generality, we can assume that the given circle has radius 1 and hence its area is π. We need to construct a square of side $\sqrt{\pi}$. But π is not algebraic over $\mathbb{Q}$ (the proof of this is not given in this book) and hence π is not algebraic over $\mathbb{Q}$. Therefore, $[\mathbb{Q}(\sqrt{\pi}) : \mathbb{Q}]$ is not finite. Thus, $\sqrt{\pi}$ is not constructible. ◀

Theorem 16.5.7. Trisecting the angle is impossible; that is, there exists an angle which cannot be trisected with ruler and compass alone.

Proof: The adjoining figure indicates that an angle θ can be constructed if and only if a segment of length $|\cos\theta|$ can be constructed. Now, 60° is a constructible angle and we shall show that it cannot be trisected. If 60° can be trisected, then the number $\cos 20^\circ$ is constructible from $\mathbb{Q}$. Let $a = \cos 20^\circ$. Since

$$\cos 3\theta = 4\cos^3\theta - 3\cos\theta \quad \text{for any } \theta,$$

we get that $\cos 60^\circ = 4\cos^3 20^\circ - 3\cos 20^\circ$ and hence

$$8a^3 - 6a - 1 = 0 \quad \left(\text{since } \cos 60^\circ = \frac{1}{2}\right).$$

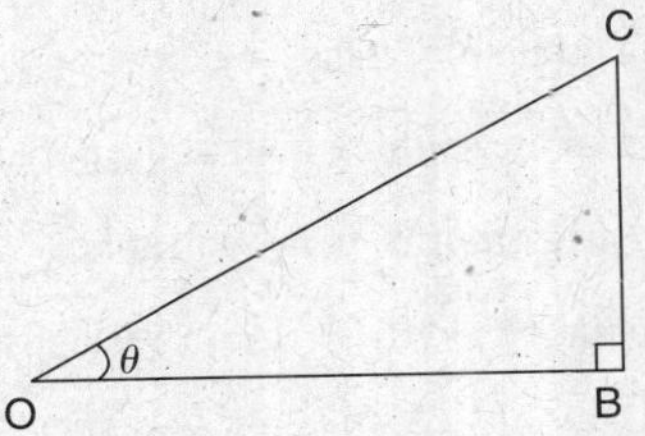

Now, the polynomial $8x^3 - 6x - 1$ is irreducible over $\mathbb{Q}$ and has a root a and therefore

$$[\mathbb{Q}(a) : \mathbb{Q}] = 3 \neq 2^r \quad \text{for any } r \in \mathbb{Z}.$$

This implies that a is not constructible. Thus, 20° is not constructible and hence 60° cannot be trisected. ◀

Theorem 16.5.8. It is impossible to construct a regular 7-gon by ruler and compass alone.

Proof: Suppose, if possible, that a regular polygon of 7 sides (regular 7-gon) is constructible. Then, the angle $(\pi/7)$ is constructible. Let $a = 2\cos(2\pi/7)$. Then, a is a constructible number. Put $\theta = (2\pi/7)$. Then,

$$\sin 4\theta = -\sin 3\theta$$

and hence $8\cos^3\theta + 4\cos^2\theta - 4\cos\theta - 1 = 0$. Therefore,

$$a^3 + a^2 - 2a - 1 = 0.$$

Then, a is a root of the polynomial

$$x^3 + x^2 - 2x - 1$$

which is irreducible over $\mathbb{Q}$. Therefore,

$$[\mathbb{Q}(a) : \mathbb{Q}] = 3 \neq 2^r \quad \text{for any } r \in \mathbb{Z},$$

which is a contradiction to our assumption that a regular 7-gon is constructible. Thus, a regular 7-gon is not constructible. ◀

EXERCISES 16

1. Prove that the identity map is the only automorphism of the field $\mathbb{R}$ of real number and hence the Galois group of $\mathbb{R}/\mathbb{Q}$ is trivial.
2. Prove that the Galois group of $\mathbb{C}/\mathbb{R}$ is a cyclic group of order 2.
3. Find the Galois group of the splitting field of the polynomial $x^4 - x^2 + 1$ over $\mathbb{Q}$.
4. Prove that the symmetric group S_5 is not solvable.
5. Prove that the polynomial $x^7 - 10x^5 + 15x + 5$ is not solvable by radicals over $\mathbb{Q}$.
6. Prove that $x^3 + x^2 - 2x - 1$ is irreducible over $\mathbb{Q}$.
7. Prove that it is impossible to construct a regular 9-gon using ruler and compass alone.
8. Prove that a regular 17-gon is constructible with ruler and compass alone.

Answers/Hints to Selected Even-Numbered Exercises

CHAPTER 1

EXERCISE 1(A)

2. (i) $\{0, 1, 2, 3, 4, 5\}$
   (ii) $\{(-1, 0) \cup (0, 1)\} \cap \mathbb{Q}$
   (iii) $\{\varnothing, \{a\}, \{c\}, \{a, c\}\}$
   (iv) $\{012, 013, \ldots, 019, 023, \ldots, 029, \ldots, 089, 123, 124, \ldots, 129, 134, \ldots, 139, \ldots\}$
   (v) $\{(0, 0)\}$
   (vi) $\varnothing$, the empty set.

4. $X = \{1, 2, 3\}$, $\mathbb{P}(X) = \{\varnothing, \{1\}, \{2\}, \{3\}, \{1, 2\}, \{2, 3\}, \{1, 3\}, X\}$

6. $X = \{1, 2, 3, \ldots, 100\}$

   $A = \{1, 4, 9, 16, 25, 36, 49, 64, 81, 100\}$ and $B = \{1, 3, 5, \ldots, 99\}$

   $C_i = \{i, i + 1, i + 2, i + 3, i + 4\}$ for each $1 \leq i \leq 96$
   (i) $A \cap B = \{1, 9, 25, 49, 81\}$
   (ii) $A \cup B \cup C_2 = \{1, 2, 3, 4, 5, 6, 7, 9, 11, 13, 15, 16, 17, \ldots\}$
   (iii) $\bigcup_{i=1}^{96} C_i = X$ and $X \cap A = A$
   (iv) $B \cap (\bigcup_{i=20}^{25} C_i) = \{21, 23, 25, 27, 29\}$
   (v) $X - (A \cup B) = \{2, 6, 8, 10, 12, 14, 18, 20, 22, 24, 26, 28, \ldots\}$
   (vi) $X - (\bigcup_{i=6}^{90} C_i) = \{1, 2, 3, 4, 5, 95, 96, 97, 98, 99, 100\}$
   (vii) $\varnothing$
   (viii) $A - B = \{4, 16, 36, 64, 100\}$

8. Straight forward verifications.

10. (i) $A \oplus B = (A - B) \cup (B - A) = (B - A) \cup (A - B) = B \oplus A$

(ii) $(A \oplus B) \oplus C = [((A - B) \cup (B - A)) - C] \cup [C - ((A - B) \cup (B - A))]$

$= \dots$

$= (A \cap B \cap C) \cup ((A - B) - C) \cup ((B - C) - A) \cup ((C - A) - B)$

The other equality is by symmetry.

12. (i) True; since $A \subseteq X \cap Y \Leftrightarrow A \subseteq X$ and $A \subseteq Y$.

(ii) False; for, if $X = \{1, 2\}$ and $Y = \{3, 4\}$, then the set $A = \{2, 3\}$ belongs $\mathbb{P}(X \cup Y)$, but A is in neither $\mathbb{P}(X)$ nor $\mathbb{P}(Y)$.

(iii) False; for the empty set $\varnothing \in \mathbb{P}(X - Y)$, but $\varnothing \notin \mathbb{P}(X) - \mathbb{P}(Y)$.

(iv) True; since $X \in \mathbb{P}(X) = \mathbb{P}(Y) \Rightarrow X \subseteq Y$.

EXERCISE 1(B)

2. (i) Clearly, $A = B \Rightarrow A - C = B - C$.

But $A - C = B - C \nRightarrow A = B$; for, if A and B are any subsets of C, then $A - C = \varnothing = B - C$.

(ii) $A \cap C = B \cap C \nRightarrow A = B$; for, if C is any subset of $A \cap B$, then $A \cap C = C = B \cap C$.

(iii) $A \cup C = B \cup C \nRightarrow A = B$; for, if A and B are any subsets of C, then $A \cup C = C = B \cup C$.

(iv) $(A - B) \times (C - D) \subseteq (A \times C) - (B \times D)$. The other inclusion may not be true; for, take $A = \{1, 2\}$, $B = \{3, 4\}$, $C = \{2, 3\}$ and $D = \{2, 5\}$. Then, $(1, 2) \in (A \times C) - (B \times D)$ and $(1, 2) \notin (A - B) \times (C - D)$, since $2 \notin C - D$.

(v) $(A \times C) \cup (B \times D) \subseteq (A \cup B) \times (C \cup D)$. The other inclusion may not be true; for, in (iv) above, $(2, 5) \in (A \cup B) \times (C \cup D)$, but $(2, 5) \notin (A \times C) \cup (B \times D)$.

4. For each $a \in A$, there are exactly m b's in B such that $(a, b) \in A \times B$. Since A is n elements, $A \times B$ has nm elements. Also, the number of relations from A to B is 2^{nm}.

6. The number of functions from A into B is m^n.

8. If $f(a) = f(b)$, then $a = b$ (since $f|\{a, b\}$ is an injection). Therefore, f is an injection if $f|Z$ is an injection for any subset Z of X. The converse is clear. This statement is not valid for surjections. The map $f: \mathbb{R} \to \mathbb{R}^+ \cup \{0\}$

given by $f(a) = |a|$ is a surjection, but the restriction of f to (0, 1) is not a surjection.

10. Yes; the number of such functions g is equal to the number of functions from $X - A$ into Y.

12. $f^{-1}[-2, 8) = (-\sqrt{8}, \sqrt{8})$;

    $f^{-1}(-\infty, 0] = \{0\}; f^{-1}(-1, 1) = (-1, 1)$;

    $f^{-1}(\mathbb{Z}) = \{\pm\sqrt{n} : 0 \leq n \in \mathbb{Z}\}$.

14. (1) If $g : B \to A$ such that $g \circ f = I_A$, then, for any $x, y \in A$,

$$f(x) = f(y) \Rightarrow g(f(x)) = g(f(y)) \Rightarrow x = y$$

and hence f is an injection, conversely suppose that f is an injection. Choose $a_0 \in A$ and define $g : B \to A$ by

$$g(b) = \begin{cases} a & \text{if } b = f(a), a \in A \\ a_0 & \text{if } b \notin f(A). \end{cases}$$

Then, g is a function and $g \circ f = I_A$.

(2) If $f \circ h = I_B$, then, for any $b \in B, f(h(b)) = b$ and $h(b) \in A$ and hence f is a surjection. Conversely, suppose that $f : A \to B$ is a surjection. For each $b \in B, f^{-1}(\{b\})$ is a nonempty subset of A and choose an element $a_b \in f^{-1}(\{b\})$ for each $b \in B$. Then, $h : B \to A$ defined by $h(b) = a_b$ is a function and $f(h(b)) = b$ for all $b \in B$.

16. Straight forward verification.

18. Define $f : \mathbb{Z} \to \mathbb{R}$ by $f(n) = |n|$ for all $n \in \mathbb{Z}$. If B is the interval (0, 1) in $\mathbb{R}$, then $f^{-1}(B) = \emptyset$ and $f(f^{-1}(B)) = f(\emptyset) = \emptyset \subset B$.

20. (i) $a \in A \Rightarrow f(a) \in f(A) \Rightarrow a \in f^{-1}(f(A))$.

    Therefore, $A \subseteq f^{-1}(f(A))$ and hence $f(A) \subseteq f(f^{-1}(f(A)))$.

$$f(f^{-1}(B)) \subseteq B \quad \text{for all } B \subseteq Y$$

and hence $f(f^{-1}(f(A))) \subseteq f(A)$. Thus, $f(f^{-1}(f(A))) = f(A)$.

(ii) Proof is similar to (i).

(iii) Clear.

22. For any a and b in X,

$$f(a) = f(b) \Rightarrow (g \circ f)(a) = (g \circ f)(b) \Rightarrow a = b$$

(since $g \circ f$ is an injection). Therefore, f is an injection. If f is a surjection also, then f is a bijection and, for any x and y in Y, choose a and $b \in X$ such that $f(a) = x$ and $f(b) = y$. Now,

$$g(x) = g(y) \Rightarrow g(f(a)) \Rightarrow g(f(b)) \Rightarrow a = b \Rightarrow x = y.$$

Thus, g is an injection.

24. (i) $n!$

(ii) 0

(iii) $mC_n \cdot n!$

26. (1) $\Rightarrow$ (2): Clearly, $A \subseteq f^{-1}(f(A))$. If $x \in f^{-1}(f(A))$, then $f(x) \in f(A)$ and hence $f(x) = f(a)$ for some $a \in A$. Since f is an injection $x = a \in A$. Thus, $f^{-1}(f(A)) = A$.

(2) $\Rightarrow$ (3): Clearly, $f(A_1 \cap A_2) \subseteq f(A_1) \cap f(A_2)$.

$$\begin{aligned} x \in f(A_1) \cap f(A_2) &\Rightarrow x = f(a_1) = f(a_2) \quad \text{for some } a_i \in A_i \\ &\Rightarrow a_1 \in f^{-1}(f(\{a_2\})) = \{a_2\} \text{ (by (2))} \\ &\Rightarrow a_1 = a_2 \in A_1 \cap A_2 \quad \text{and} \quad x = f(a_1) \in f(A_1 \cap A_2) \end{aligned}$$

Thus, $f(A_1) \cap f(A_2) = f(A_1 \cap A_2)$.

(3) $\Rightarrow$ (1): For any a and $b \in X$,

$$\begin{aligned} f(a) = f(b) &\Rightarrow f(a) \in f(\{a\}) \cap f(\{b\}) = f(\{a\} \cap \{b\}) \\ &\Rightarrow a = b \text{ (otherwise, } \{a\} \cap \{b\} = \varnothing) \end{aligned}$$

28. Consider $(f_1 \circ f_2 \circ \cdots \circ f_n) \circ (f_n^{-1} \circ f_{n-1}^{-1} \circ \cdots \circ f_1^{-1})$.

30. (i) $\Rightarrow$ (ii): If f has a left inverse, then f is an injection (by Ex. 14 of Exercise 1(B)). Suppose that $f(X) \neq Y$. Choose $x_1 \neq x_2 \in X$ and define

$$g_1, g_2 : Y \to X \text{ by } g_1(y) = \begin{cases} x & \text{if } y = f(x),\ x \in X \\ x_1 & \text{if } y \notin f(X) \end{cases}$$

$$\text{and} \quad g_2(y) = \begin{cases} x & \text{if } y = f(x), x \in X \\ x_2 & \text{if } y \notin f(X) \end{cases}.$$

Then, $g_1 \neq g_2$ (since $Y - f(X) \neq \varnothing$) and $g_1 \circ f = I_X = g_2 \circ f$, which is a contradiction to the uniqueness of the left inverse of f.

(ii) $\Rightarrow$ (i) and (ii) $\Rightarrow$ (iii) are clear.

(iii) $\Rightarrow$ (ii): If f has a right inverse, then f is a surjection. (by Ex. 14 of Exercise 1(B)). Suppose a and $b \in X$ such that $f(a) = f(b) = y_0$, say. For each $y \neq y_0$ in Y, choose $x_y \in X$ such that $f(x_y) = y$. Define $h_1, h_2 : Y \to X$ by

$$h_1(y) = \begin{cases} x_y & \text{if } y \neq y_0 \\ a & \text{if } y = y_0 \end{cases} \quad \text{and} \quad h_2(y) = \begin{cases} x_y & \text{if } y \neq y_0 \\ b & \text{if } y = y_0 \end{cases}.$$

Then, $f \circ h_1 = I_Y = f \circ h_2$ and hence, by (iii), $h_1 = h_2$ so that $a = b$. Thus, f is an injection also.

32. If a, b and c are three distinct elements of X, define f and $g : X \to X$ by

$$g(x) = \begin{cases} a & \text{if } x = b \\ b & \text{if } x = a \\ x & \text{otherwise} \end{cases} \quad \text{and} \quad f(x) = \begin{cases} a & \text{if } x = c \\ c & \text{if } x = a \\ x & \text{otherwise} \end{cases}$$

Then, $(g \circ f)(a) = g(c) = c \neq b = f(b) = (f \circ g)(a)$ and hence $g \circ f \neq f \circ g$.

34. Define g and $h : \mathbb{P}(X) \to \mathbb{P}(\mathbb{P}(X))$ by

$$g(A) = \mathbb{P}(A) \text{ and } h(A) = \{\{a\} : a \in A\} \quad \text{for any } A \in \mathbb{P}(X).$$

Then, $(f \circ g)(A) = A = (f \circ h)(A)$; that is, g and h are distinct right inverses of f (note that $\varnothing \in g(A)$ and $\varnothing \notin h(A)$).

36. (i) Follows from 10 of Exercise 1(B).

(ii) Suppose that η is a bijection and $A \neq X$. Let $y_1 \neq y_2 \in Y$. Choose $x_0 \in X - A$ and define f and $g : X \to Y$ by

$$f(x) = y_1 \quad \text{for all } x \in X \text{ and } g(x) = \begin{cases} y_1 & \text{if } x \in A \\ y_2 & \text{if } x \notin A \end{cases}.$$

Then, $f(x_0) = y_1 \neq y_2 = g(x_0)$. But $\eta(f) = \eta(g)$ and $f \neq g$. Therefore, η is a bijection implies that $A = X$.

38. This follows from Ex. 37.

EXERCISE 1(C)

2. Let $X = \{a, b, c\}$ and

$$R = \{(a, b), (b, a)\}$$
$$S = \{(a, a), (b, b), (c, c), (a, b), (b, c)\}$$
$$T = \{(a, b), (b, c), (a, c)\}.$$

Then, R is symmetric, but neither reflexive on X nor transitive. S is reflexive, but neither transitive nor symmetric. T is transitive, but neither symmetric nor reflexive.

4. Easy verification.

6. Since $a + b = b + a$, $((a, b), (a, b)) \in R$ and hence R is reflexive on X.

$a + d = b + c \Rightarrow c + b = d + a$. Therefore, R is symmetric.

$$\begin{aligned}((a, b), (c, d)) \in R \text{ and } ((c, d), (e, f)) \in R \\ &\Rightarrow a + d = b + c \text{ and } c + f = d + e \\ &\Rightarrow a + d + f = b + c + f = b + d + e \\ &\Rightarrow a + f = b + e \Rightarrow ((a, b), (e, f)) \in R\end{aligned}$$

Therefore, R is transitive. Thus, R is an equivalence relation on X.

8. (i) $(x, y) \in R \Rightarrow (a \in R(x) \Leftrightarrow (a, x) \in R \Leftrightarrow (a, y) \in R \Leftrightarrow a \in R(y)$ and hence $R(x) = R(y))$

$$R(x) = R(y) \Rightarrow x \in R(y) \Rightarrow (x, y) \in R$$

(ii) $R(x) \cap R(y) \neq \varnothing \Leftrightarrow (a, x) \in R$ and $(a, y) \in R$ for some $a \in X$

$$\Leftrightarrow (x, y) \in R$$

(iii) is trivial.

10. Straight forward verifications. In Exercise 5 above, X/R is bijective with the set $\mathbb{Q}$ of rational numbers.

12. Suppose that the given conditions are satisfied.

$$(a, b) \in R \Rightarrow (a, b) \quad \text{and} \quad (b, b) \in R \Rightarrow (b, a) \in R.$$

Therefore, R is symmetric. Also,

$$(a, b) \in R \quad \text{and} \quad (b, c) \in R \Rightarrow (c, a) \in R \Rightarrow (a, c) \in R.$$

Thus, R is transitive also. The converse is trivial.

14. (i) $\{\mathbb{Q} + r : r \in \mathbb{R}\}$

(ii) $\{\mathbb{Z} + r : r \in \mathbb{Q}\}$

(iii) R is not reflexive and hence not an equivalence relation.

(iv) $\{\{(a, b) \in \mathbb{R}^2 : a^2 + b^2 = r\} : 0 \leq r \in \mathbb{R}\}$.

(v) is not equivalence relation (it is not symmetric).

(vi) is not reflexive, since $(1, 1) \notin \mathbb{R}$

(vii) is not reflexive, since $(2^{1/4}, 2^{1/4}) \notin \mathbb{R}$

(viii) is not transitive.

(ix) $\{\{A \oplus B : B \in F\} : A \in \mathbb{P}(X)\}$, where F is the class of all finite subsets of X.

(x) $\{r\mathbb{Q}^+ : r \in \mathbb{R}^*\}$

EXERCISE 1(D)

2. (i) If $|X| = |Z|$ and $h : X \to Z$ is a bijection, then $f \circ h^{-1}$ is an injection of Z into Y. By Theorem 1.4.5, $|Y| = |Z|$.

   (ii) If $|X| = |Y|$ and $f(X) \subseteq A \subseteq Y$, then

$$|X| = |f(X)| \leq |A| \leq |Y| = |X|$$

   and hence $|A| = |Y|$.

   (iii) $X \subseteq A \subseteq Y \Rightarrow |X| \leq |A| \leq |Y|$

4. If $g : \mathbb{Z}^+ \to X$ is a bijection, then $f \circ g : \mathbb{Z}^+ \to Y$ is a surjection. By Theorem 1.4.5, Y is at most countable.

6. Follows from Corollary 1.4.8.

8. $|X| < |\mathbb{P}(X)| < |\mathbb{P}(\mathbb{P}(X))|$

10. $|\mathbb{P}(X)| = |2^X| = |2^Y| = |\mathbb{P}(Y)|$, where $2 = \{0, 1\}$.

12. $\mathscr{C} = \{C : C$ is a circle with radius for $1/n$ some $n \in \mathbb{Z}^+\}$.

14. Define $f: \mathbb{Z}^+ \to \mathbb{Z}^+ - I_n$ by $f(a) = a + n$. Then, f is a bijection.

16. Let P_n be the set of polynomials over $\mathbb{Q}$ of degree $< n$. Then, $P_n \simeq \mathbb{Q}^n$ and the set P of all polynomials over $\mathbb{Q}$ is the unions of all P_n, $n \in \mathbb{Z}^+$. $|P_n| = |\mathbb{Q}^n| = |\mathbb{Q}| = |\mathbb{Z}^+|$ and $|P| = |\bigcup_{n \in \mathbb{Z}^+} P_n| = |\mathbb{Z}^+|$.

18. If A and T are the sets of algebraic numbers and transcendental numbers, respectively, then $A \cup T = \mathbb{R}$. If T is countable, than $A \cup T$ $(= \mathbb{R})$ is countable, since A is countable.

20. The given set is bijective with $\mathbb{Q} \times \mathbb{Q} \simeq \mathbb{Z}^+$.

CHAPTER 2

EXERCISE 2(A)

2. Apply second principle of induction (Theorem 2.1.4) to prove that $2n + 1 < 2^{n-1}$ for all $5 \leq n \in \mathbb{Z}^+$.

4. This is trivial when $n = 1$ or $r = n$. Use Theorem 2.1.3 and the identity $\binom{n}{r} + \binom{n}{r+1} = \binom{n+1}{r+1}$ for all $1 \leq r < n$.

6. If $p_1, p_2, \ldots, p_n$ are all the primes, then, by Theorem 2.1.6, there exists a prime p $(= p_i)$ dividing $p_1 p_2 \cdots p_n + 1$, so that p divides 1, which is an absurd.

8. Let A be the set of all positive integers which cannot be expressed in the required form. If A is nonempty, then A has smallest member, say a. Since a cannot be prime and $a > 1$ (note that $1 \notin A$), $a = bc$, where $1 < b < a$ and $1 < c < a$. By the smallest property of a, b and $c \notin A$ and hence $a \notin A$, a contradiction. Therefore, A is empty.

10. Use Corollary 2.1.3 and Ex. 8 above.

12. Let $a = p_1^{r_1} p_2^{r_2} \cdots p_t^{r_t}$ and $b = p_1^{s_1} p_2^{s_2} \cdots p_t^{s_t}$, where p_i's are distinct primes and $r_i \geq 0$, $s_i \geq 0$.

    By interchanging p_i's if necessary, we can assume that $r_i \leq s_i$ for $1 \leq i \leq k$ and $s_i \leq r_i$ for $k < i < t$.

    Then, l.c.m. $\{a, b\} = p_1^{s_1} \cdots p_k^{s_k} p_{k+1}^{r_{k+1}} \cdots p_t^{r_t}$

    and g.c.d. $\{a,b\} = p_1^{r_1} \cdots p_k^{r_k} p_{k+1}^{s_{k+1}} \cdots p_t^{s_t}$.

    Now, $ab = \prod_{i=1}^{t} p_i^{r_i+s_i} =$ l.c.m. $\{a, b\} \cdot$ g.c.d. $\{a, b\}$.

14. Suppose that $n > 1$ and n is not a prime. Let p be the least prime dividing n. Then, $pm = n$ for some $m > 1$. Any prime dividing m should divide n and hence $p \leq m$. Now, $p^2 \leq pm = n$ and hence $p \leq \sqrt{n}$.

16. Follows from the fact that a positive integer n divides a and b if and only if n divides b and d (since $a = bc + d$ and $a - bc = d$).

18. Use Theorem 2.1.8.

20. $1273 = 969 + 304$

    $969 = 3 \cdot 304 + 57$

    $304 = 5 \cdot 57 + 19$

    $57 = 3 \cdot 19$.

    Therefore, g.c.d. $\{969, 1273\} = 19 = 304 - 5 \cdot 57$

    $$= 304 - 5 \cdot (969 - 3 \cdot 304)$$
    $$= 16 \cdot 304 - 5 \cdot 969$$
    $$= 16(1273 - 969) - 5 \cdot 969$$
    $$= 16 \cdot 1273 - 21 \cdot 969; \; x = -21$$

    and $y = 16$

22. Use Theorem 2.1.8.

24. As in Ex. 12 above,

$$c = \frac{a}{(a, b)} = \prod_{i=k+1}^{t} p_i^{r_i - s_i}$$

and $$d = \frac{b}{(a, b)} = \prod_{i=1}^{k} p_i^{s_i - r_i}.$$

Since $p_1, \ldots, p_k, \ldots, p_t$ are all distinct primes, $(c, d) = 1$.

EXERCISE 2(B)

2. (i) $\{x : 3x \equiv 5 \pmod 7\}$ (since 5 divides 15, 25 and 35)
   $= \{x : x \equiv 5 \cdot 5 \pmod 7\}$ (since $5 \cdot 3 \equiv 1 \pmod 7$)
   $= \{x : x \equiv 4 \pmod 7\} = 4 + 7\mathbb{Z}$
   (ii) $4 + 7\mathbb{Z}$
   (iii) The empty set, by Theorem 2.2.6
   (iv) $7 + 8\mathbb{Z}$
   (v) $12 + 17\mathbb{Z}$
   (vi) Empty set, by Theorem 2.2.6.

4. Use Theorem 2.2.8 for (i) and (ii) and, for others, use Theorem 2.2.9 (note that $m_7 = 5$, $m_{11} = 10$, $m_{17} = 12$, $m_{19} = 2$, $m_{23} = 7$ and $m_{29} = 3$).

6. (i) For any $i \neq j \in \{0, 1, \ldots, n-1\}$, $i \not\equiv j \pmod n$ and, since $a_i \in i + n\mathbb{Z}$, $a_i \equiv i \pmod n$ and hence $a_i \equiv a_j \pmod n$.
   (ii) Clear.
   (iii) For any $a \in \mathbb{Z}$, choose i such that $a = qn + i$, $0 \leq i \leq n - 1$ and then $a \equiv i \equiv a_i \pmod n$ and hence $a \equiv a_i \pmod n$.

8. Let $m, m + 1, \ldots, m + (n - 1)$ be n consecutive integers, where $m \in \mathbb{Z}$. Choose $0 \leq i \leq n - 1$ such that $m \equiv i \pmod n$. Then, $m + 1 \equiv i + 1 \pmod n$, $\ldots$, $m + k \equiv i + k \pmod n$ for all k. This implies that $\{m, m + 1, \ldots, m + (n - 1)\}$ is a transversal for congruence modulo n.

10. If $n = a_1 a_2 \cdots a_r$ and $m = a_{\sigma(1)} a_{\sigma(2)} \cdots a_{\sigma(r)}$, where σ is any rearrangement of $\{1, 2, \ldots, r\}$, then 3 divides $n \Leftrightarrow$ 3 divides $\sum_{i=1}^{r} a_i = \sum_{i=1}^{r} a_{\sigma(i)} \Leftrightarrow$ 3 divides m.

12. 3 divides $12x\,527846531 \Leftrightarrow$ 3 divides $1 + 2 + x + 5 + 2 + 7 + 8 + 4 + 6 + 5 + 3 + 1$
    $\Leftrightarrow$ 3 divides $44 + x$
    $\Leftrightarrow x \in \{1, 4, 7\}$.

Similarly, 9 divides the number $\Leftrightarrow$ 9 divides $44 + x$

$$\Leftrightarrow x = 1.$$

14. (i) $a \equiv b \pmod{m, \text{mod } n} \Leftrightarrow m$ and n divide $a - b$

$$\Leftrightarrow mn \text{ divides } a - b \text{ (since } (m, n) = 1)$$

(ii) $c \equiv a, d \equiv a \pmod n \Rightarrow c \equiv d \pmod n$

Similarly, $c \equiv b \pmod m$ and $d \equiv b \pmod m \Rightarrow c \equiv d \pmod m$.

Therefore, by (i), $c \equiv d \pmod{mn}$.

EXERCISE 2(C)

2. Let $a, b \in \mathbb{Q}$, $a < b$ and $A = \{r \in \mathbb{Q} : a < r < b\}$.

   Define $f : A \to \mathbb{Q}^+$ by $f(r) = \frac{r-a}{b-r}$

   and $g : \mathbb{Q}^+ \to A$ by $g(s) = \frac{a+sb}{1+s}$.

   Then, $f \circ g = I_{\mathbb{Q}^+}$ and $g \circ f = I_A$ and hence f is a bijection. Choose $c \in \mathbb{Q}$ such that $a < c < b$. Then, $(a, b) = (a, c) \cup c \cup (c, b) \simeq \mathbb{Q}^- \cup \{0\} \cup \mathbb{Q}^+ = \mathbb{Q}$ (since $(a, c) \simeq \mathbb{Q}^+ \simeq \mathbb{Q}^-$ and $(c, b) \simeq \mathbb{Q}^+$).

4. Imitate the proof of Ex. 2 above to prove that $(a, b) \simeq \mathbb{R} \simeq (c, d)$ and $(a, b)_{\mathbb{Q}} \simeq \mathbb{Q} \simeq (c, d)_{\mathbb{Q}}$.

6. Let $\epsilon > 0$ be given. Choose $n_0 \in \mathbb{Z}^+$ such that $(2/\epsilon) < n_0$. Then, for any $n, m \geq n_0$, we have

$$\left|\frac{1}{n} - \frac{1}{m}\right| \leq \frac{1}{n} + \frac{1}{m} \leq \frac{1}{n_0} + \frac{1}{n_0} < \frac{\epsilon}{2} + \frac{\epsilon}{2} = \epsilon.$$

   Thus, $\{1/n\}$ is a Cauchy sequence.

8. Clearly, $\sim$ is symmetric and reflexive on CS($\mathbb{Q}$). If $|a_n - b_n| \to 0$ and $|b_n - c_n| \to 0$, then

$$0 \leq |a_n - c_n| \leq |a_n - b_n| + |b_n - c_n| \to 0.$$

   Therefore, $\sim$ is transitive also.

10. Let $r \in \mathbb{R}$. For each $n \in \mathbb{Z}^+$, choose $a_n \in \mathbb{Q}$ such that $r - (1/2n) < a_n < r + (1/2n)$. Then, $a_n \to r$ and $\{a_n\} \in$ CS($\mathbb{Q}$). If $\{b_n\} \in$ CS($\mathbb{Q}$) such that $b_n \to r$, then

$$|a_n - b_n| \leq |a_n - r| + |r - b_n| \to 0$$

   and hence $\{a_n\} \sim \{b_n\}$.

12. $\{\tilde{a}\} = \{\tilde{b}\} \Rightarrow \{a\} \sim \{b\} \Rightarrow |a - b| = 0 \Rightarrow a = b.$

14. $|(a_n + b_n) - (a'_n + b'_n)| \leq |a_n - a'_n| + |b_n - b'_n| \to 0$

$$|a_n b_n - a'_n b'_n| = |a_n(b_n - b'_n) + (a_n - a'_n)b'_n|$$

$$\leq |a_n||b_n - b'_n| + |a_n - a'_n||b'_n| \to 0,$$

since we can find real numbers K and K' such that $|a_n| \leq K$ and $|b'_n| \leq K'$ for all n.

EXERCISE 2(D)

2. If $a < b$, then $b < a$ is not possible. The number of pairs (a, b) with $a < b$ can be atmost $\frac{n^2-n}{2} = \frac{n(n-1)}{2}$. Therefore, the number of partial orders on X can be at most $2^{\frac{n(n-1)}{2}}$.

4. For any $a, b \in X$, the set $\{a, b\}$ has least element, that is, $a \leq b$ or $b \leq a$.

6. Let each $(X_i, \leq)$ be well ordered. Consider

$$\varnothing \neq A \subseteq X = X_1 \times X_2 \times \cdots \times X_n.$$

Put $A_1 = \{x_1 \in X_1 : (x_1, x_2, \ldots, x_n) \in A$ for some $x_i \in X_i, i \geq 2\}$.

Then, $\varnothing \neq A_1 \subseteq X_1$ and A_1 has a least element, say a_1. Next, put

$$A_2 = \{x_2 \in X_2 : (a_1, x_2, x_3, \ldots, x_n) \in A \quad \text{for some } x_i \in X_i, i \geq 3\}$$

Then, $\varnothing \neq A_2 \subseteq X_2$ and A_2 has a least element, say a_2.

Again, consider $A_3 = \{x_3 \in X_3$ $(a_1, a_2, x_3, x_4, \ldots, x_n) \in A$ for some $x_i \in X_i$, $i \geq 4\}$. Continue this procedure to obtain $a_1, a_2, a_3, \ldots, a_n$. Then, $(a_1, a_2, \ldots, a_n)$ will be the least element in A.

EXERCISE 2(E)

2. Let A be an $m \times n$ matrix and B be an $r \times s$ matrix. AB is defined if and only if $n = r$, and BA is defined if and only if $s = m$. Therefore, AB and BA are defined if and only if B is an $n \times m$ matrix (or B^t is an $m \times n$ matrix).

4. Let A, B and C be $m \times n$, $r \times s$ and $t \times u$ matrices, respectively.

$$A(BC) \text{ is defined} \Leftrightarrow BC \text{ is defined and } A(BC) \text{ is defined}$$

$$\Leftrightarrow s = t \text{ and } n = r$$

$$\Leftrightarrow AB \text{ and } (AB)C \text{ are defined.}$$

Let $A = (a_{ij})$, $B = (b_{ij})$ and $C = (c_{ij})$ and suppose that $A(BC)$ (and hence $(AB)C$) is defined. For any $1 \leq i \leq m$ and $1 \leq j \leq u$,

$$\begin{aligned}\text{the } ij^{\text{th}} \text{ entry in } A(BC) &= \sum_{p=1}^{n} a_{ip}\left(\sum_{q=1}^{n} b_{pq}c_{qj}\right)\\ &= \sum_{q=1}^{n}\sum_{p=1}^{n} a_{ip}(b_{pq}c_{qj})\\ &= \sum_{q=1}^{n}\left(\sum_{p=1}^{n} a_{ip}b_{pq}\right)c_{qj}\\ &= \text{the } ij^{\text{th}} \text{ entry in } (AB)C.\end{aligned}$$

Thus, $A(BC) = (AB)C$. Similarly, others can be proved.

6. Let $A = (a_{ij})$ and $S_a = (s_{ij})$. Then, $s_{ij} = a$ or 0 according as $i = j$ or $i \neq j$. An ij^{th} entry in S_aA is

$$\sum_{r=1}^{n} s_{ir}a_{rj} = aa_{ij}, \text{ for any } 1 \leq i, j \leq n.$$

Therefore, $S_aA = aA$ and similarly, $AS_a = aA$.

8. (i) Let $A = (a_{ij})$ and $B = (b_{ij})$. Then, AB and $(AB)^t$ are $m \times r$ and $r \times m$ matrices, respectively. For any $1 \leq i \leq r$ and $1 \leq j \leq m$,

$$\begin{aligned} ij^{\text{th}} \text{ entry in } (AB)^t &= ji^{\text{th}} \text{ entry in } AB\\ &= \sum_{s=1}^{n} a_{js}b_{si}\\ &= \sum_{s=1}^{n} b'_{is}a'_{sj}\\ &= ij^{\text{th}} \text{ entry in } B^tA^t,\end{aligned}$$

where b'_{is} and a'_{sj} are the is^{th} and sj^{th} entries in B^t and A^t, respectively.

(ii) If B_1 and B_2 are such that $AB_1 = I$ and $B_2A = I$, then $B_1 = IB_1 = (B_2A)B_1 = B_2(AB_1) = B_2I = B_2$.

(iii) $(AB)(B^{-1}A^{-1}) = A(BB^{-1})A^{-1} = AIA^{-1} = AA^{-1} = I$ and, similarly $(B^{-1}A^{-1})(AB) = I$.

(iv) If $AB = I = BA$, then $A^tB^t = (BA)^t = I^t = \text{I} = B^tA^t$ and hence $(A^t)^{-1} = B^t = (A^{-1})^t$.

10. Straight-forward verification.

12. If $A = (a_{ij})$ is a skew-symmetric matrix, then $A = -A^t$ and hence $a_{ii} = -a_{ii}$, so that $a_{ii} = 0$.

14. Let $C = \frac{1}{2}(A + A^t)$ and $D = \frac{1}{2}(A - A^t)$. Then, C is a symmetric matrix and D is a skew-symmetric matrix. Also, $A = C + D$.

EXERCISE 2(F)

2. Use induction on the order of the matrix (if A is an upper triangular matrix, then $\det A = a_{11}A_{11}$).

4. Subtract x_1 times the $(r - 1)^{\text{th}}$ column from the rth column to obtain

$$\det A = \det\begin{pmatrix} 1 & 0 & 0 & \cdots & 0 \\ 1 & x_2 - x_1 & x_2(x_2 - x_1) & \cdots & x_2^{n-2}(x_2 - x_1) \\ \vdots & \vdots & \vdots & \vdots & \vdots \\ 1 & x_n - x_1 & x_n(x_n - 1) & \cdots & x_n^{n-2}(x_n - x_1) \end{pmatrix}$$

and now expand with respect to the 1st row, factor out $x_j - x_1$ and apply induction on n.

6. $A^2 = A \Rightarrow (\det A)^2 = \det A \Rightarrow \det A = 0$ or $\det A = 1$. If A is nonsingular, then $\det A \neq 0$.

8. $A^m = O_{n\times n} \Rightarrow (\det A)^m = 0 \Rightarrow \det A = 0$.

10. Follows from Theorems 2.6.2 and 2.6.8.

12. By Ex. 11, $\det A = -\det A$ and hence $\det A = 0$.

14. (i) $\det(AB) = \det A \cdot \det B = \det B \cdot \det A = \det(BA)$.
    (ii) $\det(A \cdot A^t) = \det A \cdot \det A^t = \det A \cdot \det A$.
    (iii) $\det(ABA^{-1}) = \det A \cdot \det B \cdot \det A^{-1}$
    $= \det A \cdot \det B \cdot (\det A)^{-1} = \det B$.

CHAPTER 3

EXERCISE 3(A)

2. In Exercise 1(A), $\mathbb{Z}_4 = \{0, 1, 2, 3\}$ and, in Exercise 1(F), $S = \{1, i, -1, -i\}$

0 is the identity	1 is the identity
$1 +_4 1 = 2$	$i \cdot i = -1$
$1 +_4 1 +_4 1 = 3$	$i \cdot i \cdot i = -i$

Both the tables are same, if we rename $+_4$ by $\cdot$, 0 by 1, 1 by i, 2 by -1, and 3 by $-i$.

4. The table in (3) is given by

*	a	b	c	d	e	f
a	d	e	a	c	a	b
b	e	f	d	c	f	d
c	a	d	c	a	d	c
d	c	c	a	e	d	b
e	a	f	d	d	b	d
f	b	d	c	b	d	a

$$(b*(d*a))*(c*(b*a)) = (b*c)*(c*e) = d*d = e$$
$$((a*b)*c)*d = (e*c)*d = d*d = e$$
$$(a*b)*(c*d) = e*a = a$$

6. Let S_3 be the set of bijections of $\{1, 2, 3\}$ onto itself and o be the composition of mappings in S_3. Then, o is associative, but not commutative; for consider f and g defined by $f(1) = 1, f(2) = 3, f(3) = 2$ and $g(1) = 2$, $g(2) = 1$ and $g(3) = 3$. Then, $(f \circ g)(1) = 3$ and $(g \circ f)(1) = 2$ and hence $f \circ g \neq g \circ f$.

8. No, the statement is false. Consider the operation * on $\{a, b\}$ given by the table

*	a	b
a	b	a
b	a	a

Here, $(a*a)*b = b*b = a$

and $a*(a*b) = a*a = b$.

Therefore, * is not associative. But it is commutative, since $a*b = b*a$.

10. Yes, the statement is true.

It is trivial when S has exactly one element. To prove the converse, define $a*b = b$ for all a and $b \in S$. If * is commutative, then $a = b$ for all a, $b \in S$ and hence $|S| = 1$.

12. The number of binary operations on an n-element set is n^{n^2}. The number of commutative binary operations on an n-element is $n^{\frac{n^2+n}{2}}$.

    $\therefore$ The number of noncommutative binary operations on an n-element set is $n^{n^2} - n^{\frac{n^2+n}{2}}$ and, on a 3-element set, it is $3^9 - 3^6 = 3^6 \times 26$.

14. By definition of $+$, $e + x = x = x + e$ for all $x \in S'$.

EXERCISE 3(B)

2. (i) All elements in $\mathbb{Z}_{10}$
   (ii) 1, 3, 7, 9
   (iii) 1, 5, 7, 11, 13, 17, 19, 23, 25, 29, 31, 35
   (iv) 1, 3
   (v) All bijections
   (vi) All elements in $\mathbb{P}(X)$
   (vii) All elements in $\mathbb{R}^+$
   (viii) 1 only.

4. Let $e = \begin{pmatrix} 1 & 0 \\ 0 & 1 \end{pmatrix}$, $a = \begin{pmatrix} -1 & 0 \\ 0 & 1 \end{pmatrix}$, $b = \begin{pmatrix} -1 & 0 \\ 0 & -1 \end{pmatrix}$ and $c = \begin{pmatrix} 1 & 0 \\ 0 & -1 \end{pmatrix}$. Then, e is the identity in $G = \{e, a, b, c\}$, $a^2 = e = b^2 = c^2$, $ab = c = ba$, $ac = b = ca$, $bc = a = cb$.

6. $x * x = x = e * x \Rightarrow x = e$.

8. The map $\overline{e} : X \to G$ defined by $\overline{e}(x) = e$ for all $x \in X$ is the identity element in G^X and, for any $f \in G^X$, the map $-f : X \to G$, defined by $(-f)(x) = -f(x)$ for all $x \in X$, is the inverse of f in G^X.

10. $\mathbb{Z}_n = \{0, 1, 2, \ldots, n - 1\}$ and 0 has no inverse in $(\mathbb{Z}_n, \cdot_n)$ and hence $(\mathbb{Z}_n, \cdot_n)$ is not a group.

12. For any $0 < m < p$, $(m, p) = 1$ and hence there exist a and $b \in \mathbb{Z}$ such that $am + bp = 1$. If we choose $r \in \mathbb{Z}^+$ such that $a = qp + r$, $0 < r < p$, then $rm + qp + bp = 1$ and hence $rm \equiv 1 \pmod{p}$, so that r is the inverse of m in $(G_p, \cdot_p)$.

14. Yes, take $p = 5$ in Ex. 12 above.

16. (1) $\Rightarrow$ (3): If n is not a prime, then $n = ab$ for some $0 < a < b$ and $0 < b < n$ and hence $a \cdot_n b = 0$, a contradiction to (1).

(3) $\Rightarrow$ (2): Follows from Ex. 12 above.

(2) $\Rightarrow$ (4): (2) $\Rightarrow$ (1) $\Rightarrow$ (3) $\Rightarrow$ (4)

(4) $\Rightarrow$ (1): $0 < a < n$ and $0 < b < n \Rightarrow (a, n) = 1 = (b, n)$

$$\Rightarrow (ab, n) = 1 = a \cdot_n b \neq 0$$

18. This is an n-element group and is equal to $\{e, a, a^2, \ldots, a^{n-1}\}$, where a is the rotation by angle $\frac{2\pi}{n}$ about the origin.

20. $\bar{0}$ is the identity and, for any $r \in \mathbb{Q}$, $-\bar{r}$ is the inverse of $\bar{r}$ in $\mathbb{Q}/\mathbb{Z}$.

EXERCISE 3(C)

2. Let S be any finite set with more than one element and define $x * y = y$ for all x and $y \in S$. Then, $(S, *)$ is a finite semigroup which satisfies the left cancellation law and does not satisfy the right cancellation law.

4. Imitate the proof of Theorem 3.3.2.

6. Let $(S, \cdot)$ be a finite semigroup and $a \in S$. Then, $\{a^n : n \in \mathbb{Z}^+\} \subseteq S$ and hence there exist $m < n$ in $\mathbb{Z}^+$ such that $a^m = a^n$. Put $\alpha = n - m \in \mathbb{Z}^+$. Then,

$$a^n = a^{m+\alpha} = a^m$$
$$a^{2m} = a^m \cdot a^m = a^m \cdot a^{m+\alpha} = a^{2m+\alpha}$$

By induction on k, $a^{km} = a^{km+\alpha}$ for all $k \in \mathbb{Z}^+$. Also, $a^{km+2\alpha} = a^{km+\alpha} \cdot a^\alpha = a^{km} \cdot a^\alpha = a^{km+\alpha} = a^{km}$.

By induction on r,

$$a^{km+r\alpha} = a^{km} \quad \text{for all } k \text{ and } r \in \mathbb{Z}^+.$$

In particular, $a^{\alpha m+m\alpha} = a^{\alpha m}$; that is, $(a^{\alpha m})^2 = a^{\alpha m}$.

8. Let G be a finite group and $|G| = 2m$. For any $a \in G$, let $A_\alpha = \{a, a^{-1}\}$. Then, $|A_a| \leq 2$ for all $a \in G$. Also, $A_a \cap A_b = \varnothing$ or $A_a = A_b$ for each pair of all elements a and b in G. $\{A_a : a \in G\}$ is a partition of G and hence

$$2m = |G| = \Sigma|A_a|.$$

Since $A_e = \{e\}$, there must be atleast one more $a \neq e$ such that $|A_a| = 1$; that is, $a = a^{-1}$ or $a^2 = e$ and $a \neq e$.

10. For each $a \in G$, $\{a^n : n \in \mathbb{Z}\}$ is a subset of G and hence finite, so that, there exist $m < n$ in $\mathbb{Z}$ such that $a^m = a^n$ or $a^{n-m} = e$ and $n - m \in \mathbb{Z}^+$. Therefore, for each $a \in G$, there exist a positive integer n_a such that $a^{n_a} = e$. If n is the product of all n_a, $a \in G$, then $a^n = e$ for all $a \in G$.

12. See Ex. 8 of Exercise 3(B).

14. If $|G| = 1$, then $G = \{e\}$. If $|G| = 2$, then $G = \{e, a\}$, where $a^2 = e$. Let $|G| = 3$. If $e \neq a \in G$, then $a^2 \neq e$ (otherwise, $a^2 = e$ implies that there exists $b \in G - \{e, a\}$ such that ab, b, e, a are distinct elements of G) and hence $\{e, a, a^2\} = G$. Let $|G| = 4$. If $a^2 = e$ for all $a \in G$, then $a = a^{-1}$ for all $a \in G$ and hence $ab = (ab)^{-1} = b^{-1}a^{-1} = ba$ for all a and $b \in G$. Therefore, we can assume that there exists $a \in G$ such that $a^2 \neq e$. If $a^3 = e$, then there exists $b \in G - \{e, a, a^2\}$ such that ba is an element of G other than e, a, a^2 and b so that $|G| > 4$. Thus, $a^3 \neq e$ and hence $G = \{e, a, a^2, a^3\}$, which is abelian. Next, let $|G| = 5$. We can prove that, for any $e \neq a \in G$, $G = \{e, a, a^2, a^3, a^4\}$ which is abelian.

EXERCISE 3(D)

2.

$\cdot_{11}$	1	2	3	4	5	6	7	8	9	10
1	1	2	3	4	5	6	7	8	9	10
2	2	4	6	8	10	1	3	5	7	9
3	3	6	9	1	4	7	10	2	5	8
4	4	8	1	5	9	2	6	10	3	7
5	5	10	4	9	3	8	2	7	1	6
6	6	1	7	2	8	3	9	4	10	5
7	7	3	10	6	2	9	5	1	8	4
8	8	5	2	10	7	4	1	9	6	3
9	9	7	5	3	1	10	8	6	4	2
10	10	9	8	7	6	5	4	3	2	1

4.

$+_2$	0	1
0	0	1
1	1	0

$+_2$	0	1	2
0	0	1	2
1	1	2	0
2	2	0	1

$+_4$	0	1	2	3
0	0	1	2	3
1	1	2	3	0
2	2	3	0	1
3	3	0	1	2

*	*e*	*a*	*b*	*c*
e	*e*	*a*	*b*	*c*
a	*a*	*e*	*c*	*b*
b	*b*	*c*	*e*	*a*
c	*c*	*b*	*a*	*e*

$+_5$	0	1	2	3	4
0	0	1	2	3	4
1	1	2	3	4	0
2	2	3	4	0	1
3	3	4	0	1	2
4	4	0	1	2	3

6. The bijections on a 3-element set form a nonabelian group with 6 elements. Any group with less than 6 elements is abelian (see Ex. 14 of Exercise 3(C)).

8. $\mathbb{Z}_n$, the group of integers modulo n.

10. Put $A = a$, $B = b$, $AB = c$ and $\begin{pmatrix} 1 & 0 \\ 0 & 1 \end{pmatrix} = e$. Then, $a^2 = b^2 = c^2 = -e$, $BA = -c$, $a^4 = b^4 = c^4 = e$ and $Q_8 = \{A^n B^m : n, m \in \mathbb{Z}\} = \{e, -e, a, -a, b, -b, c, -c\}$.

CHAPTER 4

EXERCISE 4(A)

2. Recall that, for any $n \times n$ matrices A and B, $|AB| = |A||B|$.
   (i) No
   (ii) Yes
   (iii) Yes
   (iv) Yes

(v) Yes

(vi) Yes

(vii) No

(viii) Yes.

4. If $G = <a>$, then $G = <a^{-1}>$ also and therefore $a = a^{-1}$ so that $a^2 = e$ and $|G| \leq 2$.

6. $\mathbb{Z}_n = <d> \Leftrightarrow ad \equiv 1 \pmod{n}$ for some $a \in \mathbb{Z}$
$\Leftrightarrow 1 = ad + bn$ for some a and $b \in \mathbb{Z}$
$\Leftrightarrow (d, n) = 1.$

8. $x, y \in C_a \Rightarrow ax = xa$ and $ay = ya$
$\Rightarrow x^{-1}a = ax^{-1}$ and $axy = xay = xya$
$\Rightarrow x^{-1}$ and $xy \in C_a$

10. $Z(G) = \bigcap_{a \in G} C_a$

12. Any subgroup of $\mathbb{Z}_n$ is of the form $<d>$ for some divisor d of n.

14. $x \mapsto axa^{-1}$ is a bijection of H onto aHa^{-1}.

16. Follows from Theorem 4.1.1.

EXERCISE 4(B)

2. First note that any cyclic group is at most countable.
   (i) No, since $\mathbb{R} - \{0\}$ is uncountable.
   (ii) $(\mathbb{P}(X), +)$ is cyclic $\Leftrightarrow |\mathbb{P}(X)| \leq 2$ (since $A + A = \emptyset$)
   $\Leftrightarrow |X| \leq 1$
   (iii) No, since $\mathbb{Q}_8$ is not abelian and a cyclic group is always abelian.
   (iv) No, if $\mathbb{Q}^+ = <a>$, then $\frac{1}{a} = a^n$ for some $n \in \mathbb{Z}$ and hence $a = 1$, so that $\mathbb{Q}^+ = <1> = \{1\}$, which is absurd.
   (v) No, since $\mathbb{R}^+$ is uncountable.
   (vi) Yes, since $\cos\frac{2\pi}{n} + i\sin\frac{2\pi}{n}$ generates.

4. (i) $<7> = \mathbb{Z}_{18}$, since $(7, 18) = 1$
   (ii) In $\mathbb{Z}_{20}$, $<5> = \{0, 5, 10, 15\}$
   (iii) In $\mathbb{Z}_{12}$, $<3> = \{0, 3, 6, 9\}$
   (iv) In $\mathbb{Z}_{16}$, $<3> = \mathbb{Z}_{16}$, since $(3, 16) = 1$

(v) In $\mathbb{C}-\{0\}$, $<2> = \{2^n : n \in \mathbb{Z}\}$

(vi) In $(\mathbb{R}^+, \cdot)$, $<\sqrt{2}> = \{2^{\frac{n}{2}} : n \in \mathbb{Z}\}$

(vii) In $(\mathbb{R}, +)$, $<\sqrt{2}> = \{n\sqrt{2} : n \in \mathbb{Z}\}$

(viii) In $\mathbb{C}-\{0\}$, $<\sqrt{-2}> = \{(i\sqrt{2})^n : n \in \mathbb{Z}\}$

(ix) In any group G, $<e> = \{e\}$.

(x) In $(3\mathbb{Z}, +)$, $<12> = \{12n : n \in \mathbb{Z}\}$.

6. Not necessary.

8. Let G be an abelian group, a and $b \in G$ and $O(a) = m = p_1^{r_1} \cdots p_k^{r_k} p_{k+1}^{r_{k+1}} \cdots p_u^{r_u}$ and $O(b) = n = p_1^{s_1} \cdots p_k^{s_k} p_{k+1}^{s_{k+1}} \cdots p_u^{s_u}$, where $p_1, \ldots, p_u$ are distinct primes and r_i and $s_i \in \mathbb{Z}$ such that $0 \le r_i \le s_i$ for $1 \le i \le k$ and $0 \le s_i \le r_i$ for $k < i \le u$. Then,

$$\text{l.c.m.}\{m, n\} = p_1^{s_1} \cdots p_k^{s_k} p_{k+1}^{r_{k+1}} \cdots p_u^{r_u}.$$

Put $m_1 = p_1^{r_1} \cdots p_k^{r_k}$ and $n_1 = p_{k+1}^{s_{k+1}} \cdots p_u^{s_u}$.

Let $c \in a^{m_1}$ and $d \in b^{n_1}$. Then,

$$O(c) = \frac{O(a)}{m_1} = p_{k+1}^{r_{k+1}} \cdots p_u^{r_u}$$

$$\text{and} \quad O(d) = \frac{O(b)}{n_1} p_1^{s_1} \cdots p_k^{s_k}.$$

$O(c)$ and $O(d)$ are relatively prime and therefore, by Ex. 7, G has a cyclic subgroup of order $O(c) \cdot O(d) = \text{l.c.m.}\{m, n\}$.

10. $x \in <a> \cap <b> \Rightarrow x = a^r = b^s$ for some $0 \le r < O(a)$ and $0 \le s < O(b)$

$$\Rightarrow \frac{O(a)}{(O(a), r)} = O(x) = \frac{O(b)}{(O(b), s)} \quad \text{(by Theorem 4.2.6)}$$

$\Rightarrow O(x)$ divides $O(a)$ and $O(b)$

$\Rightarrow O(x) = 1 \Rightarrow x = e$.

12. $16 \cdot_{20} 4 = 4$, $16 \cdot_{20} a = a$ for all $a \in G = \{4, 8, 12, 16\}$. Therefore, 16 is the identity in $(G, \cdot_{20})$. This is a cyclic group generated by 8 and also by 12 (12 is the inverse of 8).

14. No.

16. Let $G = <a>$, $H = <b>$, $O(a) = m$ and $O(b) = n$. Suppose that $(m, n) = 1$. Then, $\alpha m + \beta n = 1$ for some α and β in $\mathbb{Z}$. Let $(x, y) \in G \times H$. Then, $x = a^r$ and $y = b^s$ for some $0 \le r < m$ and $0 \le s < n$. Put $t = \beta rn + \alpha sm$.

Then,

$$
\begin{aligned}
& a^{t-r} = a^{t-r(\alpha m+\beta n)} = a^{m(\alpha s-\alpha r)} = e \\
\text{and} \quad & b^{t-s} = b^{t-s(\alpha m+\beta n)} = b^{n(\beta r-\beta s)} = e \\
\text{and hence} \quad & (x, y) = (a^r, b^s) = (a^t, b^t) = (a, b)^t.
\end{aligned}
$$

Thus, (a, b) is a generator of $G \times H$ and hence $G \times H$ is a cyclic group. Conversely, suppose that $G \times H$ is cyclic and $G \times H = <(c, d)>$. Then, $O(c, d) = |G \times H| = mn$. If $r =$ l.c.m.$\{m, n\}$, then $(c, d)^r = (c^r, d^r) = (e, e)$ and hence $O(c, d)$ divides r; that is, mn divides r. This is possible only when $r = mn$ or, equivalently $(m, n) = 1$.

18. $<a^n> \cap <a^m> = <a^r>$, where $r \equiv$ l.c.m.$\{m, n\}$ modulo 24.

20. $\{0\}$, <2>, <3>, <4>, <6>, <8>, <12> and $\mathbb{Z}_{24}$ are all the subgroups of $\mathbb{Z}_{24}$.

    $\{0\}$, <2>, <3>, <5>, <6>, <10>, <15> and $\mathbb{Z}_{30}$ are all the subgroups of $\mathbb{Z}_{30}$.

22. $\{0\}$, <5>, <25>, <125> and $\mathbb{Z}_{625}$ are all the subgroups of $\mathbb{Z}_{625}$.

24. Any infinite cyclic group has infinitely many subgroups. Therefore, if G has finitely many subgroups, then $<a>$ is finite for any $a \in G$ and, since $G = \bigcup_{a \in G} < a >$, G is finite.

26. $O(a) = O(a^{-1})$ for any $a \in G$. Also, if $O(a)$ and $O(b)$ divide n, then $O(ab)$ divides l.c.m. $\{O(a), O(b)\}$ and hence $O(ab)$ divides n.

28. If $O(a) = m$, then $12 < m \leq 24$ and hence $O(a) = 24$ and a is a generator of G.

30. $O(a) = O(a^{-1})$ and $O(ab)$ is a divisor of l.c.m.$\{O(a), O(b)\}$.

EXERCISE 4(C)

2. Define $f : [0, 1) \rightarrow \frac{\mathbb{R}}{\mathbb{Z}} (= \{x + \mathbb{Z} : x \in \mathbb{R}\})$ by

$$f(r) = r + \mathbb{Z}.$$

If $0 \leq r \leq s < 1$ and $f(r) = f(s)$, then $s - r \in \mathbb{Z}$ and hence $s - r = 0$. Therefore, f is an injection. For any $x \in \mathbb{R}$, let $[x]$ be the largest integer $\leq x$ and put $r = x - [x]$. Then, $0 \leq r < 1$ and

$$f(r) = r + \mathbb{Z} = x - [x] + \mathbb{Z} = x + \mathbb{Z} \text{ (since } [x] \in \mathbb{Z}).$$

Thus, f is a surjection also.

4. $Ha^{-1} = Hb^{-1} \Leftrightarrow a^{-1}b \in H \Leftrightarrow aH = bH.$

6. If a, b and c are the transposions (1 2), (2 3) and (3 1), respectively, then

$$aH = bH = cH = \{a, af, af^{-1}\} = Ha = Hb = Hc$$
$$\text{and} \quad eH = fH = f^{-1}H = H = He = Hf = Hf^{-1}$$

8. $f(x) = x = g(x) \Rightarrow f^{-1}(x) = x$ and $f(g(x)) = f(x) = x$

Therefore, A_x is a subgroup of $S(X)$. Since the identity $e \notin A_{x,y}$, $A_{x,y}$ is not a subgroup of $S(X)$.

10. $xyA = xzA \Rightarrow (xy)^{-1}\, xz \in A \Rightarrow y^{-1}z \in A \Rightarrow yA = zA$

$Ayx = Azx \Rightarrow (yx)\,(zx)^{-1} \in A \Rightarrow yz^{-1} \in A \Rightarrow Ay = Az$

12. xA is a subgroup $\Rightarrow e \in xA \Rightarrow x \in A \Rightarrow xA = A.$

14. If A and B are subgroups of G and x and $y \in G$ such that $xA = yB$, then $x \in xA = yB$ and hence $xB = yB$ so that $xA = xB$ and $A = B$.

EXERCISE 4(D)

2. $A \cap B$ is a subgroup of A and hence $|A \cap B|$ is a divisor of $|A|$. If $|A|$ is prime and $A \cap B \neq \{e\}$, then $|A \cap B| = |A|$ and hence $A \cap B = A$ so that $A \subseteq B$.

4. $i_G(A) = A \Leftrightarrow xA = yA$ for all x and $y \in G - A \Leftrightarrow x^{-1}y \in A$

6. For any $x \in G$, $x(A \cap B) = xA \cap xB$. If $x_1A, \ldots, x_nA$ are all the distinct left cosets of A in G and $y_1B, \ldots, y_mB$ are all the distinct left cosets of B in G, then $\{x_iA \cap y_jB : x_iA \cap y_jB \neq \varnothing\}$ are all the distinct left cosets of $A \cap B$ in G and hence $i_G(A \cap B) \leq n \cdot m = i_G(A) \cdot i_G(B)$.

8. AB is a subgroup of G and A and B are subgroups of AB. For any $x = ab$, $a \in A, b \in B$,

$$x(A \cap B) = xA \cap xB = bA \cap aB.$$

If $G = AB$ in Ex. 6 above, then $x_iA \cap y_jB \neq \varnothing$ for all i and j and hence $i_{AB}(A \cap B) = i_{AB}(A) \cdot i_{AB}(B)$.

10. $O(b) = 1$ or 31.

12. By Ex. 8 in Exercise 3(C), there is $a \neq e$ in G such that $a^2 = e$. Let $A = \{e, a\}$. Then, $i_G(A) = n$. If $b \neq a$ and $O(b) = 2$, then $b \notin A$ and 2 divides $i_G(A) = n$, which is a contradiction, since n is odd.

14. The group of bijections on a 3-element set is a nonabelian group of order 6, see Ex. 14 in Exercise 3(C).

EXERCISE 4(E)

2. $b \in B$ and $x \in A \cap B \Rightarrow bxb^{-1} \in B \cap A$ (since A is normal in G).

4. Let $X = \{1, 2, 3\}$ and $f : X \to X$ be defined by $f(1) = 2, f(2) = 3$ and $f(3) = 1$. Then, $\{e\}$, $\{e, f, f^{-1}\}$ and $S(X)$ are the only normal subgroups of $S(X)$.

6. For any $x \in G$ and $a \in A$, $xaA = xA$ and hence $Axa = Ax$ so that $xax^{-1} \in A$.

8. Let $G = D_4$, the group of symmetries a square (see Example 3.2.8). Let σ be the clock-wise rotation about the centre of the square through an angle $\frac{\pi}{2}$ and d be the reflection about the diagonal D. Let $A = \{e, d\}$ and $B = \{e, d, \sigma^2, \sigma^2 d\}$. Then, B is a normal subgroup of G (since $i_G(B) = 2$), A is a normal subgroup of B and A is not normal in G (since $\sigma^{-1}\alpha\sigma \notin A$).

10. Let $|A| = m$ and $i_G(A) = n = \left|\frac{G}{A}\right|$. Since $(m, n) = 1$, there exist integers α and β such that $\alpha m + \beta n = 1$. Clearly $x^m = e$ for all $x \in A$. Also,

$$x^m = e \Rightarrow xA = (xA)^{\alpha m + \beta n} = x^{m\alpha} A \cdot (xA)^{n\beta} = A \Rightarrow x \in A.$$

12. $a \in A$ and $x \in C(A) \Rightarrow xax^{-1} = a \in A$.

14. $C(A) \subseteq N(A)$. Equality may not hold.

16. For any $a \in A$ and $x \in G$, $xa \cdot xa \in A$ and hence $xax \in Aa^{-1} = A$. So that $xax^{-1} = (xax)(x^{-1})^2 \in AA = A$.

18. Straight forward verification.

20. Let $f \in \mathbb{Z}(S(X))$ and $x \in X$. If $f(x) \neq x$, then choose $y \in X$, such that $y \neq x$ and $y \neq f(x)$ (since $|X| \geq 3$). Define $g : X \to X$ by $g(x) = y$, $g(y) = x$ and $g(z) = z$ for all $z \in X - \{x, y\}$. Then, $(f \circ g)(x) = f(y)$ and $(g \circ f)(x) = f(x) \neq f(y)$ and hence $f \circ g \neq g \circ f$.

22. $(x_1, x_2)(e_1, y_2)(x_1, x_2)^{-1} = (x_1 e_1 x_1^{-1}, x_2 y_2 x_2^{-1}) = (e_1, x_2 y_2 x_2^{-1}) \in \{e_1\} \times G_2$.

EXERCISE 4(F)

2. $A = \{e, a^3, a^6, a^9, a^{12}\}$. $\left|\frac{G}{A}\right| = \frac{|G|}{|A|} = 3$. Let $A_1 = a^5 A$ and $A_2 = a^7 A$. Then, $\frac{G}{A} = \{A, A_1, A_2\}$.

$\cdot$	A	A_1	A_2
A	A	A_1	A_2
A_1	A_1	A_2	A
A_2	A_2	A	A_1

4. See Ex. 16 in Exercise 4(E). $(xA)^2 = x^2A = A$ for all $x \in G$. Therefore, G/A is abelian.

6. (i) is clear, by Lagrange's theorem.
   (ii) $(xA)^n = A$ and hence $x^n \in A$.
   (iii) If $O(a) = m$, then $(aA)^m = a^mA = A$ and hence $O(aA)$ divides m.

8. If $A = \langle x_1, \ldots, x_n \rangle$ and $\frac{G}{A} = \langle y_1A, \ldots, y_mA \rangle$, then $G = \langle \{x_iy_j : 1 \leq i \leq n$ and $1 \leq j \leq m\}$.

10. Let $X = \{a, b, c\}$. Then, $\{e\}$, $N = \{e, (a\ b\ c), (c\ b\ a)\}$ and $S(X)$ are all the normal subgroups of $S(X)$. $S(X)/N$ has only two members N and $N(a, b)$.

CHAPTER 5

EXERCISE 5(A)

2. If $f(ab^{-1}) = f(a)f(b)^{-1}$ for all $a, b \in G$, then $f(e) = f(aa^{-1}) = f(a)f(a)^{-1} = e'$, $f(b^{-1}) = f(eb^{-1}) = f(e)f(b)^{-1} = e'f(b)^{-1} = f(b)^{-1}$ and $f(ab) = f(a(b^{-1})^{-1}) = f(a)(f(b)^{-1})^{-1} = f(a)f(b)$ for all a and $b \in G$. The converse is clear.

4. G is abelian $\Leftrightarrow ab = ba$ for all a and $b \in G$

$$\Leftrightarrow (ab)^{-1} = b^{-1}a^{-1} = a^{-1}b^{-1} \quad \text{for all } a, b \in G$$
$$\Leftrightarrow f(ab) = f(a)f(b)$$

6. For any homomorphism $f : \mathbb{Z} \to \mathbb{Z}_2$, $f(n) = nf(1)$ for all $n \in \mathbb{Z}$ and hence

$$f(1) = 0 \Rightarrow f(n) = 0 \quad \text{for all } n \in \mathbb{Z}$$
$$\text{and} \quad f(1) = 1 \Rightarrow f(n) = 0 \text{ for all even } n \text{ and}$$
$$f(n) = 1 \text{ for all odd } n$$
$$(\text{since } 1 + 1 = 0 \text{ in } \mathbb{Z}_2)$$

8. Let $f : \mathbb{Z} \to \mathbb{Z}$ be a nontrivial endomorphism. Then, $f(n) = nf(1)$ for all $n \in \mathbb{Z}$ and hence $f(1) \neq 0$ so that

$$f(n) = f(m) \Rightarrow nf(1) = mf(1) \Rightarrow n = m.$$

10. If $f : \mathbb{Z} \to \mathbb{R}$ is a homomorphism and $f(1) = a$, then $f(n) = nf(1) = na = f_a(n)$ for all $n \in \mathbb{Z}$ and hence $f = f_a$.

12. If $O(a) = n$, then $f(a)^n = f(a^n) = f(e) = e'$ and hence $O(f(a))$ is finite and is a divisor of n (by Theorem 4.2.4 (3)).

14. If $|G| = p$, a prime, then $|\ker f|$ divides $|G| = p$ and hence $|\ker f| = 1$ or p, so that $\ker f = \{e\}$ or G.

16. f is a homomorphism $\Rightarrow aabb = f(1 + 1, 1 + 1) = f(1, 1)f(1, 1) = abab$

$$\Rightarrow ab = ba$$

f is a homomorphism $\Leftrightarrow ab = ba$.

18. Use induction on n to prove that $f(n + m) = f(n)f(m)$ (that is, $a^{n+m} = a^n b^m$) for all $n, m \in \mathbb{Z}$. $\ker f = \{0\}$ if a is of infinite order and $\ker f = n\mathbb{Z}$ if $O(a) = n$.

20. If $f: Q_8 \to \mathbb{Z}_2$ is a homomorphism such that $f(i) = 0$ and $f(j) = 1$, then $f(a)$ can be evaluated for any $a \in Q_8$.

22. Follows from the fact that $ab > 0 \Leftrightarrow (a > 0$ and $b > 0)$ or $(a < 0$ and $b < 0)$ for any $a, b \in \mathbb{R} - \{0\}$. Also, for any $x, y \in G$, $xy = 1 \Leftrightarrow x = 1 = y$ or $x = -1 = y$.

24. If $f: \mathbb{Z}_8 \to \mathbb{Z}_{24}$ is defined by $f(a) = 3a$ for all $a \in \mathbb{Z}_8$, then f is a monomorphism.

26. If f is an isomorphism of $(\mathbb{Z}, +)$ onto $(\mathbb{Z}, +)$, then $f(n) = nf(1)$ for all $n \in \mathbb{Z}$ and $f(1) = 1$ or -1 (otherwise f is not surjective) and hence f is either identity map or the inverse map.

28. $(f + g)(x) = f(x) + g(x)$ for all x. The trivial homomorphism is the identity in $\mathrm{Hom}(G, H)$. Also, if $f \in \mathrm{Hom}(G, H)$, then $-f \in \mathrm{Hom}(G, H)$ and $-f$ is the inverse of f.

30. $g \circ f$ is an injection $\Rightarrow f$ is an injection.

    $g \circ f$ is a surjection $\Rightarrow g$ is a surjection.

EXERCISE 5(B)

2. In Ex. 1, define $f: G \to G/N \times G/M$ by $f(x) = (xN, xM)$. Then, $\ker f = N \cap M$ and hence, by the Fundamental Theorem of Homomorphisms, $G/N \cap M \cong f(G)$ which is a subgroup of $G/N \times G/M$. If $G = \mathbb{Z}$, $N = n\mathbb{Z}$ and $M = m\mathbb{Z}$, then $N \cap M = k\mathbb{Z}$, where $k = \text{l.c.m.}\{n, m\}$. Recall that $\mathbb{Z}_k \cong \mathbb{Z}/k\mathbb{Z}$, $\mathbb{Z}_n \cong \mathbb{Z}/n\mathbb{Z}$ and $\mathbb{Z}_m \cong \mathbb{Z}/m\mathbb{Z}$.

4. If $\mathbb{Z}_{27}$ is a homomorphic image of $\mathbb{Z}_{72}$, then $\mathbb{Z}_{27}$ is isomorphic to a quotient of $\mathbb{Z}_{72}$ and hence there must be a subgroup A of $\mathbb{Z}_{72}$ whose index is 27. This is impossible, since 27 is not a divisor of 72.

6. If $f: G \to G'$ is an epimorphism, then $G/\ker f \cong G'$ and hence $|G'| = |G/\ker f| = i_G(\ker f)$, which is a divisor of $|G|$.

8. (i) $\Rightarrow$ There is an epimorphism $f: \mathbb{Z}_n \to \mathbb{Z}_m$

   $\Rightarrow \mathbb{Z}_n/k \cong \mathbb{Z}_m$ for some subgroup k of $\mathbb{Z}_n$

   $\Rightarrow i_{\mathbb{Z}_n}(k) = m \Rightarrow m$ divides $|\mathbb{Z}_n| = n \Rightarrow (b)$

   (ii) $\Rightarrow n = md$ for some $d \in \mathbb{Z}^+$

   $\Rightarrow$ The map $f: \mathbb{Z}_m \to \mathbb{Z}_n$ defined by $f(a) = ad$ is a monomorphism

   $\Rightarrow f$ is an isomorphism of $\mathbb{Z}_m$ onto $f(\mathbb{Z}_m) \subseteq \mathbb{Z}_n \Rightarrow (c)$

   (iii) $\Rightarrow$ There is a monomorphism $f: \mathbb{Z}_m \to \mathbb{Z}_n$

   $\Rightarrow \mathrm{O}(1) = m$ in $\mathbb{Z}_m$ and $\mathrm{O}(f(1)) = m$ in $\mathbb{Z}_n$

   $\Rightarrow m$ divides $|\mathbb{Z}_n| = n \Rightarrow (b)$

   (iv) $\Rightarrow md = n$ for some $d \in \mathbb{Z}^+$

   $\Rightarrow$ The map $f: \mathbb{Z}_n \to \mathbb{Z}_m$, defined by $f(a) = r$, where $a = qm + r$, $0 \leq r < m$, is an epimorphism $\Rightarrow (a)$

10. Follows from Ex. 8 above.

EXERCISE 5(C)

2. Let $f: G \to G'$ be a homomorphism of groups. In Ex. 1 above, take $G_1 = G/\ker f$, $G_2 = G'$, $f_2 = f$ and $f_1 : G \to G_1$ be defined by $f_1(a) = a(\ker f)$. Then, there exists a homomorphism $g : G_1 \to G_2$ such that $g \circ f_1 = f_2$; that is, $g \circ f_1 = f$ and g is a monomorphism and f_1 is an epimorphism.

4. Define $\alpha(A) = f^{-1}(A)$ for any subgroup A of G' and $\beta(B) = f(B)$ for any subgroup B of G containing $\ker f$. Then, $(\alpha \circ \beta)(B) = B$ and $(\beta \circ \alpha)(A) = A$ and hence α is a bijection of the set of subgroup of G' onto the set of subgroups of G containing $\ker f$.

6. Refer Example 3.2.8. Let a be the clock-wise rotation about the centre of the square through an angle $\pi/2$ and b be the reflection about the diagonal D_1. Then, $\mathrm{O}(a) = 4$, $\mathrm{O}(b) = 2$ and $aba = b$. The group G_8 is of order 8 and $G_8 = \{e, a, a^2, a^3, b, ab, a^2b, a^3b\}$. It can be verified that $N_1 = \{e\}$, $N_2 = \{e, a, a^2, a^3\}$, $N_3 = \{e, a^2, b, a^2b\}$, $N_4 = \{e, a^2, ab, ab, a^3b\}$, $N_5 = \{e, a^2\}$ and $N_6 = G_8$ are all the normal subgroups of G_8 and hence G_8/N_i, $1 \leq i \leq 6$, are all (up to isomorphism) the homomorphic images of G_8. Also, $\mathbb{Z}(G_8) = \{e, a^2\} = N_5$ and N_2, N_3, N_4, N_5 and G_8 are all subgroups of G_8 containing $\mathbb{Z}(G_8)$ and hence N_2/N_5, N_3/N_5, N_4/N_5, N_5/N_5, and G_8/N_5 are all the subgroups of $G_8/\mathbb{Z}(G_8)$.

8. (i) $x \in f(A) \cap f(B) \Rightarrow f(a) = x = f(b)$ for some $a \in A$ and $b \in B$

$$\Rightarrow ab^{-1} \in \ker f \subseteq A \cap B \Rightarrow a = (ab^{-1})b \in B$$
$$\Rightarrow x = f(a) \in f(A \cap B)$$

(ii) $x \in A \ker f \Rightarrow x = ab,\ a \in A,\ b \in \ker f \Rightarrow f(x) = f(ab) = f(a) \in f(A)$

$$\Rightarrow x \in f^{-1}(f(A)) \Rightarrow f(x) = f(a),\ a_1 \in A$$
$$\Rightarrow a_1^{-1}x \in \ker f \Rightarrow x = a_1(a_1^{-1}x) \in A \ker f.$$

(iii) Define $g : G \to G'/[G', G']$ by $g(x) = f(x)[G', G']$ since f is an epimorphism, so is g and $\ker g = [G, G]$. Therefore, $G/[G, G] \cong G'/[G', G']$.

10. (i) If $x_1A, \ldots, x_nA$ are all the distinct left cosets of A in G_1, then verify that $f(x_1)f(A), \ldots, f(x_n)f(A)$ are all the distinct left cosets of $f(A)$ in G'.

(ii) If $A = f^{-1}(A')$ and $f(x_1)A', \ldots, f(x_n)A'$ are all the distinct left cosets of A' in G', then $x_1A, \ldots, x_nA$ are all the distinct left cosets of A in G.

12. If $|G| = p$ and $e \neq a \in G$, then $O(a)$ divides p and $O(a) = p$ and hence $G = \langle a \rangle \cong \mathbb{Z}_p$.

14. $A_1/A_1 \cap N \cong A_1N/N = A_2N/N \cong A_2/A_2 \cap N.$

EXERCISE 5(D)

2. $2\mathbb{Z}$ is a proper subgroup of $(\mathbb{Z}, +)$ and $f: \mathbb{Z} \to 2\mathbb{Z}$, defined by $f(a) = 2a$, is an isomorphism.

4. $T_a(A) \subseteq A$, where T_a is the inner automorphism of G corresponding to any $a \in G$.

6. If x and $a \in G$ and $f(a) = a$ for all $f \in \mathrm{Aut}(G)$, then $f(xax^{-1}) = f(x)f(a)f(x)^{-1} = T_{f(x)}(a) = a = T_x(a) = xax^{-1}$ for all $f \in \mathrm{Aut}(G)$.

8. For Ex. 7, $x \mapsto x^{-1}f(x)$ is an injection of G into G; G being finite $G = \{y^{-1}f(y) : y \in G\}$. In addition, suppose that $f^2 = I_G$. Then, for any $a \in G$,

$$a^{-1}f(a) = f^2(a^{-1}f(a)) = f(f(a^{-1})a) = f((a^{-1}f(a))^{-1})$$

and therefore, by Ex. 7, $f(x) = x^{-1}$ for all $x \in G$. Now,

$$xy = (y^{-1}x^{-1})^{-1} = f(y^{-1}x^{-1}) = f(y^{-1})f(x^{-1})f(x^{-1}) = yx,$$
for any $x, y \in G$.

10. $f(aba^{-1}b^{-1}) = f(a)f(b)f(a)^{-1}f(b)^{-1}$

12. If $f: G \to G'$ is a homomorphism and $G = \langle a \rangle$, then $f(G) = \langle f(a) \rangle$.

14. For any $0 < i < p, f_i : \mathbb{Z}_p \to \mathbb{Z}_p$ defined by $f(a) = a^i$, is an automorphism of $\mathbb{Z}_p$ and $\text{Aut}(\mathbb{Z}_p) = \{f_i : 0 < i < p\}$, which is isomorphic to the multiplicative groups $\mathbb{Z}_p - \{0\}$ modulo p.

16. $(m, n) = 1 \Leftrightarrow \alpha m + \beta n = 1$ for some $\alpha, \beta \in \mathbb{Z}$

$$\Leftrightarrow 1 \equiv \alpha m \pmod{n}$$

If $(m, n) = 1$ and $a^m = b^m$, then $a^{\alpha m} = b^{\alpha m}$ and hence $a = b$, so that f_m is automorphism. Conversely, if $f_m \in \text{Aut}(G)$, then $\langle a \rangle = G = \langle f_m(a) \rangle = \langle a^m \rangle$ and hence $a = a^{\alpha m}$ for some $\alpha \in \mathbb{Z}$ so that $1 \equiv \alpha m \pmod{n}$.

18. $\text{Aut}(\mathbb{Z}_2 \times \mathbb{Z}_2) \cong S_3$

20. $\text{Aut}(\mathbb{Z}_2 \times \mathbb{Z}_3) \cong \text{Aut}(\mathbb{Z}_6) \cong \mathbb{Z}_2$

$\text{Aut}(\mathbb{Z}_2) = \{e\}, \text{Aut}(\mathbb{Z}_3) \cong \mathbb{Z}_2$

CHAPTER 6

EXERCISE 6(A)

2. (i) Yes, since $\mathbb{Z}_n$ is finite.
   (ii) No, since $f : \mathbb{Z} \to \mathbb{Z}$ defined by $f(a) = 2a$ is an injection, but not a surjection.
   (iii) Yes, since $\mathbb{Z}_n$ is finite.
   (iv) Yes, since $|X| = n \to |S(X)| = n!$ and hence $S(X)$ is finite.
   (v) Yes, since $|X| = n \to |\mathbb{P}(X)| = 2^n$
   (vi) Yes, refer the proof of Theorem 6.1.1.

4. $|X| = n < \infty \Rightarrow |S(X)| = 2^n$

   If X is infinite and $Y \subseteq X \Rightarrow S(Y) \subseteq S(X)$ and hence $2^n < |S(X)|$ for all $n \in \mathbb{Z}^+$, so that $S(X)$ is infinite.

6. Refer proof of Theorem 6.1.2.

8. Use Theorem 6.1.3.

10. Define $f : \mathbb{R} \to \mathbb{R}$ by $f(a) = \begin{cases} a+1 & \text{if } a \in \mathbb{Z} \\ a & \text{if } a \notin \mathbb{Z} \end{cases}$.

EXERCISE 6(B)

2. $O(a)$ divides $|S_n| = n!$ for any $a \in S_n$

4. $O(a) = 8 = O(b)$ and $O(c) = 4$

6. No, otherwise there is an epimorphism $\alpha : S_8 \to S_{12}$ and hence $|S_8| \geq |S_{12}|$, which is an absurd.

8. $$a\,b\,a^{-1} = \begin{pmatrix} 1 & 2 & 3 & 4 & 5 & 6 & 7 & 8 \\ 7 & 4 & 6 & 5 & 3 & 1 & 8 & 2 \end{pmatrix}$$

$$b\,c\,b^{-1} = \begin{pmatrix} 1 & 2 & 3 & 4 & 5 & 6 & 7 & 8 \\ 5 & 6 & 7 & 3 & 2 & 1 & 8 & 4 \end{pmatrix}$$

$$c\,a\,c^{-1} = \begin{pmatrix} 1 & 2 & 3 & 4 & 5 & 6 & 7 & 8 \\ 4 & 1 & 8 & 7 & 2 & 5 & 3 & 6 \end{pmatrix}$$

10. $$S_4 = \left\{ \begin{matrix} e, \begin{pmatrix} 1 & 2 & 3 & 4 \\ 2 & 1 & 4 & 3 \end{pmatrix}, \begin{pmatrix} 1 & 2 & 3 & 4 \\ 4 & 3 & 2 & 1 \end{pmatrix}, \begin{pmatrix} 1 & 2 & 3 & 4 \\ 3 & 4 & 1 & 2 \end{pmatrix}, \begin{pmatrix} 1 & 2 & 3 & 4 \\ 2 & 3 & 1 & 4 \end{pmatrix}, \\ \begin{pmatrix} 1 & 2 & 3 & 4 \\ 3 & 1 & 2 & 4 \end{pmatrix}, \begin{pmatrix} 1 & 2 & 3 & 4 \\ 2 & 4 & 3 & 1 \end{pmatrix}, \begin{pmatrix} 1 & 2 & 3 & 4 \\ 4 & 1 & 3 & 2 \end{pmatrix}, \begin{pmatrix} 1 & 2 & 3 & 4 \\ 3 & 2 & 4 & 1 \end{pmatrix}, \\ \begin{pmatrix} 1 & 2 & 3 & 4 \\ 4 & 2 & 1 & 3 \end{pmatrix}, \begin{pmatrix} 1 & 2 & 3 & 4 \\ 1 & 3 & 4 & 2 \end{pmatrix}, \begin{pmatrix} 1 & 2 & 3 & 4 \\ 1 & 4 & 2 & 3 \end{pmatrix} \end{matrix} \right\}$$

EXERCISE 6(C)

2. (i) Yes, (1 4 2 6 5 3 8 7)
   (ii) No
   (iii) Yes, (1 2 3 4 5 6 7 8)
   (iv) No
4. (i) 12
   (ii) 2
   (iii) 6
   (iv) 7.
6. Let $|f|$ = Supp f and f and g be disjoint.

$$i \notin |f| \cup |g| \Rightarrow f(i) = i = g(i) \Rightarrow f(g(i)) = i \Rightarrow i \notin |fg|.$$
$$i \in |f| \Rightarrow f(i) \neq i \Rightarrow g(i) = i \Rightarrow f(g(i)) = f(i) \neq i \Rightarrow i \in |fg|.$$
$$i \in |g| \Rightarrow g(i) \neq i \Rightarrow f(g(i)) \neq f(i) = i \Rightarrow i \in |fg|.$$

8. $f(i) = i \Rightarrow f^m(i) = i$ for all $m \in \mathbb{Z}$
10. Supp $f = \phi \Leftrightarrow i \notin$ Supp f for any $i \in I_n \Leftrightarrow f(i) = i$ for all i.

12. Follows from the observation that $i \in \text{Supp}\, f$ if and only if $f(i) \in \text{Supp}\, f$ and that f is a 3-cycle if and only if $\text{Supp}\, f = \{i, f(i), f^2(i)\}$ for some i. This cannot be extended for any r-cycle. For example, if $f = (1\ 2) \circ (3\ 4)$, then $\text{Supp}\, f = \{1, 2, 3, 4\}$, which is a 4-element set, but f is not a 4-cycle.

14. $(gfg^{-1})^n = gf^n g^{-1}$ and hence $f^n = e \Leftrightarrow (gfg^{-1})^n = e$ for any $n \in \mathbb{Z}^+$.

16. If $a = (i_1\, i_2\, i_3 \cdots i_{2m+1})$, then

$$a^2 = (i_1\, i_3\, i_5 \cdots i_{2m+1}\, i_2\, i_4 \cdots i_{2m})$$

17. No; if $a = (i_1\, i_2\, i_3 \cdots i_{2m})$, then

$$a^2 = (i_1\, i_3 \cdots i_{2m-1}) \circ (i_2\, i_4 \cdots i_{2m}).$$

18. If a is an r-cycle and $1 \leq s < r$, then $O(a) = r$ and $O(a^s) = \dfrac{O(a)}{(s, O(a))} = \dfrac{r}{(s, r)}$ and hence, if $(s, r) = 1$, then $O(a^s) = r$.

20. If $a = (i_1\, i_2 \cdots i_r)$ and $b = (j_1 j_2 \cdots j_s)$, then $f \circ a \circ f^{-1} = (f(i_1) f(i_2) \cdots f(i_r))$ and $f \circ b \circ f^{-1} = (f(j_1) f(j_2) \cdots f(j_s))$. Also, $f(i_t) = f(j_u) \Leftrightarrow i_t = j_u$.

22. Partitions of 4 are $\{1, 1, 1, 1\}$, $\{1, 1, 2\}$, $\{2, 2\}$, $\{1, 3\}$, $\{4\}$ and of 5 are $\{1, 1, 1, 1, 1\}$, $\{1, 1, 1, 2\}$, $\{1, 2, 2\}$ $\{1, 1, 3\}$ $\{2, 3\}$, $\{1, 4\}$, $\{5\}$.

24. Follows from Ex. 20 and 23 above.

26. Use Ex. 25 and the facts that $(2\ 3)\,(1\ 2)\,(2\ 3) = (1\ 3)$, $(3\ 4)\,(1\ 3)\,(3\ 4) = (1\ 4)$, … to prove that s_n is generated by $(1\ 2)$, $(2\ 3)$, …, $(n-1, n)$. If $a = (1\ 2\ 3 \ldots n)$ and $b = (1\ 2)$, then $a^k\, b\, a^{-k} = (k\, k+1)$ for all $1 \leq k \leq n-1$. Therefore, S_n is generated a and b.

28. See Ex. 20 above.

30. Let $|G| = 6$. If G is abelian, then $G \cong \mathbb{Z}_6$. Suppose that G is not abelian. Then, $O(a) = 3$ for some $a \in G$ ($O(a) = 2$ for all $a \Rightarrow G$ is abelian and $O(a) = 6$ for some $a \Rightarrow G \cong \mathbb{Z}_6$). Choose $b \in G - \{e, a, a^2\}$. Then, we can check that $b^2 \neq a$ and $b^2 \neq a^2$. Also, $b^2 \notin \{b, ab, a^2b\}$. Since e, a, a^2, b, ab, a^2b are all distinct elements, we have $G = \{e, a, a^2, b, ab, a^2b\}$. Therefore, $b^2 = e$ and $O(b) = 2$. Since $\langle a \rangle$ is of index 2, $\langle a \rangle$ is normal in G and hence $b\,a\,b^{-1} = e$, a or a^2. If $b\,a\,b^{-1} = e$, then $a = e$ and, if $b\,a\,b^{-1} = a$, then $ba = ab$ and G is abelian. Therefore, $b\,a\,b^{-1} = a^2$. Thus, $G = \langle a, b \rangle$, $a^3 = e = b^2$, $b\,a\,b = a^2$. Also, if $f = (1\ 2\ 3)$ and $g = (1\ 2)$, then $f^3 = e = g^2$ and $g\,f\,g = f^2$ in S_3. If $\alpha : S_3 \to G$ is defined by; $\alpha(f) = a$, $\alpha(g) = b$, $\alpha(f^2) = a^2$, $\alpha(e) = e$, $\alpha(fg) = ab$ and $\alpha(f^2g) = a^2b$, then α is an isomorphism.

EXERCISE 6(D)

2. (i) $f = (1\ 3\ 2\ 4) \circ (6\ 7\ 8\ 9)$ and $\text{CI}(f) = 4 + 4 - 2 = 6$
   (ii) $f = (1\ 3\ 6\ 9\ 5) \circ (2\ 4\ 7\ 8)$ and $\text{CI}(f) = 5 + 4 - 2 = 7$
   (iii) $f = (2\ 8\ 3\ 9\ 5\ 4) \circ (6\ 7)$ and $\text{CI}(f) = 6 + 2 - 2 = 6$
   (iv) $f = (2\ 3\ 6\ 8\ 9\ 4\ 5\ 7)$ and $\text{CI}(f) = 8 - 1 = 7$

4. If $f = a_1 \circ a_2 \circ \cdots \circ a_s$, then $f^{-1} = a_s^{-1} \circ \cdots \circ a_2^{-1} \circ a_1^{-1}$ and length of a_i is equal to that of a_i^{-1}. Also, a_i's are pair-wise disjoint if and only a_i^{-1}'s are so. Therefore, $\text{CI}(f) = \text{O}(a_1) + \cdots + \text{O}(a_s) - s = \text{CI}(f^{-1})$.

6. If $f = a_1 \circ a_2 \circ \cdots \circ a_s$, then $g \circ f \circ g^{-1} = (g \circ a_1 \circ g^{-1}) \circ (g \circ a_2 \circ g^{-1}) \circ \cdots \circ (g \circ a_s \circ g^{-1})$

   a_i is an r-cycle $\Leftrightarrow g \circ a_i \circ g^{-1}$ is an r-cycle
   a_i's are disjoint $\Leftrightarrow g \circ a_i \circ g^{-1}$'s are disjoint.

   Therefore, $\text{CI}(f) = \sum_{i=1}^{s} \text{O}(a_i) - s = \sum_{i=1}^{s} \text{O}(g \circ a_i \circ g^{-1}) - s = \text{CI}(g \circ f \circ g^{-1})$.

8. Let $f = a_1 \circ a_2 \circ \cdots \circ a_s$ and a_i's be disjoint cycles.

   Then, $\text{CI}(f) = \sum_{i=1}^{s} \text{O}(a_i) - s = \sum_{i=1}^{s} (\text{O}(a_i) - 1)$. Since $\text{O}(a_i) - 1 \geq 0$,

   $\text{CI}(f) = 1 \Leftrightarrow \text{O}(a_i) - 1 = 1$ for some i and $\text{O}(a_j) - 1 = 0$ for all $j \neq i$
   $\Leftrightarrow \text{O}(a_i) = 2$ and $\text{O}(a_j) = 1$ for all $j \neq i$
   $\Leftrightarrow f = a_1$ and a_1 is a transposition.

10. All transpositions in S_6; that is, $(i\ j)$, $1 \leq i < j \leq 6$.

12. $A_1 = A_2 = \{e\}$, $A_3 = \{e, a, a^2\}$, where $a = (1\ 2\ 3)$. Therefore, A_n is abelian for $n < 4$. Also, A_4 is not abelian (since $(1\ 2\ 3) \circ (2\ 3\ 4) = (1\ 2) \circ (3\ 4)$ and $(2\ 3\ 4) \circ (1\ 2\ 3) = (1\ 3) \circ (2\ 4)$) and hence A_n is not abelian for all $n \geq 4$ (since A_4 is a subgroup of A_n).

14. $|S_n| = n!$ and $|A_n| = \frac{1}{2}n!$ and hence $|S_n - A_n| = |A_n|$.

16. If $f = (1\ 2\ 3\ 4\ 5) \circ (6\ 7\ 8)$, then $\text{O}(f) = 5 \cdot 3 = 15$.

18. If $f = a_1 \circ \cdots \circ a_r$, a_i's are transpositions, then $f^2 = a_1 \circ \cdots \circ a_r \circ a_1 \circ \cdots \circ a_r$.

20. If $A_n \subsetneqq H \subsetneqq S_n$, then $|H|$ divides $n!$ and $\frac{1}{2}n!$ divides $|H|$ and $\frac{1}{2}n! < |H| < n!$ and hence $1 > \frac{n}{|H|} > 2$ which is an absurd. Since $\frac{n!}{|H|}$ is an integer.

22. Let A be a subgroup of S_n and $f \in A$ be odd. Then, $A_n \cup A_n f = S_n$ and $A_n \cap A_n f = \emptyset$, since A_n is of index 2. From this we get that A is the disjoint

union of $(A_n \cap A)$ and $(A_n f \cap A)$. Also, $g \mapsto gf$ is a bijection of $A_n \cap A$ onto $A_n f \cap A$. Therefore, $|A| = 2(A_n \cap A)$, so that $|A|$ is even.

24. If $O(f) = 5$ in S_6, then f cannot be a product of any two or more disjoint cycles; for, if $f = f_1 \circ f_2$ and f_1 and f_2 are disjoint cycles of length > 1, then $O(f) =$ l.c.m. of $|f_1|$ and $|f_2| = 2$ or 6.

26. This is an example to prove that the converse of the Langrange's Theorem is false.

28. In S_4, $C((2\ 4\ 1)) = \{e, (2\ 4\ 1)\}$

 and $C((1\ 2) \circ (3\ 4)) = \{e, (1\ 2) \circ (3\ 4), (1\ 3) \circ (2\ 4), (1\ 4) \circ (2\ 3)\}$.

30. $O(1\ 2\ 3\ 4) = 4$ and 3 does not divide 4.

32. 16, since 21 is the least positive integer n such that $f^{21} = e$.

34. All 7-cycles only.

36. $m = 3$ or 7 or 9.

38. A_n has only two cosets A_n and $A_n f$, where f is any odd permutation and $A_n f$ is the set of all odd permutations.

40. ${}^5C_3 \cdot 2 = 20$.

42. 0, since any 4-cycle is add.

44. Let $f = f_1 \circ f_2 \circ \cdots \circ f_r$, where f_i's are disjoint cycles. Then, $O(f) =$ l.c.m. $\{|f_1|, \ldots, |f_r|\}$. If $O(f)$ is odd, then $|f_i|$ is odd and hence f_i is even for all i. Therefore, f is an even permutation.

46. Since $|D_n| = 2_n$ and $O(a) = n$, $<a>$ is of index 2 and hence normal and $|D_n / <a>| = 2$.

48. If N is a normal subgroup of S_n, then $N \cap A_n$ is a normal subgroup of A_n and, since A_n is simple, $N \cap A_n = \{e\}$ or A_n and therefore $A_n \subseteq N$ so that $N = A_n$ or S_n. Since N is proper, $N = A_n$.

CHAPTER 7

EXERCISE 7(A)

2. Yes, θ is effective.

4. $\ker \theta = \{a \in G : axa^{-1}H = xH \text{ for all } x \in G\}$, which is not necessarily $\{e\}$ and hence θ is not necessarily effective.

6. Let $X = G/H$. Then, $|X| = n$. Define $\theta : G \times X \to X$ by $\theta(a, xH) = axa^{-1}H$. Then, θ is an action of G on X and hence $f_\theta : G \to S(X)\ (= S_n)$ is a homomorphism whose kernel is $\bigcap_{x \in G} xHx^{-1} = H$. Therefore, $G/H \cong f_\theta(G) \subseteq S_n$.

8. Each element of G maps edges (or faces or vertices or diagonals) onto edges (or faces or vertices or diagonals, respectively).

10. Straight forward verification.

12. If θ is the given action of G on X, then ker θ is a normal subgroup of G contained in H.

14. Let H be a subgroup of G and $\frac{|G|}{|H|} = p$. Let $X = \{xH : x \in G\}$. Then, $|X| = p$ and G acts on X by left translation; that is, $\theta(a, xH) = axH$. Then, ker θ is a normal subgroup of G and ker $\theta = \bigcap_{x \in G} xHx^{-1} \subseteq H$; $G/\ker\theta$ is isomorphic to a subgroup of S_p and hence $|G/\ker\theta|$ is a divisor of $p!$. If q is a prime dividing $|G/\ker\theta|$ then q divides $p!$ and hence $q = p$. Also, p^2 does not divide $p!$ and hence $|G/\ker\theta| = p = \frac{|G|}{|H|}$. Therefore, $|\ker\theta| = |H|$. Since ker $\theta \subseteq H$, we get that $H = \ker\theta$, which is normal in G.

16. If $|G| = 2695$ and H is a subgroup of 539 in G, then H is of index 5 and 5 is the least prime dividing $|G|$. By Ex. 14 above, H is normal in G.

18. If $|G|$ is odd and 3 divides $|G|$, then 3 is the least prime dividing $|G|$. Now use Ex. 14 above.

EXERCISE 7(B)

2. See answer for Ex. 6 in Exercise 7(A).

4. Its kernel is $\bigcap_{x \in G} xHx^{-1}$.

6. Let $\mathcal{T}(G)$ be the set of all subgroups of G and let G act on $\mathcal{T}(G)$ by conjugation (that is, $\theta(a, K) = aKa^{-1}$). Then, the orbit of H is the set of conjugates of H in G and the stabilizer of H is the normalizer of H in G. Now use Corollary 7.2.1.

8. We can assume that the action is nontrivial. By Theorem 7.2.7, St(x) is a maximal subgroup of G for each $x \in X$. If $N \subseteq$ St(x) for all $x \in X$, then the induced action of N on X is trivial. Suppose that $N \not\subseteq$ St(x) for some $x \in X$. Then, since N is normal NSt(x) is a subgroup containing St(x) properly and hence NSt(x) $= G$. Now, for any $y \in X$, there exists $g \in G$ such that $gx = y$ (since the action of G on X is primitive, it is transitive by Corollary 7.2.2).

Since $g = G = N\text{St}(x)$, $g = ab$ for some $a \in N$ and $b \in \text{St}(x)$. Now, $y = gx = abx = ax$, $a \in N$. Thus, the induced action of N on G is trivial.

10. Let the action θ of G on X is doubly transitive and R be an equivalence relation on X which is compatible with θ. If $R \neq \Delta_X$, then there exist $x \neq y \in X$ such that $(x, y) \in R$ and now, for any $z \in X$, there exists $a \in G$ such that $ax = x$ and $ay = z$ and hence $(x, z) \in R$. This proves that $R = X \times X$. Thus, the action is primitive. The converse is not in general. For, consider the following example.

    Let N be a proper nontrivial normal subgroup of a group G such that N is a maximal subgroup of G (for example, take $G = \mathbb{Z}$ and $N = p\mathbb{Z}$ for some prime p). Let θ be the action of G on G/N by left translation. Then, θ is primitive (since $\text{St}(xN) = N$ for any $x \in G$); but θ is not doubly transitive (choose $e \neq x \in N$ and $y \in G - N$. Then, there is no $a \in G$ such that $aeN = xN$ and $aeN = yN$).

EXERCISE 7(C)

2. If $C(x) = \{axa^{-1} : a \in G\}$, then $|C(x)| = 1 \Leftrightarrow axa^{-1} = x$ for all $a \in G$ $\Leftrightarrow x \in Z(G)$.

4. Let e, $a = (1\ 2\ 3)$, $b = (1\ 3\ 2)$, $c = (2\ 3)$, $d = (3\ 1)$ and $s = (1\ 2)$ be all the elements of S_3 (see Worked Exercise 6.2.1). $C(e) = \{e\}$, $C(a) = \{a, b\}$ and $C(s) = \{s, c, d\}$ are all the conjugate classes in S_3 and $\text{Cent}(a) = \{e, a, b\}$ and $\text{Cent}(s) = \{e, s\}$ and, by the class equation 7.3.2,

$$6 = |S_3| = i(\text{Cent}(a)) + i(\text{Cent}(s)) + 1 = 2 + 3 + 1.$$

    Note that $Z(S_3) = \{e\}$.

6. For any $a \in G$, $|C(a)| = i_G(\text{St}(a))$ and hence $|C(a)|$ divides $|G|$.

8. Let $i_G(H) = n$ and $X = \{xH : x \in H\}$. Then, G acts on X and $f : G \to S(X)$, defined by $f(a)(xH) = axH$, is a homomorphism whose Kernel is $K = \bigcap_{x \in G} xHx^{-1}$. Then, K is a normal subgroup of G, $K \subseteq H$ and $G/K \cong$ a subgroup of $S(X)$. Since $S(X)$ is finite, $|G/K| = i_G(K)$ is finite.

10. Let $|N| = n$ and $|G/N| = m$ and $(n, m) = 1$. Choose integers α and β such that $\alpha n + \beta m = 1$. For any $a \in G$,

$$\begin{aligned} \text{O}(a) \text{ divides } n &\Rightarrow a^n = e \quad \text{and} \quad \text{O}(aN) \text{ divides } m \\ &\Rightarrow a^{\alpha n} = e \quad \text{and} \quad a^m \in N \\ &\Rightarrow a = a^{\alpha n + \beta m} = e(a^m)^\beta \in N. \end{aligned}$$

12. $|N| = 3$, $N \nsubseteq Z(G)$. Choose $x \in N$ such that $x \notin Z(G)$. Let G act on N by conjugation. Then, the orbit $C(x) = \{axa^{-1} : a \in G\}$ is a subset of N,

since N is normal and $x \in N$. Since $x \notin Z(G)$, $|C(x)| > 1$. Since $e \notin C(x)$ (otherwise $x = e \in Z(G)$), it follows that $1 < |C(x)| < |N| = 3$ and hence $|C(x)| = 2$ and hence the stabiliser St(x) is of index 2.

14. As in Worked Exercise 7.3.1, let $\mathbb{Z}_6$ act on the set X of all possible necklaces. We have $|X| = 2^6$. Let us compute $X_a = \{x \in X : ax = x\}$ for all $a \in \mathbb{Z}_6 = \{0, 1, 2, 3, 4, 5\}$. Clearly, $|X_0| = |X| = 2^6$. Let $b_1, b_2, b_3, b_4, b_5, b_6$ be the beads. The action of 1 transforms the beads $b_1, b_2, \ldots, b_6$, in this order, to $b_2, b_3, \ldots, b_6, b_1$. In order that these are same necklaces, $b_1, b_2, \ldots, b_6$ must be add of same colour and hence $|X_1| = 2$, since we are given with two colours. The action of 2 will transform $b_1\, b_2\, b_3\, b_4\, b_5\, b_6$ to $b_3\, b_4\, b_5\, b_6\, b_1\, b_2$; for these to be same necklaces, b_1, b_3, b_5 must be of one colour and b_2, b_4, b_6 must also be of one colour. Thus, $|X_2| = 2 \cdot 2 = 4$. Similarly, we can see that $|X_3| = 8$, $|X_4| = 4$, $|X_5| = 2$ (note that $X_a = X_{a^{-1}}$ for any $a \in \mathbb{Z}_6$).

    Therefore, by Theorem 7.3.7, the number of different necklaces is

$$\frac{1}{|\mathbb{Z}_p|}\sum_{a\in\mathbb{Z}_6}|X_a| = \frac{1}{6}(2^6+2+4+8-4+2)$$
$$=14.$$

16. If G acts on $\mathbb{P}(G)$ by conjugation, then $C(S)$ is the orbit of S and $N(S)$ is the stabiliser of S.

18. Use Ex. 14 in Exercise 7(A).

20. By Worked Exercise 7.3.1, the number of different necklaces is

$$\frac{5}{11}(5^{11-1}+11-1) = \frac{5}{11}(5^{10}+10) = 4{,}338{,}925.$$

EXERCISE 7(D)

2. (i) $<3> = \{0, 3, 6, 9, 12, 15, 18, 21\}$ is the only Sylow 2-subgroups. $<8> = \{0, 8, 16\}$ is the only Sylow 3-subgroup $\{0\}$ is the only Sylow p-subgroup for prime $p \neq 2, 3$.

   (ii) Refer Worked Exercise 6.2.1. $\{e, a, b\}$ is the only Sylow 3-subgroup and $\{e, c\}$, $\{e, d\}$ and $\{e, s\}$ are all the Sylow 2-subgroups $\{e\}$ is the Sylow p-subgroup for any prime $p \neq 2, 3$.

   (iii) Refer Worked Exercise 6.4.2. $<a_1>$, $<b_1>$, $<c_1>$ and $<d_1>$ are the only Sylow 3-subgroups. $\{e, p, q, r\}$ is the only Sylow 2-subgroup.

4. Let $|G| = 56$. Then, $n_7 = 1 + 7k$ should divide $|G| = 2^3 \cdot 7$ and hence n_7 should divide 2^3, so that $n_7 = 1$ or 8. Any subgroup of order 7 is cyclic and hence generated by each of the six nonidentity elements. If $n_7 = 8$, then there are $6 \cdot 8 = 48$ elements of order 7 and the remaining 8

elements should form a unique Sylow 2-subgroup, which turns out to be normal and hence G is not simple. If $n_7 = 1$, then the unique Sylow 7-subgroup is normal and hence G is not simple.

6. Follows from the fact that, if A is a Sylow p-subgroup, then xAx^{-1} is also a Sylow p-subgroup for any p.

8. Let G be a group of order $p^r \cdot n$. Then, $n_p = 1 + pk$ for some $k \geq 0$ and n_p should divide the index of any Sylow p-subgroup. Therefore, n_p should divide n which is impossible, unless $k = 0$ (since $n < p$). Therefore, $n_p = 1$ and hence there exists unique Sylow p-subgroup which is normal in G.

10. Use Ex. 9 above and Ex. 14 of Exercise 7(A).

12. Let $|G| = 225 = 3^2 \cdot 5^2$. Then, $n_5 = 1 + 5k$ should divide 9 and hence $n_5 = 1$ so that there is a unique Sylow 5-subgroup which is normal in G.

14. $323 = 19 \cdot 17$ and use Worked Exercise 7.4.2.

16. (i) Routine verification.
 (ii) If $a = (i_1\, i_2 \cdots i_r)$ and $b = (j_1 j_2 \cdots j_r)$ are r-cycles, define $f \in S_n$ by $f(i_k) = j_k$ for $1 \leq k \leq r$ and $f(i) = i$ for all $i \neq i_k$. Then, $f \circ a \circ f^{-1} = b$ (by (*i*)).
 (iii) Use (i) and (ii) and Theorem 6.3.5.
 (iv) use (iii).

18. Let $n > 2$, $f = (1\ 2)$ and $g = (2\ 3)$. Then, $f \circ g \neq g \circ f$.

20. Follows from Theorem 7.4.2 and the fact that $|G| = |N| \cdot |G/N|$.

22. If $|G| = p^n$, then $n_p = 1 + pk$ should divide n, which is possible only when $n_p = 1$.

24. $\mathbb{Z}_7[x]$, the additive group of polynomials over $\mathbb{Z}_7$.

26. Let S be a normal (and hence unique) Sylow p-subgroup of G. Then, $f(S)$ ($\cong S/S \cap \ker f$) is a p-subgroup and hence $f(S)$ is contained in a Sylow p-subgroup. Thus, $f(S) \subseteq S$.

28. If $|G|$ is square-free, then $|G| = p_1 p_2 \cdots p_r$, where $p_1, \cdots, p_r$ are distinct primes. If A_i is the Sylow p_i-subgroup of G, then $A_i \cap A_j = \{e\}$ for $i \neq j$ and $G = A_1 A_2 \cdots A_r$. Now, $G \cong A_1 \times \cdots \times A_r \cong \mathbb{Z}_{p_1} \times \cdots \times \mathbb{Z}_{p_r} \cong \mathbb{Z}_{|G|}$.

30. $30 = 2 \cdot 3 \cdot 5$, $A_2 = \{0, 15\}$, $A_3 = \{0, 10, 20\}$ and $A_5 = \{0, 6, 12, 18, 24\}$ are respectively Sylow 2, 3 and 5-subgroups of $\mathbb{Z}_{30}$. For primes $p \neq 2, 3, 5$, $\{0\}$ is the Sylow p-subgroup.

CHAPTER 8

EXERCISE 8(A)

2. Since there is no element of order 4 in $\mathbb{Z}_2 \times \mathbb{Z}_2$, $\mathbb{Z}_4$ is not isomorphic with $\mathbb{Z}_2 \times \mathbb{Z}_2$.

4. Use Corollary 8.1.2.

6. $A \times B \cong AB = \mathbb{Z}_{10}$ (Use Ex. 4 above)

8. No, otherwise $\mathbb{Z} \times \mathbb{Z} \cong \mathbb{Z}$ and hence any nonzero subgroup of $\mathbb{Z} \times \mathbb{Z}$ is of finite index (since, it is so in $\mathbb{Z}$). Note that $\mathbb{Z} \times \{0\}$ is of infinite index in $\mathbb{Z} \times \mathbb{Z}$.

10. Use Corollary 8.1.4 and the facts that $\mathbb{Z}/m\mathbb{Z} \cong \mathbb{Z}_m$ and $m\mathbb{Z} = (p_1^{r_1}\mathbb{Z}) \cap (p_2^{r_2}\mathbb{Z}) \cap \cdots \cap (p_n^{r_n}\mathbb{Z})$.

12. By Corollary 6.4.6, the group of symmetries of a square is indecomposable. Since $n\mathbb{Z} \cap m\mathbb{Z} = [n, m]\mathbb{Z}$, where $[n, m]$ is the l.c.m. of n and m, the intersection of any two nonzero subgroups of $\mathbb{Z}$ is not zero. Therefore, $\mathbb{Z}$ is indecomposable. If A is a nonzero subgroup of $(\mathbb{Q}, +)$, then $A \cap \mathbb{Z}$ is a nonzero subgroup of $\mathbb{Z}$ and hence $A \cap B \neq \{0\}$ for any nonzero subgroups A and B of $(\mathbb{Q}, +)$. Therefore, $(\mathbb{Q}, +)$ is indecomposable.

14. Since $|A \cap B|$ is a common divisor of $|A|$ and $|B|$, we get that $A \cap B = \{e\}$. Also, $|AB| = |A| \cdot |B| = |G|$ and hence $AB = G$. Now, use Corollary 8.1.4.

16. If S_3 is decomposable, then $S_3 \cong A \times B$, where A and B are subgroups of order 2 and 3, respectively, and hence $S_3 \cong \mathbb{Z}_2 \times \mathbb{Z}_3 \cong \mathbb{Z}_6$, which is not true.

18. Straight forward verification.

20. (i) The map $f: G_1 \times G_2 \to G_2 \times G_1$, defined by $f(a, b) = (b, a)$ is an isomorphism.
    (ii) $f: (G_1 \times G_2) \times G_3 \to G_1 \times (G_2 \times G_3)$, defined by $f((a_1, a_2), a_3) = (a_1, (a_2, a_3))$ is an isomorphism.
    (iii) If $f: G_1 \to G_2$ is an isomorphism, then $g: G_1 \times G_3 \to G_2 \times G_3$, defined by $g(a, b) = (f(a), b)$ is an isomorphism.
    (iv) $G_1 \times \{e\}$ and $\{e\} \times G_2$ are subgroups of $G_1 \times G_2$ and are isomorphic to G_1 and G_2, respectively.
    (v) $\mathbb{Z} \times \mathbb{Z}$ is not cyclic and $\mathbb{Z}$ is cyclic.

EXERCISE 8(B)

2. Use Ex. 16 of Exercise 4(B).

4. Follows from $a^m = e \Leftrightarrow a_i^m = e$ for all $i \Leftrightarrow O(a_i)$ divides m for all $1 \leq i \leq n$.

6. Let $a = (a_1, \ldots, a_n) \in G_1 \times \cdots \times G_n$. Then, $a \in \mathbb{Z}(G) \Leftrightarrow a_i b_i = b_i a_i$ for all $b_i \in G_i$ and for all $1 \leq i \leq n \Leftrightarrow a_i \in \mathbb{Z}(G_i)$ for all i.

8. $(x_1, x_2)(a_1, a_2)(x_1, x_2)^{-1} = (x_1 a_1 x_1^{-1}, x_2 a_2 x_2^{-1}) \in N_1 \times N_2$ if $a_i \in N_i$. Define $f: G_1 \times G_2 \to G/N_1 \times G/N_2$ by $f(x_1, x_2) = (x_1 N_1, x_2 N_2)$. Then, f is an epimorphism and $\ker f = N_1 \times N_2$.

10. If G is an infinite cyclic group, then $G \cong \mathbb{Z}$ and, since $\mathbb{Z}$ is not a p-group, G is so.

12. Use Ex. 4 above.

14. Use Ex. 2 above.

16. Use Ex. 14 above and Ex. 12 of Exercise 8(A).

18. Since $G = A_1 A_2 \cdots A_r$, where $A_i = \langle a_i \rangle$, we can use Theorem 8.1.1.

20. Theorem 8.3.1 can be used.

EXERCISE 8(C)

2. Since $\mathbb{P}(X) \cong \mathbb{Z}_2^5 = \mathbb{Z}_2 \times \mathbb{Z}_2 \times \mathbb{Z}_2 \times \mathbb{Z}_2 \times \mathbb{Z}_2$, the invariants of $\mathbb{P}(X)$ are 2, 2, 2, 2, 2.

4. $G \cong \mathbb{Z}_4 \times \mathbb{Z}_4 \times \mathbb{Z}_4 \times \mathbb{Z}_4$.

CHAPTER 9

EXERCISE 9(A)

2. (i) 14

   (ii) 3

   (iii) $\begin{pmatrix} 4 & 4 \\ 4 & 4 \end{pmatrix}$

   (iv) 78

   (v) 0

   (vi) $\{2\}$.

4. (i) $\Rightarrow$ (ii) for any $x \in R$, $x = x + 0 = x + (0 \cdot 0) = (x + 0) \cdot (x + 0) = x^2$ and $x = x + (x \cdot 0) = (x + x) \cdot (x + 0) = x^2 + x^2 = x + x$ and hence $x = 0$.

   (ii) $\Rightarrow$ (iii) $\Rightarrow$ (i) are trivial. Also (ii) $\Leftrightarrow$ (iv) is clear.

6. This follows from $px = 0$ for all $x \in \mathbb{Z}_p$.

8. Straight forward verification. Note that -1 is the new additive identity and 0 is the new multiplicative identity and that $-a - 1 - 1$ is the additive inverse of a.

10. $a = 0 \Rightarrow ab = 0 = ba$.

    $a \neq 0 \Rightarrow ab = ba, \quad$ since $a(ba) = (ab)a$

12. $ab = 0 \Rightarrow ba = (ba)^n = b(ab)^{n-1}a = b \circ a = 0$.

14. The ring of 2×2 matrices over $\mathbb{Z}_2$.

EXERCISE 9(B)

2. If a is a unit, then $ab = 0 \Rightarrow b = a^{-1}(ab) = a^{-1}(0) = 0$.

4. a is a unit $\Rightarrow a = a^2a^{-1} = aa^{-1} = 1$.

   a is nilpotent $\Rightarrow a^n = 0, n \in \mathbb{Z}^+ \Rightarrow a = a^n = 0$.

6. Let $n = p_1^{a_1} \cdots p_r^{a_r}$, where p_i's are distinct primes and $a_i \in \mathbb{Z}^+$. Then, $C(n) = p_1 \cdots p_n$. Any $0 \leq x < n$ is a nilpotent if and only if $C(n)$ divides x and hence x is of the form $C(n)a$, $0 \leq a < \frac{n}{C(n)}$. Thus, the number of nilpotents in $\mathbb{Z}_n$ is $n/C(n)$.

8. Straight forward verification.

10. $(ab)(ab) = a(ba)b = a(ab)b = a^2b^2 = ab$ and hence ab is an idempotent. $a + b$ need not be an idempotent. For example, the unity 1 is an idempotent and $1 + 1$ is not in $\mathbb{Z}_3$.

12. Clear.

14. If n is the least positive integer such that $a^n = 0$, then $n > 1$ (since $a \neq 0$), $a^{n-1} \neq 0$ and $a(a^{n-1}) = 0$ and hence a is a zero divisor.

16. We can suppose that $a \neq 0$. If $a^n = 0$ and $b = -a$, then $b^n = 0$ for some $n > 1$ and $(1 + a)(1 + b + \cdots - b^{n-1}) = (1 - b)(1 + b + \cdots + b^{n-1}) = 1$ and hence $1 + a$ is a unit.

18. Consider $(axa - ax)^2 = axa \cdot axa + ax \cdot ax - axa \cdot ax - ax \cdot axa = axaxa + axax - axax - axaxa = 0$ and hence $axa - ax$ is a nilpotent, so that $axa - ax = 0$; i.e., $axa = ax$. Similarly, $axa = xa$.

EXERCISE 9(C)

2. $(a + b)^{3r} = a^{3r} + b^{3r}$ (since 3 divides ${}^{3r}C_s$ for any $0 < s < 3r$).

4. If R is a finite ring, then $O(a)$ is finite in the group $(R, +)$ and, if m is the l.c.m. $\{O(a) : a \in R\}$ then $ma = 0$ for all $a \in R$ and hence char(R) is finite.

6. $\mathbb{Z}_5$.

8. $(a + b)^p = a^p + b^p = a + b$ and $(a \cdot b)^p = a^p b^p = ab$. Also, $a \in A \Rightarrow a = a^p \Rightarrow a(1 - a^{p-1}) = 0 \Rightarrow a = 0$ or $a^{p-1} = 1 \Rightarrow a = 0$ or a is a unit.

10. Direct verification.

12. Let $ab = 1$ and $R = \{a_1, a_2, \ldots, a_n\}$. Consider $Ra = \{a_1a, a_2a, \ldots, a_na\} \subseteq R$. For any i and j,

$$a_ia = a_ja \Rightarrow (a_i - a_j)a = 0 \Rightarrow a_i - a_j = (a_i - a_j)ab = 0$$

Therefore, Ra and R have the same number of elements and hence $R = Ra$. In particular, $1 = ca$ for some $c \in R$. Now, $b = 1b = (ca)b = c(ab) = c$ and hence $ba = 1$.

EXERCISE 9(D)

2. Direct verification.

4. S is a subring of $(\mathbb{Z}_n, +_n, \cdot_n) \Leftrightarrow S$ is a subgroup of $(\mathbb{Z}_n, +_n) \Leftrightarrow S = \langle m \rangle$ for some divisor m of n.

6. Let $A = \{x_1x_2 \cdots x_n : n > 0 \text{ and } x_1, x_2, \ldots, x_n \in X\}$.

   Then, $(X) = \{a_1 + a_2 + \cdots + a_n : a_i \text{ or } -a_i \in A\}$.

8. Let X be an infinite set and R be the set of all finite subsets of X. Then, $(R, +, \cap)$ is a ring without unity. For any finite subset A of X, $\mathbb{P}(A)$ is a subring of R and A is the unity in $\mathbb{P}(A)$. (Here, $+$ is the symmetric difference of sets.)

10. $S \cup T$ is a subring $\Rightarrow S \cup T$ is a subgroup of $(R, +) \Rightarrow S \subseteq T$ or $T \subseteq S$.

12. Follows from Theorem 4.1.2.

14. Consider $(a + b)^2 + (a + b) \in c(R)$. Therefore, $ab + ba \in c(R)$. $a(ab + ba) = (ab + ba)a$ and hence $a^2b = ba^2$. Since $(a^2 + a)b = b(a^2 + a)$, we get $ab = ba$.

EXERCISE 9(E)

2. (i) Yes; since $\mathbb{Z}_n$ is finite, any injection of $\mathbb{Z}_n$ into $\mathbb{Z}_n$ is a bijection.
   (ii) No, there is no homomorphism of $\mathbb{Z}_2$ into $\mathbb{Z}_3$.
   (iii) Yes; since $\mathbb{Z}_n$ will be isomorphic to a subgroup of $\mathbb{Z}_m$ and we can use Lagrange's Theorem.
   (iv) Yes; if $f: \mathbb{Z}/m\mathbb{Z} \to \mathbb{Z}/n\mathbb{Z}$ is defined by $f(a + m\mathbb{Z}) = a + n\mathbb{Z}$, then f is a well-defined epimorphism.
   (v) Yes, if R is not trivial.
   (viii) Yes, namely the zero map and the identity map.

EXERCISE 9(F)

2. The ring of all 2×2 matrices over $\mathbb{Z}_3$.

4. No; consider $\begin{pmatrix} 0 & 0+1 \\ 0+1 & 1 \end{pmatrix} \cdot \begin{pmatrix} 1 & 1+0 \\ 1+0 & 0 \end{pmatrix} = \begin{pmatrix} 1 & 0 \\ 2 & 1 \end{pmatrix}$.

8. 0 is the only nilpotent and 0 and 1 are the only idempotents.

10. Direct verification.

12. The set of all scalar matrices; that is, matrices $A = (a_{ij})$ such that $a_{ij} = 0$ for all $i \neq j$ and $a_{ii} = a_{jj}$ for all i and j.

14. $\mathbb{R}$

16. If $A = (a_{ij})$ is such that $BA = B = AB$ for all $B \in M_n(R)$, then prove that $a_{ij} = 1$ or 0 according as $i = j$ or $i \neq j$.

EXERCISE 9(G)

2. (i) $\mathbb{R} \times \mathbb{R}$ is not integral domain and hence not a field, since $(1, 0) \cdot (0, 1) = (0, 0)$
   (ii) $\mathbb{Q}^n$ is not an integral domain, not a field.

(iii) $\mathbb{Z}[\sqrt{2}]$ is an integral domain, but not a field.
(iv) $\mathbb{Z}[i]$ is an ID, but not a field.
(v) $\mathbb{Z}_5[i]$ is a field.
(vi) $\mathbb{Z}_4[i]$ is not an ID.
(vii) $\mathbb{Z}_3[i]$ is a field.
(viii) $\mathbb{Z}_2[i]$ is a field.
(ix) $\mathbb{Z}_5 \times \mathbb{Z}_3$ is not an ID.
(x) $\mathbb{Z}_{19}$ is a field.

4. R is a field and Boolean ring $\Rightarrow R$ is an ID and $a^2 = a \Rightarrow a(a-1) = 0$ $\Rightarrow a = 0$ or $a = 1$ for all $a \in R \Rightarrow R = \{0, 1\}$, converse is clear.

6. Direct verification.

8. $\mathbb{Z}_n[i]$ is finite.

12. No.

14. $(R - \{0\}, \cdot)$ is a finite semigroup satisfying the cancellation laws. By Theorem 3.3.5, $(R - \{0\}, \cdot)$ is a group. Therefore, R is a field.

16. If F is a finite field, then F is an ID and $\text{char}(F) \neq 0$. Therefore, by Theorem 9.7.4, $\text{char}(F)$ is prime.

18. Imitate Ex. 14 above.

20. Let n be the least positive integer such that $na = 0$. Then, $(n-1)a \neq 0$ and $(nb) \cdot (n-1)a = (n-1)b \cdot na = 0$ and hence $nb = 0$ for all $b \in R$. Thus, $\text{char}(R) > 0$.

CHAPTER 10

EXERCISE 10(A)

2. (i) $\mathbb{Z}$ is not an ideal of $\mathbb{Q}$, since $\frac{1}{2} \cdot 1 \notin \mathbb{Z}$.
(ii) $\mathbb{Q}$ is not an ideal of $\mathbb{R}$, $\sqrt{2} \cdot 1 \notin \mathbb{Q}$.
(iii) No, $\mathbb{Z}$ is a subring of $\mathbb{Q}$, but not an ideal.
(iv) True.
(v) True.
(vi) True, $\{0\}$ is an ideal in any ring $\mathbb{R}$.
(vii) No, if $\mathbb{R}$ is finite and $|\mathbb{R}| = n$, then there are at most 2^n ideals in $\mathbb{R}$.

(viii) $\mathbb{Z}_4$ has exactly three ideals, namely $\{0\}$, $\{0, 2\}$ and $\mathbb{Z}_4$.

4. If R is a division ring, I is a nonzero ideal of R and $0 \neq a \in I$, then $1 = aa^{-1} \in I$ and hence $I = R$, so that R has only two ideals. The converse is not true; for, $M_2(\mathbb{R})$, the set of 2×2 matrices forms a ring with exactly two ideals, but it is not a division ring.

6. (i)
$$\begin{aligned} x \in (I + J)^* &\Leftrightarrow x(a + b) = 0 \text{ for all } a \in I, b \in J. \\ &\Leftrightarrow xa = 0 = xb \quad \text{for all } a \in I, b \in J \\ &\Leftrightarrow x \in I^* \cap J^* \\ &\Leftrightarrow x \in (I \cup J)^*. \end{aligned}$$

(ii) Clear.

(iii) Since $I \cap J \subseteq I$ and J, $I^* \subseteq (I \cap J)^*$ and $J^* \subseteq (I \cap J)^*$ and hence $I^* + J^* \subseteq (I \cap J)^*$.

8. Let I be an ideal and $p \in I$, where p is a prime. Then, $p\mathbb{Z} \subseteq I$. If $p\mathbb{Z} \neq I$, then there exists $a \in I$ such that $a \notin p\mathbb{Z}$; that is, p does not divide a and hence $(a, p) = 1$. Then, there exist integers α and β such that $1 = \alpha a + \beta p \in I$ and hence $I = \mathbb{Z}$.

10. (i) $r(I \cap J) \subseteq r(I)$ and $r(J)$

$$\begin{aligned} a \in r(I) \cap r(J) &\Rightarrow a^n \in I \text{ and } a^m \in J, \text{ for some } n, m \in \mathbb{Z}^+ \\ &\Rightarrow a^{n+m} \in I \cap J, n + m \in \mathbb{Z}^+ \\ &\Rightarrow a \in r(I \cap J). \end{aligned}$$

(ii) $I \subseteq J$ and $a \in r(I) \Rightarrow a^n \in I \subseteq J \Rightarrow a \in r(J)$.

(iii) $r(I) \subseteq r(I + J)$ and $r(J) \subseteq r(I + J)$, since I and $J \subseteq I + J$. Therefore, $r(I) + r(J) \subseteq r(I + J)$.

(iv) $I + J \subseteq r(I) + r(J)$. Hence, $r(I + J) \subseteq r(r(I) + r(J))$.

$$\begin{aligned} a \in r(r(I) + r(J)) &\Rightarrow a^n = x + y, n \in \mathbb{Z}^+, x \in r(I) \text{ and } y \in r(J) \\ &\Rightarrow a^n = x + y, x^r \in I, y^s \in J \\ &\Rightarrow a^{n(r+s)} = (x + y)^{r+s} \in I + J \\ &\Rightarrow a \in r(I + J). \end{aligned}$$

12. (i) $24\mathbb{Z}$

(ii) $3\mathbb{Z}$

(iii) $3\mathbb{Z}$

(iv) $6\mathbb{Z}$

(v) $\{0\}$

(vi) $2\mathbb{Z}$

22. Let I be an ideal in $M_n(R)$ and J be the set of all 11th (first row and first column) entries in members of I. Then prove that $I = M_n(J)$, by establishing that, if $A = (a_{ij})$ and E_{ij}'s are the matrix units, then

$$E_{1r}\left(\sum a_{is}E_{is}\right)E_{j1} = a_{rj}E_{11}$$

If R is simple, then $J = \{0\}$ or R and hence $I = \{0\}$ or $M_n(R)$. Thus, $M_n(R)$ is a simple ring.

EXERCISE 10(B)

2. (i) R has unity $\Rightarrow R/I$ has unity. The converse fails.
   (ii) True.
   (iii) True.
   (iv) False; for $\mathbb{Z}$ is an ID and $\mathbb{Z}/6\mathbb{Z}$ is not an ID.
   (v) False.
   (vi) False; for $R/R = \{0\}$.
   (vii) True; for $R/\{0\} = R$.

4. Suppose that the multiplication of additive cosets of S are well defined. Then, for any $a \in S$ and $r \in R$, $a + S = 0 + S$ and hence $ar + S = 0r + S = S$ which implies that $ar \in S$. Similarly, $ra \in S$. Thus, S is an ideal of R. Converse is trivial.

6.

	Idempotents	Nilpotents	Units
$\mathbb{Z}/I;\ I = 6\mathbb{Z}$	$I, 1 + I, 3 + I, 4 + I$	I	$1 + I; 5 + I$
$\mathbb{Z}/I;\ I = 8\mathbb{Z}$	$I, 1 + I$	$I, 2 + I, 4 + I, 6 + I$	$1 + I, 3 + I, 7 + I, 5 + I$
$\mathbb{Z}/I;\ I = 7\mathbb{Z}$	$I, 1 + I$	I	$a + I, 1 \leq a \leq 6.$

8. (i) $\Leftrightarrow$ (ii): n is prime $\Leftrightarrow (a, n) = 1$ for all $1 \leq a < n \Leftrightarrow ab \equiv 1 \pmod{n}$ for some b, for each $1 \leq a < n \Leftrightarrow a + n\mathbb{Z}$ is a unit for all $1 \leq a < n$.

   (ii) $\Leftrightarrow$ (iii) is clear, since $\mathbb{Z}/n\mathbb{Z}$ is finite.

10. $n\mathbb{Z}$, $n \geq 0$, are the only ideals of $\mathbb{Z}$.

12. If $I = 2\mathbb{Z}$ in $\mathbb{Z}$, then $(0 + I)(0 + I) = 4\mathbb{Z} \subsetneqq 2\mathbb{Z} = 0 + I$

14. If $(n, m) = 1$, then $n\mathbb{Z} \cap m\mathbb{Z} = nm\mathbb{Z}$ and $n\mathbb{Z} + m\mathbb{Z} = \mathbb{Z}$ and hence $\mathbb{Z}_n \cong \mathbb{Z}/n\mathbb{Z} = n\mathbb{Z} + m\mathbb{Z}/n\mathbb{Z} \cong m\mathbb{Z}/n\mathbb{Z} \cap m\mathbb{Z} = m\mathbb{Z}/nm\mathbb{Z}$.

16. The ideals of $\mathbb{Z}/n\mathbb{Z}$ are of the form $d\mathbb{Z}/n\mathbb{Z}$, where d is a divisor of n and the homomorphic images of $\mathbb{Z}/n\mathbb{Z}$ are of the form $\mathbb{Z}/d\mathbb{Z}$, where d is a divisor of n.

EXERCISE 10(C)

2. Direct verification.

EXERCISE 10(D)

2. Let $P = \bigcap_{\alpha\in\Delta} P_\alpha$. Then, clearly P is a proper ideal of R. If $a \notin P$ and $b \notin P$, then there exist α and $\beta \in \Delta$ such that $a \notin P_\alpha$ and $b \notin P_\beta$. Choose $\gamma \in \Delta$ for which $P_\gamma \subseteq P_\alpha$ and $P_\gamma \subseteq P_\beta$. Then, a and $b \notin P_\gamma$ and hence $ab \notin P_\gamma$, so that $ab \notin P$.

4. Use Ex. 2 and 3 above.

6. Clearly prime $\Rightarrow$ primary. $4\mathbb{Z}$ is primary and not prime in $\mathbb{Z}$.

8. Note that $1 \in \sqrt{I} \Leftrightarrow 1 \in I$. Since I is proper, $\sqrt{I}$ is a proper ideal of R and $I \subseteq \sqrt{I}$. Also, $ab \in \sqrt{I} \Rightarrow (ab)^n \in I$ for some $m > 0 \Rightarrow a^n b^n \in I$ $\Rightarrow a^n \in I$ or $b^n \in \sqrt{I} \Rightarrow a \in \sqrt{I}$ or $b \in \sqrt{I}$. Thus, $\sqrt{I}$ is a prime ideal. Also, if $I \subseteq P$ and P is a prime ideal, then $\sqrt{I} \subseteq \sqrt{P} = P$.

10. $ab \equiv 0 \pmod{p^n} \Rightarrow p^n$ divides $ab \Rightarrow p$ divides a or $b \Rightarrow p^n$ divides a^n or $b^n \Rightarrow a^n \equiv 0 \pmod{n}$ or $b^n \equiv 0 \pmod{n}$.

12. Let R be regular and I be an ideal of R. If $a \in \sqrt{I}$, then $a^n \in I$ and choose $b \in R$ such $aba = a$. Now, $abab = ab$ and $ab = (ab)^n = a^n b^n \in I$ and hence $a = (ab)a \in I$. Therefore, $\sqrt{I} = I$. For the converse, if $a \in R$, then aRa is an ideal and $a^3 \in aRa$ and hence $a \in \sqrt{aRa} = aRa$.

14. (i) $\Rightarrow$ (ii): Let P be the unique prime ideal of R. Then, $P = N$, the set of nilpotents in R. If a is a nonunit in R, use the Zorn's lemma to get a prime ideal $(= p)$ containing a and hence $a \in N$.

    (ii) $\Rightarrow$ (iii): $ab = 0$ and $b^n \neq 0$ for all $n > 0 \Rightarrow b$ is not nilpotent $\Rightarrow b$ is a unit (by (ii)) $\Rightarrow a = (ab)b^{-1} = 0$. Thus, $\{0\}$ is primary. If $a \neq 0$ is a nonunit, then $a^n = 0$ (by (ii)) for some $n > 1$. If n is least such that $a^n = 0$, then $a^{n-1} \cdot a = 0$ and $a^{n-1} \neq 0$ and hence a is a zero divisor.

    (iii) $\Rightarrow$ (i): $N =$ radical of $\{0\}$. If $\{0\}$ is primary, then N is a prime ideal. If P is any prime ideal, then $N \subseteq P$. Also, $0 \neq a \in P \Rightarrow a$ is a nonunit $\Rightarrow a$ is a zero divisor (by (iii)) $\Rightarrow ab = 0$ for some $b \neq 0 \Rightarrow a^n = 0$ for some $n \Rightarrow a \in N$. Therefore, $P \subseteq N$ and $P = N$. Thus, N is the only prime ideal of R.

16. If $m > 1$ and m^2 divides n, then $m^2r = n$ for some $r \geq 1$. Now, for any $a \in R$, $(mra)^2 = m^2r^2a = n(ra) = 0$ and hence $mra \in N(R) = \{0\}$, so that $mra = 0$, which is a contradiction to the least property of $n = \text{char}(R)$.

EXERCISE 10(E)

2. $9\mathbb{Z}$

4. Use Ex. 3 and the fact that a nonzero ideal of $\mathbb{Z}$ is prime if and only if it is maximal.

6. Let $X = \{x_1, x_2, \ldots, x_n\}$ and M be a maximal ideal in $(\mathbb{P}(x), +, \cap)$. Then, $\sum_{i=1}^{n} \{x_i\} = X \notin M$ and hence $\{x_i\} \notin M$ for some i and, since $\{x_i\} \cap \{x_j\} = \varnothing \in M$, $\{x_j\} \in M$ for all $j \neq i$. Let

$$M_i = \{A \subseteq X : x_i \notin A\}.$$

Then, M_i is a maximal ideal of $\mathbb{P}(X)$ and $M \subseteq M_i$ so that $M = M_i$.

8. Let $x \in X$ and define $\alpha : \mathscr{C}(X, \mathbb{R}) \to \mathbb{R}$ by $\alpha(f) = f(x)$ for any $f \in \mathscr{C}(X, \mathbb{R})$. α is an epimorphism of rings and $\ker\alpha = M_x$. Therefore, $\mathscr{C}(X, \mathbb{R})/M_x \cong \mathbb{R}$. Since $\mathbb{R}$ is a field, so is $\mathscr{C}(X, \mathbb{R})/M_x$. Thus, M_x is a maximal ideal.

10. Consider the map $f: \mathbb{Z}[i] \to \mathbb{Z}_2$ defined by $f(a + bi) = 0$ if $a - b$ is even and, $= 1$ if $a - b$ is odd. Prove that f is an epimorphism of rings and $\ker f = I$. Therefore, $\mathbb{Z}[i]/I \cong \mathbb{Z}_2$ and hence I is a maximal ideal and the quotient has exactly 2 elements.

12. I is not a prime ideal and $|\mathbb{Z}[i]/I| = 25$.

14. Use the fact that $R/\ker f \cong S$ and that $I \mapsto f(I)$ is a one-to-one correspondence between the set of ideals of R containing $\ker f$ and the set of ideals of S. Also, $I \subseteq J \Rightarrow f(I) \subseteq f(J)$.

16. R/P becomes a finite commutative ring without zero divisors and hence R/P is a field and P is a maximal ideal.

18. M/M^n is the only prime ideal of R/M^n; for, if Q is a prime ideal of R/M^n, then $Q = P/M^n$ for some prime ideal P of R containing M^n and hence M, so that $P = M$.

20. (i) For any $0 \neq a \in R$, $1 - ra$ is a nonunit $\Leftrightarrow 1 - ra \in M$ for some max ideal $M \Leftrightarrow a \notin M$, for some $M \Leftrightarrow a \notin J(R) \Leftrightarrow J(R) = \{0\} \Leftrightarrow$ R is semisimple.

    (ii) If $0 \neq a \in R$, then $aba = a$ for some $b \in R$ and hence $a(1 - ba) = 0$, so that $1 - ba$ is a nonunit.

    (iii) Let $J(R)$ be the Jacobson radical of R. Then,

$$\begin{aligned} a \in J(R) &\Rightarrow a + I \in M/I \text{ for all maximal ideals } M \text{ of } R \\ &\Rightarrow a + I \in J(R/I) \Rightarrow a + I = I \Rightarrow a \in I. \end{aligned}$$

EXERCISE 10(F)

2. Since $\text{char}(\mathbb{Z}_n) = n = \text{char}(R)$, $n(a, r) = (0, 0)$ for all $(a, r) \in \mathbb{Z}_n \times R$ and n is the least such positive integer. Therefore, $\text{char}(\mathbb{Z}_n \times R) = n$.

4. The field of quotients of R is $\mathbb{Q}$ itself, since any nonzero $\frac{m}{n}$ in $\mathbb{Q}$ can be expressed as $\frac{m}{2} \cdot \left(\frac{n}{2}\right)^{-1}$ and $\frac{m}{2}$ and $\frac{n}{2} \in R$.

6. If F is a field, consider the homomorphism $f: \mathbb{Z} \to F$ defined by $f(n) = n \cdot 1$, where 1 is the unity in F. Then, $\ker f = \{0\}$ or $p\mathbb{Z}$ for some prime p (if $p = \text{char}(F)$). Then, $\mathbb{Z}$ or $\mathbb{Z}_p$ ($= \mathbb{Z}/p\mathbb{Z}$) is isomorphic to a subring of F and hence $\mathbb{Q}$ or $\mathbb{Z}_p$ is isomorphic to a subfield of F.

8. (ii) $\mathbb{Q}$

   (iii) $\mathbb{Z}_{79}$

10. If $|F| = 9$, then $\text{char}(F) = 3$ and $\mathbb{Z}_3$ is the prime subfield of F.

12. Let R be an integral domain and $f: R \to R$ an automorphism. Define f^*: $F \to F$ by $f^*\left(\frac{a}{b}\right) = \frac{f(a)}{f(b)}$ for any $\frac{a}{b}$ in the field F of quotients of F. It can be checked that f^* is well defined and is an automorphism of F. Also, $f^*/R = f$.

14. By Ex. 13, R can be embedded in an integral domain S and hence R can be embedded in the field of quotients of S.

16. Refer the construction of field of quotients.

18. For any prime ideal P of R disjoint with S, $S^{-1}P = \{\frac{a}{s} : a \in P, s \in S\}$ is a prime ideal of $S^{-1}R$ and any prime ideal of $S^{-1}R$ is of this form. It can be checked that $P \mapsto S^{-1}P$ is a one-to-one correspondence between the prime ideals of R disjoint with S and the prime ideals of $S^{-1}R$.

CHAPTER 11

EXERCISE 11(A)

2. (i) True, since $\mathbb{Z}_2$ is an ID.
   (ii) False, since x has no inverse in $R[x]$, for any ring R.
   (iii) False, since $\mathbb{Z}_6$ is not an ID.
   (iv) False, $R[x]$ is always infinite.
   (v) False, since 1 and -1 are the only units in $\mathbb{Z}[x]$.
   (vi) True, since $2^2 = 0$ in $\mathbb{Z}_4$ and hence in $\mathbb{Z}_4[x]$.
   (vii) False
   (viii) False.

4. Direct verification.

6. Let $f \in \mathbb{R}[x]$. Then, f is a unit $\Rightarrow fg = 1, g \in \mathbb{R}[x] \Rightarrow \deg f + \deg g = 0$ $\Rightarrow \deg f = 0 = \deg g \Rightarrow f$ is a nonzero constant polynomial.

8. $2^6 - 1 = 63$.

10. $I = xR[x]$.

12. First observe that, if $a^n = 0$, then $(1 - a)(1 + a + \cdots + a^{n-1}) = 1$ and hence $1 - a$ is a unit. Therefore, if u is a unit and a is a nilpotent in any commutative ring R, then $u - a = u(1 - u^{-1}a)$ is a unit. Prove the result by induction on the degree of $f(x)$. The result is trivial if $\deg f(x) = 0$. Let $\deg f(x) = n > 0$ and assume the result for all polynomials of degree less than n. Let $f(x) = a_0 + a_1x + \cdots + a_nx^n$. Suppose that $f(x)$ is a unit in $R[x]$. Then, $f(x)g(x) = 1$ for some $g(x) = b_0 + b_1x + \cdots + b_mx^m \in R[x]$. Then, we have (1) $a_nb_m = 0$, (2) $a_{n-1}b_m + a_nb_{m-1} = 0$, (3) $a_{n-2}b_m + a_{n-1}b_{m-1} + a_nb_{m-2} = 0$, and, in general $\sum\limits_{i+j=k} a_ib_j = 0$ for $k > 0$ and $a_0b_0 = 1$. Thus, a_0 is a unit and $a_0^{-1} = b_0$. By multiplying (2) with a_n, we get $a_n^2\, b_{m-1} = 0$. Also, by multiplying (3) with a_n^2, we get $a_n^3b_{m-2} = 0$ and, in general $a_n^ib_{m-i+1} = 0$. In particular, $a_n^{m+1}b_0 = 0$ and hence $a_n^{m+1} = 0$. Therefore, a_n is nilpotent and so is a^nx^n and hence $f(x) - a^nx^n$ is a unit and is of degree less than n. By induction hypothesis, we get that $a_{n-1}, a_{n-2}, \ldots, a_1$ are nilpotents and a_0 is a unit in R. Converse is clear from our first observation.

14. Clearly, $I[x]$ is a proper ideal of $R[x]$. Suppose that I is a prime ideal of R. Let $f(x) = a_0 + a_1x + \cdots + a_nx^n$ and $g(x) = b_0 + b_1x + \cdots + b_mx^m \in R[x] - I[x]$. Let i be the least such that $a_i \notin I$ and j is the least such that $b_j \notin I$.

Then, $a_ib_j \notin I$, $a_r \in I$, $b_s \in I$ for all $r < i$ and $s < j$. Then, the coefficient of x^{i+j} in $f(x)g(x)$ is not in I and hence $f(x)g(x) \notin I[x]$. Thus, $I[x]$ is a prime ideal of $R[x]$. Converse is clear.

16. Define $\alpha : R[x] \rightarrow R$ by $\alpha(f(x))$ = the constant term in $f(x)$. Then, α is an epimorphism of rings and its kernel is $<x>$. Thus, $R[x]/<x> \cong R$.

EXERCISE 11(B)

2. (i) 2
   (ii) 1
   (iii) 0
   (iv) 4
   (v) 4
4. The Kernel of f_a is an ideal of $F[x]$ and any nonzero ideal of $F[x]$ is infinite.
6. $(1 + 4x)(1 + 4x) = 1$ in $\mathbb{Z}_8[x]$.
8. $f(a) = g(a) \Leftrightarrow a$ is a root of the polynomial $f(x) - g(x)$.
10. Use Fermat's Theorem 4.4.7.

EXERCISE 11(C)

2. Let P be a prime ideal and I be an ideal of R such that $P \subseteq I$. Choose a and $b \in R$ such that $P = aR$ and $I = bR$. Then, $a = bc$ for some $c \in R$ and hence $bc \in P$, so that $b \in P$ or $c \in P$. If $b \in P$, then $I = P$. If $c \in P$, then $c = ad$ and hence $a = bad$.
4. Let $I = \{f(x) \in F[x] : a \text{ is a root of } f(x)\}$. We can assume that $I \neq \{0\}$. If $g(x)$ is a nonzero polynomial of least degree in I, then we can prove that I is the ideal generated by $g(x)$ in $F[x]$ (use division algorithm).
6. Every element of $\mathbb{Z}_3$ is a root of $x^3 - x$ and hence of $(x^3 - x)f(x)$ for any $f(x) \in \mathbb{Z}_3[x]$.
8. By Euler's Theorem 4.4.6, $a^{p-1} \equiv 1 \pmod p$ for all $1 \leq a \leq p - 1$ and hence each $1 \leq a \leq p - 1$ is a root of $x^{p-1} - 1 \in \mathbb{Z}_p[x]$. Also, each $1 \leq a \leq p - 1$ is a root of $f(x) = (x - 1)(x - 2) \ldots (x - (p - 1)) - (x^{p-1} - 1)$. Therefore, $f(x)$ has $p - 1$ roots and its degree is $p - 2$. Therefore, $f(x)$ must be the zero polynomial.
10. $\mathbb{Z}[x]/<x>$ ($\cong \mathbb{Z}$) is an integral domain, but not a field and hence $<x>$ is a prime ideal but not a maximal ideal in $\mathbb{Z}[x]$.

12. The evaluation map $f_i : \mathbb{R}[x] \to \mathbb{C}$, defined by $f_i(g(x)) = g(i)$, is an epimorphism and its kernel is $<1 + x^2>$.

14. Let $f(x) = a_0 + a_1x + \cdots + a_{n-1}x^{n-1} + x^n$, $a_i \in \mathbb{Z}$ and $a = \frac{r}{s}$, where r and $s \in \mathbb{Z}$, $(r, s) = 1$ and $s > 0$. Suppose that $f(a) = 0$. Then, $a_0 + a_1\left(\frac{r}{s}\right) + \cdots + a_{n-1}\left(\frac{r}{s}\right)^{n-1} + \left(\frac{r}{s}\right)^n = 0$ and hence $s^n a_0 + s^{n-1}a_1 r + \cdots + sa_{n-1}r^{n-1} + r^n = 0$. If $s > 1$ and p is a prime dividing s, then p should divide r^n and hence p divides r which is a contradiction to the assumption that $(r, s) = 1$. Therefore, $s = 1$ and $a \in \mathbb{Z}$.

EXERCISE 11(D)

2. Follows from Theorem 11.4.2.

4. $1 + x^2$ is irreducible over $\mathbb{Z}_3$ and hence $\mathbb{Z}_3[x]/<1 + x^2>$ is a field and has exactly 9 elements.

6. a is a unit in $F[x]$ and hence $f(x)$ and $af(x)$ are associates.

8. $1 + 2x^2 + x^4$ has no root in $\mathbb{Z}_3$, but it is reducible in $\mathbb{Z}_3[x]$, since $1 + 2x^2 + x^4 = (1 + x^2)(1 + x^2)$.

10. Follows from the fact that R is an ID if and only if $R[x]$ is an ID.

12. Consider the epimorphism $f(x, y) \mapsto f(x, -x)$ of $F[x, y]$ onto $F[x]$. Its Kernel is $<x + y>$. Therefore, $F[x, y]/<x + y> \cong F[x]$. Similarly, $F[x, y]/<x + y> \cong F[y]$.

14. For each $n \in \mathbb{Z}^+$, $(1 + 2x^n)(1 + 2x^n) = 1$ in $\mathbb{Z}_4[x]$ and hence $1 + 2x^n$ is a unit in $\mathbb{Z}_4[x]$ for all $n \in \mathbb{Z}^+$. Also, $(2x^n)^2 = 0$ in $\mathbb{Z}_4[x]$ and hence $2x^n$ is a nilpotent in $\mathbb{Z}_4[x]$ for all $n \in \mathbb{Z}^+$.

CHAPTER 12

EXERCISE 12(A)

2. (i) $1, -1$
   (ii) $1, -1, i, -i$
   (iii) $\pm 1, \pm(1+\sqrt{2}), \pm(1-\sqrt{2})$
   (iv) ± 1
   (v) $\mathbb{R} - \{0\}$
   (vi) $1, 2, 3, 4$

4. $(2+\sqrt{3})(2-\sqrt{3})=1$ and $(2+\sqrt{3})\sqrt{3}=3+2\sqrt{3}$

6. $1 + 2x + 3x^2$, $2 + 4x - x^2$, $3 - x + 2x^2$, $-3 + x + 5x^2$, $-2 + 3x + x^2$, $-1 - 2x - 3x^2$

8. Use induction on n.

10. Let $a = bu$, where u is a unit. Then, $a = bc \Rightarrow bu = bc \Rightarrow u = c$.

EXERCISE 12(B)

2. Refer Worked Exercise 11.3.1.

4. R is a PID and S is a homomorphic image of R, then $S \cong R/I$ for some ideal I of R and any ideal of R/I is of the form $<x>/I$ for some $x \in R$ such that $I \subseteq <x>$ and then $x + I$ generates $<x>/I$. Therefore, R/I and hence S is a PID.

6. If p is an odd prime, then the ideal $\langle 2, \sqrt{-p} \rangle$ generated by 2 and $\sqrt{-p}$ in $\mathbb{Z}[\sqrt{-p}]$ is not a principal ideal.

8. Let I be a nonzero ideal of $\mathbb{Z}[\sqrt{-2}]$ and $A = \{a^2 + 2b^2 : 0 \neq a + b\sqrt{-2} \in I\}$. Choose $x = a + b\sqrt{-2} \in I$ such that $a^2 + 2b^2$ is least in A. Then, I is the ideal generated by x in $\mathbb{Z}[\sqrt{-2}]$.

10. If I is an ideal of $S^{-1}R$ and $J = \{a \in R : \frac{a}{1} \in I\}$, then, J is an ideal of R and hence $J = <a_0> = a_0R$ for some $a_0 \in J$. Now, $\left\langle \frac{a_0}{1} \right\rangle = I\left(\frac{a}{s} \in I \Rightarrow \frac{a}{1} = \frac{a}{s} \cdot \frac{s}{1} \in I \Rightarrow a \in J \Rightarrow a = a_0 r, r \in R \text{ and } \frac{a}{s} = \frac{a_0}{1} \cdot \frac{r}{s} \right)$.

12. See Ex. 4 above and consider $R/\{0\} = R$.

14. M is a maximal ideal of a PID if and only if $M = <p>$ for some irreducible element. Also, $<p> = <q> \Leftrightarrow p$ and q are associates.

16. R is not an integral domain and hence not PID.

18. Let P be a nonzero ideal of R. Suppose that P is primary. Then, the radical $r(P)$ is prime and hence maximal ideal in R, so that $r(P) = <p>$ for some prime element p. If $p \in P$, then $P = <p>$. Suppose $p \notin P$. Choose least n such that $p^n \in P$. Then prove that $P = <p^n>$. Conversely, suppose that $P = <p^n>$, where p is a prime in R and $n \in \mathbb{Z}^+$. Let a and $b \in R$ such that $ab \in P$ and $a \notin P$. Then, $p^n | ab$ and $p^n \nmid a$. Let m be the largest integer such that $p^m | a$. Then, $p^n c = a$ for some $c \in \mathbb{R}$ such that $p \nmid c$. Also, $m < n$ and $ab = p^m(cb)$, since $p^{m+1} | ab$, it follows that $p | b$ and $b^n \in <p^n> = P$. Thus, P is a primary ideal.

EXERCISE 12(C)

2. Let $a_1, a_2, \ldots, a_n$ be elements of R. If some $a_i = 0$, then 0 is the unique l.c.m. of $a_1, a_2, \ldots, a_n$. Suppose that $a_i \neq 0$ for each i, we can write

$$a_1 = p_1^{r_1} p_2^{r_2} \cdots p_t^{r_t} \quad \text{and} \quad a_2 = p_1^{s_1} p_2^{s_2} \cdots p_t^{s_t}$$

for some distinct prime elements $p_1, p_2, \ldots, p_t$ and nonnegative integers r_i, s_i. Then, $\prod_{i=1}^{t} p_i^{\max\{r_i, s_i\}}$ is the l.c.m. of $\{a_1\ a_2\}$. We can extend this to $a_1, a_2, \ldots, a_n$ using induction.

4. Follows from the observations that, for any $s \in S$, $\frac{1}{s}$ is a unit in $S^{-1}R$ and hence $\frac{a}{s}$ is prime (irreducible) if and only if $\frac{a}{1}$ is prime (irr) in $S^{-1}R$. Also, if a is prime in R, then $\frac{a}{1}$ is prime in $S^{-1}R$.

6. Any nonzero nonunit a can be written as a product $p_1 p_2 \ldots p_n$, where p_i's are primes and hence $a \sim p^n$. Let I be a nonzero proper ideal in R. Then, $p^n \in I$ for some $n \in \mathbb{Z}^+$. Let n be the least such that $p^n \in I$. Then, $<p^n> \subseteq I$. Also, $0 \neq a \in I \Rightarrow a$ is a nonunit $\Rightarrow a = p^m, n \leq m \Rightarrow a = p^m . p^{m-n} \in <p^n>$. Thus, $I = <p^n>$.

8. Note that $<a> \subseteq <b> \Leftrightarrow b$ divides a. Also, if $a = p_1^{r_1} p_2^{r_2} \cdots p_n^{r_n}$, where p_i's are distinct primes and r_i's are nonnegative integers, then there are at most finitely many principal ideals containing $<a>$. Thus, there cannot be a strictly increasing infinite sequence of principal ideals.

10. These follow from the fact that any two nonzero elements a and b can be written as $a = p_1^{r_1} p_2^{r_2} \cdots p_n^{r_n}$, and $b = p_1^{s_1} \cdots p_n^{s_n}$, where p_i's are distinct primes and r_i and s_i are nonnegative integers and that

$$(a, b) = \prod_{i=1}^{n} p_i^{\min\{r_i, s_i\}}$$

$$\text{and} \quad [a, b] = \prod_{i=1}^{n} p_i^{\max\{r_i, s_i\}}$$

EXERCISE 12(D)

2. There are no nonzero nonunits in any field.

4. Use Theorem 12.4.4 and Corollary 12.4.5 or the fact that $f(x)$ of degree ≤ 3 is reducible over a field F if and only if $f(x)$ has a root in F.

(i) $f(x) = 15 - 9x^2 + 6x^3 + 2x^4$ is primitive, 3 is a prime, $3|15$, $3|9$, $3|6$, $3\nmid 2$ and $3^2\nmid 15$. Therefore, by Corollary 12.4.5, $f(x)$ is irreducible over $\mathbb{Z}$.

(ii) There are no roots of $3 + 2x^2 + x^3$ in $\mathbb{Q}$ and hence it is irreducible.

(iii) $4 + 2x + x^3$ has no roots in $\mathbb{Z}_5$ and hence it is irreducible.

(iv) Let $f(x) = 1 + x^2 + x^5 \in \mathbb{Z}_2[x]$. Suppose that $f(x)$ is reducible, since $f(x)$ has no root in $\mathbb{Z}_2$, $f(x)$ has no linear factors and hence there exists an irreducible factor of degree 2 of $f(x)$. $1 + x + x^2$ is the only quadratic irreducible polynomial over $\mathbb{Z}_2$. Therefore, $1 + x + x^2$ should divide $f(x)$. If $1 + x^2 + x^5 = (1 + x + x^2)(a + bx + cx^2 + dx^3)$, then by comparing the coefficients, we get that $a = 1 = b = c = d$, which is a contradiction. Thus, $f(x)$ is irreducible over $\mathbb{Z}_2$.

(v) $9 - x^3$ has no root in $\mathbb{Z}_3$ and hence $9 - x^3$ is irreducible over the field $\mathbb{Z}_{31}$.

(vi) If $f(x) = 1 + x^3 + x^6$, then $f(x + 1) = 3 + 9x + 18x^2 + 21x^3 + 15x^4 + 6x^5 + x^6$, which is irreducible over $\mathbb{Q}$ (by the Eisenstein's criterion) and hence $f(x)$ is irreducible over $\mathbb{Q}$.

(vii) By the Eisenstein's criterion, $5 + 10x + 15x^3 + 2x^5$ is irreducible over $\mathbb{Q}$ and $\mathbb{Z}$ (it is primitive).

(viii) $2 + 2x + x^4$ is irreducible over $\mathbb{Z}$ and $\mathbb{Q}$.

(ix) Since 4 is a root of $9 - x^3$ in $\mathbb{Z}_{11}$, $9 - x^3$ is reducible over $\mathbb{Z}_{11}$.

(x) This is irreducible over $\mathbb{Q}$, by the Eisenstein's criterion.

6. Use Eisenstein's criterion.

8. x^2, $1 + x^2$, $1 + x + x^2$ and $x + x^2$ are the only polynomials of degree 2 over $\mathbb{Z}_2$ and among these $1 + x + x^2$ is the only irreducible polynomial.

10. $a + bx + x^2$, $a, b \in \mathbb{Z}_p$, are all the monic polynomials of degree 2 in $\mathbb{Z}_p[x]$ and these are p^2 in number. Among these the reducible polynomials are of the form $(x - a)(x - b)$, where a and $b \in \mathbb{Z}_p$. Since $(x - a)(x - b)$ and $(x - b)(x - a)$ are the same polynomials, the number of reducible polynomials is $pC_2 + p(= \frac{p(p-1)}{2} + p)$. Thus, the number of irreducible monic polynomials of degree 2 over $\mathbb{Z}_p$ is

$$p^2 - \left(\frac{p(p-1)}{2} + p\right) = \frac{p(p-1)}{2}.$$

EXERCISE 12(E)

2. $\mathbb{Z}[x]$ is not a PID and hence not an Euclidean domain.

4. $R[x]$ is an Euclidean domain $\Rightarrow R[x]$ is a PID $\Rightarrow R$ is a field.

6. $g(-a) = g((-1)a) = g(-1)g(a) = g(a)$ (by Theorem 12.5.1 (2)).

8. For any $a+b\sqrt{3} \in \mathbb{Z}(\sqrt{3})$, define $g(a+b\sqrt{3}) = |a^2 - 3b^2|$ note that $g(a+b\sqrt{3}) = 0$ if and only if $(a+b\sqrt{3}) = 0$. Verify that g is a gauge function for $\mathbb{Z}[\sqrt{3}]$ with respect to which $\mathbb{Z}(\sqrt{3})$ is an Euclidean domain. Use the technique of the proof of Worked Exercise 12.5.3.

EXERCISE 12(F)

2. $2 = (1 + i)(1 - i)$ and 2 divides neither $1 + i$ nor $1 - i$

   $17 = (4 + i)(4 - i)$ and 17 divides neither $4 + i$ nor $4 - i$.

4. This is true in any Euclidean domain and $\mathbb{Z}[i]$ is a Euclidean domain.

6. Let $g(a)$ be least in $A = \{g(b) : b$ is a nonzero nonunit in $R\}$. Let $x \in R$. Then, $x = qa + r$ for some q and $r \in R$ such that $r = 0$ or $g(r) < g(a)$. If $r = 0$, then a divides x. If $r \neq 0$, then, by the least property of $g(a)$, r is a unit and $x - r = qa$ and hence a divides $x + (-r)$ and $-r$ is a unit in R.

8. g.c.d.$\{-3 + 11i, 8 - i\} = 2 + i$ (use the Euclidean algorithm).

10. $x = 1$ and $y = -3i$.

CHAPTER 13

EXERCISE 13(A)

2. (i), (ii), (iii), (iv) are $\mathbb{R}$-submodules of $\mathbb{R}^4$ and others are not.

4. $\mathbb{Z}$ is a faithful $\mathbb{Z}$-module and, for any nonzero proper ideal I of a ring R with unity, R/I is an R-module, which is not faithful, since $\text{Ann}(R/I) = I$.

6. Follows from $P \subseteq P + Q$ and $P \cap Q \subseteq Q$.

8. Let $M = \mathbb{Z}_2 \times \mathbb{Z}_2$. Then, M is a $\mathbb{Z}$-module. Let $N = \{(0, 0), (1, 1)\}$, $P = \mathbb{Z}_2 \times \{0\}$ and $Q = \{0\} \times \mathbb{Z}_2$. Then, N, P, Q are $\mathbb{Z}$-submodules of M.

Here, we have $N \cap (P + Q) = N \cap M = N$, $N \cap P = \{(0, 0)\}$ and $N \cap Q = \{(0, 0)\}$ and hence $(N \cap P) + (N \cap Q) = \{(0, 0)\} \neq N \cap (P + Q)$.

10. If $0 \neq x \in N_0$, then $<x>$ is a nonzero R-submodule and hence $N_0 \subseteq <x> \subseteq N_0$, so that $N_0 = <x>$.

12. Clearly, θ_N is an R-congruence on M. If θ is any R-congruence on M and $N = \theta(0)$, then N is an R-submodule of M and $\theta_N = \theta$.

EXERCISE 13(B)

4. If f and g are R-endomorphisms of M, it can be verified that $f + g$ and $f \circ g$ are also R-endomorphisms of M.

6. Direct verification.

8. If f and $g \in M$, then $f + g \in M$ and $af \in M$ for all $a \in \mathbb{R}$. Also, $(f + g)' = f' + g'$ and $(af)' = (af)'$ and hence $f \mapsto f'$ is an $\mathbb{R}$-homomorphism of M into $\mathbb{R}^{\mathbb{R}}$.

10. If $f : M \to N$ is an R-isomorphism and $x_1, \ldots, x_n \in M$, then $M = <x_1, \ldots, x_n> \Leftrightarrow N = <f(x_1), \ldots, f(x_n)>$.

12. For any $a \in \mathbb{Q}$, define $f_a : \mathbb{Q} \to \mathbb{Q}$ by $f_a(r) = ar$. Then, $f_a \in \text{End}_{\mathbb{Z}}(\mathbb{Q})$ and $a \mapsto f_a$ is a ring isomorphism of $\mathbb{Q}$ onto $\text{End}_{\mathbb{Z}}(\mathbb{Q})$. (If $f \in \text{End}_{\mathbb{Z}}(\mathbb{Q})$ and $f(1) = a$. Then, verify that $f(n) = an$ for all $n \in \mathbb{Z}$ and $m\, f\left(\frac{n}{m}\right) = f(n) = an$ and hence $f = f_a$.)

14. Define $f : M \to M/A \times M/B$ by $f(x) = (x - A, x + B)$. Then, f is an $\mathbb{R}$-homomorphism and $\ker f = A \cap B$. Also, let $(x + A, y + B) \in M/A \times M/B$, $x, y \in M$. Then, $x - y \in M = A + B$ and hence $x - y = a + b$ for some $a \in A$ and $b \in B$. Put $z = x - a = y + b$. Then, $z - x \in A$ and $z - y \in B$ and hence $f(z) = (z + A, z + B) = (x + A, y + B)$. Thus, f is an epimorphism. By Theorem 13.2.4, $M/A \cap B \cong M/A \times M/B$.

16. Clearly, θ_f is an R-congruence and $\theta_f(0) = \ker f$. For any $s = f(z)$, $z \in M$,

$$\{x \in M : f(x) = s = f(z)\} = z + \ker f.$$

EXERCISE 13(C)

2. No in (i) and (iv).

   Yes in (ii) and (iii).

4. For each $i \in I$, let $N_i = \{\alpha \in R^I : |\alpha| \subseteq \{i\}\}$. Then, each N_i is an R-submodule of $R^{(I)}$. Define $f_i : N_i \to R$ by $f_i(\alpha) = \alpha(i)$. Then, f_i is an R-isomorphism and hence $N_i \cong R$, as R-modules. Also, $g : R^{(I)} \to \bigoplus_{i \in I} N_i$, defined by $g(\alpha)(i) = f_i^{-1}(\alpha(i))$, is an R-isomorphism.

6. Let $N = \prod_{i \in I} M_i$ be the direct product and $\Pi_i : N \to M_i$ be the ith projection. Then, each Π_i is an R-epimorphism. By Theorem 13.3.1 (2), there exists a R-homomorphism $\Pi : N \to M$ such that $p_i \circ \Pi = \Pi_i$. Since Π_i is surjective, so is p_i. Thus, p_i is an R-epimorphism.

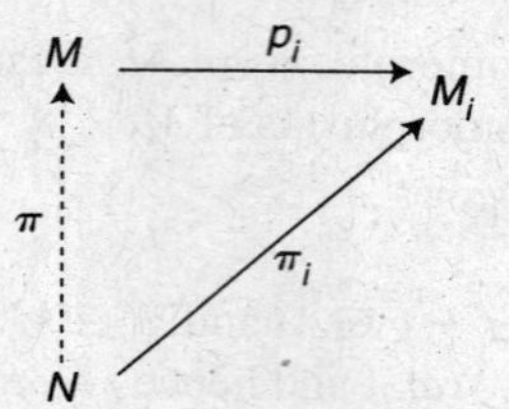

8. Let $N = \{\alpha \in \prod_{i \in I} M_i : |\alpha| \text{ is finite}\}$. Then, N is an R-module ($\mathbb{R}$-submodule of $\prod_{i \in I} M_i$). Define $g_i : M_i \to N$ by $g_i(x_i)(j) = x_i$ or 0 according as $j = i$ or $j \neq i$. Then, g_i is an R-monomorphism. By Definition 13.3.3 (2), there exists homomorphism $g : M \to N$ such that $g \circ f_i = g_i$. Since g_i is an injection, f_i is also an injection. Thus, each f_i is a R-monomorphism.

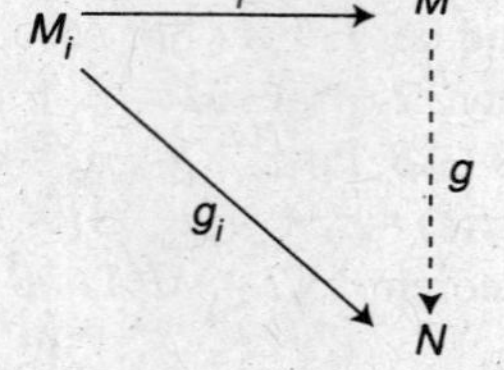

10. Let each f_i be an R-isomorphism. Then, for any $x, y \in \prod_i M_i$,

$$\begin{aligned} f(x) = f(y) &\Rightarrow f(x)(i) = f(y)(i) \text{ for all } i \in I \\ &\Rightarrow f_i(x(i)) = f_i(y(i)) \\ &\Rightarrow x(i) = y(i) \text{ for all } i \in I \Rightarrow x = y. \end{aligned}$$

Thus, f is an injection. Also, for any $y \in \prod_i N_i$, choose $x_i \in M_i$ such that $f_i(x_i) = y(i)$ for each $i \in I$. Then, $f(\alpha) = y$, where $\alpha \in \prod_i M_i$ is defined by $\alpha(i) = x_i$. Thus, f is an R-isomorphism.

EXERCISE 13(D)

2. M is precisely the external direct sum $\bigoplus_{x \in X} M_x$ where M_x is the $\mathbb{R}$-module $\mathbb{R}$ for each $x \in X$. Since $\mathbb{R}$ is simple as an $\mathbb{R}$-module (by Example 13.4.1 (1)), so is each M_x. Therefore, M is a completely reducible $\mathbb{R}$-module. M is simple $\Leftrightarrow M \cong \mathbb{R} \Leftrightarrow X = \{x\}$ (if $x \neq y \in X$, then M_x can be treated as a nonzero proper submodule of M).

4. Since N is proper, $M/N \neq \{0\}$. By Theorem 13.4.3, $M = N \oplus (\bigoplus_{\alpha \in I} M_\alpha)$, where $\{M_\alpha\}_{\alpha \in I}$ is a nonempty (since $M \neq N$) family of simple R-modules. Then, $M/N \cong \bigoplus_{\alpha \in I} M_\alpha$ and hence M/N is completely irreducible.

6. If $M = \bigoplus_{\alpha \in \Delta} M_\alpha$ and $M_\alpha = \bigoplus_{i \in I_\alpha} M_{\alpha_i}$, for each $\alpha \in \Delta$, where each M_{α_i} is simple, then $M \cong \bigoplus_{\alpha \in \Delta,\, i \in I_\alpha} M_{\alpha_i}$ and hence M is completely reducible.

EXERCISE 13(E)

2. Any element of M can be written as $r_1x_1 + \cdots + r_nx_n$, $x_1, \ldots, x_n \in B$ and $r_1, \ldots, r_n \in R$. Also, since B is linearly independent, this expression is unique.

4. If B_i is a basis for M_i, then prove that $B_1 \times B_2 \times \cdots \times B_n$ is a basis for $M_1 \times M_2 \times \cdots \times M_n$.

6. Let I be an ideal of $\mathbb{Z}$. Then, $I = n\mathbb{Z}$ for some $n \geq 0$ and $\{n\}$ is a basis for I as a $\mathbb{Z}$-module.

8. Let M be an R-module. Let $N = \bigoplus_{x \in M} R_x$, where R_x is the R-module R. Then, N is a free R-module and define $f: N \to M$ by $f(\alpha) = \sum_{x \in M} \alpha(x)x$. Then, f is an R-epimorphism and hence M is a homomorphic image of the free R-module N.

10. If $f: M \to N$ is an R-isomorphism and $B \subseteq M$, then B is a basis for $M \Leftrightarrow f(B)$ is a basis for N.

12. If B is a basis and $B \subsetneq A \subseteq M$, then choose $y \in A$ such that $y \notin B$. Then, $y = r_1x_1 + \cdots + r_nx_n$ for some $x_i \in B$. Then, $r_1x_1 + \cdots + r_nx_n + (-1)y = 0$ and hence A is linearly dependent.

14. Let B be a basis for M. For each $x \in B$, choose $y_x \in M$ such that $f(x) = y_x + N$. Any $y \in F$ can be uniquely expressed as $y = r_1x_1 + \cdots + r_nx_n$,

where $x_i \in B$ and $r_i \in R$ and now define $g(y) = r_1 y_{x_1} + \cdots + r_n y_{x_n}$. Then, $g : F \to M$ is a homomorphism and

$$g(y) + M = \sum_{i=1}^{w} r_i (y_{x_i} + M) = \sum_{i=1}^{n} r_i f(x_i) = f\left(\sum_i r_i x_i\right) = f(y).$$

EXERCISE 13(F)

2. Let $e_4 = (0, 0, 0, 1, 0)$ and $e_5 = (0, 0, 0, 0, 1)$. Then, $\{e_1, e_2, e_3, e_4, e_5\}$ is a basis of F^5.

4. The matrices of D with respective to B and C are respectively

$$\begin{pmatrix} 0 & 0 & 0 & 0 & 0 \\ 1 & 0 & 0 & 0 & 0 \\ 0 & 2 & 0 & 0 & 0 \\ 0 & 0 & 3 & 0 & 0 \\ 0 & 0 & 0 & 4 & 0 \end{pmatrix} \text{ and } \begin{pmatrix} 0 & 0 & 0 & 0 & 0 \\ 1 & 0 & 0 & 0 & 0 \\ -1 & 2 & 0 & 0 & 0 \\ -1 & -1 & 3 & 0 & 0 \\ -1 & -1 & -1 & 4 & 0 \end{pmatrix}.$$

The matrices of transformations from B to C and C to B are respectively

$$\begin{pmatrix} 1 & 0 & 0 & 0 & 0 \\ -1 & 1 & 0 & 0 & 0 \\ 0 & -1 & 1 & 0 & 0 \\ 0 & 0 & -1 & 1 & 0 \\ 0 & 0 & 0 & -1 & 1 \end{pmatrix} \text{ and } \begin{pmatrix} 1 & 0 & 0 & 0 & 0 \\ 1 & 1 & 0 & 0 & 0 \\ 1 & 1 & 1 & 0 & 0 \\ 1 & 1 & 1 & 1 & 0 \\ 1 & 1 & 1 & 1 & 1 \end{pmatrix}.$$

CHAPTER 14

EXERCISE 14(A)

2. $[K : \mathbb{Q}] = 4$, since $[K : \mathbb{Q}(\sqrt{p})] = 2$ and $[\mathbb{Q}(\sqrt{p}) : \mathbb{Q}] = 2$

4. $[F_n : \mathbb{Q}] = [F_n : F_{n-1}][F_{n-1} : F_{n-2}] \ldots [F_1 : \mathbb{Q}] = 2^n$

   and $[\mathbb{R} : \mathbb{Q}] > [F_n : \mathbb{Q}] = 2^n$ for all n.

6. No, since $\text{char}(\mathbb{Z}_p) = p$ and $\text{char}(\mathbb{Q}) = 0$

8. For any field F, $\text{char}(F) = O(1)$ in the group $(F, +)$.

EXERCISE 14(B)

2. $\{a_0 + a_1 r + a_2 r^2 + a_3 r^3 : a_0, a_1, a_2, a_3 \in \mathbb{Q}\}$, where $r^4 - 2 = 0$
4. Infinite (see Ex. 4 in Exercise 14(A)).
6. (i) $x^2 - 5$

   (ii) $23 - 10x + x^2$

   (iii) $7 - 10x + x^2$

   (iv) $-1 + 2x^2 + x^4$

   (v) $625 - 58x^2 + x^4$

   (vi) $x^2 - 6$

8. $F \subseteq K \subseteq E \Rightarrow [E : K][K : F] = (E : F) = p$, a prime

   $\Rightarrow [E : K] = 1$ or $[K : F] = 1$

   $\Rightarrow K = E$ or $K = F$.

10. Since $a^2 \in F(a)$, we have $F \subseteq F(a^2) \subseteq F(a)$ and hence $n = [F(a^2) : F] \leq [F(a) : F] = m$, say. Let $f(x) = a_0 + a_1 x + \cdots + a_n x^n =$ be the minimal polynomial of a^2 over F. Then, $a_0 + a_1 a^2 + \cdots + a_n a^{2n} = 0$ and hence a is a root of the polynomial $g(x) = a_0 + a_1 x^2 + \cdots + a_n x^{2n}$. This implies that m divides $2n$ since m is odd, m divides n and hence $m \leq n$. Thus,

$$[F(a^2) : F] = n = m = [F(a) : F].$$

Also since $F(a^2) \subseteq F(a)$, it follows that $F(a^2) = F(a)$.

EXERCISE 14(C)

2. Let $F = \bigcup_{n=1}^{\infty} F_n$. a and $b \in F \Rightarrow a \in F_n$ and $b \in F_m$ for some n and $m \Rightarrow$ a and $b \in F_n$ or F_m, according as $m \leq n$ or $n \leq m \Rightarrow a \pm b$ and $a \cdot b \in F_n \Rightarrow a \pm b$ and $a \cdot b \in F$. Thus, F is a field and F_n is a subfield of F for each n.

4. If K is algebraically closed, then, for any $a \in K$, $x^n - a$ has all the roots in K (by Theorem 14.3.1) and hence $n < |K|$ for all $n \in \mathbb{Z}^+$. Therefore, K is infinite.

6. Use the argument given in Ex. 2 above and the fact that, for any α and $\beta \in \Delta$, $F_\alpha \subseteq F_\beta$ or $F_\beta \subseteq F_\alpha$.

8. Let a be a root of $p(x)$ in E. Let b be another root of $p(x)$ in some extension K of F. Then, $F[x]/\langle p(x)\rangle \cong F(a)$ and $F[x]/\langle p(x)\rangle \cong F(b)$ and

hence $F(a) \cong F(b)$. Therefore, we can treat $f(x)$ as a polynomial over $F(a)$ as well as over $F(b)$. Also, E is a splitting field of $f(x)$, treated as a polynomial over $F(a)$. If L is a splitting of $f(x)$, treated as a polynomial over $F(b)$, then by the uniqueness of the splitting field, $E \cong L$. Since $f(x)$, as a polynomial over F, splits in E, we get that $L \subseteq E(b)$. Since L contains all the roots of $f(x)$, in particular, $b \in L$, we have $E(b) \subseteq L$. Therefore, $E(b) = L$. Also,

$$\begin{aligned}[E(b) : F] &= [L : F] = [L : F(b)][F(b) : F] \\ &= [E : F(a)][F(a) : F] = [E : F]\end{aligned}$$

and, since $E \subseteq E(b)$, we have $E(b) = E$ and, in particular, $b \in E$. Thus, E contains all the roots of $p(x)$.

10. Consider $f(x) = x^p - 1 = (x - 1)(x^{p-1} + x^{p-2} + \cdots + x + 1)$. The splitting field of $f(x) \in \mathbb{Q}[x]$ is same as the splitting field of $g(x) = 1 + x + \cdots + x^{p-1} \in \mathbb{Q}[x]$. If a is a primitive p^{th} root of unity, that is, $a = e^{2\pi i/p}$, and E is the splitting field of $f(x)$, then

$$E = \mathbb{Q}(1, a, a^2, \ldots, a^{p-1}) = \mathbb{Q}(a).$$

Since $g(x)$ is irreducible in $\mathbb{Q}$, we have $[E : \mathbb{Q}] = p - 1$.

EXERCISE 14(D)

2. Use induction on n.

4. Use induction on m.

6. Use Theorem 14.4.4(2).

EXERCISE 14(E)

2. $F = GF(2^3)$. $F^* = F - \{0\}$ is a cyclic group of order 7 and hence any $1 \neq a \in F^*$ generates F^*.

4. $1 + x + x^2$ is an irreducible polynomial over $\mathbb{Z}_2$ and hence $\mathbb{Z}_2[x]/\langle 1 + x + x^2\rangle$ is a field with four elements, say F. Then, $F = \{0, 1, a, 1 + a\}$, where $1 + a + a^2 = 0$. Then, $x^2 + ax + 1$, $x^3 + x^2 + ax + (1 + a)$ and $x^4 + ax^3 + ax^2 + (1 + a)x + 1$ are irreducible polynomials of degrees 2, 3 and 4, respectively over F.

6. Let F be a finite field. Then, by Theorem 14.5.1, $|F| = p^n$, where $n \in \mathbb{Z}^+$ and $p = \text{char}(F)$. If $p = 2$, then the map $\alpha : F \to F$ defined by $\alpha(a) = a^2$ is an injection and hence a bijection, so that any $x \in F$ can be expressed as $x = a^2 = a^2 + 0^2$ for some $a \in F$. Next, suppose that $p > 2$; that is,

p is an odd prime. Let $x \in F$ and $S = \{x - a^2 : a \in F\}$ and $T = \{b^2 : b \in F\}$. Then, S and T are subsets of F, $|S| = (p^n - 1)/2 + 1 = \frac{p^n+1}{2}$ and $|T| = \frac{p^n+1}{2}$ (note that $a \neq -a$ for any $a \in F$, since char$(F) \neq 2$). Since $|S \cup T| \leq p^n$ and $|S \cup T| = |S| + |T| - |S \cap T|$, it follows that $S \cap T$ is nonempty and hence there exists $x - a^2 = b^2 \in T \cap S$; that is, $x = a^2 + b^2$ for some a and $b \in F$.

8. Since char$(F) = p$, the map $\alpha : F \to F$ defined by $\alpha(x) = x^p$ is an epimorphism of the ring F (recall that $x^{|F|} = x$ for all $x \in F$ and $|F| = p^n$). Since F is finite and $\alpha : F \to F$ is a surjection, we get that α is an injection and hence, for each $a \in F$, there exists unique $x \in F$ such that $a = \alpha(x) = x^p$.

CHAPTER 15

EXERCISE 15(A)

2. We have $1 + z + z^2 = 0$. Consider $a = \sqrt{2} + z$. Then, $a^2 = 2 + 2\sqrt{2}\,z - (1 + z) = 1 + (2\sqrt{2} - 1)z \in \mathbb{Q}(a)$ and hence $(2\sqrt{2} - 1)z \in \mathbb{Q}(a)$. Therefore, $(2\sqrt{2} - 1)^3 = ((2\sqrt{2} - 1)z)^3 \in \mathbb{Q}(a)$; that is, $8 - 1 - 6\sqrt{2}\,(2\sqrt{2} - 1) \in \mathbb{Q}(a)$ and hence $\sqrt{2} \in \mathbb{Q}(a)$. Since $(\sqrt{2} - 1)z \in \mathbb{Q}(a)$, it follows that $z \in \mathbb{Q}(a)$. Thus, $\mathbb{Q}(\sqrt{2}, z) \subseteq \mathbb{Q}(a)$. Clearly, $a \in \mathbb{Q}(\sqrt{2}, z)$ and hence $\mathbb{Q}(a) \subseteq (\sqrt{2}, z)$.

4. Let K be a finite extension of a finite field F and $|F| = p^n$, where char$(F) = p =$ char(K). Then, K is also finite and $|K| = p^m$ for some $m \in \mathbb{Z}^+$. We have $K = F(a)$ for some $a \in K$. Hence, $a^{p^m} = a$. Therefore, a is a separable element and hence $F(a)$ $(=K)$ is a separable extension of F.

6. $x^4 - 2$ is an irreducible polynomial over $\mathbb{Q}$ and $2^{1/4}$ is a root of $x^4 - 2$ belonging to $\mathbb{Q}(2^{1/4})$; but $2^{1/4}i$, which is also a root of $x^4 - 2$ does not belong to $\mathbb{Q}(2^{1/4})$. Therefore, $\mathbb{Q}(2^{1/4})$ is not a normal extension of $\mathbb{Q}$.

8. Use Theorem 15.1.5.

EXERCISE 15(B)

2. Clearly the identity map and the map $c : \mathbb{C} \to \mathbb{C}$, defined by $c(a + ib) = a - ib$, are in $G(\mathbb{C}/R)$. Also, if $f \in G(C/\mathbb{R})$, then $f(i)^2 = f(i^2) = f(-1) = 1$ and hence $f(i) = \pm i$ so that, for each $a + ib \in \mathbb{C}$, $f(a + ib) = a + ib$ or $a - ib$. Therefore, $f =$ Id, or c. Thus, $G(\mathbb{C}/\mathbb{R})$ is a two element group and cyclic.

4. Let $\sqrt[3]{2} = u$. Then, $u^3 = 2$ and $u \in \mathbb{R}$. We are given that σ is the automorphism of $E = \mathbb{Q}(\omega, u)$ such that $\sigma(r) = r$ for all $r \in \mathbb{Q}$, $\sigma(\omega) = \omega^2$ and $\sigma(u) = u\omega$. Clearly, $\sigma \circ \sigma = \text{Id}$ and hence $H = \{\text{Id}, \sigma\}$ is a subgroup of $G(E/\mathbb{Q})$. Let E_H be the fixed field of H. Since $\sigma(u\omega^2) = \sigma(u)\sigma(\omega)^2 = u\omega^2$ and hence $u\omega^2 \in E_H$, $\mathbb{Q}(u\omega^2) \subseteq E_H$. On the other hand, let $a \in E_H$. That is, $\sigma(a) = a$. Observe that $\{1, \omega\}$ is a basis of $\mathbb{Q}(u, \omega)$ over $\mathbb{Q}(u)$ and that $\{1, u, u^2\}$ is a basis of $\mathbb{Q}(u)$ over $\mathbb{Q}$. Therefore, $\{1, u, u^2, \omega, \omega u, \omega u^2\}$ is a basis of $\mathbb{Q}(u, \omega)(= E)$ over $\mathbb{Q}$ and hence we can write

$$a = r_0 + r_1 u + r_2 u^2 + r_3 \omega + r_4 \omega u + r_5 \omega u^2$$

for some $r_i \in \mathbb{Q}$. Now,

$$\begin{aligned} a = \sigma(a) &= r_0 + r_1 u\omega + r_2 u^2\omega^2 + r_3\omega^2 + r_4\omega^3 u + r_5\omega^4 u^2 \\ &= r_0 + r_1 u\omega + r_2 u^2(-1 - \omega) + r_3(-1 - \omega) + r_4 u + r_5 \omega u^2 \\ &= (r_0 - r_3) + r_4 u - r_2 u^2 - r_3\omega + r_1\omega u + (-r_2 + r_5)\omega u^2 \end{aligned}$$

From this, it follows that $r_0 = r_0 - r_3$, $r_1 = r_4$, $r_2 = -r_2$, $r_3 = -r_3$, $r_4 = r_1$, and $r_5 = -r_2 + r_5$ and hence $r_3 = 0 = r_2$, $r_1 = r_4$ and r_0 and r_5 are arbitrary. Therefore,

$$a = r_0 + r_1(u + \omega u) + r_5\omega u^2 = r_0 - r_1\omega^2 u + r_5(\omega^2 u)^2$$

Thus, $a \in \mathbb{Q}(\omega^2 u)$. Therefore, $E_H = \mathbb{Q}(\omega^2 u)$.

6. Since $0 = a^5 - 1 = (a - 1)(1 + a + a^2 + a^3 + a^4)$ and $a \neq 1$, it follows that a is a root of the polynomial $f(x) = 1 + x + x^2 + x^3 + x^4$ which is irreducible over $\mathbb{Q}$. Therefore, $[\mathbb{Q}(a) : \mathbb{Q}] = 4$. The roots of $x^5 - 1$ are $1, a, a^2, a^3$ and a^4 which are in $\mathbb{Q}(a)$. Therefore, $\mathbb{Q}(a)$ is the splitting field of $x^5 - 1$ over $\mathbb{Q}$. Thus, $\mathbb{Q}[a]$ is a normal extension of $\mathbb{Q}$ and the Galois group $G(\mathbb{Q}(a)/\mathbb{Q})$ is of order $[\mathbb{Q}(a) : \mathbb{Q}] = 4$. Observe that $\{1, a, a^2, a^3\}$ is a basis of $\mathbb{Q}(a)$ over $\mathbb{Q}$ and hence any element of $\mathbb{Q}(a)$ can be uniquely expressed as

$$r_0 + r_1 a + r_2 a^2 + r_3 a^3, r_i \in \mathbb{Q}$$

and the four elements of $G(\mathbb{Q}(a)/\mathbb{Q})$ are Id, f, g and h, where $f(a) = a^2$, $g(a) = a^3$ and $h(a) = a^4$, which form a cyclic group with f (or g) as a generator. Thus, $G(\mathbb{Q}(a)/\mathbb{Q}) \cong \mathbb{Z}_4$.

EXERCISE 15(C)

2. It is known that the Klein four-group is the group of order 4 in which each element is of order ≤ 2 and is isomorphic to $\mathbb{Z}_2 \times \mathbb{Z}_2$. We prove that the Galois group of $x^4 + 1 \in \mathbb{Q}[x]$ is of order 4 and $f^2 = \text{Id}$ for each f in this group. Let $F = \mathbb{Q}(a)$, where $a = e^{i\pi/4}$. Since a, a^3, a^5 and a^7 are the roots of $x^4 + 1$, it follows that F is the splitting field of

$x^4 + 1$ over $\mathbb{Q}$. Also, $[F : \mathbb{Q}] = 4$, since $x^4 + 1$ is irreducible over $\mathbb{Q}$. char($\mathbb{Q}$) = 0 and hence F is a normal separable extension of $\mathbb{Q}$. Therefore, $G(F/\mathbb{Q})$ is of order $[F : \mathbb{Q}] = 4$. Note that any element b of F can be expressed as $b = r_0 + r_1 a + r_2 a^2 + r_3 a^3$ with $r_i \in \mathbb{Q}$, and any $f \in G(F/\mathbb{Q})$ is determined by its value at a; since $f(a)$ must be a root of $x^4 + 1$, the four elements of $G(F/\mathbb{Q})$ are Id, f, g and h, where $f(a) = a^3$, $g(a) = a^5$ and $h(a) = a^7$. Also, note that $f^2(a) = f(f(a)) = f(a^3) = f(a)^3 = a^9 = a$, $h^2(a) = a^{49} = a$ and $g^2(a) = a^{25} = a$ and hence $f^2 = \text{Id} = g^2 = h^2$. Thus, $G(F/\mathbb{Q}) \cong \mathbb{Z}_2 \times \mathbb{Z}_2$.

4. Let $E_m = \mathbb{Q}(a)$, $G =$ The Galois group $G(E_m/\mathbb{Q})$. Then, $|G| = [E_m : \mathbb{Q}] = 2^m$. Choose subgroup H_{m-1} of order 2^{m-1} in G (Use Sylow Theorem I). Then, H_{m-1} is of index 2 in G and hence normal in G. Let E_{m-1} be fixed field of H_{m-1}. Then, E_m is a normal extension of E_{m-1}. Also, $\mathbb{Q} \subseteq E_{m-1} \subseteq E_m$, $[E_m : E_{m-1}] = 2$ and

$$\left|G\left(\frac{E_{m-1}}{\mathbb{Q}}\right)\right| = \frac{\left|G\left(\frac{E_m}{\mathbb{Q}}\right)\right|}{\left|G\left(\frac{E_m}{E_{m-1}}\right)\right|} = 2^{m-1}$$

Now, choose a subgroup H_{m-2} of order 2^{m-2} in $G(E_{m-1}/\mathbb{Q})$ and let E_{m-2} be the fixed field of H_{m-2}. This process can be continued to construct the required fields $\mathbb{Q} = E_0 \subset E_1 \subset \cdots \subset E_{m-1} \subset E_m = \mathbb{Q}(a)$.

6. Note that the only nonidentity automorphism in $G(\mathbb{Q}(a)/\mathbb{Q})$ is the f for which $f(a) = a^2$ (and hence $f(a^2) = a$, since $a^3 = 1$). Therefore, $G(\mathbb{Q}(a)/\mathbb{Q})$ is the two element group, which has only two subgroups, namely {Id} and the whole group and the corresponding fixed fields are $\mathbb{Q}(a)$ and $\mathbb{Q}$, respectively.

CHAPTER 16

EXERCISE 16

2. If $f \in G(\mathbb{C}/\mathbb{R})$, then $f(i)^2 = f(i^2) = f(-1) = -1$ and hence $f(i) = i$ or $-i$. Therefore, $G(\mathbb{C}/\mathbb{R}) = \{\text{Id}, f\}$, where $f(a + bi) = a - bi$ for any $a + bi \in \mathbb{C}$.

4. Refer Worked Exercise 16.3.3.

6. If $f(x) = x^3 + x^2 - 2x - 1$, then $f(x - 2) = x^3 + 7x^2 + 14x + 7$. Now, use the Eisenstein's criterion.

Index